PROCEEDINGS OF THE
INTERNATIONAL CONGRESS OF
MATHEMATICIANS
1958

*The publication has been aided
by a grant to the
International Congress of Mathematicians
by* UNESCO

PROCEEDINGS OF THE INTERNATIONAL CONGRESS OF MATHEMATICIANS

14–21 AUGUST 1958

EDITED BY

J. A. TODD, F.R.S.

LECTURER IN MATHEMATICS IN THE
UNIVERSITY OF CAMBRIDGE

CAMBRIDGE
AT THE UNIVERSITY PRESS
1960

CAMBRIDGE UNIVERSITY PRESS
Cambridge, New York, Melbourne, Madrid, Cape Town,
Singapore, São Paulo, Delhi, Mexico City

Cambridge University Press
The Edinburgh Building, Cambridge CB2 8RU, UK

Published in the United States of America by Cambridge University Press, New York

www.cambridge.org
Information on this title: www.cambridge.org/9781107622661

First published 1960
First paperback edition 2013

A catalogue record for this publication is available from the British Library

ISBN 978-1-107-62266-1 Paperback

CONTENTS

Preface   *page* vii

Officers of the Congress and Members of Committees and
    Sub-committees   ix

List of Donors   xii

Scientific Programme   xiv

Secretary's Report   xliii

Report of the Inaugural Session   xlviii

Report of the Closing Session   lv

The Work of the Fields Medallists   lvii

One-hour Addresses   1

Half-hour Addresses   279

Index   573

PREFACE

This volume contains the official record of the International Congress of Mathematicians held in Edinburgh from 14 to 21 August 1958, and the text of the addresses given by invitation of the Programme Committee. In accordance with a decision already announced, the short communications made by members at the Congress are not included in the *Proceedings*, but the names of the communicators and the titles of their papers will be found in the section giving the scientific programme. Summaries of these communications, if received in time, were printed in the volume of abstracts issued to members during the Congress.

J. A. T.

CAMBRIDGE
December 1958

OFFICERS OF THE CONGRESS AND MEMBERS OF COMMITTEES AND SUB-COMMITTEES

PATRON

H.R.H. THE PRINCE PHILIP, DUKE OF EDINBURGH

HOSTS

The City of Edinburgh
The University of Edinburgh
The Royal Society of Edinburgh
The Royal Society

CONGRESS COMMITTEE

Chairman: The Lord Provost of the City of Edinburgh

Vice-Chairman: The Principal of the University of Edinburgh

COUNCILLOR TOM CURR, representing the City Council of Edinburgh

PROFESSOR A. C. AITKEN, representing the University of Edinburgh

PROFESSOR SIR EDMUND WHITTAKER,* representing The Royal Society of Edinburgh

PROFESSOR W. V. D. HODGE, representing The Royal Society

PROFESSOR J. H. C. WHITEHEAD, representing the London Mathematical Society

DR R. SCHLAPP, representing the Edinburgh Mathematical Society

PROFESSOR H. DAVENPORT, representing the National Committee for Mathematics

J. MACPHERSON AND F. SMITHIES, Joint Secretaries

OFFICERS OF THE CONGRESS

President: W. V. D. HODGE
Treasurer: A. L. IMRIE
Secretary: F. SMITHIES
Secretary of Executive Committee: R. SCHLAPP
Advisory Secretary: J. MACPHERSON

* Died March 1956; succeeded by Dr W. L. Edge.

EXECUTIVE COMMITTEE

W. V. D. Hodge (*Chairman*), A. C. Aitken (*Vice-Chairman*), D. A. Allan,
R. Bell (replaced by J. G. Dunbar), E. T. Copson, Sir J. Erskine,
W. H. M. Greaves (replaced by D. E. Rutherford), R. Ll. Gwilt
(replaced by K. K. Weatherhead), A. L. Imrie, I. A. Johnson-
Gilbert (replaced by D. M. Weatherstone), N. Kemmer, M. H. A.
Newman, R. A. Rankin, John Reid, F. Smithies, W. V. Stevens,
C. H. Stewart, Sir John Storrar (from May 1957 replaced by
W. Borland), G. Temple, Sir Edmund Whittaker (replaced by
W. L. Edge), E. M. Wright.
R. Schlapp, P. Heywood (*Joint Secretaries*).

SUB-COMMITTEES

[The Chairman and Secretary of the Executive Committee and the
Congress Secretary were *ex officio* members of all sub-committees.]

Accommodation. C. H. Stewart (*Chairman*), N. Kemmer, G. MacKenzie,
L. D. Macmillan, J. MacPherson, John Reid, D. M. Weatherstone.

Entertainments. I. A. Johnson-Gilbert (replaced by D. M. Weatherstone)
(*Chairman*), E. T. Copson, H. M. Jamieson, John Reid, W. V.
Stevens.

Excursions. W. H. M. Greaves (replaced by D. E. Rutherford) (*Chairman*), Mrs Aitken, D. A. Allan, E. T. Copson, R. P. Gillespie,
D. Russell, Mrs Schlapp. L. M. Brown (*Secretary*).

Finance. R. Bell (replaced by J. G. Dunbar) (*Chairman*), A. C. Aitken,
Sir John Erskine, R. Ll. Gwilt (replaced by K. K. Weatherhead),
A. L. Imrie, I. A. Johnson-Gilbert (replaced by D. M. Weather-
stone), M. H. A. Newman, R. A. Rankin, D. E. Rutherford,
W. V. Stevens, C. H. Stewart.

Proceedings. J. A. Todd (*Chairman*), M. F. Atiyah, H. Davenport,
P. Du Val, H. Heilbronn, K. A. Hirsch, N. Kemmer, D. V. Lindley,
E. A. Maxwell, D. Rees, H. P. F. Swinnerton-Dyer, A. G. Walker,
J. H. Williamson, E. C. Zeeman.

Registration and Reception. R. A. Rankin (*Chairman*), I. M. H. Ethering-
ton, T. S. Graham. P. Heywood (*Secretary*).

Scientific Programme. M. H. A. Newman (*Chairman*), A. C. Aitken,
M. S. Bartlett, Miss M. L. Cartwright, J. L. B. Cooper, H. Davenport, P. Hall, N. Kemmer, M. J. Lighthill, E. A. Maxwell, D. G.

Northcott, W. W. Rogosinski, A. G. Walker, J. H. C. Whitehead, M. V. Wilkes. J. A. Green (*Secretary*).

Travel Grants. A. G. Walker (*Chairman*), A. Fletcher, A. L. Imrie, T. J. Willmore.

Book Exhibition. T. A. A. Broadbent (*Chairman*), C. B. Allendoerfer (U.S.A.), E. Bompiani (Italy), G. Choquet (France), E. Kamke (Germany), L. Locher-Ernst (Switzerland), A. Thin.

LIST OF DONORS

Albright & Wilson Ltd.
Associated Electrical Industries Ltd.
Babcock & Wilcox Ltd.
William Baird & Co. Ltd.
Bank of Scotland
Barr & Stroud Ltd.
J. Bibby & Sons Ltd.
Boots Pure Drug Co. Ltd.
Wilmot Breeden Ltd.
Bristol Aeroplane Co. Ltd.
British Aluminium Co. Ltd.
British Mathematical Colloquium
British Paints Ltd.
British Ropes Ltd.
British Tabulating Machine Co. Ltd.
The Rt. Hon. The Viscount Bruce, C.H., M.C., F.R.S.
Bruce Peebles & Co. Ltd.
Burmah Oil Co. Ltd.
The Cementation Co. Ltd.
Central Electricity Authority
Colvilles Ltd.
Commercial Bank of Scotland, Ltd.
T. & A. Constable Ltd.
Courtaulds Ltd.
Coventry Gauge and Tool Co. Ltd.
Cow & Gate Ltd.
Currys Ltd.
Decca Record Co. Ltd.
De Havilland Holdings Ltd.
Dominion Insurance Co. Ltd.
English Electric Co. Ltd.
Ferranti Ltd.
Fisons Ltd.
General Electric Co. Ltd.
The Great Universal Stores Ltd.

Gestetner Ltd.
James Grant & Co. (West) Ltd.
Grants, North Bridge, Edinburgh
Arthur Guinness Son & Co. Ltd.
IBM United Kingdom Ltd.
Imperial Chemical Industries Ltd.
Jenners (Princes Street) Ltd.
Johnson, Matthey & Co. Ltd.
Johnson & Phillips Ltd.
Laporte Industries Ltd.
Joseph Lucas Ltd.
McDougall's Educational Co. Ltd.
Wm. Arnott McLeod & Co. Ltd.
McVitie & Price Ltd.
Metropolitan-Vickers Electrical Co. Ltd.
The Monotype Corporation Ltd.
Morgan Crucible Co. Ltd.
Morrison & Gibb Ltd.
Mullard Ltd.
Munro & Miller Ltd.
Murphy Radio Ltd.
Michael Nairn & Co. Ltd.
National Bank of Scotland Ltd.
National Cash Register Co. Ltd.
National Coal Board
North British and Mercantile Insurance Co. Ltd.
North British Distillery Co. Ltd.
North British Rubber Co. Ltd.
Oliver & Boyd Ltd.
Pergamon Press Ltd.
Pillans & Wilson Ltd.
Pye Ltd.
Joseph Rank Ltd.
Henry Robb Ltd.
Rolls-Royce Ltd.
Captain H. K. Salvesen
Harold Samuel, Esq.
Sangamo Weston Ltd.

Scottish Agricultural Industries Ltd.
Scottish Brewers Ltd.
Scottish Equitable Life Assurance Society
Scottish Life Assurance Co. Ltd.
Scottish Widows' Fund and Life Assurance Society
Shell Petroleum Co. Ltd.
South of Scotland Electricity Board
Standard Life Assurance Company
J. G. Thomson & Co. Ltd.
Vickers Ltd.
Geo. Wimpey & Co. Ltd.

SCIENTIFIC PROGRAMME

ONE-HOUR ADDRESSES BY INVITATION OF THE ORGANIZING COMMITTEE

NOTE. *An asterisk indicates that no manuscript has been received for publicatton.*

ALEXANDROV, A. D. Modern development of surface theory.

BOGOLYUBOV, N. N. and VLADIMIROV, V. S. On some mathematical problems of quantum field theory.

CARTAN, H. Sur les fonctions de plusieurs variables complexes: les espaces analytiques.

CHEVALLEY, C. La théorie des groupes algébriques.

EILENBERG, S. Applications of homological algebra in topology.*

FELLER, W. Some new connections between probability and classical analysis.

GÅRDING, L. Some trends and problems in linear partial differential equations.

GROTHENDIECK, A. The cohomology theory of abstract algebraic varieties.

HIRZEBRUCH, F. Komplexe Mannigfaltigkeiten.

KLEENE, S. C. Mathematical logic: constructive and non-constructive operations.

LANCZOS, C. Extended boundary value problems.

PONTRYAGIN, L. S. Оптимальные процессы регулирования. (Optimal processes of regulation.)

ROTH, K. F. Rational approximations to algebraic numbers.

SCHIFFER, M. Extremum problems and variational methods in conformal mapping.

STEENROD, N. E. Cohomology operations and symmetric products.*

TEMPLE, G. Linearization and delinearization.

THOM, R. Des variétés triangulées aux variétés différentiables.

UHLENBECK, G. E. Some fundamental problems in statistical physics.

WIELANDT, H. Entwicklungslinien in der Strukturtheorie der endlichen Gruppen.

HALF-HOUR ADDRESSES BY INVITATION OF THE ORGANIZING COMMITTEE

SECTION I

BETH, E. W. Completeness results for formal systems.

KREISEL, G. Ordinal logics and the characterization of informal concepts of proof.

MARKOV, A. A. Неразрешимость проблемы гомеоморфии. (Insolubility of the problem of homeomorphy.)

SECTION II

DEURING, M. Neuere Ergebnisse über algebraische Funktionenkörper.*

HIGMAN, G. Lie ring methods in the theory of finite nilpotent groups.

LINNIK, YU. V. О проблеме делителей и родственных ей бинарных аддитивных проблемах. (On divisor problems and some related binary additive problems.)

ROQUETTE, P. Some fundamental theorems on abelian function fields.

SHIMURA, G. Fonctions automorphes et correspondances modulaires.

SECTION III

ARNOLD, V. I. Некоторые вопросы приближения и представления функций. (Some questions of approximation and representation of functions.)

BERS, L. Spaces of Riemann surfaces.

GRAUERT, H. Die Riemannschen Flächen der Funktionentheorie mehrerer Veränderlichen.

HEINS, M. Functions of bounded characteristic and Lindelöfian maps.

LIONS, J. L. Problèmes mixtes abstraits.

MENCHOFF, D. E. О сходимости тригонометрических рядов. (The convergence of trigonometric series.)

MINAKSHISUNDARAM, S. Hilbert algebras.

SZ.-NAGY, B. Spectral sets and normal dilations of operators.

SECTION IV

BOTT, R. An application of the Morse theory to the topology of Lie groups.

KOSIŃSKI, A. On some problems connected with the topology of manifolds.

MOORE, J. C. Homology structure of group-like spaces.*

PAPAKYRIAKOPOULOS, C. D. The theory of 3-dimensional manifolds since 1950.

SECTION V

CHERN, S. S. Differential geometry and integral geometry.

MATSUSAKA, T. The polarization of algebraic varieties, and some of its applications.

MILNOR, J. W. and KERVAIRE, MICHEL A. Bernoulli numbers, homotopy groups, and a theorem of Rohlin.

NAGATA, M. On the fourteenth problem of Hilbert.

NIJENHUIS, A. Geometric aspects of formal differential operations on tensor fields.

SAMUEL, P. Relations d'équivalence en géométrie algébrique.

SEGRE, B. On Galois geometries.

WANG, H. C. Some geometrical aspects of coset spaces of Lie groups.

SECTION VI

CHUNG, K. L. Continuous parameter Markov chains.

GNEDENKO, B. V. О предельных теоремах теории вероятностей. (Limit theorems of probability theory.)

RÉNYI, A. Probabilistic methods in number theory.

SAVAGE, L. J. Recent tendencies in the foundations of statistics.

SECTION VII

LEHMER, D. H. Discrete variable methods in numerical analysis.

RUTISHAUSER, H. Survey of experiments on the solution of linear systems.*

WIJNGAARDEN, A. VAN. Summation of series.*

SECTION VIII

HOFMANN, J. E. Über eine Euklid-Bearbeitung, die dem Albertus Magnus zugeschriebenwird.

KUREPA, G. Some principles of mathematical education.

SHORT COMMUNICATIONS

NOTE. *These communications are not printed in the Proceedings.*
See Preface, p. vii

SECTION I

BAUMSLAG, G., BOONE, W. G. and NEUMANN, B. H. Unsolvable problems about elements and subgroups of groups.

BETH, E. W. A proof of Craig's Lemma by means of semantic tableaux.

BODIOU, G. Implication stricte et détermination des lois de probabilité.

CHURCH, A. Consistency proof for recursive arithmetic.

COPELAND, A. H. Sr. Induction and observational check.

CURRY, H. B. The tectonic property for calculuses.

GANDY, R. O. Denumerable models in set theory.

GOODSTEIN, R. L. Primitively recursive algebraic and transcendental numbers.

HARROP, R. The finite model property and subsystems of classical propositional calculus.

KALMÁR, L. Problems concerning the conductivity states of multipoles.

KAPUANO, I. Solution du problème restreint du continu posé par Lusin.

KEMENY, J. G. Undecidable problems of elementary number theory.

KUREPA, G. On regressive functions.

LASKI, J. G. Naming real numbers.

LÖB, M. H. A simplified normal form for recursively enumerable predicates.

MOSTOWSKI, A. A semantical characterization of invariant and absolute formulas.

NOVIKOV, P. S. Some algorithmic questions of algebra.

PORTÉ, J. Derived rules and non-normal matrices in the propositional calculus.

RIGUET, J. Junction of multipoles and local structures.

ROBINSON, A. On a problem of Tarski's.

ROSE, A. The use of universal decision elements as flip-flops.

b

TARJAN, R. On instrumentation in logical problems.

TARSKI, A. An extension of the Löwenheim–Skolem theorem to a second-order logic.

SECTION II

ALBERT, A. A. and PAIGE, L. J. Jordan algebras with 3 generators.

ALDERSON, H. Logarithmetics and direct products of plain quasi-groups.

ALLING, N. L. On ordered divisible groups.

ALMEIDA-COSTA, A. Idéaux demi-premiers, μ- et π-systèmes d'idéaux.

AMEMIYA, I. and HALPERIN, I. Non-associative regular rings and von Neumann's co-ordinatization theorem.

AMITSUR, S. A. Cohomology groups of two fields and simple algebras.

ASPLUND, E. Inverses of non-singular matrices (a_{ij}) satisfying $a_{ij} = 0$ for $i + r < j$.

BATEMAN, P. T. On the non-vanishing of $L(1, \chi)$ for real residue-characters.

BAUMSLAG, G. A theory of free WR π-groups.

BEHRENS, E. A. Treue Moduln mit distributivem Untermodulverband.

BENNETT, A. A. A tensor algebra.

BERGER, R. Differentialmoduln zweistufiger diskreter Bewertungsringe in algebraischen Funktionenkörper.

BLACKBURN, N. A special class of p-groups.

BOHUN-CHUDYNIV, V. On semi-associative and semi-commutative algebras.

BOHUN-CHUDYNIV, V. Methods of constructing orthogonal and unitary rectangular matrices.

BRAUER, R. On groups of even order.

BRITTON, J. L. The word problem for groups.

BRUCK, R. H. Normal endomorphisms of a loop.

CARTER, R. W. On a class of finite soluble groups.

CHALK, J. H. H. Some ternary and quaternary Pellian equations.

CHÂTELET, F. Points rationnels sur une surface cubique.

CROISOT, R. Théorie noethérienne des anneaux, des demi-groupes et des modules dans le cas non commutatif.

DAVENPORT, H. Some recent results in analytic number theory.

DELANGE, H. Sur certaines fonctions arithmétiques.

DILWORTH, R. P. Generalized direct-union representation theorems for algebras.

DIVINSKY, N. J. On simple semi-radical and radical algebras.

DUPARC, H. J. A. On pseudo-primes of order 2.

DWINGER, P. Homomorphisms of α-complete Boolean algebras.

ERDŐS, P. and SCHERK, P. On a question of additive number theory.

ESTERMANN, T. The representation of a number as a sum of three squares.

FAITH, C. C. A note on algebraic division ring extensions.

FEIT, W. On a class of groups related to Frobenius groups.

FELSCHER, W. Jordan–Hölder theory and modularly ordered sets.

FINDLAY, G. D. and LAMBEK, J. On modules of quotients.

FORT, M. K. and HEDLUND, G. A. Minimal coverings of pairs by triples.

FRÖHLICH, A. Central extensions of finite algebraic number fields.

FUCHS, L. Multiplications on an abelian group.

GLODEN, A. Système trigrade normal et quadrilatère inscrit à côtés et diagonales en nombres entiers.

GODWIN, H. J. The determination of fields with a given subfield.

GOLDIE, A. W. The structure of rings with maximum condition.

GONSHOR, H. Algebras arising from genetics.

GRELL, H. Zur Theorie der Ringerweiterungen von Integritäts-bereichen mit eingeschränkter Minimalbedingung.

GROOT, J. DE. Automorphism groups of rings.

GRUENBERG, K. W. The Engel elements of a soluble group.

GUNDLACH, K. B. Dirichletreihen zur Hilbertschen Modulgruppe.

HALE, V. W. D. Triple algebras satisfying the entropic law.

HASELGROVE, C. B. Numerical integration and diophantine approximation.

HENRIKSEN, M. A representation for a class of Archimedean lattice-ordered algebras.

HERRMANN, O. Über einen Fundamentalbereich verallgemeinerter Hilbertscher Modulgruppen.

HERTZIG, D. On simple algebraic groups.

HUGHES, N. J. S. Jordan-Hölder-Schreier refinement theorem.

JENNER, W. E. Complex multiplication in a class of Riemann matrices.

JONES, B. W. Spinor-genera of quadratic forms.

KANOLD, H. J. Ein Satz über zahlentheoretische Funktionen.

KANTZ, G. Ein Hauptidealringkriterium für Ringe mit eindeutiger Primelementzerlegung.

KASCH, F. Zur Mahlerschen Vermutung über S-Zahlen.

KEMPERMAN, J. H. B. On small sumsets in a group.

KNESER, M. and TAMAGAWA, T. Another formulation and proof of Siegel's theorems on quadratic forms.

KOCHENDÖRFFER, R. A theorem on Sylow groups.

KRASNER, M. Prolongement analytique dans les corps valués complets.

KRUSKAL, J. B. Theory of well-partial-ordering.

KÜPISCH, H. Über 'generalized uniserial' Algebren.

LAMAN, G. On automorphisms of transformation groups of polynomial algebras.

LAMPRECHT, E. Zur Klassifikation der Differentiale algebraischer Funktionenkörper.

LEHMER, D. H. and LEHMER, E. Density of primes having various specified properties.

LESIEUR, L. Théorie noethérienne des anneaux, des demi-groupes et des modules dans le cas non commutatif.

LINT, J. H. VAN. Linear relations for modular forms.

LOONSTRA, F. Extensions of homomorphisms.

ŁOŚ, J. Sur l'existence de sous-groupes λ-purs.

LÖWIG, H. F. J. On families of algebras.

MARCZEWSKI, E. Un schème algébrique des notions d'indépendance.

MORAN, S. Associative operations on groups.

MORDELL, L. J. Integer solutions of two simultaneous quadratic equations.

NÉRON, A. L. Valeur asymptotique du nombre des points rationnels de hauteur bornée sur une courbe elliptique.

NEUBÜSER, J. Über homogene Gruppen.

OLIVEIRA, J. T. DE. Homomorphisms of modular systems.

OSBORN, J. M. Loops with the weak inverse property.

PATTERSON, E. M. Automorphisms and derivations of degree 2 or 3.

PEREMANS, W. Holomorphs and complete groups.

PICCARD, S. Deux problèmes de la théorie générale des groupes.

POST, K. A. Rank functions on semigroups.

POTTER, H. S. A. Derivatives of matrix functions.

RADO, R. Graphs without triangles, having any transfinite chromatic number.

RAŠAJSKY, N. B. Sur le système d'équations en involution de Darboux-Lie.

READ, R. C. Enumeration of superposed graphs.

RÉDEI, L. Die zweistufig nichtkommutativen endlichen Ringe.

RIBENBOIM, P. Inversion de séries de Dirichlet formelles.

RIBENBOIM, P. Un théorème d'existence pour le prolongement des valuations de Krull.

ROSSER, J. B. Real roots of real Dirichlet L-series, II.

SAWYER, D. B. A note on asymmetric approximation.

SCHMIDT, W. On the Jacobi algorithm.

SCHNEIDER, H. Matrix algebras relatively bounded in norm.

SCHWARZ, S. On dual semigroups.

SHAFAREVICH, S. I. Группа эллиптических кривых. (A group of elliptic curves.)

SHANKS, M. E. Remarks on field topologies.

SHEPHARD, G. C. Convex bodies.

SIERPIŃSKI, W. Quelques hypothèses sur les nombres premiers.

SINGER, J. A class of groups associated with Latin squares.

SUPRUNENKO, D. A. Локально нильпотентные подгруппы вещественной полной линейной группы. (Locally nilpotent subgroups of the real full linear group.)

TAMARI, D. Embedding of semi-groups in groupoids of operators and their extension to groups.

THOMA, E. Die unitären Darstellungen der Bewegungsgruppe des R^2.

TURÁN-SÓS, V. On diophantine approximation.

VINOGRADOV, I. M. Некоторое усовершенствование метода оценки тригонометрических сумм. (Some refinements of the method of estimating trigonometric sums.)

WATSON, G. L. Integral quadratic forms.

ZASSENHAUS, H. J. Invariant bilinear forms on Lie algebras.

ZULAUF, A. A series depending on the zeros of Riemann's ζ-function.

SECTION III

AGMON, S. On coercive integro-differential forms.

AKUTOWICZ, E. J. On extrapolating a positive definite function from a finite interval.

ALEXIEWICZ, A. Two-norm spaces.

ALFSEN, E. M. Non-linear integration.

ALTEN, H.-W. Über die Nullstellen der Legendreschen Funktionen erster und zweiter Art.

ARSCOTT, F. M. The ellipsoidal wave equation—a three-parameter eigenvalue problem.

ARSOVE, M. G. Some phenomena encountered in the study of Pincherle bases.

BAKER, I. N. Fix-points and iterates of entire functions.

BARRY, P. D. The minimum modulus of certain integral functions of zero order.

BASS, J. Fonctions d'autocorrélation et sommes trigonométriques.

BAUER, H. Über die Existenz von Massen mit vorgegebenem Bild.

BERG, L. Taubersche Sätze für Potenzreihen.

BIERNACKI, M. Sur le second théorème de la moyenne du calcul intégral.

BONSALL, F. F. The iteration of operators mapping a positive cone into itself.

BORWEIN, D. Multiplication of $(C, -\kappa)$-summable series.

BOSANQUET, L. S. The summability of Laplace–Stieltjes integrals.

BRELOT, M. Sur la convergence en théorie du potentiel.

BREMERMANN, H. J. Characterization of the Šilov boundary of arbitrary domains in C^n and applications.

BREMERMANN, H. J. Über die Cartan–Thullensche Theorie in komplexen Banach-Räumen unendlicher Dimension.

BROUSSE, P. Dérivée normale de l'intégrale de Poisson généralisée.

BROWDER, F. Regularity of solutions of elliptic boundary-value problems.

BROWN, L., SHIELDS, A. and ZELLER, K. Representation of zero as a sum of exponentials.

CALABI, E. Elliptic solutions of the equation prescribing the Hessian.

CAMERON, R. H. A complex sequential Wiener integral or Feynman integral.

CAMPBELL, R. Détermination effective de toutes les moyennes de Cesàro d'ordre entier pour les séries de polynômes orthogonaux comprenant ceux de Laguerre et de Hermite.

CASTRO, A. DE. On the existence of periodic solutions of differential equations.

CATTABRIGA, L. A boundary value problem for non-linear fourth order parabolic equations.

CESARI, L. Recent results in surface area theory.

CESARI, L. Existence of periodic solutions of weakly nonlinear Lipschitzian differential systems.

CILIBERTO, C. Problemi di Mayer–Lagrange.

CIMMINO, G. On some types of boundary conditions for solutions of partial differential equations.

CODDINGTON, E. A. Unitary equivalence of some self-adjoint dilations of ordinary differential operators.

COLLATZ, L. Tschebyscheffsche Approximation mit gebrochenen Funktionen.

COLLINGWOOD, E. F. On the cluster sets of arbitrary functions.

COLMEZ, J. Spectres continus d'un opérateur auto-adjoint.

CONTI, R. Differential systems with linear boundary conditions.

CORDES, H. O. On local boundary conditions and maximal operators.

DENJOY, A. Approximation des nombres réels par ceux d'un corps fourni par un groupe fuchsien.

DOLBEAULT, P. Formes méromorphes et courants associés.

DOLINSKY, R. Neuere Untersuchungen über Nullstellen der Abschnitte von Potenzreihen.

DUFF, G. F. D. A mixed problem for linear differential equations.

ENDL, K. Über Verfahren vom Hausdorffschen Typus und ein Momentenproblem.

EVANS, W. B. Convergence properties of infinite exponentials.

EVERITT, W. N. Integrable-square solutions of ordinary self-adjoint differential equations.

FELL, J. M. G. Dual spaces of C^*-algebras.

FIORENZA, R. Sul problema della derivata obliqua per le equazioni di tipo ellittico.

FISHEL, B. Boundary values of distributions.

FLEMING, W. H. Generalized surfaces.

FLETT, T. M. Some generalizations of Tauber's theorem.

FØLNER, E. Generalized almost periodic functions in groups.

FOX, C. Fractional integration and chain transforms.

FRANK, E. Continued fraction expansions for the ratios of certain generalized hypergeometric functions.

FREUD, G. On smooth functions.

GAGLIARDO, E. Propriétés de certaines classes de fonctions de n variables.

GEHRING, F. W. Harmonic functions and Tauberian theorems.

GHAFFARI, A. On the behaviour in the large of the integral curves of a non-linear differential equation.

GHIKA, A. Transformations de certains espaces vectoriels métrisables et espaces de Banach en espaces hilbertiens.

GLODEN, R. F. Séries de Fourier et polynômes.

HAJÓS, G. On the definition of surface area in elementary cases.

HÄLLSTRÖM, G. J. af. Zur Wertverteilung pseudomeromorpher Funktionen.

HAMMER, P. C. Basic theory of the oscillation of functions.

HARTMAN, S. Beitrag zur Theorie des Massringes mit Faltung.

HAYES, W. D. The quasiconservative oscillator.

HAYMAN, W. K. Picard values of functions and their derivatives.

HILLE, E. Green's transforms and singular boundary value problems.

HIRSCHMAN, I. I. On multiplier transforms.

HORVÁTH, J. Basic sets of polynomial solutions for partial differential equations.

ILIEFF, L. Einige Sätze über die analytische Nichtforsetzbarkeit der Potenzreihen.

JURKAT, W. B. On term by term differentiation almost everywhere.

KAZARINOFF, N. D. Asymptotic theory of second-order differential equations with two simple turning points.

KENNEDY, P. B. On the growth of multivalent functions.

KEOGH, F. R. The length of level curves in bounded star-shaped domains.

KNOBLOCH, H. W. Connections between formal and convergent solutions of linear differential equations.

KÖNIG, H. Eine Klasse von linearen Transformationen.

KÖTHE, G. M. Semireflexive locally convex spaces.

KOOSIS, P. New proof of a theorem of Riesz.

KRABBE, G. L. A family of non-spectral operators which satisfy the spectral theorem.

KUIPERS, L. Generalized Legendre associated functions.

KULTZE, R. Zur Theorie Fredholmscher Endomorphismen in nuklearen Räumen.

KUREPA, S. A cosine functional equation in linear vector spaces.

KURZWEIL, J. Continuous dependence on a parameter and generalized ordinary differential equations.

LAMBE, C. G. Approximate values of Lamé functions.

LAVRENTIEV, M. Jr. Cauchy problem for elliptic equations.

LAX, P. D. Translation-invariant spaces.

LEHRER, Y. On Appell's functions.

LEHTO, O. A generalization of Picard's theorem.

LEICHTWEISS, K. Selbstadjungierte lineare normierte Räume.

LEJA, F. Sur les moyennes arithmétiques, géométriques et harmoniques des distances mutuelles des points d'un ensemble.

LEKKERKERKER, C. G. On the convergence of infinite sums and integrals.

LELONG, P. Sur les propriétés métriques locales d'un ensemble analytique complexe.

LEMPERT, K. Perturbation theory for linear operators in Hilbert space.

LEWIN, L. Some unsolved problems concerning polylogarithms and associated functions.

LOHWATER, A. J. The cluster set of functions defined in a circle.

ŁOJASIEWICZ, S. Sur l'identification des fonctions avec des distributions.

ŁOJASIEWICZ, S. Division d'une distribution par une fonction analytique des variables réelles.

LOMBARDI, L. The gradient-method in the calculus of variations.

LORENTZ, G. G. and ZELLER, K. Rearrangements of series and analytic sets.

LUXEMBURG, W. On the theory of function norms.

MACINTYRE, A. J. On $\cos x \cdot \cos (x/2) \cdot \cos (x/3)$ ....

MAIER, W. Studien zu Riemanns Zeta-funktion.

MALLIAVIN, P. Théorème de Stone–Weierstrass–Bernstein sur un espace localement compact.

MARCHENKO, V. A. Разложение по собственным функциям несамо-сопряженных сингулярных дифференциальных операторов. (The expansion in proper functions of non-self-adjoint singular differential operators.)

MARCUS, S. Sur la mesurabilité des ensembles projectifs.

MAUDE, R. Hyper-Dirichlet series as functions of two variables.

MEHDI, M. R. Summability factors for generalized absolute summability.

MEULENBELD, B. Properties of the generalized Legendre associated functions.

MIKOLÁS, M. Differentiation and integration of complex order in the theory of trigonometrical series and generalized zeta-functions.

MIRANDA, C. Il teorema del massimo modulo per le equazioni lineari ellittiche.

MITCHELL, J. M. An integral theorem for a class of harmonic vectors of three variables.

MYRBERG, L. J. Eine Bemerkung zum Picardschen Satz.

NASH, J. Parabolic equations.

NASTOLD, H. J. Meromorphe Schnitte komplexanalytischer Vektor-raumbündel und Anwendungen.

NEVILLE, E. H. The standard elliptic integrals of the third kind.

NICKEL, K. Fehlerabschätzungen bei Systemen parabolischer Differentialgleichungen.

NICOLESCU, M. Sur quelques problèmes concernant les opérateurs linéaires dans une algèbre normée.

NIRENBERG, L. Inequalities for derivatives.

NITSCHE, J. C. C. Solutions of non-linear elliptic equations.

NOBLE, M. E. Some convergence criteria for Fourier series.

OBRECHKOFF, N. Sur quelques questions de la sommation absolue des séries divergentes.

ORLICZ, W. Über Funktionen von verallgemeinerter beschränkter Variation.

PEIXOTO, M. M. On structural stability.

PESCHL, E. F. Über eine Verallgemeinerung des Schwarzschen Lemmas auf Lösungen von gewissen elliptischen Differential-gleichungen.

PETERSEN, G. M. Matrices and norms.

PETROVSKY, I. G. and LANDIS, E. M. О числе предельных циклов уравнения с рациональной правой частью. (On the number of limiting cycles of an equation with a rational right-hand side.)

PEYOVITCH, T. Propriétés asymptotiques d'une équation différentielle.

PLESSIS, N. DU. Some honest operators.

POPOV, S. B. Sur l'équation de Bessel généralisée.

PRASAD, B. N. The second theorem of consistency for absolute Riesz summability.

RANKIN, R. A. The Schwarzian derivative and uniformization.

RÉNYI, C. On multiple zeros of derivatives of periodic entire functions.

REVUZ, A. Mesures sur les espaces topologiques ordonnés.

RIVLIN, T. J. and SHAPIRO, H. S. Unique approximation by polynomials in several real variables.

RIZZONELLI, P. On integro-differential equations of 'Faltung' type.

ROSSUM, H. VAN. On the Padé table of functions related to Bessel functions.

RUBEL, L. A. Minorization and Fourier transforms.

RUDIN, W. Lacunary trigonometric series.

RUNCK, P. O. Über das Konvergenzverhalten der Interpolationsformeln von Lagrange und Hermite mit äquidistanten Knoten.

SAN JUAN LLOSA, R. Sur la caractérisation de la somme des séries divergentes moyennant les lois formelles du calcul.

SARD, A. Images of critical sets.

SCHÄFFER, J. J. Function spaces with translations.

SCHOTTLAENDER, S. Reihenentwicklung nach Laguerreschen Polynomen.

SCHRÖDER, J. Funktionalanalytische Behandlung von Iterationsverfahren.

SEBASTIÃO E SILVA, J. Deux généralisations successives de la notion de distribution.

SEGAL, I. E. Group-invariant analysis in function-space.

SEIBERT, P. Boundary problems of Riemann covering surfaces.

SHAGINYAN, A. L. Некоторые задачи теории приближений в комплексной области. (Some problems in the theory of approximations in the complex domain.)

SOBOLEV, S. L. Теоремы вложения для абстрактных функций множеств. (Embedding theorems for abstract functions of sets.)

SONNENSCHEIN, J. Une classe de procédés de sommation.

SYER, F. J. D. Complete explicit solution of a very general equation.

SZEKERES, G. Regular growth of real functions.

TAKAHASHI, R. Les fonctions sphériques dans les groupes de Lorentz généralisés.

TAMMI, O. On schlicht domains bounded by a circle.

TAYLOR, S. J. On sharpening the Lebesgue density theorem.

TEODORESCU, N. V. Dérivées spatiales et opérateurs différentiels linéaires généralisés.

THIELMAN, H. P. Functions whose second differences converge to zero.

THRON, W. J. Convergence regions for continued fractions.

TILLMANN, H. G. Runge's approximation theorem and the theory of analytic functionals.

TORNEHAVE, H. On multiply almost periodic entire functions.

TREJO, C. A. On Zygmund's weak type of quasi-linear operations.

TRJITZINSKY, W. J. Metric theory in spaces having a measure.

ULUÇAY, C. On a general principle in the theory of analytic functions.

URSELL, H. D. Inequalities between sums of powers.

VALA, K. E. Sur les produits tensoriels des espaces de Hilbert.

VERMES, P. Summability of power series at unbounded sets of isolated points.

VINCENSINI, P. Sur une dégénérescence géométrique.

WALTER, W. Über ganze Lösungen der Differentialgleichung $\Delta^p(u) = f(u)$.

WAŻEWSKI, T. Sur l'allure asymptotique des intégrales des équations différentielles ordinaires.

WEISS, G. Applications of a theorem of Marcinkiewicz.

WERMER, J. Polynomial approximation in C^n.

WILLIAMSON, J. H. On constructions of Wiener–Pitt and Šreĭder.

WOLF, F. The essential spectrum of a partial elliptic differential operator.

YOSIDA, K. On the differentiability of a semi-group of operators.

YOUNG, L. C. Partial area.

ZLÁMÁL, M. Über das gemischte Problem für die Gleichung $\epsilon\alpha(t)u_{tt} + \beta(t)u_t = a(x)u_{xx}$ bei kleinem ϵ.

SECTION IV

ADAMS, J. F. On the non-existence of elements of Hopf invariant 1.

AIGNER, A. Die Häufigkeitsverteilung gewisser Typen von endlichen Graphen.

ALEXANDROV, P. S. Theory of bicompact spaces.

ANDERSON, R. D. Superplexes.

AQUARO, G. Uniform structures and compactifications.

AUBERT, K. E. On the maximal ideal method in topological algebra.

BARCUS, W. D. On the suspension of a loop space.

BASS, R. W. Global topological structure of arbitrary dynamical systems.

BAUER, F. W. Tangentialstrukturen.

BING, R. H. Some consequences of the approximation theorem for surfaces.

CARTWRIGHT, M. L. Discontinuous infinite minimal sets in the plane.

COPELAND, A. H. Jr, and BROWN, E. H. The groups of homotopy classes of maps between certain polyhedra.

CSÁSZÁR, Á. Structures topologiques générales.

CURTIS, M. L. and FORT, M. K. Singular homology of one-dimensional spaces.

DEDECKER, P. Cohomologie à coefficients non-abéliens.

DOLD, A. Die geometrische Realisierung eines schiefen kartesischen Produktes.

HARROLD, O. G. The Schoenflies problem in three-space.

HIRSCH, G. Homology operations.

HIRSCH, M. W. Immersions in manifolds.

ISBELL, J. R. Embeddings and inverse limits.

KAN, D. M. An axiomatization of the homotopy groups.

KATĚTOV, M. On the co-dimension of a set in a topological space.

KELDYSH, L. V. Открытые отображения компактов.
(Open mappings of compacta.)

KERVAIRE, M. A. Non-parallelizability of the n-sphere for $n > 7$.

KURATOWSKI, K. Some problems connected with cohomotopy groups
of open subsets of Euclidean space.

MCDOWELL, R. H. and GROOT, J. DE. Extension of homeomorphisms
on metric spaces.

MORREY, C. B. The analytic embedding of analytic manifolds and
related topics.

PUPPE, D. Ein Satz über semi-simpliziale Monoidkomplexe.

REMMERT, R. Kohärente Bildgarben.

RUDIN, M. E. Triangulating a tetrahedron.

SALZMANN, H. Topologische projektive Ebenen.

SMIRNOV, YU. M. О бесконечномерных пространствах.
(Infinite-dimensional spaces.)

SPANIER, E. H. Infinite symmetric products and function spaces.

STEIN, K. Konvergenzsätze über analytische Mengen.

SWAN, R. G. The homology of cyclic products.

WHYBURN, G. T. Uniform convergence of mappings.

WILLE, R. J. Rigid continua.

ZEEMAN, E. C. The relation between the Leray and Serre spectral
sequences of a continuous map.

SECTION V

ABELLANAS, P. Cohomology on an algebraic surface.

ACZÉL, J. Anwendung der Methode der Funktionalgleichungen in
der Theorie der geometrischen Objekte.

ANDRÉ, J. On a characterization of finite translation-planes.

BACHMANN, F. Teilebenen hyperbolischer projektivmetrischer Ebenen.

BEHARI, R. Some properties and applications of Eisenhart's generalized Riemann space.

BENZ, W. Über Kreisgeometrien mit eingeschränkter Möbiusgruppe.

BILINSKI, S. Über die Ordnungszahl der Klassen Eulerscher Polyeder.

BLANUŠA, D. Skew regular polyhedrons and polytopes.

BLUM, R. On Riemann spaces of conform class 1.

BOER, J. H. DE. Remark on the multiplicities of connected components in the case of an algebraic correspondence.

BURAU, W. Neue geometrische Betrachtungsweisen in der projektiven Invariantentheorie.

CALAPSO, R. Sulle coppie di congruenze W aderenti ad una stessa superficie.

CANTONI, L. Sulle trasformazioni puntuali tra spazi ordinari che posseggono un'unica congruenza di curve caratteristiche.

CATTANEO, G. I. Basic vector fields, affine transformations and applications.

DEMBOWSKI, P. Free projective planes and quadrangle transitivity.

DOLBEAULT-LEMOINE, SIMONE. Réductibilité de variétés plongées.

EELLS, J. On characteristic classes and curvature.

FERNÁNDEZ, G. Differential properties of the projective plane as a surface of Euclidean four-dimensional space.

GAETA, F. Sur certains algorithmes tensoriels qui remplacent l'élimination.

GAUTHIER, L. The geometry of the curves of genus 3 over a field of characteristic 3.

GODEAUX, L. Construction de surfaces projectivement canoniques.

GRAY, J. W. Contact structures.

GROSJEAN, P. V. The 'semi-metric geometry' and the unified field theory.

GUGGENHEIMER, H. Axioms and postulates in the foundations of geometry.

HACHTROUDI, M. Espaces de Riemann harmoniquement isotropes.

HAEFLIGER, A. Singularités des sections des espaces fibrés différentiables.

HAIMOVICI, A. Sur certains invariants dans un espace à connexion affine.

HAIMOVICI, M. Sur la décomposition des systèmes différentiels extérieurs.

HOHENBERG, F. Komplexe Erweiterung der gewöhnlichen Schraublinie.

HOPF, H. and KATSURADA, Y. Some congruence theorems for closed hypersurfaces in Riemann spaces.

KARZEL, H. Verallgemeinerte absolute Geometrien.

KAWAGUCHI, A. On connections in higher-order spaces.

KLEE, V. Some characterizations of convex polyhedra.

KLINGENBERG, W. Zur Struktur kompakter Riemannscher Mannigfaltigkeiten.

LAUGWITZ, D. On a conjecture of H. Weyl concerning the 'problem of space'.

LICHNEROWICZ, A. Sur les transformations analytiques des variétés kählériennes compactes.

LINGENBERG, R. Über Gruppen, in denen der allgemeine Satz von den drei Spiegelungen gilt.

MARCHIONNA, E. Sur les surfaces complémentaires arithmétiquement normales.

MARCHIONNA TIBILETTI, C. Un complément à un théorème d'existence des fonctions algébriques.

MARKUS, L. and CALABI, E. Relativistic space forms.

MATHEEV, A. Géométrie du champ de 'vecteurs' dans un espace cayleyen.

MEDEK, V. Lineare Systeme von projektiven Verwandschaften.

MOTZKIN, T. S. Face dimensions of convex sets.

MURRE, J. P. Quadratic and birational transformations.

NASH, J. The imbedding problem for Riemannian manifolds.

PAPY, G. L. A. Espaces différentiels.

PINL, M. Minimal surfaces with constant Gaussian curvature.

RIZZA, G. B. Holomorphic deviation for the $2p$-dimensional sections of complex analytic manifolds.

SOMMER, F. Probleme der komplexen Differentialgeometrie.

STEIN, E. Generalized Segre varieties.

TITS, J. Sur les géométries définies par des axiomes d'incidence.

VARGA, O. Über verallgemeinerte nichteuklidische Geometrien.

VEN, A. J. H. M. VAN DE. On the normal bundle of an algebraic variety in a projective space.

VIDAL ABASCAL, E. A generalization of integral invariants.

VINCENSINI, P. Sur une transformation de l'ensemble des corps convexes de E_n.

VOSS, K. Flachpunkte auf negativ gekrümmten Flächen.

WILLMORE, T. J. A generalization of Walker's operation of torsional derivation.

YANO, K. Harmonic and Killing vector fields in compact orientable Riemannian spaces with boundary.

ZOBEL, A. Condition calculus on an open variety.

SECTION VI

ANDERSEN, E. S. Fluctuations of sums of random variables.

BARANKIN, E. W. and KATZ, M. Jr. Sufficient statistics of minimal dimension.

BARTLETT, M. S. Quasi-stationary behaviour of some non-linear biological models.

BHARUCHA-REID, A. T. and RUBIN, H. Generating functions and the operator theory of Markov processes with a denumerable state space.

BIRNBAUM, Z. W. Inequalities for stochastic processes with known covariance function.

BOSE, R. C. and KUEBLER, R. R. Construction of error detecting and error correcting codes for the symmetric binary channel.

CHARNES, A. and COOPER, W. W. Extremal principles for traffic flows in a network.

CHERNOFF, H. Sequential design of experiments.

COHN, R. M. Randomness and overlap of order statistics.

COX, D. R. A special renewal problem.

DOSS, S. On a sufficient statistic for a family of Laplacian variables taking values in an abelian group.

DUGUÉ, D. Intégration aléatoire.

EISENHART, C. A test for extreme residuals.

GARREAU, G. A. The influence of electronic computers on applied statistics.

GILLIS, J. The Ehrenfest distribution.

GYIRES, B. On generalized infinite stochastic Toeplitz matrices.

HARRIS, T. E. A stationary measure for the multiplicative process.

HENNEQUIN, P. L. Processus en cascade à n dimensions et problème de moments.

HOTELLING, H. A multiple-stage experiment for finding the best variety.

JACOBS, K. n-person Markov processes.

KAMPÉ DE FÉRIET, J. Intégrales aléatoires d'équations aux dérivées partielles et problème abstrait de Cauchy.

KATZ, L. Probability distribution of components of a random mapping.

KENDALL, D. G. The rate of convergence problem for denumerable Markov chains and processes.

KRICKEBERG, K. Necessary conditions in the theory of convergence of martingales.

KRIEZIS, P. On the unification of theories of probability.

LUKACS, E. Extensions of a theorem of Marcinkiewicz.

MIHOC, G. Fonctions d'estimation pour les paramètres d'une chaîne à liaisons complètes.

MOURIER, E. Problèmes d'estimation et tests d'hypothèses pour des fonctions aléatoires.

MUILWIJK, J. A class of growth models.

NARAYANA, T. A method of estimating the ED 50 for quantal data.

NOETHER, G. E. On a class of paired comparison procedures.

OLIVEIRA, J. T. DE. Estimation by the minimum discrepancy method.

PARZEN, E. An approach to time series analysis.

PRIESTLEY, M. B. On the analysis of stationary processes with mixed spectra.

PROKHOROV, Y. V. Усиленная устойчивость сумм и неограниченно-делимые распределения. (Strong stability of sums and infinitely divisible distributions.)

RAMAKRISHNAN, A. Some limiting stochastic operations.

RANKIN, B. Computable probability spaces.

REICH, E. On the mixed-type Markov process of Takács.

REIERSØL, O. A contribution to the algebra of analysis of variance.

ROY, S. N. 'Normal' multivariate confidence bands.

SARYMSAKOV, T. A. О некоторых новых результатах по цепям Маркова. (On some new results on Markov chains.)

SERAZHDINOV, S. Кн. Локальная предельная теорема для цепей Маркова с непрерывным временем. (A local limit theorem for Markov chains with continuous time.)

SEVASTIANOV, B. A. Ветвящиеся случайные процессы для частиц, диффундирующих в ограниченной области с поглощающими границами. (Branching random processes for particles diffusing in a bounded region with an absorbing boundary.)

SMITH, C. A. B. Transfinite ordinals and two-person games.

STATULEVICHUS, V. A. Предельные теоремы для неоднородных цепей Маркова. (Limit theorems for non-homogeneous Markov chains.)

SUPPES, P. Some mathematical and statistical problems which arise in learning-theory.

SYSKI, R. A note on congestion theory.

TAKÁCS, L. On a sojourn time problem.

TEICHER, H. On the mixture of distributions.

VILLEGAS, C. On the estimation of the coefficients of a linear functional relationship.

VINCZE, L. On some joint limiting distributions in the theory of order statistics.

ZAREMBA, S. K. On the choice of parameter values in estimates of the spectral density.

ZOUTENDIJK, G. Markov runs in an arbitrary set.

SECTION VII

ABLOW, C. M. and PERRY, C. L. Iterative solutions of the Dirichlet problem for $\Delta u = u^2$.

ABRAMOWITZ, M. Heat transfer in laminar flow.

ACKROYD, R. T. and BALL, J. M. Application of spherical and cylindrical means to reactor theory.

ACKROYD, R. T. and PERKS, M. A. Accuracy and scope of variational and difference solutions of the diffusion equations of reactor theory.

ADLER, G. Des types nouveaux des problèmes aux limites de l'équation de la chaleur.

AGOSTINELLI, C. Sur les tourbillons sphériques en magnétohydrodynamique.

ALBASINY, E. L. The numerical solution of nonlinear heat conduction problems on digital computers.

ALT, F. L. Heat flow in the presence of moisture.

BARTLETT, C. C. and NOBLE, B. The computation of eigenvalues for certain mixed boundary value problems.

BEL, L. Sur les discontinuités des dérivées secondes des potentiels de gravitation.

BENDAT, J. S. Optimum time variable filters.

BIELECKI, A. Sur une généralisation de la notion de stabilité.

BITZADZE, A. V. Об общей смешанной задаче для уравнений смешанного типя. (On the general mixed problem for equations of mixed type.)

BOGOLYUBOV, N. N. and VLADIMIROV, V. S. Об аналитическом продолжении обобщенных функций. (On analytic continuation of generalized functions.)

BONNOR, W. B. Gravitational radiation.

BOUIX, M. Application des distributions aux problèmes des limites et des sources dans les équations aux dérivées partielles.

BRADISTILOV, G. Relation entre le nombre des mouvements périodiques et asymptotiques autour de la position d'équilibre de l'n-pendule multiple situé dans un plan.

BRESSAU, A. Sur les sollicitations lagrangiennes dépendant même des vélocités et provenant d'un potentiel géneralisé.

BRUN, V. An application of a 'carpenter's curve' to Simpson formulas.

CADE, R. Conditions of equilibrium for an electric double layer in fluid.

CAPRIOLI, L. Sulle guide d'onda con pareti assorbenti.

CAPRIZ, G. Successive approximations in problems of elastic stability.

CARR, J. W. III. Application of the Weissinger truncation.

CARTER, M. D. and PACK, D. C. A theorem on compressible flows with circular sector hodographs.

CATTANEO, C. General relativity, relative mass, momentum, energy and gravitational force in a general system of reference.

CHAMBERS, L. G. Propagation of electromagnetic waves in a waveguide containing a moving medium.

CLENSHAW, C. W. The numerical calculation of the Chebyshev expansions of mathematical functions.

DAS, S. C. Streamlines in plane Bingham body under compression.

DIAZ, J. B. and PAYNE, L. E. Mean value theorems in the theory of elasticity.

DOUGLAS, A. S. On the solution of elliptic partial differential equations by difference methods.

DRESSLER, R. F. Unsteady non-linear waves in sloping channels.

DUNGEY, J. W. Propagation of a pulse across a magnetic field in a collision-free plasma.

EICKER, F. Über eine Rand- und Mittelwertaufgabe der Poissonschen Differentialgleichung.

FAIRTHORNE, R. A. Canonical representation of scripts and vocabularies.

FAULKNER, F. D. Variational problems of rocket trajectories.

FAURE, R. Synchronisation des systèmes mécaniques. Divers types de solutions périodiques.

GARIBALDI, A. C. On the variational principles of Gauss and others.

GILBERT, C. Gravitation and the principle of stationary action.

GILLES, D. C. The treatment of singularities in the numerical solution of partial differential equations.

GOODWIN, E. T. A problem of absorption with chemical reaction.

GROSSWALD, E. Transformations used in the mechanical integration of systems of differential equations.

HELLIWELL, J. B. The flow past a closed body in a high subsonic stream.

HENRICI, P. On the speed of convergence of cyclic and quasicyclic Jacobi methods for computing eigenvalues of hermitian matrices.

HERRMANN, H. Optimales Programmieren für elektronische Analogie-Rechenmaschinen.

HERSCH, J. Un aspect des équations aux différences.

HOCHSTRASSER, U. W. Finding the smallest eigenvalues of matrices.

HOLT, M. Supersonic flow past a double wedge of finite span.

HYMAN, M. A. Zeros of general functions.

HYMAN, M. A. Non-iterative numerical solution of boundary-value problems.

JACOB, C. Propriétés de la fonction de Green modifiée.

KRUSKAL, M. D. The gyration of a charged particle.

KUNTZMANN, J. Evaluation d'erreur dans les problèmes différentiels de conditions initiales.

KYNCH, G. J. Viscosity of suspensions.

LADYZHENSKAYA, O. A. Исследование стационарных и неста-ционарных течений вязкой жидкости. (Investigation of stationary and non-stationary flow of a viscous liquid.)

LADYZHENSKAYA, O. A. and FADDEEV, D. D. К теории возму-щений непрерывного спектра. (Perturbation theory of a con-tinuous spectrum.)

LANGER, R. E. Solutions of a fourth-order differential equation of a hydrodynamic type.

LESLIE, D. C. M. Lifetime of an earth satellite.

LIPTON, S. The evaluation of the bi-variate normal integral over a particular non-rectangular region.

LOVASS-NAGY, V. Generalization of the method of symmetrical components and its application to mathematical investigation of polyphase electrical systems.

LUDFORD, G. S. S. The magneto-hydrodynamic shock.

MACKIE, A. G. Some exact solutions of Chaplygin's equation which represent flow past bodies.

MAGNUS, W. The discriminant of Hill's equation.

MAYNE, A. J. Rapidly converging iterative processes for general types of equation.

MERSON, R. H. Applications of the method of differential correction to the analysis of earth satellite observations.

MEYER, R. E. Non-uniform shock propagation in a stratified gas.

MICHEL, J. G. L. Use of feed-back set-ups on differential analysers.

MISHCHENKO, E. F. Асимптотическая теория релаксационных колебаний. (Asymptotic theory of relaxation oscillations.)

MITCHELL, A. R. Some inherent difficulties in the solution of two-point boundary problems by finite difference methods.

MITROPOLSKY, YU. A. Некоторые вопросы асимптотического интегрирования нелинейных дифференциальных уравнений. (Some questions on the asymptotic integration of non-linear differential equations.)

MUIR, W. W. A step-by-step method for the solution of boundary value problems.

NARDINI, R. Sui fronti d'onda nella magneto-elasticità.

NORTON, H. J. The theory and practice of the iterative method for determining eigenvalues.

O'BEIRNE, T. H. Automatic subtabulation to tenths with certain desk machines.

OLEYNIK, O. A. О задаче Коши для квазилинейных гиперболических уравнений. (On Cauchy's problem for quasi-linear hyperbolic equations.)

ORLOFF, C. Application pratique des accords numériques (spectres mathématiques) à l'analyse numérique.

PAYNE, L. E. and WEINBERGER, H. F. Approximation in exterior mixed boundary value problems.

PAYNE, L. E. and WEINBERGER, H. F. A Rellich identity for second-order systems.

PEARCE, R. P. Generation, convection and dissipation of vorticity in the atmosphere.

PEKERIS, C. L. Approximate solutions of the Schrödinger wave equation for the helium atom.

PEKERIS, C. L. and ALTERMAN, Z. A method of solving the non-linear differential equations of atmospheric tides.

PEKERIS, C. L. and RABINOWITZ, P. Numerical evaluation of the subdominant eigenvalues and eigenvectors of the matrix equation $Ax + \lambda Bx = 0$.

PHAM, M. Q. Le principe de Fermat en relativité générale.

PHAM, T. H. Sur la méthode des singularités en relativité générale.

PIGNEDOLI, A. New researches on the motion of relativistic energy particles in electric and magnetic superposed fields.

PINI DE SOCIO, M. Propagazione di onde non sinusoidali in un gas ionizzato soggetto a un campo magnetico.

POLLAK, H. O. and GILBERT, E. N. Amplitude distribution of impulse noise.

PROUSE, J. On the solution of the mixed problem for non-linear hyperbolic partial differential equations by finite differences.

RADOJČIĆ, M. An axiomatic deduction of the special theory of relativity.

RADOK, J. R. M. Method of functional extrapolation for the numerical integration of ordinary differential equations.

REICHARDT, H. Ausstrahlungsbedingungen für die Wellengleichung.

xl SCIENTIFIC PROGRAMME

REIERSØL, O. Methods of deriving recurrence formulas.

REISSIG, R. Ljapunowsche Funktionen als Kriterium für das D-Verhalten dynamischer Systeme.

REY PASTOR, M. J. Calcul d'un type de réacteur hétérogène.

REZA, F. Generalized linear systems.

ROBINSON, A. and KIRKBY, S. Flutter derivatives of a wing-tailplane combination.

ROBINSON, P. D. Wave-functions for the hydrogen atom in spheroidal co-ordinates.

RODRIGUEZ, A. E. Approximate distribution functions and a linearized form of the Born–Green equation for monatomic fluids.

RÓZSA, P. On some new methods of linear algebra and their applications.

SAIBEL, E. Heat conduction in thin films.

SALZER, H. E. Optimal points for numerical differentiation.

SANDERS, J. Elasticity, hydrodynamics and function-theory.

SCHIELDROP, E. B. Some remarks concerning the general principle of least action.

SELMER, E. S. Numerical integration by non-equidistant ordinates.

SETH, B. R. Poro-plastic deformation.

SHÜ, S. S. Applications of Jacobi's expansion and its generalization to boundary-value problems of ordinary differential equations.

SITNIKOV, K. A. Инварианты однородной и изотропной турбулентности в вязкой сжимаемой жидкости. (Invariants of homogeneous and isotropic turbulence in a viscous compressible liquid.)

SOUTHARD, T. H. Approximation and tabulation of Weierstrass elliptic functions.

STALLMANN, F. Über die numerische Behandlung der konformen Abbildung durch das Iterationsverfahren von Schwarz und Neumann.

STEWARD, G. C. The cardinal points in plane kinematics.

STOKER, J. J. Elastic stability of thick plates and shells.

STUART, J. T. On the application of perturbation theory to non-linear problems of hydrodynamic stability.

SYNGE, J. L. Elasticity in general relativity.

SZABÓ, J. J. Application de la théorie matricielle au calcul des ponts à poutres multiples.

SZEBEHELY, V. G. A generalization of the problem of two centres of gravitation.

TAUB, A. H. Rotational flows in relativistic hydrodynamics.

TCHEN, C. M. Mathematical structure of correlation functions for interacting particles.

THOMSON, J. Y. A finite difference solution of the compressible boundary layer equations.

TRANTER, C. J. On some series containing Bessel functions.

ULAM, S. M. Dynamics of infinite systems of mass particles.

VACCA, M. T. Tourbillon magnéto-hydrodynamique élicoïdal.

VARGA, R. S. Over-relaxation for non-negative primitive matrices.

VEKUA, I. N. О некоторых геометрических и механических приложениях теории обобщенных аналитических функций. (Some geometrical and mechanical applications of the theory of generalized analytic functions.)

VOGEL, T. M. Systèmes dynamiques à hérédité non linéaire et à mémoire totale.

WATSON, J. Some aspects of the non-linear theory of hydrodynamic stability.

WILSON, S. B. L. A finite difference method for wave problems.

ZIN, G. Analytical foundations of electromagnetic theory.

SECTION VIII

ALLENDOERFER, C. B. Teaching mathematics on television.

BUNT, L. N. H. An investigation into the possibility of teaching probability and statistics in Dutch secondary schools.

BURCKHARDT, J. J. Zwei griechische Ephemeriden.

CALVERT, H. R. The need for the sector as a computing and drawing instrument.

CASSINA, U. A study of the present state of teaching the elements of geometry in Italy.

DRENCKHAHN, F. Elementargeometrie in Unterricht: vom Gestaltlichen aus und in didaktischen Experimenten.

DUREN, W. L. Jr. The reform of college mathematics teaching in the United States.

EISELE-HALPERN, C. The theory of probability in Charles S. Peirce's logic and history of science.

GEARY, A. Diploma in Technology: a new departure in technical education in England.

HALL, J. A. P. Robert Recorde, 1510?–1558.

I.C.M.I., FIRST TOPIC. Mathematical instruction up to the age of fifteen years, reported by Professor H. F. Fehr, Columbia University, New York.

I.C.M.I., SECOND TOPIC. The scientific bases of mathematics in secondary education, reported by Professor H. Behnke, Münster.

I.C.M.I., THIRD TOPIC. Comparative study of methods of initiation into geometry, reported by Professor H. Freudenthal, Utrecht.

ILIĆ-DAJOVIĆ, M. Sur la réalisation d'un enseignement moderne de la géométrie élémentaire.

JAEGER, A. Contributions to the American undergraduate instruction for the prospective mathematician.

MAY, K. O. Stimulating undergraduate research.

McCONNELL, J. Sir Edmund Whittaker's philosophy of science.

NESS, W. Beispiel und Gegenbeispiel in der Mathematik.

PIENE, K. Statistics in secondary schools.

PRICE, G. BALEY. National Science Foundation Summer Schools.

STORER, W. O. Symbolism and the rules of operations in school mathematics.

TANNER, R. C. H. 'Equal and unequal'.

TRICOMI, F. G. Quo vadimus?

TUCKER, A. W. The work of the (American) Commission of Mathematics.

VAUGHAN, H. E. The U.I.C.S.M. Secondary School Programme.

SECRETARY'S REPORT

PREPARATIONS

The invitation to the International Congress of Mathematicians to meet in Edinburgh in 1958 was sponsored by the City of Edinburgh, the University of Edinburgh, the Royal Society and the Royal Society of Edinburgh; it was conveyed to the Amsterdam Congress in September 1954 by Professor W. V. D. Hodge, and was unanimously accepted.

After informal discussions on the procedure to be adopted, the first (and only) meeting of the Congress Committee was held in Edinburgh on 29 April 1955; this Committee contained representatives of the four sponsoring bodies and of the London Mathematical Society, the Edinburgh Mathematical Society and the British National Committee for Mathematics. It was announced at this meeting that H.R.H. the Duke of Edinburgh had graciously consented to become Patron of the Congress.

The Congress Committee appointed the principal officers of the Congress and set up an Executive Committee to supervise the detailed arrangements, keeping only certain formal and ceremonial matters in its own hands.

The first meeting of the Executive Committee was held on 5 October 1955. It was decided to set up a series of sub-committees to be responsible for the detailed work of organization; membership lists of these are given earlier in the present volume. Most of the work of preparation was done by these sub-committees, in particular by their chairmen and secretaries, and by the other officers of the Congress. In fact, only two further meetings of the Executive Committee were held, on 3 October 1956 and on 6 January 1958, and the remaining work of the Congress Committee was handled by correspondence.

It would take up too much space to describe in detail the work of the various sub-committees, whose functions are clearly indicated by their titles. Suffice it to say that a great deal of hard work was done by a great many people, both mathematicians and others, associated with these sub-committees, and that the Congress owes them an immense debt of gratitude. Mathematicians from all parts of the country contributed to their labours; for instance, much of the work on the scientific programme was done in Manchester, and travel grants for invited speakers were mainly dealt with in Liverpool.

FINANCE

The City Chamberlain of the City of Edinburgh (Mr A. L. Imrie) acted as Treasurer of the Congress, and the organisers owe a great debt to him, and to Dr C. H. Stout and Mr D. R. Ritchie of his staff, for the assistance which they gave.

There were four sources of revenue: (1) donations, both in the form of direct grants and by placing facilities at our disposal without charge, from a number of bodies, especially those represented on the Congress Committee; (2) donations from industrial and other organizations, and from private individuals (see List of Donors on p. xii); (3) a subvention from the International Mathematical Union for (a) organizational expenses, (b) travel grants, (c) publication of *Proceedings*; (4) membership fees of £5 for full members and £2 for associate members.

THE CONGRESS

The Congress itself may be said to have begun with the opening of registration at 2 p.m. on 13 August 1958, and an informal social gathering took place in the headquarters the same evening. The Congress headquarters were situated in the Edinburgh University Union, occupying a central position relative to the rooms where most of the lectures took place.

The inaugural session took place in the McEwan Hall at 10 a.m. on Thursday, 14 August 1958, and the closing session in the same hall at 2.30 p.m. on Thursday, 21 August. Detailed accounts of these sessions will be found at the end of this Report.

The Congress was attended by 1658 full members and 757 associate members; this is the largest total number for any International Congress of Mathematicians, though the number of full members is slightly less than at Harvard (1700) in 1950.

Universities and scientific organizations throughout the world were invited to appoint official delegates to the Congress. Similar invitations were sent to industrial organizations who had made substantial contributions to the funds of the Congress. The number of delegates appointed was 582, representing 308 organizations.

SCIENTIFIC PROGRAMME

The scientific programme of the Congress consisted of invited lectures, lasting either for an hour or for half an hour, and of offered communications, lasting for fifteen minutes each. The numbers of these delivered were as follows: 19 one-hour lectures, 37 half-hour lectures, and 604 fifteen-minute communications. The half-hour lectures and fifteen-minute communications were allocated to sections according to subject; there were originally eight sections, but they were later subdivided, the final list being:

I.	Logic and Foundations	VB.	Differential Geometry
IIA.	Algebra	VI.	Probability and Statistics
IIB.	Theory of Numbers	VIIA.	Applied Mathematics
IIIA.	Classical Analysis	VIIB.	Mathematical Physics
IIIB.	Functional Analysis	VIIC.	Numerical Analysis
IV.	Topology	VIII.	History and Education
VA.	Algebraic Geometry		

Even with this degree of subdivision there were often several sessions in progress in a single section at the same time.

In Section VIII a number of special sessions were arranged by the International Commission on Mathematical Instruction. These were devoted to reports and discussions on three prepared topics, namely:

(i) Mathematical instruction up to the age of fifteen years.

(ii) The scientific bases of mathematics in secondary education.

(iii) Comparative study of methods of initiation into geometry.

Most of the sessions were held in rooms in the University of Edinburgh; some of the invited lectures, however, took place in George Heriot's School, in the Heriot-Watt College, or in Moray House. Mathematicians from all parts of the world acted as chairmen of the sessions.

SOCIAL EVENTS AND ENTERTAINMENTS

The informal social gathering on the evening of Wednesday, 13 August, has already been mentioned. On the afternoon of Thursday, 14 August, the Lord Provost, Magistrates and Council of the City of Edinburgh held a garden party for members of the Congress in the grounds of Lauriston Castle. On the evening of Saturday, 16 August, an informal dance took place in the McEwan Hall; a team from the Scottish Country Dance Society was present to give demonstrations. Instead of a banquet, as at previous Congresses, a Congress Reception was given; this was held in the Royal Scottish Museum on the evening of Wednesday, 20 August.

On the evening of Friday, 15 August, members of the Congress were able to choose from three entertainments: (i) a chamber music recital in the Freemasons' Hall; (ii) an evening of Scottish song and dance in the Music Hall; (iii) a programme of Scottish films at the Gateway Theatre.

EXCURSIONS

The main Congress excursions were held on Sunday, 17 August. The most popular was the steamer cruise from Glasgow down the Clyde and round the island of Bute, ending at Gourock; this was a full-day excursion. The alternative was an afternoon excursion by coach to Loch Lubnaig and Loch Earn.

On the afternoon of Tuesday, 19 August, a wide variety of excursions was available to members of the Congress; scenic, historic, artistic and technological interests were all catered for. Sight-seeing bus tours of the City of Edinburgh were provided on two evenings during the Congress.

Special excursions were provided for associate members on three mornings; most of these were to various firms, factories and workshops, but there were also trips down the Royal Mile and to the Royal Botanic Garden, and visits to Hopetoun House and Lennoxlove. In addition, all associate members were invited by the Royal Zoological Society of Scotland and the Royal Society of Edinburgh to visit the Royal Scottish Zoological Gardens on the morning of Friday, 15 August.

EXHIBITIONS

An exhibition of current mathematical books was on view during the Congress on the premises of Messrs James Thin, South Bridge, Edinburgh; the books were selected by an international sub-committee under the chairmanship of Professor T. A. A. Broadbent.

An exhibition of school text-books was arranged by the International Commission on Mathematical Instruction, and was shown at Moray House during the Congress.

The Monotype Corporation arranged an exhibition of mathematical typography at the Heriot-Watt College.

Other exhibitions of mathematical books were on view in the Scottish National Library and the Edinburgh University Library.

ACKNOWLEDGEMENTS

A separate list of industrial firms who gave financial assistance to the Congress is printed earlier in this volume, and some other donations have been mentioned in § 2 of this Report.

Besides those services by individuals and organizations that have been indicated, explicitly or implicitly, earlier in this Report, there are others that should not be passed over unnoticed. Some of these are mentioned below; doubtless many more have been inadvertently omitted. To all these the warmest thanks of the organizers of the Congress are due for their willing help and co-operation.

In addition to the general entertainments described in §5, official delegates to the Congress were hospitably entertained at various functions by H.M. Government, by the City of Edinburgh and the University of Edinburgh, and by the Company of Merchants of the City of Edinburgh.

Secretarial assistance was provided to many officers of the Congress by the institutions to which they belonged; special mention may be made of the Universities of Edinburgh, Glasgow, Cambridge, Manchester and Liverpool, and of the Royal Naval College at Greenwich.

Folders in which members could carry Congress papers were presented jointly by B.O.A.C. and B.E.A.

Messrs Oliver and Boyd presented each member of the Congress with a recently published illustrated booklet entitled *Presenting Edinburgh*.

The Edinburgh Festival Society Ltd. performed the onerous task of arranging the accommodation of Congress members in hotels, boarding houses, private lodgings and University hostels.

Messrs T. and A. Constable Ltd. printed the booklets supplied to Congress members on their registration.

The Congress badge was designed by Mr Walter Pritchard and manufactured by Messrs H. W. Miller.

The Scottish Tourist Board, British Railways, Scottish Omnibuses Ltd. and the Edinburgh City Transport Department took part in arrangements for excursions and other transport for Congress members.

Sir Henry Lunn Ltd. were official travel agents for the Congress.

The Post Office and the Commercial Bank of Scotland made special facilities available for Congress members.

The co-operation of the Edinburgh City Police, the Automobile Association, the Royal Automobile Club and the Royal Scottish Automobile Club in arranging sign-posting and in controlling the movement of traffic near the Congress headquarters must also be acknowledged.

Students from Edinburgh and other universities acted as stewards at headquarters, in the lecture rooms and elsewhere; they were supervised by Dr M. F. Atiyah and Dr J. C. Polkinghorne.

I.B.M. United Kingdom Ltd. provided office equipment on loan for use in the Congress office.

Finally, the devoted work of Mrs Fenton and Miss Watson, who assisted in the Congress office, should be acknowledged.

REPORT OF THE
INAUGURAL SESSION

The inaugural session of the Edinburgh Congress took place in the McEwan Hall on the morning of Thursday, 14 August 1958. The Right Honourable Ian Johnson-Gilbert, Lord Provost of the City of Edinburgh, and Chairman of the Congress Committee, presided over the meeting.

The Lord Provost opened the proceedings by welcoming the Congress on behalf of the City of Edinburgh. His address was followed by other speeches of welcome: these were given by Sir Edward V. Appleton, Vice-Chancellor and Principal of the University of Edinburgh, and Vice-Chairman of the Congress Committee, on behalf of the University; by Sir David Brunt on behalf of the Royal Society; and by Professor N. Feather on behalf of the Royal Society of Edinburgh.

Professor J. F. Koksma, Secretary of the Amsterdam Congress of 1954, then spoke as follows:

My Lord Provost, Ladies and Gentlemen: Professor Schouten, President of the Amsterdam Congress of 1954, being prevented by reasons of health from coming to Edinburgh, to his and our deep regret, has asked me to transmit a message to you, a message which contains a proposal. I think the best thing I can do is to read you the letter he wrote to me on the 5th of August. I should only like to add the remark that Professor Schouten does not mention any motive for his proposal, presumably for the trivial reason that in the eyes of all of us such a mention would be superfluous. After having heard his letter you will all, I am sure, agree with Professor Schouten's views.

Epe, August 5, 1958

Dear Colleague,

It is the custom that the President of the last Congress proposes the name of the person to be elected by the Congress as its President.

As I cannot attend the Congress, I beg you to bring my best wishes for the Edinburgh Congress, and to propose Professor W. V. D. Hodge as President.

Yours sincerely,

J. A. SCHOUTEN

The proposal that Professor W. V. D. Hodge be elected President of the Congress was then put to the meeting, and carried by acclamation.

Professor Hodge then read to the meeting the following message from H.R.H. The Prince Philip, Duke of Edinburgh, Patron of the Congress:

Buckingham Palace

When preparations for this International Mathematical Congress began, more than three years ago, I was invited to accept the office of Patron. I gladly accepted this invitation recognizing, as I do, the quite essential part that mathematics has to play in the modern world. This is the age of applied science and many practical details of our lives, our transport, our communications, our engineering, our agriculture as well as our explorations of nearer and farther space, are governed more and more by technology which itself rests on a mathematical basis.

It seems to me most fitting that a Mathematical Congress should meet in Edinburgh, for this is the birthplace of John Napier's logarithms, that indispensable tool of the technical man all the world over.

Friendship between nations grows from personal friendship between individuals. Therefore a congress such as this has a wider significance and can do much to deepen and enrich international amity.

I wish this Edinburgh Congress all success in its labours. I trust that many new friendships will be made as well as old ones renewed and that, when you leave Edinburgh, you will take with you happy memories of this ancient, famous and beautiful city.

PHILIP

August 1958

Professor Hodge then proposed that the following reply be sent to His Royal Highness:

H.R.H. The Duke of Edinburgh,
Balmoral Castle

The mathematicians assembled in Edinburgh for the International Congress of Mathematicians thank Your Royal Highness most warmly for the gracious message which you have sent as Patron. They are proud to have as Patron one whose great interest in all branches of science is known throughout the world, and they send respectful greetings to Your Royal Highness.

W. V. D. HODGE
President

This proposal was carried unanimously.

Professor Hodge then gave his Presidential Address to the Congress. The address was as follows:

PRESIDENTIAL ADDRESS

I am most grateful to Professor Koksma for the kind words he has used about me, and I am deeply honoured by the manner in which you have accepted me as your President.

When the idea of holding this Congress in Edinburgh was first mooted, it was the hope of all British mathematicians that we should have as our President Sir Edmund Whittaker, that great figure in the life of the City and University of Edinburgh, so much respected and loved in the mathematical world. But in March 1956 he passed away. I count it not the least of my claims to be President of this Congress that I was one of those fortunate enough to receive their first introduction to higher mathematics in Sir Edmund's classes.

The preparations for this Congress have been onerous, and I would first like to pay tribute to all who have helped to make it possible for us to bring our plans to maturity. It has been a most moving and pleasant experience to find that so many were willing to contribute both their time and their money to the enterprise. Our four hosts, the City of Edinburgh, the University of Edinburgh, the Royal Society, and the Royal Society of Edinburgh, have proved far from merely formal sponsors. Each in its own way has contributed practical help to an extent which cannot easily be measured, and the goodwill which they have shown to us throughout has been quite indispensable. Next, I should like to pay my tribute to all those individuals, mathematicians and others, throughout the country, who have laboured long for the success of the Congress. If I do not name them individually it is simply because the list is too long. I should also like to thank the various institutions to which these people are attached, who have so generously allowed them to use their facilities for the work of the Congress. And, finally, we are most grateful to the International Mathematical Union and the many learned societies, industrial organizations, and individuals who have contributed most generously to the cost of this enterprise.

At the Harvard Congress of 1950 Professor Veblen referred to the difficulties encountered by the organizers of International Congresses, caused by the ever increasing number of people professionally engaged in the study of mathematics, and at Amsterdam in 1954 Professor Schouten spoke of the same problem. As you can well imagine, the organizers of the present Congress have had to face this problem once again. I should like to take up a little of your time by giving my own personal reflexions on this matter.

The International Congresses of Mathematicians, which are held every four years, serve a number of purposes. The most important is to get together the leaders in all branches of mathematics so that they may discuss their common problems and exchange ideas on them. In saying this, I wish to emphasize the phrase 'all branches of mathematics'. In recent years there has been a steady growth in the number of symposia held, many with the support of the International Mathematical Union. These symposia have done excellent work in advancing research in special fields. But this is not enough. It is essential for the well-being of mathematics that there should be periodic gatherings attended by representatives of all branches of the subject, and this for several reasons: in my personal opinion, the most important reason is that gatherings such as this serve as an invaluable safeguard against the dangers of excessive specialization.

The problem of specialization is a difficult one. Mathematics is now so vast that few can hope to cover the whole range, and much of our progress has been due to the efforts of men and women who have devoted their lives to work in a narrow field of research. Most of us must continue to work in specialized fields, and with good fortune we can make our contribution to mathematics as a whole in this way. But there are dangers in this. There is always the risk that we may come to regard our own special problems as all-important; and to regard mathematics simply as a system of conclusions drawn from definitions and postulates that must be consistent, but otherwise may be created at the free will of the mathematician. As Professor Courant has justly remarked: 'If this description were accurate, mathematics could not attract any intelligent person. It would be a game with definitions, rules, and syllogisms without motive or goal.... Only under the discipline of responsibility to the organic whole, only guided by intrinsic necessity, can the free mind achieve results of scientific value.'

I believe that mathematicians are now much less likely to fall into this danger than they were some time ago. But over-specialization also produces a practical difficulty. As we all know from our own experience, in order to make progress in our own field we must know what is going on in other fields, and what new techniques are being developed elsewhere in mathematics. The problem we are faced with is simply that of maintaining contact with all the main developments going on in mathematics while working intensively in our own specialized field. Some solution of this problem is essential, and International Congresses can go a long way towards giving the required answer. These Congresses provide an opportunity for periodic stocktaking, and the opportunities they provide for surveying the whole field of mathematics are a way of counteracting the evils of excessive specialization, and of determining the 'intrinsic necessity' to which Professor Courant refers: they may thus vitally influence the whole course of mathematics in the succeeding years.

The organizers of this Congress have planned our meetings so as to pass under review all the main developments in mathematics and to try to get things into perspective. In the main, we have followed the traditional divisions into sections, but we have, surely not before time, given topology a section to itself, and we have somewhat changed the emphasis in the sections dealing with applied mathematics. But the one-hour speakers are not assigned to sections. They have been picked as a team so that a continuous spectrum will be presented and they have been asked to make their lectures broad surveys of recent developments. In this way it is our hope that their contributions will present a general survey of all that is important in modern mathematics, and that when our *Proceedings* are published, they will form a focus from which many of the developments of mathematics in the next few years may begin.

Over one week, it is not possible to cover the whole range of mathematics, and at the same time to deal adequately with the wide applications of mathematics to other fields of intellectual endeavour. No mathematician can be indifferent to the ever growing number of applications of mathematics to the various sciences—physical, biological, and social—and in industry the present generation has witnessed with pride the revolution brought about by the introduction of statistical methods, and by the spectacular development since the war of the science of computing. We should like to include in the business of this Congress a thorough study of all the applications of mathematics to Science and Technology. But factors of time and space make this a practical impossibility, and our business is primarily concerned with that abstract science of mathematics whose laws govern so much of our knowledge. Hence most of our work will be concerned with pure mathematics. But not all. In our sections dealing with applied mathematics we have endeavoured to overcome, to some degree, the limitations of time-table, by inviting some distinguished exponents of other sciences to talk to us about their mathematical problems: we are at least establishing contact with them on ground which gives hope of fruitful cooperation.

In one instance, we have gone further. The youngest child of mathematics, the science of computing, perhaps because of its youth, has presented the mother science with many fascinating new problems, and we have considerably enlarged the amount of space given to this subject. Practical considerations have, however, forced us to confine ourselves to the mathematical side of this science, leaving the engineering side for other Congresses.

You will see that we have again included a section dealing with history and education in mathematics. Mathematics has a great history, and mathematicians should know something of it. The problems of mathematical education are many, and the International Commission on Mathematical Instruction has devoted much time to some of these problems. The meetings of the Commission form part of the work of the section on Education. The work of the Commission will be concerned with three problems of importance in the field of mathematical education which have been selected for special study during the last few years; but, at the same time, there are other problems in mathematical education, particularly on the higher levels, which concern us all. It is part of our duty to see that our pupils who go on to walks of life outside the academic field understand that mathematics is an integral part of world culture; not only a pillar of the technological civilization of today, but an essential item in the intellectual equipment of the good citizen. To achieve this state, it is first necessary that the training we give our young men and women should be aimed at developing this understanding of principles and encouraging their interest, instead of crushing it beneath a mass of technicalities; and secondly, that we should be prepared to take the trouble to give accounts of our work to the mathematically educated layman.

Another respect in which our Congresses differ from symposia lies in the fact that membership of a Congress is open to all mathematicians, while that of a symposium is by invitation, and is therefore confined to those who have been or are making a name for themselves in their particular field. Hence Congresses offer almost the only opportunity for many young mathematicians to meet and listen to the leaders in their subject. We welcome the large number of young people who are attending a Congress for the first time. Many will be here just to listen, but they will be able to meet and discuss

problems with more mature mathematicians. Others, and the number of
them is very large, are presenting papers and it is to be expected that,
amongst the 650 papers offered, a number will attract attention from the
more senior of us, and may prove a foretaste of great things to come. On this
occasion we do not propose to publish these short papers; it is better that
they should follow the normal channels of publication, but, in the expectation
that many of the papers read at this Congress will excite considerable
interest, we have made arrangements for a number of small discussion rooms
to be available where groups can get together informally and discuss their
ideas more fully.

Believing, as I do, that we have provided for those essential needs of
mathematicians which only an International Congress can satisfy, I wish you
all a profitable and enjoyable week in Edinburgh.

Professor H. Hopf, chairman of the Fields Medals Committee, then
read the report of the Committee, which was as follows:

My Lord Provost! Ladies and Gentlemen! As on the occasion of the last
three International Congresses of Mathematicians, so at this Congress two
Fields Medals are to be awarded. It is already a tradition that the recipients
of the medals are *young* mathematicians. This is not expressly prescribed in
the memorandum of the donor, the late Professor Fields. It is only said that
the awards should be made 'in recognition of work already done, and *as an
encouragement for further achievement on the part of the recipients*'—and this
has been interpreted to imply that the recipients should be young. However, the other day, a friend of mine made the remark that when one looks
at the present situation in mathematics and the developments in recent years,
one feels that it is the old rather than the young who need encouragement.
But even the point of this *bon mot* persuades us again to applaud and reward
youth. Thus the Committee on the Fields Medals 1958 agreed, from the
beginning, to keep to the tradition of awarding the medals to mathematicians
of the younger generation.

This Committee on the Fields Medals, which was set up by the Organizing
Committee of the International Congress of Mathematicians, Edinburgh,
consists of eight members, namely: Chandrasekharan, Bombay; Friedrichs,
New York; Hall, Cambridge (England); Hopf, Zürich; Kolmogoroff, Moscow;
Schwartz, Paris; Siegel, Göttingen; and Zariski, Cambridge (U.S.A.). Each
of us first wrote down his own list of nominees. The combined list contained
thirty-eight names. Let me address these words to the thirty-six who will
not be named here: 'The Committee on the Fields Medals wishes to express
its sincere appreciation and admiration for the work you have done. The high
quality and the great variety of your achievements augur well for the future
of mathematics. These very attributes have created considerable trouble for
our Committee; again and again have we regretted that more than two
medals could not be presented.'

The great variety within mathematics is due not only to the multiplicity
of the branches of mathematics, but also to the diversity of the general tasks

that face a mathematician in any branch. A task which is particularly fundamental, is: *to solve old problems*; and another, no less fundamental, is: *to open
the way to new developments*. Our Committee is glad to have found two
young mathematicians who have done unusually good work, one in each of
these directions. As Chairman of the Committee on the Fields Medals 1958,
I have the honour and pleasure to announce that the Committee has decided
to award the Medals to

KLAUS FRIEDRICH ROTH, of the University of London, for solving
a famous problem of number theory, namely, the determination of
the exact exponent in the Thue–Siegel inequality;

and to

RENÉ THOM, of the University of Strasbourg, for creating the theory
of 'Cobordisme' which has, within the few years of its existence, led
to the most penetrating insight into the topology of differentiable
manifolds.

Detailed reports on the work of the laureates will be given in a special
session; Professor Davenport will speak on Dr Roth's work, and I on
Professor Thom's.

May I now ask Dr Roth and Professor Thom to come forward to receive the
Medals from the hands of the Lord Provost of Edinburgh?

After reading the report, Professor Hopf introduced Dr K. F. Roth
and Professor R. Thom to the Lord Provost, who presented the medals
to the two prize-winners.

The Lord Provost then declared the Inaugural Session closed, and
the proceedings terminated with the playing of the National Anthem
of the United Kingdom.

REPORT OF THE CLOSING SESSION

The closing session of the Edinburgh Congress took place in the McEwan Hall on the afternoon of Thursday, 21 August 1958. Professor W. V. D. Hodge, President of the Congress, was in the chair.

Professor Hodge made the following statement about the 1962 Congress and the procedure to be followed in deciding where the 1966 Congress should be held:

PRESIDENT'S CLOSING REMARKS

Before passing to the business of this meeting, may I, on behalf of all members of the Congress, express our sincere sympathy with the delegation from the Netherlands in the tragic accident which occurred to one of its members on Tuesday. It is the fervent hope of all of us that Professor van Wijngaarden will make a good recovery from his injuries.

As those who were present at the International Congress in Amsterdam will remember, a committee consisting of representatives of the International Mathematical Union and of the organizers of the 1958 Congress was appointed to consider the location of the Congress of 1962. This committee consisted of Professors Hopf, Chandrasekharan and MacLane representing the Union, and Dr Smithies and myself representing this Congress.

The committee has discussed with the representatives of a number of countries the possibilities of holding the next Congress in a number of places. I am authorized by the committee to say that while for reasons of a technical nature it is not possible to make any announcement today of the name of the host country for 1962, the prospects of holding a Congress in that year amount to a certainty. In order to remove any element of mystery from this statement, I will explain that one country represented here is very anxious to be our host but is unable to issue a formal invitation until certain consultations are completed at home; while another country has generously expressed its willingness to await the conclusion of these consultations and has promised to issue an invitation if, but only if, the first country finds itself unable to do so.

The necessary consultations will be completed in a few months. The joint committee, therefore, undertakes to make an announcement by 1 January 1959. This announcement will be sent to the Adhering Organizations of the International Mathematical Union and subsequently published in *International Mathematical News*. If any country which is not a member of the International Mathematical Union wishes to be informed directly, will its representatives please send to Professor Eckmann, Secretary of the Union, the name and address of the organization which should be informed.

Another matter concerning future Congresses must now be decided. I should like to propose that a committee of five, three to be appointed by the Executive Committee of the Union and two by the organizers of the 1962 Congress, be appointed to consider the location of the Congress of 1966.

The proposals made by Professor Hodge were unanimously approved.

Professor H. Hopf made a brief report on the work done at the Assembly of the International Mathematical Union, which had met at St Andrews from 11 to 13 August 1958.

Dr F. Smithies, Secretary of the Congress, made some formal announcements; he also reminded members of the decision taken earlier not to include the texts or abstracts of fifteen-minute communications in the published *Proceedings of the Congress*.

Professor B. Jessen then addressed the Congress as follows:

The opportunity has been given to me of saying some words at this closing session of our congress. I shall try to express the feelings of gratitude to our hosts that I am sure are shared by all members of the congress.

We must all be happy that our British colleagues, when inviting the congress to meet in Britain, chose Edinburgh as the meeting place, thus making it possible for us to enjoy for a while the special atmosphere of this ancient and beautiful city and to become acquainted also with other parts of Scotland. We have reason to be most grateful to the city of Edinburgh, to its famous university, and to its people for the hearty hospitality with which they have received us.

I believe that only those who have tried it quite know what it means to organize a congress of the size of our international mathematical congresses. It must be an enormous amount of work that the organizing committee and its helpers over a long period of time have put into the planning of the congress. I must express the admiration that I am sure we all feel for the way in which everything has been arranged. You certainly have done an excellent job. I wish that we could thank you all individually, including all the young people who have been around to help us with the many practical problems that invariably arise. But you have been very shy about it, and have not even printed the names of the organising committee in the membership list.

It is certainly not possible at the present stage to sum up the results of this congress. Through the choice of the invited speakers and through the large number of communications of other members the congress has presented a picture of mathematics today with all its trends. But the international congresses have another purpose, which I believe is just as important, that of promoting the fellowship between mathematicians of all countries. This fellowship has its roots in our common love for our science, to whose growth we all try to contribute. It is the responsibility of each generation to take care that this fellowship is maintained and strengthened, and extended to the new generation. It increases our joy in our work and, like the similar fellowship among scientists of other fields, sets an example for international collaboration. I am sure that, through the way in which this congress has been prepared, it has also admirably served its purpose in this respect.

The song of Auld Lang Syne has as its theme friendship and kindness. Our British colleagues and friends have at this congress given us both in full measure. Thank you.

Professor Hodge then declared the Edinburgh Congress closed.

THE WORK OF THE FIELDS MEDALLISTS

THE WORK OF K. F. ROTH

By H. DAVENPORT

On the three previous occasions on which Fields Medals have been presented, the addresses on the achievements of the recipients have been given either by the Chairman or by a member of the awarding Committee. On this occasion, Professor Siegel was to have spoken about the work of Dr Roth, but as he is unfortunately unable to be present the duty has devolved on me. It is a pleasant duty, in that it requires me to pay tribute to the work of a colleague and friend.

Dr Roth's greatest achievement is by now well known to mathematicians generally; it is his solution, in 1955, of the principal problem concerning approximation to algebraic numbers by rational numbers.

If α is any irrational number, whether algebraic or not, there are infinitely many rational numbers p/q such that

$$\left| \frac{p}{q} - \alpha \right| < \frac{1}{q^2};$$

for example the convergents to the continued fraction for α. It is therefore natural to attempt to characterize irrational numbers in terms of the exponents μ for which there are infinitely many approximations satisfying

$$\left| \frac{p}{q} - \alpha \right| < \frac{1}{q^\mu}.$$

For convenience, I denote by $\bar{\mu} = \bar{\mu}(\alpha)$ the upper bound of such exponents μ. Obviously $\bar{\mu}(\alpha) \geqslant 2$.

The problem is: what can be said about the value of $\bar{\mu}(\alpha)$ when α is *algebraic*? In 1844 Liouville showed, in a very simple manner, that $\bar{\mu}(\alpha) \leqslant n$ if α is an algebraic number of degree n. In fact, if α is a root of the irreducible equation $f(x) = 0$, where $f(x)$ has integral coefficients (not all 0), then on the one hand

$$\left| f\left(\frac{p}{q} \right) \right| \geqslant \frac{1}{q^n},$$

and on the other hand it is easily seen that

$$\left|f\left(\frac{p}{q}\right)\right| = \left|f\left(\frac{p}{q}-\alpha+\alpha\right)\right| < c\left|\frac{p}{q}-\alpha\right|,$$

where c depends only on α. Comparison of these inequalities leads to the result. If $n = 2$ we get $\bar{\mu}(\alpha) = 2$; thus quadratic irrationals are about as badly approximable as any irrational number can be.

There are simple considerations which suggest that Liouville's result is far from being best possible. But it was not until 1908 that this was proved; in that year the Norwegian mathematician Axel Thue showed that $\bar{\mu}(\alpha) \leqslant \frac{1}{2}n+1$. In 1921 Siegel made further very substantial progress, and obtained $\bar{\mu}(\alpha) < 2\sqrt{n}$ approximately, the precise result being a little better than this. In 1947 Dyson improved Siegel's inequality to $\bar{\mu}(\alpha) \leqslant \sqrt{(2n)}$.

In all this work, extending over a period of 40 years, the basic idea was the use of polynomials in two variables. Suppose $f(x_1, x_2)$ is a polynomial with integral coefficients, of degree r_1 in x_1 and r_2 in x_2, and suppose p_1/q_1 and p_2/q_2 are two rational approximations to α. Then

$$\left|f\left(\frac{p_1}{q_1}, \frac{p_2}{q_2}\right)\right| \geqslant \frac{1}{q_1^{r_1}q_2^{r_2}},$$

provided of course that $\qquad f\left(\frac{p_1}{q_1}, \frac{p_2}{q_2}\right) \neq 0.$

Suppose further that $f(\alpha, \alpha) = 0$ and that the Taylor expansion of $f(x_1, x_2)$ in powers of $x_1-\alpha$ and $x_2-\alpha$ has all its 'early' coefficients zero, a condition which can be made precise in various ways. Then one can obtain an upper bound for

$$\left|f\left(\frac{p_1}{q_1}, \frac{p_2}{q_2}\right)\right|$$

in terms of $\qquad \left|\frac{p_1}{q_1}-\alpha\right| \quad \text{and} \quad \left|\frac{p_2}{q_2}-\alpha\right|,$

and the principle is to combine this with the previous lower bound in such a way as to establish that p_1/q_1 and p_2/q_2 cannot both be very good approximations to α. Finally, p_1/q_1 and p_2/q_2 are chosen in a suitable way from the infinite sequence of approximations.

The proof of the existence of a polynomial $f(x_1, x_2)$ with all the desired properties is a difficult matter, and the condition that

$$f\left(\frac{p_1}{q_1}, \frac{p_2}{q_2}\right) \neq 0$$

is particularly troublesome. No explicit construction for such a polynomial has yet been found. During the course of their work, the four
mathematicians I have mentioned developed methods of great subtlety
and interest, and other important ideas which are relevant to the problem
were contributed by Gelfond, Mahler and Schneider.

In 1955 Roth finally solved the problem: he proved that $\overline{\mu}(\alpha) = 2$ for
any algebraic number α. The achievement is one that speaks for itself;
it closes a chapter, and a new chapter will now be opened. Roth's
theorem settles a question which is both of a fundamental nature and
of extreme difficulty. It will stand as a landmark in mathematics for
as long as mathematics is cultivated.

It is not my intention to describe or analyse Dr Roth's proof, particularly as he will be speaking about it himself. My own impression of
his proof is that it is a structure, inevitably of some complexity, every
part of which fits into its proper place and carries its proper share of the
total load. As you have probably anticipated from my description of
previous work, it uses polynomials in an arbitrarily large number of
variables, instead of in two variables. It had indeed long been realized
that this would be necessary, but the difficulties in the way had appeared
to be quite insuperable.

I turn now to another achievement of Dr Roth, which seems to me to
be also of the first magnitude, though the problem to which it relates is
perhaps of less universal interest. Let

$$n_1, \ n_2, \ n_3, \ \ldots$$

be a sequence of natural numbers, and suppose that no three of the
numbers are in arithmetic progression; in other words,

$$n_i + n_j \neq 2n_k$$

unless $i = j = k$. It was conjectured by Erdős and Turán in 1935
(though the conjecture is believed to be older) that such a sequence must
have zero density, that is, the number $N(x)$ of terms not exceeding x
must satisfy

$$\frac{N(x)}{x} \to 0 \quad \text{as} \quad x \to \infty.$$

This problem was the subject of several interesting and ingenious papers,
but it resisted all attempts at solution for a long time. The conjecture
was proved by Dr Roth in 1952, and his proof is one of great interest and
originality. He first considers a set of numbers $n_1, n_2, \ldots, n_r \leqslant x$ with the
property in question, for which r is a maximum, and proves that such

a set, if dense, would have to have considerable regularity of distribution. This regularity is of two kinds: regularity of distribution in position and regularity of distribution among the residue-classes to any modulus. Then he applies the analytic method developed by Hardy and Littlewood for problems of an additive character, and the features of regularity prove to be just sufficient to give the estimates necessary for the method to succeed. The final conclusion is that $N(x) < cx/(\log\log x)$, where c is an absolute constant. I can recall no other instance, of comparable importance, in which the Hardy–Littlewood method has been used to elucidate the additive properties of an unknown sequence, instead of a special sequence such as the kth powers or the primes.

There are other achievements of Dr Roth which stand out by their originality and novelty, and it is with reluctance that I pass over them. Those I have outlined already will, I am sure, satisfy you that the recognition which has come to him is well merited.

The Duchess, in *Alice in Wonderland*, said that there is a moral in everything if only you can find it. It is not difficult to find a moral in Dr Roth's work. It is that the great unsolved problems of mathematics may still yield to direct attack, however difficult and forbidding they appear to be, and however much effort has already been spent on them.

THE WORK OF R. THOM

By H. HOPF

René Thom wurde 1923 in Montbéliard geboren. Er absolvierte von 1943 bis 1946 die Ecole Normale Supérieure in Paris und ging danach an die Universität Strasbourg, an der er heute 'Professeur sans Chaire' ist. In Strasbourg entstand seine Thèse 'Espaces fibrés en sphères et carrés de Steenrod' (*Ann. Sci. Ecole norm. sup.* (3), 69), mit der er 1951 an der Universität Paris zum Doktor promovierte. Im Jahre 1954 erschien die durch die Thèse schon vorbereitete Abhandlung 'Quelques propriétés globales des variétés différentiables' (*Comm. Math. Helvet.* 28), in der die Theorie des *Cobordismus* begründet wird. Heute, also nach vier Jahren, darf man sagen, daß seit langer Zeit nur wenige Ereignisse die Topologie, und durch die Topologie weitere Zweige der Mathematik, so stark beeinflußt haben wie das Erscheinen dieser Arbeit. Ich will versuchen, die Grundidee und die Grundzüge der Theorie des Cobordismus hier kurz zu beschreiben.

Man betrachtet k-dimensionale kompakte unberandete orientierte Mannigfaltigkeiten $A, B, \ldots$; man setzt nicht voraus, daß sie zusammenhängend sind (eine 1-dimensionale A ist also die Vereinigung endlich vieler zueinander fremder geschlossenen Linien, eine 2-dimensionale A die Vereinigung endlich vieler fremder geschlossener Flächen usw.). Mit $A + B$ wird die Vereinigung zueinander fremder Exemplare von A und B, mit $-A$ die zu A homöomorphe, aber entgegengesetzt orientierte Mannigfaltigkeit, mit $A - B$ die Summe $A + (-B)$ bezeichnet. $A.B$ ist, für Mannigfaltkgkeiten beliebiger Dimensionen, das cartesische Produkt. Daneben betrachtet man, für jedes k, $(k+1)$-dimensionale kompakte berandete orientierte Mannigfaltigkeiten $U, V, \ldots$, deren Ränder k-dimensionale Mannigfaltigkeiten sind. Wenn zu der k-dimensionalen A eine $(k+1)$-dimensionale U existiert, deren Rand ein Exemplar von A ist, so sagt man: 'A ist berandend', und man schreibt: $A \sim 0$; wenn $A - B$ berandend ist, so sagt man: 'A und B sind cobordant', und man schreibt: $A \sim B$. Die Relation $\sim$ definiert die 'Cobordismusklassen' der Mannigfaltigkeiten $A, B, \ldots$. Diese Klasseneinteilung ist verträglich mit der Addition; die Klassen jeder Dimension k bilden bezüglich der Addition eine Abelsche Gruppe Ω^k; ihr 0-Element ist die Klasse der berandenden A. Die Klasseneinteilung ist auch verträglich mit der cartesischen Multiplikation; so werden die Gruppen Ω^k mit $k = 0, 1, 2, \ldots$ zu einem Ring

$$\Omega = \Omega^0 + \Omega^1 + \ldots + \Omega^k + \ldots$$

verschmolzen. Das ist die '*Thomsche Algebra*'.

Man wird sich hier fragen: 'Ist das nicht alles ganz trivial? Kann denn eine so primitive Definition die Grundlage von neuen und interessanten Einsichten bilden?' Aber dieselbe Frage hat man sich auch gestellt, als Hurewicz 1935 die Homotopiegruppen definiert hatte; und als man dann die Konsequenzen dieser primitiven Definition sah, da gestand mancher Mathematiker: 'Das hätte ich nicht entdecken können —das wäre mir zu einfach gewesen.' Es bedurfte in der Tat eines genialen Mathematikers wie Hurewicz, sich durch das anscheinend zu Einfache nicht abschrecken zu lassen, sondern zu sehen, wie tief diese einfachen Begriffe in das Wesen der Probleme eindringen. Ich bin der Ansicht, daß Hurewicz einer der größten Geometer unserer Zeit gewesen ist—sein früher Tod wird auch während dieses Kongresses schmerzlich fühlbar sein—und ich persönlich habe kaum ein höheres Lob an einen Mathematiker zu vergeben, als ihn in die Nähe von Hurewicz zu stellen.

Nun, Thoms Entdeckung des Cobordismus und der Algebra Ω erinnert mich immer wieder—bei aller inhaltlichen Verschiedenheit—an die Entdeckung der Homotopiegruppen durch Hurewicz.

Eine der gar nicht trivialen Einsichten, die Thom offenbar von Anfang an hatte, war die, daß der Begriff des Cobordismus besonders gut der Untersuchung der *differenzierbaren* Mannigfaltigkeiten angepaßt ist. Auf diese werden wir uns jetzt beschränken. Eine differenzierbare Mannigfaltigkeit läßt sich immer mit einer Riemannschen Metrik versehen, also mit einer Metrik, die im unendlich-Kleinen euklidisch ist; so kommen die orthogonalen Gruppen ins Spiel. Thom verbindet mit den orthogonalen Gruppen $SO(n)$ gewisse topologische Komplexe $M(SO(n))$; unter Benutzung tiefgehender und erst in den letzten Jahren entwickelter Hilfsmittel (Eilenberg, MacLane, Cartan, Serre) zeigt er, daß die Gruppen Ω^k mit gewissen Homotopiegruppen der Räume $M(SO(n))$ isomorph sind, und es gelingt ihm, für viele Fälle diese Gruppen zu berechnen oder wenigstens recht präzise Aussagen über sie zu machen. Die Hauptergebnisse sind die folgenden: Es ist

$$\Omega^0 = Z, \quad \Omega^1 = \Omega^2 = \Omega^3 = 0, \quad \Omega^4 = Z, \quad \Omega^5 = Z_2, \quad \Omega^6 = \Omega^7 = 0,$$

wobei Z die unendlich zyklische, Z_2 die Gruppe der Ordnung 2 bezeichnet. Für alle k, die $\not\equiv 0 \bmod 4$ sind, ist Ω^k endlich. Für $k = 4m$ ist Ω^k direkte Summe von $\pi(m)$ Gruppen Z, wobei $\pi(m)$ die Anzahl der Partitionen (Zerlegungen in positive Summanden) von m ist, und einer endlichen Gruppe. Die erwähnten endlichen Gruppen sind schwer zu bestimmen; befreit man sich aber von ihnen dadurch, daß man Ω durch die 'schwache' Thomsche Algebra Ω' ersetzt, die aus Ω entsteht, indem man rationale Zahlen als Koeffizienten der Elemente von Ω zuläßt (sodaß also Ω' das Tensorprodukt von Ω und dem Körper der rationalen Zahlen ist), dann läßt sich nicht nur die additive, sondern sogar die multiplikative Struktur der Algebra genau angeben: Ω' ist die Algebra der Polynome mit rationalen Koeffizienten in Variablen $P^0, P^2, ..., P^{2m}, ...$, wobei P^{2m} durch den komplexen projektiven Raum von $2m$ komplexen (also $4m$ reellen) Dimensionen repräsentiert wird. Hieraus folgt: jede Mannigfaltigkeit M besitzt ein ganzzahliges Vielfaches nM, das mit einer wohlbestimmten ganzzahligen linearen Verbindung von cartesischen Produkten komplexer projektiver Räume gerader Dimensionen cobordant ist.

Alles dies dient natürlich in erster Linie dazu, diejenigen Eigenschaften von Mannigfaltigkeiten zu untersuchen, welche Invarianten der Cobordismusklassen sind. Es sind dies in erster Linie diejenigen Eigenschaften, die mit den schwer zu untersuchenden 'Pontrjaginschen

Zahlen' der Mannigfaltigkeiten zusammenhängen. Ein soeben erwähnter Satz zeigt, daß es darauf ankommt, einerseits die betrachtete Mannigfaltigkeit in der Algebra Ω darzustellen, und andererseits die Pontrjaginschen Zahlen der Räume P^m zu ermitteln; die Rolle, welche diese aus der klassischen algebraischen Geometrie stammenden Räume hier für die Topologie beliebiger (differenzierbarer) Mannigfaltigkeiten spielen, ist höchst bemerkenswert.

In diesen Rahmen gehört die bisher wichtigste Auswirkung der Theorie des Cobordismus; sie bildet einen der Pfeiler, auf denen F. Hirzebruch's Theorie des Riemann–Roch'schen Satzes ruht—eine Theorie, die man ihrerseits zu den bedeutendsten mathematischen Fortschritten der letzten Jahre zählen darf (F. Hirzebruch, *Neue topologische Methoden in der algebraischen Geometrie*, Springer-Verlag, 1956).

Eine andersartige Anwendung der Cobordismus-Theorie hat J. Milnor gemacht: er hat die aufsehenerregende Tatsache bewiesen, daß die Sphäre S^7 mehrere verschiedene differenzierbare Strukturen tragen kann, womit zum ersten Mal gezeigt worden ist, daß es zwei differenzierbare Mannigfaltigkeiten gibt, die sich zwar eineindeutig und stetig, aber nicht eineindeutig und differenzierbar aufeinander abbilden lassen; an der Spitze von Milnors Beweis steht der Satz von Thom, daß $\Omega^7 = 0$, daß also jede 7-dimensionale (geschlossene orientierbare) Mannigfaltigkeit 'berandend' ist (*Ann. Math.* (2), 64, 1956).

Ich kann hier nicht mehr auf andere unter den vielen interessanten Resultaten von Thom eingehen, weder aus den oben erwähnten, noch aus den Arbeiten, die ich nicht erwähnt habe. Aber eine Bemerkung prinzipieller Art will ich noch machen:

Die Topologie befindet sich, wie viele andere Zweige der Mathematik, heute in einem Stadium kräftiger und konsequenter Algebraisierung; dieser Prozeß hat außerordentlich klärende, vereinfachende und vereinheitlichende Erfolge gezeitigt und zu unerwarteten neuen Resultaten geführt; dabei ist es nicht etwa so, daß die Algebra nur die Hilfsmittel zur Behandlung topologischer Probleme liefert, es zeigt sich vielmehr, daß die meisten Probleme selbst eine Seite von ausgesprochen algebraischem Charakter besitzen. Die großen Erfolge dieser Entwicklung bringen aber, wie mir scheint, auch eine gewisse Gefahr mit sich, nämlich die Gefahr einer Störung des mathematischen Gleichgewichtes, indem eine Tendenz entsteht, den geometrischen Inhalt der topologischen Probleme und Situationen ganz zu vernachlässigen; diese Vernachlässigung aber würde eine Verarmung der Mathematik bedeuten. Gerade im Hinblick auf diese Gefahr finde ich, daß Thoms Leistungen etwas außerordentlich Ermutigendes und Erfreuliches an sich haben: auch Thom beherrscht und benutzt natürlich die modernen algebraischen Methoden und sieht die algebraischen Seiten seiner Probleme, aber seine grundlegenden Ideen, von deren großartiger Einfachheit ich vorhin gesprochen habe, sind von durchaus geometrisch-anschaulicher Natur. Diese Ideen haben die Mathematik wesentlich bereichert, und alles deutet darauf hin, daß die Wirkung Thomscher Ideen—mögen sie nun in den schon bekannten oder in noch ungeschriebenen Arbeiten zum Ausdruck kommen—noch lange nicht erschöpft ist.

ONE HOUR ADDRESSES

MODERN DEVELOPMENT OF
SURFACE THEORY

By A. D. ALEXANDROV

I will try to give in my lecture an outline of a general theory of surfaces as it has been developed during the past decade by a group of Russian geometers, U. F. Borisov, V. A. Zalgaller, A. V. Pogorelov, U. G. Reshetnak, I. J. Backelman, V. V. Streltsov and myself. This theory arose as a natural generalization of the theory of convex surfaces, the systematic presentation of which was given in my book *The Intrinsic Geometry of Convex Surfaces*, published just 10 years ago. Now this general theory has grown into an extensive branch of geometry with its own concepts, problems, methods and numerous results.

It would be hopeless to try to give here more than a general idea of the theory, so that all details and many results even of a fundamental character must be omitted. In constructing the foundations of the theory my aim was to define and to study the most general surfaces which allow of concepts and results analogous to those of classic Gaussian theory of surfaces. There are, first of all, two basic concepts of Gaussian theory; that of the intrinsic metric of a surface and that of its curvature. We accept an integral point of view according to which the metric is determined not by means of a line-element but by means of the distances between points measured *in* the surface, and the curvature is considered as a set-function, so that we mean integral curvature of a domain on the surface instead of the curvature at a point.

Let a surface S possess the property that any two of its points x, y can be joined by a curve $\widehat{xy}$ which lies in S and has a finite length $s(\widehat{xy})$. We define the intrinsic distance as

$$\rho_s(x,y) = \inf_{\widehat{xy} \subset S} s(\widehat{xy}). \tag{1}$$

It is evident that it satisfies the usual conditions imposed upon the general concept of a metric:

(1) $\rho(x,y) = 0$ if and only if $x = y$;

(2) $\rho(x,y) + \rho(y,z) \geqslant \rho(z,x).$

Thus the surface becomes a metric space with the metric ρ_s.

There are two points of view; one may consider a surface as a figure in the space, being interested in its special shape; or the surface may be considered as a metric space with the intrinsic metric ρ. In this case we speak of the intrinsic geometry of the surface, while in the first case we speak of its external geometry.

Corresponding to these two points of view there are two concepts of the curvature of a surface. The external curvature is measured by means of the area of the spherical image and the intrinsic curvature is measured by means of the excesses of geodesic triangles, the excess of a triangle T with the angles α, β, γ being, by definition,

$$\delta(T) = \alpha + \beta + \gamma - \pi.$$

As we are going to consider surfaces with a definite, i.e. finite, curvature we speak of surfaces with bounded curvature. Thus the objects of the purely intrinsic theory are two-dimensional metric manifolds with bounded intrinsic curvature (M.B.C.) and the objects of the external theory are surfaces with bounded curvature (S.B.C.).

The intrinsic and the external theories are not independent; there exists a close connection between them and, first of all, the well-known theorem by Gauss which asserts the equality of the intrinsic and the external curvatures for, at least, sufficiently regular surfaces. Thus we have sketched a certain programme; to give the strict definitions of manifolds of bounded curvature and of surfaces of bounded external curvature, to study their properties and to establish the connection between the intrinsic and the external properties of these surfaces.

A somewhat different, and in some respects even more general, approach to the theory of surfaces may be based upon the concept of parallel translation, which is closely connected with the concept of curvature because of the well-known Gauss–Bonnet theorem. The parallel translation of a vector along a curve on a surface can be defined both in intrinsic and external terms by means of the Levi-Civita construction. If we follow this trend of ideas the objects of the theory are the metric manifolds and the surfaces where parallel translation of vectors is defined for a sufficiently ample set of curves. Such surfaces were studied recently by Borisov, and I will give an account of his results in the last part of my lecture.

1. The definition of M.B.C.

1.1. The intrinsic metric.
The length of a curve is defined in any metric space in the usual manner; it is the supremum of the sums of the

distances between successive points of the curve. If any two points of a set S in a metric space can be joined by a curve which lies in S and has finite length, we call the set metrically connected. Now we say that the metric of a space is intrinsic in itself, or, simply, intrinsic, provided that the space is metrically connected and the distance between any two points is equal to the infimum of the lengths of curves joining these points.

The introduction of this concept is justified by the following theorem. Let S be a metrically connected set in a metric space R. Then if we define the distance
$$\rho_s(x, y) = \inf_{\widehat{xy} \subset S} s(\widehat{xy}),$$
the metric thus introduced in S proves to be intrinsic in the above sense. Accordingly we speak of the intrinsic metric induced in S by the metric of the surrounding space R. The definition of the metric of a surface given above is a particular instance of this general definition. Thus, our theorem being applied, we see that this metric is intrinsic in itself.

An M.B.C. must be a surface considered from the intrinsic point of view. Therefore it is natural that our first postulate in the definition of an M.B.C. should be the following one. An M.B.C. is a two-dimensional metric manifold with a metric intrinsic in itself.

1.2. The angle. In order to formulate the condition of the boundedness of the curvature by means of the excesses of triangles we have to define a triangle and an angle. (These definitions will be valid for any metric space.) We define, first, the shortest line or segment xy as a curve joining the points x, y and having the length equal to the distance $\rho(x, y)$ between them. Then it is evident what is understood by a triangle or a polygon. We note that in any manifold and even in any locally compact space with an intrinsic metric each point has a neighbourhood any two points of which can be joined by a segment.

The definition of an angle is given as follows. Let L, M be two curves with the common initial point O. Take the variable points $X \in L$, $Y \in M$ $(X, Y \neq O)$ and construct the plane triangle $O'X'Y'$ with sides equal to the distances OX, OY, XY. Let $\gamma(XY)$ be the angle of this triangle at the vertex O'. We define the upper angle between the curves L and M as
$$\bar{\alpha}(LM) = \overline{\lim_{X, Y \to O}} \gamma(XY).$$
This angle always exists.

Further, we say that there exists a definite angle between the curves L, M provided that the limit of the angle $\gamma(XY)$ exists, and in this case we define the angle
$$\alpha(LM) = \lim_{X, Y \to O} \gamma(XY).$$

Making use of the upper angle, which always exists, we define the excess of a triangle T as $$\delta(T) = \bar{\alpha} + \bar{\beta} + \bar{\gamma} - \pi,$$

$\bar{\alpha}$, $\bar{\beta}$, $\bar{\gamma}$ being the upper angles between the sides of T.

1.3. The condition of the boundedness of the curvature. Now we are ready to formulate the second and last postulate which defines an M.B.C. Each point has a neighbourhood U such that the sum of the excesses of pairwise non-overlapping triangles lying in U is uniformly bounded above: $$\Sigma\,\delta(T) < N.$$

N does not depend upon the set of the triangles and depends upon the neighbourhood U only.

Thus, briefly speaking, an M.B.C. is a 2-manifold with an intrinsic metric and with uniformly bounded sums of the excesses of non-overlapping triangles, at least in certain neighbourhoods which cover the manifold. Sometimes we speak of a metric of bounded curvature, which is preferable in the case where we have to consider not only one but many metrics given in the same manifold, i.e. when the manifolds are topologically mapped on to one and the same manifold.

1.4. Curvature. The definition of curvature is quite natural and runs as follows. We define the positive and the negative parts of the curvature of an open set G as the upper and the lower bounds of the sums of the excesses of the pairwise non-overlapping triangles lying in G $$\omega^+(G) = \sup \Sigma\,\delta(T), \quad \omega^-(G) = \inf \Sigma\,\delta(T).$$

The curvature itself is defined as $$\omega(G) = \omega^+(G) + \omega^-(G),$$

and the absolute curvature $$\Omega(G) = \omega^+ - \omega^-.$$

After that one can prolong these set-functions on to the ring of Borel sets by following the routine of measure theory. Then the fundamental fact is that these set-functions prove to be totally additive.

Our conditions concerning the excesses of triangles seem to be, in a certain respect, the minimum one has to suppose in order to have the possibility of defining the curvature as a totally additive set-function.

Zalgaller has given an abstract construction of a measure (non-negative totally additive set-function) which covers the definition of Lebesgue's measure, the above definition of the curvature and of many other set-functions which occur in geometry provided the definition starts from certain magnitudes ascribed to such elementary sets as the area of rectangles in the case of Lebesgue's measure or the excesses of triangles in the case of curvature.

1.5. Some other concepts. We define, further, such concepts as the area of a domain, the direction of a curve at a point, and the integral geodesic curvature (the bend) of a curve. For example, two curves are said to have the same definite direction at their common initial point provided the upper angle between them is equal to zero. The angle between two curves depends upon their directions only, i.e. it is the same for all pairs of curves with the same directions. The set of directions at a given point is isometric, with respect to its angular metric, to the set of generators of a cone.

2. The study of M.B.C. by means of approximation

2.1. With the exception of direct methods the first and most fruitful method in the theory of M.B.C. is that of approximation of general M.B.C. by means of polyhedra. First of all we have the following fundamental theorem: Let an intrinsic metric ρ given in a manifold M be the limit of a uniformly convergent sequence of metrics ρ_n with uniformly bounded positive parts of curvatures. Then ρ is a metric of bounded curvature also, and the curvatures of the metrics ρ_n weakly converge to the curvature ω of ρ in the sense that for any continuous function $f(x)$ different from zero on a compact set only,

$$\int f(x)\,\omega_n(dM) \to \int f(x)\,\omega(dM).$$

In particular, the limit of Riemannian metrics with uniformly bounded positive parts of curvature, i.e. $\int_{K>0} K\,dS$, is a metric of bounded curvature.

2.2. The simplest M.B.C. are manifolds with polyhedral metrics, or, in short, polyhedra. A polyhedron is such a manifold with an intrinsic metric, each point of which has a neighbourhood isometric to a cone. This descriptive definition is equivalent to a constructive one: a polyhedron is a manifold with intrinsic metric constructed of plane triangles, or in other words it allows of a subdivision into triangles isometric to plane ones. The curvature of a polyhedron is concentrated in its vertices, i.e. in such points the neighbourhoods of which do not reduce to pieces of the plane. The whole angle θ around such a point is different from 2π and is connected with the curvature ω of the point by the equation

$$\omega = 2\pi - \theta.$$

The curvature of a polyhedron is the sum of the curvatures at the vertices and its positive part is the sum extended over the vertices with the whole angle $\theta < 2\pi$. The convergence theorem above implies that the limit of polyhedral metrics with uniformly bounded positive parts of curvatures is a metric of bounded curvature.

The converse theorem exists in the following form: Any metric of bounded curvature is a limit of a sequence of polyhedral metrics with uniformly bounded absolute curvatures. Or in a more exact form, let P be a compact polygon in an M.B.C., R, and let ρ be the intrinsic metric induced in P by the metric of R. There exists a sequence of polyhedra P_n with uniformly bounded absolute curvatures, and of mappings of these on to P, such that the metrics determined in P by these mappings uniformly converge to ρ. And the positive and the negative parts of their curvatures weakly converge to the corresponding parts of the curvature of the metric ρ. We say that the convergence is regular.

2.3. If we combine this result with the previous convergence theorem, we get a new definition of an M.B.C. as a manifold which is, at least locally, the limit of polyhedra with uniformly bounded positive parts of curvatures. Polyhedra being, obviously, the limits of Riemannian manifolds, an M.B.C. proves to be the limit of Riemannian manifolds with uniformly bounded positive parts of their integral curvatures. In other words, the class of M.B.C. is the closure of the class of Riemannian or of polyhedral manifolds in the sense of uniform convergence of metrics under the condition of the uniform boundedness of positive parts of curvatures, or, what proves to be the same, the boundedness of absolute integral curvatures.

2.4. The above theorems provide the foundations of a method of studying M.B.C. by means of approximation by polyhedra. This method is applied to the study of some fundamental properties of M.B.C. For instance, we define the area of a polygon P in an M.B.C. as the limit of the areas of polyhedra regularly convergent to P.

In order to ensure a standard application of this method we have to supply ourselves with a number of general theorems concerning the convergence of various magnitudes associated with an M.B.C. and figures in it, such as polygons, curves, angles, area, integral geodesic curvature, etc. In fact, we have such theorems at our disposal.

Suppose, now, we are given a problem concerning M.B.C. Then we formulate it for polyhedra, and attempt to solve it for them. Polyhedra being the objects of elementary geometry, the problem reduces to one of a rather intuitive character. And if we succeed in solving the problem for polyhedra it only remains for us to apply suitable convergence theorems in order to obtain the general result. Most of the concrete results in the theory of M.B.C. have been obtained in this way.

3. Analytic characterization of M.B.C.

3.1. Probably the most important result obtained by means of this method is the following theorem by Reshetnak (1953). The metric of each M.B.C. may be determined by means of a line-element of the form

$$ds^2 = \lambda(dx^2 + dy^2), \tag{1}$$

where $\log \lambda$ is the difference of two subharmonic functions, and conversely, any metric determined by such a line-element, with the same condition for λ, is a metric of bounded curvature.

More exactly the first part of the theorem may be expressed as follows: Let P be a polygon in an M.B.C., R, homeomorphic to a circle. Then, by means of a map of P on to a domain D of the xy-plane, one can introduce in P co-ordinates x, y in such a way that the length of any curve in P which is the image of a broken line L in D is equal to

$$s = \int_L \sqrt{\{\lambda(dx^2 + dy^2)\}}. \tag{2}$$

And if we put $z = x + iy$, $\lambda(x, y) = \lambda(z)$ is representable by means of the following formula

$$\log \lambda(z) = -\frac{1}{\pi} \int_D \log|z - \zeta| \, \omega(dE_\zeta) + h(z), \tag{3}$$

where $\omega(E_\zeta)$ is the curvature of the set in P corresponding to $E_\zeta \subset D$; the integral is understood in the Lebesgue–Radon sense, and $h(z)$ is a suitably chosen harmonic function in D. Since $\omega = \omega^+ + \omega^-$, $\omega^+ \geq 0$, $\omega^- \leq 0$, the well-known integral representation of subharmonic functions implies that $\log \lambda$ is the difference of two such functions.

This theorem generalizes the well-known fact that the line-element of a regular surface may always be represented in the conformal form (1) and λ is connected with Gaussian curvature by the formula

$$\Delta \log \lambda = -2K\lambda. \tag{4}$$

If we consider this formula as a Poisson equation for $\log \lambda$ we just get the solution in the form (3) with $\omega(dE_\zeta) = K\lambda \, d\xi \, d\eta$.

3.2. Reshetnak's theorem adds to the two above definitions of an M.B.C. (i.e. the initial axiomatic one and the constructive one) a third one, the analytic definition. It opens the way for applications of analytic methods to the study of M.B.C. But so far nobody has followed this way to any considerable extent. Almost all results in the theory so far have been obtained by means of geometric methods.

4. Geometrical methods and some results of the theory of M.B.C.

4.1. There are two geometric methods in the theory of M.B.C., that of approximation by polyhedra and the other one which I call the method of cutting and gluing. It is based upon 'the theorem of gluing'. As a polyhedron is constructed or glued up of triangles, so, more generally, an M.B.C. may be constructed of pieces of given M.B.C., for example, of polygons cut out of any M.B.C., by means of gluing them together along segments of their boundaries. The possibility of such a construction under certain conditions imposed upon the boundaries of the glued pieces is ensured by a theorem which I call the 'theorem of gluing'.

In the simplest case when the glued pieces are polygons the theorem reduces to the following statement: Suppose we are given a complex of polygons cut out of some M.B.C.; suppose the complex is a manifold R with a boundary (possibly void) and that the identified segments of the sides of the polygons have equal lengths. Then if we define for any two points $x, y \in R$ the distance as the greatest lower bound of the lengths of curves joining x, y (the lengths being defined by metrics which are already given in each polygon) then R turns out to be an M.B.C.

In the case of more general domains than polygons an additional condition is necessary. It concerns the integral geodesic curvatures of the boundaries, for these curvatures as segment functions should be of bounded variation. For instance the conditions are fulfilled provided we have pieces of regular surfaces bounded by curves with piecewise continuous geodesic curvature.

4.2. The method of cutting and gluing is as old as geometry itself. The ancient proofs of Pythagoras's theorem as well as many other ancient proofs in elementary geometry consist just in cutting certain figures into suitable pieces and rearranging, or, let me say, gluing those pieces together so as to make the statement obvious. We apply just the same idea to our far more general and abstract figures.

4.3. In order to show that we do not only have general definitions let me mention a few results which were obtained by means of our methods and which were new for regular surfaces as well. I formulate these results in ordinary terms of differential geometry in order to avoid further definitions of certain concepts of the general theory.

(1) Consider surfaces S in a space of constant curvature K_0. Let them be homeomorphic to a circle and have prescribed the perimeter p and the positive part of their relative curvature, i.e.

$$\int_{K > K_0} K \, dS,$$

S being the area and K the Gaussian curvature. We put the question: what is the upper bound for the area of such surfaces? The answer is that it is the area of the circular cone with the same prescribed data (provided such a cone does exist, which is ensured by a simple condition).

(2) Consider the same surfaces in the same space and suppose they have non-positive relative curvature, i.e. $K \leqslant K_0$, so that for any domain, $\omega \leqslant K_0 S$. We ask, once again, about the maximum of the area. The answer is that the maximum is attained by the surfaces isometric to a circle with the same perimeter. It is worth mentioning that Reshetnak succeeded in proving that such an isometric inequality for any small circle on a surface is not only necessary but also sufficient for the Gaussian curvature of the surface to be $\leqslant K_0$.

(3) Consider now a curve L on a surface S homeomorphic to a circle. Let ω^+ be the positive part of the curvature of S, and s and τ the length and the integral of the absolute value of the geodesic curvature of the line L. The following results hold.

(3a) Let $\omega^+ + \tau < \pi$ and the distance between the ends of L be r. Then

$$s \leqslant \frac{r}{\cos \frac{1}{2}(\omega^+ + \tau)},$$

and the estimation is sharp. The equality is attained in the case of an isosceles triangle, r being its base and s the sum of two other sides.

(3b) Under the condition $\omega^+ + \tau < 2\pi$, the curve L either has no multiple points and it is necessarily so provided $\omega^+ + \tau < \pi$, or may be divided into two branches without such points. In the latter case it consists of a loop (i.e. a curve without multiple points and with coincident ends) and of one or two branches each of which has neither multiple points nor common points with the domain bounded by the loop.

(3c) Under the same condition $\omega^+ + \tau < 2\pi$ the length of the curve does not exceed a certain constant which depends on $\omega^+ + \tau$ and the size (the diameter, or the perimeter) of the surface. In particular, p being the perimeter of the surface,

$$s \leqslant \frac{p}{1 + \cos \frac{1}{2}(\omega^+ + \tau)}, \quad \text{if} \quad \omega^+ + \tau \leqslant \pi,$$

and
$$s \leqslant \frac{p}{\sin \frac{1}{2}(\omega^+ + \tau)}, \quad \text{if} \quad \pi < \omega^+ + \tau < 2\pi.$$

These estimations are sharp.

The above theorems are cited as examples of numerous concrete results that are obtained in our theory. I could give many other examples, some similar to the above and some of quite different type, but there is not time.

5. The surfaces which are M.B.C.

5.1. The question arises as to what are the surfaces, in ordinary Euclidean space E^3, which, from the intrinsic point of view, are M.B.C., and whether or not it is possible to embed any M.B.C., at least locally, in E^3. According to the beautiful Nash–Kuiper theorem, any Riemannian manifold allows of such an embedding, and even in the large provided it is orientable and compact. Any M.B.C. can be approximated by Riemannian manifolds, and this makes it obvious enough that one can extend the Nash–Kuiper result to an M.B.C. But the Nash–Kuiper embedding is too arbitrary, and it does not ensure that deep connection between the external and the intrinsic properties of the surface which is characteristic for more regular embeddings of Riemannian manifolds. The most fundamental of these connections being Gauss's theorem, we define a Gaussian embedding as one which preserves the validity of Gauss's theorem in at least a suitably generalized form. Thus our problem consists in finding Gaussian embeddings.

5.2. But in order to approach the solution of the problem it is necessary, at first, to determine and to study the classes of surfaces among which the realization of a given M.B.C. may be sought. Hence, our first problem is to determine and to study sufficiently general surfaces which are M.B.C. from the viewpoint of their intrinsic metric and which allow of a generalization of Gauss's theorem. After that the second problem is to prove, if possible, that an M.B.C., possibly subject to certain additional conditions, can be realized in E^3 as a surface of the given class. The third problem consists of a deeper study of the dependence of the external properties of the surface upon the intrinsic ones. In particular there is the question concerning the dependence of the degree of regularity of the surface upon that of its intrinsic metric. The special importance of the last problem for the general theory consists in the fact that its positive solution makes one sure that the solutions of problems of embedding and bending of surfaces obtained in the scope of general theory give the solution of corresponding problems in terms of differential geometry provided the surfaces have sufficiently regular intrinsic metric.

5.3. The surfaces R.D.C. (representable by differences of convex functions). As far as our first problem is concerned, the following results have been obtained. I studied the surfaces which, at least locally, allow of a representation by an explicit equation of the form $z = f(x, y)$, where x, y, z are Cartesian co-ordinates and f is the difference of two convex functions. In short these are R.D.C. surfaces. Any polyhedron which allows of a local representation by the equation $z = f(x, y)$, any convex surface, and any surface with first derivatives subject to the Lipschitz condition are R.D.C. surfaces. It is not difficult to verify that an R.D.C. surface can be approximated by regular surfaces with uniformly bounded absolute curvatures. Therefore, our convergence theorem for M.B.C. being applied, we see that these surfaces are M.B.C.

To establish the connection between the intrinsic and the external properties of these surfaces was not so easy. A generalized Gauss's theorem exists, but the exact definitions of the spherical image and of the external curvature require some consideration because of the absence of either tangent or supporting plane at an arbitrary point. I will not give such details here.

The connection between the intrinsic and the external geometry is not exhausted by Gauss's theorem. For instance, we have an intrinsic concept of an angle and of a direction of a curve. For R.D.C. surfaces they prove to be equivalent to the corresponding external concepts. The existence of an intrinsic direction of a curve at its initial point O proves to be equivalent to the existence of the ordinary tangent line, and the angle between two curves proves to be equal to the angle between their tangents measured on the tangent cone of the surface at the point O.

5.4. Surfaces with generalized second derivatives. Backelman has studied the surfaces which allow of a parametric representation $x(u, v)$, $y(u, v)$, $z(u, v)$ with continuous first derivatives and with generalized second derivatives summable by squares. He succeeded in extending to such surfaces all basic results of ordinary differential geometry provided the second derivatives which occur in its formulae are understood as generalized ones.

5.5. Smooth S.B.C. Backelman's surfaces are not, in general, R.D.C. Still they are included in a class of surfaces studied by Pogorelov. These are smooth surfaces with bounded external curvature. The exact definition is the following. The surface S is supposed to have at each point a tangent plane which continuously depends upon the point. Therefore the spherical image is defined. Let now $F_1, ..., F_n$ be closed sets on S, pairwise without common points, and let $\sigma(F_i)$ be the areas of their spherical images. The condition of the boundedness of the external curvature requires that $\Sigma \sigma(F_i)$ be uniformly bounded for all such systems of closed sets F_i.

First Pogorelov proves that his surfaces are M.B.C. Now as far as Gauss's theorem is concerned, there is no difficulty in defining absolute external curvature as sup $\Sigma \sigma(F_i)$. But it proved to be far more difficult to define curvature with suitable sign and to prove Gauss's theorem for it. Pogorelov's considerations are rather subtle. First he divides the points of the surface into two classes: the ordinary points and the non-ordinary ones, the first being characterized by the following property. An ordinary point A has such a neighbourhood U that no point $X \in U$ has the same spherical image as A. It is proved that the non-ordinary points are in a certain sense negligible.

The ordinary points are classified according to their indices, i.e. according to the number and the sense of the circuits of the spherical image of a closed simple curve surrounding the point. The sign of the index is ascribed to the point, and the positive part of the curvature is defined as the area of the spherical image of the set of all positive points, provided the multiplicity of the spherical image is taken into consideration. The negative part of the curvature is defined similarly. Then Pogorelov proves that these set-functions are equal to the corresponding intrinsic magnitudes.

Some additional results are worth mentioning. The surfaces with everywhere positive curvature are convex, and the surfaces whose positive and negative curvature both vanish are developable. If a surface has a locally one-to-one spherical image then it is convex, provided the spherical map preserves orientation, otherwise it is of negative curvature.

5.6. Backelman has noted lately that Pogorelov's proof of the theorem that his surfaces are M.B.C. does not make use of the smoothness of the surfaces and therefore allows of an immediate generalization to the surfaces with the following property.

Consider a surface S, and divide it into small pieces S_i. Consider the spherical image of those points of an S_i at which S_i has a supporting plane, with the exception of the points which lie upon the boundary of S_i. Let $\sigma^+(S_i)$ be the area of this spherical image. The condition imposed upon the surface is: the sums $\Sigma\sigma^+(S_i)$ are required to be uniformly bounded for all subdivisions of S into pieces S_i. It is obvious that in this construction we just catch the positive part of the external curvature of the surface. Hence the surfaces subject to the above condition may be characterized as the surfaces of bounded positive external curvature, quite analogously to our definition of M.B.C. where the condition of boundedness is imposed just upon the positive part of the curvature.

The simple repetition of Pogorelov's proof for smooth surfaces—for Pogorelov just makes use of the above construction—leads to the result: a surface with bounded positive part of the external curvature is an M.B.C. But up to now this is essentially all that is known about such surfaces. In particular we have neither a proof of Gauss's theorem nor even the definition of curvature for such surfaces. Their study is our next problem.

I believe that these surfaces form that general class of surfaces among which we have to expect the local realization of abstract M.B.C. They include all the above classes of surfaces, such as, for instance, the R.D.C. surfaces.

6. Embedding problems

6.1. As far as the embedding problems are concerned we have no general results except those for convex surfaces. First of all there is my old theorem on the embedding of manifolds into a space $R^3_{K_0}$ of constant curvature K_0. In the case of a compact manifold it reads as follows. Let M be an M.B.C. homeomorphic to a sphere and let its curvature ω for any domain G be subject to the condition $\omega(G) \geqslant K_0 S(G)$, S being the area. Then there exists in $R^3_{K_0}$ a convex surface isometric to M. According to Pogorelov's theorem this surface is unique up to a motion or reflection.

6.2. The problem of the regularity of the embedding provided the metric of M is representable by means of a line-element ds^2 has been solved to the following extent. Suppose the coefficients of the line-element have first derivatives subject to the Lipschitz condition. Then the surface S is smooth and realizes not only the metric but the line-element ds^2 itself. That is (in the case of Euclidean space) there exists a parametrization u, v of M such that the vector-function $\mathbf{x}(u, v)$ representing the surface S satisfies the equation $d\mathbf{x}^2 = ds^2$. In other words, the embedding solves not only the generalized problem but also the classical problem itself.

The same property of regularity takes place in the small, i.e. for a convex surface realizing any domain in an M.B.C., provided the curvature is subject to the inequalities:

$$C > \frac{\omega(G)}{S(G)} > K_0 + \epsilon \quad (\epsilon = \text{const} > 0).$$

6.3. Pogorelov established stronger regularity of a convex surface with prescribed line-element, provided the latter is at least five times differentiable. Pogorelov's theorem reads as follows: If the line-element of a convex surface S in $R^3_{K_0}$ is $k + 1$ times ($k \geqslant 4$) differentiable (analytic) and has curvature $K > K_0$ everywhere, then S is k times differentiable (analytic). The result is valid for any convex surface in a space.

6.4. Notwithstanding their strength, these results seem to me far from being sufficient. If a line-element is $k + 1$ times differentiable the surface proves to be k times differentiable (or we have somewhat stronger results if a Lipschitz condition is implied). But in my theorem above $k = 1$ and in Pogorelov's theorem $k \geqslant 4$. Meanwhile, the ordinary formulae of differential geometry imply second derivatives, or third ones if we write the Peterson–Codazzi equations not in an integral but in their ordinary form. Therefore the most interesting and important problem consists in finding the minimal conditions which ensure two times or three times differentiability of the surface. This problem, however, still remains unsolved.

6.5. I would like to mention that Pogorelov succeeded recently in proving theorems on the embedding of a manifold into a three-dimensional Riemannian space (the curvature of the manifold being sufficiently great). The theorem of regularity and that of the uniqueness of the embedding are also established. A short exposition was published in *Vestnik* of Leningrad University, 1957, N 7, and the full details are given in a booklet published later by Kharkov University Press.

7. The surfaces with parallel translation of vectors

7.1. As was mentioned at the beginning of my lecture there is a different approach to the theory of general surfaces based upon the concept of parallel translation of a vector. This idea has been recently realized by Borisov.

The definition of parallel translation was given by Levi-Civita. Consider a smooth surface S and a line $L \subset S$ joining the points A, B. Take a vector $\mathbf{a}$ tangent to S at the point A. Take, now, successive points $A = X_0, X_1, \ldots, X_n = B \in L$. If we project the vector $\mathbf{a}$ onto the tangent plane P_1 at the point X_1, we get a vector $\mathbf{a}_1$ at X_1; then project this vector onto the tangent plane P_2 at the point X_2 and so on. At the end we get a vector $\mathbf{a}_n$ at the point B. Suppose that the vector $\mathbf{a}_n$ tends to a certain limit $\mathbf{b}$ provided the points X_i are taken to be more and more dense upon the line L. Then we say that $\mathbf{b}$ is the result of parallel translation of $\mathbf{a}$ along the curve L.

This is the external definition of parallel translation by means of projections. A somewhat different definition may be given when, instead of projecting a vector from one tangent plane P_i onto another one P_{i+1}, we revolve the first plane P_i around the line of intersection of P_i and P_{i+1} and transfer in this way a vector given in P_i into the plane P_{i+1}. This operation being applied, we have parallel translation by the development of tangent planes.

Both definitions are equivalent for regular curves on regular surfaces, but, in general, they are not. Borisov has proved the following simple theorem which seems to be the more interesting because of the simplicity of its result. The lengths of vectors $\mathbf{a}_n$ converge to a certain limit, provided we use the projection, if and only if the curve has the following property. The sum of the squares of the angles between the normals to the surface at the successive points $X_1, \ldots, X_n$ tends to zero when the points are distributed more and more densely on the curve. If the curve has such a property the parallel translation along it is unique for any vector, i.e. not only the lengths of the vectors $\mathbf{a}_n$, but the vectors themselves with their directions have a definite limit. Under the same condition for the curve the parallel translation by developing of tangent planes is unique also, and gives the same result. But the converse statement is not true, which shows, in particular, that the two definitions of parallel translations are not equivalent.

In the following we understand by a parallel translation the operation defined by means of projections.

7.2. Borisov proves, further, that the uniform convergence of the parallel translation on a compact set of rectifiable curves (on a given surface) is equivalent to the condition that $\theta^2/\rho \to 0$ uniformly as $\rho \to 0$,

ρ being the distance between any two points and θ the angle between the normals at them.

Borisov studies the surfaces subject to this condition. First of all he proves that the parallel translation has an intrinsic meaning. Its intrinsic definition may be given for Borisov's surfaces as follows. A vector a is translated along a geodesic, i.e. the shortest line, if the angle between the vector and the line remains unchanged. The translation along a curve is defined by means of parallel translation along inscribed geodesic broken lines with the natural passage to the limit. Borisov proves that this intrinsic parallel translation is defined for any rectifiable curve and is equivalent to the external parallel translation as it has been defined above.

It is necessary to note that this is not so very simple, for we must have, at first, an intrinsic co-ordinate-free, purely metric definition of a vector in the surface and of the angle between the vector and the shortest line. These definitions are based upon a concept of angle which is somewhat more general than that used in the theory of M.B.C. Two curves are said to have the same direction at a point O if the angle between them is equal to zero. The concept of a direction being thus introduced, we have immediately the concept of a vector.

We shall not mention Borisov's other results with the exception of the Gauss–Bonnet theorem. Borisov proves that for any domain G with rectifiable boundary L homeomorphic to a circle and having spherical image within a hemisphere the Gauss–Bonnet formula holds. The rotation of a vector under parallel translation along L is equal to the area of the spherical image of G defined as a certain curvilinear integral along the spherical image of L. This result seems to me the more interesting in that Borisov's surfaces are not, generally speaking, manifolds of bounded curvature, so that their curvature is not a totally additive set-function.

ON SOME MATHEMATICAL PROBLEMS
OF QUANTUM FIELD THEORY

By N. N. BOGOLYUBOV AND V. S. VLADIMIROV

In this talk, which has been prepared by Vladimirov and myself, I shall be concerned with some mathematical problems, which arose in connection with the quantum field theory.

In our opinion, modern quantum field theory is of great interest for mathematics as a source of new mathematical ideas. Already the usual, non-relativistic, quantum mechanics has essentially influenced the development of a new branch of mathematics—the theory of generalized functions or distributions. It is sufficient to mention that the well-known delta-functions were first introduced by Dirac in his quantum mechanical investigations, and that he actually made use of the notion of the improper, or weak, transition to the limit as the way of defining such functions.

In dealing with these functions, physicists did not focus appropriate attention on the question of rigorous mathematical foundation of their arguments. Because of such a lack of mathematical 'rules of behaviour' mathematicians were for a long time very sceptical as to the possibility of introducing the concept of delta functions, and of analogous non-conventional functions, into their world.

Now, however, after the work of Sobolev[1] and Schwartz[2] the theory of generalized functions has become a fully recognized mathematical theory. Its applications to other branches of mathematics are still expanding. Quantum field theory deals with generalized functions which in Schwartz's terminology are called tempered distributions.

Let us explain here briefly what these generalized functions are. A theoretical physicist would say something like this—the singular or generalized functions are defined merely by prescribing the rules for integrating them with sufficiently regular functions. Unlike ordinary conventional functions they are not characterized by giving their values for different values of their arguments.

Let us now formalize this intuitive way of presentation and formulate one of many possible rigorous definitions. Consider Schwartz's space S of infinitely differentiable functions $F(x)$; $x = (x_1, ..., x_n)$, such that $F(x)$ and any of its partial derivatives decrease at infinity more rapidly than any power of $1/|x|$. Consider linear functionals $L(F)$ in this space and introduce a convention of representing them in the form

$$L(F) = \int K(x)\,F(x)\,dx.$$

Then the $K(x)$ are called generalized functions. The generalized functions introduced in such a way possess many very convenient properties. For example, they can be infinitely differentiated; they always have Fourier transforms and so on. But all this good behaviour disappears when we try to perform non-linear operations on them.

Even such an apparently elementary operation as the multiplication of two generalized functions is, in general, meaningless. We wish to stress that this fact is very closely connected with the appearance of the so-called 'divergences' in quantum field theory. Let us make only a few remarks on this subject. The main problem of the theory is to determine the scattering matrix S, the elements of which may be used to calculate the probabilities of different processes occurring in particle collisions. For the determination of the S-matrix its elements are usually expanded into the perturbation theory series in a small parameter characterizing the interaction strength.

In the case of electrodynamics this parameter can actually be considered as small (it is equal to $1/137$) and therefore perturbation theory was first applied in detail to the problems of quantum electrodynamics. As is well known, in quantum field theory the nth term of the expansion of the S-matrix may be represented as an integral of the so-called time-ordered product $T\{\mathscr{L}(x_1)\ldots\mathscr{L}(x_n)\}.$

Here $x = (x^0, x^1, x^2, x^3)$, x^0 is the time, $\mathscr{L}(x)$ is the operator representing the interaction Lagrangian.

A purely formal 'definition' of the time-ordered product is the following: it is the product of the operators ordered with the time increasing from right to left. The explicit form for any matrix element of such an expression leads to products of the type

$$\Pi\mathscr{D}^{(c)}(x_i - x_j), \tag{1}$$

where $\mathscr{D}^{(c)}(x)$ are generalized functions defined by integral representations of the form

$$\mathscr{D}^{(c)}(x) = \mathscr{P}\!\left(\frac{\partial}{\partial x}\right)\int \frac{e^{ipx}}{p^2 - m^2}\,dp. \tag{2}$$

Here $\mathscr{P}\,(\partial/\partial x)$ is a certain polynomial, the integrals are taken over the whole 4-space of the momentum variables $p = (p^0, p^1, p^2, p^3)$, and

$$p^2 = (p^0)^2 - \sum_{1\leqslant\alpha\leqslant3}(p^\alpha)^2,\quad px = p^0 x^0 - \sum_{1\leqslant\alpha\leqslant3}p^\alpha x^\alpha.$$

By formally evaluating the expressions involving the products (1) by means of Fourier transforms, we are led to the well-known divergences due to the insufficiently rapid decrease of the integrands with increase of p.

In quantum field theory a special subtraction formalism has been developed by means of which the divergences from every expansion term of the S-matrix can be removed. In fundamental papers by Dyson[4] and Salam[5] a rather general form was given to this technique, which however still had the form of a recipe, as in the corresponding considerations one had to deal with meaningless divergent expressions. An interesting property of the products of type (1) was noticed by Parasiuk[6].

In trying to attach some meaning to them, Parasiuk has considered the approximating sequences of regular functions $\mathscr{D}_M^{(c)}(x)$, which tend to $\mathscr{D}^{(c)}(x)$ as $M \to \infty$ in the improper sense. It turned out that the product

$$\Pi \mathscr{D}_M^{(c)}(x_i - x_j)$$

does not have, in general, even an improper limit; such a limit exists, however, in a more restricted sense. In fact, if one introduces a more restrictive definition of the local improper limit in the space of points $(x_1, ..., x_n)$ excluding the possibility of any pair of arguments x_i, x_j coinciding, then this limit exists. More exactly, this limit exists on the subspace of those trial functions which, together with their partial derivatives, vanish for coinciding arguments. Thus the problem of completing the definition of the product (1) is reduced to that of extending the linear continuous functional defined at first only on the above-mentioned subspace to the whole space of trial functions.

This problem is not as arbitrary as it might seem at first sight. This is due to the fact that the extension must be done in such a way as not to violate the basic physical requirements for the S-matrix, i.e. the most general conditions such as covariance, unitarity and causality. These physical conditions have been formulated mathematically, in [3], in the form of relations to be satisfied by the terms of the S-matrix expansion.

The problem of extending the notion of product, taking account of the above conditions, has been solved in a paper written together with Parasiuk[7]. The resulting algorithm turned out, in fact, to be essentially equivalent to the Dyson–Salam subtraction procedure. It should be emphasized that the problem of removing the divergences was solved only in the framework of perturbation theory. The convergence of the perturbation theory series itself is not yet proved. Moreover, some rather convincing arguments for its divergence have been put forward.

Leaving aside these more delicate questions, we will turn our attention to more outstanding features. It was convenient to speak about the small parameter in quantum electrodynamics, when investigating the interaction of electrons with photons. However, for strong interactions, e.g. nucleon-meson interaction, the corresponding parameter characterizing the interaction strength equals approximately 15. Therefore, it is quite hopeless to obtain even an asymptotic approximation by means of only a few terms of the perturbation theory series. As for the problem of the regularization of exact equations in quantum field theory, it is still rather far from its solution. In particular, it becomes necessary to define the multiplication of generalized functions of more general type than (2). This circumstance determines the great interest of any attempts to go beyond the framework of perturbation theory.

Recently a new trend characterized by an interesting and profound approach has appeared. Independently of any special variety of meson theory, these investigations start only from the most fundamental hypotheses of present-day quantum field theory such as covariance, unitarity and causality of the scattering matrix, and the properties of the energy spectrum. As a result of these investigations various so-called dispersion relations† have been obtained between quantities which can be determined from the scattering experiments.

Dispersion relations, therefore, can be experimentally verified. There are no indications yet that they contradict experimental facts. But the experimental investigations become more and more exact. Let us imagine the situation which would arise if it were shown that some dispersion relations are not fulfilled. This would mean that the most fundamental hypotheses of present-day theory need a radical reconsideration, since only they are taken into account in the derivation of the dispersion relations. In this connection it is quite understandable that great attention is being paid to the question of a rigorous proof of the dispersion relations. One must be sure that the violation of dispersion relations implies, in fact, the incorrectness of the basic axioms of the present-day theory.

Moreover, it became clear recently that the dispersion relations are a rather effective means for obtaining information on the interactions of elementary particles. They may also be used to obtain approximate equations for the determination of the probabilities of different collision processes which occur at not too high energies.

The dispersion relations are exact integral relations between the real

† Cf. [8], where further references are contained.

and imaginary parts of matrix elements of the scattering matrix. The
principal means for obtaining dispersion relations is Cauchy's theorem,
for the application of which it is necessary to know the analyticity
properties of the matrix elements in their complex arguments. Since the
matrix elements are generalized functions, the problem of analytical
extension of generalized functions arises.

In order to explain this, let us simplify somewhat the real situation.
Suppose that two generalized functions $f_r(t)$ and $f_a(t)$ of one variable t
are given, and that in virtue of the causality conditions $f_r(t)$ vanishes for
$t < 0$ (retarded function) and $f_a(t)$ vanishes for $t > 0$ (advanced function).
Then it is clear that their Fourier transforms $\tilde{f}_r(E)$ and $\tilde{f}_a(E)$ admit
analytic extension into the upper or lower half-planes, respectively. Now
we make use of the so-called energy-spectrum condition, mathematically
expressed as the coincidence of the functions $\tilde{f}_j(E)$ $(j = r, a)$ in the energy
range $|E| < \mu$. From this condition follows the existence of a single
analytic function $\tilde{f}(z)$ holomorphic throughout the plane of the complex
variable z, except for the cut

$$-\infty < \operatorname{Re} z \leqslant -\mu, \quad \mu \leqslant \operatorname{Re} z < \infty, \; \operatorname{Im} z = 0.$$

Since $f_j(t)$ are generalized functions, the function $\tilde{f}(z)$ so constructed is
bounded polynomially (cf. [9]) in the domain where

$$|\operatorname{Im} z| \geqslant \delta > 0.$$

Denoting by n the degree of the majorant polynomial and applying
Cauchy's theorem with an approximate contour of integration one may
obtain the integral representation for the function $\tilde{f}(z)$,

$$\tilde{f}(z) = \frac{(z-E_0)^{n+1}}{2\pi i}\left(\int_{-\infty}^{-\mu} + \int_{\mu}^{\infty}\right)\frac{[\tilde{f}_r(E')-\tilde{f}_a(E')]}{(E'-z)(E'-E_0)^{n+1}}dE'$$
$$+ \sum_{k=0}^{n}\frac{1}{k!}\tilde{f}^{(k)}(E_0)(z-E_0)^k,$$

where $|E_0| < \mu$. Passing to the real axis we may easily obtain the
required dispersion relation.

Of course, the real situation is much more complicated, because the
matrix elements depend upon many variables. So, for instance, for
functions $\mathscr{F}_r(x)$ and $\mathscr{F}_a(x)$ dependent upon one four-vector

$$x = (x^0, x^1, x^2, x^3)$$

and vanishing outside the advanced $(x \gtrsim 0)$ and retarded $(x \lesssim 0)$ light-
cones respectively we come across the following difficulty. In order to
carry out the above scheme for obtaining dispersion relations applicable

to this case, it is necessary to establish the analyticity properties of the functions

$$\mathscr{F}_j(E, \mathbf{e}) = \int \exp\left[i(Ex^0 - \{E^2 - \mu^2\}^{\frac{1}{2}} \mathbf{ex})\right] \mathscr{F}_j(x)\, dx, \quad |\mathbf{e}| = 1, \ (j = r, a)$$

(3)

in the complex domain of the energy variable E.

It can be seen, however, from the given formula that these functions may be analytically extended into the corresponding half-planes only under the non-physical condition $\mu^2 < 0$.† Moreover, the expressions (3) themselves are defined only on two intervals of the real axis

$$-\infty < E < -\mu, \quad \mu < E < \infty.$$

For the analytical extension of these functions we must resort to complicated artificial methods. A welcome circumstance is the so-called energy-spectrum condition according to which the functions $\mathscr{F}_j(p)$ coincide in a certain domain of the momentum variables.

A general purely mathematical problem which arises here belongs to the following type. In the space of n four-vectors $(x_1, \ldots, x_n)$ generalized functions $\mathscr{F}_j(x_1, \ldots, x_n)$ are given which vanish outside the corresponding combinations of the retarded and advanced light-cones. Let, further, their Fourier transforms $\mathscr{F}_j(p_1, \ldots, p_n)$ coincide in a certain domain $\mathscr{G}_0$ of the momentum variables $(p_1, \ldots, p_n)$. The question is whether there exists in the space of $4n$ complex variables

$$z = (z_1, \ldots, z_n) = (p_1 + iq_1, \ldots, p_n + iq_n)$$

a function $\mathscr{F}(z)$ holomorphic in a certain domain $\mathscr{G}$ and coinciding for real z from the domain $\mathscr{G}_0$ with the functions $\mathscr{F}_j(p_1, \ldots, p_n)$. Since the actual form of the functions $\mathscr{F}_j$ is unknown, the intersection of all such domains is of interest (if, of course, it is not empty).

We understand by analytical extension of the generalized function $\tilde{f}(p)$ from the real domain $\mathscr{G}_0$ into the complex domain $\mathscr{G}$ a holomorphic function $\tilde{f}(p + iq)$ in the domain $\mathscr{G}$ which possesses the properties:

(1) for every fixed q the function $\tilde{f}(p + iq)$ is a generalized function with respect to p, for those p for which $p + iq \in \mathscr{G}$;

(2) $\tilde{f}(p + iq)$ is convergent in the improper sense in the domain $\mathscr{G}_0$ to $\tilde{f}(p)$ if $q \to 0$. This means that for any trial function $\phi(p)$ the support of which is confined to $\mathscr{G}_0$ the limit

$$\int \tilde{f}(p + iq)\, \phi(p)\, dp \to \int \tilde{f}(p)\, \phi(p)\, dp$$

---

† In connection with this let us note the investigations by Schwartz[10] and Lions[11], where the association between the support of the generalized function and the analyticity properties of its Laplace transformation has been studied.

for $q \to 0$ exists. Thus, we see that the questions under consideration are concentrated on two intensively developing branches of mathematics— the theory of generalized functions and that of functions of several complex variables.

Let us give some definite results, which apart from their own interest may play the role of lemmas in the consideration of more complicated cases. We point out first an important theorem of the so-called 'edge of the wedge theorem' type.

Theorem 1. Let $\mathscr{F}_r(x)$ and $\mathscr{F}_a(x)$ be two generalized functions which vanish outside the advanced and retarded light-cones respectively. Let their Fourier transforms $\mathscr{F}_j(p)$ $(j = r, a)$ coincide in the domain $\mathscr{G}_0$. Then there exists a function $\mathscr{F}(z)$ of four complex variables

$$z = (z^0, z^1, z^2, z^3)$$

holomorphic in a domain $\mathscr{G}$, the real section of which is $\mathscr{G}_0$. This function is such that for all real $z = p$ in $\mathscr{G}_0$

$$\mathscr{F}(p) = \mathscr{F}_r(p) = \mathscr{F}_a(p).$$

The theorem stated above follows directly from the theorem proved in 1956 by one of the authors of this report ([8], Mathematical Appendix). This theorem states:

Let the functions $\mathscr{F}_j(x)$ $(j = r, a)$ possess the retarded and advanced properties respectively and let their Fourier transforms coincide in a sphere of radius η with centre at the origin. Then there exists a simultaneous analytical extension $\mathscr{F}(z)$ of the functions $\mathscr{F}_j(p)$ into the domain

$$|z_0| < 0.18\eta, \quad |\mathbf{z}| < 0.18\eta. \tag{4}$$

Furthermore, for sufficiently large indices r and q one may construct the 'universal' functions $H_j(z, p)$ with the properties:

(a) $H_j(z, p)$ as a function of p belongs to the class $C(r, q; 4)$ (cf. [3]) while as a function of z it is holomorphic in the domain (4).

(b) In the domain (4) $\mathscr{F}(z)$ admits the integral representation

$$\mathscr{F}(z) = \sum_{j=r,a} \int \mathscr{F}_j(p) H_j(z, p) \, dp. \tag{5}$$

We outline briefly a scheme for proving the last theorem. Introducing the cut-off factor $-\epsilon x_0^2 - \epsilon \mathbf{x}^2$ into the exponent we define the 'smoothed' Fourier transforms $\mathscr{F}_j(z, \epsilon)$ which are entire analytic functions of z and construct the function

$$\mathscr{F}(z, \epsilon) = \begin{cases} \mathscr{F}_r(z, \epsilon) - T(z, \epsilon) & (\operatorname{Im} z_0 > 0), \\ \mathscr{F}_a(z, \epsilon) - T(z, \epsilon) & (\operatorname{Im} z_0 < 0), \end{cases}$$

where
$$T(z, \epsilon) = \frac{1}{2\pi i} \int_{-R}^{R} \frac{\mathscr{F}_r(t, \mathbf{z}, \epsilon) - \mathscr{F}_a(t, \mathbf{z}, \epsilon)}{t - z_0} dt.$$

This function is holomorphic inside the circle of radius R.

Further, a set of z is defined for which $T(z, \epsilon)$ tends uniformly to zero if $\epsilon \to +0$. Making use of these facts and applying Cauchy's theorem we determine the domain of uniform convergence of the function $\mathscr{F}(z, \epsilon)$ for $\epsilon \to +0$ to the analytical function $\mathscr{F}(z)$. At the same time formula (5) is obtained.

Later Theorem 1 was proved anew by Bremermann, Oehme and Taylor[12] using other methods where the analogous theorem was called the 'edge of the wedge theorem'; a third proof was given by Jost and Lehmann[13]. It should be emphasized that there are at least three methods of proving the theorems such as Theorem 1 and its various generalizations and modifications.

(1) The method of the theory of generalized functions using different ways of parametrization. The basic ideas of this method in its simplest form have been illustrated by the scheme of the proof of Theorem 1 given above. This method was further developed and improved in our papers[14, 15, 16].

(2) The method of Bremermann, Oehme and Taylor[12] in which the basis for the proofs is the theory of functions of several complex variables: the expansion in series and construction of the envelope of holomorphy of domains. By an envelope of holomorphy of the domain $\mathscr{G}$ (cf., for instance, Bochner and Martin[17], chap. IV) we understand the intersection of the domains of holomorphy of all functions which are holomorphic in $\mathscr{G}$. If the envelope of holomorphy of the domain coincides with the domain itself, then we shall call this domain a domain of holomorphy.

(3) The method of Jost and Lehmann[13], which was later developed and improved by Dyson[18], the basis of which is the construction of an integral representation for generalized functions the Fourier transforms of which vanish outside the light-cone, whereas the functions themselves vanish in a given region of the type

$$a - \sqrt{(\mathbf{p}^2 + b^2)} < p^0 < \sqrt{(\mathbf{p}^2 + c^2)} - a. \tag{6}$$

Let us review Dyson's considerations. Let the function $f(x)$ vanish outside the light-cone. Then the function $\tilde{f}(p)$ admits a representation

$$\tilde{f}(p) = \mathscr{F}(p^0, p^1, p^2, p^3, 0, 0) = \mathscr{F}(\hat{p}), \tag{7}$$

where $\mathscr{F}(p)$ is a function of six variables $p = (p^0, p^1, \ldots, p^5)$ satisfying the six-dimensional wave equation

$$\square_6 \mathscr{F}(p) = 0.$$

Using the representation of a solution of the wave equation and taking into account (7) we obtain the representation for the function $\tilde{f}(p)$,

$$\tilde{f}(p) = \int d\Sigma_\alpha \left[\mathscr{F}(r), \frac{\partial}{\partial r^\alpha} \mathscr{D}(r - \hat{p}) \right], \tag{8}$$

where the integration is made over an arbitrary space-like surface in the six-dimensional space, and where $\mathscr{D}(p)$ is the odd invariant function in the six-dimensional space

$$\mathscr{D}(p) = \frac{1}{2\pi^2} \mathscr{E}(p^0) \delta'(p^2).$$

To satisfy the condition that $\tilde{f}(p)$ vanishes in the region (6) a corresponding space-like surface is chosen in representation (8) (from the so-called admissible hyperboloids in the six-dimensional space). This leads to the final representation

$$\tilde{f}(p) = \int d_4 u \int dk^2 \mathscr{E}(p^0 - u^0) \delta[(p - u)^2 - k^2] \overline{\psi}(u, k^2), \tag{9}$$

where the function $\overline{\psi}(u, k^2)$ vanishes everywhere except for those (u, k) which lie in a domain defined uniquely by the parameters a, b and c. We shall not write out the particular form of this domain. One can find the details in the original paper by Dyson[18].

Representation (9) is unique. The inverse statement also holds: any function represented in the form (9) vanishes in the region (6), whereas its Fourier-transform vanishes outside the light-cone.

Let us give some examples of the particular evaluation of the domains $\mathscr{G}$ under different assumptions about the domains $\mathscr{G}_0$.

(1) Let $\mathscr{G}_0$ be a slab $|p^0| < m$. Then $\mathscr{G}$ is a set of points $z = p + iq$ satisfying the inequality[8]

$$|\mathbf{q}| < |\mathrm{Im}\, \{(p_0 + iq_0)^2 - m^2\}^{\frac{1}{2}}|. \tag{10}$$

This result has been obtained also in the paper[12]. It was proved that the domain (10) does not admit further analytical extension, i.e. it is a domain of holomorphy. The proof is based on the construction of the envelope of holomorphy of the domains of the special form:

$$[p + iq : p^0 + iq^0 \in \mathscr{B}, |\mathbf{q}| < \mathscr{V}(p_0, q_0), |\mathbf{p}| < \infty], \tag{11}$$

the so-called semitubes. In Bremermann's paper[19] it was proved that the envelope of holomorphy of the semitube (11) is a domain of the same form where the function $\mathscr{V}$ is to be replaced by its smallest super-harmonic majorant.

(2) Let $\mathscr{G}_0$ be the exterior of the hyperboloid

$$(p^0)^2 < \mathbf{p}^2 + m \quad (m \gtrless 0).$$

Then $\mathscr{G}$ is the whole space with the exception of the points at infinity and the analytical hypersurface

$$(z^0)^2 - (z^1)^2 - (z^2)^2 - (z^3)^2 = \rho \quad (\max(0, m) \leqslant \rho < \infty).$$

Therefore, further analytical extension of the domain obtained is impossible (cf. Behnke and Thullen[20]). Moreover, the function $\mathscr{F}(z)$ admits the representation

$$\mathscr{F}(z) = \Sigma \mathscr{P}_k(z)\, \phi_k\{(z^0)^2 - (z^1)^2 - (z^2)^2 - (z^3)^2\}, \tag{12}$$

where the sum has a finite number of terms. In the representation (12) $\mathscr{P}_k(z)$ are polynomials, while the functions $\phi_k(w)$ are bounded polynomially. For $m > 0$, this result may also be obtained using the Jost–Lehmann–Dyson method if we put in (6) $a = 0$, $b^2 = c^2 = m$.

(3) If $\mathscr{G}_0$ is the domain

$$c_1 - (\gamma_1 \mathbf{p}^2 + \sigma^2)^{\frac{1}{2}} < p^0 < c_2 + (\gamma_2 \mathbf{p}^2 + \delta^2)^{\frac{1}{2}} \quad (\gamma_i \geqslant 0),$$

then the domain $\mathscr{G}$ is a set of points $p + iq$ satisfying the conditions: either $q^2 > 0$ or $q^2 \leqslant 0$,

$$c_1 - \{\gamma_1 e_2(|\mathbf{p}| - |q|) + \sigma^2\}^{\frac{1}{2}} + |q| < p^0 < c_2 + \{\gamma_2 e_2(|\mathbf{p}| - |q|) + \delta^2\}^{\frac{1}{2}} - |q|,$$

where $e_2(\xi) = \xi^2$ if $\xi > 0$ and $e_2(\xi) = 0$ if $\xi < 0$. In this example the domain $\mathscr{G}$ is, in fact, wider but is defined by more complicated conditions (cf. for example, papers [14] and [16]).

These results enable one to prove the following theorem which is the basic one for proving the dispersion relations for elastic particle scattering.

Theorem 2. Let four generalized functions

$$\mathscr{F}_{ij}(x_1, x_2, x_3, x_4) \quad (i, j = r, a),$$

of four 4-vectors be given, invariant with respect to the transformations of the inhomogeneous orthochronous Lorentz group. Suppose that these generalized functions satisfy the following conditions:

$$\mathscr{F}_{rr} = 0, \quad \text{if} \quad x_1 \lesssim x_3 \quad \text{or} \quad x_2 \lesssim x_4;$$

$$\mathscr{F}_{ra} = 0, \quad \text{if} \quad x_1 \lesssim x_3 \quad \text{or} \quad x_2 \gtrsim x_4;$$

$$\mathscr{F}_{ar} = 0, \quad \text{if} \quad x_1 \gtrsim x_3 \quad \text{or} \quad x_2 \lesssim x_4;$$

$$\mathscr{F}_{aa} = 0, \quad \text{if} \quad x_1 \gtrsim x_3 \quad \text{or} \quad x_2 \gtrsim x_4.$$

Suppose, further, that their Fourier transforms

$$\mathscr{F}_{ij}(p_1, p_2, p_3, p_4)$$

defined, evidently, on a manifold

$$p_1 + p_2 + p_3 + p_4 = 0 \tag{13}$$

satisfy the conditions:

$\mathscr{F}_{rj} = \mathscr{F}_{aj}$, if $p_1^2 < (\mathscr{M} + \mu)^2$ and $p_3^2 < \gamma^2 \mu^2$ $(j = r, a)$;

$\mathscr{F}_{ir} = \mathscr{F}_{ia}$, if $p_2^2 < (\mathscr{M} + \mu)^2$ and $p_4^2 < \gamma^2 \mu^2$ $(i = r, a)$;

$\mathscr{F}_{ij} = 0$, if $(p_1 + p_3)^2 < (\mathscr{M} + \mu)^2$ or $p_1^0 + p_3^0 < 0$ $(i, j = r, a)$.

We suppose that $\gamma > 1$, $\mathscr{M} + \mu \geqslant \gamma \mu$.

Let V and τ_0 be any fixed numbers satisfying the inequalities

$$V < \tau_0 \leqslant \mu^2, \quad \tau_0 \geqslant -\mu(2\mathscr{M} + \mu).$$

Then there exist a sufficiently small positive number ρ and a generalized function

$$\phi(z_1, z_2, \ldots, z_5; \xi)$$

of the real variable ξ with the properties:

(1) $\phi(z_1, \ldots, z_5; \xi)$ is analytic with respect to $(z_1, \ldots, z_5)$ in the domain

$$\left. \begin{array}{l} |z_1 - \mathscr{M}^2| < \rho, \quad |z_2 - \mathscr{M}^2| < \rho, \quad |z_3 - \tau| < \rho, \quad |z_4 - \tau| < \rho, \\ |z_5 + 4\Delta^2| < \rho/\xi, \end{array} \right\} \tag{14}$$

where the real parameters τ and Δ^2 independently vary in the intervals

$$V \leqslant \tau \leqslant \tau_0, \quad 0 \leqslant \Delta^2 \leqslant \gamma \mu^2 \left(1 - \frac{\mu^2 - \tau_0}{2\mu(\mathscr{M} + \mu)} \right) - \tau_0. \tag{15}$$

(2) $\phi(z_1, \ldots, z_5; \xi) = 0$ if $\xi < (\mathscr{M} + \mu)^2$.

(3) For real $(p_1, \ldots, p_4)$ from the manifold (13) for which the quantities

$$z_i = p_i^2 \ (i = 1, \ldots, 4), \quad z_5 = (p_1 + p_2)^2, \quad \xi = (p_1 + p_3)^2$$

satisfy the inequalities (14) and $p_1^0 + p_3^0 \geqslant 0$ we have a representation of the form

$$\mathscr{F}_{ij}(p_1, \ldots, p_4) = \phi[p_1^2, p_2^2, p_3^2, p_4^2, (p_1 + p_2)^2; (p_1 + p_3)^2].$$

The upper limit for Δ^2 in (15) may be increased. To do this we have to turn our attention to the numerical solution of a system of three algebraic equations with three unknowns (cf. the details in [16]). So, for instance, for meson-nucleon scattering ($\gamma = 3$, $\tau_0 = \mu^2$, $\mathscr{M} = 7\mu$) formula (15) yields

$$\Delta^2 \leqslant 2\mu^2, \tag{16}$$

while the numerical calculation gives

$$\Delta^2 \leqslant 2 \cdot 56 \mu^2.$$

The estimate (16) has been obtained also in the paper by Bremermann, Oehme and Taylor using the above-mentioned semitube method. In this paper attention is drawn to the fact that the interval (16) may be extended by means of further analytical extension.

Recently Lehmann obtained the result[21]

$$\Delta^2 < \frac{8}{3} \frac{2\mathcal{M} + \mu}{2\mathcal{M} - \mu} \mu^2 \sim 3\mu^2,$$

using the method of the Jost–Lehmann–Dyson integral representations. This result was also given in a footnote in [12].

The examples given by Bremermann *et al.*[12] show that an arbitrary increase of the upper limit for Δ^2 in Theorem 2 is impossible. Thus, for instance, a theorem of the type of Theorem 2 for nucleon-nucleon scattering may be proved only for the non-physical mass ratio

$$\mu > (\sqrt{2} - 1)\mathcal{M}, \qquad (17)$$

if one takes into consideration only the causality conditions and spectrum. However, there exist also the symmetry conditions of the Green functions. It is an open question whether these conditions make further extension possible or not. At any rate, for the nucleon vertex function the example given by Jost[22] shows that even the use of the symmetry conditions does not provide any dispersion relations if

$$\mu < \left(\frac{2}{\sqrt{3}} - 1 \right) \mathcal{M}$$

which is actually satisfied for real mass ratios.

In conclusion, let us dwell upon the papers devoted to the study of the analyticity properties of the vacuum expectation values of scalar fields. In Källén and Wightman's paper[23] were established those analyticity properties of the vacuum expectation values of three scalar fields

$$\mathcal{F}^{\mathcal{A}\mathcal{B}\mathcal{C}}(x - x', x' - x'') = \langle 0 \, | \, \mathcal{A}(x) \, \mathcal{B}(x') \, \mathcal{C}(x'') | \, 0 \rangle,$$

which are due to local commutativity, the spectrum properties and Lorentz invariance. It was found that all six different vacuum expectation values which may be obtained by the commutation of three operators $\mathcal{A}(x)$, $\mathcal{B}(x')$ and $\mathcal{C}(x'')$ are the boundary values of the same analytical function of three complex variables the domain of holomorphy of which appears to be limited by seven analytical hypersurfaces. It was proved that further analytical extension of the domain obtained is impossible if only the above-mentioned properties are used. The methods employed in the theory of functions of several complex variables, particularly the analytical extensions, are widely used in the proofs of this paper. Moreover, the Bargmann–Hall–Wightman theorem[24] is essentially used. It states:

Let f, a function of n complex four-vectors $z_1, \ldots, z_n$, be analytic in the tube

$$-\infty < \operatorname{Re} z_j < \infty, \quad \operatorname{Im} z_j^0 > |\operatorname{Im} \mathbf{z}_j| \quad (j = 1, \ldots, n) \qquad (18)$$

and invariant with respect to the homogeneous orthochronous Lorentz group. Then f is a function of only the scalar products $z_j z_k$ $(j, k = 1, \ldots, n)$. This function is analytic on the complex manifold of the corresponding scalar products, when $(z_1, \ldots, z_n)$ vary over the domain (18).

It is interesting to note that, though domain (18) does not contain any real point, Lorentz invariance implies analyticity in a certain region of the real points. In fact, as Jost has shown[25], this region consists of points $(x_1, \ldots, x_n)$ possessing the following property: any convex domain of the form

$$\xi = \sum_{k=1}^{n} \lambda_k x_k \quad \left(\lambda_k \geqslant 0, \quad \sum_{k=1}^{n} \lambda_k = 1 \right)$$

consists entirely of space-like vectors. The example constructed by Jost in[25] shows that this domain cannot be extended without additional assumptions.

The results given play an essential role in establishing the connection between the analyticity properties of the vacuum expectation values and the principal hypotheses of the theory (cf. Wightman[26], Dyson[27] and the papers mentioned above by Källén and Wightman[23], Hall and Wightman[24] and Jost[25]).

REFERENCES

[1] Соболев, С. Л. Méthode nouvelle à résoudre le problème de Cauchy pour les équations linéaires hyperboliques normales. *Матем. сб.* 1 (43), 39–71 (1936).

[2] Schwartz, L. *Théorie des distributions*, 1–2. Paris 1950–51.

[3] Боголюбов, Н. Н. и Ширков, Д. В. Вопросы квантовой теории поля, 1, II, *Успехи физ. наук*, 55, 149–214 (1955); 57, 1–92 (1955).

[4] Dyson, F. J. *S*-matrix in quantum electrodynamics. *Phys. Rev.* 75, 1736–1745 (1949).

[5] Salam, A. Divergent integrals in renormalizable field theories. *Phys. Rev.* 84, 426–431 (1951).

[6] Парасюк, О. С. Умножение причинных функций при несовпадающих аргументах, *Известия Ак. Наук СССР, сер. математич.* 20, no. 6 (1956).

[7] Bogoliubow, N. N. und Parasiuk, O. S. Über die Multiplikation der Kausalfunktionen in der Quantentheorie der Felder. *Acta math.* 97, 227–266 (1957).

[8] Боголюбов, Н. Н., Медведев, Б. В. и Поливанов, М. К. *Вопросы теории дисперсионных соотношений.* М., Гостехиздат, 1958.

[9] Боголюбов, Н. Н. и Парасюк, О. С. Об аналитическом продолжении обобщенных функций. *ДАН СССР,* 109, 717–719 (1956).

[10] Schwartz, L. Transformation de Laplace des distributions. *Medd. Lunds Univ. mat. Semin. (Supplementband),* 196–206 (1952).

[11] Lions, J. L. Supports dans la transformation de Laplace. *J. Analyse Math.* 1–2, 369–380 (1952–3).

[12] Bremermann, H. J., Oehme, R. and Taylor, J. G. A proof of dispersion relations in quantized field theories. *Phys. Rev.* 109, 2178–2190 (1958).

[13] Jost, R. and Lehmann, H. Integraldarstellung kausaler Kommutatoren. *Nuovo Cimento,* 5, 1598–1610 (1957).

[14] Боголюбов, Н. Н. и Владимиров, В. С. Об аналитическом продолжении обобщенных функций. *Изв. Ак. Наук, сер. математич.* 22, 15–48 (1958).

[15] Боголюбов, Н. Н. и Владимиров, В. С. Одна теорема об аналитическом продолжении обобщенных функций, *Научные доклады высшей школы,* 3, 26–35 (1958).

[16] Владимиров, В. С. Об определении области аналитичности. *Изв. Ак. Наук, сер. математич,* 23, 275–294 (1959).

[17] Бохнер, С. и Мартин, У. Т. *Функции многих комплексных переменных,* ИИЛ, Москва, 1951.

[18] Dyson, F. J. Integral representations of causal commutators. *Phys. Rev.* 110, 1460–1464 (1958).

[19] Bremermann, H. J. Die Holomorphiehüllen der Tuben- und Halbtubengebiete. *Math. Annalen,* 127, 406–423 (1954).

[20] Behnke, H. und Thullen, P. *Theorie der Funktionen mehrerer komplexer Veränderlichen.* Berlin, 1934.

[21] Lehmann, H. Analytic properties of scattering amplitudes as functions of momentum transfer. *Nuovo Cimento,* 10, 579–589 (1958).

[22] Jost, R. Ein Beispiel zum Nucleon-Vertex. *Helv. phys. acta,* 31, 263–273 (1958).

[23] Källén, C. and Wightman, A. S. The analytic properties of the vacuum expectation value of a product of three scalar local fields (preprint, 1958).

[24] Hall, D. and Wightman, A. S. A theorem on invariant analytic functions with applications to relativistic quantum field theory. *K. danske vidensk. selsk (Mat.-fys. Medd.),* 31, no. 5 (1957).

[25] Jost, R. Eine Bemerkung zum CTP-Theorem. *Helv. phys. acta,* 30, 409–416 (1957).

[26] Wightman, A. S. Quantum field theory in terms of vacuum expectation values. *Phys. Rev.* 101, 860–866 (1956).

[27] Dyson, F. J. Connection between local commutativity and regularity of Wightman Functions (preprint, 1958).

SUR LES FONCTIONS DE PLUSIEURS VARIABLES COMPLEXES: LES ESPACES ANALYTIQUES

Par HENRI CARTAN

Je voudrais résumer ici quelques résultats obtenus depuis trois ou quatre ans dans la théorie des espaces analytiques.

1. Notions préliminaires

Les fonctions analytiques de plusieurs variables complexes n'ont été étudiées, pendant longtemps, que dans les ouverts des espaces numériques C^n (C^n désigne l'espace dont les points ont pour coordonnées n nombres complexes). A une époque relativement récente on a abordé l'étude systématique des *variétés analytiques* (complexes); la notion de variété analytique est aujourd'hui familière à tous les mathématiciens. En gros, une variété analytique de dimension (complexe) n est un espace topologique séparé, au voisinage de chaque point duquel on s'est donné un ou plusieurs systèmes de 'coordonnées locales' (complexes), en nombre égal à n, le passage d'un système de coordonnées locales à un autre s'effectuant par des transformations holomorphes. Pour tout ouvert U d'une variété analytique X, on a la notion de *fonction holomorphe* dans U (f est holomorphe dans U si, au voisinage de chaque point de U, f peut s'exprimer comme fonction holomorphe des coordonnées locales); les fonctions holomorphes dans U forment un anneau $\mathscr{H}(U)$. Notons que la notion de fonction holomorphe a un caractère local: pour qu'une f continue dans U soit holomorphe dans U, il faut et il suffit que la restriction f_i de f à chacun des ouverts U_i d'un recouvrement ouvert de U soit holomorphe dans U_i. Ceci conduit à considérer, en chaque point $x \in X$, l'anneau $\mathscr{H}_x$ des 'germes' de fonctions holomorphes au point x (anneau qui est la limite inductive des anneaux $\mathscr{H}(U)$ relatifs aux ouverts U contenant x). La connaissance, pour chaque point $x \in X$, de l'anneau $\mathscr{H}_x$, détermine entièrement la structure de variété analytique de X. D'une façon précise, supposons donné, pour chaque point x d'un espace topologique séparé X, un sous-anneau $\mathscr{H}_x$ de l'anneau des germes de fonctions continues (à valeurs complexes) au point x (nous dirons alors que X est un espace *annelé*); pour qu'il existe sur X une structure de variété analytique de dimension n telle que, pour chaque x, $\mathscr{H}_x$ soit précisément l'anneau des germes de fonctions holomorphes au point x, il faut et il suffit que X puisse être recouvert par des ouverts U_i jouissant de la propriété suivante: il existe un homéomorphisme ϕ_i de U_i sur un ouvert A_i de $\mathbf{C}^n$, de manière que, pour chaque point $x \in U_i$, l'homéomorphisme ϕ_i transporte l'anneau $\mathscr{H}_x$ sur l'anneau des germes de fonctions holomorphes au point $\phi_i(x)$ de l'espace $\mathbf{C}^n$. Cette condition exprime, en somme, que ϕ_i définit un *isomorphisme* de U_i, comme espace annelé, sur l'ouvert A_i muni de sa structure naturelle d'espace annelé. Ainsi les ouverts de $\mathbf{C}^n$, munis de leur structure naturelle d'espace annelé, constituent des *modèles locaux* pour les variétés analytiques de dimension n.

Mais la notion de variété analytique n'est pas suffisamment générale. Prenons un exemple: une variété algébrique, plongée sans singularité dans un espace projectif complexe, peut bien être considérée comme une variété analytique; mais une variété algébrique plongée avec singularités dans un espace projectif ne peut pas rentrer dans le cadre trop étroit des variétés analytiques. Cet exemple suggère la nécessité d'élargir la notion de variété analytique, et explique pourquoi la notion plus générale d'"espace analytique" a récemment acquis droit de cité en Mathématiques. Les espaces analytiques sont, en quelque sorte, des variétés analytiques pouvant admettre certaines singularités internes. Etant donné l'importance qu'a prise récemment la notion d'espace analytique, nous allons entrer dans quelques détails.

2. Sous-ensembles analytiques

On va utiliser une catégorie de 'modèles' plus étendue que la catégorie des ouverts de $\mathbf{C}^n$. Avant de la définir avec précision, il nous faut rappeler une définition et quelques résultats classiques. Soit A un ouvert de $\mathbf{C}^n$; on dit qu'une partie M de A est un *sous-ensemble analytique* de A si M est *fermé* dans A et si, pour chaque point $x \in M$, il existe un ouvert U contenant x et contenu dans A, et un système fini de fonctions f_i holomorphes dans U, de telle manière que $M \cap U$ soit exactement l'ensemble des points de U où s'annulent simultanément les fonctions f_i; en bref, un sous-ensemble analytique de A est un sous-ensemble fermé qui, au voisinage de chacun de ses points, peut se définir par des équations holomorphes. La structure locale des ensembles analytiques est bien connue depuis Weierstrass;† si x est un point d'un ensemble analytique M, M est, au voisinage de x, réunion d'un nombre fini d'ensembles analytiques M_i dont chacun est 'irréductible' au point x, et les M_i sont

_____

† Voir par exemple [27], et [7], Exposé 14.

entièrement déterminés au point de vue local: si l'on a deux décompositions de M en composantes irréductibles au point x, ces deux décompositions coïncident, à l'ordre près, dans un voisinage assez petit de x. De plus, on peut donner une description locale d'un ensemble analytique M *irréductible* au point x: il est possible de choisir, au voisinage de x, des coordonnées locales $x_1, ..., x_n$ dans l'espace ambiant, nulles au point x, et un entier $p \leqslant n$, de manière que soient vérifiées les conditions suivantes. L'application f de M dans $\mathbf{C}^p$ définie par

$$(x_1, ..., x_n) \to (x_1, ..., x_p)$$

est ce qu'on appelle un 'revêtement ramifié' de degré k au voisinage de x: cela veut dire que f applique tout voisinage assez petit de x dans M *sur* un voisinage de 0 dans $\mathbf{C}^p$, et que l'image réciproque d'un point de $\mathbf{C}^p$ assez voisin de 0 se compose 'en général' de k points distincts de M (voisins de x), et possède en tout cas au plus k points; l'entier p s'appelle la dimension (complexe) de M au point x (et en fait, la dimension topologique de M, au voisinage de x, est égale à $2p$). D'une façon plus précise, il est possible d'exclure de $\mathbf{C}^p$, au voisinage de 0, un sous-ensemble analytique R dont toutes les composantes irréductibles en 0 sont de dimension $< p$, de manière que la restriction de f à $M - f^{-1}(R)$ fasse de $M - f^{-1}(R)$ un revêtement (véritable) à k feuillets de $\mathbf{C}^p - R$, du moins dans des voisinages assez petits de $x \in M$ et de $0 \in \mathbf{C}^p$. Les coordonnées de chacun des k points de $M - f^{-1}(R)$ que f transforme en un point donné $(x_1, ..., x_p)$ sont des fonctions *holomorphes* de $x_1, ..., x_p$. On voit que $M - f^{-1}(R)$ est, au voisinage de chacun de ses points, une sous-variété analytique (sans singularités) de dimension p dans l'espace ambiant $\mathbf{C}^n$; de plus cette variété est *connexe*: d'une façon précise, il existe un système fondamental de voisinages ouverts de x, dans l'espace ambiant, qui coupent $M - f^{-1}(R)$ suivant un ensemble connexe.

Soit M un sous-ensemble analytique au voisinage d'un point $x \in \mathbf{C}^n$; nous avons dit que M est réunion, au voisinage de x, de ses composantes irréductibles au point x. On dit que M est de dimension $\leqslant p$ au point x si toutes les composantes irréductibles de M au point x ont une dimension $\leqslant p$; et l'on dit que M est purement p-dimensionnel au point x si toutes les composantes de M au point x ont la dimension p.

Soit à nouveau M un sous-ensemble analytique d'un ouvert $A \subset \mathbf{C}^n$. On dit que M est de dimension $\leqslant p$ si M est de dimension $\leqslant p$ en chacun de ses points; et que M est purement p-dimensionnel si M est purement p-dimensionnel en chacun de ses points. Un point $x \in M$ est dit *régulier* si M est, au voisinage de ce point, une sous-variété de l'espace ambiant;

l'ensemble des points réguliers de M est ouvert dans M et dense dans M, et l'ensemble des points non-réguliers, ou *singuliers*, de M, est un sous-ensemble analytique.† Si M est purement p-dimensionnel, l'ensemble des points réguliers de M est une sous-variété de dimension p en chacun de ses points, et l'ensemble des points singuliers de M est un sous-ensemble analytique de dimension $\leqslant p-1$.

Tout cela est bien classique. Mais un résultat récent de Lelong[22], dont de Rham[33] vient de donner une autre démonstration, concerne l'intégration des formes différentielles sur les sous-ensembles analytiques et fournit une information précieuse sur la nature des singularités d'un tel ensemble. Soit M un sous-ensemble analytique purement p-dimensionnel d'un ouvert $A \subset \mathbf{C}^n$; soit M' l'ouvert de M formé des points réguliers de M; il est bien connu que, au voisinage de chaque point de $M-M'$, M' est de volume $(2p)$-dimensionnel fini, et par suite toute forme différentielle ω de degré $2p$, définie dans A et à support compact, possède une intégrale $\int \omega$ étendue à M'. Ainsi M définit un *courant* de dimension $2p$ dans l'ouvert A. Lelong et de Rham montrent que *ce courant est fermé*; cela revient à dire que, pour toute forme différentielle ϖ de degré $2p-1$, à support compact, l'intégrale $\int d\varpi$ étendue à M' est nulle (ce résultat vaut si la forme ϖ est de classe C^1). Ce théorème exprime, en somme, que l'ensemble $M-M'$ des points singuliers de M est négligeable comme courant de dimension (réelle) $2p-1$; on savait seulement que sa dimension comme espace topologique est $\leqslant 2p-2$.

3. Espaces analytiques

Etant maintenant un peu familiarisés avec les sous-ensembles analytiques, nous pouvons aborder la définition générale d'un *espace analytique*.‡ Soit M un sous-ensemble analytique d'un ouvert $A \subset \mathbf{C}^n$; M possède une structure annelée naturelle: on attache à chaque point $x \in M$ l'anneau $\mathscr{H}_x$ des germes de fonctions induits sur M par les germes de fonctions holomorphes de l'espace ambiant $\mathbf{C}^n$. Les espaces annelés ainsi attachés aux divers sous-ensembles analytiques (pour tous les ouverts $A \subset \mathbf{C}^n$, et pour toutes les valeurs de n) sont nos 'modèles'. Par définition, un 'espace analytique' est un espace annelé X, dont la topologie est séparée, et qui satisfait à la condition suivante: chaque

† Voir [8], Exposé 9.

‡ Il y a essentiellement deux définitions possibles d'un espace analytique: celle dite de Behnke–Stein (voir [4]), et celle dite de Cartan–Serre (voir [7], Exposé 13; [8], Exposé 6; et [35]). L'équivalence des deux définitions n'a été démontrée que tout récemment par Grauert et Remmert (*C.R. Acad. Sci., Paris*, 245, 918–921 (1957)). Nous donnons ici la définition 'de Cartan–Serre'.

point de X possède un voisinage ouvert U qui est *isomorphe* (comme espace annelé) à l'un des modèles qu'on vient de définir. Etant donnés deux espaces analytiques X et X', une application $\phi: X \to X'$ sera dite analytique (ou holomorphe) si c'est un 'morphisme d'espaces annelés', ce qui signifie ceci: ϕ est une application continue telle que, pour tout point $x \in X$ et tout germe $f \in \mathscr{H}_{\phi(x)}$, le germe composé $f \circ \phi$ appartienne à $\mathscr{H}_x$. En particulier, les fonctions holomorphes (scalaires) sur X ne sont autres que les fonctions continues qui, en chaque point $x \in X$, appartiennent à l'anneau $\mathscr{H}_x$. L'anneau $\mathscr{H}_x$ est ainsi l'anneau des germes de fonctions holomorphes au point x.

On voit que les espaces analytiques forment une 'catégorie' avec morphismes, au sens technique de ce terme; et les variétés analytiques forment une sous-catégorie de la catégorie des espaces analytiques (c'est d'ailleurs une sous-catégorie 'pleine', dans le jargon des spécialistes; autrement dit, les applications analytiques d'une variété analytique X dans une autre X' sont les mêmes, que l'on considère X et X' comme des variétés analytiques ou comme des espaces analytiques).

On a une notion évidente de *sous-espace analytique* d'un espace analytique X; c'est un sous-ensemble fermé Y de X, tel que Y puisse être défini, au voisinage de chacun de ses points y, en annulant un nombre fini de fonctions de l'anneau $\mathscr{H}_y$ de l'espace ambiant X. On munit Y d'une structure annelée en associant à chaque point $y \in Y$ l'anneau des germes induits sur Y par les éléments de $\mathscr{H}_y$; et on voit facilement que cet espace annelé Y est un espace analytique.

Toutes les notions qui ont été définies sur les modèles et qui ont un caractère local, invariant par isomorphisme, se transportent aux espaces analytiques. Par exemple, un espace analytique X peut, en un de ses points x, être *irréductible* ou non; en tout cas, x possède un voisinage dans lequel X est réunion d'un nombre fini de sous-espaces analytiques irréductibles au point x. On a la notion d'espace purement n-dimensionnel, et celle d'espace de dimension $\leqslant n$. Un espace analytique X, purement n-dimensionnel, est de dimension topologique $2n$; si X est de dimension $\leqslant n$, il est de dimension topologique $\leqslant 2n$. Un point $x \in X$ est *uniformisable* s'il possède un voisinage ouvert isomorphe à une variété analytique; l'ensemble des points uniformisables est un ouvert partout dense, et son complémentaire est un sous-espace analytique. Si X est purement 1-dimensionnel, tous les points de X sont uniformisables. Tous ces faits sont bien connus.

On a aussi la notion *globale* d'irréductibilité: X est globalement irréductible si X n'est pas la réunion de deux sous-espaces analytiques

X' et X'' tous deux distincts de X. Si X n'est pas irréductible, on appelle *composante irréductible* de X (au sens global) tout sous-espace analytique Y tel que X soit réunion de Y et d'un sous-espace analytique $Y' \neq X$. On démontre que X est la réunion de ses composantes irréductibles, et que celles-ci forment une famille *localement finie* (i.e. chaque point de x possède un voisinage qui ne rencontre qu'un nombre fini de composantes irréductibles). Les composantes irréductibles de X ne sont pas autre chose que les adhérences des composantes connexes de l'ensemble des points uniformisables de X. Tout espace irréductible est purement dimensionnel.

On a une catégorie intermédiaire entre celle des variétés analytiques et celle des espaces analytiques: c'est celle des espaces analytiques *normaux*. Ce sont les espaces annelés dont les modèles sont les sous-ensembles analytiques normaux des ouverts d'un espace $\mathbf{C}^n$. Pour qu'un espace analytique X soit normal, il faut et il suffit que, pour chaque point $x \in X$, l'anneau $\mathscr{H}_x$ soit intègre et intégralement clos; en particulier, X est irréductible en chaque point x (car ceci équivaut à dire que l'anneau $\mathscr{H}_x$ est intègre). Toute variété analytique est un espace normal. Si X est un espace analytique quelconque, l'ensemble des $x \in X$ tels que $\mathscr{H}_x$ soit intègre et intégralement clos est un ouvert partout dense, et son complémentaire est un sous-espace analytique (Oka).† De plus, on peut attacher canoniquement à l'espace X un espace analytique normal $\tilde{X}$ (dit 'normalisé' de X), dont les points sont en correspondance biunivoque avec les composantes irréductibles de X en chacun de ses points, de telle manière que l'application naturelle $\tilde{X} \to X$ soit holomorphe et 'propre' (i.e. l'image réciproque de tout compact de X est un compact de $\tilde{X}$). L'espace normalisé $\tilde{X}$ jouit de la propriété universelle suivante: toute application holomorphe surjective $Y \to X$, où Y est un espace analytique normal, se factorise d'une seule manière en $Y \to \tilde{X} \to X$, où l'application $Y \to \tilde{X}$ est holomorphe.

Signalons deux propriétés importantes des espaces analytiques normaux: les composantes irréductibles (au sens global) d'un espace normal ne sont autres que ses composantes connexes; les points non-uniformisables d'un espace normal de dimension n forment un sous-espace analytique de dimension $\leqslant n-2$.‡

4. Etude géométrique d'une application analytique $X \to Y$

Soit f une application analytique d'un espace analytique X dans un espace analytique Y. Il est évident que l'image réciproque $f^{-1}(Y')$ d'un

---

† Voir [25], et [8], Exposé 10.   ‡ Voir [8], Exposé 11, théorème 2.

sous-espace analytique Y' de Y est un sous-espace analytique de X. En revanche, on sait bien que l'image directe d'un sous-espace analytique de X n'est pas, en général, un sous-espace analytique de Y. Déjà $f(X)$ n'est pas nécessairement fermé dans Y; et il n'est même pas vrai, en général, que chaque point de $f(X)$ possède dans Y un voisinage ouvert U tel que $f(X) \cap U$ soit un sous-ensemble analytique de U. Un contre-exemple classique est le suivant: X est l'espace $\mathbf{C}^2$ (coordonnées x_1, x_2), Y est l'espace $\mathbf{C}^2$ (coordonnées y_1, y_2), et f est l'application

$$(x_1, x_2) \to (x_1, x_1 x_2);$$

l'image $f(X)$ se compose de tous les points de Y, sauf ceux pour lesquels $y_1 = 0$, $y_2 \neq 0$; quel que soit l'ouvert U contenant l'origine, $U \cap f(X)$ n'est pas fermé dans U.

Une analyse subtile du comportement d'une application holomorphe $f: X \to Y$ a conduit Stein et Remmert à d'importants résultats,[†] dont je voudrais mentionner quelques-uns. Considérons, pour chaque point $x \in X$, la 'fibre' $f^{-1}(f(x))$, qui est un sous-espace analytique contenant x; soit $d(x)$ la plus grande des dimensions de ses composantes irréductibles au point x; pour chaque entier k, soit X_k l'ensemble des points de X où $d(x) \geqslant k$. On a $X = X_0 \supset X_1 \supset \ldots \supset X_k \supset \ldots$, et l'on montre que les X_k sont des *sous-espaces analytiques*; de plus, chaque fibre de l'application $X_k \to Y$ induite par f est de dimension $\geqslant k$ en chacun de ses points. Si X est de dimension finie, on définit une suite de sous-ensembles analytiques

$$\varnothing = X(-1) \subset X(0) \subset \ldots \subset X(r) \subset \ldots,$$

avec $X(r) = X$ pour r grand. Les $X(r)$ se définissent par récurrence descendante sur r: à chaque composante irréductible A de $X(r+1)$, on associe le sous-espace A_k (avec $k = \dim(A) - r$), et $X(r)$ est réunion des A_k. On démontre alors ceci: tout point $x \in X(r) - X(r-1)$ possède dans $X(r)$ un voisinage dont l'image par f est un sous-ensemble analytique purement r-dimensionnel au point $f(x)$. A partir de là, on démontre le résultat fondamental de Remmert: *si l'application analytique $f: X \to Y$ est propre* (c'est-à-dire, répétons-le, si l'image réciproque de tout compact de Y est un compact de X), *alors l'image $f(X)$ est un sous-espace analytique de Y, et la dimension de $f(X)$ est égale au plus petit des entiers r tels que $X(r) = X$*; de plus, si X est (globalement) irréductible, il en est de même de $f(X)$.

Dans les démonstrations des résultats précédents, l'on fait un usage essentiel d'un théorème de Remmert et Stein,[‡] qui sert dans maintes

† Voir [29], [32], [38].   ‡ Voir [28], ainsi que [8], Exposés 13 et 14 de Stein.

circonstances, et se formule ainsi: *soit Z un sous-ensemble analytique, de dimension < n, d'un espace analytique Y; et soit A un sous-ensemble analytique purement n-dimensionnel de l'ouvert Y − Z; alors l'adhérence de A dans Y est un sous-ensemble analytique purement n-dimensionnel de Y.*

Le théorème de Remmert s'applique notamment lorsque X est un espace analytique *compact*, car f est alors automatiquement propre: l'image $f(X)$ est alors toujours un sous-ensemble analytique de Y. En particulier, supposons que Y soit l'espace projectif (complexe) P_n; pour toute application analytique $f: X \to P_n$, l'image $f(X)$ est un sous-ensemble *algébrique* de P_n, d'après le célèbre théorème de Chow (lequel est d'ailleurs une conséquence immédiate du théorème de Remmert–Stein, comme je l'avais signalé dans la conférence que j'ai faite au Congrès de Harvard en 1950). A titre d'application,† considérons k fonctions méromorphes f_i sur un espace analytique *compact* X; elles définissent une application analytique $f: X \to (P_1)^k$. A vrai dire, ceci n'est pas tout à fait correct, à cause des points d'indétermination des f_i; mais on voit facilement qu'on peut 'modifier' l'espace X de façon que l'application f soit partout définie: d'une façon précise, il existe un espace analytique compact X' et une application analytique $\pi: X' \to X$ qui définit un isomorphisme des corps de fonctions méromorphes $K(X)$ et $K(X')$, et qui jouit de la propriété que les $f_i \circ \pi = g_i$ n'ont pas points d'indétermination sur X'. Les g_i définissent donc une application analytique $g: X' \to (P_1)^k$, dont l'image est un sous-espace algébrique. Si les f_i sont analytiquement dépendantes (c'est-à-dire si les différentielles df_i sont linéairement dépendantes), il en est de même des g_i, donc le rang de l'application g est $< k$, et l'image $g(X')$ est distincte de $(P_1)^k$; il existe donc un polynôme non identiquement nul qui s'annule sur l'image de g, autrement dit les f_i sont *algébriquement dépendantes* (théorème de Thimm, démontré ainsi par Remmert). De la même manière, on montre que si n désigne la dimension de l'espace analytique compact X, le corps $K(X)$ des fonctions méromorphes est une extension algébrique simple d'un corps de fractions rationnelles à k variables, avec $k \leqslant n$ (théorème annoncé tout d'abord par Chow).

5. Quotients d'espaces analytiques

Les questions précédentes sont en rapport étroit avec le problème, étudié par Stein[39], des espaces quotients d'espaces analytiques. Soit X un espace analytique, que nous supposerons *normal*; soit R une relation d'équivalence *propre* sur X (ceci signifie que R satisfait à l'une des trois conditions suivantes, équivalentes:

(i) le R-saturé de tout compact de X est compact;

(ii) l'espace quotient X/R est localement compact et l'application $p\colon X \to X/R$ est propre;

(iii) R désignant le graphe de la relation d'équivalence, l'application de projection $R \to X$ est propre).

Sur l'espace quotient X/R on a une structure annelée évidente: à tout ouvert $U \subset X/R$ on associe l'anneau $\mathscr{H}(U)$ des fonctions continues f sur U, telles que $p \circ f$ soit holomorphe dans $p^{-1}(U)$; cet anneau s'identifie à celui des fonctions holomorphes dans $p^{-1}(U)$ et constantes sur les classes d'équivalence. Le problème se pose de savoir si X/R, muni de cette structure annelée, est un espace analytique normal. Il faut, bien entendu, faire des hypothèses convenables sur la relation R. Nous ferons désormais l'hypothèse suivante:

(H) chaque point $z \in X/R$ possède un voisinage ouvert U tel que, pour tout couple de fibres distinctes de $p^{-1}(U)$, il existe une application holomorphe de $p^{-1}(U)$ dans un espace analytique, constante sur les fibres, et prenant des valeurs distinctes sur les deux fibres en question. (Par exemple, il en est bien ainsi lorsque les fonctions de l'anneau $\mathscr{H}(U)$ séparent les points de U.)

Avec l'hypothèse (H), il n'est pas encore certain que X/R soit un espace analytique normal, mais c'est presque vrai. D'une façon précise, l'application propre $p\colon X \to X/R$ admet une factorisation $X \xrightarrow{f} Y \xrightarrow{g} X/R$, où Y est un espace analytique normal, f une surjection holomorphe (et propre), et g un morphisme d'espaces annelés jouissant de la propriété suivante: les fibres de g sont des ensembles finis, et il existe un sous-ensemble fermé 'mince' A de X/R tel que $g^{-1}(z)$ soit réduit à un seul point lorsque $z \notin A$ (on dit qu'une partie fermée A de X/R est 'mince' si tout point de A possède un voisinage ouvert U tel qu'il existe une fonction de $\mathscr{H}(U)$ nulle sur $A \cap U$ et non identiquement nulle). De plus, une telle factorisation est *unique* 'à un isomorphisme près', et g induit un isomorphisme des espaces annelés $Y - g^{-1}(A)$ et $(X/R) - A$.

L'espace Y est ainsi déterminé à un isomorphisme près par la relation d'équivalence R (supposée satisfaire à (H)); on peut l'appeler le *quotient analytique* de X relativement à R.

Le théorème précédent, qui résulte des travaux de Stein, possède d'intéressantes applications, comme on le verra tout à l'heure.

6. Classification des espaces analytiques

Proposons-nous d'abord de voir comment un espace analytique X se comporte vis-à-vis des fonctions holomorphes. Soit $\mathscr{H}(X)$ l'anneau des fonctions holomorphes dans X (tout entier); cet anneau peut se réduire aux constantes, par exemple si X est *compact* et connexe. Considérons, pour un espace analytique X, les propriétés suivantes:

(a) Les éléments de $\mathscr{H}(X)$ séparent les points de X (autrement dit, pour tout couple de points distincts x, x', il existe une f holomorphe dans X et telle que $f(x) \neq f(x')$).

(b) Pour chaque point $x \in X$, il existe un système fini de $f_i \in \mathscr{H}(X)$ qui est 'séparant' au point x (on entend par là que, pour l'application $f: X \to \mathbf{C}^n$ définie par les n fonctions f_i, x est un point isolé de la fibre $f^{-1}[f(x)]$).

(c) Tout sous-ensemble analytique compact de X est fini.

Il est facile de voir que (a) entraîne (b). D'autre part, il est presque immédiat que (b) entraîne (c).

Grauert[14] a démontré le résultat surprenant que voici: si X est irréductible, la condition (b) entraîne que X est *réunion dénombrable de compacts* (on sait que Calabi et Rosenlicht[6] avaient donné l'exemple d'une variété analytique, connexe, qui n'est pas réunion dénombrable de compacts). Grauert a aussi montré que si X, irréductible et de dimension n, satisfait à (b), il existe un système de n fonctions $f_i \in \mathscr{H}(X)$ qui est 'séparant' en tout point $x \in X$.

A côté des propriétés (a), (b), (c), il est une propriété d'une nature différente: on dit que X est *holomorphiquement convexe* si, pour tout compact $K \subset X$, l'ensemble $\hat{K}$ des $x \in X$ tels que l'on ait

$$|f(x)| \leqslant \sup_K |f| \quad \text{pour toute} \quad f \in \mathscr{H}(X)$$

est compact. Il revient au même de dire que, pour tout sous-ensemble infini et discret de X, il existe une $f \in \mathscr{H}(X)$ qui n'est pas bornée sur cet ensemble. Il est trivial que tout espace analytique compact est holomorphiquement convexe; en revanche, un espace compact ne satisfait à (a), (b) ou (c) que s'il est fini.

Rappelons le théorème connu: pour qu'un domaine étalé sans ramification dans $\mathbf{C}^n$ soit un domaine d'holomorphie, il faut et il suffit qu'il soit holomorphiquement convexe (Oka[26] pour le cas général des domaines à une infinité de feuillets).

Pour un espace X holomorphiquement convexe, les conditions (b) et (c) sont équivalentes, comme on le voit facilement. De plus Grauert a prouvé (ce qui est beaucoup plus difficile) que (*b*) et (*a*) sont équivalentes pour un X holomorphiquement convexe[14]. Une *variété analytique* X qui est holomorphiquement convexe et satisfait à l'une des conditions équivalentes (*a*), (*b*), (*c*) est une *variété de Stein*, et réciproquement. Dans le cas général, un espace analytique X qui est holomorphiquement convexe et satisfait à (*a*), (*b*), (*c*) est appelé par Grauert un espace *holomorphiquement complet*. Il est immédiat que tout sous-espace analytique d'un espace holomorphiquement complet est holomorphiquement complet. C'est pour les espaces holomorphiquement complets que les théorèmes fondamentaux de la théorie des faisceaux analytiques cohérents, établis antérieurement pour les variétés de Stein, sont valables; mais ceci est un autre sujet, qui nous entraînerait trop loin.

Soit X un espace holomorphiquement convexe, que nous supposerons normal; nous ne faisons sur X aucune des hypothèses (*a*), (*b*), ou (*c*). Considérons, dans X, la relation d'équivalence R que voici: x et x' sont R-équivalents si $g(x) = g(x')$ pour toute $g \in \mathscr{H}(X)$. Il est clair que la condition (H) du no. 5 est remplie; on peut donc appliquer ici le théorème de ce numéro. Soit alors Y le 'quotient analytique' de X relativement à la relation R; on voit tout de suite que la surjection holomorphe $f : X \to Y$ définit un isomorphisme des anneaux de fonctions holomorphes $\mathscr{H}(X)$ et $\mathscr{H}(Y)$; puisque l'application f est propre, l'espace Y est, comme X, holomorphiquement convexe. De plus il est évident que Y satisfait à la condition (*b*), donc Y est holomorphiquement complet, et en particulier Y satisfait à (*a*); il en résulte que l'application $Y \to X/R$ est un isomorphisme d'espaces annelés. Ainsi *l'espace quotient X/R, muni de sa structure annelée, est un espace normal, holomorphiquement complet.* On le notera X^*; il est naturellement attaché à X (Remmert[31] le nomme le 'noyau' de X). On en déduit notamment: tout espace analytique irréductible et holomorphiquement convexe est réunion dénombrable de compacts (cf.[31]): si X est normal, cela tient au fait que X^* est réunion dénombrable de compacts d'après Grauert, et que l'application $X \to X^*$ est propre; si X n'est pas normal, on considère le normalisé $\tilde{X}$.

On peut considérer d'autres classes d'espaces analytiques. Nous dirons que X est *projectivement complet* ('analytiquement complet' dans la terminologie de Grauert et Remmert) si X est holomorphiquement convexe et si, pour tout $x \in X$, il existe une application holomorphe de X dans un espace projectif P_k qui est 'séparante' au point x. On peut démontrer que tout espace analytique normal X, holomorphiquement convexe, possède un plus grand quotient qui est projectivement complet: c'est un espace normal Y, projectivement complet, muni d'une surjection holomorphe et propre $f: X \to Y$, qui jouit de la propriété universelle suivante: toute application holomorphe de X dans un espace projectivement complet Z se factorise (nécessairement d'une seule manière) en

$$X \xrightarrow{f} Y \xrightarrow{g} Z,$$ où g est holomorphe.

Nous dirons qu'un espace analytique X est *algébriquement complet* s'il est holomorphiquement convexe et si, pour tout $x \in X$, il existe une application holomorphe de X dans un espace algébrique (non nécessairement projectif) qui est séparante au point x. Tout espace analytique normal X, holomorphiquement convexe, possède un plus grand quotient algébriquement complet, qui jouit d'une propriété universelle vis-à-vis des applications holomorphes de X dans les espaces algébriquement complets.

De là résulte en particulier ceci: tout espace analytique normal et compact X possède un plus grand quotient qui soit une variété algébrique projective (normale); tout espace analytique normal et compact possède un plus grand quotient qui soit une variété algébrique (non nécessairement projective).

Les résultats précédents, dont la démonstration sera publiée ailleurs, constituent une généralisation d'une situation bien connue: tout tore complexe possède un plus grand quotient qui est une variété abélienne (c'est-à-dire un tore complexe satisfaisant aux conditions de Riemann).

7. Problèmes de plongement

Il s'agit de 'réaliser' certains espaces analytiques comme sous-espaces d'espaces particulièrement simples, tels que les espaces numériques C^n et les espaces projectifs P_n. Dans chaque cas, le sens du mot 'réaliser' a besoin d'être précisé.

Si l'on veut réaliser un espace analytique X dans un espace C^n, le moins que l'on puisse exiger est de trouver une application holomorphe et injective $f: X \to C^n$. Or ceci n'est possible que si les fonctions holomorphes sur X séparent les points de X (condition (a) du no. 6). Remmert[31] a montré que cette condition nécessaire (a) est aussi suffisante, du moins si l'on suppose que X est réunion dénombrable de compacts (ce qui est automatiquement le cas lorsque X est irréductible). D'une façon précise, si X est purement k-dimensionnel, satisfait à (a) et est réunion dénombrable de compacts, il existe une application holomorphe et injective de X dans C^{2k+1}.

On peut être plus exigeant, en demandant une application $f: X \to C^n$ qui soit non seulement holomorphe et injective, mais *propre*. Ceci impose à X de satisfaire à (a) et d'être holomorphiquement convexe; autrement dit, X doit être holomorphiquement complet. Remmert[31] montre que, réciproquement, tout X holomorphiquement complet qui est réunion dénombrable de compacts peut être plongé dans un C^n par une application f holomorphe, injective et propre; alors l'image $f(X)$ est un sous-espace analytique Y de C^n; mais il faut prendre garde que cette 'réalisation' Y de X ne respecte pas nécessairement les structures annelées. Cependant, lorsque X est une véritable variété analytique (variété de Stein), on peut réaliser X comme sous-variété analytique d'un espace C^n (avec un n qui ne dépend que de la dimension de X).

Je voudrais maintenant dire quelques mots des plongements dans l'espace projectif (l'image étant alors un sous-ensemble algébrique). On a établi ces dernières années une série de théorèmes qui garantissent l'existence de tels plongements. Sans entrer dans le détail (ce qui nécessiterait toute une conférence), rappelons seulement le théorème fondamental de Kodaira[20]: une variété analytique compacte sur laquelle existe une forme de Kähler à périodes rationnelles est isomorphe à une variété algébrique plongée sans singularités dans un espace projectif.

Soit X une variété analytique dans laquelle un groupe discret d'automorphismes G *opère proprement* (ce qui signifie que, pour tout compact $K \subset X$, les $s \in G$ tels que sK rencontre K sont en nombre fini). Considérons l'espace quotient X/G muni de sa structure annelée (cf. no. 5); on montre[10] que c'est un espace analytique normal (mais ce n'est pas, en général, une variété analytique, à cause de l'existence de points fixes pour les transformations de G); plus généralement, si X est un espace analytique normal et si G est un groupe discret opérant proprement dans X, X/G est un espace analytique normal. Cela étant, si X/G est *compact*, il est naturel de se demander si X/G peut être réalisé comme sous-espace analytique (donc algébrique) d'un espace projectif. Effectivement, il en est toujours ainsi lorsque X est un *domaine borné* d'un espace numérique C^n; le plongement projectif de X/G peut alors être obtenu au moyen d'un système fini de formes automorphes (séries de Poincaré) d'un même poids;† la variété algébrique, image du plongement, est 'projectivement normale'.

Mais les cas les plus intéressants, dans la théorie des fonctions automorphes, sont ceux où l'espace X/G *n'est pas compact*; alors il ne peut être question de réaliser X/G comme variété algébrique dans un espace projectif. On peut néanmoins se proposer de chercher une application analytique $f: X/G \to P_n$ qui soit injective et définisse un isomorphisme

† Voir [10] et [2].

de l'espace analytique X/G sur un 'ouvert de Zariski' d'une variété algébrique projective V (c'est-à-dire sur le complémentaire, dans V, d'un sous-ensemble algébrique W). On sait maintenant que ceci est possible dans la théorie des fonctions modulaires de Siegel[36, 37]. D'une façon précise, soit X l'espace de Siegel, formé des (n, n)-matrices symétriques complexes $z = x + iy$ telles que y soit définie-positive; le groupe symplectique réel $Sp(n, R)$, formé des $(2n, 2n)$-matrices réelles

$$M = \begin{pmatrix} a & b \\ c & d \end{pmatrix},$$

où a, b, c, d sont des (n, n)-matrices telles que ${}^t MJM = J$, avec

$$J = \begin{pmatrix} 0 & 1_n \\ -1_n & 0 \end{pmatrix},$$

opère dans X par $z \to (az + b)(cz + d)^{-1}$; dans ce groupe de transformations de X, on considère le sous-groupe discret G défini par les matrices à coefficients entiers. Le quotient $X/G = V_n$ est un espace analytique normal, non compact. Satake[34] a montré comment on peut compactifier V_n en définissant une topologie convenable sur la réunion de $V_n, V_{n-1}, \dots$, V_1, V_0, et il a de plus défini une structure annelée sur ce compactifié V_n^*, structure qui induit, bien entendu, les structures d'espace analytique des sous-espaces V_n (ouvert dans V_n^*), V_{n-1}, etc. Puis Baily[3] a prouvé que l'espace annelé V_n^* est effectivement un espace analytique normal, ainsi que l'avait conjecturé Satake, et a de plus montré que V_n^* peut se réaliser comme variété algébrique dans un espace projectif,

$$V_{n-1}^* = V_n^* - V_n$$

s'identifiant à une sous-variété algébrique de V_n^*. Le plongement projectif peut être obtenu par des formes automorphes d'un même poids convenable (il s'agit de formes automorphes pour les puissances du facteur d'automorphie $\det(cz + d)^2$). L'existence d'un tel plongement permet de prouver que toute fonction méromorphe dans X et invariante par G s'exprime comme quotient de deux formes automorphes de même poids (du moins si $n \geqslant 2$). On peut compléter ces résultats, et montrer que l'algèbre graduée des formes automorphes des divers poids est engendrée par un nombre fini d'éléments (comme algèbre sur le corps complexe).†
D'autre part, tous ces résultats s'étendent au cas de n'importe quel sous-groupe du groupe symplectique qui est 'commensurable' au groupe modulaire; les variétés algébriques projectives qui sont ainsi attachées à ces groupes sont des 'revêtements ramifiés' les unes des autres.

† Voir [12], Exposé 17.

8. Application à la théorie des variétés analytiques réelles

Il est superflu de rappeler la définition d'une variété analytique réelle (abstraite); les modèles sont ici les ouverts de l'espace numérique réel R^n, et les changements de coordonnées locales sont analytiques-réels. Nous ne considérerons que les variétés analytiques-réelles qui peuvent être recouvertes au moyen d'une famille dénombrable de compacts, ce qui revient à supposer l'existence d'une base dénombrable d'ouverts pour la topologie.

Soit V une variété analytique-réelle de dimension n; les résultats de Whitney[41] permettent d'affirmer l'existence d'une application injective et propre $f: V \to R^{2n+1}$, indéfiniment différentiable et de rang n en tout point, dont l'image est une sous-variété de R^{2n+1} qu'on peut même supposer analytique. La question était restée ouverte de savoir si l'on peut exiger en outre que le plongement f soit *analytique*; en d'autres termes, toute variété analytique-réelle (abstraite) peut-elle être réalisée, au sens de la structure analytique, comme sous-variété analytique d'un espace numérique réel? Il y a un an à peine, une réponse positive a été donnée par Morrey[24] dans le cas où V est compacte; auparavant, Malgrange[23] avait donné une réponse affirmative pour toute variété analytique V, compacte ou non, mais sous la restriction de l'existence d'un ds^2 analytique sur V (la méthode de Malgrange reposait sur la théorie des équations elliptiques). Grauert[18] vient de prouver enfin que toute variété analytique-réelle (sans aucune autre restriction que l'hypothèse d'une base dénombrable d'ouverts) peut se réaliser comme sous-variété analytique d'un espace numérique; ce résultat est obtenu par des méthodes analytiques-complexes, et c'est à ce titre qu'il en est question ici. Voici quelques détails au sujet de cet important théorème.

On sait que toute variété analytique-réelle V peut être plongée comme sous-espace fermé d'une variété analytique complexe X, de manière que X soit une 'complexification' de V: ceci signifie que chaque point $x \in V$ possède un voisinage ouvert U (dans X) dans lequel on a un système de coordonnées locales complexes tel que les points de $V \cap U$ soient précisément les points à coordonées réelles, celles-ci servant de coordonnées locales pour V. Si n est la dimension réelle de V, n est donc la dimension complexe de X. Grauert montre qu'ainsi plongée dans X, V possède un système fondamental de voisinages ouverts qui sont des variétés de Stein; il est impossible de donner ici une idée de la démonstration, fort délicate, et qui met en œuvre la théorie des faisceaux analytiques cohérents et celle des fonctions plurisousharmoniques (introduites par Lelong il y a plus de dix ans). A ce propos, il est bon de noter qu'une condition nécessaire et suffisante pour qu'une variété analytique-complexe X, connexe et holomorphiquement convexe, soit une variété de Stein, est qu'il existe sur X une fonction 'strictement plurisousharmonique'.

Revenons à la variété analytique-réelle V, plongée dans une variété de Stein X qui en est une complexification. Appliquons à X le théorème de plongement de Remmert (no. 7); ceci donne un plongement analytique-réel de V dans un espace numérique réel. On pourrait aussi, sans utiliser le théorème de Remmert, procéder comme suit: le fait que V possède un système fondamental de voisinages ouverts qui sont des variétés de Stein entraîne que les théorèmes fondamentaux de la théorie des faisceaux analytiques cohérents sont applicables à la variété analytique-réelle V[11]; on sait alors que l'anneau des fonctions analytiques-réelles, sur V, est assez riche pour fournir une application analytique de V dans un espace $\mathbf{R}^k$, dont le rang soit égal en tout point de V à la dimension de V; d'où l'existence d'un ds^2 analytique sur V, et l'on peut appliquer le théorème de Malgrange.

L'existence d'un plongement analytique propre de V dans un $\mathbf{R}^k$ permet d'appliquer à V le théorème d'approximation de Whitney[40]: toute fonction p fois continûment différentiable sur V peut être arbitrairement approchée par des fonctions *analytiques* sur V, l'approximation s'entendant au sens de la convergence uniforme de la fonction et de chacune de ses dérivées d'ordre $\leqslant p$; et l'on peut même exiger une convergence de plus en plus rapide à l'infini. De là résulte évidemment que si une variété analytique-réelle est réalisable différentiablement dans un espace $\mathbf{R}^k$, elle est aussi réalisable analytiquement dans le même $\mathbf{R}^k$. Toute variété analytique-réelle V de dimension n peut donc être analytiquement réalisée dans $\mathbf{R}^{2n+1}$.

D'autre part, le fait que la théorie des faisceaux analytiques cohérents s'applique à toute variété analytique-réelle V a des conséquences agréables, telles que celles-ci: toute sous-variété analytique W de V peut être définie globalement par des équations analytiques $f_i = 0$, en nombre fini (les f_i étant analytiques dans V tout entière); de plus, toute fonction analytique sur W est induite par une fonction analytique dans V; la cohomologie réelle de V peut se calculer au moyen des formes différentielles analytiques, etc....[11]

Notons que tous ces résultats nécessitent l'usage des méthodes analytiques-complexes, qui semblent ainsi commander toute étude approfondie de l'analytique-réel. Ceci est confirmé par le fait que la seule notion de *sous-ensemble analytique-réel* (d'une variété analytique-réelle V) qui ne conduise pas à des propriétés pathologiques doit se référer à l'espace complexe ambiant: il faut considérer les sous-ensembles fermés E de V tels qu'il existe une complexification X de V et un sous-ensemble analytique-complexe E' de X, de manière que $E = E' \cap V$. On démontre[11] que ce sont aussi les sous-ensembles de V qui peuvent être définis globalement par un nombre fini d'équations analytiques. La notion de sous-ensemble analytique-réel a ainsi un caractère essentiellement *global*, contrairement à ce qui avait lieu pour les sous-ensembles analytiques-complexes.

Bruhat et Whitney[5] viennent d'étudier ces sous-ensembles analytiques-réels d'une variété analytique-réelle V. Ils prouvent notamment que si E est un sous-ensemble analytique de V, il existe une famille localement finie de sous-ensembles analytiques irréductibles (globalement) E_i telle que $E_i \not\subset E_j$ pour $i \neq j$, et que E soit la réunion des E_i; cette famille est uniquement déterminée à l'ordre près. De plus, si E est irréductible et de dimension p, tout sous-ensemble analytique de E, distinct de E, a toutes ses composantes irréductibles de dimension $\leqslant p - 1$ (c'est là un résultat qui semble naturel; néanmoins il serait faux si l'on avait adopté, pour la notion de sous-ensemble analytique, la définition de caractère local à laquelle on songe naturellement).

9. Espaces fibrés analytiques†

Nous nous bornerons, pour simplifier l'exposition, au cas des *fibrés principaux*. Considérons d'abord le cas analytique-complexe: on a un espace analytique X, un groupe de Lie complexe G, et l'on considère les fibrés analytiques principaux (localement triviaux au sens analytique-complexe) ayant pour base X et pour groupe structural G; deux tels fibrés P et P' sont isomorphes s'il existe un isomorphisme de l'espace analytique P sur l'espace analytique P', qui soit compatible avec les opérations du groupe G et qui induise l'application identique de la base X. On sait que les classes de fibrés isomorphes sont en correspondance biunivoque avec l'ensemble de cohomologie $H^1(X, \mathbf{G}^a)$, $\mathbf{G}^a$ désignant le faisceau des germes d'applications *holomorphes* de X dans G. On pourrait aussi considérer les classes de fibrés *topologiques* principaux, qui sont en correspondance biunivoque avec les éléments de $H^1(X, \mathbf{G}^c)$, $\mathbf{G}^c$ désignant le faisceau des germes d'applications *continues* de X dans G. On a une application naturelle

$$* \qquad H^1(X, \mathbf{G}^a) \to H^1(X, \mathbf{G}^c)$$

† Voir les travaux de Grauert[15, 16, 17], ainsi que l'exposition qui en est faite dans [9].

définie par l'injection $G^a \to G^c$; elle n'est, en général, ni injective ni surjective. Cependant Grauert a démontré que si X est un espace *holo-morphiquement complet* (cf. no. 6), l'application * est *bijective*; autrement dit, si deux fibrés analytiques principaux de base X et de groupe structural G sont topologiquement isomorphes, ils sont analytiquement isomorphes; et tout fibré topologique principal, de base X et de groupe G, peut être muni d'une structure de fibré analytique principal, compatible avec sa structure de fibré topologique. Ce résultat important est établi par des méthodes fort difficiles, et qu'il ne semble pas possible de simplifier substantiellement dans l'état actuel des Mathématiques. Les démonstrations font d'ailleurs intervenir simultanément d'autres problèmes. En voici quelques-uns, que nous formulons seulement dans un cas particulier pour simplifier l'exposé: toute application continue $X \to G$ est-elle homotope à une application holomorphe? Si deux applications holomorphes $X \to G$ sont homotopes dans l'espace des applications continues, le sont-elles dans l'espace des applications holomorphes? Si une application holomorphe $Y \to G$ (où Y désigne un sous-espace analytique de X) est prolongeable en une application continue $X \to G$, est-elle prolongeable en une application holomorphe $X \to G$? Toutes ces questions reçoivent une réponse affirmative lorsque l'espace X est holomorphiquement complet. Si l'on ne fait pas cette hypothèse, on a des réponses partielles lorsque le groupe G est résoluble (Frenkel[13]).

D'une manière générale, lorsque X n'est pas holomorphiquement complet, la classification des fibrés analytiques de base X est un problème fort intéressant mais sur lequel on ne sait encore que peu de choses. La classification des fibrés vectoriels a été faite par Grothendieck[19] dans le cas où X est la droite projective complexe, et par Atiyah[1] lorsque X est une courbe algébrique de genre 1.

Je voudrais encore dire quelques mots des fibrés analytiques-réels. On peut vérifier que les méthodes de Grauert sont susceptibles, moyennant des modifications adéquates, d'être appliquées aux fibrés analytiques-réels, compte tenu du fait que la théorie des faisceaux analytiques cohérents s'applique maintenant aux variétés analytiques-réelles sans aucune restriction (grâce au théorème de plongement de Grauert). On peut alors montrer que tous les résultats énoncés plus haut pour le cas où X est un espace analytique holomorphiquement complet et G un groupe de Lie complexe, sont valables lorsque X est une variété analytique-réelle et G un groupe de Lie réel. En particulier, *la classification analytique des fibrés principaux coïncide avec leur classification topologique.*

10. Conclusion

Cet aperçu de résultats récents dans la théorie des espaces analytiques est forcément incomplet. Je regrette notamment de n'avoir même pas mentionné la toute nouvelle théorie des 'déformations de structures complexes'; mais c'est un sujet qui apparaît déjà assez vaste pour nécessiter une conférence à lui seul.† J'ai dû aussi laisser de côté le problème du prolongement des sous-ensembles analytiques (complexes), auquel Rothstein a apporté de si intéressantes contributions, ainsi que le problème analogue du prolongement des faisceaux analytiques cohérents. Je signale enfin un problème qui mérite de retenir l'attention: sur un espace analytique général, on n'a pas encore de théorie satisfaisante des formes différentielles; si on considère un point non-uniformisable et que l'on réalise un voisinage de ce point par un sous-ensemble analytique d'un ouvert d'un espace $\mathbf{C}^n$, il y a certainement lieu de considérer d'autres 'formes différentielles' que celles qui sont induites par les formes différentielles de l'espace ambiant $\mathbf{C}^n$.

BIBLIOGRAPHIE

[1] Atiyah, M. F. Vector bundles over an elliptic curve. *Proc. Lond. Math. Soc.* (3), 7, 414–452 (1957).

[2] Baily, W. L. On the quotient of an analytic manifold by a group of analytic homeomorphisms. *Proc. Nat. Acad. Sci., Wash.*, 40, 804–808 (1954).

[3] Baily, W. L. Satake's compactification of V_n. *Amer. J. Math.* 80, 348–364 (1958).

[4] Behnke, H. und Stein, K. Modifikation komplexer Mannigfaltigkeiten und Riemannscher Gebiete. *Math. Ann.* 124, 1–16 (1951).

[5] Bruhat, F. et Whitney, H. Quelques propriétés fondamentales des ensembles analytiques-réels. *Comment Math. Helv.* 33, 132–160 (1959).

[6] Calabi, E. and Rosenlicht, M. Complex analytic manifolds without countable base. *Proc. Amer. Math. Soc.* 4, 335–340 (1953).

[7] Cartan, H. Séminaire 1951–52, École Normale Supérieure, Paris.

[8] Cartan, H. Séminaire 1953–54, École Normale Supérieure, Paris.

[9] Cartan, H. Espaces fibrés analytiques. *Symposium de Topologie, Mexico,* 1956, 97–121.

[10] Cartan, H. Quotient d'un espace analytique par un groupe d'automorphismes. Algebraic Geometry and Algebraic Topology. *A Symposium in Honor of S. Lefschetz*, 90–102. Princeton, 1957.

[11] Cartan, H. Variétés analytiques réelles et variétés analytiques complexes. *Bull. Soc. Math. Fr.* 85, 77–99 (1957).

[12] Cartan, H. Séminaire 1957–58, École Normale Supérieure, Paris.

[13] Frenkel, J. Cohomologie non abélienne et espaces fibrés. *Bull. Soc. Math. Fr.* 85, 135–220 (1957).

[14] Grauert, H. Charakterisierung der holomorph-vollständiger komplexen Räume. *Math. Ann.* 129, 233–259 (1955).

† Voir surtout [21], où l'on trouvera une bibliographie de la question.

52 HENRI CARTAN

[15] Grauert, H. Approximationssätze für holomorphe Funktionen mit Werten in komplexen Räumen. *Math. Ann.* 133, 139–159 (1957).

[16] Grauert, H. Holomorphe Funktionen mit Werten in komplexen Lieschen Gruppen. *Math. Ann.* 133, 450–472 (1957).

[17] Grauert, H. Analytische Faserungen über holomorph-vollständigen Räumen. *Math. Ann.* 135, 263–273 (1958).

[18] Grauert, H. On Levi's problem and the imbedding of real-analytic manifolds. *Ann. Math.* (2), 68, 460–472 (1958).

[19] Grothendieck, A. Sur la classification des fibrés holomorphes sur la sphère de Riemann. *Amer. J. Math.* 79, 121–138 (1957).

[20] Kodaira, K. On Kähler varieties of restricted type. *Ann. Math.* (2), 60, 28–48 (1954).

[21] Kodaira, K. and Spencer, D. C. On deformations of complex analytic structures. *Ann. Math.* (2), 67, 328–466 (1958).

[22] Lelong, P. Intégration sur un ensemble analytique complexe. *Bull. Soc. Math. Fr.* 85, 239–261 (1957).

[23] Malgrange, B. Plongement des variétés analytiques-réelles. *Bull. Soc. Math. Fr.* 85, 101–112 (1957).

[24] Morrey, Ch. B. The analytic imbedding of abstract real-analytic manifolds. *Ann. Math.* (2), 68, 159–201 (1958).

[25] Oka, K. Sur les fonctions analytiques de plusieurs variables, VIII. *J. Math. Soc. Japan*, 3, 204–214, 259–278 (1951).

[26] Oka, K. Sur les fonctions analytiques de plusieurs variables, IX. Domaines finis sans point intérieur. *Jap. J. Math.* 23, 97–155 (1953).

[27] Osgood, W. F. *Lehrbuch der Funktionentheorie.* (Leipzig, 1928–32.)

[28] Remmert, R. und Stein, K. Über die wesentlichen Singularitäten analytischer Mengen. *Math. Ann.* 126, 263–306 (1953).

[29] Remmert, R. Projektionen analytischer Mengen. *Math. Ann.* 130, 410–441 (1956).

[30] Remmert, R. Meromorphe Funktionen in kompakten komplexen Räumen. *Math. Ann.* 132, 277–288 (1956).

[31] Remmert, R. Sur les espaces analytiques holomorphiquement séparables et holomorphiquement convexes. *C.R. Acad. Sci., Paris*, 243, 118–121 (1956).

[32] Remmert, R. Holomorphe und meromorphe Abbildungen komplexer Räume. *Math. Ann.* 133, 328–370 (1957).

[33] de Rham, G. Seminar on several complex variables. *Inst. Adv. Study*, mimeographed Notes, 1957–58.

[34] Satake, I. On the compactification of the Siegel space. *J. Indian Math. Soc.* 20, 259–281 (1956).

[35] Serre, J. P. Géométrie algébrique et géométrie analytique. *Ann. Inst. Fourier*, 6, 1–42 (1955–6).

[36] Siegel, C. L. Einführung in die Theorie der Modulfunktionen n-ten Grades. *Math. Ann.* 116, 617–657 (1939).

[37] Siegel, C. L. Symplectic geometry. *Amer. J. Math.* 65, 1–86 (1943).

[38] Stein, K. Analytische Abbildungen allgemeiner analytischer Räume. *Colloque de Topologie, Strasbourg*, avril 1954.

[39] Stein, K. Analytische Zerlegungen komplexer Räume. *Math. Ann.* 132, 63–93 (1956).

[40] Whitney, H. Analytic extensions of differentiable functions defined in closed sets. *Trans. Amer. Math. Soc.* 36, 63–89 (1934).

[41] Whitney, H. The self-intersections of a smooth n-manifold in $2n$-space. *Ann. Math.* (2), 44, 220–246 (1945).

LA THÉORIE DES GROUPES ALGÉBRIQUES

Par C. CHEVALLEY

1. La notion de groupe algébrique

La notion de groupe algébrique repose sur celle de variété algébrique, de la même manière que celle de groupe topologique dépend de la notion d'espace topologique. Il ne saurait être question d'exposer ici avec précision la définition des variétés algébriques; nous allons cependant indiquer les caractères essentiels de ce type d'objets mathématiques. Une variété algébrique peut être définie par les données d'un ensemble U, son ensemble de points, et d'un ensemble de fonctions, son corps de fonctions rationnelles; ces fonctions sont des applications de parties de U dans un corps algébriquement clos K, qu'on appelle le corps des constantes: par ailleurs, elles forment un corps relativement à des opérations d'addition et de multiplication qui jouissent des propriétés suivantes: si des fonctions rationnelles u et v sont définies en un point x, $u+v$ et uv sont également définies en x et y prennent les valeurs $u(x)+v(x)$ et $u(x)\,v(x)$ respectivement; de plus, si u est définie en x et $u(x) \neq 0$, u^{-1} est définie en x; les applications constantes de U dans K sont des fonctions rationnelles, et forment un corps isomorphe à K. Une variété algébrique est munie d'une topologie dont la famille d'ouverts est engendrée par les ensembles de définition des fonctions rationnelles; cette topologie est d'un type très particulier, puisque les ouverts non vides y sont tous denses dans l'espace entier. Pour tout point x de la variété, il existe un voisinage $A(x)$ de x et un certain nombre de fonctions rationnelles $u_1, ..., u_n$, partout définies sur $A(x)$, telles que l'application

$$x' \to (u_1(x'), ..., u_n(x'))$$

soit une bijection de $A(x)$ sur un sous-ensemble algébrique de l'espace numérique K^n (une partie de K^n est dite algébrique si elle se compose de tous les points dont les coordonnées satisfont à un certain système d'équations algébriques); de plus, toute fonction rationnelle u sur U s'exprime comme fraction rationnelle en $u_1, ..., u_n$, et même comme polynôme si elle est définie en tout point de $A(x)$; on dit alors que $u_1, ..., u_n$ forment un système de coordonnées en x sur U. Une application f d'une variété V dans une variété U est dite rationnelle (on dit alors aussi que f est un morphisme) si f est continue et possède la propriété suivante: si $y \in V$ et si $u_1, ..., u_n$ est un système de coordonnées sur

U en $f(y)$, il y a un système de coordonnées $(v_1, \ldots, v_m)$ en y sur V tel que, au voisinage de y, les $u_i(f(y))$ s'expriment comme polynômes en les $v_j(y)$.

Si U et U' sont des variétés, le produit cartésien $U \times U'$ possède une structure de variété qui possède la propriété suivante: si $u_1, \ldots, u_m$ (resp. $u_1', \ldots, u_{m'}'$) forment un système de coordonnées sur U (resp. U') en un point x_0 (resp. x_0'), les fonctions $(x, x') \to u_i(x)$, $(x, x') \to u_j(x')$ forment un système de coordonnées en (x_0, x_0') sur $U \times U'$.

Un *groupe algébrique* est un groupe G qui est muni d'une structure de variété algébrique et qui possède la propriété suivante: l'application $(s, t) \to st^{-1}$ est un morphisme de $G \times G$ dans G.

Par exemple, le groupe $GL(n)$ des matrices inversibles de degré n à coefficients dans K est un groupe algébrique (les fonctions rationnelles sur le groupe étant les fonctions de matrices qui peuvent s'exprimer comme fonctions rationnelles des coefficients des matrices).

Un autre exemple se construit comme suit. Soit $\mathbf{C}$ le corps des complexes, et soit n un entier > 0; soit P un sous-groupe discret de $\mathbf{C}^n$ engendré par $2n$ points linéairement indépendants sur le corps $\mathbf{R}$ des réels. Le groupe $\mathbf{C}^n/P = Z$ est alors un groupe commutatif compact; il est muni d'une structure de groupe de Lie complexe de dimension complexe n. Soit maintenant L le corps des fonctions méromorphes sur $\mathbf{C}^n$ qui admettent tous les éléments du groupe P comme périodes. A chaque fonction de L est associée une fonction définie sur une partie de Z, à savoir l'image par l'application canonique $\mathbf{C}^n \to \mathbf{C}^n/P$ de l'ensemble des points où la fonction donnée est holomorphe. En général, le corps L ne contiendra que les constantes. Cependant, si le groupe P satisfait à certaines conditions (qui s'expriment au moyen de la théorie des matrices de Riemann), le corps L contiendra 'assez de fonctions' dans le sens que, pour deux points distincts de Z, on pourra trouver une fonction de L qui soit définie en ces deux points et y prenne des valeurs distinctes. On montre alors que l'on peut munir le groupe Z d'une structure de groupe algébrique en définissant les fonctions rationnelles comme étant les fonctions sur Z définies par les éléments de L. Les groupes algébriques ainsi définis s'appellent les variétés abéliennes; il vaudrait sans doute mieux les appeler groupes abéliens, n'était la confusion fâcheuse qui tend à se produire entre ce sens du mot abélien et le sens usuel de 'commutatif'. Il importe de noter que les divers groupes algébriques qu'on peut obtenir ainsi à partir des divers groupes P satisfaisant aux conditions requises ne sont pas tous isomorphes entre eux, alors que les groupes de Lie complexes dont ils proviennent sont tous isomorphes.

Soit G un groupe algébrique, et soit H un sous-groupe de G qui est une partie fermée de G (au sens de la topologie définie ci-dessus sur la variété G). On montre alors que la composante connexe H_0 de l'élément neutre dans H est un sous-groupe distingué d'indice fini de H, et est une partie fermée de G; de plus, H_0 peut être muni d'une structure de groupe algébrique, une fonction rationnelle sur H_0 coïncidant localement (au voisinage de chaque point de son ensemble de définition) avec la restriction à H_0 d'une fonction rationnelle sur G. Un groupe H_0 défini de cette manière s'appelle un sous-groupe algébrique de G.

Par exemple, le groupe des éléments de déterminant 1 de $GL(n)$ est un sous-groupe algébrique de $GL(n)$; il en est de même du groupe des matrices orthogonales de déterminant 1 contenues dans $GL(n)$, et, si n est pair, du groupe des matrices symplectiques contenues dans $GL(n)$. Citons encore le groupe des matrices triangulaires (a_{ij}) $(a_{ij} = 0$ si $i < j)$, le groupe des matrices triangulaires unipotentes $(a_{ij} = 0$ si $i < j, a_{ii} = 1)$, le groupe des matrices diagonales.

Les sous-groupes algébriques des variétés abéliennes sont eux-mêmes des variétés abéliennes.

2. Caractérisation des deux types de groupes algébriques

Les groupes algébriques qui sont isomorphes à des sous-groupes fermés de groupes du type $GL(n)$ sont appelés les groupes linéaires; les propriétés de ces groupes sont très différentes de celles des variétés abéliennes définies plus haut. Il est remarquable que le fait pour un groupe algébrique G d'être du type linéaire ou du type abélien puisse se reconnaître par le seul examen de la variété du groupe, abstraction faite de sa loi de composition.

Commençons par les groupes linéaires. On dit qu'une variété algébrique U est affine s'il existe un système de coordonnées sur U valable sur toute la variété, c'est-à-dire d'une manière plus précise si U est isomorphe à une sous-variété fermée d'un espace numérique K^n. Ceci étant, on montre qu'une condition nécessaire et suffisante pour qu'un groupe algébrique soit linéaire est que sa variété soit une variété affine. La condition est évidemment nécessaire. Si elle est satisfaite, on montre qu'on peut former des sous-espaces vectoriels de dimensions finies de l'algèbre des fonctions partout définies sur le groupe (c'est-à-dire des fonctions qui s'expriment comme polynômes en les coordonnées d'un système de coordonnées valable sur toute la variété) qui sont invariants par les opérations de translation du groupe; ces espaces fournissent des représentations linéaires du groupe qui permettent de construire un isomorphisme du groupe avec un groupe linéaire.

Passons maintenant aux groupes du type abélien. Ces groupes n'ont été introduits jusqu'ici que dans le cas où le corps de base est celui des complexes; c'est Weil qui a donné la généralisation de ces groupes au cas d'un corps de base de caractéristique quelconque. La définition des variétés abéliennes repose sur la notion de variété complète qui généralise en géométrie algébrique celle des espaces compacts (on observera que les variétés abéliennes introduites plus haut sont, du point de vue des groupes de Lie, des groupes compacts). Considérons une variété U et un morphisme f d'une sous-variété ouverte V' d'une variété V dans U; le graphe de f est une partie Γ de la variété produit $U \times V$; prenons son adhérence $\overline{\Gamma}$. Bien que l'application f ne soit pas définie aux points de $V - V'$, il peut se produire que, si $y \in V - V'$, il y ait un ou des points $x \in U$ tels que $(x, y) \in \overline{\Gamma}$; on dit alors que x est une valeur d'adhérence de f en y. La variété U étant donnée, si pour tous choix de V, V' et f et pour tout $y \in V - V'$ il existe au moins une valeur d'adhérence de f en y, on dit que la variété U est *complète*. Ceci étant, si K est le corps des complexes, on peut montrer que les variétés abéliennes comme définies plus haut sont exactement tous les groupes algébriques qui sont des variétés complètes. Il est donc naturel, pour un corps de base quelconque, d'appeler variétés abéliennes les groupes algébriques qui sont des variétés complètes. Ces groupes sont nécessairement commutatifs; leur étude est à certains égards plus difficile que celle des groupes linéaires; le simple fait qu'ils soient commutatifs oblige par exemple à aller chercher beaucoup plus profondément les éléments de structure propres à les caractériser. Cette étude a été cependant poussée très loin dans les travaux de Weil et de ses successeurs; nous y reviendrons.

Tandis qu'un groupe linéaire admet une représentation linéaire fidèle, un groupe du type abélien n'admet aucune représentation linéaire non triviale; un groupe ne peut donc être à fois linéaire et du type abélien sans se réduire à son élément neutre.

Par contre, on peut montrer que les groupes algébriques les plus généraux peuvent se construire à partir des groupes linéaires et des groupes du type abélien. Indiquons d'abord que, si G est un groupe algébrique et H un sous-groupe algébrique de G, l'ensemble G/H des classes (à droite ou à gauche) de G suivant H peut être muni de manière naturelle d'une structure de variété; l'application naturelle $f: G \to G/H$ est un morphisme, et tout morphisme de G dans une variété qui est constant sur chaque classe suivant H se décompose en l'application f

suivie d'un morphisme de G/H. Si H est de plus distingué, G/H est un groupe, et, muni de la structure de variété dont on vient de parler, un groupe algébrique. Ceci étant, on peut montrer que tout groupe algébrique G admet un sous-groupe algébrique distingué H et un seul tel que H soit linéaire et G/H complet (i.e. du type abélien); il existe diverses démonstrations de ce théorème, dont deux publiées, l'une par Barsotti et l'autre par Rosenlicht. On a encore peu de renseignements sur le problème réciproque, à savoir le problème de déterminer les groupes algébriques admettant un sous-groupe distingué linéaire H donné admettant comme quotient un groupe donné de type abélien. Cependant, des cas importants ont été étudiés par Rosenlicht, Serre et Lang; cette question très importante est liée aux généralisations de la théorie du corps de classes aux variétés algébriques.

3. Variétés abéliennes

3.1. Décomposition en variétés simples. Isogénies. Une variété
abélienne A est dite simple si elle ne se réduit pas à son élément neutre et si elle n'a aucun sous-groupe fermé connexe non trivial autre que la variété tout entière. On est naturellement conduit à chercher à décomposer une variété abélienne quelconque en produit de variétés abéliennes simples. Cependant, il se produit ici un phénomène analogue à celui que l'on rencontre dans la théorie des groupes de Lie semi-simples, un groupe de Lie semi-simple étant seulement localement, mais en général pas globalement, isomorphe à un produit de groupes simples. Pour en arriver à un énoncé exact, on introduit la notion d'isogénie de variétés abéliennes. On appelle en général homomorphisme d'une variété abélienne A dans une variété abélienne B une application de A dans B qui est à la fois un homomorphisme de groupes et un morphisme de variétés. On dit que A est isogène à B s'il existe un homomorphisme surjectif de noyau fini de A dans B; cette relation, évidemment réflexive et transitive, se trouve être aussi symétrique. Ceci étant, on montre que toute variété abélienne A est isogène à un produit de variétés abéliennes simples; le nombre des facteurs du produit qui sont isogènes à une variété abélienne simple quelconque ne dépend que de A.

3.2. L'anneau des endomorphismes. Soit A une variété abélienne; un élément de structure d'importance fondamentale que l'on peut attacher à A est son anneau d'endomorphismes $\mathfrak{A}(A)$; il se compose de tous les homomorphismes de A dans A, avec les lois de composition

définies par les formules $(\alpha+\beta)(x) = \alpha(x)+\beta(x)$, $(\alpha\beta)(x) = \alpha(\beta(x))$. Parmi les endomorphismes figurent notamment les multiplications par les entiers naturels, à savoir les applications $x \to nx$; l'étude de ces applications fournit des renseignements sur les points d'ordre fini de A; réciproquement, le fait que les endomorphismes transforment en lui-même l'ensemble des points d'ordres finis fournit des renseignements précieux sur l'anneau $\mathfrak{A}(A)$.

Avant d'indiquer les résultats auxquels on arrive dans le cas général, nous considérerons d'abord ce qui se passe dans le cas où A est de dimension 1 et où K est le corps des complexes. La variété A est alors une courbe elliptique; elle se met sous la forme C/P, où P est un sous-groupe discret de rang 2 de C; les fonctions rationnelles sur A sont les fonctions elliptiques admettant les points de P comme périodes. Un endomorphisme de A est la transformation de A induite par une opération de la forme $z \to \alpha z$ dans C, α étant un nombre complexe tel que $\alpha P \subset P$: ces opérations sont ce qu'on appelle les multiplications complexes attachées à la courbe elliptique A; on sait que les nombres α tels que $\alpha P \subset P$ forment un anneau qui ou bien se réduit à l'anneau des entiers rationnels ou bien est un sous-anneau de l'anneau des entiers d'un corps imaginaire quadratique L; de plus, dans ce dernier cas, la théorie de la multiplication complexe est très intimement associée à la théorie du corps de classes sur L. Ceci indique que, les résultats généraux relatifs à l'anneau $\mathfrak{A}(A)$ pour une variété abélienne quelconque A une fois obtenus, on peut espérer que ces résultats puissent servir de base à une généralisation de la théorie arithmétique de la multiplication complexe à des corps de nombres plus généraux que les corps imaginaires quadratiques. Cet espoir a été brillamment réalisé par les travaux de Shimura et Taniyama.

Soit A une variété abélienne quelconque; à tout endomorphisme α de A se trouve associé un entier $\nu(\alpha)$ défini comme suit: si α n'est pas surjectif, $\nu(\alpha) = 0$; dans le cas contraire, α définit un isomorphisme du corps $F(A)$ des fonctions rationnelles sur A sur un sous-corps $F'(A)$ de lui-même, et $\nu(\alpha)$ est alors le degré de l'extension $F(A)/F'(A)$. Si cette extension est séparable, ce qui se produit toujours dans le cas où ν est la multiplication par un entier premier à la caractéristique de K, $\nu(\alpha)$ est égal à l'ordre du noyau de α. Un premier résultat fondamental de la théorie affirme que, si α est la multiplication par un entier $k > 0$, on a $\nu(\alpha) = k^{2\dim A}$; il en résulte facilement que, si k est premier à la caractéristique, les points d'ordres diviseurs de k forment un groupe fini isomorphe au produit de 2 dim A groupes cycliques d'ordre k. On en déduit que, si l est un nombre premier distinct de la caractéristique, les points d'ordres diviseurs de puissances de l forment un groupe qui est le produit de $2 \dim A$ exemplaires du groupe additif des entiers l-adiques réduits modulo l. Soit M_l ce groupe: il est clair que les endomorphismes de A opèrent sur le groupe M_l, ce qui donne une représentation de l'anneau $\mathfrak{A}(A)$ dans l'anneau des endomorphismes de M_l. Or on voit facilement que ce dernier est isomorphe à l'anneau des matrices de degré $2 \dim A$ à coefficients dans l'anneau des entiers l-adiques, anneau que nous désignerons par $\mathscr{M}_l$. Utilisant la représentation ainsi obtenue de $\mathfrak{A}(A)$ dans $\mathscr{M}_l$, Weil obtient les résultats suivants. Le groupe additif de $\mathfrak{A}(A)$ admet un ensemble fini de générateurs, et n'admet pas de torsion; il en résulte que l'anneau $\mathfrak{A}(A)$ se plonge dans une algèbre de dimension finie $\mathfrak{A}_{\mathbf{Q}}(A)$ sur le corps $\mathbf{Q}$ des rationnels. Cette algèbre est semi-simple. D'une manière plus précise, si A est une variété abélienne simple, $\mathfrak{A}(A)$ n'admet aucun diviseur de zéro $\neq 0$, de sorte que $\mathfrak{A}_{\mathbf{Q}}(A)$ est dans ce cas un corps gauche. Ce corps n'est d'ailleurs pas quelconque: il admet un antiautomorphisme $\alpha \to \alpha'$ involutif tel que l'on ait $\sigma(\alpha\alpha') > 0$ pour tout $\alpha \neq 0$, σ étant la trace. Dans le cas général, il y a des variétés abéliennes simples $B_1, \ldots, B_m$, non isogènes entre elles, telles que A soit isogène au produit de n_1 fois la variété $B_1, \ldots, n_m$ fois la variété B_m. L'algèbre $\mathfrak{A}_{\mathbf{Q}}(A)$ se décompose alors en m algèbres simples, dont la i-ième est l'algèbre des matrices de degrè n_i à coefficients dans $\mathfrak{A}_{\mathbf{Q}}(A_i)$.

Nous avons déjà signalé l'application de la théorie des variétés abéliennes aux généralisations de la multiplication complexe. Une autre application (c'est celle pour laquelle la théorie a été édifiée) se rapporte à la théorie des corps de fonctions algébriques d'une variable sur un corps fini k à q éléments. Soit C une courbe; l'anneau des correspondances entre la courbe C et elle-même est isomorphe à l'anneau des endomorphismes de la jacobienne J de la courbe C, de la manière suivante: si Γ est une correspondance entre C et C (que nous supposons dépourvue de composantes de la forme $x \times C$, ce qui signifie qu'il n'y a pas de point de C auquel tous les points de C correspondent), les points qui correspondent à un point x de C, affectés de multiplicités convenables, forment un diviseur $\mathfrak{d}(x)$ sur la courbe C. Par linéarité, on déduit de l'application $x \to \mathfrak{d}(x)$ une application $\mathfrak{a} \to \mathfrak{d}(\mathfrak{a})$ du groupe des diviseurs $\mathfrak{a}$ de C dans lui-même (si $\mathfrak{a} = \Sigma a_i x_i$, on a $\mathfrak{d}(\mathfrak{a}) = \Sigma a_i \mathfrak{d}(x_i)$); on montre que cette application transforme les diviseurs principaux en diviseurs principaux, donc définit un endomorphisme du groupe des classes de diviseurs de C, c'est-à-dire du groupe sous-jacent à la jacobienne J de C dans lui-même; on montre que cet endomorphisme est un morphisme de la variété J; c'est donc un endomorphisme de J, dont on montre qu'il ne dépend que de la classe de Γ. On montre que l'application ainsi obtenue de l'anneau des classes de correspondances dans $\mathfrak{A}(J)$ est un isomorphisme du premier de ces anneaux sur le second. Or, supposons la courbe C définie par une équation $F(x, y) = 0$ à coefficients dans le corps k; il y a alors une correspondance remarquable Γ entre C et elle-même, à savoir celle qui fait correspondre à tout point (x, y) le point (x^q, y^q). La puissance n-ième de cette correspondance associe au point (x, y) le point (x^{q^n}, y^{q^n}); les points de C qui se correspondent à eux-mêmes au moyen de cette correspondance sont ceux dont les coordonnées appartiennent à l'extension unique k_n de degré n du corps k. Par ailleurs, Γ définit un endomorphisme α' de J; Weil a déduit des propriétés de l'anneau d'endomorphismes de la jacobienne des renseignements sur la croissance en fonction de n du nombre des points fixes de la correspondance Γ, c'est-à-dire du nombre des points de C à coordonnées dans k_n; c'est la méthode par laquelle il a pu établir l'hypothèse de Riemann relative à la fonction ζ définie par la courbe C.

3.3. Variétés abéliennes et variétés de Picard. La théorie des variétés abéliennes trouve d'autres applications dans la théorie des diviseurs sur une variété algébrique complète U quelconque; il s'agit ici de généraliser la relation qui existe entre une courbe et sa jacobienne. Les diviseurs sur U sont les combinaisons formelles à coefficients entiers d'hypersurfaces tracées sur U; ils forment un groupe $\mathfrak{D}(U)$. On appelle famille algébrique de diviseurs sur U une loi qui fait correspondre à tout point t d'une variété non singulière T (dite variété des paramètres) un diviseur $D(t)$ sur U, loi qui doit satisfaire à certaines conditions qui permettent de qualifier d'algébrique l'application $t \to D(t)$. Un diviseur est dit algébriquement équivalent à 0 s'il appartient à une famille algébrique de diviseurs qui contient aussi le diviseur 0; les diviseurs algébriquement équivalents à 0 forment un sous-groupe $\mathfrak{D}_0$ de $\mathfrak{D}$. Dans le cas d'une courbe, $\mathfrak{D}_0$ n'est autre que le groupe des diviseurs de degré 0, i.e. des combinaisons formelles de points dans lesquelles la somme des coefficients est 0. Parmi les diviseurs algébriquement équivalents à 0 figurent les diviseurs principaux, i.e. les diviseurs de fonctions sur la variété; ils forment un sous-groupe $\mathfrak{D}_p$ de $\mathfrak{D}_0$. Dans le cas d'une courbe, $\mathfrak{D}_0/\mathfrak{D}_p$ n'est autre que la jacobienne de la courbe. On établit que, dans le cas général, le groupe $\mathfrak{D}_0/\mathfrak{D}_p$ peut encore être muni d'une structure de variété qui en fait une variété abélienne P, et qui satisfait à la condition suivante: si $t \to D(t)$ est une famille algébrique de diviseurs de $\mathfrak{D}_0$, et si on désigne par $ClD(t)$ la classe de $U(t)$ modulo $\mathfrak{D}_p$, $t \to ClD(t)$ est un morphisme de la variété T dans la variété P. La variété P ainsi obtenue s'appelle la variété de Picard de la variété U. La construction de la variété de Picard a fait l'objet de nombreux travaux récents, dus notamment à Matsusaka, Samuel, Néron et Weil.

Les notations étant comme ci-dessus, le théorème de Néron–Severi affirme que le groupe $\mathfrak{D}/\mathfrak{D}_0$ admet un ensemble fini de générateurs; Lang a donné récemment de ce théorème une élégante démonstration basée sur l'étude des points d'ordres finis de la variété de Picard.

Par ailleurs, les résultats obtenus relatifs à la variété de Picard permettent d'étudier les extensions abéliennes non ramifiées du corps $F(U)$ des fonctions sur la variété U, en utilisant la théorie classique des extensions kummériennes. Une extension cyclique de degré diviseur de n (n étant un entier premier à la caractéristique) du corps $F(U)$ s'obtient par extraction de la racine n-ième d'une fonction u sur U. Comme en théorie des nombres algébriques, pour que l'extension ainsi obtenue soit non ramifiée, il faut et suffit que le diviseur de la fonction u soit de la forme nD, où D est un diviseur; par ailleurs, si D est lui-même principal, l'extension est triviale. On est donc conduit à étudier les éléments d'ordres diviseurs de n dans le groupe $\mathfrak{D}/\mathfrak{D}_p$. Or le groupe des éléments d'ordres diviseurs de n dans $\mathfrak{D}_0/\mathfrak{D}_p$ est connu par la théorie de la variété de Picard; par ailleurs, le groupe des éléments d'ordres diviseurs de n dans $\mathfrak{D}/\mathfrak{D}_0$ est un groupe fini dont l'ordre est borné en fonction de n. On conçoit donc qu'on puisse obtenir des renseignements assez précis sur la théorie des extensions abéliennes non ramifiées de $F(U)$; et de fait, Lang est parvenu à généraliser aux extensions abéliennes des corps de fonctions algébriques une grande partie de la théorie du corps de classes pour les corps de nombres algébriques.

La théorie de la variété de Picard P d'une variété quelconque s'applique en particulier au cas où U est elle-même une variété abélienne A. On trouve que P est alors isogène à A, sans lui être nécessairement isomorphe. Cartier a réussi tout récemment à montrer que la relation ainsi établie entre une variété abélienne A et sa variété de Picard P est une relation réciproque; en d'autres termes, la variété de Picard de P est isomorphe à A. On arrive ainsi à une relation de dualité entre variétés abéliennes qui rappelle, par ses propriétés formelles, la relation de dualité entre les groupes commutatifs finis et leurs groupes de caractères.

4. Les groupes linéaires

Le développement de la théorie des groupes linéaires algébriques a comporté deux phases assez distinctes, qui se différencient l'une de l'autre aussi bien par la nature des problèmes traités que par celle des méthodes employées.

La première phase a été marquée par le souci de s'inspirer de la théorie générale des groupes de Lie. Cette dernière ne fournissait à la théorie des variétés abéliennes qu'un invariant de nature triviale, puisque l'algèbre de Lie d'un groupe commutatif analytique ne dépend que de la dimension de ce groupe. Il n'en est plus ainsi dans le cas des groupes linéaires; cela se conçoit, puisqu'un sous-groupe analytique du groupe linéaire complet $GL(n, \mathbf{C})$ est entièrement déterminé par son algèbre de Lie. Aussi a-t-on cherché d'abord à exploiter au maximum les renseignements sur un groupe algébrique que l'on peut tirer de l'étude de son algèbre de Lie: c'est la voie dans laquelle s'est engagé le fondateur de la théorie, Maurer; ses travaux ont été repris et complétés par ceux de Chevalley.

On aperçoit aisément une méthode pour construire des sous-groupes algébriques du groupe $GL(n, \mathbf{C})$. Ce dernier opère en effet de manière naturelle sur les espaces de tenseurs de diverses espèces construits sur l'espace vectoriel $\mathbf{C}^n$, et les composantes du transformé $s(T)$ d'un tenseur T par une opération s de $GL(n, \mathbf{C})$ s'expriment comme fonctions rationnelles des coefficients de la matrice s. On dit qu'un tenseur T est un *invariant* d'un sous-groupe G de $GL(n, \mathbf{C})$ si on a $s(T) = T$ pour tout $s \in G$, et un *semi-invariant* si $s(T)$ est de la forme $a(s) T$ (pour $s \in G$), $a(s)$ étant un scalaire. Ceci étant, si on se donne un certain nombre de tenseurs $T_1, \ldots, T_h$ d'espèces quelconques, l'ensemble des éléments s de $GL(n, \mathbf{C})$ qui admettent chacun des T_i comme semi-invariant est évidemment un groupe algébrique. On peut montrer que tout sous-groupe algébrique G de $GL(n, \mathbf{C})$ peut se définir de la manière précédente. L'idée de la démonstration est la suivante. Les fonctions polynômes sur $GL(n, \mathbf{C})$ (celles qui peuvent s'exprimer comme des polynômes en les coefficients d'une matrice de $GL(n, \mathbf{C})$) forment une algèbre P; celles d'entre elles qui sont homogènes d'un degré déterminé d sont des tenseurs. Le groupe $GL(n, \mathbf{C})$ opère sur l'algèbre P. Par ailleurs, un sous-groupe algébrique G de $GL(n, \mathbf{C})$ est défini par un certain idéal $\mathfrak{a}$ de P, qui se compose des fonctions nulles sur G. On voit tout de suite qu'une condition nécessaire et suffisante pour qu'un élément s appartienne à G est que s transforme $\mathfrak{a}$ en lui-même; il suffit même que s transforme en lui-même l'espace des fonctions de $\mathfrak{a}$ d'un degré inférieur à un certain entier d (pourvu que d soit assez grand). La condition pour que $s \in G$ s'exprime donc par la condition que s transforme en eux-mêmes certains sous-espaces vectoriels de certains espaces tensoriels; or on voit facilement que cette condition peut encore se traduire par la condition que s doit laisser semi-invariants certains tenseurs en nombre fini. Utilisant ce théorème, on peut montrer que, si G est un sous-groupe algébrique de $GL(n, \mathbf{C})$ et H un sous-groupe algébrique et distingué de G, le groupe algébrique G/H est isomorphe à un groupe linéaire. Par ailleurs, il convient de noter que le résultat précédent et sa démonstration s'étendent sans modification au cas où le corps $\mathbf{C}$ est remplacé par un corps quelconque.

Soit G un sous-groupe quelconque de $GL(n, \mathbf{C})$; il y a alors un plus petit groupe algébrique G^* contenant G: il se compose des opérations qui admettent comme semi-invariants tous les semi-invariants de G; on l'appelle la coque algébrique du groupe G. On montre que, si G est un groupe analytique, G^* contient G comme sous-groupe distingué et a même groupe des commutateurs que G, de sorte que G^*/G est commutatif; d'une manière plus précise, on a $G^* = G$ toutes les fois que G est lui-même le groupe des commutateurs d'un groupe analytique. Par ailleurs, on montre que le groupe G^* est engendré par les coques algébriques des sous-groupes à un paramètre du groupe analytique G (i.e. des groupes de la forme $\{\exp tX\}$, où X est une matrice fixe). Or les résultats cités plus haut sur la détermination d'un groupe par ses semi-invariants permettent de déterminer explicitement la coque algébrique G^* d'un groupe à un paramètre G. Supposons que G se compose des opérations $\exp tX$. La matrice X peut se représenter de manière unique comme somme d'une matrice nilpotente N et d'une matrice semi-simple D (i.e. d'une matrice réductible à la forme diagonale) qui commute avec N. Ceci étant, pour qu'une matrice X' appartienne à l'algèbre de Lie de G^*, il faut et suffit qu'elle puisse se mettre sous la forme $cN + D'$ où c est un scalaire et D' une matrice qui se met sous la forme $P(D)$, P étant un polynôme qui doit posséder la propriété suivante: si $d_1, \ldots, d_n$ sont les racines caractéristiques de D, on doit avoir $\Sigma z_i P(d_i) = 0$ toutes les fois que $z_1, \ldots, z_n$ sont des entiers tels que $\Sigma z_i d_i = 0$. Les matrices X' qui possèdent cette propriété s'appellent les répliques de X; pour qu'une sous-algèbre $\mathfrak{g}$ de l'algèbre de Lie de $GL(n, \mathbf{C})$ soit l'algèbre de Lie d'un sous-groupe algébrique de $GL(n, \mathbf{C})$, il faut et suffit que toute réplique de toute matrice appartenant à $\mathfrak{g}$ appartienne encore à $\mathfrak{g}$.

Les résultats précédents fournissent des renseignements sur les points d'un groupe linéaire algébrique irréductible G à coefficients dans un sous-corps donné K de **C**. Nous supposerons que G est défini sur K, c'est-à-dire que l'idéal des fonctions polynômes nulles sur G est engendré par des polynômes à coefficients dans K; il revient au même de dire que l'algèbre de Lie de G est engendrée par des matrices à coefficients dans K. Dans ce cas, on peut montrer que G contient 'assez' de points à coefficients dans K, en ce sens que toute fonction polynôme sur $GL(n, \mathbf{C})$ qui est nulle sur l'ensemble des points de G à coefficients dans K est identiquement nulle sur G. Le principe de la démonstration consiste à utiliser les théorèmes cités plus haut pour se ramener au cas où G est commutatif; dans ce cas, on fabrique des points de G à coefficients dans K en partant de points s à coefficients dans une extension algébrique galoisienne de K et en formant le produit de s par ses conjugués relativement à K; comme ces conjugués commutent entre eux, leur produit est à coefficients dans K. On notera que la question des points à coordonnées dans un corps donné K se pose encore de la même manière quand **C** est remplacé par un corps algébriquement clos de caractéristique quelconque; mais, dans ce cas, Rosenlicht a montré que le théorème d'existence d'assez de points à coordonnées dans K n'est en général plus vrai, bien qu'il le soit encore si on suppose le corps K parfait.

Les résultats relatifs à la correspondance entre les groupes algébriques et leurs algèbres de Lie tombent malheureusement en défaut dès que la caractéristique p du corps de base cesse d'être nulle. On peut bien encore associer à tout groupe algébrique de dimension n une algèbre de Lie, qui est aussi de dimension n; cette algèbre de Lie est d'ailleurs une p-algèbre au sens de Jacobson, c'est-à-dire une algèbre de Lie dans laquelle est définie une opération de puissance p-ième, liée à l'opération de crochet par certaines identités; la présence de cette opération résulte du fait que la puissance p-ième d'une dérivation d'un corps de caractéristique p est encore une dérivation. Mais, alors que les théorèmes de la théorie classique qui permettent d'inférer les propriétés de l'algèbre de Lie à partir de celles du groupe restent en général vrais en caractéristique p, il n'en est pas de même des théorèmes allant dans l'autre direction. Citons par exemple le fait que plusieurs sous-groupes irréductibles d'un même groupe algébrique G peuvent avoir la même algèbre de Lie, et que des groupes algébriques non résolubles (comme le groupe linéaire unimodulaire à 2 variables en caractéristique 2) peuvent avoir des algèbres de Lie résolubles. Il résulte de là que les méthodes inspirées de la théorie des groupes de Lie perdent toute leur efficacité dans l'étude des groupes linéaires en caractéristique p, tout comme dans le cas des variétés abéliennes en caractéristique 0; elles ont dû être remplacées par des méthodes directes, qui s'appuyent ici encore fortement sur la géométrie algébrique.

Les premiers résultats relatifs à la théorie des groupes linéaires en caractéristique p furent ceux de Kolchin relatifs aux groupes résolubles; Kolchin a notamment démontré que le théorème de Lie, qui affirme que les matrices d'un groupe linéaire analytique résoluble complexe peuvent être simultanément réduites à la forme triangulaire reste vrai pour les groupes algébriques résolubles irréductibles sur des corps algébriquement clos de caractéristique quelconque; le résultat était d'autant plus frappant que l'on savait à l'époque que l'énoncé correspondant pour les algèbres de Lie est faux.

C'est au mémoire fondamental de Borel que la théorie des groupes linéaires algébriques sur un corps algébriquement clos K de caractéristique quelconque doit l'aspect de doctrine harmonieuse et cohérente qu'elle revêt aujourd'hui. Soit G un groupe linéaire algébrique irréductible sur K. Il est clair que G lui-même ne saurait être une variété complète sans se réduire à son élément neutre. Mais G peut néanmoins avoir des espaces homogènes complets, c'est-à-dire des sous-groupes fermés H tels que la variété G/H soit complète. C'est ainsi que l'espace projectif de dimension $n-1$, qui est une variété complète, est un espace homogène du groupe linéaire général à n variables. Cependant, on peut montrer qu'un groupe linéaire résoluble ne peut avoir aucun espace homogène complet; d'une manière plus précise, si un groupe linéaire résoluble opère rationnellement sur une variété complète U quelconque, il y a au moins un point de U qui est invariant par les opérations du groupe. La démonstration se fait en se ramenant d'abord au cas où le groupe est commutatif, puis, de là, au cas où le groupe est de dimension 1; or on peut montrer qu'un groupe algébrique irréductible de dimension 1 est isomorphe soit au groupe additif du corps de base soit au groupe multiplicatif des éléments $\neq 0$ du corps de base; dans chacun de ces cas, il est facile de montrer directement que le groupe n'admet aucun espace homogène complet non trivial. Il résulte de là que, si un groupe linéaire irréductible G opère dans une variété complète U, tout sous-groupe résoluble irréductible de G est contenu dans le groupe d'isotropie d'au moins un point de U. Par ailleurs, le groupe G admet au moins un espace homogène complet tel que les groupes d'isotropie des points de cet espace soient résolubles. Pour le montrer, il suffit de plonger G dans le groupe linéaire complet $GL(n, K)$ et d'observer que le théorème est vrai pour $GL(n, K)$. Or le quotient de $GL(n, K)$ par le groupe des matrices triangulaires n'est autre que la variété des drapeaux de l'espace vectoriel

K^n sur K, c'est-à-dire la variété composée des suites $(L_1, \ldots, L_n)$ où L_i est un sous-espace de dimension i de K^n et $L_i \subset L_{i+1}$ $(1 \leqslant i \leqslant n-1)$; cette variété est complète. On démontre en même temps par cette méthode que tout sous-groupe résoluble irréductible de $GL(n, K)$ est conjugué à un sous-groupe de groupe triangulaire, ce qui fournit une nouvelle démonstration du théorème de Lie–Kolchin. Si G est un groupe linéaire algébrique irréductible quelconque, on appelle maintenant groupes de Borel les sous-groupes résolubles irréductibles maximaux de G; ce sont aussi les éléments minimaux de l'ensemble des sous-groupes fermés H tels que G/H soit complet; on montre que ce sont aussi des sous-groupes résolubles maximaux (sans condition d'irreductibilité) de G. Les groupes de Borel d'un groupe G sont tous conjugués entre eux dans G.

On appelle tore un groupe linéaire qui est isomorphe au produit d'un certain nombre de fois le groupe multiplicatif du corps de base par lui-même; la raison de cette terminologie est que ces groupes jouent dans la théorie des groupes algébriques un rôle très analogue à celui que jouent les groupes toroïdaux dans la théorie des groupes compacts. Dans le cas du groupe linéaire général, le groupe des matrices diagonales est un tore maximal, et tout autre tore contenu dans le groupe est conjugué à un sous-groupe du groupe des matrices diagonales. Si G est un groupe linéaire algébrique quelconque, les tores maximaux de G sont encore tous conjugués entre eux; leur dimension commune fournit un invariant important du groupe G, que l'on appelle son rang. On appelle enfin groupes de Cartan de G les centralisateurs des tores maximaux de G; tout groupe de Cartan est le produit direct d'un tore maximal et d'un groupe formé d'éléments unipotents (une matrice est dite unipotente si elle est la somme de la matrice unité et d'une matrice nilpotente).

Les résultats se précisent encore dans le cas où G est un groupe semi-simple, c'est-à-dire où G n'admet aucun sous-groupe résoluble distingué de dimension > 0. Dans ce cas, on démontre que les tores maximaux sont leurs propres centralisateurs, de sorte que les groupes de Cartan sont des tores. Chaque tore maximal T n'est contenu que dans un nombre fini de groupes de Borel $B_1, \ldots, B_N$; ces groupes sont permutés de manière simplement transitive entre eux par les opérations du normalisateur N du groupe T; le groupe $N/T = W$ est un groupe fini; dans le cas classique, l'importance de ce groupe dans la théorie des groupes semi-simples avait été reconnue par Weyl; aussi l'appelle-t-on le groupe de Weyl du groupe G. Soit B l'un des groupes B_i; il est le produit semi-direct du groupe T et d'un groupe B^u formé de matrices unipotentes. De plus, alors que la structure des groupes unipotents généraux en caractéristiques $p > 0$ semble être fort complexe, les groupes B^u qui proviennent de groupes semi-simples sont relativement peu compliqués: B^u est en effet le produit semi-direct d'un certain nombre de groupes H_i de dimension 1, isomorphes au groupe additif du corps K, dont les normalisateurs contiennent T. Si H_i est l'un de ces groupes, dont on suppose fixée une paramétrisation au moyen des éléments de K, le transformé par un élément t de T du point de paramètre θ sur H_i est le point de paramètre $\alpha_i(t) \theta$, $\alpha_i(t)$ étant un homomorphisme du groupe T dans le groupe multiplicatif du corps K. Les divers homomorphismes de T que l'on obtient de cette manière sont appelés les racines du groupe G (relativement à T); le groupe de Weyl permute entre elles les racines. On retrouve ainsi tous les éléments de structure qui ont permis à Killing et à Cartan de donner dans le cas classique la classification complète des groupes de Lie semi-simples; on peut alors procéder à cette classification dans le cadre plus général de la théorie des groupes semi-simples sur un corps algébriquement clos de caractéristique quelconque; ceci fait, on constate qu'il y a exactement autant de types que dans la théorie classique, de sorte que la classification est entièrement indépendante de la caractéristique du corps de base. Il convient de noter cependant que l'existence de groupes des divers types prévus par la classification ne peut encore être établie qu'à partir de la connaissance de l'existence de ces groupes dans le cas classique. Pour passer du cas classique au cas de caractéristique p, on utilise un procédé qui permet d'associer à tout groupe semi-simple complexe G qui soit son propre groupe adjoint et à tout corps K (pas nécessairement algébriquement clos) un groupe de matrices à coefficients dans K; appliqué aux corps finis K, ce procédé permet de construire des groupes finis qui fournissent certains groupes simples finis. Parmi les groupes finis simples ainsi obtenus, ceux qui correspondent aux groupes simples classiques ou aux groupes exceptionnels de type G_2 étaient connus depuis les travaux de Dickson.

Soit G un groupe linéaire irréductible, et soit B un groupe de Borel de G. L'espace homogène complet G/B ne dépend que du quotient de G par son radical; les variétés complètes que l'on obtient ainsi possèdent des propriétés intéressantes. Le groupe B, étant un sous-groupe de G, opère dans G/B; le théorème de Bruhat affirme que l'espace G/B se décompose en un nombre fini d'orbites pour le groupe B. Chacune de ces orbites est une sous-variété (non fermée en général) de G/B, isomorphe à un espace numérique K^ν. Dans le cas classique, on obtient de cette manière une décomposition cellulaire de l'espace compact G/B en cellules de dimensions paires; il en résulte immédiatement que, pour toute dimension m, le groupe d'homologie entière pour la dimension m de G/B est un groupe libre de rang égal au nombre des cellules de dimension m. Or le nombre de ces cellules peut en principe se calculer dès que l'on connaît la manière dont le groupe de Weyl opère sur les racines. On trouve ainsi des formules relatives à l'homologie de G/B qui ont été également obtenues par voie transcendante par Bott. Par ailleurs, on sait que c'est encore un problème ouvert que de généraliser aux variétés définies sur les corps de caractéristique $p > 0$ les notions que la topologie fournit à l'étude des variétés algébriques complexes; ce problème est intimement lié aux questions d'analyse diophantienne par les conjectures de Weil. Dans le cas des variétés G/B, on peut montrer que la notion d'anneau de classes d'équivalence rationnelle, due à Chow, fournit la solution du problème: on trouve en effet que, pour une caractéristique quelconque, l'anneau de Chow de la variété G/B est isomorphe à l'anneau de cohomologie de la variété complexe qui correspond au groupe complexe homologue de groupe G.

SOME NEW CONNECTIONS BETWEEN PROBABILITY AND CLASSICAL ANALYSIS

By WILLIAM FELLER†

Recent research has revealed the intimate relationship between potential theory and Markov processes, and has supplied new examples of the fertility of a probabilistic approach to problems of classical analysis. Choquet's work on capacities, the Beurling–Deny theory of general potentials, Doob's probabilistic approach to the Dirichlet problem, and Hunt's basic results concerning potentials and Markov processes are closely related, despite the diversity of formal appearances and methods. I had hoped in this address to discuss the interconnections between these theories, but the task proved too overwhelming for the limited time and my own limitations. I am therefore compelled unashamedly to restrict this talk to some related aspects of my own work. I propose to describe the two boundaries and topologies induced by the annihilators of certain operators; to discuss their justification, their use for an abstract theory of so-called boundary value problems, and their connection with an invariant theory of operators of local character (which generalize the ill-defined notion of differential operators).

I shall not endeavor to develop a theory or even to state results in a precise form. Rather I shall try to explain the background and the purpose of the theory by means of a few simple examples, preferably using classical harmonic functions. Although everything will be interpreted probabilistically, the main emphasis is purely analytic.

In the sequel D will always denote a topological space and $\mathbf{C}$ the familiar Banach space of continuous functions in D with

$$\|f\| = \sup_{p \in D} |f(p)|.$$

Unless otherwise stated all operators will act on continuous functions.

1. Boundaries induced by positive operators

1.1. Harmonic functions. It is convenient to start with a probabilistic interpretation of harmonic functions by means of an *ad hoc* constructed random walk. In § 2 we shall approach the Laplacian more directly and more naturally.

† Research in part supported by the Office of Ordnance Research, U.S. Army.

Let D be the open unit disc in the plane and for each point $p \in D$ let $D_p \subset D$ be the greatest open disc centered at p and contained in D. We shall study the operator T from C to C for which $Tf(p)$ *equals the arithmetic mean* of f over D_p. Thus

$$Tf = \int_D K(., q) f(q) \, dq,$$ (1)

where $K(p, q)$ equals $|D_p|^{-1}$ or 0 according as $q \in D_p$ or $q \bar{\in} D_p$. Similarly, T^n is induced by a kernel $K^{(n)}$.

This T determines a discrete *random walk* with arbitrary initial position $p \in D$ in which $K^{(n)}(p, .)$ is the density of the random position $Q_n \in D$ after n steps. We obtain a well-defined measure in the space of all sequences $\{Q_n\}$ $(Q_0 = p, Q_n \in D)$ and it can be shown that almost all sequences are convergent to a point of the boundary circle. For our purposes a less refined purely analytical statement will suffice. For any set $A \subset D$ the probability that $Q_n \in A$ equals $T^n \chi(p)$, where χ is the characteristic function of A. It is easily verified that $T^n f \to 0$ for each $f \in C$ vanishing at the boundary. Therefore $T^n \chi \to 0$ for each compact A, and this is equivalent to the statement that Q_n approaches the boundary in probability.

This result can be rendered more precise as follows. Let $\Gamma \subset B$ be a set of the boundary circle B, and u_Γ the harmonic function determined, in the classical sense, by the boundary values 1 on Γ and 0 on $B - \Gamma$. Then $u_\Gamma(p)$ *is the probability that* Q_n *approaches* Γ as $n \to \infty$.

Harmonic functions appear in this context because they are eigenfunctions satisfying $\phi = T\phi$. The set $\mathfrak{P}$ of all solutions of this equation such that $0 \leqslant \phi \leqslant 1$ is a convex set and the harmonic functions u_Γ coincide with its *extremals*. The sets

$$\Gamma_\epsilon = \{q \in D \mid u_\Gamma(q) > 1 - \epsilon\}$$ (2)

are a system of deleted neighborhoods of Γ. If χ is the characteristic function of Γ_ϵ then $T^n \chi \to u_\Gamma$, and hence

$$u_\Gamma(p) = \lim_{n \to \infty} \mathrm{prob}\, \{Q_n \in \Gamma_\epsilon\}.$$ (3)

From this it follows easily that $u_\Gamma(p)$ is the probability of an actual asymptotic approach to Γ, but for our purposes the weaker statement (3) fully suffices. [An alternative proof is given in the next section.]

The point to be emphasized is that this set-up is not of an analytic nature but can be carried out abstractly for a large class of operators in an arbitrary topological space D. No boundary need be defined *a priori*, and it is natural to *define* a boundary in such a way that the sets Γ_ϵ become neighborhoods of the corresponding boundary sets. Again, D may possess a boundary which does not admit of the interpretation of Γ_ϵ as neighborhoods, and it may be necessary to introduce a new boundary appropriate to the study of the transformation T. The simplest example is obtained by mapping the unit disc D conformally onto a domain $\tilde{D}$ whose boundary $\tilde{B}$ is of a complicated structure with prime ends, etc. Under this conformal map T and its random walk are carried over to $\tilde{D}$, the new eigenfunctions are again harmonic, but obviously the convergence properties of the random walk remain true only if we replace the 'natural' boundary of $\tilde{D}$ by the boundary and topology induced by the conformal map. This, of course, is a probabilistic version of the now familiar observation due to Martin[17] that the study of harmonic functions in complicated domains requires the introduction of an appropriate boundary.

We pass to a more interesting example of a different kind.

1.2. Relativization and isomorphisms.
For an arbitrary (not necessarily bounded) $\psi > 0$ harmonic in the disc D we define a new transformation by

$$T_\psi f = \psi^{-1} T(f\psi). \tag{4}$$

Its kernel is given by

$$K_\psi(p, q) = \psi^{-1}(p) K(p, q) \psi(q). \tag{5}$$

A function $v > 0$ satisfies $v = T_\psi v$ if and only if $\phi = v\psi$ satisfies $\phi = T\phi$ (is harmonic). We have thus a 1-1 correspondence between the positive eigenfunctions of T and T_ψ with 1 and ψ corresponding to ψ^{-1} and 1, respectively.

Operators of this form will be called *similar* to T. Clearly similarity is transitive, symmetric and reflexive. A closely related transformation has been used by Brelot[1] for harmonic functions. We shall see that the notion of similarity is exceedingly useful, and has its counterpart in similar semigroups and differential equations. Here we use it to illustrate the notion of the boundary induced by T_ψ and to derive a new proof for our interpretation of the harmonic function u_Γ.

Denote by $\mathfrak{P}$ and $\mathfrak{P}_\psi$, respectively, the sets of positive eigenfunctions of T and T_ψ bounded by 1. For simplicity of exposition consider the case $\psi = u_\Gamma$ where ψ is an extremal of $\mathfrak{P}$. The mapping $u\psi \leftrightarrow \phi$ establishes a 1-1 correspondence between $\mathfrak{P}_\psi$ and the subset of $\mathfrak{P}$ of elements such that $\phi \leqslant u_\Gamma$; this correspondence preserves extremals. Now the structure of the set $\mathfrak{P}$ of harmonic functions is best described in terms of the 'natural' boundary of D. Since relations between T, $\mathfrak{P}$, and $D \cup B$ with the natural topology will carry over to T_ψ, $\mathfrak{P}_\psi$ and $D \cup \Gamma$, the set Γ plays for T_ψ the role that B plays for T and may be considered the *boundary induced by* T_ψ.

The same conclusion may be reached probabilistically. From $\psi^{-1} = T_\psi \psi^{-1}$ it follows that $T_\psi^n f \to 0$ for each $f \in \mathbf{C}$ vanishing along Γ. Hence if χ is the characteristic function of a neighborhood U of Γ we have $T_\psi^n \chi \to 1$: thus in the random walk associated with T_ψ the paths converge in probability to the boundary set Γ.

This remark leads to a new proof of the interpretation of u_Γ given in the preceding section. Using (5) it is seen that the relation $T_\psi^n \chi \to 1$ may be rewritten as

$$\int_U K^{(n)}(p,q)\,\psi(q)\,dq \to \psi(p). \tag{6}$$

Now the neighborhood U of Γ may be chosen so small that in it $\psi = u_\Gamma$ is arbitrarily close to 1 and we conclude that in the T-random walk $\mathbf{P}\{Q_n \in U\} \to u_\Gamma(p)$ for each neighborhood of Γ. This is a slightly weakened version of the interpretation of $u_\Gamma(p)$ as *probability of an asymptotic approach to* Γ.

Given this interpretation of u_Γ we see that in the T-random walk K_ψ represents the *conditional* transition probability density given the event that the paths converge to Γ. Probabilistically, then, *the T_ψ-random walk is obtained from the T-random walk by conditioning or relativization*: in the T-walk we pay attention only to paths converging to Γ. More precisely, let $\mathfrak{S}$ be the set of all sequences $\{Q_n\}$, $(Q_0 = p, Q_n \in D)$, with the measure $\mathbf{P}$ induced by T, and let $\mathfrak{S}_\Gamma$ be the subset of sequences converging to Γ. Then the T_ψ-walk assigns zero probability to $\mathfrak{S} - \mathfrak{S}_\Gamma$ and probability $\mathbf{P}\{\mathfrak{A}\} \div \mathbf{P}\{\mathfrak{S}\} = \mathbf{P}\{\mathfrak{A}\}\,\psi^{-1}(p)$ to the subsets $\mathfrak{A} \subset \mathfrak{S}$.

1.3. Abstract construction; restricted and total boundaries. We pass to the extreme case where D is the set of integers $1, 2, \ldots$, with the discrete topology. This has the advantage that no preconceived intuitive notion of boundary obscures the view. The boundaries may be of a complicated topological structure and the present case will clearly reveal the features and problems of the most general set-up.

$\mathbf{C}$ is now the space of bounded functions and we write $\mathbf{f} \in \mathbf{C}$ as a column matrix with elements $f(i)$. We consider an operator T defined by a matrix Π with elements $\Pi(i,j)$ so that in matrix notation $T\mathbf{f} = \Pi\mathbf{f}$. The matrix Π is supposed to be substochastic, that is, its elements are ≥ 0, its row sums ≤ 1. We denote by $\mathfrak{P}$ the set of all eigenvectors $\boldsymbol{\phi}$ such that $\boldsymbol{\phi} = \Pi\boldsymbol{\phi}$

PROBABILITY AND CLASSICAL ANALYSIS

and $0 \leqslant \phi \leqslant 1$, and by $\mathfrak{P}^{\infty}$ the set of all (possibly unbounded) solutions $\phi > 0$ of $\phi = \Pi\phi$. To avoid trivialities we shall assume: (i) each ϕ is strictly positive, (ii) $\mathfrak{P}$ contains at least two independent vectors. [Condition (i) eliminates the nuisance of partitioned matrices requiring words rather than thoughts; (ii) eliminates empty and single-point boundaries.]

Again Π may be interpreted as the matrix of transition probabilities in a *random walk* (Markov chain), the row defects $1 - \Sigma\Pi(., j)$ accounting for the possibility of a termination of the process. For an arbitrary initial position $i \in D$ we have a probability measure on the set $\mathfrak{S}$ of all terminating or infinite sequences of integers $\{Q_n\}$. The subset $\mathfrak{S}^{(n)}$ of sequences of length $\geqslant n$ has probability $\Sigma_j \Pi^n(i, j)$, and hence the probability of the set $\mathfrak{S}^{(\infty)}$ of infinite sequences is given by the ith element of $\bar{\phi} = \lim \Pi^n 1$. Note that $\bar{\phi}$ is the *maximal* element of $\mathfrak{P}$.

We proceed to introduce a *restricted* or $\mathfrak{P}$-*boundary* B, and a *total* or $\mathfrak{P}^{\infty}$-*boundary* $B^{\infty} \supset B$. We begin with the extremely simple special case where $\phi = \Pi\phi$ has only *finitely* many independent solutions.

(a) *The $\mathfrak{P}$-boundary.* Let $\mathfrak{P}$ be spanned by N non-zero vectors $\phi^{(1)}, ..., \phi^{(N)}$. These can be chosen as extremal elements of $\mathfrak{P}$, which amounts to saying that $\|\phi^{(k)}\| = 1$ and $\bar{\phi} = \phi^{(1)} + ... + \phi^{(N)}$. For fixed k and $\epsilon > 0$ put $\Gamma_{\epsilon}^{(k)} = \{i \mid \phi^{(k)}(i) > 1 - \epsilon\}$. As $\epsilon \to 0$ we get a nest of non-empty sets with empty intersection; from $\bar{\phi} \leqslant 1$ we conclude that for fixed $\epsilon > \frac{1}{2}$ the sets $\Gamma_{\epsilon}^{(j)}$ and $\Gamma_{\epsilon}^{(k)}$ are non-overlapping $(j \neq k)$.

The *restricted boundary* B consists of N points $\beta^{(1)}, ..., \beta^{(N)}$ such that $\Gamma_{\epsilon}^{(k)}$ is a deleted neighborhood of $\beta^{(k)}$. We can extend the definition of each $\phi \in \mathfrak{P}$ to $D \cup B$ by putting $\phi^{(j)}(\beta^{(k)}) = 1$ or 0 according as $j = k$ or $j \neq k$. Then each ϕ is continuous in $D \cup B$, and the 'Dirichlet problem' is soluble: to prescribed boundary values there corresponds exactly one $\phi \in \mathfrak{P}$.

Finally with an obvious notation, $\Pi^n(i, \Gamma_{\epsilon}^{(k)}) \to \phi^{(k)}(i)$, as $n \to \infty$ for each fixed ϵ. From this one deduces that in the random walk starting at i the sample sequences $\{Q_n\}$ converge with probability $\phi^{(k)}(i)$ to $\beta^{(k)}$; with probability $1 - \bar{\phi}(i)$ they terminate; and the probability of no convergence is 0.

(b) *The $\mathfrak{P}^{\infty}$-boundary* can be introduced directly, but it is more convenient to use the similarity transformation introduced in § 1.2. A matrix Π_{ψ} is *similar* to Π if either $\Pi_{\psi} = \Pi$ or

$$\Pi_{\psi}(i, j) = \psi^{-1}(i)\,\Pi(i, j)\,\psi(j), \tag{7}$$

where $\psi \in \mathfrak{P}^{\infty}$. We recall that similarity is transitive, reflexive, and symmetric and that the mapping $\phi \leftrightarrow v$ where $v(i) = \phi(i)\,\psi^{-1}(i)$ establishes a 1-1 correspondence between $\mathfrak{P}^\infty$ and $\mathfrak{P}^\infty_\psi$.

To see the relation between B and the corresponding restricted boundary B_ψ induced by Π_ψ consider the typical case $\psi = \phi^{(1)} + \phi^{(2)}$. To $\phi^{(3)}, \ldots, \phi^{(N)}$ and to each unbounded $\phi \in \mathfrak{P}^\infty$ there correspond unbounded vectors in $\mathfrak{P}^\infty_\psi$ and the boundary B_ψ reduces to two points whose neighborhoods coincide with the neighborhoods of $\beta^{(1)}$ and $\beta^{(2)}$. For the interpretation of this Π_ψ in terms of *conditional probabilities* see § 1.2.

In general, if B_ψ is the restricted boundary induced by Π_ψ we shall identify points of B and B_ψ with coinciding systems of neighborhoods. With this identification we define the *total boundary B^∞ induced by Π as the union of the boundaries B_ψ as ψ ranges over* $\mathfrak{P}^\infty$. All *similar matrices induce the same total boundary*, and if $\mathfrak{P}^\infty$ is spanned by M independent vectors, then B^∞ contains exactly M points.

Note that $D \cup B^\infty$ need not be compact. (No compactification seems natural or desirable for our purposes.)

(c) *The maximal ideal boundaries.* When $\mathfrak{P}$ is not spanned by denumerably many elements the extremal elements of $\mathfrak{P}$ do not correspond to points of the prospective boundary, but rather to sets of positive capacity. No satisfactory definition of points and neighborhoods is known. Now both $\mathfrak{P}$ and $\mathfrak{P}^\infty$ have a lattice structure similar to that of harmonic functions, and the correspondence between $\mathfrak{P}^\infty$ and $\mathfrak{P}^\infty_\psi$ is a lattice isomorphism. This makes it possible to define points of B and B^∞ by maximal ideals in $\mathfrak{P}$ and $\mathfrak{P}^\infty$, respectively (see [5]). Unfortunately these boundaries are absurdly large. For example, sample sequences converge to sets rather than to points; each $\phi \in \mathfrak{P}$ has continuous boundary values which is at variance with the desirable model of harmonic function in a disc D with the natural topology. That the lattices $\mathfrak{P}$ and $\mathfrak{P}^\infty$ are isomorphic to lattices of continuous functions on some Hausdorff spaces is, of course, well known (see, for example, Kadison [15]). To us the main point is that this Hausdorff space appears as a *boundary* of D and is useful as such.

Our maximal ideal boundaries serve well for an orientation and as a guide, but the introduction of a less clumsy and more appropriate boundary is an open, and promising, problem.

[*Postscript.* A partial solution has now been obtained by J. L. Doob [19] who uses Martin's original construction. The Martin boundary is sufficiently small for sample sequences to converge toward points. In the finite case, however, this boundary may be bigger than ours; its neighborhoods are larger and this could lead to difficulties in connection with boundary value problems and non-minimal semigroups.]

2. Semigroups and differential equations

2.1. Orientation. We turn to the more interesting study of a family of positive operators $\{T(t)\}$, $t \geqslant 0$, from $\mathbf{C}$ to $\mathbf{C}$ with $\|T(t)\| \leqslant 1$, and with the semigroup property $T(t+s) = T(t)\,T(s)$. The probabilistic counterpart to a fixed $T(t)$ is a (possibly terminating) random walk with jumps taking place at times $t, 2t, \ldots$. To the whole semigroup there corresponds a random motion (Markov process) in D with continuous time: the sample space $\mathfrak{S}$ is the space of functions Q defined in an interval $0 \leqslant t < \tau \leqslant \infty$ such that $Q(t) \in D$, and $Q(0) = p \in D$ is a given point. The semigroup induces a **P**-measure in $\mathfrak{S}$ such that for each Borel set $A \subset D$ we have $\mathbf{P}\{Q(t) \in A\} = T(t)\,\chi(p)$ where χ is the characteristic function of A. Needless to say with the **P**-measure almost all paths are reasonably regular. In particular, they are continuous when the semigroup is generated by a differential operator ('diffusion processes').

If a path Q is defined only within a finite interval $0 \leqslant t < \tau$ (that is, if the process terminates at time τ) then, except on a null set of paths, as $t \uparrow \tau$ either $Q(t) \to q \in D$ or $Q(t)$ has no point of accumulation in D. In the first case we say that the process terminates at q, in the second that it *terminates 'at the boundary'*. However, it remains to justify this expression.

For this purpose we shall have to introduce a boundary B induced by the *generator* Ω of the semigroup. It will be seen that all operators $T(t)$, $t > 0$, *induce the same boundary* $\pi \subset B$, and that the process *either terminates at $B - \pi$ or approaches π asymptotically* without reaching it.

We say that the semigroup is *generated by* Ω if for a dense set of 'smooth' $f \in \mathbf{C}$

$$\lim_{t \to 0} t^{-1}\{T(t)\,f - f\} = \Omega f \in \mathbf{C} \tag{8}$$

in the sense of pointwise convergence. According to this definition, introduced in[11], Ω may generate many semigroups; the infinitesimal generator in the sense of Hille[13] is a contraction $\Omega|\Sigma$ of Ω obtained by imposing *lateral conditions* (see §4.3).

Putting $T(t)\,f = u(t, .)$ the function u will satisfy the functional equation $u_t = \Omega u$ with the 'initial condition' $u(0, .) = f$. (This is literally true for smooth f, and in an operational sense for all f.) In classical terms we are concerned with 'solving' this equation. We consider first a family of 'similar' generators and then the relation between generators with the same annihilators.

2.2. The Laplacian. Isomorphisms. We return to harmonic functions and take as example the familiar *heat equation* $u_t = \Delta u$ in the unit disc D. Among the positive semigroups generated by Δ there exists a *minimal semigroup*. In classical terms $u(t, .) = T(t)f$ is the solution of $u_t = \Delta u$ with initial condition $u(0, .) = f$ and *boundary condition* that u vanishes at the circle B. Associated with it is the *Wiener process* in D terminating at B. Physically the process represents a homogeneous diffusion in D with 'absorbing boundaries', or heat conduction with zero temperature at the boundary.

For this minimal semigroup $\|T(t)\| < 1$ and therefore the boundary π induced by $T(t)$ is empty, but we are concerned with the boundary induced by harmonic functions, that is, the *annihilators* of the generator Δ. In this connection, of course, we adhere to the natural topology of the closed disc. As in § 1.1 let u_Γ be the harmonic function in D determined by the boundary values 1 on the set $\Gamma \subset B$ and 0 on $B - \Gamma$. For the process starting at the point $p \in D$ we have the following analogue to the results of § 1.1.

$u_\Gamma(p)$ *is the probability that the process terminates at the boundary set* Γ. The probability that this happens before time t is $v(t, p)$ where v is the solution of $v_t = \Delta v$ with zero initial values and boundary values 1 on Γ and 0 on $B - \Gamma$.

This assertion becomes plausible on observing that for bounded harmonic ψ obviously
$$T(t)\,\psi = \psi - v(t, .), \tag{9}$$

where $v_t = \Delta v$ and v has zero initial values and boundary values coinciding (almost everywhere) with those of ψ. Now $T(t)\,1(p) = \mathbf{P}\{Q(t) \in D\}$ is the probability that the process does not terminate before it. Together with (9) this implies the assertion for the particular case $u_\Gamma = 1$, or $\Gamma = B$, at least if we accept as fact that the process does not terminate in the interior of D. We shall only indicate how the general assertion may be reduced to this particular case by a generalization of the method of *similarity transformations* or isomorphisms, introduced in § 1.2.

For positive harmonic ψ we define a new semigroup of positive operators from $\mathbf{C}$ to $\mathbf{C}$ by
$$T_\psi(t)\,f = \psi^{-1}T(t)\,(f\psi). \tag{10}$$

If $0 \leqslant f \leqslant 1$ then $0 \leqslant T_\psi(t)f \leqslant \psi^{-1} \leqslant T(t)\,\psi \leqslant 1$ and hence $\|T_\psi(t)\| \leqslant 1$. A glance at (8) shows that the semigroup $\{T_\psi\}$ is generated by the operator Δ_ψ defined by
$$\Delta_\psi f = \psi^{-1}\Delta(f\psi). \tag{11}$$

For simplicity of exposition we now restrict the consideration to $\psi = u_\Gamma$. First note that ψ^{-1} is unbounded near $B - \Gamma$ and close to 1 near Γ, and that $T_\psi(t)^{-1} \leqslant \psi^{-1}$. From this it is easy to conclude that *the random process corresponding to* $\{T_\psi\}$ *terminates at* Γ. Our interpretation of u_Γ is obtained from this by adapting the argument following (6). As at the end of § 1.2 we remark that the kernel of the T_ψ-semigroup represents the conditional transition probability densities of the T-process given that the path terminates at Γ. In the sense explained above the T_ψ-*process is therefore simply the restriction of the T-process to the paths terminating at* Γ.

Returning to the analytical relationship between the T- and the T_ψ-semigroup note that $\Delta_\psi \phi = 0$ if and only if $\phi\psi$ is harmonic. The positive (possible unbounded) annihilators of Δ and Δ_ψ stand in a 1-1 correspondence (which is a lattice isomorphism). The bounded annihilators of Δ_ψ correspond to the harmonic functions dominated by $\psi = u_\Gamma$, and thus the set Γ is the appropriate boundary for $\{T_\psi\}$ just as the circle B is for $\{T\}$. In short we find the same situation as in § 1.3.

The circle B, *induced by the harmonic functions, represents the 'total' boundary for the family of all similar semigroups* $\{T_\psi(t)\}$ *[or all differential equations* $u_t = \Delta_\psi u$]. *The boundary induced by the bounded annihilators of* Δ_ψ *corresponds to the subset* $\Gamma \subset B$.

That Γ is the appropriate boundary for the T_ψ-semigroup reflects the fact that the common *range of the transformations* $T_\psi(t)$ *is characterized by the side condition that* $T_\psi(t)f$ *vanishes along* Γ just as $T(t)f$ vanishes along B.

We have here the simplest example of a 'boundary condition' and see that it relates to our boundary rather than the 'natural' one. In the terminology of classical differential equations, Δ_ψ is a differential operator with coefficients singular along $B - \Gamma$, and boundary conditions can be imposed only along Γ. How vague and unsatisfactory such descriptions can be is known from the simple Sturm–Liouville theory in one dimension.

2.3. The active boundary. Singular operators.

A new phenomenon may be described in connection with the operator $\Omega = \omega\Delta$ in the disc D where $\omega > 0$ is continuous in D but may tend to zero or infinity near the circle B. This operator has the same annihilators as the Laplacian Δ and it is interesting to compare the semigroups generated by $\omega\Delta$ with those generated by Δ. In classical terms we are concerned with the integration of the parabolic differential equation $u_t = \omega\Delta u$ which may be singular owing to the behavior of ω near B. We describe here the main features of the theory carried out in[6] for denumerable spaces (the Kolmogoroff differential equations) by methods of much wider applicability.

There exists a uniquely determined *minimal* positive semigroup $\{T(t)\}$ generated by $\omega\Delta$ and for it $\|T(t)\| \leqslant 1$. However, for an appropriate choice of ω it may happen that $T(t)\,1 = 1$: the minimal semigroup is in this case *unique*; by contrast to the heat equation no boundary conditions need or can be imposed in this case, and the induced random process (diffusion) does not terminate at a finite time.

With an arbitrary $\omega > 0$ let $\{T(t)\}$ be the minimal semigroup generated by $\omega\Delta$ in the unit disc D. Then the following is true.

(a) *The passive boundary π.* There exists a set $\pi \subset B$ of the unit circle (determined up to a null set) such that for each $t > 0$ the eigenfunctions ϕ of $\phi = T(t)\,\phi$ such that $0 \leqslant \phi \leqslant 1$ coincide with the positive harmonic functions (the annihilators of $\omega\Delta$) dominated by u_π, the harmonic function with boundary values 1 on π and 0 on $B - \pi$. In this sense *the boundary induced by each $T(t)$ coincides with π.* In the case $\omega = 1$ (the heat equation) π is empty. For any set $\Gamma \subset \pi$ the value $u_\Gamma(p)$ is the probability in the $T(t)$-process starting at the point $p \in D$ that Γ is *asymptotically approached as $t \to \infty$*; the probability of reaching Γ at a finite time is zero.

(b) *The active boundary $A = B - \pi$.* For each set $\Gamma \subset A$ the value $u_\Gamma(p)$ *is the probability that the process will terminate at Γ.* The probability that this occurs before time t equals $v(t, p)$ where v is a solution of $v_t = \omega\Delta v$ with zero initial values and boundary values 1 on Γ and 0 on $A - \Gamma$.

No boundary conditions can be imposed along the passive boundary. This summary explains the relations between minimal *semigroups generated by operators with the same annihilators.*

In analytic terms we may characterize the active and passive boundaries as follows. For $\lambda \geqslant 0$ the bounded positive solutions of

$$\lambda\phi - \omega\Delta\phi = 0 \tag{12}$$

form a convex set $\mathfrak{P}_\lambda$ endowed by a lattice structure similar to that of harmonic functions. Now there exists a 1-1 correspondence (lattice isomorphism) between the $\mathfrak{P}_\lambda$ for $\lambda > 0$ on one hand, and the harmonic functions dominated by u_A on the other hand. Thus the *active boundary is induced by each $\mathfrak{P}_\lambda$ for $\lambda > 0$.* An alternative interpretation may be given in terms of the resolvent $(\lambda - \omega\Delta)^{-1}$ and a discrete random walk associated with it.

3. The adjoint boundary

3.1. Duality. So far we have restricted consideration to operators acting on functions. Actually the study of an operator T from C to C cannot be separated from the study of the adjoint operator T^* which takes *measures* into measures. In probability T^* is the primary notion, although we are compelled to take T as the basic tool of the theory of Markov processes. The reasons are discussed in[7]; see also Dynkin[3].

To avoid new notations let D be an open domain of the plane and suppose that T is of the form (1) with an arbitrary *positive* kernel. The adjoint transformation acts on all measures, but it is convenient to consider only absolutely continuous measures and treat T^* as an operator on densities. Let then $\mathbf{L}$ be the Banach space of integrable functions in D with the usual norm

$$N(u) = \int_D |u|\, dq. \tag{13}$$

Then T^* as an operator from $\mathbf{L}$ to $\mathbf{L}$ carries $u \in \mathbf{L}$ into

$$T^*(u) = \int_D u(p)\, K(p, .)\, dp. \tag{14}$$

It is positive and $N(T^*) \leqslant 1$. The transformation (14) remains meaningful for all $u > 0$, although the integral may diverge.

In principle the construction of a boundary induced by T^* should follow the method used for T, but fortunately *an extremely simple trick will save us the trouble of a repeated construction*.

We start from the set $\mathfrak{P}^*$ of all finite eigenfunctions $u > 0$ of $u = T^*u$. They need not be integrable, but for simplicity we shall suppose that each $u \in \mathfrak{P}^*$ is strictly positive and continuous in D.

For an arbitrary $u \in \mathfrak{P}^*$ define a kernel $\tilde{K}$ by

$$\tilde{K}(p,q) = u(q)\, K(q,p)\, u^{-1}(p), \tag{15}$$

where $p \in D$, $q \in D$. Clearly

$$\int_D \tilde{K}(p,q)\, dq = 1 \tag{16}$$

for each p, and thus $\tilde{K}$ represents the kernel of a new transformation $\tilde{T}$ from C to C.

If $v \in \mathfrak{P}^*$ then $\phi = vu^{-1}$ is a continuous eigenfunction of $\tilde{T}\phi = \phi$, not necessarily bounded. Conversely, to each positive eigenfunction of $\tilde{T}\phi = \phi$ there corresponds the element $v = \phi u \in \mathfrak{P}^*$. This establishes a 1-1 correspondence between $\mathfrak{P}^*$ and the set $\mathfrak{P}^\infty$ of positive eigenfunctions of $\tilde{T}$. (These sets are endowed with lattice structures and the correspondence is a lattice isomorphism, but we omit these details.) In § 1.3 we have seen that $\tilde{T}$ induces a *total boundary* relative to the set $\mathfrak{P}^\infty$ of all positive eigenfunctions of $\tilde{T}\phi = \phi$. If in (15) we replace u by another element of $\mathfrak{P}^*$ then $\tilde{T}$ is replaced by a *similar* operator (as defined in § 1.3) and the total boundary remains unchanged. This justifies the

Definition. The adjoint boundary B^* induced by T is the total boundary induced by the operator $\tilde{T}$ acting from C to C. It is independent of the choice of $u \in \mathfrak{P}^*$.

Probabilistic interpretation. Consider first the case where $N(u) = 1$ and interpret u as the stationary probability density of the position (at any time) in a Markov chain with transition probability densities K. This process is defined for all integral values of the time parameter from $-\infty$ to ∞. In[16] Kolmogoroff pointed out that in this process $\tilde{K}(p, q)$ represents the conditional probability density of the position q at time n given that at time $n + 1$ the position is p. Moreover, the same relationship exists between the higher order transition probability densities of the K-chain and the $\tilde{K}$-chain. In other words, for the K-chain, $\tilde{K}$ represents the transition probability densities in the negative time direction: the $\tilde{K}$-chain is obtained from the K-chain by *reversing the time direction*.

Roughly speaking, then, the boundary induced by T represents the directions towards which the sample sequences can converge, and the adjoint boundary the directions from which the process can originate. This description applies to, and becomes more concrete in connection with, continuous time parameter processes and leads to an interpretation of the boundary conditions for differential equations.

Analytically there is no change in the situation when the integral (13) diverges, and it is therefore annoying that Kolmogoroff's intuitive interpretation of $\tilde{K}$ breaks down. However, as Derman[2] pointed out, it may be salvaged for non-integrable u by considering a whole family of chains.

3.2. Relations between the two topologies.

It is natural to ask whether and how the topology of $D \cup B$ induced by T is related to the topology of $D \cup B^*$ induced by T^*. A first answer is that every imaginable situation can arise. For the familiar symmetric operators of analysis the two spaces are identical. Still simpler is the other extreme, where B and B^* are disjoint and have disjoint neighborhoods. Most vexing are the intermediate cases. Among the examples given in[5] there appears the following configuration:

The boundary B consists of m points $\beta^{(1)}, \ldots, \beta^{(m)}$ and B^* consists of n points $\gamma^{(1)}, \ldots, \gamma^{(n)}$. Each deleted neighborhood of B is a deleted neighborhood of B^* and vice versa. However, each neighborhood of $\gamma^{(1)}$ contains a neighborhood of $\beta^{(1)} \cup \beta^{(2)}$ and each neighborhood of $\beta^{(1)}$ or $\beta^{(2)}$ is contained in a neighborhood of $\gamma^{(1)}$. Roughly speaking, the point $\gamma^{(1)}$ is the same as the set $\beta^{(1)} \cup \beta^{(2)}$. Similarly, two points of B may be equivalent to the union of the three points of B^*, etc.

These phenomena lead to many new problems connecting topological and analytical problems and are interesting in connection with boundary conditions; see the end of § 4.3.

4. Background and program

4.1. The problems. Given a topological space D an important problem of probability theory is to find the most general Markov process on D. (Hunt's beautiful results in [14] permit us to reformulate this in terms of *potentials*.) Space does not permit us here to analyze why and how this problem is reduced to that of finding semigroups of operators from **C** to **C**. Anyhow, the following slightly more general problem is of obvious interest in itself.

Find all operators Ω generating [in the sense of (8)] *positive semigroups* $\{T(t)\}$ *from* **C** *to* **C**. We have omitted the restriction $\|T(t)\| \leqslant 1$, which becomes more and more untenable even for probability theory and excludes semigroups of interest in potential theory, in diffusion (with creation of masses), and heat conduction. [We note in passing that our method of isomorphisms cannot be fully exploited as long as one adheres to the conventional Banach norm. It would be interesting and highly desirable to reformulate the whole theory free from restriction to normed spaces utilizing the *Köthe–Mackey* concept of dual spaces.]

Given a generator Ω we face the problem of *finding all positive semigroups generated by it and to discover the analytic and probabilistic relations among them*. The first part leads to a strict formulation of the vague and unsatisfactory notion '*boundary conditions*' for differential equations. The problem is to construct all possible lateral conditions; in this form it is analogous to the construction of self-adjoint contractions in Hilbert space, but it leads to new angles.

Finally, it is important to *find the generator of the adjoint semigroup*. (In special cases this amounts to finding the physicist's Fokker–Planck, or continuity, equation.)

This point of view links together operators which classically seemed worlds apart. For example, when D is the set of integers we are led to

6

infinite systems of ordinary differential equations that can be treated precisely as partial differential equations; in fact, we obtain the boundary conditions with an analogue of '*normal derivatives*' at the boundary in a form which is applicable also, say, to harmonic functions in a domain with non-differentiable boundary. Again, when D is the real line, the semigroup for which $u(t, .) = T(t)f$ is harmonic in the half plane $t > 0$ is generated by a Riesz potential, and Elliott[4] has shown that its restriction to finite intervals is closely related to a second-order differential operator.

It is possible now to see, in rough outlines, the way to a rather general solution of the problems stated. Space does not permit us to go into details, and we proceed instead to indicate how the first problem is connected with an intrinsic theory of differential operators (or their analogue) in arbitrary spaces.

4.2. Operators of local character. The concept of a differential operator is defined only in special spaces and depends on a co-ordinate system. We replace it by the more meaningful and more general notion of an operator of local character. The positive semigroups generated by an operator of local character present an obvious analytic interest. (The corresponding Markov processes are the only ones whose path functions are continuous with probability one; see Ray[18].) They are a natural generalization of the second-order elliptic differential operators in Euclidean spaces: on one hand, such operators share the main properties of the Laplacian and generate positive semigroups. On the other hand, the derivation of the diffusion equation based on the local character condition (which I regret having introduced in [12] under the misleading name of Lindeberg condition) shows that no differential operator of higher order shares this property.

For an operator Ω of local character to generate a positive semigroup such that $\|T(t)\| \leqslant 1$ it is necessary, and very likely also sufficient, that it have the following *positive maximum property*: for each f in the (local) domain of Ω such that f attains a positive local maximum at the point p one has $\Omega f(p) \leqslant 0$. Dropping the norm condition $\|T(t)\| \leqslant 1$ leads to the *weak maximum condition* which requires $\Omega f(p) \leqslant 0$ only at points p such that $f(p) = 0$ and $f \leqslant 0$ in a neighborhood of p.

Operators of local character with the maximum property promise to be a fertile analogue in topological spaces of the Laplacian and general elliptic operators. It is a challenging problem (now within reach) to find a canonical expression for the operators of this class. A complete answer exists only in one dimension[9], but unpublished results of McKean point toward a solution in Euclidean spaces.

The notion of *induced boundaries* plays an important role in this connection. To explain it, consider an operator Ω of local character with the maximum property in a one-dimensional interval I. It is easily seen that Ω can have at most two independent annihilators. We may suppose, if necessary, that the domain of Ω has been extended so that Ω possesses the maximal number of annihilators compatible with its definition. We say then that the point $p \in I$ is regular if in a neighborhood N of p there exist two independent annihilators, that is, if the local topology induced by Ω agrees with the given topology. At a point p near which there exists only one annihilator, Ω induces a topology in which each interval has but one boundary point (is half open). This is reflected both in the analytic and the probabilistic properties of the corresponding processes. For example, all paths starting at a deficient point go in one direction (see Dynkin[3]). Similarly, in an arbitrary topological space D it will be desirable to avoid singularities and pathologies by requiring that in the interior of D the *local topology* induced by Ω agrees with the given topology.

An indication of the general character of our operators is obtained from the *canonical form* of an operator Ω of local character with the positive maximum property in an open interval I without singular points. It is given by the following theorem[9]. The interval I can be parametrized by a 'canonical scale' x in such a way that each f in the domain of Ω has one-sided derivatives with respect to x and they are continuous except for jumps. (We denote them indiscriminately by f'.) Moreover, there exist two Borel measures m and γ in I, the m-measure of each interval being positive, such that for each f in the domain of Ω,

$$\Omega f . dm = df' - f d\gamma \tag{17}$$

in the sense that the integrals are equal. Of course, if m and γ are absolutely continuous this canonical form reduces to $\Omega f = af'' - cf$. However, this traditional form requires artificial restrictions on the coefficients, whereas (17) is of an intrinsic nature and its theory is simpler and more flexible. Thus, in the equation of the vibrating string our m represents the mechanical mass and γ an attractive elastic force[10]. The form (17) now covers cases where either the mass or the force are concentrated at individual points, which are usually treated by artificial passages to the limit, whereas the generalized form of Ω makes available once and for all the basic existence and expansion theorem.

4.3. Lateral conditions. A *conditioning set* Σ for Ω is a set such that for $f \in C$ and each $\lambda > \lambda_0$ there exists exactly one solution $F \in \Sigma$ of

$$\lambda F - \Omega F = f. \tag{18}$$

Using the Hille–Yosida theory it can be shown that each semigroup generated by Ω corresponds to a conditioning set. For example, the minimal semigroups generated by $\Omega = \omega \Delta$ (§ 2.3) correspond to restricting the domain of Ω to functions vanishing at the active boundary. The problem 'to find a semigroup generated by the differential operator Ω restricted by the conditioning set Σ' is the rigorous formulation of the vague and not always soluble problem of 'integrating the differential equation $u_t = \Omega u$ with the boundary condition $u \in \Sigma$'.

For simplicity consider an operator Ω such that for $\lambda > 0$ the equation $\lambda \xi - \Omega \xi = 0$ admits exactly m independent solutions $\xi \in C$. Then the *active boundary* A induced by Ω consists of m points $\beta^{(1)}, \dots, \beta^{(m)}$. For example, if Ω is a second-order differential operator in one dimension, say of the form (17), $m \leqslant 2$. All F in the domain of Ω will be continuous in $D \cup A$. The solution $F^{\min}$ of (18) corresponding to the *minimal* semigroup generated by Ω satisfies the boundary condition $F^{\min}(\beta^{(j)}) = 0$. Every other solution is of the form

$$F = F^{\min} + \sum_{j=1}^{m} \Phi^{(j)} \xi^{(j)}, \tag{19}$$

where $\xi^{(j)}$ is the solution of $\lambda \xi - \Omega \xi = 0$ such that $\xi^{(j)}(\beta^{(k)}) = 0$ or 1 according as $j \neq k$ or $j = k$, and where $\Phi^{(j)}$ is a certain functional of f.

Our problem now consists in determining these arbitrary functionals in their dependence on λ in such a way that *the range Σ of the transformation $f \to F$ is independent of λ*, and to describe all possible such ranges (conditioning sets). In [6] and [8] this problem is solved by a method of wide applicability. It is interesting that even in the case of differential equations the lateral condition $F \in \Sigma$ need not be of local character; hence the classical notion of boundary condition is too restrictive.

The *adjoint* semigroup is generated by an operator Ω^* acting on measures. It is noteworthy that even for a differential operator Ω the adjoint Ω^* need not be of local character. It appears, however, that for the minimal semigroup Ω^* shares the local character of Ω.

It is customary to consider (in Euclidean spaces) only differential operators Ω which are symmetric or so nearly symmetric that Ω and Ω^* induce the same boundary. New and interesting phenomena occur for truly unsymmetric operators. An abstract formulation of many classical problems is to find semigroups generated by Ω whose adjoint is generated by a given formal adjoint Ω^* of Ω.

Suppose that the minimal semigroup generated by Ω has an adjoint generated by Ω^*. Let the *adjoint boundary* consist of n points $\gamma^{(1)}, \ldots, \gamma^{(n)}$ corresponding to n independent solutions $\eta^{(k)}$ of $\lambda\eta - \Omega^*\eta = 0$. For the semigroup corresponding to (19) to be generated by Ω^* it is necessary and sufficient that each functional $\Phi^{(j)}$ is a linear combination of the $\eta^{(k)}$, so that the general solution of our problem involves mn free parameters to be determined in such a way that the range of the transformation (19) be independent of λ.

The solution of this problem given in [6] shows clearly the abstract generalization of the *normal derivatives* appearing in the mixed boundary value problem for harmonic functions. (One advantage of the abstract approach is to derive the boundary conditions in a form that is always meaningful, whereas the classical normal derivatives impose a regularity condition on the boundary.)

The general solution seems also to indicate that the topological question whether $\beta^{(j)}$ and $\gamma^{(k)}$ have disjoint neighborhoods is related to the behavior of the inner product of $\xi^{(j)}$ with $\eta^{(k)}$ as $\lambda \to \infty$ (see § 3.2). However, the precise way in which the relations between the two topologies induced by Ω are reflected in the analytical properties of Ω is an open and promising problem.

REFERENCES

[1] Brelot, M. Le problème de Dirichlet. Axiomatique et frontière de Martin. *J. Math. pures appl.* (9), 35, 297–335 (1956).

[2] Derman, C. A solution to a set of fundamental equations in Markov chains. *Proc. Amer. Math. Soc.* 5, 332–334 (1954).

[3] Dynkin, E. B. The infinitesimal operators of Markov processes. [In Russian, English summary.] *Teoria Verojatnostei i le Primenenija*, 1, 38–60 (1956).

[4] Elliott, J. The boundary value problems and semi-groups associated with certain integro-differential operators. *Trans. Amer. Math. Soc.* 75, 300–331 (1954).

[5] Feller, W. Boundaries induced by non-negative matrices. *Trans. Amer. Math. Soc.* 83, 19–54 (1956).

[6] Feller, W. On boundaries and lateral conditions for the Kolmogorov differential equations. *Ann. Math.* (2), 65, 527–570 (1957).

[7] Feller, W. The general diffusion operator and positivity preserving semigroups in one dimension. *Ann. Math.* (2), 60, 417–436 (1954).

[8] Feller, W. Generalized second order differential operators and their lateral conditions. *Ill. J. Math.* 1, 459–504 (1957).

[9] Feller, W. On the intrinsic form for second order differential operators. *Ill. J. Math.* 2, 1–18 (1958).

[10] Feller, W. On the equation of the vibrating string. *J. Math. Mech.* **8**, 339–348 (1959).

[11] Feller, W. The parabolic differential equation and the associated semi-groups of transformations. *Ann. Math.* (2), **55**, 468–519 (1952).

[12] Feller, W. Zur Theorie der stochastischen Prozesse. *Math. Ann.* **113**, 113–160 (1936).

[13] Hille, E. and Phillips, R. S. *Functional Analysis and Semi-groups.* Revised edition. Amer. Math. Soc. Colloquium Publ. no. 31 (1957).

[14] Hunt, G. A. Markoff processes and potentials. I–III. *Ill. J. Math.* **1**, 44–93 (1957); **2**, 151–213 (1958).

[15] Kadison, R. V. A representation theory for commutative topological algebra. *Mem. Amer. Math. Soc.* no. 7 (1951).

[16] Kolmogoroff, A. Zur Theorie der Markoffschen Ketten. *Math. Ann.* **112**, 155–160 (1935).

[17] Martin, R. S. Minimal positive harmonic functions. *Trans. Amer. Math. Soc.* **49**, 137–172 (1941).

[18] Ray, D. Stationary Markov processes with continuous paths. *Trans. Amer. Math. Soc.* **82**, 452–493 (1956).

[*Added in proof*

[19] Doob, J. L. Discrete potential theory and boundaries. To appear in the *J. Math. Mech.*]

SOME TRENDS AND PROBLEMS IN LINEAR PARTIAL DIFFERENTIAL EQUATIONS

By LARS GÅRDING

The theory of partial differential equations is a very large and very loosely connected part of present-day mathematics. It has, however, a core which has grown out of the boundary value problems of classical physics and now qualifies as a mathematical theory in its own right. The present trend or drift of the subject is very clear. There is an increased interest in general problems connected with this core, and this interest is due to some very successful applications of the theory of linear topological spaces which have been made recently. After the pioneering work by Schauder[1],† the interaction with the theory of linear spaces has been rather slow in coming, but is now producing good results in the hands of a generation that learnt both subjects at the same time. When a problem about partial differential operators has been fitted into the abstract theory, all that remains is usually to prove a suitable inequality and much of our new knowledge is, in fact, essentially contained in such inequalities. But the abstract theory is not only a tool, it is also a guide to general and fruitful problems.

In my exposition I shall often refer to a comprehensive survey[2] of problems in the theory of partial differential equations written by Petrowsky in 1946.

1. Operators with constant coefficients

We shall write linear differential operators in the form

$$A(x, D) = \Sigma A_\alpha(x) D_\alpha \quad (|\alpha| \leqslant m),$$

where x is a point in R^n, $D_k = \partial/\partial x^k$, $D_\alpha = D_{\alpha_1} \dots D_{\alpha_k}$, $A_\alpha(x)$ is permutation symmetric in α and $|\alpha| = k$ is the length of the index α, i.e. the order of D_α. The order of A is supposed to be m and we write

$$PA(x, D) = \Sigma A_\alpha(x) D_\alpha \quad (|\alpha| = m),$$

for the principal part of A. The two corresponding characteristic polynomials are

$$A(x, \zeta) = \Sigma A_\alpha(x) \zeta_\alpha \quad (|\alpha| \leqslant m),$$

and

$$PA(x, \zeta) = \Sigma A_\alpha(x) \zeta_\alpha \quad (|\alpha| = m),$$

† N.B. Superior figures in the text refer to the bibliography. The figure [0] means that the result is not published.

where $\zeta = (\zeta_1, ..., \zeta_n)$ has complex components and $\zeta_\alpha = \zeta_{\alpha_1} ... \zeta_{\alpha_k}$. A hypersurface $s(x) = 0$ is said to be characteristic with respect to A at a point x if $PA(x, s_x) = 0$, $(s_x = \operatorname{grad} s)$. A manifold of arbitrary dimension is said to be free if no characteristic hyperplane is tangent to it.

Let us start with operators $A(D)$ with constant coefficients. It is natural to ask if they have interesting properties in common. We shall see that the answer is yes. The guide to fruitful problems in this connection has been the theory of distributions (Soboleff[3], Schwartz[3]) rather than classical physics, and this shows that a new theory does not have to solve old problems to motivate its existence. It is quite enough to state and solve new ones. Classically, a natural domain for a differential operator A is a set of smooth functions defined in an open set $\mathcal{O}$, e.g. the space $\mathcal{E} = \mathcal{E}(\mathcal{O})$ of infinitely differentiable functions and its subspace $\mathcal{D} = \mathcal{D}(\mathcal{O})$ of functions with compact supports. If these spaces are equipped with the Schwartz topologies, any A will map them continuously into themselves. Via the integral

$$\langle h, g \rangle = \int h(x) g(x) \, dx \quad (g \in \mathcal{D}),$$

any h in $\mathcal{E}$ and, in fact, any locally integrable function gives rise to a continuous linear functional on $\mathcal{D}$, and, if $h \in \mathcal{E}$, we have

$$\langle Ah, g \rangle = \langle h, \check{A}g \rangle, \quad \check{A}(D) = A(-D).$$

Let $\mathcal{D}'$ be the space of distributions, i.e. the dual of the space $\mathcal{D}$. The same equality serves to define $Ah \in \mathcal{D}'$ when $h \in \mathcal{D}'$, provided we let $\langle f, g \rangle$ denote the value of $f \in \mathcal{D}'$ at $g \in \mathcal{D}$. In this way we have imbedded $\mathcal{E}$ into $\mathcal{D}'$, we have extended the notion of the integral and we have extended A from $\mathcal{E}$ to $\mathcal{D}'$. In many ways, the elements f of $\mathcal{D}'$ behave like functions. They can be studied locally by restriction to $\mathcal{D}(V)$, where V is a small open set contained in $\mathcal{O}$, and they are limits of functions in $\mathcal{D}$. Taking $\mathcal{D}'$ as a domain of definition gives new perspectives on classical results and leads to new interesting problems.

Let us consider A on $\mathcal{D}$. Taking the Fourier–Laplace transform

$$(\mathcal{F}f)(\zeta) = \int e^{-x\zeta} f(x) \, dx,$$

we see that $\mathcal{F}Af(\zeta) = A(\zeta)\mathcal{F}f(\zeta)$, i.e. $\mathcal{F}$ turns A into multiplication by its characteristic polynomial. An immediate consequence is that f vanishes if Af vanishes provided $A \neq 0$, which we now assume. A more precise result is the inequality

$$\int |f(x)|^2 \, dx \leqslant c \int |Af(x)|^2 \, dx \quad (f \in \mathcal{D}), \tag{1}$$

where c depends only on the support of f (Malgrange[4], Hörmander[5]). Applied to $\check{A}$ it shows among other things that the mapping $\check{A}\colon \mathscr{D} \to \mathscr{E}$ has a continuous inverse. By the abstract theory, in this case the Hahn–Banach theorem, this implies that

$$A\mathscr{D}' \supset \mathscr{E}'. \tag{2}$$

The space on the right consists of all distributions with compact supports in $\mathcal{O}$. In particular, the equation $AF = \delta$ has solutions $F \in \mathscr{D}'(R^n)$. These are elementary solutions of A and were obtained long ago by summations of the integral

$$(2\pi i)^{-n}\int e^{\zeta x}A(\zeta)^{-1}d\zeta \quad (\mathrm{Re}\,\zeta = 0),$$

in various special cases by, for example, Fredholm[6], Zeilon[7] and Herglotz[8]. There is also a method which applies to the general case (Hörmander[0]). Deeper results of the same nature as (2) are $A\mathscr{E} = \mathscr{E}$ and $A\mathscr{D}_b' = \mathscr{D}_b'$ (Malgrange[4]), which hold when $\mathcal{O}$ is convex; $\mathscr{D}_b'$ is the space of distributions of finite order. Ehrenpreis[9] has shown that $A\mathscr{D}' = \mathscr{D}'$ when $\mathcal{O} = R^n$, and Łojasiewicz[10] and Hörmander[11] that $A\mathscr{S}' = \mathscr{S}'$ where $\mathscr{S}'$ is the space of tempered distributions. It would be interesting to know, for example, if $A\mathscr{D}' = \mathscr{D}'$ for an arbitrary convex $\mathcal{O}$ and to clarify the role of the convexity. This leads to interesting uniqueness properties for solutions of $Au = 0$ (Malgrange[4]).

Inequalities of the type (1) can serve as a basis for comparison between operators. An operator A is said to majorize another, B, if (1) holds in a bounded region with Bf on the left side. This concept is due to Hörmander[5] who proved that A majorizes B if and only if $\Sigma\,|B^{(\alpha)}(\zeta)|^2$ is majorized by a constant times $\Sigma\,|A^{(\alpha)}(\zeta)|^2$ for all ζ with $\mathrm{Re}\,\zeta = 0$, $(A^{(\alpha)}(\zeta) = \partial^\alpha A(\zeta), \partial^k = \partial/\partial\zeta_k)$. If majorization is taken with respect to functions which do not vanish at the boundary (in which case $c\int |f(x)|^2\,dx$ should be added to the right side of (1)), it turns out that A majorizes only operators of the form $c_1 A + c_2$ where c_1 and c_2 are constants, provided A is not a polynomial in one variable.

Let us now consider solutions u of $Au = 0$. A solution of the form polynomial times $e^{\zeta x}$ where $A(\zeta) = 0$ will be called exponential. Malgrange[4] has shown that if $\mathcal{O}$ is convex, the exponential solutions span all solutions in $\mathscr{E}$ and in $\mathscr{D}'$. Still more precise results have been obtained by Ehrenpreis[0] for entire functions. Simple examples show that, in general, $Au = 0$ has more solutions in $\mathscr{D}'$ than in $\mathscr{E}$. On the other hand, Weyl[12] (see also Caccioppoli[13]) has shown that any locally integrable solution of $Au = 0$ is harmonic (Weyl's lemma). This led Schwartz to the following question. When is it true that $Au = 0$ has the same solutions in $\mathscr{D}'$ and $\mathscr{E}$, i.e. when is every solution u in $\mathscr{D}'$ an infinitely differentiable function? An operator with this property is called hypoelliptic. Hörmander[5] proved that A is hypoelliptic if and only if $A(\zeta) \to \infty$ with $\operatorname{Im} \zeta$, uniformly when $\operatorname{Re} \zeta$ stays bounded. This class includes the parabolic operators and also the elliptic ones, which are characterized by the stronger property that the solutions of $Au = 0$ are analytic functions of the real variables x. That this is equivalent to the classical algebraic condition that $PA(\zeta) \neq 0$ when $\operatorname{Re} \zeta = 0$ and $\zeta \neq 0$ had been proved by Petrowsky[14]. Let $x = (x', x'')$ be a partition of the co-ordinates x into sets x' and x'' with n' and $n'' = n - n'$ elements. Generalizing our present problem we may ask when it is true that every solution $u \in \mathscr{D}'$ of $Au = 0$ is hypoelliptic in x' in the sense that $\int u(x', x'') \phi(x'') \, dx''$ is infinitely differentiable when $\phi \in \mathscr{D}$. A necessary and sufficient condition is that $A = \Sigma A_\alpha(D') D_\alpha''$, where A_0 is hypoelliptic and dominates the other A_α in the strict sense so that $A_\alpha(\zeta')/A_0(\zeta') \to 0$ as $\operatorname{Im} \zeta' \to \infty$, uniformly when $\operatorname{Re} \zeta'$ is bounded (Gårding and Malgrange[39]). This shows, for example, that when A is the wave operator $\square = D_1^2 - D_2^2 - \ldots - D_n^2$, any solution u is infinitely differentiable in $x^2, \ldots, x^n$. Another consequence is that any solution u of $Au = 0$ gives rise to distributions on noncharacteristic hyperplanes which are restrictions of u. In particular, it makes sense to speak of the Cauchy data of u on such a hyperplane. An important, but somewhat neglected, problem is to characterize these data. It includes as a special case the characterization of the boundary values of an analytic function in a strip, where, of course, many results are known, e.g. the Paley Wiener theorem[15, 3]. For a general A, some results have been obtained by Ehrenpreis[0]. The subject of partial ellipticity should also offer things of interest.

The hyperbolic operators can be characterized as those for which $AF = \delta$ has a solution F whose support can be contained in some proper half-cone with its vertex at the origin. Let $\sigma x \leqslant 0$ be a half-space which has at most the origin in common with the support of F. We say that A is hyperbolic σ. An equivalent condition is that $PA(\sigma) \neq 0$ and that $A(t\sigma + \zeta) \neq 0$ when ζ is purely imaginary and t is large enough (Gårding[16]). It follows that A is hyperbolic $-\sigma$, that PA is hyperbolic σ (but not conversely!) and that the real algebraic surface $PA(\zeta) = 0$ has m sheets. If these do not intersect, except at the origin, A is said to be strictly hyperbolic. In any case, the elementary solution F is uniquely determined. Cauchy's problem for $Au = 0$ is correctly set with Cauchy data in $\mathscr{E}$ or $\mathscr{D}'$ on the plane $\sigma x = 0$ and the solution will be in $\mathscr{E}$ and $\mathscr{D}'$

respectively in the whole space. The value of u at a point y with $\sigma y > 0$ depends only on the Cauchy data in the intersection of $\sigma x = 0$ with $y - \operatorname{supp} F$. To take a classical example, the wave operator $\square$ is hyperbolic $\sigma = (1, 0, \ldots, 0)$. The support of F is in this case the cone $x^1 \geqslant 0$, $xx = x^1x^1 - x^2x^2 - \ldots - x^nx^n \geqslant 0$ if n is odd and its boundary $x^1 \geqslant 0$, $xx = 0$ if n is even. This is the precise formulation of Huygens's principle. The support of F has been determined by Petrowsky[17] for the case when $A = PA$ is homogeneous and strictly hyperbolic. He found other cases of Huygens's principle with lacunas or 'holes' in the support, and suggested a similar investigation when $A = PA$ is not necessarily strictly hyperbolic. The discovery that in this case the dimension of the support may be much less than the dimension of its convex hull (Gårding[18]) has not diminished the interest of this problem. A successful treatment will probably not be easy and cannot avoid the topology of the complex hypersurface $A(\zeta) = 0$. Clearly, lacunas are an exceptional phenomenon, but it is interesting to note that they occur also for wave operators with variable lower terms. This has been proved by Stellmacher[19], who at the same time disproved Hadamard's conjecture[20] that the only second-order hyperbolic operators with lacunas are the wave operators in even dimensions.

2. Analytic coefficients

By Cauchy–Kovalevskaja's theorem we know that Cauchy's problem can be solved locally for analytic data on an analytic non-characteristic hypersurface S provided the coefficients of the operator $A = A(x, D)$ are analytic. The local character of this theorem leads to questions of analytic continuation. These have been attacked recently by Leray[23], who proposes to show that all the singularities of the solution lie on (complex) characteristics issuing from the singularities of the data or on characteristics which are tangent to S. His generalization[23] of the Laplace transform, which replaces the cycles $\operatorname{Re} \zeta = $ constant by compact cycles, shows promise of being an instrument which can single out natural classes of elementary solutions for operators with constant and polynomial coefficients and give information about them which is detailed enough to solve, for example, the problem of the lacunas.

Elliptic operators A with analytic coefficients are of special interest. John[24] has shown that they have analytic local elementary solutions, and from this it follows easily that if $u \in \mathscr{D}', f$ is analytic, and $Au = f$, then u is analytic. This completes classical results to the effect that all sufficiently differentiable solutions of analytic elliptic equations, even non-linear, are analytic (Petrowsky[14], see also Friedman[25]). Recently, Malgrange[4] has extended Runge's theorem to analytic elliptic operators. He shows that every solution u in $\mathscr{E}(\mathcal{O})$ of $Au = 0$ is a limit of solutions in $\mathscr{E}(\mathcal{O}')$ where $\mathcal{O}'$ is any open set which contains $\mathcal{O}$, provided $\mathcal{O}$ has no holes, i.e. $\mathcal{O}' - \mathcal{O}$ has no components which are compact relative to $\mathcal{O}$. Naturally, A has to be elliptic in $\mathcal{O}'$. The theorem is also true for analytic manifolds. Holmgren[21] proved long ago, using Cauchy-Kovalevskaja's theorem, that a smooth solution of Cauchy's problem for a free hyper-surface is unique. This type of uniqueness problem can be stated more generally as follows. Let $Au = 0$ and let u vanish of infinite order on a k-dimensional analytic manifold M. When is it true that u vanishes in a neighbourhood of M? The cases $k = n - 1$ and $k = 0$ correspond to Cauchy's problem with zero data and the case when u has an infinite zero at a point respectively. Perfecting the technique of Holmgren, John[22] has shown that it is sufficient that M be free and that domains of uniqueness can be obtained by deforming M through free manifolds keeping its boundary fixed. If M is linear and A has constant coefficients, this result is best possible. In fact, then $Au = 0$ has non-trivial solutions vanishing of infinite order on any given characteristic hyperplane (Hörmander[5]).

3. Non-analytic coefficients

The extension of Holmgren's uniqueness theorem to operators with non-analytic coefficients presents considerable difficulties for operators which are not hyperbolic or parabolic. If this could be achieved for elliptic operators, Malgrange's generalization of Runge's theorem would hold in the non-analytic case. At the time of Petrowsky's report, the only significant result was due to Carleman[26, 27] who had shown unique-ness for a first-order real system in two variables with non-multiple characteristics. He had also proved that if the system is elliptic, a solu-tion with a zero of infinite order must vanish identically. The first result has now been extended by Calderón[28] to, for example, Cauchy's pro-blem with data on a plane $\sigma x = $ constant for operators A with real principal parts and no multiple characteristics in the sense that $PA(x, t\sigma + \zeta)$ has no multiple zeros in t when ζ is real. The requirements on the coefficients are very mild. The proof uses the Zygmund–Calderón singular operators[29] which are an efficient substitute for the Fourier transform. Counter-examples to uniqueness have been found by Landis[30], de Giorgi[31] and Pliś[32]. Carleman's second result has been extended to elliptic operators A of the second order with real principal part by Aronszajn[33] (see also Cordes[34]) who extended an inequality by Heinz[35] and Müller[36] modelled on an inequality by Carleman[27]. Aronszajn's result implies uniqueness of Cauchy's problem, which has been proved in another way by Landis[37]. Hörmander[38] has proved that inequalities of the Carleman model do not work when the multiplicity of the characteristics exceeds two. The whole situation is, however, far from clear and we have no real idea of the mechanism of uniqueness since necessary conditions are not explored. The general uniqueness problem stated above is unexplored in the non-analytic case.

For operators whose coefficients are not infinitely differentiable, $\mathcal{D}'$ is no longer a natural domain of definition since $A = \Sigma(-D_\alpha) A_\alpha$ does not map $\mathcal{D}$ into itself, and, if the coefficients are only continuous, $\check{A}$ need not even have a sense. One way out of this difficulty is to consider bilinear forms

$$\int \Sigma A_{\alpha\beta}(x) D_\alpha f(x) D_\beta g(x) \, dx, \tag{3}$$

rather than differential operators. When the coefficients $A_{\alpha\beta}$ are sufficiently smooth and f or g belongs to $\mathcal{D}$ we can write such a form as

$$\int (\Sigma(-D)_\beta A_{\alpha\beta}(x) D_\alpha f(x)) g(x) \, dx,$$

or
$$\int f(x) \Sigma(-D)_\alpha A_{\alpha\beta}(x) D_\beta g(x) \, dx,$$

but (3) has a sense also when the coefficients are only locally integrable. In classical physics, forms of the type (3) with $f = g$ appear as expressions for various kinds of energy and are prior to differential operators. We shall find it convenient to associate with (3) double differential operators

$$A = A(x, \bar{x}, D, \bar{D}) = \Sigma A_{\alpha\beta}(x, \bar{x}) D_\alpha \bar{D}_\beta, \tag{4}$$

where $\bar{x}$ is a new set of variables and $A_{\alpha\beta}(x, x) = A_{\alpha\beta}(x)$. Then the integrand equals
$$[Af(x) g(\bar{x})]_{x=\bar{x}}.$$

Following Hörmander[5] we abuse this notation by shortening it to $Af(x) g(x)$. Accordingly, the integral (3) shall be written as $\langle Af, g \rangle$. For instance, taking $A = \Sigma D_k \bar{D_k}$, we obtain the Dirichlet integral. We assume that the total order n of A is independent of x and $\bar{x}$ and define PA in the obvious way. The characteristic polynomial of A is defined to be

$$A(x, \bar{x}, \zeta, \bar{\zeta}) = \Sigma A_{\alpha\beta}(x, \bar{x}) \zeta_\alpha \bar{\zeta_\beta},$$

where $\bar{\zeta}$ is the conjugate of ζ. We say that A is elliptic or hyperbolic if $A(x, x, D, -D)$ is elliptic or strongly hyperbolic respectively. Hyperbolic double differential operators have been introduced by Gårding[40]. Here I only want to point out that some of the work that has been done on the regularity of solutions of elliptic equations (see Nirenberg[41]) can be stated very conveniently in terms of double differential operators. Put

$$|D^k f(x)|^2 = \Sigma \, |D_\alpha f(x)|^2 \quad (|\alpha| \leqslant k)$$

and

$$|D^k f, \mathcal{O}| = \left[\int_{\mathcal{O}} |D^k f(x)|^2 \, dx \right]^{\frac{1}{2}}, \tag{5}$$

where $\mathcal{O}$ is open, and put with Laurent Schwartz

$$|D^{-k} f, \mathcal{O}| = \sup |\langle f, g \rangle| \, |D^k g, \mathcal{O}|^{-1} \quad (g \in D(\mathcal{O})). \tag{6}$$

If the value $+\infty$ is permitted, all these norms have a sense for any $f \in \mathscr{D}'(\mathcal{O})$. Let $0 \leqslant k \leqslant m$ and suppose that A has the (double) order k; $m - k$, i.e. that the sum in (4) runs over $|\alpha| \leqslant k$ and $|\beta| \leqslant m - k$. Then if $|D^k f, \mathcal{O}| < \infty$ we can compute

$$|D^{k-m} A f, \mathcal{O}| = \sup |\langle A f, g \rangle| \, |D^{m-k} g, \mathcal{O}|^{-1} \quad (g \in \mathscr{D}(\mathcal{O})).$$

Suppose now that A is uniformly elliptic, that $k < m$ and that the double order of $A - PA$ does not exceed k; $m - k - 1$. Suppose also that $\mathcal{O}' \subset \mathcal{O}$ has compact closure in $\mathcal{O}$. Then one can show that

$$|D^{k+1} f, \mathcal{O}'| \leqslant c |D^{k+1-m} A f, \mathcal{O}| + c \, |D^k f, \mathcal{O}| \tag{7}$$

with some finite constant c, provided the coefficients of A are bounded and those of PA are Lipschitz continuous or, more generally, Lipschitz continuous in the norm $(\int |a|^n dx)^{1/n}$ (Gårding[0]). In particular, if $Af = 0$ and $|D^k f, \mathcal{O}| < \infty$, then $|D^{k+1} f, \mathcal{O}'| < \infty$. Iterations of this inequality lead to various generalizations of Weyl's lemma. It has numerous variants involving norms

$$\left[\int |D^k f(x)|^p \, dx \right]^{1/p} \quad (1 \leqslant p \leqslant \infty),$$

and Hölder continuity. Very sharp results in this direction, which have important applications to non-linear equations, are known for $n = 2$ (see Nirenberg[41]). They have recently been extended to general n by Nash[42] and de Giorgi[43]. The latter proves that if $A = \Sigma A_{jk} D_j \overline{D}_k$ $(A_{jk} = A_{kj})$ is real and uniformly elliptic, any solution f of $Af = 0$ with $|D^1 f, \mathcal{O}| < \infty$ is Hölder continuous. The proof uses a contraction property of Dirichlet's integral which has been explored systematically by

Beurling and Deny[44]. Nash has got corresponding results also for parabolic operators. It would be interesting to have some idea of continuity requirements which are necessary for inequalities of the type (7).

Weyl's lemma has been extended to hypoelliptic operators with variable coefficients by Hörmander[45] and Malgrange[46].

4. Boundary problems

Let $\mathcal{O} \subset R^n$ be open and bounded and let $H = H(\mathcal{O})$ be the Hilbert space of all square integrable functions in $\mathcal{O}$. Let $A = \Sigma A_\alpha D_\alpha$ be a differential operator with smooth coefficients and let $\bar{A} = \Sigma(-D)_\alpha \overline{A_\alpha}$ be its formal adjoint. With the domain $\mathscr{D}$ they become densely defined operators from H to H. One of the problems posed by Petrowsky was to find correctly set boundary problems for arbitrary differential operators. A somewhat abstract answer to this problem has been found by Vishik[47]. He interprets a closed operator B between A and $\bar{A}^*$ as giving a boundary problem if the range of B is H and B^{-1} is bounded. It is easy to see that such a B exists if and only if A and $\bar{A}$ have bounded inverses. Hence the inequality (1) shows that Petrowsky's question can be given a positive answer for operators with constant coefficients. For sufficiently small regions Hörmander[5] has extended the inequality (1) to a general class of operators with variable coefficients, those of principal type and with real principal parts. Still, Petrowsky's question remains open if one adds that the boundary problem should be specified in a more classical fashion, e.g. so that the domain of B is given by equations $Qf = 0$ on the boundary of $\mathcal{O}$, when Q runs through a set of differential operators.

One of the achievements of the last decade is the extension of Dirichlet's problem to general elliptic equations and systems which are positive definite in a certain sense, the so-called strongly elliptic operators (Vishik[48]). I will try to give a brief sketch of the theory. The method is closely connected with Dirichlet's principle and can be considered as a variant of the method of orthogonal projection by H. Weyl[12]. Let $\mathcal{O}$ be bounded and define $|D^k f| = |D^k f, \mathcal{O}|$ as before. Let H^k be all functions for which this norm is finite and let H_0^k be the closure of $\mathscr{D}(\mathcal{O})$ in H^k. It is easy to see that $|D^k f|^2$ and $\int \Sigma |D_\alpha f(x)|^2 dx$, $(|\alpha| = k)$, are equivalent square norms in H_0^k. Let H_0^{-k} be the dual of H_0^k with respect to the duality

$$\langle f, g \rangle = \int f(x) g(x) \, dx.$$

It consists of all $f \in \mathscr{D}'(\mathcal{O})$ for which the norm

$$|D^{-k} f| = \sup |\langle f, g \rangle| \, |D^k g|^{-1} \quad (g \in \mathscr{D}),$$

is finite.

Let $\qquad A = \Sigma A_{\alpha\beta} D_\alpha \overline{D}_\beta \quad (|\alpha| = |\beta| = m)$ \hfill (8)

be a double differential operator of (double) order $m; m$ with constant coefficients whose real part

$$\mathrm{Re}\, A = \Sigma \tfrac{1}{2}(A_{\alpha\beta} + \bar{A}_{\beta\alpha})\, D_\alpha \overline{D}_\beta$$

is elliptic. Changing the sign of A we may assume that $\mathrm{Re}\, A \geqslant 0$, i.e. $\mathrm{Re}\, A(\zeta, \bar\zeta) \geqslant 0$ when $\mathrm{Re}\, \zeta = 0$. Put $(Af, g) = \langle Af, \bar{g}\rangle$. A Fourier transformation shows that $(\mathrm{Re}\, Af, f)$ and $|D^m f|^2$ are equivalent square norms on $\mathscr{D}$. As remarked by Laurent Schwartz[0], this means that A is a linear homeomorphism

$$H_0^m \to H_0^{-m}. \tag{9}$$

In particular, the equation $Au = f \in H = H^0$ has a unique solution u in H_0^m. Weyl's lemma shows that when f is smooth, then u is smooth and satisfies

$$\Sigma A_{\alpha\beta} D_\alpha(-D)_\beta u = f.$$

Hence we have solved an inhomogeneous variant of Dirichlet's problem in an almost trivial way. The non-trivial part is the verification that when f is smooth and $\mathcal{O}$ is smoothly bounded, then the solution is smooth in the closure of $\mathcal{O}$. This result can be stated as follows. If $\mathcal{O}$ is bounded by a smooth hypersurface of dimension $n-1$, the mapping (9) induces linear homeomorphisms

$$H^{m+k} \cap H_0^m \to H_0^{k-m} \quad (0 \leqslant k \leqslant m), \tag{10}$$

$$H^{m+k} \cap H_0^m \to H^{k-m} \quad (k \geqslant m). \tag{11}$$

Observe that both sides decrease as k increases. The last result is due to Ladyženskaja[49] and Guseva[50], the first one to Lions[51]. For $m = 2$, mixed boundary problems have been treated similarly by Soboleff and Vishik[71]. The smoothness properties of (9) expressed by (10) and (11) have a local character. In fact, both sides of these formulas could be taken relative to some smoothly bounded subset $\mathcal{O}'$ of $\mathcal{O}$ which has a part of its boundary in common with the boundary of $\mathcal{O}$ and then A and $\mathcal{O}$ do not have to be smooth outside $\mathcal{O}'$.

Some reflection shows that if B has order $m; m$ and sufficiently small coefficients, then $A + B$ still gives a homeomorphism (9). More generally, it has been shown (see, for example, Gårding[52]) that a uniformly elliptic operator A of order $m; m$ with uniformly continuous principal part and bounded lower terms gives rise to a Fredholm mapping (9). By definition, such a mapping has a closed range whose codimension is finite and equal to the dimension of the nullspace. If the coefficients of A and the boundary of $\mathcal{O}$ are sufficiently smooth, A also induces Fredholm mappings (10) and (11). It is probable that a uniformly elliptic operator A of order $m + k$; $m - k$ ($0 \leqslant k \leqslant m$), with a uniformly continuous principal part and bounded lower terms gives a Fredholm mapping (10) when $\mathcal{O}$ is smoothly bounded. It would be interesting to investigate the smoothness properties of (9) near lower-dimensional parts of the boundary of $\mathcal{O}$. The behaviour of the elements of H_0^k near such parts has been studied by Soboleff[53] in his classical paper on the polyharmonic equation.

Consider a strongly elliptic operator A of the form (8) and suppose that it is real, i.e. that $A = \operatorname{Re} A$. Changing the sign of A, we may also suppose that $A \geqslant 0$. When $m = 1$, $|D^m f|^2$ and $(Af, f) + (f, f)$ are trivially equivalent square norms of H^m, but when $m > 1$, this is no longer true. This concept is due to Aronszajn[54] who obtained conditions for coerciveness when A is locally positive in the sense that $Af\bar{f} \geqslant 0$ for all f. His results were generalized by Schechter[55] and Agmon[56]. They are part of a general and as yet incomplete theory of majorization of real double differential operators on various classes $\mathscr{C}$ of functions. When the operators are squares and $\mathscr{C} = \mathscr{D}$, it reduces to Hörmander's theory[5].

A variety of boundary problems for strongly elliptic operators have been explored in a general way by, for example, Vishik[47], Aronszajn[57], Browder[58] and Lions[59]. Much less has been done for general elliptic operators and systems. Important results on the regularity properties of boundary problems for elliptic and hypoelliptic operators have been obtained by Hörmander[60], but so far we have insufficient knowledge of what the natural boundary problems are except, of course, in special cases like the Cauchy–Riemann equations. One cannot expect to find very close analogues of the Riemann mapping theorem, but it should be possible to find interesting classes of non-linear boundary problems which include it (cf. Beurling[61]).

The theory of Cauchy's problem for hyperbolic operators and systems with variable coefficients is at present in good shape. Petrowsky's fundamental work[62] has been simplified by Leray[63] and extended to the case when the data are distributions. Both these authors use the Cauchy–Kovalevskaja theorem and an approximation via operators with analytic coefficients, but the existence theorems can also be made to follow directly from certain inequalities of the Friedrichs–Lewy type (Gårding[40]). It remains to study the properties of the solutions, in particular the propagation of discontinuities. A start has been made by Leray[23], Courant and Lax[64] and Lax[65].

The work that has been done on, for example, Dirichlet's problem for elliptic operators, and Cauchy's problem for hyperbolic operators has

led to new problems and has given us new techniques, but on the whole
the results are straightforward extensions from the classical case $m = 2$
and they are of limited physical interest. The boundary problems that
occur in practice in connection with, for example, the wave equation
are mostly of the mixed type. In these, the boundary has a space-like
part C_2 carrying two data and a characteristic or time-like part C_1
carrying one datum. A large number of papers have been written about
this problem (see, for example, recent contributions by Duff[66]) but
a systematic theory is still missing. It would be interesting to have it
cast in the form of appropriate inequalities. Mixed problems for general
parabolic type equations in cylindrical domains have attracted much
interest. In particular, Ladyzenskaja[67] and Lions[68] have developed
a powerful variant of the method of orthogonal projection adapted to
this case. Still, I feel that these problems are less significant than the
true mixed problems where the boundary is more arbitrary. Such problems for hyperbolic operators of higher order in two variables occur in
gas dynamics and linear cases have recently been successfully treated by
Campbell and Robinson[69] (see also Thomée[70]). It would be interesting
to have an extension to several variables (cf. Friedrichs[77]).

5. Spectral theory

Let H be the Hilbert space of all square integrable functions in an open
set $\mathcal{O}$ with the scalar product $(f, g) = \int f(x)\overline{g(x)}\,dx$ and let B be a self-
adjoint extension of a differential operator A with domain of definition
$\mathcal{D}$ and smooth coefficients. It follows from a theorem of Gelfand and
Kostiutjenko[72] (see also Berezanskij[73] and Maurin[74]) that B has a
complete set of eigendistributions of A. When A is elliptic these are
smooth functions. The summability properties of the corresponding
eigenfunction expansion are determined by the kernels $e(\lambda, x, y)$ of the
projections E_λ associated with B,

$$E_\lambda f(x) = \int e(\lambda, x, y) f(y)\,dy.$$

When A is elliptic and B is bounded from below, the spectral function
e is a smooth Carleman kernel. It turns out that A alone determines the
properties of e for large λ. In fact, if e' is a spectral function belonging
to another self-adjoint extension B' of A in another region $\mathcal{O}'$, then the
difference $e' - e$ is so stongly oscillating in $\mathcal{O} \cap \mathcal{O}'$ as $\lambda \to \infty$ that a sufficiently high Riesz mean of it tends to zero like any negative power of λ
(Bergendal[75]). For special operators A and B, Lewitan[76] has obtained
many sharp estimates of e and of the convergence or summability $E_\lambda f \to f$

($\lambda \to \infty$). It would be interesting to extend them to the general case. In particular, it should be possible to prove convergence when f is locally smooth and satisfies appropriate boundary conditions. It would also be interesting to remove the restriction that B is bounded from below, as has been done by Lewitan[76] in special cases, and to investigate the role of the ellipticity for the local character of the convergence.

6. Conclusion

I cannot go here into many important parts of the subject like difference equations, the theory of systems and applications to quantum mechanics and differential geometry. My subject has been the general theory of partial differential operators. It has grown out of classical physics, but has no really important applications to it. Still, physics remains its main source of interesting problems. I have the feeling that general talks of the kind I have given are perhaps less useful than periodical surveys of unsolved problems in physics which seem to require new mathematical techniques. Such surveys will of course not be news to specialists, but they will provide many mathematicians with worthwhile problems. Planned efforts to further the interaction between physics and mathematics have been rare. They ought to be one of the major concerns of the international mathematical congresses.

REFERENCES

[1] See, for example: Schauder, J., Équations du type elliptique, problèmes linéaires. *Enseign. math.* 35, 126–139 (1936). Das Anfangswertproblem einer quasilinearen hyperbolischen Differentialgleichung zweiter Ordnung in beliebiger Anzahl von unabhängigen Veränderlichen. *Fundam. Math.* 24, 213–246 (1935). And Leray, J .and Schauder, J., Topologie et équations fonctionnelles. *Ann. sci. Éc. norm. sup.*, Paris, (3), 51, 45–78 (1934).

[2] Petrowsky, I. G. On some problems in the theory of partial differential equations. *Usp. Mat. Nauk,* 1, nos. 3–4 (13–14), 44–70 (AMS translations no. 12) (1946).

[3] Soboleff, S. L. Méthode nouvelle à résoudre le problème de Cauchy pour les équations linéaires hyperboliques. *Mat. Sbornik,* 1 (43), 39–71 (1936). Schwartz, L. *Théorie des distributions,* I, II. Paris, 1951.

[4] Malgrange, B. Existence et approximation des solutions des équations aux dérivées partielles et des équations de convolution. *Ann. Inst. Fourier,* 6, 271–354 (1955–6).

[5] Hörmander, L. On the theory of general partial differential operators. *Acta Math.* 94, 160–248 (1955).

[6] Fredholm, I. Sur les équations de l'équilibre d'un corps solide élastique. *Acta Math.* 23, 1–42 (1900).

[7] Zeilon, N. Sur les intégrales fondamentales des équations à caractéristiques réelles de la physique mathématique. *Ark. Mat. Astr. Fys.* 9, 1–70 (1913).

[8] Herglotz, G. Über die Integration linearer, partieller Differentialgleichungen
     mit konstanten Koeffizienten. I–III. *Ber. sächs. Ges. (Akad.) Wiss.*
     78, 93–126, 297–318 (1926); 80, 69–114 (1928).

[9] Ehrenpreis, L. Solutions of some problems of division. III. *Amer. J. Math.*
     78, 685–715 (1956).

[10] Łojasiewicz, S. Division d'une distribution par une fonction analytique
     de variables réelles. *C.R. Acad. Sci., Paris*, 246, 683–686 (1958).

[11] Hörmander, L. On the division of distributions by polynomials. To be
     published in *Ark. Mat.*

[12] Weyl, H. The method of orthogonal projection in potential theory. *Duke
     Math. J.* 7, 411–444 (1940).

[13] Caccioppoli, R. Sui teoremi di existenza di Riemann. *Ann. Scuola Norm.
     Sup. Pisa*, 6, 177–187 (1937).

[14] Petrowsky, I. G. Sur l'analyticité des solutions des systèmes d'équations
     différentielles. *Mat. Sbornik*, 5 (47), 3–68 (1939).

[15] Paley, R. E. A. C. and Wiener, N. *Fourier transforms in the complex
     domain. Amer. Math. Soc. Coll. Publ.* 19, 1–14 (1934).

[16] Gårding, L. Linear hyperbolic partial differential equations with constant
     coefficients. *Acta Math.* 35, 1–62 (1950).

[17] Petrowsky, I. G. On the diffusion of waves and the lacunas for hyperbolic
     equations. *Mat. Sbornik*, 17 (59), 289–370 (1945).

[18] Gårding, L. The solution of Cauchy's problem for two totally hyperbolic
     linear differential equations by means of Riesz integrals. *Ann. Math.*
     (2), 48, 785–826 (1947).

[19] Stellmacher, K. Eine Klasse huyghenscher Differentialgleichungen und
     ihre Integration. *Math. Ann.* 130, 219–233 (1955).

[20] Hadamard, J. *Le problème de Cauchy.* Paris, 1932.

[21] Holmgren, E. Über Systeme von linearen partiellen Differentialgleichungen.
     Öfv. K. Vetensk-Akad. Förh. 58, 91–103 (1901).

[22] John, F. On linear partial differential equations with analytic coefficients.
     Unique continuation of data. *Commun. Pure Appl. Math.* 2, 209–253 (1949).

[23] Leray, J. Uniformisation de la solution du problème linéaire analytique
     de Cauchy près de la variété qui porte les données de Cauchy. *Bull. Soc.
     Math. Fr.* (1957). La solution unitaire d'un opérateur différentiel linéaire.
     C.R. Acad. Sci., Paris, 245, 2146–2152 (1957).

[24] John, F. The fundamental solution of linear elliptic differential equations
     with analytic coefficients. *Commun. Pure Appl. Math.* 3, 273–304 (1950).

[25] Friedman, A. On the regularity of the solutions of non-linear elliptic and
     parabolic systems of partial differential equations. *J. Math. Mech.* 7,
     43–60 (1958).

[26] Carleman, T. Sur un problème d'unicité pour les systèmes d'équations aux
     dérivées partielles à deux variables indépendantes. *Ark. Mat. Astr. Fys.*
     26 B, no. 17, 1–9 (1939).

[27] Carleman, T. Sur les systèmes linéaires aux dérivées partielles du premier
     ordre à deux variables. *C.R. Acad. Sci., Paris*, 197, 471–474 (1933).

[28] Calderón, A. P. Uniqueness in the Cauchy problem for partial differential
     equations. *Amer. J. Math.* 80, 16–36 (1958).

[29] Calderón, A. P. and Zygmund, A. Singular integral operators and differ-
     ential equations. *Amer. J. Math.* 79, 901–921 (1957).

[30] Landis, E. M. An example of non-uniqueness of the solution of Cauchy's
     problem. *Mat. Sbornik*, 27 (69), 319–323 (1950).

[31] de Giorgi, E. Un esempio di non-unicità della soluzione del problema di Cauchy, relativo ad una equazione differenziale lineare a derivate parziali di tipo parabolico. *R. Mat. Appl.*, ser. 5, 14, 382–387 (1955).

[32] Pliś, A. The problem of uniqueness for the solution of a system of partial differential equations. *Bull. Acad. Pol. Sci.* 2, 55–57 (1954).

[33] Aronszajn, N. Sur l'unicité du prolongement des solutions des équations aux dérivées partielles elliptiques du second ordre. *C.R. Acad. Sci., Paris*, 242, 723–725 (1956).

[34] Cordes, H. O. Über die Bestimmtheit der Lösungen elliptischer Differentialgleichungen durch Anfangsvorgaben. *Nachr. Akad. Wiss. Göttingen (Math.-Phys. Kl.)*, IIa, no. 11, 239–253 (1956).

[35] Heinz, E. Über die Eindeutigkeit beim Cauchyschen Anfangswertproblem einer elliptischen Differentialgleichung zweiter Ordnung. *Nachr. Gött.* 1, 1–12 (1955).

[36] Müller, C. On the behavior of the solutions of the differential equation $\Delta u = F(x, u)$ in the neighborhood of a point. *Commun. Pure Appl. Math.* 8, 505–515 (1956).

[37] Landis, E. M. On some properties of solutions of elliptic equations. *Dokl. Akad. Nauk*, 107, 640–643 (1956).

[38] Hörmander, L. On the uniqueness of the Cauchy problem. To be published.

[39] Gårding, L. and Malgrange, B. Opérateurs différentiels partiellement hypoelliptiques. *C.R. Acad. Sci. Paris*, 247, 2083–2085 (1958).

[40] Gårding, L. *Cauchy's problem for hyperbolic equations.* Lecture notes, University of Chicago, 1957.

[41] Nirenberg, L. Estimates and existence of solutions of elliptic equations. *Commun. Pure Appl. Math.* 9, 509–530 (1956).

[42] Nash, J. Parabolic equations. *Proc. Nat. Acad. Sci., Wash.*, 43, 754–758 (1957).

[43] Giorgi, E. de. Sulla differenziabilità e l'analiticità delle estremali degli integrali multipli regolari. *Mem. R. Accad. Sci. Torino*, (3), 3, 25–43 (1957).

[44] Beurling, A. and Deny, J. Espaces de Dirichlet. I. *Acta Math.* 99, 203–224 (1958).

[45] Hörmander, L. On interior regularity of the solutions of partial differential equations. *Commun. Pure Appl. Math.* 11, 197–218 (1958).

[46] Malgrange, B. Sur une classe d'opérateurs différentiels hypoelliptiques. *Bull. Soc. Math. Fr.* 85, 3 (1957).

[47] Vishik, M. I. On general boundary problems for elliptic differential equations. *Trudy Mosk. Mat. Ob.* 1, 187–246 (1952).

[48] Vishik, M. I. On strongly elliptic systems of differential equations. *Mat. Sbornik*, 29 (71), 615–676 (1951).

[49] Ladyženskaja, O. A. On the closure of the elliptic operator. *Dokl. Akad. Nauk*, 79, 723–725 (1951).

[50] Guseva, O. V. On boundary problems for strongly elliptic systems. *Dokl. Akad. Nauk*, 102, 1069–1072 (1955).

[51] Lions, J.-L. See Magenes, E. and Stampacchia, G, I problemi al contorno per le equazioni differenziali di tipo ellittico. *Ann. Scuola Norm. Sup. Pisa*, ser. III, 12, 247–358 (1958).

[52] Gårding, L. Dirichlet's problem for linear elliptic partial differential equations. *Math. Scand.* 1, 55–72 (1953).

[53] Soboleff, S. L. Sur un problème limite pour les équations polyharmoniques. *Mat. Sbornik*, 2 (44), 745–757 (1937).

[54] Aronszajn, N. On coercive integro-differential quadratic forms. *Tech. Rep. Proc. Conf. Partial Diff. Eq. Univ. of Kansas*, no. 14, 94–106 (1954).

[55] Schechter, M. Coerciveness of linear partial differential operators for functions satisfying zero Dirichlet-type boundary data. *Commun. Pure Appl. Math.* 11, 153–174 (1958).

[56] Agmon, S. The coerciveness problem for integro-differential forms. *Tech. note 5, contract AF* 61(052)–04. Jerusalem, 1958.

[57] Aronszajn, N. Applied functional analysis. *Proc. Int. Congr. Math.* 2, 123–127 (1950).

[58] Browder, F. On the regularity properties of solutions of elliptic differential equations. *Commun. Pure Appl. Math.* 9, 351–361 (1956).

[59] Lions, J.-L. Problèmes aux limites en théorie des distributions. *Acta Math.* 94, 13–153 (1955).

[60] Hörmander, L. On the regularity of the solutions of boundary problems. *Acta Math.* 99, 225–264 (1958).

[61] Beurling, A. An extension of the Riemann mapping theorem. *Acta Math.* 90, 117–130 (1953).

[62] Petrowsky, I. G. Über das Cauchysche Problem für Systeme von partiellen Differentialgleichungen. *Mat. Sbornik*, 2 (44) 815–868 (1937).

[63] Leray, J. *Hyperbolic differential equations.* The Institute for Advanced Study, Princeton, N.J., 1953.

[64] Courant, R. and Lax, P. The propagation of discontinuities in wave motion. *Proc. Nat. Acad. Sci., Wash.*, 42, 872–876 (1956).

[65] Lax, P. Asymptotic solutions of oscillatory initial value problems. *Duke Math. J.* 24, 627–646 (1957).

[66] Duff, G. F. D. A mixed problem for normal hyperbolic linear partial differential equations of second order. *Canad. J. Math.* 9, 141–160 (1957).

[67] Ladyženskaja, O. A. On the solution of non-stationary operator equations. *Mat. Sbornik*, 39 (81), 491–524 (1956).

[68] Lions, J.-L. Problèmes mixtes abstraits. *Proc. Int. Congr. Math. Edinburgh*, 389–397, (1960).

[69] Campbell, L. L. and Robinson, A. Mixed problems for hyperbolic partial differential equations. *Proc. Lond. Math. Soc.* (3), 5, 129–147 (1955).

[70] Thomée, V. Estimates of the Friedrichs–Lewy type for mixed problems in the theory of linear hyperbolic differential equations in two independent variables. *Math. Scand.* 5, 93–113 (1957).

[71] Soboleff, S. L. and Vishik, M. I. A general form of boundary problem for elliptic partial differential equations. *Dokl. Akad. Nauk*, 121, 521–523 (1958).

[72] Gelfand, I. M. and Kostiučenko, A. G. Expansion in eigenfunctions of differential and other operators. *Dokl. Akad. Nauk, SSSR*, 103, 349–352 (1955).

[73] Berezanskij, Yu M. Eine Verallgemeinerung des Bochnerschen Satzes auf die Eigenfunctionsentwicklung partieller Differentialoperatoren. *Dokl. Akad. Nauk, SSSR*, 110, 893–896 (1956).

[74] Maurin, K. Entwicklung positiv definiter Kerne nach Eigendistributionen. Differenzierbarkeit der Spektralfunktion eines hypoelliptischen Operators. *Bull. Acad. Pol. Sci.*, 6, 149–155 (1958).

[75] Bergendal, G. On spectral functions belonging to an elliptic differential operator with variable coefficients. *Math. Scand.* 5, 241–254 (1957).

[76] Lewitan, B. M. Some problems in the spectral theory of self-adjoint differential operators. *Uspechi Mat. Nauk*, 11, 117–144 (1956).

[77] Friedrichs, K. O. Symmetric positive linear differential equations. *Commun. Pure Appl. Math.* 11, 333–418 (1958).

THE COHOMOLOGY THEORY OF
ABSTRACT ALGEBRAIC VARIETIES

By ALEXANDER GROTHENDIECK

It is less than four years since cohomological methods (i.e. methods of Homological Algebra) were introduced into Algebraic Geometry in Serre's fundamental paper[11], and it seems already certain that they are to overflow this part of mathematics in the coming years, from the foundations up to the most advanced parts. All we can do here is to sketch briefly some of the ideas and results. None of these have been published in their final form, but most of them originated in or were suggested by Serre's paper.

Let us first give an outline of the main topics of cohomological investigation in Algebraic Geometry, as they appear at present. The need of a theory of cohomology for 'abstract' algebraic varieties was first emphasized by Weil, in order to be able to give a precise meaning to his celebrated conjectures in Diophantine Geometry[20]. Therefore the initial aim was to find the '*Weil cohomology*' of an algebraic variety, which should have as coefficients something 'at least as good' as a field of *characteristic* 0, and have such formal properties (e.g. duality, Künneth formula) as to yield the analogue of Lefschetz's 'fixed-point formula'. Serre's general idea has been that the usual 'Zariski topology' of a variety (in which the closed sets are the algebraic subsets) is a suitable one for applying methods of Algebraic Topology. His first approach was hoped to yield at least the right Betti numbers of a variety, it being evident from the start that it could not be considered as the Weil cohomology itself, as the coefficient field for cohomology was the ground field of the variety, and therefore not in general of characteristic 0. In fact, even the hope of getting the 'true' *Betti numbers* has failed, and so have other attempts of Serre's[12] to get Weil's cohomology by taking the cohomology of the variety with values, not in the sheaf of local rings themselves, but in the sheaves of Witt-vectors constructed on the latter. He gets in this way modules over the ring $W(k)$ of infinite Witt vectors on the ground field k, and $W(k)$ is a ring of characteristic 0 even if k is of characteristic $p \neq 0$. Unfortunately, modules thus obtained over $W(k)$ may be infinitely generated, even when the variety V is an abelian variety[13]. Although interesting relations must certainly exist between these cohomology groups and the 'true ones', it seems certain now that the Weil cohomology has to be defined by a completely different approach. Such an approach was recently suggested to me by the *connections between sheaf-theoretic cohomology and cohomology of Galois groups on the one hand, and the classification of unramified coverings of a variety on the other* (as explained quite unsystematically in Serre's tentative Mexico paper[12]), and by Serre's idea that a 'reasonable' algebraic principal fiber space with structure group G, defined on a variety V, if it is not locally trivial, should become locally trivial on some covering of V *unramified* over a given point of V. This has been the starting-point of a definition of the Weil cohomology (involving both 'spatial' and Galois cohomology), which seems to be the right one, and which gives clear suggestions how Weil's conjectures may be attacked by the machinery of Homological Algebra. As I have not begun these investigations seriously as yet, and as moreover this theory has a quite distinct flavor from the one of the theory of algebraic coherent sheaves which we shall now be concerned with, we shall not dwell any longer on Weil's cohomology. Let us merely remark that the definition alluded to has already been the starting-point of a theory of cohomological dimension of fields, developed recently by Tate[18].

The second main topic for cohomological methods is the *cohomology theory of algebraic coherent sheaves*, as initiated by Serre. Although inadequate for Weil's purposes, it is at present yielding a wealth of new methods and new notions, and gives the key even for results which were not commonly thought to be concerned with sheaves, still less with cohomology, such as Zariski's theorem on 'holomorphic functions' and his 'main theorem'—which can be stated now in a more satisfactory way, as we shall see, and proved by the same uniform elementary methods. The main parts of the theory, at present, can be listed as follows:

(*a*) General finiteness and asymptotic behaviour theorems.

(*b*) Duality theorems, including (respectively identical with) a cohomological theory of residues.

(*c*) Riemann–Roch theorem, including the theory of Chern classes for algebraic coherent sheaves.

(*d*) Some special results, concerning mainly abelian varieties.

The third main topic consists in the *application of the cohomological methods to local algebra*. Initiated by Koszul and Cartan–Eilenberg in connection with Hilbert's 'theorem of syzygies', the systematic use of these methods is mainly due again to Serre. The results are the *characterization* of regular local rings as those whose global cohomological dimension is finite, the clarification of *Cohen–Macaulay's equidimensionality theorem* by means of the notion of *cohomological codimension*[23], and especially the possibility of giving (for the first time as it seems) a *theory of intersections*, really satisfactory by its algebraic simplicity and its generality. Serre's result just quoted, that regular local rings are the only ones of finite global cohomological dimension, accounts for the fact that only for such local rings does a satisfactory theory of intersections exist. I cannot give any details here on these subjects, nor on various results I have obtained by means of a *local duality theory*, which seems to be the tool which is to replace differential forms in the case of unequal characteristics, and gives, in the general context of commutative algebra, a clarification of the notion of residue, which as yet was not at all well understood. The motivation of this latter work has been the attempt to get a global theory of duality in cohomology for algebraic varieties admitting arbitrary singularities, in order to be able to develop intersection formulae for cycles with arbitrary singularities, in a nonsingular algebraic variety, formulas which contain also a 'Lefschetz formula mod. p'[8]. In fact, once a proper local formalism is obtained, the global statements become almost trivial. As a general fact, it appears that, to a great extent, the 'local' results already contain a global one; more precisely, global results on varieties of dimension n can frequently be deduced from corresponding local ones for rings of Krull dimension $n+1$.

We will therefore turn now to giving some main ideas in the second topic, that is, the cohomology theory of algebraic coherent sheaves. First, I would like, however, to emphasize one point common to *all* of the topics considered (except *perhaps* for (d)), and in fact to all of the standard techniques in Algebraic Geometry. Namely, that the natural range of the notions dealt with, and the methods used, are not really algebraic varieties. Thus, we know that an *affine* algebraic variety with ground field k is determined by its co-ordinate ring, which is an arbitrary finitely generated k-algebra without nilpotent elements; therefore, any statement concerning affine algebraic varieties can be viewed also as a statement concerning rings A of the previous type. Now it appears that most of such statements make sense, and are true, if we assume only A to be a commutative ring with unit, provided we sometimes submit it to some mild restriction, as being noetherian, for instance. In the same way, most of the results proved for the local rings of algebraic geometry, make sense and are true for arbitrary noetherian local rings. Besides, frequently when it seemed at first sight that the statement only made sense when a ground field k was involved, as in questions in which differential forms are considered, further consideration of the matter showed that this impression was erroneous, and that a better understanding is obtained by replacing k by a ring B such that A is a finitely generated B-algebra. Geometrically, this means that instead of a single affine algebraic variety V (as defined by A) we are considering a 'regular map' or 'morphism' of V into another affine variety W, and properties of the variety V then are generalized to properties of a morphism $V \to W$ (the 'absolute' notion for V being obtained from the more general 'relative' notion by taking W reduced to a point). On the other hand, one should not prevent the rings having nilpotent elements, and by no means exclude them without serious reasons. Now just as arbitrary commutative rings can be thought of as a proper generalization of *affine* algebraic varieties, one can find a corresponding suitable generalization of arbitrary algebraic varieties (defined over an arbitrary field). This was done by Nagata[9] in a particular case, yet following the definition of schemata given by Chevalley[4] he had to stick to the irreducible case, and with no nilpotent elements involved. The principle of the right definition is again to be found in Serre's fundamental paper[11], and is as follows. If A is any commutative ring, then the set Spec (A) of all prime ideals of A can be turned into a topological space in a classical way, the closed subsets consisting of those prime ideals which contain a given subset of A. On the other hand, there is a sheaf of rings defined in a natural way on Spec (A), the fiber of this sheaf at the point p being the local ring A_p. More generally, every module M over A defines a sheaf of modules on Spec (A), the fiber of which at the point p is the localized module M_p over A_p. Now we call *pre-schema* a topological space X with a sheaf of rings $\mathcal{O}_X$ on X, called its *structure sheaf*, such that every point of X has an open neighborhood isomorphic to some Spec (A). If X and Y are two pre-schemas, a *morphism* f from X into Y is a continuous map $f \colon X \to Y$, together with a corresponding homomorphism $f^* \colon \mathcal{O}_Y \to \mathcal{O}_X$ for the structure sheaves, submitted to the one condition: if $x \in X$, $y = f(x)$, then the inverse image by $f^* \colon \mathcal{O}_{Y,y} \to \mathcal{O}_{X,x}$ of the maximal ideal in $\mathcal{O}_{X,x}$ is the maximal ideal in $\mathcal{O}_{Y,y}$. If X and Y are the prime spectra of rings A and B, then it can be shown that the morphisms from X into Y correspond exactly to the ring homomorphisms of B into A, as they should.

As was explained before, if we consider morphisms $f: X \to S$ with a fixed pre-schema S, S plays the part of a ground-field. Besides, if $S = \mathrm{Spec}\,(A)$, then X is a pre-schema *over* S if and only if the sheaf of rings $\mathcal{O}_X$ is a sheaf of A-algebras. In the category of all pre-schemata over a given S, there exists a product (which corresponds to the tensor product of algebras over a commutative ring A). Using this, one can define objects like pre-schemata of groups, etc., over a fixed ground-pre-schema S. One can also use products in order to introduce a mild separation condition on pre-schemata, suggested by the usual condition in the definition of algebraic varieties, so that one gets what can be called a *schema*. A schema is called *noetherian* if it is the union of a finite number of open sets isomorphic to the prime spectra of *noetherian* rings. A schema X over another S is called of *finite type over S* (or the morphism $f: X \to S$ is called of finite type) if, for every affine open set U in S, $f^{-1}(U)$ is the union of a finite number of affine open sets, corresponding to *finitely generated* algebras over the ring of U. Most of the notions and results of usual Algebraic Geometry can now be stated and proved in this new context, provided essentially that in some questions one sticks to noetherian schemata and to morphisms which are of finite type. Let us only remark here that the notion of a *complete* variety yields the notion of a *proper morphism* (which in the case of Algebraic Geometry over the complex numbers is the same as proper in the usual topological sense): the morphism $f: X \to S$ is called proper if it is of finite type, and if for every noetherian schema Y over S, the projection $X \times_S Y \to Y$ is a closed map. One can define also projective and quasi-projective morphisms, corresponding to the notions of projective and quasi-projective varieties. Many properties on general morphisms can be reduced to the case of projective or quasi-projective ones, using a suitable generalization of Chow's well-known lemma.

A sheaf of modules F on the schema X is called *quasi-coherent* if on each affine open set U in X, it can be defined by a module M over the ring A of U. If X is noetherian, then F is coherent (in the general sense of[11]) if and only if it is quasi-coherent, and the modules M are finitely generated. Quasi-coherent and coherent algebraic sheaves behave as nicely under the usual sheaf-theoretic operations (tensor products, sheaves of homomorphisms, direct and inverse images, and derived functors of the previous ones) as could possibly be expected from them.

We are now in a position to state results in the proper context. We shall limit ourselves, however, to (*a*) and (*b*). The Riemann–Roch theorem, proved independently by Washnitzer[19] and myself[2] in abstract algebraic geometry, will be exposed by Hirzebruch at this Congress. Let us only remark that the present formulation of this theorem, as suggested by Hirzebruch's formula, is substantially stronger than the latter, because as usual a statement about a *complete variety* is replaced by one about a *proper morphism*. As for the 'special results' alluded to in (*d*), due to Barsotti, Cartier, Rosenlicht and Serre, let us only state that one knows about all one wants and expects on the cohomology of an abelian variety, and a great deal of the relations between the cohomology of a variety and its Albanese variety. In particular, one gets by cohomological methods the absence of 'torsion' in an abelian variety[1, 3a, 13] and the biduality of abelian varieties[3]. As yet, there has been no question of stating such results in the general context of a schema over another, although this is certainly a reasonable one.

The main results in the cohomology theory of morphisms of schemata are the following (Theorems 1 and 2 being rather straightforward adaptations of Serre's results[11]).

Theorem 1. Let F be a quasi-coherent sheaf on the affine schema X, defined by a module M over the co-ordinate ring A of X. Then $H^i(X, F) = 0$ for $i > 0$, and $H^0(X, F) = M$.

Of course, the cohomology groups are taken in the general sense of cohomological algebra in abelian categories[5, 6]. It is important for technical reasons not to take as a *definition* of cohomology the Čech cohomology as was done in[11]; in virtue of Leray's spectral sequence for a covering, Theorem 1 implies that the cohomology groups of a quasi-coherent sheaf on a schema X can be computed by the Čech method, but this should be considered as an accidental phenomenon. There is a converse to Theorem 1[17], to the effect that if X is a noetherian pre-schema for which $H^1(X, F) = 0$ for every coherent subsheaf F of $\mathcal{O}_X$, then X is affine. This can be shown in this general case by a suitable adaptation of Serre's proof.

Let us recall that if f is a continuous map from a space X into another Y, then for every abelian sheaf F on X, one defines the direct image $f_*(F)$ of F by f to be the sheaf on Y the sections of which on an open set $U \subset Y$ are the sections of F on $f^{-1}(U)$. Taking the right derived functors of the functor f_*, we get the 'higher direct images' $R^q f_*(F)$ of F by f. The sheaf $R^q f_*(F)$ is the sheaf associated to the pre-sheaf

$$U \to H^q(f^{-1}(U), F),$$

and is well known as entering in the beginning term of Leray's spectral sequence for the continuous map f and the sheaf F. In case f is a morphism of schemata and F is an algebraic sheaf, the higher direct images are algebraic sheaves, which are quasi-coherent whenever F is quasi-coherent and f of finite type. In this case, the group of sections of $R^q f_*(F)$ on an *affine* open set U of Y is *identical* with $H^q(f^{-1}(U), F)$, as follows at once from Leray's spectral sequence, for instance. Theorem 1 is easily generalized to a property of affine morphisms $f: X \to Y$, i.e. morphisms for which the inverse image of an affine open set is affine (which in fact is a property local with respect to Y): if f is affine, then for every quasi-coherent sheaf F on X, we have $R^q f_*(F) = 0$ for $q > 0$, and the converse holds if X is noetherian.

The next theorem is concerned with *projective morphisms*. Let Y be a pre-schema, and let $\mathscr{S}$ be a quasi-coherent sheaf of graded algebras on Y. For the sake of simplicity, we suppose that $\mathscr{S}$ has only positive degrees, and is generated (as a sheaf of $\mathscr{O}_Y$-algebras) by $\mathscr{S}^1$. Then by generalizing well-known constructions, one can define a pre-schema X over Y (in fact a schema if Y is one itself), and on X an algebraic sheaf locally isomorphic to $\mathscr{O}_X$, denoted by $\mathscr{O}_X(1)$. More generally, for every quasi-coherent sheaf $\mathscr{M}$ of graded $\mathscr{S}$-modules, one defines a quasi-coherent sheaf $F(\mathscr{M})$ on X, the functor $\mathscr{M} \to F(\mathscr{M})$ being exact and compatible with the usual operations such as tensor products and $\mathscr{T}or_i$, $\mathscr{H}om$ sheaves and $\mathscr{E}xt^i$. Taking $\mathscr{M} = \mathscr{S}$, one gets $\mathscr{O}_X$, and shifting degrees by n in S one gets sheaves $\mathscr{O}_X(n)$, which can (by what we just stated) be obtained as n-fold tensor products of $\mathscr{O}_X(1)$ with itself. The definitions are such that, if f is the projection $f: X \to Y$, then we have a natural homomorphism (compatible with f) of $\mathscr{M}^0$ into $F(\mathscr{M})$, and hence also from $\mathscr{M}^n$ into $F(\mathscr{M}(n)) = F(\mathscr{M})(n)$, where for every algebraic sheaf G on X, we denote by $G(n)$ the tensor product $G \otimes_{\mathscr{O}_x} \mathscr{O}_X(n)$. The formalism just sketched, and in particular getting a pre-schema over Y from a graded quasi-coherent sheaf $\mathscr{S}$ of algebras on Y, should be considered as the natural general description of the 'blowing-up' process: this is obtained by taking a coherent subsheaf $\mathscr{J}$ of $\mathscr{O}_Y$, taking the subsheaves $\mathscr{J}^n$ of $\mathscr{O}_Y$ generated by this sheaf of ideals, and taking for $\mathscr{S}$ the direct sum of the sheaves $\mathscr{J}^n$, which is a graded sheaf of algebras in an obvious way. A pre-schema X over Y is called *projective over* Y (and the morphism $f: X \to Y$ is called a *projective morphism*) if X can be obtained, by the process alluded to above, from a sheaf $\mathscr{S}$ such that $\mathscr{S}^1$ is a finitely generated sheaf (i.e. a coherent sheaf if Y is assumed noetherian). Thus, if Y is an affine noetherian schema which is of integrity (i.e. irreducible and $\mathcal{O}_Y$ without nilpotent elements) then the schemata over Y which are projective over Y, and for which the projection $f\colon X \to Y$ is a birational equivalence, are those which can be obtained from Y by blowing up a coherent sheaf of ideals on Y (so that we get 'practically all' birational proper morphisms by the standard blowing up process). Let us also remark that if we take for $\mathcal{S}$ the symmetric algebra of a locally free sheaf of modules $\mathcal{V}$ on Y, then the corresponding X over Y should be looked at as the fiber space with projective spaces as fibers, which corresponds to the 'vector bundle' defined by the sheaf $\mathcal{V}'$ dual to $\mathcal{V}$. (In fact, all 'geometric' constructions of Algebraic Geometry can be carried over in the context of schemata.) The main facts about the cohomological theory of projective morphisms are stated in the following theorem:

Theorem 2. Let $f\colon X \to Y$ be a projective morphism, defined by a graded sheaf of algebras $\mathcal{S}$ on Y. Y is supposed noetherian, and we suppose that $\mathcal{S}^1$ is generated as a sheaf of modules by r generators (locally). Then the following statements hold for every coherent sheaf F on X:

    (i) $F(n)$ is 'generated by its sections' for n large, provided we restrict to the points of X which lie above an *affine* open set of X;

    (ii) $R^q f_*(F(n)) = 0$ for $q > 0$, n large;

    (iii) $R^q f_*(F) = 0$ for $q > r$;

    (iv) The sheaves $R^q f_*(F)$ are coherent.

The first statement (i) implies also that *every* coherent sheaf F on X can be obtained from a suitable quasi-coherent sheaf of graded modules over $\mathcal{S}$.

The next two theorems are proved by first dealing with the case of a projective morphism (N.B. projective morphisms are proper). Then the statements reduce to statements on graded modules over a polynomial ring $A[X_1, ..., X_n]$, with A noetherian, and can be proved by an easy *decreasing induction* on the dimension i in the cohomology. This explains why a complete statement and a simple proof of these results, even restricted to $i = 0$ (i.e. when not concerned with genuine cohomology), could not be achieved by non-cohomological methods. The case of a general morphism is then reduced to the case of a projective one by Chow's lemma, as in [7].

Theorem 3. Let $f\colon X \to Y$ be a proper morphism, Y being noetherian. Then, for every coherent sheaf F on X the sheaves $R^q f_*(F)$ are coherent.

Theorem 4. Let $f\colon X \to Y$ and F be as before, let y be a point of Y, then $R^q f_*(F)_y$ is a finitely generated module over the local ring $\mathcal{O}_y$, and the m_y-adic completion of this module is naturally isomorphic to

$$\varprojlim_k H^q(f^{-1}(y), F \otimes_{\mathcal{O}_y} (\mathcal{O}_y / m_y^k))$$

(where m_y is the maximal ideal of $\mathcal{O}_y$).

This should be considered as the complete statement of Zariski's main result on 'holomorphic functions'[21]. Here, the lim group should be considered as the 'holomorphic cohomology' of X along the fiber $f^{-1}(y)$, with coefficients in the sheaf F. From Theorem 4, one gets a result which is global with respect to Y: if Y' is a closed subset of Y, and $X' = f^{-1}(Y')$, then the holomorphic cohomology of X along X' (with coefficients in F) is the ending of a spectral sequence of cohomological type, the beginning term of which is $E_2^{p,q} = H^p(Y/Y', R^q f_*(F))$ (where the second member denotes the holomorphic cohomology of Y along Y').

Theorem 4 yields at once Zariski's connectedness theorem, in the following general form. Let $f: X \to Y$ be a proper morphism, Y no-etherian. Then by Theorem 3 the direct image $f_*(\mathcal{O}_X)$ is a *coherent* sheaf of $\mathcal{O}_Y$-algebras $\mathscr{B}$ on Y. Let $y \in Y$, then $\mathscr{B}_y$ is an $\mathcal{O}_y$-algebra which is a finitely generated module. It follows at once from Theorem 4 (with $q = 0$) that the set of connected components of the fiber $f^{-1}(y)$ is in one to one correspondence with the maximal ideals of $\mathscr{B}_y$ (which of course are finite in number). If, for instance, X and Y are of integrity, f surjective, then the field K of Y is a subfield of the field L of X, and $\mathscr{B}_y$ is a subring of L which is integral over the subring $\mathcal{O}_y$ of K. If the integral closure of $\mathcal{O}_y$ in K has only one maximal ideal (we then say that y is a 'unibranch point' of Y) and if the algebraic closure of K in L is purely inseparable over K, then it follows at once that $\mathscr{B}_y$ has only one maximal ideal. Therefore, the fiber of f at y is connected. (N.B. no analytic irreducibility of $\mathcal{O}_y$ was needed.)

We can state things in a more geometric way, by using the fact that the coherent sheaf $f_*(\mathcal{O}_X) = \mathscr{B}$ of $\mathcal{O}_Y$-algebras defines in a natural way a pre-schema Y' over Y (characterized by the condition that Y' be affine over Y and the direct image of $\mathcal{O}_{Y'}$ should be $\mathscr{B}$). It follows from the Cohen–Seidenberg theorems that Y' is also proper over Y, and that the fibers in the projection $Y' \to Y$ are finite. (Conversely, it follows from Theorems 3 and 4 that any Y' over Y having these properties can be defined by a coherent sheaf of algebras on Y. We shall say that such a pre-schema Y' over Y is *integral over* Y.) Using the fact that we have an isomorphism $B \xrightarrow{\sim} f^*(\mathcal{O}_X)$, one sees that $f: X \to Y$ can be factored in a natural way as $X \xrightarrow{f'} Y' \xrightarrow{g} Y$, where now f' is such that $f'_*(\mathcal{O}_X) = \mathcal{O}_{Y'}$. Zariski's connectedness theorem can now be stated in the following way: the connected components of $f^{-1}(y)$ are in a one to one correspondence with the elements of $g^{-1}(y)$, or equivalently: the fibers of f' are connected. (This canonical factorization of f was suggested by work of Stein on analytic spaces.)

Using Theorems 3 and 4 and the connectedness theorem, one obtains also, by global and cohomological methods, the most general statement of Zariski's 'main theorem' in commutative algebra, which we will state here in the geometric language:

Theorem 5. Let $f: X \to Y$ be a quasi-projective morphism from a schema X into a noetherian schema Y. The points in X which are isolated in their fiber $f^{-1}(f(x))$ form an open set U. There exists a schema Y' integral over Y, and an isomorphism of U onto an open subset U' of Y', such that the restriction of f to U is identical to the compositum $U \to Y' \to Y$.

This statement is somewhat more general than the usual purely local statement. It should be remarked that the local rings of X and Y may contain nilpotent elements. The proof of Theorem 5 is obtained by first reducing to the case when X is proper over Y. Then Y' is the schema over Y constructed above with the sheaf $f_*(\mathcal{O}_X)$, and the fact that the canonical factorization of f induces an isomorphism from U onto an open subset of Y' is easily deduced from the connectedness theorem. The fact that there should exist cohomological proofs of Zariski's connectedness theorem and of his 'main theorem' was first conjectured by Serre. That these results hold also for general schemata, not only those of algebraic geometry, yields results such as the following (which I understand is due to Chow): let A be a noetherian local integrity domain which is normal (or more generally, 'unibranch'); then the algebraic projective space over the residue field k of A defined by the graded k-algebra $G(A) = \Sigma m^n / m^{n+1}$ ($m =$ maximal ideal of A) is connected.

It remains to say a few words on the duality theorems, although time does not permit us to give precise statements. Let us recall first Serre's original statement [22]: if X^n is a non-singular projective variety defined over an algebraically closed field k, E a vector bundle over X, then there is a natural duality between $H^i(X, \mathcal{O}_X(E))$ and $H^{n-i}(X, \Omega_X^n(E'))$, where $\mathcal{O}_X(E)$ is the sheaf of germs of regular sections of E, and $\Omega_X^n(E')$ the sheaf of germs of regular n-differential forms of X with values in the dual vector-bundle E'. This duality is obtained by cup-product, using the fact that $H^n(X, \Omega_X^n)$ itself is naturally isomorphic to the ground-field k. Various generalizations of this result, and applications to Poincaré duality (including a Lefschetz formula) were given in [8]. However, the case of singular varieties was not taken into account, so that the statements on Poincaré duality were necessarily restricted to non-singular cycles and non-singular intersections of such cycles. Moreover, the proper algebraic foundations for the *covariant* features of the cohomology of non-singular varieties, as suggested by Poincaré duality, remained unclear. By developing a general theory of residues, this situation can now be considered as (potentially) completely clarified. To simplify matters, let us consider schemata of finite type over an arbitrary ground-field k, i.e. algebraic spaces defined over k, the local rings of X being allowed, however, to have nilpotent elements (which is technically of the highest importance). Then, on such an X, there is a canonically defined complex of algebraic quasi-coherent sheaves $\mathscr{K}_X$, with positive degrees, and a differential operator of degree -1. In dimension i, $\mathscr{K}_X$ is the direct sum of sheaves $\mathscr{D}(X/Y)$, where Y runs over the set of all closed irreducible subsets of X of dimension i. The sheaf $\mathscr{D}(X/Y)$ is the extension to X of a *constant* sheaf on Y, corresponding to a module over the local ring $\mathcal{O}_{X/Y}$ of Y in X, which should be called the *dual module* of this local ring. In general, if A is a locality, over the ground-field k (that is, a local ring of a finitely generated k-algebra) there is a well-defined A-module $D(A)$, called its dual (with respect to k), which can be obtained in the following way: take a subfield L of A separable over k such that the residue field of A be algebraic over L, then

$$D(A) = \mathrm{Hom}_L(A, \Omega^p(L)),$$

where Hom denotes the *continuous* homomorphisms, and where $\Omega^p(L)$ is the vector space (of dimension 1) over L of the highest dimensional differential forms of L with respect to k. (The fact that the second member does not depend on the choice of L is not trivial, and is obtained as a consequence of an alternative cohomological definition of $D(A)$ as a direct limit of Ext modules.) The sheaf $\mathscr{K}_X$ is an *injective* sheaf of $\mathcal{O}_X$-modules. The definition of its differential operator is a rather subtle one. The complex of sheaves $\mathscr{K}_X$ should be called the *residue complex* of X (or the complex of '*generalized Cousin data*' of X). It turns out that if X is non-singular and separable over k, then $\mathscr{K}_X$ is an (injective) *resolution* of the sheaf Ω_X^n of germs of differential forms of highest degree on X. In any case, if $\dim X = n$, then the sheaf of cycles of degree n in $\mathscr{K}_X$, denoted by ω_X, should be considered as playing the part of differential forms of degree n on X. In fact, there is a natural homo-morphism (defined by the well-known Kähler process) of Ω_X^n into ω_X; if X is normal and separable over k, then ω_X is nothing else but the sheaf of germs of differential forms on X which are regular at simple points. On the other hand, if X is Cohen–Macaulay (i.e. its local rings are Cohen–Macaulay), then $\mathscr{K}_X$ is a resolution of ω_X (and conversely). This shows that in general the sheaves $H_i(\mathscr{K}_X)$ are not zero even for $i \neq n$ (N.B. they never are for $i = n$), however, these sheaves are always coherent (although $\mathscr{K}_X$ itself is much too big to be coherent).

Now let F be an arbitrary coherent sheaf on X, we write

$$H_i(X, F) = H_i(\mathrm{Hom}_{\mathcal{O}_X}(X; F, \mathscr{K}_X)).$$

Thus the *homology* of X with coefficients in F is defined as a *contravariant* functor in F, which moreover is a 'homological functor' because the complex $\mathscr{K}_X$ is injective, i.e. $F \rightarrow \mathrm{Hom}_{\mathcal{O}_X}(X; F, \mathscr{K}_X)$ is exact. It has all the properties one expects from homology, in particular, it is *covariant* with X with respect to *proper* morphisms (the sheaf $\mathscr{K}_X$ itself behaving covariantly with X), the cohomology of X operates on the homology by *cap-product*. Relative homology and cohomology groups of various types can also be defined. If X is Cohen–Macaulay, then the above definition gives

$$H_i(X, F) \simeq \mathrm{Ext}_{\mathcal{O}_X}^{n-i}(X; F, \omega_X). \tag{1}$$

This relation is replaced by a spectral sequence in the general case. If X is Cohen–Macaulay and moreover F is locally free, then the above relation yields

$$H_i(X, F) \simeq H^{n-i}(X, F' \otimes \omega_X) \tag{2}$$

so that in this case, homology can be expressed as cohomology. In the general case, this relation is replaced by a spectral sequence, analogous to Cartan's spectral sequence for an arbitrary topological space, connecting its homology and its cohomology (and yielding Poincaré duality in case the space is a variety).

Using the definitions of H_n and ω_X, we see that (if $\dim X = n$) there is a canonical element in $H_n(X, \omega_X)$, called the *canonical homology class* of X; as homology is contravariant in the argument, it follows that there is a canonical homology class in $H_n(X, \Omega_X^n)$, and as homology is co-variant with the space, it follows that for each p-cycle Z in X, there is a homology class $\gamma_X(Z)$ in $H_p(X, \Omega_X^p)$. The formation of these classes is compatible with direct images of cycles and homology classes. More-over, if X is non-singular and separable over k, then $\gamma(Z)$ can be regarded as an element of $H^{n-p}(X, \Omega_X^{n-p})$ in virtue of (2). Using some simple facts of local character connecting the intersection and inverse images of cycles with some partially defined multiplicative structure in $\mathcal{K}_X$ (for non-singular X), we get that formation of characteristic classes is compatible with products and inverse images (which solves the difficulties encountered in [8]).

Serre's duality theorem is now generalized in the following way. Suppose F is a coherent sheaf on X with complete support, then by cap-product there is a natural pairing

$$H_i(X, F) \times H^i(X, F) \to H_0(X, \mathcal{O}_X/\mathcal{J}),\tag{3}$$

where $\mathcal{J}$ is any coherent sheaf of ideals on X such that $\mathcal{J}.F = 0$. We can take $\mathcal{J}$ such that $\mathcal{O}_X/\mathcal{J}$ has complete support, i.e. is the sheaf of local rings of an algebraic subspace Y of X. Then the second member of (3) is nothing else but $H_0(Y, \mathcal{O}_Y)$, and using the sum of residues one gets a natural k-homomorphism

$$H_0(X, \mathcal{O}_X/J) = H_0(Y, \mathcal{O}_Y) \to k.\tag{3a}$$

Using (3) and (3a), one gets a pairing

$$H_i(X, F) \times H^i(X, F) \to k.\tag{4}$$

This pairing is a duality: thus for complete supports, homology and cohomology are dual to each other. Moreover, in this duality, the direct image of homology is transposed to the inverse image of cohomology (as is needed for the formalism of Poincaré duality).

We have sketched the statement of results for schemata of finite type over a field k. More general results hold in fact, and the duality theorem can be stated for any coherent sheaf F on a noetherian schema X, such that the support of F is complete (i.e. proper over some Artin ring A, which need not be a field). Combining this with Theorem 4, one gets an equivalent statement, concerned with the homological properties of an arbitrary proper morphism (the basis Y being noetherian).

To conclude this talk, I should state some unsolved questions. In fact it is perhaps too early to do so, as the available new techniques have not been tried seriously enough to know whether or not they are able to solve these questions. The following two (which are probably related) will perhaps show some resistance. For the sake of brevity, we state them in the context of Algebraic Geometry.

Problem A. (Kodaira's vanishing theorem.) Let V be a non-singular projective variety, L a negative line bundle on V (i.e. such that some negative multiple of L defines a projective embedding of V). If V is of dimension n, is it true that $H^i(X, \mathcal{O}_X(L)) = 0$ for $0 \leqslant i < n$?

It is not difficult to show, using duality, that this problem is equivalent to the following one, which no longer involves cohomology: Let V and L be as before, and let W be a non-singular hyperplane section of V (with respect to some projective embedding of V). Then is it true that every regular $(n-1)$-form, on W, with coefficients in L, is the residue of a rational differential n-form on V, with coefficients in L, having a divisor $\geqslant -W$?

Problem B. Let f be a birational proper morphism of a non-singular variety X into another Y. Is it true that $R^q f_*(\mathcal{O}_X) = 0$ for $q > 0$?

Using Leray's spectral sequence, this is equivalent to the problem: if $f: X \to Y$ is as above, is it true that $H^p(X, \mathcal{O}_X) \xrightarrow{\sim} H^p(Y, \mathcal{O}_Y)$ for every p? More generally, it would follow that if E is a vector bundle on Y, F its inverse image on X, then $H^p(X, \mathcal{O}_X(F)) \xrightarrow{\sim} H^p(Y, \mathcal{O}_Y(E))$ ('birational invariance of cohomology'). It would imply that the arithmetic genus of a complete non-singular variety is a birational invariant (this is only known in the classical case when k is the field of complex numbers and X projective, by using symmetry $h^{0,p} = h^{p,0}$), more generally that (f being as before) the Todd classes of Y are the direct images by f of the Todd classes of X (as is seen by applying the Riemann–Roch formula[2]). The answer to Problem B is not known even for non-singular projective varieties over the field of complex numbers. It should be remarked that it is essential for the statement to be true, that both X and Y should be non-singular.

There are some other questions in which the cohomological methods and the heuristic insight provided by the point of view of schemata will probably prove helpful. The most important at present seem to be the following ones.

Problem C. Let X be a noetherian schema, Y a complete sub-schema. State conditions under which it is possible to 'blow down' Y to a point, i.e. to find a proper morphism of X into a schema X', mapping Y into a single point y and inducing an isomorphism of $X - Y$ onto $X' - (Y)$.

According to Grauert, this problem is closely related to problem A. Moreover, it is connected with the theory of holomorphic functions of Zariski, a necessary (yet not sufficient) condition for the blowing-down to be possible being the following (in virtue of Theorem 4): the ring of holomorphic functions of X along Y must be noetherian and of Krull dimension equal to dim X.

Problem D. Let G be a schema of groups over the schema Y (G of finite type over Y), let H be a closed sub-schema of groups of G. Prove the existence of the schema G/H over X (defined by the usual universal mapping properties).

Problem E. Let X be a schema proper over another Y (Y noetherian). Prove the existence of a schema of abelian groups of Y, playing the part of a relative Picard variety with respect to the determination of locally free rank one sheaves on products $X \times_Y Z$ (Z being a variable 'parameter schema' over Y).

We shall not give here the precise definition of a 'relative Picard schema', but only remark that if this schema exists then it behaves in the simplest conceivable way with respect to change of base-space Y (which should be looked at as the analogue of a change of a base-field). Moreover, the fact that we do admit nilpotent elements in the local rings of the schemata Z provides a great amount of supplementary information on the Picard variety (especially its infinitesimal structure), which does not seem to have been obtained as yet even in the classical case. These remarks still hold if we consider the more general following situation: Given a schema of groups G over X (G of finite type over X), for every Z over Y, we consider on $X \times_Y Z$ the inverse image G^Z of G by the projection $X \times_Y Z \to X$ and the isomorphism classes of schemata over $X \times_Y Z$ which are principal under G^Z. This leads to a definition of a generalized relative Picard variety of X, with respect to G, which should be a schema over Y, and a schema of abelian groups over Y if G is abelian. It is hoped that a general existence theorem of such generalized relative Picard schema exists and will be proved in the future.

As a quite general fact, it is believed that a better insight in any part of even the most classical Algebraic Geometry will be obtained by trying to re-state all known facts and problems in the context of schemata. This work is now begun, and will be carried on in a treatise on Algebraic Geometry which, it is hoped, will be written in the following years by J. Dieudonné and myself, and which is expected to give a systematic account of all the questions touched upon in this talk.

REFERENCES

[1] Barsotti, I. Abelian varieties over fields of positive characteristic. *R.C. Circ. mat. Palermo*, (2), 5, 1–25 (1956).

[2] Borel, A. and Serre, J. P. Le théorème de Riemann–Roch. *Bull. Soc. Math. Fr.* 86, 97–136 (1958).

[3] Cartier, P. Thèse. Paris, 1959. (To appear.)

[3a] Cartier, P. *Séminaire Bourbaki*. (May, 1958).

[4] Chevalley, C. and Cartan, H. *Séminaire École Normale Supérieure* (1955–56).

[5] Godement, R. Topologie algébrique et théorie des faisceaux. *Act. Sci. Ind.* no. 1252. Hermann, Paris, 1958.

[6] Grothendieck, A. Sur quelques points d'algèbre homologique. *Tôhoku Math. J.* 9, nos. 2–3, 119–221 (August, 1957).

[7] Grothendieck, A. Sur les faisceaux algébriques et les faisceaux analytiques cohérents. In *Séminaire Cartan*, 1956–57 (exposé no. 2).

[8] Grothendieck, A. Théorèmes de dualité pour les faisceaux algébriques cohérents. *Séminaire Bourbaki* (Mai, 1957).

[9] Nagata, M. A general theory of algebraic geometry over Dedekind domains. *Amer. J. Math.* 78, 78–116 (1956).

[10] Rosenlicht, M. Extensions of vector groups by abelian varieties. *Amer. J. Math.* 80, 685–714 (1958).

[11] Serre, J. P. Faisceaux algébriques cohérents. *Ann. Math.* (2), 61, 197–278 (1955).

[12] Serre, J. P. Sur la topologie des variétés algébriques en caractéristique *p*. *Symposium de Topologie Algébrique, Mexico* (August, 1956).

[13] Serre, J. P. Quelques propriétés des variétés abéliennes en caractéristique *p*. *Amer. J. Math.* 80 (1958).

[14] Serre, J. P. Sur la dimension homologique des anneaux et modules noethériens. *Proc. Int. Symp. Algebraic Number Theory, Tokyo-Nikko*, pp. 175–189 (1955).

[15] Serre, J. P. Théories des intersections. Cours professé au Collège de France 1957/58, rédigé par P. Gabriel. (To appear in mimeographed notes.)

[16] Serre, J. P. Groupes algébriques et corps de classes, d'après un cours professé au Collège de France. (To appear in *Act. Sci. Ind.* Hermann, Paris.)

[17] Serre, J. P. Sur la cohomologie des variétés algébriques. *J. Math. pures appl.* 36, 1–16 (1957).

[18] Tate, J. Groups of Galois type and cohomological dimension of fields. (To appear.)

[19] Washnitzer, A. Geometric syzygies. *Amer. J. Math.* 81, 171–248 (1959).

[20] Weil, A. Number of solutions of equations in finite fields. *Bull. Amer. Math. Soc.* 55, 497–508 (1949).

[20a] Weil, A. Abstract versus classical algebraic geometry. *Proc. Int. Congr. Math. Amst.* 3, 550–558 (1954).

[21] Zariski, O. Theory and applications of holomorphic functions on algebraic varieties over arbitrary groundfields. *Mem. Amer. Math. Soc.* no. 5 (1951).

[22] Zariski, O. Scientific report on the second summer Institute. III. Algebraic sheaf theory. *Bull. Amer. Math. Soc.* 62, 117–141 (1956).

[23] Auslander, M. and Buchsbaum, R. A. Codimension and multiplicity. *Ann. Math.* (2), 68, 625–657 (1958).

KOMPLEXE MANNIGFALTIGKEITEN

Von F. HIRZEBRUCH

1. Einleitung

Vor etwa 10 Jahren veröffentlichten Ehresmann und Hopf Arbeiten[6,14], in denen untersucht wurde, ob gegebene differenzierbare Mannigfaltigkeiten eine komplexe Struktur zulassen. Dabei führten sie die fast-komplexen Mannigfaltigkeiten als *Verallgemeinerung* der komplexen ein. *Beispiele* für komplexe Mannigfaltigkeiten werden durch die algebraischen Mannigfaltigkeiten geliefert. Damit sollen in diesem Vortrag kompakte komplexe Mannigfaltigkeiten gemeint sein, die sich holomorph und singularitätenfrei in einen komplexen projektiven Raum einbetten lassen.

Von komplexen Mannigfaltigkeiten selbst wird kaum die Rede sein. Es soll einerseits über Resultate von Milnor[16] über fast-komplexe Mannigfaltigkeiten berichtet werden und andererseits über die neue Formulierung des Satzes von Riemann–Roch für algebraische Mannigfaltigkeiten beliebiger Dimension durch Grothendieck[2]. Milnors Untersuchungen stehen in engem Zusammenhang mit der Thomschen Cobordisme-Theorie[21]. Grothendieck hat den in [13] bewiesenen Satz von Riemann–Roch weiter verallgemeinert. Seine allgemeinere Formulierung ermöglichte einen rein algebraischen Beweis für Grundkörper beliebiger Charakteristik. Bindeglied zwischen den Überlegungen von Milnor und Grothendieck ist für diesen Vortrag die Theorie der Chernschen Klassen. Ich beginne daher damit, daß ich an die Definition der fast-komplexen Mannigfaltigkeiten und der Chernschen Cohomologieklassen einer solchen Mannigfaltigkeit erinnere.

Falls nichts Gegenteiliges gesagt wird, sollen Mannigfaltigkeiten immer kompakt sein. Es ist für das Folgende wichtig, ausdrücklich darauf hinzuweisen, daß Mannigfaltigkeiten im allgemeinen *nicht* als *zusammenhängend* vorausgesetzt werden.

2. Fast-komplexe Mannigfaltigkeiten[1,20]

Es sei X eine $2n$-dimensionale differenzierbare Mannigfaltigkeit. Der Raum T_x der Tangentialvektoren an X im Punkte $x \in X$ ist ein $2n$-dimensionaler Vektorraum über dem Körper R der komplexen Zahlen. Eine fast-komplexe Struktur auf X ist ein stetiges Tensorfeld J, das jedem $x \in X$ einen R-Endomorphismus J_x von T_x mit $J_x \circ J_x = -Id$ zuordnet.

Durch J_x wird T_x wie folgt mit der Struktur eines Vektorraumes über dem Körper C der komplexen Zahlen versehen: Für $a+bi \in C$ (a, b reell) und $v \in T_x$ wird definiert

$$(a+bi)\,v = a+b(J_x v).$$

Eine fast-komplexe Mannigfaltigkeit ist eine differenzierbare Mannigfaltigkeit mit fast-komplexer Struktur. Die Aussage 'Die Vektoren $v_1, \ldots, v_p \in T_x$ sind linear-unabhängig über C' ist für eine fast-komplexe X sinnvoll.

3. Chernsche Klassen [1, 13, 20]

Auf einer orientierten (kompakten) differenzierbaren Mannigfaltigkeit X gibt es immer Vektorfelder mit endlich vielen Singularitäten (Nullstellen). Jeder Singularität ist eine ganze Zahl als Vielfachheit zugeordnet. Ein alter Satz von Hopf besagt, daß die Anzahl der Singularitäten eines Vektorfeldes (gezählt mit den richtigen Vielfachheiten) unabhängig von dem Vektorfeld ist: Diese Anzahl ist immer gleich der Eulerschen Charakteristik $e(X)$. Unter einem p-Feld auf X verstehen wir ein p-tupel von Vektorfeldern auf X. Nun sei X mit einer fast-komplexen Struktur versehen. Ein Punkt $x \in X$ heißt Singularität eines gegebenen p-Feldes, wenn die p-Vektoren des p-Feldes in x als Vektoren von T_x linear-abhängig über C sind. Es sei n die komplexe Dimension von X und $1 \leqslant p \leqslant n$. Die Singularitätenmenge eines p-Feldes ist im allgemeinen ein $2(p-1)$-dimensionaler Zyklus. Die (ganzzahlige) Homologieklasse dieses Zyklus ist unabhängig von der Wahl des p-Feldes. Sie heißt Chernsche Homologieklasse der p-Felder. Die Chernsche Homologieklasse der 1-Felder ist 0-dimensional. Zu ihrer Definition ist eine fast-komplexe Struktur nicht erforderlich.

Die fast-komplexe Mannigfaltigkeit X ist in natürlicher Weise orientiert, da jeder Tangentialraum T_x als Vektorraum über C orientiert ist. Also ist der Chernschen Homologieklasse der p-Felder durch den Poincaréschen Dualitätssatz, der ja eine kanonische Isomorphie zwischen der ganzzahligen r-dimensionalen Homologiegruppe und der ganzzahligen Cohomologiegruppe $H^{2n-r}(X, Z)$ herstellt, eine Cohomologieklasse $c_{n-p+1} \in H^{2(n-p+1)}(X, Z)$ zugeordnet. Damit sind die Chernschen Cohomologieklassen c_i ($1 \leqslant i \leqslant n$) von X definiert ($c_i \in H^{2i}(X, Z)$).

Wir haben vorstehend die Hindernis-Definition der Chernschen Klassen angedeutet und dabei die Homologiesprache verwendet. Natürlich ist die Cohomologiesprache für die Hindernistheorie besser geeignet: Man trianguliere X. Es gibt dann immer singularitätenfreie

p-Felder über dem $(2n - 2p + 1)$-dimensionalen Skelett, und das Hindernis gegen singularitätenfreie Fortsetzung auf das $(2n - 2p + 2)$-Skelett ist eine Cohomologieklasse der Dimension $2(n - p + 1)$ mit Koeffizienten in der $(2n - 2p + 1)$-ten Homotopiegruppe der Stiefelschen Mannigfaltigkeit aller p-tupel komplex-linear-unabhängiger Vektoren des $\mathbf{C}^n$, die bekanntlich den kompakten homogenen Raum $\mathbf{U}(n)/\mathbf{U}(n - p)$ als Deformationsretrakt hat. $\pi_{2n-2p+1}(\mathbf{U}(n)/\mathbf{U}(n - p))$ ist die erste nicht verschwindende Homotopiegruppe von $\mathbf{U}(n)/\mathbf{U}(n - p)$. Sie ist unendlich-zyklisch. Man kann c_i auch definieren als das erste Hindernis gegen die Reduktion der Strukturgruppe des komplexen Tangentialbündels auf $\mathbf{U}(i - 1)$.

4. Chernsche Zahlen

X sei weiterhin fast-komplex und habe die komplexe Dimension n. Der Cohomologiering $H^*(X, \mathbf{Z})$ ist ein graduierter Ring mit den Cohomologiegruppen $H^j(X, \mathbf{Z})$ als direkten Summanden, der vermöge des Poincaréschen Dualitätssatzes zum Homologie-Schnittring der Mannigfaltigkeit X isomorph ist. Das Einselement 1 von $H^*(X, \mathbf{Z})$ gehört zu $H^0(X, \mathbf{Z})$ und entspricht beim Poincaréschen Isomorphismus dem Grundzyklus der orientierten X. Die Gruppen $H^{2i}(X, \mathbf{Z})$ gehören zum Zentrum des Cohomologieringes. Sofern wir also nur Chernsche Klassen miteinander multiplizieren, gilt das kommutative Gesetz.

Jeder Partition ω von n soll eine ganze Zahl zugeordnet werden: $\omega = (r_1, \ldots, r_j)$ wird von natürlichen Zahlen gebildet, deren Summe gleich n ist. $c_{r_1} c_{r_2} \ldots c_{r_j}$ ist dann eine Cohomologieklasse der Dimension $2n$, die auch mit c_ω bezeichnet werde. $c_\omega[X]$ soll den Wert von c_ω auf dem Grundzyklus der orientierten X andeuten. Jeder Partition ω von n ist also die ganze Zahl $c_\omega[X]$ zugeordnet. Das sind die Chernschen Zahlen von X. Wegen des in §3 erwähnten Satzes von Hopf ist $c_n[X]$ die Eulersche Charakteristik. Die Chernschen Zahlen, die, wie Beispiele zeigen, im allgemeinen von der fast-komplexen Struktur und nicht nur von der differenzierbaren Struktur abhängen ([1], § 13.9) sind also Verallgemeinerungen der Eulerschen Charakteristik, die natürlich nichts mit der fast-komplexen Struktur zu tun hat. Die Chernschen Zahlen lassen sich auch als Schnittzahlen Chernscher Homologieklassen interpretieren. Das entspricht mehr dem Vorgehen in der algebraischen Geometrie, wo die Chernschen Klassen ersetzt werden durch gewisse Äquivalenzklassen algebraischer Zyklen bezüglich rationaler Äquivalenz (kanonische Klassen). Zur Definition der Chernschen Zahlen auf rein algebraische Weise ist die Theorie des Schnittringes der Äquivalenzklassen algebraischer Zyklen erforderlich (vgl. die Arbeiten von Chow).

5. Chernsche Zahlen und arithmetisches Geschlecht[13]

X sei eine n-dimensionale algebraische Mannigfaltigkeit (wobei wir unter 'Dimension' einer algebraischen oder fast-komplexen Mannigfaltigkeit immer die komplexe Dimension verstehen wollen). g_i sei die Dimension über C des Vektorraumes aller holomorphen Differentialformen von X vom Grade i. Das arithmetische Geschlecht $\chi(X)$ ist gegeben durch

$$\chi(X) = \sum_{i=0}^{n} (-1)^i g_i.$$

Per definitionem ist g_0 die Dimension über C des Vektorraumes der holomorphen Funktionen. Da X kompakt ist, ist g_0 gleich der Anzahl der Zusammenhangskomponenten von X. Man kann $\chi(X)$ darstellen als Linearkombination der Chernschen Zahlen von X mit rationalen Koeffizienten. Für jede Dimension n hat man ein Polynom T_n vom Gewicht n in den c_i, das diese Linearkombination angibt. Es gilt für

$$n = 1: \quad \chi(X) = \tfrac{1}{2} c_1(X],$$

$$n = 2: \quad \chi(X) = \tfrac{1}{12}(c_2 + c_1^2)[X],$$

$$n = 3: \quad \chi(X) = \tfrac{1}{24} c_2 c_1[X],$$

$$n = 4: \quad \chi(X) = \tfrac{1}{720}(-c_4 + c_3 c_1 + 3c_2^2 + 4c_2 c_1^2 - c_1^4)[X].$$

Das ist der Satz über das Toddsche Geschlecht, der in[13] bewiesen wurde und einen Spezialfall des Satzes von Riemann–Roch darstellt.

Ein mehrfach-projektiver Raum ist ein Produkt von komplexen projektiven Räumen, dessen komplexe Dimension gleich n ist. Es gibt $\pi(n)$ mehrfach-projektive Räume der Dimension n, wo $\pi(n)$ die Anzahl der Partitionen von n ist. Die $\pi(n)$ Koeffizienten des Toddschen Polynoms T_n sind zum Beispiel dadurch gegeben, daß $T_n(c_1, ..., c_n)$ auf allen mehrfach-projektiven Räumen der Dimension n den Wert 1 annehmen muß. (Die Chernschen Zahlen der mehrfach-projektiven Räume sind wohlbekannt.) Für jede fast-komplexe Mannigfaltigkeit X können wir das Toddsche Geschlecht $T(X)$ als rationale Zahl durch die Toddsche Linearkombination T_n der Chernschen Zahlen von X definieren. Für algebraische X ist $\chi(X) = T(X)$, das Toddsche Geschlecht also ganzzahlig.

6. Chernsche Zahlen und Satz von Milnor[16]

Ein System von $\pi(n)$ ganzen Zahlen, oder genauer eine Abbildung der Menge der Partitionen von n in die ganzen Zahlen, muß gewisse Bedingungen erfüllen, um als das System der Chernschen Zahlen einer n-dimensionalen algebraischen Mannigfaltigkeit X auftreten zu können. Eine notwendige Bedingung wird durch die Ganzzahligkeit des Toddschen Geschlechtes gegeben, die für $n = 1$ besagt, daß die Eulersche Charakteristik gerade sein muß, und zum Beispiel für $n = 3$, daß $c_1 c_2[X]$ durch 24 teilbar sein muß. Für $n = 1$ ist die Teilbarkeit von $c_1[X]$ durch 2 auch die einzige Bedingung: Die positiven geraden Zahlen werden durch disjunkte Vereinigungen von Riemannschen Zahlenkugeln (projektiven Geraden) erhalten, die geraden Zahlen $\leqslant 0$ durch zusammenhängende Riemannsche Flächen vom Geschlecht $\geqslant 1$. Milnor hat sich folgende Frage gestellt:

Welche Systeme von $\pi(n)$ ganzen Zahlen treten als System der Chernschen Zahlen einer fast-komplexen X der komplexen Dimension n auf?

Milnor hat gezeigt, daß man für jedes n ein 'vollständiges' System von Kongruenzen für die Chernschen Zahlen aufstellen kann, deren Gültigkeit notwendig und hinreichend dafür ist, daß ein vorgegebenes System von $\pi(n)$ ganzen Zahlen als System der Chernschen Zahlen einer n-dimensionalen fast-komplexen X auftritt. Milnor beweist nämlich folgenden

Satz. Ein System von $\pi(n)$ ganzen Zahlen tritt dann und nur dann als System der Chernschen Zahlen einer n-dimensionalen fast-komplexen X auf, wenn es als System der Chernschen Zahlen einer n-dimensionalen algebraischen Y auftritt, die einer wohlbestimmten Menge $\mathfrak{M}$ von algebraischen Mannigfaltigkeiten angehört.

Zur Menge $\mathfrak{M}$ gehören die komplexen projektiven Räume, ferner die Hyperfläche $H_{(r,t)}$ des zweifach-projektiven Raumes $P_r(C) \times P_t(C)$ vom Doppelgrad $(1,1)$ und mit $r > 1$, $t > 1$. Milnor zeigt durch eine einfache Überlegung, daß es zu jeder algebraischen bzw. fast-komplexen X ein 'Negativum' gibt, das ist eine algebraische bzw. fast-komplexe Mannigfaltigkeit X' gleichdimensional mit X, so daß jede Chernsche Zahl von X' gleich dem negativen der entsprechenden Chernschen Zahl von X ist. Wir nehmen nun in die Menge $\mathfrak{M}$ noch jeweils ein Negativum der oben erwähnten Mannigfaltigkeiten auf. $\mathfrak{M}$ wird aus den bereits in $\mathfrak{M}$ aufgenommenen Mannigfaltigkeiten durch Summenbildung (disjunktes Vereinigen) und cartesische Produktbildung erzeugt.

Da man die Chernschen Zahlen der $\mathfrak{M}$ erzeugenden Mannigfaltigkeiten berechnen kann, und da man das Verhalten der Chernschen Zahlen bei Summen- und Produktbildung kennt, kann man durch Lösen gewisser endlich vieler linearer Gleichungen entscheiden, ob ein vorgegebenes System von $\pi(n)$ ganzen Zahlen als System der Chernschen

Zahlen einer n-dimensionalen $X \in \mathfrak{M}$ auftritt. Man wird so zu dem weiter oben erwähnten vollständigen System von Kongruenzen geführt. Es ist bisher nicht gelungen, dieses vollständige System für jedes n explizit niederzuschreiben. Gewisse Kongruenzen ergeben sich aus der algebraischen Geometrie (Satz von Riemann–Roch). Denn gilt eine Kongruenz für algebraische X, dann gilt sie nach dem Satz von Milnor auch für fast-komplexe X; zum Beispiel:

Korollar. Das Toddsche Geschlecht einer fast-komplexen Mannigfaltigkeit ist eine ganze Zahl.

Bisher war nur bekannt[11, 13], daß das Toddsche Geschlecht einer n-dimensionalen fast-komplexen X multipliziert mit 2^{n-1} eine ganze Zahl ist. Für eine algebraische X kann man auch die Zahlen

$$\chi^p(X) = \sum_{q=0}^{n} (-1)^q h^{p,q}$$

durch Linearkombinationen Chernscher Zahlen ausdrücken. Das gibt weitere Kongruenzen für fast-komplexe X. So ist zum Beispiel für eine 4-dimensionale algebraische X

$$4\chi(X) - \chi^1(X) = \tfrac{1}{12}(2c_4 + c_3 c_1) [X].$$

Also ist für eine 4-dimensionale fast-komplexe X die Zahl

$$(2c_4 + c_3 c_1) [X]$$

durch 12 teilbar.

Korollar. Die Eulersche Charakteristik einer 4-dimensionalen fast-komplexen X, deren zweite Bettische Zahl verschwindet, ist durch 6 teilbar. Insbesondere ist die quaternionale projektive Ebene nicht fast-komplex (vgl. [12]).

Milnor[16] gelangt zu dem oben angegebenen Satz, indem er für fast-komplexe Mannigfaltigkeiten ein Analogon der Thomschen Cobordisme-Theorie[21] aufstellt. Milnor nennt zwei fast-komplexe Mannigfaltigkeiten gleicher Dimension c-äquivalent, wenn sie in ihren Chernschen Zahlen übereinstimmen. Die c-Äquivalenzklassen der n-dimensionalen fast-komplexen Mannigfaltigkeiten bilden bezüglich des disjunkten Vereinigens wegen der weiter oben erwähnten Existenz des Negativums eine Gruppe Γ^n. Milnor ordnet der unitären Gruppe $U(k)$ einen Raum $M(U(k))$ zu, so wie Thom es mit den orthogonalen Gruppen gemacht hat. Er zeigt, daß $\pi_{i+2k}(M(U(k)))$ für große k nur von i abhängt und zwar verschwindet diese 'stabile' Gruppe für ungerades i und ist isomorph zu Γ^n für $i = 2n$. Die Spektralsequenz von J. F. Adams (*Comm. Math. Helvet.* 32, 180–214 (1958)) wird entscheidend herangezogen.

Milnor zeigt weiter, daß $\Gamma^* = \sum_{n=0}^{\infty} \Gamma^n$ bezüglich des cartesischen Produktes einen graduierten Ring bildet, der dem graduierten Polynomring über Z in unbestimmten $x_1, x_2, x_3, \ldots$ isomorph ist. x_n entspricht dabei einem Element von Γ^n. Einen Isomorphismus von $Z[x_1, x_2, \ldots]$ auf Γ^* erhält man, indem man x_n irgendeine n-dimensionale fast-komplexe Mannigfaltigkeit X_n zuordnet, deren Chernsche Zahlen eine gewisse Bedingung erfüllen. Um diese Bedingung zu beschreiben, führen wir das Polynom $s_n(\sigma_1, \ldots, \sigma_n)$ ein, das die symmetrische Funktion $t_1^n + t_2^n + \ldots + t_n^n$ durch die elementar-symmetrischen Funktionen σ_i in den t_j ausdrückt. Dann definieren wir $s(X)$ für eine n-dimensionale fast-komplexe X als Linearkombination Chernscher Zahlen folgendermaßen:

$$s(X) = s_n(c_1, \ldots, c_n)\,[X] \quad (c_i \text{ Chernsche Klasse von } X).$$

X_n muß folgende Bedingung erfüllen: Ist $n+1$ keine Primzahlpotenz, dann $s(X_n) = \pm 1$. Ist $n+1 = q^r$ mit q Primzahl, dann $s(X_n) = \pm q$. Eine Folge von Mannigfaltigkeiten X_n läßt sich aus den oben angegebenen Erzeugenden der Klasse $\mathfrak{M}$ konstruieren, da

$$s(\mathbf{P}_n(\mathbf{C})) = n+1 \quad \text{und} \quad s(\mathbf{H}_{r,t}) = -\binom{r+t}{r} \quad (r > 1, t > 1).$$

In der Arbeit [16] erhält Milnor mit denselben Methoden auch neue Informationen über den Thomschen Cobordisme-Ring Ω^*. Er zeigt, daß Ω^* nur 2-Torsion besitzt und daß Ω^* modulo dem Ideal der Torsionselemente die Struktur eines Polynomringes besitzt.

7. Bemerkungen zum Satz von Milnor

(1) Zunächst geben wir für $1 \leqslant n \leqslant 4$ ein vollständiges System von Kongruenzen für die Chernschen Zahlen einer n-dimensionalen fastkomplexen Mannigfaltigkeit an.

$n = 1$: $\quad c_1 \equiv 0 \mod 2$.

$n = 2$: $\quad c_2 + c_1^2 \equiv 0 \mod 12$.

$n = 3$: $\quad c_2 c_1 \equiv 0 \mod 24, \quad c_3 \equiv c_1^3 \equiv 0 \mod 2$.

$n = 4$: $\quad -c_4 + c_3 c_1 + 3c_2^2 + 4c_2 c_1^2 - c_1^4 \equiv 0 \mod 720,$

$\qquad\qquad c_2 c_1^2 + 2c_1^4 \equiv 0 \mod 12, \quad -2c_4 + c_3 c_1 \equiv 0 \mod 4$.

(2) Aus den Milnorschen Überlegungen kann man folgern, daß es eine natürliche Zahl gibt, so daß jedes System von $\pi(n)$ ganzen Zahlen multipliziert mit dieser Zahl als System der Chernschen Zahlen einer

n-dimensionalen fast-komplexen Mannigfaltigkeit auftritt, und daß die kleinste natürliche Zahl γ_n mit dieser Eigenschaft gleich dem Nenner des n-ten Toddschen Polynoms ist (siehe [13]):

$$\gamma_n = \prod_q q^{[n/(q-1)]} \quad (q \text{ durchläuft alle Primzahlen}).$$

Analoges gilt für Pontrjaginsche Zahlen und differenzierbare Mannigfaltigkeiten. Die Zahl $\overline{N}_k$, nach der in [13], S. 80 gefragt wird, ist gleich dem Nenner des Polynoms L_k:

$$\overline{N}_k = \prod_q q^{[2k/(q-1)]} \quad (q \text{ durchläuft die ungeraden Primzahlen}).$$

(3) Die Frage, wann ein System von $\pi(n)$ ganzen Zahlen als System der Chernschen Zahlen einer *zusammenhängenden* fast-komplexen (bzw. algebraischen) Mannigfaltigkeit auftritt, bleibt ungelöst. Wahrscheinlich müssen dann für die Chernschen Zahlen neben den Kongruenzen noch gewisse Ungleichungen erfüllt werden.

(4) Von Milnor und dem Verfasser wurde in Gesprächen der Begriff der verallgemeinerten fast-komplexen Struktur einer differenzierbaren Mannigfaltigkeit X eingeführt und angewandt: Eine solche Struktur ist eine komplexe Struktur (zur Terminologie [1], § 7.3) einer trivialen Erweiterung des reellen Tangentialbündels von X. Unter einer trivialen Erweiterung des Tangentialbündels wird dabei seine Whitneysche Summe mit einem trivialen reellen Vektorraum-Bündel verstanden. Für verallgemeinerte fast-komplexe Mannigfaltigkeiten sind ebenfalls Chernsche Klassen und, falls die Dimension von X gerade ist, auch Chernsche Zahlen definiert. Für die im vorigen Paragraphen skizzierten Untersuchungen von Milnor ist der Begriff der verallgemeinerten fast-komplexen Mannigfaltigkeit adäquater: Γ^n ist auch isomorph zur Gruppe der c-Äquivalenzklassen der verallgemeinerten fast-komplexen Mannigfaltigkeiten der komplexen Dimension n, so daß zum Beispiel der Satz über die Ganzheit des Toddschen Geschlechtes auch für verallgemeinerte fast-komplexe Mannigfaltigkeiten richtig ist. Für eine verallgemeinerte fast-komplexe Mannigfaltigkeit X der komplexen Dimension n ist $c_n[X]$ im allgemeinen von der Eulerschen Charakteristik $e(X)$ verschieden. So besitzt zum Beispiel die Sphäre S_{2n} eine verallgemeinerte fast-komplexe Struktur mit $c_n[S_{2n}] = 0$. Aber das zweite Korollar in § 6 gilt, wenn man 'fast-komplex' durch 'verallgemeinert fast-komplex' und 'Eulersche Charakteristik' durch $c_4[X]$ ersetzt. Aus den Pontrjaginschen Klassen der quaternionalen projektiven Ebene $P_2(K)$ berechnet man, daß $c_4[P_2(K)]$ für eine verallgemeinerte fast-

komplexe Struktur nur ± 3 sein kann[12], so daß also $\mathbf{P}_2(\mathbf{K})$ keine verallgemeinerte fast-komplexe Struktur besitzt. Mit Hilfe der bekannten Einbettung von $\mathbf{P}_{n-1}(\mathbf{K})$ in $\mathbf{P}_n(\mathbf{K})$ beweist man das entsprechende für die höher-dimensionalen quaternionalen projektiven Räume: Das Normalbündel von $\mathbf{P}_{n-1}(\mathbf{K})$ in $\mathbf{P}_n(\mathbf{K})$ besitzt eine komplexe Struktur. Da es zu jedem komplexen Vektorraum-Bündel ξ ein komplexes Vektorraum-Bündel mit derselben Basis gibt, dessen Whitneysche Summe mit ξ ein triviales Bündel ist, würde nämlich aus der Existenz einer verallgemeinerten fast-komplexen Struktur auf $\mathbf{P}_n(\mathbf{K})$ die Existenz einer solchen Struktur auf $\mathbf{P}_{n-1}(\mathbf{K})$ folgen. Man erhält so den

Satz. Die quaternionalen projektiven Räume $\mathbf{P}_n(\mathbf{K})$, versehen mit der üblichen differenzierbaren Struktur, besitzen für $n \geqslant 2$ keine verallgemeinerte fast-komplexe (und damit auch keine fast-komplexe) Struktur.

Dies verallgemeinert einen in [12] angegebenen Satz.

(5) Die Existenz von fast-komplexen bzw. verallgemeinerten fast-komplexen Strukturen auf einer differenzierbaren X kann auch mit Hilfe der Theorie der (höheren) Hindernisse untersucht werden, die gewisse Cohomologieklassen von X mit den Homotopiegruppen

$$\pi_i(\mathrm{SO}(2n)/\mathrm{U}(n)), \quad i \leqslant 2n-1,$$

bzw. mit den stabilen Homotopiegruppen $\pi_i(\mathrm{SO}(2m)/\mathrm{U}(m))$, m groß, als Koeffizienten sind. Da auf Grund der Resultate von Bott[3-5] alle diese Homotopiegruppen wohlbekannt sind, kann man hoffen, daß die Hindernistheorie sich systematisch durchführen läßt. (Im fast-komplexen Fall ist nur die Gruppe $\pi_{2n-1}(\mathrm{SO}(2n)/\mathrm{U}(n))$ nicht stabil, aber auch diese Gruppe läßt sich mit Hilfe der Bottschen Resultate explizit angeben.)

8. Komplexe Vektorraum-Bündel[1, 13]

Ein komplexes Vektorraum-Bündel ξ mit der typischen Faser $\mathbf{C}^q$ hat einen Totalraum E_ξ, eine Basis B_ξ und eine (stetige) Projektion π_ξ von E_ξ auf B_ξ. Ferner ist jede Faser $\pi_\xi^{-1}(x)$, wo $x \in B_\xi$, mit der Struktur eines q-dimensionalen Vektorraumes über $\mathbf{C}$ versehen. Schließlich sind eine offene Überdeckung $\{U_j\}$ von B_ξ und q Abbildungen $s_i^{(j)} \colon U_j \to E_\xi$, $(i = 1, \ldots, q)$, für jedes j ausgezeichnet, die Schnitte sind, d.h. $\pi_\xi \circ s_i^{(j)} = \text{Identität}$, und lokale Produktdarstellungen von ξ liefern, d.h. für jedes j ist die durch

$$(x, \lambda_1, \ldots, \lambda_q) \to \sum_{i=1}^q \lambda_i s_i^{(j)}(x) \quad (\lambda_i \in \mathbf{C})$$

gelieferte Abbildung $U_j \times \mathbf{C}^q \to \pi_\xi^{-1}(U_j)$ ein Homöomorphismus.

Für ξ sind Chernsche Klassen $c_i(\xi) \in H^{2i}(B_\xi, \mathbf{Z})$ $(1 \leqslant i \leqslant q)$ analog zum speziellen Fall des komplexen Tangentialbündels einer fast-komplexen Mannigfaltigkeit definiert (vgl. § 3). In der Formulierung des Satzes von Riemann–Roch ist eine gewisse gemischt-dimensionale Cohomologieklasse von B_ξ von Bedeutung, der sogenannte Chernsche Charakter von ξ, der mit ch(ξ) bezeichnet wird und (im Falle eines endlich-dimensionalen Basisraumes B_ξ) ein Element des Cohomologie-ringes $H^*(B_\xi, \mathbf{Z}) \otimes \mathbf{Q}$ ist. (Wie üblich bezeichnet $\mathbf{Q}$ die rationalen Zahlen.) Die Definition von ch(ξ) erfolgt mit Hilfe der in § 6 angeführten Polynome s_k.

$$\mathrm{ch}\,(\xi) = q + \sum_{k=1}^{\infty} (k!)^{-1} s_k(c_1(\xi), \ldots, c_k(\xi)).$$

Dabei sind die Chernschen Klassen $c_k(\xi)$ für $k > q$ ($\mathbf{C}^q$ ist typische Faser von ξ) gleich 0 zu setzen. Die Summe ist endlich, wenn B_ξ endlich-dimensional ist.

Schließlich erinnern wir an die Definition des holomorphen Vektor-raum-Bündels. Diese erfolgt genauso wie oben. E_ξ, B_ξ sind jetzt komplexe Mannigfaltigkeiten oder komplexe Räume, π_ξ und die $s_\xi^{(j)}$ sind holomorphe Abbildungen.

9. Der Satz von Riemann–Roch[13]

Es sei X eine algebraische Kurve (kompakte Riemannsche Fläche). Ein Divisor D ist eine endliche formale Linearkombination von Punkten von X mit ganzzahligen Koeffizienten.

$$D = m_1\mathfrak{p}_1 + \ldots + m_k\mathfrak{p}_k.$$

Das Problem von Riemann–Roch ist, die Dimension des Vektorraumes über $\mathbf{C}$ aller derjenigen auf X meromorphen Funktionen f zu bestimmen, deren Divisor plus dem gegebenen Divisor D ein nicht-negativer Divisor ist, d. h. ein Divisor mit ausschließlich nicht-negativen Vielfachheiten. Dieser Vektorraum wird mit $\mathfrak{L}(D)$ bezeichnet. Seine Dimension ist immer endlich. Jedem Divisor D wird ein holomorphes Geradenbündel $\{D\}$ (Vektorraum-Bündel mit $\mathbf{C} = \mathbf{C}^1$ als Faser und X als Basis) so zuge-ordnet, daß $\mathfrak{L}(D)$ dem Vektorraum über $\mathbf{C}$ aller globalen holomorphen Schnitte von $\{D\}$ kanonisch isomorph ist. Das führt zu folgender Verall-gemeinerung des Problems von Riemann–Roch: X sei eine algebraische Mannigfaltigkeit und ξ ein holomorphes Vektorraum-Bündel über X. *Man bestimme die Dimension des Vektorraumes über $\mathbf{C}$ der globalen holo-morphen Schnitte von ξ.* Es gibt keine allgemeine Antwort auf diese Frage.

Erinnern wir zunächst an den Begriff der analytischen Garbe über einer komplexen Mannigfaltigkeit X. Einfachstes Beispiel ist die Garbe $\mathcal{O}_X$, die durch das Garbendatum beschrieben wird, das jeder offenen Menge U von X, den Ring der in U holomorphen Funktionen zuordnet. Nachdem die Garbe $\mathcal{O}_X$ eingeführt ist, kann man den gerade erwähnten Ring identifizieren mit $\Gamma(U, \mathcal{O}_X)$, dem Ring der Schnitte über U der Garbe $\mathcal{O}_X$. Eine analytische Garbe kann durch ein Garbendatum angegeben werden, das jeder offenen Menge U von X einen Modul über $\Gamma(U, \mathcal{O}_X)$ zuordnet. Jedes holomorphe Vektorraum-Bündel ξ über X bestimmt eine analytische Garbe $\mathcal{O}(\xi)$ vermittels des Garbendatums, welches jeder offenen Menge U von X den $\Gamma(U, \mathcal{O}_X)$-Modul der holomorphen Schnitte von ξ über U zuordnet. Falls ξ das triviale Geradenbündel ist, dann ist $\mathcal{O}(\xi) = \mathcal{O}_X$. Die analytischen Garben heißen auch Garben von $\mathcal{O}_X$-Moduln[18]. Homomorphismen solcher Garben sind per definitionem immer $\mathcal{O}_X$-Homomorphismen[18].

Es sei nun X wieder eine algebraische Mannigfaltigkeit und ξ ein holomorphes Vektorraum-Bündel über X. Dann bezeichnet man mit $\chi(X, \xi)$ die alternierende Summe der Dimensionen der Cohomologiegruppen von X mit Koeffizienten in $\mathcal{O}(\xi)$. Diese Cohomologiegruppen sind endlich-dimensionale Vektorräume über $\mathbf{C}$.

$$\chi(X, \xi) = \sum_{i=0}^{\infty} (-1)^i \dim_{\mathbf{C}} H^i(X, \mathcal{O}(\xi))$$

$\dim_{\mathbf{C}} H^0(X, \mathcal{O}(\xi))$ ist dabei die Zahl, nach der im Problem vom Riemann–Roch gefragt wird. $H^i(X, \mathcal{O}(\xi))$ verschwindet, wenn i größer als die komplexe Dimension von X ist. Wenn ξ das triviale Geradenbündel ist, dann ist $\chi(X, \xi)$ das in § 5 betrachtete arithmetische Geschlecht $\chi(X)$. Der Satz von Riemann–Roch besagt, daß sich $\chi(X, \xi)$ durch die Chernschen Klassen von X und ξ ausdrücken läßt. Die Chernschen Klassen von X sind natürlich diejenigen des Tangentialbündels von X.

Wir bezeichnen mit c_i die Chernschen Klassen von X und führen die totale Toddsche Klasse ein durch

$$\mathscr{T}(X) = \sum_{j=0}^{\infty} T_j(c_1, \ldots, c_j), \quad \text{wo} \quad T_0 = 1 \in H^0(X, \mathbf{Z}).$$

Die T_j sind die in § 5 betrachteten Toddschen Polynome. Wir multiplizieren den Chernschen Charakter ch (ξ) (siehe § 8), mit $\mathscr{T}(X)$ und erhalten ein (gemischt-dimensionales) Element des Cohomologieringes $H^*(X, \mathbf{Z}) \otimes \mathbf{Q}$.

$$\mathrm{ch}\,(\xi) \mathscr{T}(X) \in H^*(X, \mathbf{Z}) \otimes \mathbf{Q}.$$

Ist α ein beliebiges Element von $H^*(X, \mathbf{Z}) \otimes \mathbf{Q}$ und hat X die komplexe

Dimension n, dann wird unter $\alpha[X]$ der Wert der $2n$-dimensionalen Komponente von α auf dem Grundzyklus der orientierten X verstanden. $\alpha[X]$ ist eine rationale Zahl.

Der Satz von Riemann–Roch[13] *besagt*

$$\chi(X,\xi) = (\mathrm{ch}\,(\xi)\,\mathscr{T}(X))\,[X].$$

Die vorstehende Gleichung ist eine Verallgemeinerung der in § 5 betrachteten Übereinstimmung von Toddschem und arithmetischem Geschlecht. Wenn ξ das triviale Geradenbündel ist, dann ist nämlich $\mathrm{ch}\,(\xi) = 1$. Per definitionem ist $\mathscr{T}(X)\,[X]$ das Toddsche Geschlecht $T(X)$.

Die rationale Zahl $(\mathrm{ch}\,(\xi)\,\mathscr{T}(X))\,[X]$ ist auch für eine fast-komplexe Mannigfaltigkeit X und ein (stetiges) komplexes Vektorraum-Bündel ξ wohl-definiert. Sie werde mit $T(X,\xi)$ bezeichnet. Es ist keineswegs klar, daß diese Zahl ganz ist. Der Satz von Riemann–Roch impliziert ihre Ganzheit für eine algebraische X und ein holomorphes ξ. Den Überlegungen von [1], wo gezeigt wurde, daß $T(X,\xi)$ multipliziert mit einer geeigneten Potenz von 2 ganz ist, folgend, kann man die *Ganzheit von $T(X,\xi)$ für eine fast-komplexe X und ein stetiges ξ* beweisen. Man hat dazu die Resultate von Milnor (§ 6) heranzuziehen (mit der in § 7 (4) angegebenen Verallgemeinerung).

Ganzzahligkeitsaussagen über $T(X,\xi)$ lassen sich, wie in [1] gezeigt wird, zu Sätzen über kompakte orientierte differenzierbare Mannigfaltigkeiten umformen. Man erhält so folgenden Satz, der in [1] nur 'bis auf Potenzen von 2' bewiesen werden konnte.

Satz. Es sei X eine kompakte orientierte differenzierbare Mannigfaltigkeit, $p_i \in H^{4i}(X,\mathbf{Z})$ sei die i-te Pontrjaginsche Klasse von X und $\{\hat{A}_j(p_1,\ldots,p_j)\}$ sei die multiplikative Folge von Polynomen, die zur Potenzreihe

$$\frac{\frac{1}{2}\sqrt{z}}{\sinh\frac{1}{2}\sqrt{z}}$$

gehört. Ferner seien d ein Element von $H^2(X,\mathbf{Z})$, das $\mathrm{mod}\,2$ gleich der Stiefel–Whitneyschen Klasse $w_2 \in H^2(X,\mathbf{Z}_2)$ ist, und ξ ein komplexes Vektorraum-Bündel über X. Dann ist

$$\left(\mathrm{ch}\,(\xi)\,e^{\frac{1}{2}d}\sum_{j=0}^{\infty}\hat{A}_j(p_1,\ldots,p_j)\right)[X]$$

eine ganze Zahl.

Die hier kurz referierten Dinge sollen in einer an [1] anschließenden gemeinsamen Arbeit mit A. Borel dargestellt werden ([1], Part III). Der vorstehende Satz läßt manche Folgerungen zu, so ergibt sich der **Satz** von Bott[4], daß die Chernsche Klasse c_n eines komplexen Vektorraum-Bündels über der Sphäre S_{2n} durch $(n-1)!$ teilbar ist. Ferner hat Milnor[15,17] mit Hilfe des Satzes Ergebnisse über die stabilen Homotopiegruppen der Sphären erhalten.

10. Kohärente analytische Garben

Es sei X eine komplexe Mannigfaltigkeit. Der Begriff der kohärenten analytischen Garbe ist lokaler Natur. Zu diesen Garben gehören einmal die Garben $\mathcal{O}(\xi)$, wo ξ ein holomorphes Vektorraum-Bündel ist, und allgemeiner die analytischen Garben, die sich lokal als Kokern eines lokalen Homomorphismus $\mathcal{O}(\xi) \to \mathcal{O}(\xi')$ darstellen lassen[18], wo ξ, ξ' triviale Vektorraum-Bündel sind. Das sind alle kohärenten analytischen Garben. Ist X algebraisch, zusammenhängend, n-dimensional, dann lassen sich die kohärenten analytischen Garben kennzeichnen, als diejenigen analytischen Garben G, zu denen es holomorphe Vektorraum-Bündel $\xi_0, \xi_1, \ldots, \xi_n$ über X und globale Garbenhomomorphismen gibt, so daß man eine exakte Sequenz

$$0 \to \mathcal{O}(\xi_n) \to \mathcal{O}(\xi_{n-1}) \to \ldots \to \mathcal{O}(\xi_0) \to G \to 0 \tag{1}$$

hat[2]. Eine solche Sequenz nennt man eine Auflösung von G in Vektorraum-Bündel. Diese Auflösungen hängen mit dem Syzygien-Satz von Hilbert zusammen. Von nun an seien die auftretenden Räume X, Y, zusammenhängend.

X sei weiterhin algebraisch und G kohärent analytisch über X. Grothendieck[2] definiert den Chernschen Charakter von G mittels einer Auflösung von G in Vektorraum-Bündel wie folgt:

$$\text{ch}(G) = \sum_{i=0}^{n} (-1)^i \text{ch}(\xi_i) \in H^*(X, \mathbf{Z}) \otimes \mathbf{Q}. \tag{2}$$

Er zeigt, daß diese Definition legitim ist, d. h. daß $\text{ch}(G)$ unabhängig von der Wahl der Auflösung ist. Wichtig ist dabei die Linearität des Chernschen Charakters: Wenn $0 \to \xi' \to \xi \to \xi'' \to 0$ eine exakte Sequenz von Vektorraum-Bündeln ist, dann ist

$$\text{ch}(\xi) = \text{ch}(\xi') + \text{ch}(\xi'').$$

Für die kohärente analytische Garbe G über der algebraischen X kann man die Zahl $\chi(X, G)$ wie im Falle eines Vektorraum-Bündels definieren

$$\chi(X, G) = \sum_{i=0}^{\infty} (-1)^i \dim_{\mathbf{C}} H^i(X, G).$$

Diese Summe enthält wieder nur endlich viele Glieder, die Dimensionen der Cohomologiegruppen sind endlich. Nimmt man für G eine Auflösung

(1), dann folgt aus elementaren Eigenschaften der Eulerschen Charakteristik, daß

$$\chi(X, G) = \sum_{j=0}^{n} (-1)^j \chi(X, \xi_j). \tag{3}$$

Wegen (2), (3) besitzt der Satz von Riemann–Roch die folgende Ausdehnung auf kohärente Garben über einer algebraischen X:

$$\chi(X, G) = (\operatorname{ch}(G) \mathscr{T}(X)) [X]. \tag{4}$$

11. Direkte Bilder analytischer Garben [2, 9]

Im Satz von Riemann–Roch (4) tritt die gemischt-dimensionale Cohomologieklasse $\operatorname{ch}(G)\mathscr{T}(X) \in H^*(X, \mathbf{Z}) \otimes \mathbf{Q}$ auf. Eigentlich spielt aber in (4) nur die Komponente dieser Cohomologieklasse eine Rolle, deren Dimension gleich $2 . \dim_{\mathbf{C}} X$ ist. Grothendieck hat dem Satz von Riemann–Roch nun eine allgemeinere Formulierung gegeben, in der die gesamte Klasse $\operatorname{ch}(G)\mathscr{T}(X)$ herangezogen wird: Grothendieck betrachtet zwei algebraische Mannigfaltigkeiten X und Y, eine kohärente analytische Garbe G über X und eine holomorphe Abbildung f von X in Y. Dann hat man einen *additiven* (nicht multiplikativen) Homomorphismus

$$f_* \colon \quad H^*(X, \mathbf{Z}) \to H^*(Y, \mathbf{Z})$$

und ebenso einen Homomorphismus f_* der mit $\mathbf{Q}$ tensorierten Gruppen. f_* ist folgendermaßen definiert. Man nimmt zu einer Cohomologieklasse α von X die ihr via Poincaré-Dualität entsprechende Homologieklasse, bildet diese durch f in Y ab, erhält eine Homologieklasse von Y und definiert dann $f_*\alpha$ als die der letzten Homologieklasse via Poincaré-Dualität entsprechende Cohomologieklasse. Offensichtlich ist $\dim f_*\alpha = \dim \alpha + 2 . (\dim_{\mathbf{C}} Y - \dim_{\mathbf{C}} X)$. Grothendieck versucht nun, $f_*(\operatorname{ch}(G) \mathscr{T}(X))$ in der Form $\operatorname{ch}(?) \mathscr{T}(Y)$ darzustellen. Dazu werden die direkten Bilder analytischer Garben benutzt.

Es seien X und Y komplexe Mannigfaltigkeiten (nicht notwendigerweise kompakt), G eine analytische Garbe über X und f eine holomorphe Abbildung von X in Y. Das q-te direkte Bild von G ist eine analytische Garbe über Y, die mit $f_*^q(G)$ bezeichnet wird. Zu ihrer Definition geben wir ein Garbendatum an: Für eine beliebige offene Menge U von Y betrachten wir die Cohomologiegruppe $H^q(f^{-1}(U), G)$. Diese ist ein Modul über den in $f^{-1}(U)$ holomorphen Funktionen. Da jede in U holomorphe Funktion zu einer in $f^{-1}(U)$ holomorphen Funktion geliftet werden kann, ist $H^q(f^{-1}(U), G)$ auch ein Modul über $\Gamma(U, \mathscr{O}_Y)$, dem Ring der in U holomorphen Funktionen. Die $\Gamma(U, \mathscr{O}_Y)$-Moduln $H^q(f^{-1}(U), G)$

liefern ein Garbendatum für die analytische Garbe $f_*^q(G)$ über Y. Diese direkten Bilder treten im wesentlichen bereits in den fundamentalen Arbeiten von Leray auf. $f_*^q(G)$ ist die Nullgarbe, wenn $q > 2\dim_{\mathbb{C}} X$.

Satz. Wenn X und Y algebraische Mannigfaltigkeiten sind (vgl. Einleitung), wenn $f: X \to Y$ holomorph ist und G eine kohärente analytische Garbe über X ist, dann sind die direkten Bilder $f_*^q(G)$ kohärente analytische Garben über Y.

Dieser Satz, über den wir gleich noch einige Bemerkungen machen werden, ermöglicht die folgende Definition

$$f_! \operatorname{ch}(G) = \sum_{q=0}^{\infty} (-1)^q \operatorname{ch}(f_*^q(G)).$$

In der Tat, die $f_*^q(G)$ sind kohärent, also sind ihre Chernschen Charaktere nach § 10 wohldefiniert. (Der Chernsche Charakter der Nullgarbe ist gleich 0.)

12. Der Satz von Riemann–Roch–Grothendieck[2]

Satz. Es seien X und Y algebraische Mannigfaltigkeiten (vgl. Einleitung), f eine holomorphe Abbildung von X in Y und G eine kohärente analytische Garbe über X. Dann gilt in $H^*(Y,\mathbb{Z}) \otimes \mathbb{Q}$ die folgende Gleichung:

$$f_*(\operatorname{ch}(G)\mathcal{T}(X)) = (f_! \operatorname{ch}(G))\mathcal{T}(Y).$$

(1) Dieser Satz etabliert eine starke Kovarianz-Eigenschaft der Toddschen Polynome. Wenn zum Beispiel X aus Y durch Aufblasen einer Untermannigfaltigkeit von Y entsteht (monoidale Transformation) und f die natürliche birationale Abbildung von X auf Y ist und $G = \mathcal{O}_X$ ist, dann ist $f_*^0(\mathcal{O}_X) = \mathcal{O}_Y$ und $f_*^q(\mathcal{O}_X) = 0$ für $q > 0$. In diesem Falle besagt die Grothendiecksche Formel

$$f_*(\mathcal{T}(X)) = \mathcal{T}(Y).$$

Es entsteht die Frage, ob diese Gleichung für jede birationale holomorphe Abbildung einer algebraischen X auf eine algebraische Y gilt.

(2) Der übliche Satz von Riemann–Roch[13] (vgl. die Formel (4) von § 10) ergibt sich, wenn man für Y einen Punkt nimmt und für f die konstante Abbildung. Dann ist $\mathcal{T}(Y) = 1$, ferner ist

$$\operatorname{ch}(f_*^q(G)) = \dim_{\mathbb{C}} H^q(X,G) \cdot 1.$$

Die Kohärenz der Bildgarben besagt hier, daß die Dimensionen der $H^q(X,G)$ über $\mathbb{C}$ endlich sind. Weiter beachte man, daß f_* auf allen

Komponenten von $\mathrm{ch}\,(G)\mathscr{T}(X)$, deren Dimension kleiner ist als $2\,.\dim_{\mathbb{C}}X$, verschwindet, und daß

$$f_*(\mathrm{ch}\,(G)\,\mathscr{T}(X)) = (\mathrm{ch}\,(G)\mathscr{T}(X))\,[X]\,.1\,.$$

(3) Die wichtigste Anwendung des Grothendieckschen Satzes ist wohl der übliche Satz von Riemann–Roch, da für diesen bisher kein algebraischer Beweis bekannt war. Die allgemeinere Formulierung hat den algebraischen Beweis möglich gemacht, der zunächst für den Fall einer Projektion $f\colon X\times P\to X$ (P ein projektiver Raum), und dann für Injektionen $f\colon X\to Y$ durchzuführen ist. Im zweiten Fall spielen die monoidalen Transformationen eine entscheidende Rolle. Wie gesagt, Grothendiecks Beweis ist rein algebraisch, er ist gültig für Grundkörper beliebiger Charakteristik. Man hat in Formulierung und Beweis des Satzes die analytischen Begriffe durch die entsprechenden algebraischen zu ersetzen (holomorph durch regulär, kohärent analytisch durch kohärent algebraisch usw.), die Zariski-Tolopogie und (anstelle des Cohomologieringes) den Schnittring (Chowschen Ring) der Äquivalenzklassen algebraischer Zyklen zu verwenden. Der Satz von § 11 wird ebenfalls ein rein algebraischer Satz. In diesem Vortrag haben wir durchweg die analytische Sprache verwendet, und der hier formulierte Satz von Grothendiek hat daher a priori (selbst bei Beschränkung auf Charakteristik 0) einen anderen Inhalt als der in [2] angegebene Satz von Riemann–Roch–Grothendieck. Wegen der Serreschen (GAGA)-Korrespondenzsätze[19] zwischen analytischer und algebraischer Geometrie waren wir aber berechtigt, die analytische Sprache zu verwenden.

13. Mögliche Verallgemeinerungen

Wie weit gilt der Satz von Riemann–Roch für komplexe Mannigfaltigkeiten? Die übliche Formulierung von § 9 ist sinnvoll für kompakte komplexe Mannigfaltigkeiten und holomorphe Vektorraum-Bündel ξ. Die Vermutung, daß die Gleichung

$$\chi(X,\xi) = (\mathrm{ch}\,(\xi)\,.\,\mathscr{T}\,(X))\,[X]$$

in diesem Falle richtig ist, wird bestärkt durch das neue Resultat (§ 9), daß die rechte Seite eine ganze Zahl ist.

Um zu einer Verallgemeinerung im Sinne von Grothendieck zu kommen, ist zunächst zu untersuchen, in welchen Fällen die direkten Bilder kohärenter analytischer Garben wieder kohärent sind. Darüber haben Grauert und Remmert[7–10] wichtige Resultate erhalten (sogar im Falle komplexer Räume). Man hat folgende allgemeine *Vermutung* (Grothen-dieck–Grauert–Remmert):

Bei einer eigentlichen holomorphen Abbildung f eines komplexen Raumes X in einen komplexen Raum Y sind die direkten Bilder einer kohärenten analytischen Garbe über X kohärente analytische Garben über Y.

X und Y werden hier nicht als kompakt vorausgesetzt. 'Eigentlich' bedeutet, daß für jede kompakte Teilmenge K von Y auch $f^{-1}(K)$ kompakt ist.

Grauert und Remmert[8] haben gezeigt, daß die obige Vermutung richtig ist, wenn X analytisch-vollständig und Y holomorph-vollständig ist. Zu den analytisch-vollständigen Mannigfaltigkeiten[8] gehören sowohl die holomorph-vollständigen (Steinschen) als auch die algebraischen Mannigfaltigkeiten. Die Klasse der analytisch-vollständigen Mannigfaltigkeiten ist abgeschlossen bezüglich Produktbildung und Bildung von Untermannigfaltigkeiten.

Aus der Gültigkeit der obigen Vermutung für analytisch-vollständiges X und holomorph-vollständiges Y ergibt sich leicht ihre Gültigkeit für analytisch-vollständiges X und beliebiges Y. Das schließt den Satz von § 11 ein, der aber leichter über die algebraische Geometrie zu erhalten ist (§ 12 (3)).

Weiter muß man den Chernschen Charakter einer kohärenten analytischen Garbe über X (bzw. Y) definieren können. Dazu braucht man die Auflösung einer solchen Garbe in (endlich-viele) holomorphe Vektorraum-Bündel über zusammenhängenden relativ-kompakten Teilmengen von X (bzw. Y). Nach mündlicher Mitteilung von Remmert ist das möglich, wenn X (bzw. Y) analytisch-vollständig ist.

Der Satz von Riemann–Grothendieck scheint richtig zu sein für analytisch-vollständige Mannigfaltigkeiten X, Y, eine holomorphe eigentliche Abbildung f: X → Y und eine kohärente analytische Garbe über X.

Ob Grauert, Grothendieck oder Remmert zur Stunde ein exakter Beweis bekannt ist, weiß der Verfasser nicht.

LITERATUR

[1] Borel, A. and Hirzebruch, F. Characteristic classes and homogeneous spaces. I. *Amer. J. Math.* 80, 458–538 (1958) and II (to appear).

[2] Borel, A. and Serre, J. P. Le théorème de Riemann-Roch. *Bull. Soc. Math. France*, 86, 97–136 (1958).

[3] Bott, R. The stable homotopy of the classical groups. *Proc. Nat. Acad. Sci., Wash.*, 43, 933–935 (1957).

[4] Bott, R. and Milnor, J. W. On the parallelizability of the spheres. *Bull. Amer. Math. Soc.* 64, 87–89 (1958).

[5] Bott, R. Applications of Morse theory to the homotopy of Lie groups. *Proc. Int. Congr. Math. Edinburgh*, 423–426 (1960).

[6] Ehresmann, Ch. Sur la théorie des espaces fibrés. *Colloque International de Topologie Algébrique, Paris*, 3–15 (1949).

[7] Grauert, H. and Remmert, R. Faisceaux analytiques cohérents sur le produit d'un espace analytique et un espace projectif. *C.R. Acad. Sci., Paris*, 245, 819–822 (1957).

[8] Grauert, H. and Remmert, R. Espaces analytiquement complets. *C.R. Acad. Sci., Paris*, 245, 882–885 (1957).

[9] Grauert, H. and Remmert, R. Bilder und Urbilder analytischer Garben. *Ann. Math.* (2), 68, 393–443 (1958).

[10] Grauert, H. and Remmert, R. Komplexe Räume. *Math. Ann.* 136, 245–318 (1958).

[11] Hirzebruch, F. On Steenrod's reduced powers, the index of inertia and the Todd genus. *Proc. Nat. Acad. Sci., Wash.*, 39, 951–956 (1953).

[12] Hirzebruch, F. Über die quaternionalen projektiven Räume. *S.B. bayer. Akad. Wiss.* 301–312 (1953).

[13] Hirzebruch, F. *Neue topologische Methoden in der algebraischen Geometrie.* Springer-Verlag, Berlin–Göttingen–Heidelberg, 1956.

[14] Hopf, H. Zur Topologie der komplexen Mannigfaltigkeiten. *Studies and Essays presented to R. Courant, New York*, pp. 167–185 (1948).

[15] Milnor, J. W. On the Whitehead homomorphism *J. Bull. Amer. Math. Soc.* 64, 79–82 (1958).

[16] Milnor, J. W. On the cobordisme ring Ω^*, and a complex analogue (to appear).

[17] Milnor, J. W. and Kervaire, M. A. On Bernoulli numbers, homotopy groups, and a theorem of Rochlin. *Proc. Int. Congr. Math. Edinburgh*, 454–458 (1960).

[18] Serre, J. P. Faisceaux algébriques cohérents. *Ann. Math.* (2), 61, 197–278 (1955).

[19] Serre, J. P. Géométrie algébrique et géométrie analytique. *Ann. Inst. Fourier*, 6, 1–42 (1956).

[20] Steenrod, N. *The Topology of Fibre Bundles.* Princeton University Press, Princeton, New Jersey, 1951.

[21] Thom, R. Quelques propriétés globales des variétés différentiables. *Comment. Math. Helv.* 28, 17–86 (1954).

Zusätze bei Korrektur

(1) Die Vermutung von § 13 über die direkten Bilder wurde inzwischen von H. Grauert bewiesen.

(2) Der Satz von Riemann-Roch-Grothendieck (§ 12) hat ein differenzierbares Analogon gefunden (M. F. Atiyah and F. Hirzebruch, *Bull. Amer. Math. Soc.*, to appear).

MATHEMATICAL LOGIC: CONSTRUCTIVE AND NON-CONSTRUCTIVE OPERATIONS

By S. C. KLEENE

1. Mathematical logic

Early in the century, especially in connection with Hilbert's treatment of geometry (1899), it was being said that the theorems of an axiomatic theory express truths about whatever systems of objects make the axioms true.

In the simplest case, a system S consists of a non-empty set D (the *domain*), in which there are distinguished certain *individuals*, and over which there are defined certain n-place *functions* (or operations) taking values in D, and certain n-place *predicates* (or properties and relations), i.e. functions taking propositions as values.

The *elementary* (or *first-order*) *predicate calculus* provides a language for discussing such systems. To a preassigned list of (non-logical) *constants* for the distinguished individuals, functions and predicates, we add the *propositional connectives* $\rightarrow$ ('implies' or 'if... then...'), & ('and'), v ('or'), $\neg$ ('not'), the *universal quantifier* (a) ('for all a (in D)'), and the *existential quantifier* (Ea) ('(there) exists (an) a (in D such that)').

For example, when S is the arithmetic of the natural numbers 0, 1, 2, ..., with 0, 1, $+$, $\cdot$, $=$, $>$ in their usual senses,

$$(\alpha) \quad a = b+1, \quad (\beta) \quad (Eb)(a = b+1), \quad (\gamma) \quad a > 0,$$

$$(\delta) \quad a > 0 \rightarrow (Eb)(a = b+1), \quad (\varepsilon) \quad (a)[a > 0 \rightarrow (Eb)(a = b+1)],$$

are *formulas*. Formula (α) (containing a, b, *free*) expresses a 2-place predicate (relation), (β)–(δ) (containing a free) express 1-place predicates (properties), and (ε) (containing no variable free, i.e. a *sentence*) expresses a proposition.

When (a, b) are $(3, 2)$, (α) is true. Hence when a is 3, (β) is true, also (γ); and hence by the truth table for $\rightarrow$ (right), (δ) is true. Similarly, for any other a, (δ) is true. Hence (ε) is true. Truth tables, which in principle go back to Peirce (1885) and Frege (1891), were first fully exploited by Łukasiewicz (1921) and Post (1921), and truth definitions generally by Tarski (1933).

$A \rightarrow B$		
A ＼ B	True	False
True	True	False
False	True	True

We need one elementary technical result of logic. In any formula, the quantifiers can be advanced (step by step) to the front, preserving the truth or falsity of the proposition, or of any value of the predicate, expressed. (For example,

$$[(a) A(a)] \to (a) B(a)$$

is equivalent to $\qquad (Ea)(b)[A(a) \to B(b)].)$

The resulting formula we call a *prenex form* of the original.

I. $\begin{Bmatrix} \text{Löwenheim (1915).} \\ \text{Skolem (1920).} \end{Bmatrix}$ *If* $\begin{Bmatrix} a\ sentence\ A\ is \\ sentences\ A_0, A_1, A_2, \dots\ are \end{Bmatrix}$ *true of a given system* S, *then* $\begin{Bmatrix} it\ is \\ they\ are\ all \end{Bmatrix}$ *true of a system* S_1 *with countable domain* D_1.

Proof. Say a prenex form of A is

$$(Eb)(c)(Ed)(e)(f)(Eg) A(b,c,d,e,f,g) \tag{i}$$

(all quantifiers shown). This being true of S with domain D, there are an individual β and (by the axiom of choice) functions $\delta(c)$ and $\gamma(c,e,f)$ such that

$$(c)(e)(f) A(\beta, c, \delta(c), e, f, \gamma(c,e,f)) \tag{ii}$$

is true. Now (ii), and hence (i), will remain true if we cut down the domain (without otherwise altering the functions and predicates) from D to its least subset D_1 containing β (and the distinguished individuals of S) and closed under δ, γ (and the functions of S). The new domain D_1 is countable; indeed all its members have names in the list $t_0, t_1, t_2, \dots$, of the distinct *terms* without variables formable using β, δ, γ and the symbols for the distinguished individuals and functions of S. (We can always arrange to have at least one individual, and one function, symbol.) For the version with $A_0, A_1, A_2, \dots$, we use different symbols in the role of β, δ, γ with each prenex form.

Continuing the example, (i) will be true of a system S_1 with domain D_1 whose members are named by $t_0, t_1, t_2, \dots$, if each of the expressions $A(\beta, t_c, \delta(t_c), t_e, t_f, \gamma(t_c, t_e, t_f))$ $(c,e,f = 0,1,2,\dots)$ is true; enumerate these (or for $A_0, A_1, A_2, \dots$, the expressions arising similarly from the various prenex forms) as $A^0, A^1, A^2, \dots$.

For the next theorem we simply try in all possible ways to make $A^0, A^1, A^2, \dots$ simultaneously true. We obtain the greatest freedom to do this by interpreting each term t_i as representing a different individual, say i. Thereby we can choose the value of each expression $P(t_{c_1}, \dots, t_{c_n})$ (P an n-place predicate symbol) as true or false independently of the others. Enumerate these (without repetitions) as $Q_0, Q_1, Q_2, \ldots$. Choosing their values successively can be correlated to following a path (indicated by arrows) in the tree (right); e.g. if we choose Q_0 true, Q_1 false, Q_2 false, ..., we follow the path $VV_0 V_{01} V_{011} \ldots$. As soon as the values already chosen make any one of

$$A^0, A^1, A^2, \ldots$$

false, we are defeated for that sequence of choices, and terminate the path.

Now by König's Unendlichkeitslemma (1926) (= a classical version of Brouwer's fan theorem, 1924), if (*Case* 1) arbitrarily long finite paths exist, there is an infinite path. (We follow such a path by choosing each time an arrow belonging to arbitrarily long finite paths.) Thereby we obtain the first alternative of:

II. *Either* (1) *all of* $A^0, A^1, A^2, \ldots$ $\left(and\ hence\ \begin{cases} A \\ all\ of\ A_0, A_1, A_2, \ldots \end{cases}\right)$ *are true of some system* S_1 *with the domain* $D_1 = \{0, 1, 2, \ldots\}$, *or else* (2) *some 'Herbrand conjunction'* $A^{j_1} \& \ldots \& A^{j_m}$

$\left(and\ hence\ \begin{cases} A \\ some\ A_{k_1} \& \ldots \& A_{k_n} \end{cases}\right)$

is false of every system S.

If (*Case* 2) there is a finite upper bound $b+2$ to the lengths of paths, then for each of the 2^{b+1} ways of choosing the values of $Q_0, \ldots, Q_b$ some particular A^j will be false. The conjunction $A^{j_1} \& \ldots \& A^{j_m}$ ($m \leqslant 2^{b+1}$) of these A^j's will be false for all 2^{b+1} ways, and thus of all systems S. Likewise A itself (or the conjunction $A_{k_1} \& \ldots \& A_{k_n}$ of those $A_0, A_1, A_2, \ldots$ from which $A^{j_1}, \ldots, A^{j_m}$ arise); for were A true of an S, we would be led as under I to values of $Q_0, \ldots, Q_b$ making $A^{j_1}, \ldots, A^{j_m}$ all true. (Here we need δ and γ for only finitely many arguments, symbolized by terms occurring in $A^{j_1} \& \ldots \& A^{j_m}$, so I is reproved without using the axiom of choice.)

II includes as much of Gödel's completeness theorem for the predicate calculus (1930), and of Herbrand's theorem (1930), as we can state in *model theory*. The theory of models concerns 'mutual relations between sentences of formalized theories and mathematical systems [*models*] in which these sentences hold' (Tarski, 1954–5).

Gödel's completeness theorem (II_G) has (2_G) $\begin{cases} \neg A \\ some \neg (A_{k_1} \& \ldots \& A_{k_n}) \end{cases}$ *is provable in the predicate calculus* in place of (2), and Herbrand's theorem (II_H) gives the equivalence of (2_G) to (2).

However, *if* we agree here that a 'proof' of a sentence should be a finite linguistic construction, recognizable as being made in accordance with preassigned rules and whose existence assures the 'truth' of the sentence in the appropriate sense, we already have (II), since the verification of (2) for a given $A^{j_1} \& \ldots \& A^{j_m}$ is such a construction.

What usual proofs of Gödel's completeness theorem add is that the *proof* of $\neg A$ (or $\neg(A_{k_1} \& \ldots \& A_{k_n})$) for (2_G) can be effected in a usual *formal system* of *axioms* and *rules of inference* for the predicate calculus as given in *proof theory*.

Proof theory is a modern version of the axiomatic-deductive method, which goes back to Pythagoras (reputedly), Aristotle and Euclid. Since Frege (1879), it has been emphasized that, in order to exclude hidden assumptions, the axioms and rules of inference should be specified by referring only to the *form* of the linguistic expressions (i.e. not to the interpretations or models); hence the term 'formal system'.

With Hilbert since 1904 appeared the idea of proving in a *metatheory* or *metamathematics* theorems about formal systems (cf. Hilbert–Bernays, 1934, 1939; Kleene, 1952). Thus we can talk of proving (metamathematically) that in (2_G) there is a (formal) proof of $\neg A$.

In *Hilbert's* metamathematics it was intended that only safe ('constructive' or 'finitary') methods should be used. That certain methods outrun intuition and even consistency, the mathematical public was forced to recognize by the paradoxes in which Cantor's set theory culminated in 1895. Hilbert hoped to save 'classical mathematics' (including the usual arithmetic and analysis and a suitably restricted axiomatized set theory), which he acknowledged to outrun intuition, by codifying it as a formal system, and proving this system *consistent* (i.e. that no 'contradictory' pair of sentences C and $\neg C$ are provable in it) by finitary metamathematics. Kronecker earlier (in the 1880's), and others later, proposed rather a direct redevelopment of mathematics on a less or more wide constructive basis, such as the intuitionistic (Brouwer, 1908; Heyting, 1956) or the operative (Lorenzen, 1950, 1955).

In a model S_1 as constructed above for II, = may not express equality (identity). (For I, it will if it does for S.) But *if* $A_0, A_1, A_2, \ldots$ include the usual axioms for equality, then the relation $\{x = y$ is true of the *above* $S_1\}$ will be an equivalence relation under which the equivalence classes will constitute the domain (countably infinite *or finite*) of a *new* model S_1 *with* = *as equality* (Gödel, 1930). For our applications, we may take II to be thus strengthened.

Applying (II_G) with $\left\{ \begin{array}{l} \neg C \\ \neg C, B_0, B_1, B_2, \ldots \end{array} \right\}$ as the $\left\{ \begin{array}{l} A \\ A_0, A_1, A_2, \ldots \end{array} \right\} : (II_G')$.

In $\left\{ \begin{array}{l} \textit{the predicate calculus} \\ \textit{theories formalized by the predicate calculus with axioms } B_0, B_1, B_2, \ldots \end{array} \right\}$,
each sentence C which is true of

$\left\{ \begin{array}{l} \textit{every system S} \\ \textit{every system S which makes } B_0, B_1, B_2, \ldots \textit{ true} \end{array} \right\}$

is provable as a theorem. This confirms that the predicate calculus fully accomplishes (for 'elementary theories') what we started out by considering as the role of logic. But what is combined with this in Gödel's completeness theorem (including Löwenheim's theorem) is more than was sought, and makes the theorem as much an incompleteness theorem for axiom systems as it is a completeness theorem for logic.

Thus the Löwenheim–Skolem theorem I shows that the axioms of an axiomatic set theory have a countable model (if they have any model at all), despite Cantor's theorem holding in the theory (the Skolem 'paradox', 1922–3).

Furthermore, II entails: (II″) *If the sentences of each finite subset* $A_{k_1}, \ldots, A_{k_n}$ *of* $A_0, A_1, A_2, \ldots$ *are true of a respective system S, then there is a system* S_1, *with countable domain, of which* $A_0, A_1, A_2, \ldots$ *are all true.* This gives the following theorem, found by Skolem (1933, 1934) using another method (and partially anticipated by Tarski, 1927–8).

III. *Say the constants include* $0, +1, =$, *and suppose* $B_0, B_1, B_2, \ldots$ *are true of the system* S_0 *of the natural numbers. Then there is a system* S_1, *with countable domain, not isomorphic to* S_0 *of which* $B_0, B_1, B_2, \ldots$ *are also true.*

Proof. Let $A_0, A_1, A_2, \ldots$ be $B_0, B_1, B_2, \ldots, \neg 0 = \pi, \neg 1 = \pi, \neg 2 = \pi, \ldots$ where π is a new individual symbol. Each $A_{k_1}, \ldots, A_{k_n}$ is true of an S obtained from S_0 by interpreting π as a natural number different from each n for which $\neg n = \pi$ is among $A_{k_1}, \ldots, A_{k_n}$.

Applications of Gödel's completeness theorem to algebra were noted about 1946–7 by Tarski, Henkin and A. Robinson, and have been cultivated since. We have been supposing the number of symbols at most countably infinite, as must be the case of any language in actual use. However, Malcev (1936) extended the completeness theorem to languages with arbitrarily (possibly uncountably) many constants, and Henkin (1947) used such languages to represent the complete addition and multiplication tables, etc., of algebraic systems in the set of formulas for application of the extensions of I–II.

Returning to countable languages, we may consider ones with more than one type of variables, e.g. a second-order predicate calculus with variables ranging over a domain D of individuals and also variables ranging over a collection M of subsets of D. A *standard* model for a set of sentences $A_0, A_1, A_2, \ldots$ is one with $M = \{$the set 2^D of all subsets of $D\}$. The above results do not extend when only standard models are used, in view of the categoricity of Peano's axioms for the natural numbers (using a variable over 2^D to express induction). However, Henkin (1947, 1950) introduced the notion of a *general* model in which M may be an appropriate subset of 2^D, and with which he obtained an extension of Gödel's completeness theorem. Thus we are still unable to characterize the natural numbers, except by reading into the axioms the notion of all possible subsets, which is hardly simpler.

We have given the foregoing model theory as part of the familiar classical mathematics, and for the classical 'two-valued' form of the predicate calculus. The negative results obtain all the more from the constructive standpoints. The axiomatic method cannot provide an autonomous foundation for mathematics. The rules of the language of the axioms must (at some level) be understood, and not merely described by more axioms; and this amounts to presupposing the natural numbers intuitively.

2. Constructive and non-constructive operations

The awareness that some mathematical operations are 'constructive', and others are not (at least directly) such, must go far back in mathematical history; witness the word 'algorithm'. A computer cannot tabulate the truth or falsity of $(Ex)\, R(a, x)$, where the variables range over the natural numbers, unless for the particular R he has some theory which gives him an equivalent 'constructive' definition of $(Ex)\, R(a, x)$. Say triples $b_0,\ b_1,\ b_2$ are mapped constructively into single numbers b, with constructive inverses $(b)_0,\ (b)_1,\ (b)_2$. Such a theory is known for $R(a, x) \equiv (a)_0\,(x)_0 + (a)_1\,(x)_1 = (a)_2$, using Euclid's algorithm; but not today for $R(a, x) \equiv ((x)_0 + 1)^{(x)_3} + ((x)_1 + 1)^{(x)_3} = ((x)_2 + 1)^{(x)_3}\ \&\ (x)_3 > a$, where the value just for $a = 2$ would 'decide' Fermat's 'last theorem'.

In 1936 the claim was made, by Church first and independently by Turing and by Post, that a certain class of functions definable mathematically (in one of several equivalent ways) includes all that are 'computable' or 'effectively calculable' or 'constructively defined' (*Church's thesis*), and conversely that all the functions of this class are 'computable' (*Converse of Church's thesis*).

The definition of this class of functions is not itself constructive. It consists in specifying constructively a type of computation procedure. But a given such procedure may or may not terminate for all arguments, so as to compute a (completely defined) function. (Otherwise, by Cantor's diagonal method one could get constructively outside the class, so Church's thesis could not hold.)

The converse of Church's thesis constructively interpreted means that, whenever one has a constructive proof that the computation procedure always terminates, the function is computable. It is hardly debatable then. A possibility for skepticism remains to one who wishes computability to include constructive provability that the computation procedure always terminates, while allowing the condition that it always terminate to be understood classically; he may imagine that there might be cases when the procedure does always terminate but without there being any constructive proof of that fact.

Much work has been done, especially by Péter since 1932, on special classes of computable functions, for which classes proofs are known that all the computation procedures always terminate.

To Church's thesis itself, the only suggested counterexamples involve 'computation procedures' in which the computer is to perform steps depending on some unpredictable future state of his mind, or in which the 'procedure' is somehow to vary with the argument of the function. But for the thesis, 'computation' is intended to mean of a predetermined function independent of the computer, by only preassigned rules independent of the argument.

We shall now present (essentially) Turing's definition of the class of the 'computable' functions. (Among the equivalents that appear in the literature are the Church–Kleene λ-definable functions, 1933–5, the Herbrand–Gödel general recursive functions, 1934, and definitions using Post's canonical systems, 1943, and Markov's algorithms, 1951.)

Instead of a human computer subjected to preassigned instructions, we can speak of a *machine*. Turing's theory is about *ideal* (digital) computing machines, unhampered by finiteness of storage space or fallibility of functioning. More recently the notion of an *automaton* has been used, by von Neumann (1951); the automaton should not be finite (Kleene, 1956), but potentially infinite (Church, 1957). We want a fixed finite amount of structure (or information) to establish the computation *procedure* for a function $\phi(a)$, while an unbounded amount of space and time must be available to accomodate the argument a and the computation.

The machine or automaton shall accordingly consists of $\aleph_0$ *cells*, each adjacent to at most a given finite number of other cells; but only a finite diversity of structure shall be built into it, the rest of the infinity consisting of identical repetition. Here we use the idea from information theory that information is conveyed only when the signal is not predictable. In order to simplify our brief discussion, we can specialize to the case when the cells are $c_0, c_1, c_2, \ldots$, in the order type of the natural numbers, each c_i (except c_0) being adjacent to exactly two others c_{i-1} and c_{i+1}. The general defense of the Church–Turing thesis then requires arguing that no other arrangement of the cells (with only a finite diversity of structure) would make a function computable that is not computable in this space.

Discrete *moments* of time $0, 1, 2, \ldots$ are distinguished. *States* $s_0, \ldots, s_l$ are given, in one of which each cell shall be at each moment. At moment 0, all but a finite number of the cells shall be in the *passive* state s_0. A *table* is given which determines the state of each cell c_i at moment $t + 1$ from its state and the states of the adjacent cells (for $i = 0$, s_0 replacing the state of c_{i-1}) at moment t; the output of this table shall differ from s_0 only when an input does.

To set the problem, say of computing $\phi(a)$ for a as argument, we can take the states at $t = 0$ of the cells $c_0, c_1, c_2, \ldots$ to be

$$s_0 \underbrace{s_1 \cdots s_1}_{a \text{ times}} s_2 s_0 s_0 s_0 \cdots.$$

The answer shall be receivable by the states being

$$s_0 \underbrace{s_1 \cdots s_1}_{a \text{ times}} s_1 s_0 \underbrace{s_1 \cdots s_1}_{\phi(a) \text{ times}} s_3 s_0 s_0 s_0 \cdots$$

at a later moment $t = x$ when s_3 first occurs. (The fundamental representation of a natural number b is by b successive marks, so it can be argued that a computation problem is solved only when it is possible to present the solution in this representation.)

One may for example imagine the cells $c_0, c_1, c_2, \ldots$ as representing sheets of paper, each admitting one of finitely many symbols on each of finitely many squares, and one of them carrying as part of its state a human computer in one of finitely many states of mind (cf. Kleene, 1952).

Machines can be used similarly to compute n-place functions $\phi(a_1, \ldots, a_n)$; and they can be used to 'decide' predicates $P(a_1, \ldots, a_n)$ by computing 0 to represent truth and 1 falsity.

The behavior of a machine is completely described by its table, which can be written in code form as a natural number, its *index*.

Let $T(i, a, x) \equiv \{i$ is the index of a Turing machine M_i, which, when applied to compute for a as argument, first at moment x has computed a value $\phi_i(a)\}$.

Here $\phi_i(a)$ is an incompletely defined function of i and a, its condition of definition being $(Ex)\,T(i, a, x)$.

We can constructively decide whether a given i is the index of a machine M_i, and if so given also a and x imitate M_i's behavior for a as argument at moments $0, \ldots, x$ successively. Thus, given i, a, x, we can decide whether $T(i, a, x)$ is true or false. (So there is by Church's thesis, and in a detailed treatment of the subject we would actually construct, a machine that decides $T(i, a, x)$.)

IV. *The function*

$$\psi(a) = \begin{cases} \phi_a(a) + 1 & if \quad (Ex)\,T(a, a, x), \\ 0 & otherwise \end{cases} \tag{A}$$

is uncomputable.

Proof. Were $\psi(a)$ computable, it would be computed by a machine M_q; so for each a, (B) $\psi(a) = \phi_q(a)$ and (C) $(Ex)\,T(q, a, x)$. Substituting q for a in (C) and using (A), $\psi(q) = \phi_q(q) + 1$, which contradicts (B) with q substituted for a.

V. *The predicate* $(Ex)\,T(a, a, x)$ *is undecidable.*

Proof. Were $(Ex)\,T(a, a, x)$ decidable, we could compute $\psi(a)$ by first deciding $(Ex)\,T(a, a, x)$, and according to the answer, either imitating machine M_a applied to a as argument to compute $\phi_a(a)$ and adding 1, or writing 0. This is Church's theorem 1936, but with a different example of an absolutely undecidable predicate.

In a standard formal system N of arithmetic (or 'number theory'), each decidable predicate, such as $T(i, a, x)$, can be expressed; hence also $(Ex)\,T(a, a, x)$, by a sentence C_a (constructively obtainable from a). Now, for particular a, $(Ex)\,T(a, a, x)$ when true can be 'proved' by doing the computation that shows $T(a, a, x)$ to be true for the appropriate x. This intuitive proof is available formally in a standard N. Thus

$$(Ex)\,T(a, a, x) \rightarrow \{C_a \text{ is provable}\}. \tag{a}$$

Also we are assuming of N that only true formulas are provable in it, so

$$\{C_a \text{ is provable}\} \rightarrow (Ex)\,T(a, a, x). \tag{b}$$

Now V gives:

VI. *There is no procedure for deciding whether a given sentence is provable in a formal system N of arithmetic; briefly, N is 'undecidable'* (Church 1936).

Continuing, could we in N also prove $\neg C_a$ whenever $(Ex)\,T(a,a,x)$ is false, besides only then so

$$\{\neg C_a \text{ is provable}\} \to \neg(Ex)\,T(a,a,x), \tag{c}$$

we would be able, by searching for C_a or $\neg C_a$ among the provable sentences, to decide $(Ex)\,T(a,a,x)$. So, again from V:

VII. *In a formal system N of arithmetic, there is a sentence C_q such that C_q and $\neg C_q$ are both unprovable, though $\neg C_q$ is true (i.e. $\neg(Ex)\,T(q,q,x)$).*

This gives Gödel's famous incompleteness theorem (1931), generalized to apply to all formal systems N satisfying very general conditions, and with the 'formally undecidable' sentence C_q expressing the value, for an argument q depending on the system, of a preassigned predicate $(Ex)\,T(a,a,x)$. The above proof is indirect, the existence of q being inferred from the absurdity that $\neg C_a$ is provable for all a for which it is true. But we can make it direct, by taking as q the index of a machine M_q which, given a, searches through the proofs in N for one of $\neg C_a$, and if one is found writes 0 (but otherwise never computes a value), so

$$(Ex)\,T(q,a,x) \equiv \{\neg C_a \text{ is provable}\}. \tag{d}$$

Substituting q for a in (b)–(d), the three conclusions of VII follow.

Here we have used the feature of formal systems, essential for the purpose which they are intended to serve, that a proof of a sentence can be constructively recognized as being such (and also that C_a can be constructively found from a). Without this feature, we would have a trivial counterexample to VII by taking all the true sentences as the axioms of N. With it, by Church's thesis we conclude the existence of an M_q to any such system. Here the computability notion can be applied directly to the linguistic symbolism, or the latter can be converted to natural numbers as we have already done with machine tables (by a 'Gödel numbering').

The application of Church's thesis by which we obtain VII for all systems N can be avoided for a particular system by actually constructing the M_q for it. This in effect Gödel did in proving his theorem for a particular system before Church's thesis had appeared.

In retrospect, Skolem's theorem III on the existence of unintended models S_1 of systems of sentences $B_0, B_1, B_2, \ldots$ intended to describe the natural numbers suggests Gödel's theorem VII. (Compare the example of Euclid's fifth postulate.) Indeed, for an N based on the elementary predicate calculus, (II'_G) shows that C_q is false of such an S_1. However, III applies even when $B_0, B_1, B_2, \ldots$ are all the true sentences, unlike VII.

I do not consider that VII means we must give up the emphasis on formal systems. The reasons which make a formal system the only accurate way of saying explicitly what assumptions go into a proof are still cogent. Rather VII indicates that, contrary to Hilbert's program, the path of mathematical conquest (even within the already fixed territory of arithmetic) shall not consist solely in discovering new proofs from given axioms by given rules of inference, but also in adducing new axioms or rules. There remains the question whether mathematicians can agree on the validity of the new methods.

In VII, no sooner are we aware that $\neg C_q$ is unprovable than we also know that $\neg C_q$ is true, so we can extend N by adding $\neg C_q$ as a new axiom. This process can be repeated, finitely often, and indeed transfinitely often within the limits of structural constructiveness.

It is illuminating to consider wherein the intuitive proof of $\neg C_q$ transcends N. We only conclude the truth of $\neg C_q$ when we accept (c). By (a), (c) reduces to the consistency of N, which is expressible in N via Gödel numbering by a sentence 'Consis'. The rest of the reasoning that $\neg C_q$ is true is elementary, though tedious when executed in full detail; so we may expect (as has been confirmed by Hilbert and Bernays (1939) for the usual systems as N) that it can be formalized in N. So Consis cannot be provable in N, or $\neg C_q$ would be, contrary to VII. Thus:

VIII. *In a usual formal system N of arithmetic, the sentence* Consis *expressing the consistency of N is unprovable* (Gödel's second incompleteness theorem, 1931).

Thus a system N formalizing classical mathematics cannot be proved consistent, as Hilbert hoped, by a 'subset' of the methods formalized in N.

Gentzen (1936, 1938) gave a proof of the consistency of a system N of arithmetic, in which the method transcending N is a form of transfinite induction over the ordinal numbers $<$ Cantor's first epsilon-number ϵ_0; and other such proofs have appeared since. It is a rather subjective matter whether this should make us feel safer about N than we already feel on the basis of its axioms being true, and its rules of inference preserving truth, under an interpretation ('truth definition') that as classical mathematicians we presumably accept. By a reduction of classical to intuitionistic logic given by Kolmogorov (1925), Gödel (1932–3), Gentzen (1936) and Bernays, the consistency proof by a truth definition can even be managed intuitionistically.

Kreisel (1951–2, 1958) finds the significance of the consistency proofs using ϵ_0-induction in by-products. When a sentence $(a)\,(Eb)\,R(a,b)$ (R decidable) is proved, then $(a)\,R(a,\beta(a))$ will be true for certain functions β, including $\beta(a) = \{$the least b such that $R(a,b)$ is true$\}$, which is computable. It is clear that in a given system N only a subclass of the computable functions can thus be proved to exist; indeed Kleene (1936) gave a proof of Gödel's incompleteness theorem from this idea. Kreisel, however, extracts from Ackermann's consistency proof (1940) a different characterization (not directly from N) of this subclass of the computable functions. The possibility thus appears that some true formula $(a)\,(Eb)\,R(a,b)$ might be shown to be unprovable in N because no β for it is in this subclass.

From Church's theorem other undecidability results follow. The theory of $(Ex)\,T(a,a,x)$ can be formalized in a system N_1 consisting of finitely many axioms $B_1, ..., B_k$ adjoined to the (elementary) predicate calculus. So $(Ex)\,T(a,a,x) \equiv \{C_a$ is provable in $N_1\} \equiv \{B_1 \,\&\, ... \,\&\, B_k \to C_a$ is provable in the predicate calculus$\}$. Thence from V:

IX. *The elementary predicate calculus is undecidable* (Church, 1936a; Turing, 1936–7).

Various formal systems obtained by adjoining axioms for algebraic systems to the predicate calculus have been shown undecidable by Tarski and others using a method of Tarski (1949) (cf. Tarski *et al.* 1953).

Negative solutions to the problems of the existence of various algebraic algorithms have been obtained by Post (1947), Markov since 1947, and others; in particular, Novikov (1952, 1955) showed the word problem for groups unsolvable.

Turing (1939) introduced the notion of a function $\phi(a)$ *computable from* another function $\psi(a)$ (or predicate $Q(a)$). A simple plan under the above treatment is to print the values of ψ into the space, in this respect alone violating the demand that only a finite amount of information be incorporated, by accenting successions of $\psi(0)+1, \psi(1)+1, \psi(2)+1, ...$ cells, preceded and separated by single unaccented cells. In effect, we double the number of states from $s_0, ..., s_l$ to $s_0, ..., s_l, s_0', ..., s_l'$.

When the theory is thus relativized to a given predicate $Q(a)$, the decidable predicate $T(i,a,x)$ becomes a predicate $T^Q(i,a,x)$ decidable from Q, and IV, V assume relativized versions IV*, V*.

X. *If $R^Q(a,x)$ is decidable from Q, there is a computable function $\theta(a)$ such that* $(Ex)\,R^Q(a,x) \equiv (Ex)\,T^Q(\theta(a), \theta(a), x)$.

Proof. Given a, let $M_{\theta(a)}$ be a machine which tries to compute from Q the constant function whose value is the least x such that $R^Q(a,x)$, by testing successively $x = 0, x = 1, x = 2, ...$.

Thus $(Ex) R^Q(a, x)$ is decidable from $(Ex) T^Q(a, a, x)$ by first computing $\theta(a)$. In particular (taking $R^Q(a, x) \equiv \bar{\cdot} Q(a) \,\&\, x = x$), $Q(a)$ is decidable from $(Ex) T^Q(a, a, x)$; but by V*, not conversely. This Post (1948) expressed by saying $(Ex) T^Q(a, a, x)$ is of 'higher degree (of unsolvability)' than $Q(a)$. Predicates and functions are of the 'same degree' when each is decidable (or computable) from the other. A decidable predicate is of the lowest degree ('solvability'). Starting from say $H_{(0)}(a) \equiv a = a$, and for each n defining $H_{n+1}(a) \equiv (Ex) T^{H_{(n)}}(a, a, x)$, we obtain predicates $H_{(n)}(a)$ ($n = 0, 1, 2, \ldots$) of ascending degrees. These predicates, together with those decidable from them, turn out to be exactly the predicates (called *arithmetical* by Gödel, 1931) expressible in the usual system of arithmetic. Thus the arithmetical predicates fall into a hierarchy, first described by Kleene (1943) and Mostowski (1946) in terms of the numbers of quantifiers necessary to define them in prenex form from decidable predicates.

The hierarchy can be extended into the transfinite (Davis, Kleene, Mostowski, Post, about 1950; cf. Mostowski, 1951; Kleene, 1955). One method is to consider $H_{(n)}(a)$ as a predicate $H(n, a)$ of both variables; this is of higher degree than each $H_{(n)}(a)$, and thus is non-arithmetical. 'Contracting' $H(n, a)$ to a one-place predicate $H((a)_1, (a)_0)$, which we write $H_{(\omega)}(a)$, we can proceed as before to $H_{(\omega+1)}(a), H_{(\omega+2)}(a), \ldots$. In general, at a limit ordinal ξ of Cantor's second number class approached through an increasing sequence $\{\xi_n\}$, we consider $H_{(\xi_n)}(a)$ as a predicate of n, a, and contract.

However, we have no uniform method, or justification, for picking a particular increasing sequence $\{\xi_n\}$ for ξ. So a diversity of predicates $H_{(\xi)}$ arise, for each transfinite ξ, depending on the selections of increasing sequences. Worse than this, even for $\xi = \omega$, the use of arbitrary increasing sequences $\{\xi_n\}$ with $\lim_n \xi_n = \xi$ (above we used $\xi_n = n$) will give predicates of arbitrarily high degree. This suggests restricting the sequences $\{\xi_n\}$ to be computable, after rendering ordinals accessible to the above notion of computability by representing them in a suitable system of notations, which can be natural numbers (Church–Kleene, 1936; Kleene, 1938). This being done, the diversity in predicates at a given transfinite level ξ, which remains due to the possibility of using different computable increasing sequences, was shown by Spector (1955) to be confined always within a degree. The predicates thus definable corresponding to constructive ordinals, together with all predicates decidable (and functions computable) from them, we call *hyperarithmetical* (Kleene, 1955a).

It was noticed, about 1957, by Addison, Büchi, Grzegorczyk, Kleene,

Kuznecov and Myhill (cf. Grzegorczyk *et al.* 1958) that the hyperarithmetical predicates are exactly the predicates expressible unambiguously by a formula of the elementary predicate calculus, when the domain is the natural numbers.

Kleene (1957) formulated computability from higher-type objects, such as from the existential quantifier (Ex) considered as a functional E which operates on a predicate to produce a truth value (or on a function ψ to produce the number 0 if $(Ex)(\psi(x) = 0)$ and 1 otherwise). The hyperarithmetical functions $\phi(a_1, \ldots, a_n)$ are exactly those computable from E; thus, operating constructively, except for using a number quantifier, we obtain not merely the usual predicates of arithmetic but the hyperarithmetical predicates.

REFERENCES

Ackermann, W.
  1940. Zur Widerspruchsfreiheit der Zahlentheorie. *Math. Ann.* 117, 162–194.

Brouwer, L. E. J.
  1908. De onbetrouwbaarheid der logische principes. (The untrustworthiness of the principles of logic.) *Tijdschrift voor wijsbegeerte*, 2, 152–158.
  1924. Beweis, dass jede volle Funktion gleichmässig stetig ist. *Proc. Akad. Wet. Amst.* 27, 189–193.

Church, A.
  1933. A set of postulates for the foundation of logic (second paper). *Ann. Math.* (2), 34, 839–864.
  1936. An unsolvable problem of elementary number theory. *Amer. J. Math.* 58, 345–363.
  1936a. A note on the Entscheidungsproblem. *J. Symb. Logic*, 1, 40–41. Correction, *ibid.* 101–102.
  1941. *The Calculi of Lambda-Conversion.* Ann. of Math. Studies, no. 6. Princeton University Press, Princeton, N.J.
  1957. Application of recursive arithmetic to the problem of circuit synthesis. *Summaries of Talks Presented at the Summer Institute of Symbolic Logic in 1957 at Cornell University* (mimeographed), 1, 3–50; 3, 429.

Church, A. and Kleene, S. C.
  1933–5. See Church (1933), Kleene (1935), Church (1941).
  1936. Formal definitions in the theory of ordinal numbers. *Fundam. Math.* 28, 11–21.

Frege, G.
  1879. *Begriffsschrift, eine der arithmetischen nachgebildete Formelsprache des reinen Denkens.* Nebert, Halle.
  1891. *Funktion und Begriff.* Jena.

Gentzen, G.
  1936. Die Widerspruchsfreiheit der reinen Zahlentheorie. *Math. Ann.* 112, 493–565.

1938. Neue Fassung des Widerspruchsfreiheitsbeweises für die reine Zahlentheorie. *Forschungen zur Logik und zur Grundlegung der exakten Wissenschaften*, N.S., no. 4, 19–44. Hirzel, Leipzig.

Gödel, K.
1930. Die Vollständingkeit der Axiome des logischen Funktionenkalküls. *Monatsh. Math. Phys.* 37, 349–360.
1931. Über formal unentscheidbare Sätze der Principia Mathematica und verwandter Systeme. I. *Monatsh. Math. Phys.* 38, 173–198.
1932–3. Zur intuitionistischen Arithmetik und Zahlentheorie. *Ergebn. math. Kolloq.* Heft 4, 34–38 (for 1931–2, publ. 1933).
1934. *On Undecidable Propositions of Formal Mathematical Systems* (mimeographed). Princeton, N.J.

Grzegorczyk, A., Mostowski, A. and Ryll-Nardzewski, C.
1958. The classical and the ω-complete arithmetic. *J. Symb. Logic*, 23, 188–206.

Henkin, L.
1947. *The Completeness of Formal Systems.* Princeton University Ph.D. Thesis, Princeton, N.J.
1950. Completeness in the theory of types. *J. Symb. Logic*, 15, 81–91.

Herbrand, J.
1930. *Recherches sur la théorie de la démonstration.* Travaux de la Société des Sciences et des Lettres de Varsovie, Classe III, sciences mathématiques et physiques, no. 33.

Herbrand, J. and Gödel, K.
1934. See Gödel (1934), Kleene (1936, 1952).

Heyting, A.
1956. *Intuitionism, An Introduction.* North Holland Publ. Co., Amsterdam.

Hilbert, D.
1899. *Grundlagen der Geometrie*, 7th ed. (1930), Teubner, Leipzig and Berlin.
1904. Über die Grundlagen der Logik und der Arithmetik. *Verhand. Dritten Int. Math.-Kong. Heidelberg* 1904, 247–261 (publ. Leipzig 1905).

Hilbert, D. and Bernays, P.
1934. *Grundlagen der Mathematik*, vol. 1. Springer, Berlin.
1939. *Grundlagen der Mathematik*, vol. 2. Springer, Berlin.

Kleene, S. C.
1935. A theory of positive integers in formal logic. *Amer. J. Math.* 57, 153–173, 219–244.
1936. General recursive functions of natural numbers. *Math. Ann.* 112, 727–742.
1938. On notation for ordinal numbers. *J. Symb. Logic*, 3, 150–155.
1943. Recursive predicates and quantifiers. *Trans. Amer. Math. Soc.* 53, 41–73.
1952. *Introduction to Metamathematics.* North Holland Publ. Co. (Amsterdam), Noordhoff (Groningen), Van Nostrand (New York and Toronto).
1955. Arithmetical predicates and function quantifiers. *Trans. Amer. Math. Soc.* 79, 312–340.
1955a. Hierarchies of number-theoretic predicates. *Bull. Amer. Math. Soc.* 61, 193–213.
1956. Representation of events in nerve nets and finite automata. *Automata Studies.* Ann. of Math. Studies, no. 34, 3–41. Princeton University Press, Princeton, N.J.

Kleene, S. C.
1957. Recursive functionals of higher finite types. *Summaries of Talks Presented at the Summer Institute of Symbolic Logic in 1957 at Cornell University* (mimeographed), 1, 148–154. Errata, 3, 429,

Kolmogorov, A.
1925. Sur le principe de tertium non datur. *Rec. Math.*, *Moscou*, 32, 646–667.

König, D.
1926. Sur les correspondences multivoques des ensembles. *Fundam. Math.* 8, 114–134.

Kreisel, G.
1951–2. On the interpretation of non-finitist proofs. *J. Symb. Logic*, 16, 241–267; 17, 43–58.
1958. Mathematical significance of consistency proofs. *J. Symb. Logic*, 23, 155–182.

Lorenzen, P.
1950. Konstruktive Begründung der Mathematik. *Math. Z.* 53, 162–202.
1955. *Einführung in die operative Logik und Mathematik.* Springer, Berlin, Göttingen and Heidelberg.

Löwenheim, L.
1915. Über Möglichkeiten im Relativkalkül. *Math. Ann.* 76, 447–470.

Lukasiewicz, Jan
1921. Logika dwuwartościowa. (Two-valued logic.) *Przegląd Filozoficzny*, 23, 189–205.

Malcev, A.
1936. Untersuchungen aus dem Gebiete der mathematischen Logik. *Mat. Sbornik*, 1 (43), 323–336.

Markov, A. A.
1947. Névozmožnosť nékotoryh algorifmov v téorii associativnyh sistém. (On the impossibility of certain algorithms in the theory of associative systems.) *Dokl. Akad. Nauk*, *SSSR*, N.S., 55, 587–590.
1951. Téoriá algorifmov. (The theory of algorithms.) *Trudy Matématičéskogo Instituta iméni V. A. Steklova*, 38, 176–189.

Mostowski, A.
1946. On definable sets of positive integers. *Fundam. Math.* 34, 81–112.
1951. A classification of logical systems. *Studia Philosophica*, 4, 237–274.

Novikov, P. S.
1952. Ob algoritmíčeśkoj nérazréśimosti problémy toždéstva. (On algorithmic unsolvability of the word problem.) *Dokl. Akad. Nauk*, *SSSR*, N.S., 85, 709–712.
1955. Ob algoritmiiceśkoj nérazréśimosti problémy toždéstva slov v téorii grupp. (On the algorithmic unsolvability of the word problem in group theory.) *Trudy Mat. Inst. im. Steklov*, no. 44. Izdat. Akad. Nauk, SSSR. Moscow.

Peirce, C. S.
1885. On the algebra of logic: A contribution to the philosophy of notation. *Amer. J. Math.* 7, 180–202.

Péter, R.
1951. *Rekursive Funktionen.* Akadémiai Kiadó (Akademischer Verlag), Budapest.

Post, E.
1921. Introduction to a general theory of elementary propositions. *Amer. J. Math.* 43, 163–185.
1936. Finite combinatory processes—formulation. I. *J. Symb. Logic*, 1, 103–105.
1943. Formal reductions of the general combinatorial decision problem. *Amer. J. Math.* 65, 197–215.
1947. Recursive unsolvability of a problem of Thue. *J. Symb. Logic*, 12, 1–11.
1948. Degrees of recursive unsolvability (abstract). *Bull. Amer. Math. Soc.* 54, 641–642.
Robinson, A.
1951. *On the Metamathematics of Algebra.* North Holland Publ. Co., Amsterdam.
Skolem, T.
1920. Logisch-kombinatorische Untersuchungen über die Erfüllbarkeit oder Beweisbarkeit mathematischer Sätze nebst einem Theoreme über dichte Mengen. *Skrifter utgit av Videnskapsselskapet i Kristiania*, I. *Mathematisk-naturvidenskabelig klasse*, no. 4.
1922–3. Einige Bemerkungen zur axiomatischen Begründung der Mengenlehre. *Wissenschaftliche Vorträge gehalten auf dem Fünften Kongress der Skandinavischen Mathematiker in Helsingfors vom 4. bis 7. Juli 1922* (publ. Helsingfors, 1923), 217–232.
1933. Über die Unmöglichkeit einer vollständigen Charakterisierung der Zahlenreihe mittels eines endlichen Axiomensystems. *Norsk. mat. Foren. Skr.* ser. 2, no. 10, 73–82.
1934. Über die Nicht-charakterisierbarkeit der Zahlenreihe mittels endlich oder abzählbar unendlich vieler Aussagen mit ausschliesslich Zahlenvariablen. *Fundam. Math.* 23, 150–161.
Spector, C.
1955. Recursive well-orderings. *J. Symb. Logic*, 20, 151–163.
Tarski, A.
1927–8. See the Bemerkung der Redaktion in Skolem, 1934, p. 161.
1933. Der Wahrheitsbegriff in den formalisierten Sprachen. *Studia philosophica*, 1 (1936, tr. from Polish original, 1933). Engl. tr. in A. Tarski, *Logic, Semantics, Metamathematics*, Oxford University Press, Oxford, 1956.
1954–5. Contributions to the theory of models. *Proc. Akad. Wet. Amst.*, ser. A, 57, 572–578; 58, 56–64.
Tarski, A., Mostowski, A. and Robinson, R. M.
1953. *Undecidable Theories.* North Holland Publ. Co., Amsterdam.
Turing, A. M.
1936–7. On computable numbers, with an application to the Entscheidungsproblem. *Proc. Lond. Math. Soc.* (2), 42, 230–265. A correction, *ibid.* 43, 544–546.
1939. Systems of logic based on ordinals. *Proc. Lond. Math. Soc.* (2), 45, 161–228.
von Neumann, J.
1951. The general and logical theory of automata. *Cerebral Mechanisms in Behavior, The Hixon Symposium*, pp. 1–31 (editor, Jeffress, Lloyd A.). Wiley, New York.

EXTENDED BOUNDARY VALUE PROBLEMS

By C. LANCZOS

1. Introduction

The topic to which the following discussions are devoted is as old as the history of quantitative thinking. It was observed from the time of antiquity that continuous phenomena can be approached from the viewpoint of treating them as the limits of discontinuous happenings. This age-old problem was revitalized in our own days and became once more of topical interest on account of the sensational development of the big electronic digital computers. With the ever-increasing memory-capacity of the new machines it becomes more and more possible to tackle many of the customary types of boundary value problems in a practical way. Partial differential equations of two or even three dimensions can be coded for the big machines, and we come nearer and nearer to the state in which the physicist or the engineer may get all the relevant answers for which he is striving by putting his problems on one of the high-powered electronic machines.

In this development a very definite *approximation procedure* plays a vital role. The given partial differential equation is changed to a *difference equation* and coded as a simultaneous set of algebraic equations. If, in particular, the given differential equation is of the *linear* type—and in all the following discussions we will restrict ourselves to the domain of linear operators—the resulting set of simultaneous linear equations can be characterized by the simple matrix equation

$$Ay = b, \tag{1}$$

where the matrix A takes the place of the given linear differential operator, the vector y corresponds to the unknown function, while the given right side of the differential equation, including the given boundary conditions, is absorbed by the vector b.

That this algebraization of a problem in linear differential equations is always possible is by no means self-evident. Anybody who has ever coded such a problem for the big machines will inevitably run into some questions which cannot be answered in a trivial way. Let us consider for the sake of illustration a somewhat over-simplified but characteristic example. Given the potential equation

$$\frac{\partial^2 \phi}{\partial x^2} + \frac{\partial^2 \phi}{\partial y^2} = 0 \tag{2}$$

to be solved for a certain square-domain of the variables x and y. We set up a square grid of points and change the given differential equation into a difference equation. We observe at once that we obtain only $(n-2)^2$ equations for n^2 quantities, which demands $4n-4$ more data for a full algebraic characterization of our problem. Hence we have to add $4n-4$ boundary data. How shall we choose these data? The mathematician tells us that we will be wise if we add as boundary data the values of the

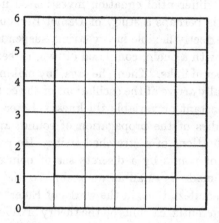

function ϕ in the grid-points along the four boundaries of our square. But we may have different ideas and tell him that we would prefer to choose our data along three sides only but omit the line 6, instead of which we want to give the functional values along the line 1. He will advise us strongly against such an idea. On the other hand, if the given differential equation happens to be

$$\frac{\partial^2 \phi}{\partial x^2} - \frac{1}{c^2}\frac{\partial^2 \phi}{\partial y^2} = 0 \tag{3}$$

he gives us exactly the opposite advice. If we try to understand these puzzling prescriptions, he refers us to the 'theory of characteristics' which, however, is not an algebraic theory, while our desire would be to understand the nature of differential equations purely from the *algebraic* point of view. This desire is not unjustified if it is true that a linear differential equation can be conceived as the limit of a set of linear algebraic equations, obtained by replacing the derivatives by difference coefficients. But then, why is it that the nature of the boundary value problem differs so completely in the elliptic and in the hyperbolic case, although algebraically they seem to be equivalent? Before we come to the general discussion of such problems, a brief glance on the historical development of the subject will not be out of place.

2. Historical survey

The close relation between continuous and discrete operators was recognized from the very beginnings of higher mathematics. But the first example of a differential equation investigated in a consistently algebraic manner is perhaps a study of Daniel Bernoulli[1] concerning the motion of a perfectly flexible heavy string, suspended between two points. He starts with a chain, composed of two, three, four, and later an arbitrary number of links. 'Then', he says, 'by making the number of links infinite, I finally arrive at the oscillations of the completely flexible chain of either constant or variable thickness.' Later Lagrange[10], in his admirable studies of the propagation of sound, applied a similar method to the vibrations of a stretched string. He replaced the continuous manifold of points by a discrete set of points whose mutual distance could be made as small as we wish. We find the same basic idea in manifold manifestations in the works of Euler[4], who based his entire theory of differential calculus on the theory of difference equations with gradually decreasing increments. Thus he derived the fundamental differential equation of variational calculus by replacing the variational integral by a sum, and the derivatives of the unknown functions by difference coefficients. In the new form the problem became an ordinary maximum–minimum problem which could be solved by the tools of elementary calculus. He was not aware that this method involves the exchange of two limit processes which demands specific justification.

However, perhaps the greatest and most consistent exponent of the algebraic method was Lord Rayleigh[12] in his investigations of acoustic and elastic vibrations (of the years 1877–94). He gained deep insight into the nature of orthogonal function systems by his algebraic method, and it was in fact this method which led him to the discovery of the fundamental properties of orthogonal expansions. Even in his time the discretization of continuous operators was performed without any qualms of conscience, as a matter of 'physical intuition'. He starts out with the following general remarks in the introductory chapter of his great researches on vibrating systems:† 'Strictly speaking, the displacements possible to a natural system are infinitely various, and

† Cf. [12], vol. 1, chap. IV, p. 91.

cannot be represented as made up of a finite number of displacements of specified type. To the elementary parts of a solid body any arbitrary displacements may be given, subject to conditions of continuity. It is only by a process of abstraction of the kind constantly practised in Natural Philosophy, that solids are treated as rigid, fluids as incompressible, and other simplifications introduced so that the position of a system comes to depend on a finite number of co-ordinates. It is not, however, our intention to exclude the consideration of systems possessing infinitely various freedom, on the contrary, some of the most interesting applications of the results of this chapter will be in that direction. But such systems are most conveniently conceived as limits of others, whose freedom is of a more restricted kind. We shall accordingly commence with a system, whose position is specified by a finite number of independent co-ordinates $\psi_1, \psi_2, \psi_3, \ldots$, etc.'

With Lord Rayleigh we come to the turn of the century and it was exactly around that time that a new epoch of mathematical rigor takes its departure, with the classical investigations of Fredholm (1900–3) concerning the theory of a certain type of functional equations, now called 'integral equations of the Fredholm type'[6]. Fredholm tackles the problem on an algebraic basis and arrives at his results by a rigorous estimation of infinite determinants. This was the first time that the algebraization of infinitesimal processes was carried through to its final consequences with full mathematical rigor. In the dazzling light of this new approach the previous algebraic attempts were put in the balance and found wanting. While the results of Fredholm remained above all reproach, it was pointed out that similar results cannot be expected if we depart from the Fredholm type of integral equations.†
The direct algebraization of differential equations was looked upon with suspicion and admitted without reservation only if the given problem could first be transformed into an integral equation of the Fredholm type.

This demand restricts, however, the type of boundary value problems admitted to an unnecessary degree. It includes only those problems which in the algebraic formulation are characterized by an $n \times n$ matrix (which means that the number of equations and unknowns must be equal), the determinant of which is different from zero. Beyond this requirement, however, the demand of the existence of a Green's function (in other words the *existence of the inverse operator*), restricts our possibilities still further.‡ Hadamard[7], in his lectures on the Cauchy

† Cf. [8], p. 1344.   ‡ Cf. [8], p. 1362; [2], p. 358.

problem, called this type of boundary value problems 'well-posed', or 'correctly set' ('un problème correctement posé'). A problem is eligible to this distinction if the following two conditions are satisfied:

(1) The given data are sufficient to obtain one and only one solution.

(2) Within a certain general class of functions the given data can be prescribed freely.

Without impinging in the least on the importance of these problems, we can hardly doubt that these requirements handicap our possibilities quite severely. In the first place, how shall we decide in a given case whether a given problem is well-posed or not? This requires a very elaborate preliminary investigation of the given differential equation. Our present knowledge goes hardly beyond the realm of second-order operators. There we have the three types of elliptic, parabolic, and hyperbolic equations, and we know that a 'well-posed' problem demands in the case of elliptic differential equations the prescription of boundary conditions, in the other two cases the prescription of initial conditions. But how much or how little we should prescribe in the case of differential equations of third or fourth or higher order for the sake of a well-posed problem, it is impossible to tell.

Apart from this difficulty, we encounter well-defined and reasonable problems which do not fall in the well-posed category. Consider the case of a conservative field of force, characterized by the equation

$$\text{grad}\,\phi = \mathbf{F}. \tag{4}$$

Here the scalar field ϕ is transformed into the vector field $\mathbf{F}$. If our aim is to obtain ϕ by observing $\mathbf{F}$, we have clearly an over-determined problem which is not solvable if $\mathbf{F}$ is freely prescribed, which is solvable, however, if $\mathbf{F}$ satisfies the compatibility condition

$$\text{curl}\,\mathbf{F} = 0. \tag{5}$$

The algebraic picture associated with this problem involves a matrix A of n rows and m columns in which $n > m$.

On the other hand, consider the differential equation

$$\text{div}\,\mathbf{E} = \rho, \tag{6}$$

where a vector field $\mathbf{E}$ is transformed into a scalar field ρ. The vector field $\mathbf{E}$ is by no means determined by this equation but we would like to know what conclusions can be drawn from the fact that ρ is given. Here we have an example of an under-determined system, algebraically characterized by a matrix A of n rows and m columns in which $n < m$.

One of the most fundamental equations of the theory of analytical functions is Cauchy's integral theorem

$$f(z) = \frac{1}{2\pi i} \oint \frac{f(\zeta)\,d\zeta}{\zeta - z}, \tag{7}$$

which determines the value of $f(z)$ inside a domain if it is given on the boundary of the domain. The corresponding theorem in the theory of the Newtonian potential is the equation

$$\phi(P) = \frac{1}{4\pi} \int_S \left(\frac{\partial \phi}{\partial n} \frac{1}{r_{PS}} - \phi \frac{\partial}{\partial n} \frac{1}{r_{PS}} \right) dS,$$

by which the function ϕ can be evaluated at the inside point P, if the values of ϕ and $\partial\phi/\partial n$ are given on the boundary S of the domain. Conceiving these problems as boundary value problems, both theorems suffer from the fact of over-determination. The potential function ϕ is uniquely determined by the boundary values $\phi(S)$ *alone*, without giving the values of $(\partial\phi/\partial n)(S)$. In the case of the analytical function $f(z)$ it is unnecessary to give $f(\zeta)$ all along the boundary; it suffices if $f(\zeta)$ is given along an arbitrary *open* portion of the boundary and we are entitled to ask the question

$$\tag{8}$$

how to obtain $f(z)$ in the inside of the domain in terms of these data. The problem is not of the well-posed type and has no elementary solution but we know that the solution exists and we are entitled to pose the problem.

We now ask quite generally the following question. Given a linear partial differential equation or any system of such equations with added boundary conditions which are chosen in an arbitrarily judicious or injudicious fashion, thus leading to an arbitrarily over-determined or under-determined system. Can we treat such a problem successfully by an algebraic method, and if so, can we transfer the results without difficulty to the field of continuous operators?

From the historical standpoint this problem is not in line with the Fredholm type of investigations since the method of determinants loses its significance if we depart from the realm of square matrices to the realm of the more general $n \times m$ matrices. It so happens, however, that the stakes of the algebraic method are much more widely set than the theory of determinants. If we follow up the later development of Fredholm's theory, we come to the classical investigations of Hilbert[9] who also employed the algebraic method for the development of the theory of integral equations (1904–10) but from an entirely different viewpoint. In Hilbert's theory the emphasis is laid on the geometry of second-order surfaces which are put in a Euclidean space of increasingly many dimensions. The principal axis transformation of these surfaces became the central item in Hilbert's theory. This geometrical method gave a second and independent rigorous foundation of the theory of integral equations, without recourse to the theory of determinants. The same theory was later put on a more analytical basis by E. Schmidt[13]. It is this principal axis theory of quadratic forms which provides the proper frame of reference for our much more general problem and which yields a suitable universal platform for the understanding of the behaviour of both well-posed and not well-posed boundary value problems.

3. The fundamental eigenvalue problem

We start with the algebraic equation (1), assuming A as a general $n \times m$ matrix, with $n \gtreqless m$. The case $n > m$ can be pictured as follows:

$$A\,y = b \qquad \begin{array}{c} \textit{over-determined} \\ n > m \\ (\mathrm{grad}\,\phi = \mathbf{F}) \end{array} \qquad (9)$$

The case $n < m$ can be pictured as follows:

$$A\,y = b \qquad \begin{array}{c} \textit{under-determined} \\ n < m \\ (\mathrm{div}\,\mathbf{E} = \phi) \end{array} \qquad (10)$$

The case $n = m$ (with the added condition $\det A \neq 0$) belongs to the usual 'well-posed' case:

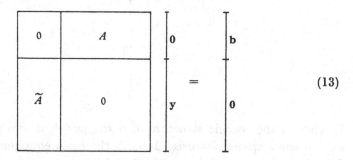

$$\qquad \qquad \qquad \qquad \qquad \qquad \qquad \text{'well-posed'} \qquad \qquad (11)$$
$$\qquad \qquad \qquad \qquad \qquad \qquad \qquad n = m$$

Our diagrams demonstrate by inspection an important feature of our problem. Usually we have the $n \times n$ case in hand and consider a matrix A as an operator which transforms a vector $\mathbf{y}$ into another vector $\mathbf{y}'$. But the general case $n \neq m$ shows that the case $n = m$ hides an important feature of our problem, viz. that the two vectors $\mathbf{y}$ and $\mathbf{b}$ belong to *two different spaces*. The vector $\mathbf{y}$ on which A operates belongs to an m-dimensional space, the vector $\mathbf{b}$ into which $\mathbf{y}$ is transformed, to an n-dimensional space. Hence a general matrix A takes a vector from one space and transforms it into a vector of another space. It is necessary that in all our following discussions we should keep the separateness of these two spaces clearly in mind.†

The fundamental first step, from which everything will follow with logical necessity, is an apparent triviality. We extend the basic equation (1) by a second equation and consider the resulting system of a single unit:

$$A\mathbf{y} = \mathbf{b}, \quad \tilde{A}.\mathbf{0} = \mathbf{0}. \qquad (12)$$

This extension is reflected in the following matrix diagram:

$$\qquad \qquad \qquad \qquad \qquad \qquad \qquad \qquad \qquad \qquad (13)$$

† Without this distinction serious misunderstandings are prone to happen; e.g. the so-called 'integral equations of the first kind', are often declared analytically inferior to the Fredholm type of equations (see, for example, [8], p. 1453), solely because the function of the left side and the transformed function of the right side do not belong to the same class of functions.

We notice that we have extended our matrix A to a new $(n+m) \times (n+m)$ square matrix S:

$$S = \begin{array}{c|c} 0 & A \\ \hline \tilde{A} & 0 \end{array} \tag{14}$$

This matrix is not only square but even *symmetric*:

$$S = \tilde{S}. \tag{15}$$

We know that symmetric square matrices have particularly desirable properties. In particular, we know that such a matrix can always be *diagonalized* by a proper orthogonal transformation. This requires the determination of the principal axes of the matrix, on the basis of the equation

$$S\mathbf{w} = \lambda \mathbf{w}. \tag{16}$$

The vector $\mathbf{w}$ has $n+m$ components which we will record in the following form:

$$\mathbf{w} = \begin{array}{|c|} \mathbf{x} \\ \\ \mathbf{y} \end{array} \tag{17}$$

In view of the specific structure of S the principal axis problem will exhibit some special features. Indeed, the basic equation (16) can be formulated in terms of the matrix A as follows:

$$A\mathbf{y} = \lambda\mathbf{x}, \quad \tilde{A}\mathbf{x} = \lambda\mathbf{y}. \tag{18}$$

We will call this system the 'shifted eigenvalue problem', because the customary eigenvalue problem† of a square matrix is defined by the equations

$$A\mathbf{y} = \lambda\mathbf{y}, \quad \tilde{A}\mathbf{x} = \lambda\mathbf{x}, \tag{19}$$

while in our case—which remains meaningful for the general case of an arbitrary $n \times m$ matrix—the position of the vectors $\mathbf{x}$ and $\mathbf{y}$ on the right sides is *reversed*.

If we multiply the second equation (18) by A, we see at once that the $\mathbf{x}$-vectors in themselves can be defined as the solution of the eigenvalue problem

$$A\tilde{A}\mathbf{x} = \lambda^2\mathbf{x}, \tag{20}$$

while multiplication of the first equation by $\tilde{A}$ shows that similarly the $\mathbf{y}$ vectors in themselves can be defined as the solution of the eigenvalue problem

$$\tilde{A}A\mathbf{y} = \lambda^2\mathbf{y}. \tag{21}$$

Since, however, the matrices $A\tilde{A}$ and $\tilde{A}A$ are in themselves respectively symmetric $n \times n$ and $m \times m$ matrices, we see that the $\mathbf{x}$ vectors themselves are mutually orthogonal to each other, and so are the $\mathbf{y}$ vectors:

$$\mathbf{x}_i \cdot \mathbf{x}_k = \delta_{ik}, \quad \mathbf{y}_i \cdot \mathbf{y}_k = \delta_{ik}. \tag{22}$$

Moreover, the vectors $\mathbf{x}_i$, if plotted as columns, fill out a full $n \times n$ space, the vectors $\mathbf{y}_j$ a full $m \times m$ space. Hence we can include the solution of the given eigenvalue problem (18) in the following matrix diagram:

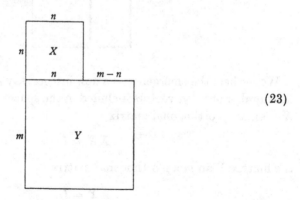

$$\tag{23}$$

Although the two spaces X and Y are separated, there is a correlation between them on the basis of the equation

$$\mathbf{x}_i = \frac{A\mathbf{y}_i}{\lambda_i}, \tag{24}$$

† Cf. [5], p. 64.

which shows that to every y_i vector a corresponding x_i vector can be found. And yet our diagram shows that this pairing between the two kinds of vector cannot hold generally if n and m are not equal. In our illustration $m > n$ and $m-n$ y_i vectors remain unpaired. But the relation (24) breaks down only for $\lambda_i = 0$. Thus we see that the eigenvalue $\lambda = 0$ must be present at least $m-n$ times. In actual fact the multiplicity of the zero eigenvalue may be still larger since it is possible that even some of the x_i vectors belong to the eigenvalue $\lambda = 0$.

It will now be our aim to *separate the zero eigenvalue from the non-vanishing eigenvalues*. We will thus bracket out all those principal axes x_i and y_i which are truly paired on the basis of a λ_i which is not zero. The matrix of these x_i vectors shall be denoted by X, the matrix of the corresponding y_i vectors by Y, while the subspaces associated with the zero eigenvalue shall be denoted by X_0 and Y_0.

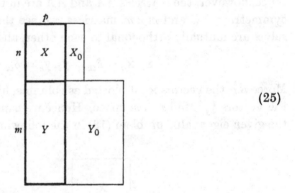

$$(25)$$

We see here the emergence of a new integer, say p, which characterizes the number of x_i, y_i vectors included in the spaces X and Y. The matrix X is an $n \times p$ orthogonal matrix

$$\tilde{X}X = I, \tag{26}$$

the matrix Y an $m \times p$ orthogonal matrix

$$\tilde{Y}Y = I. \tag{27}$$

(The products $X\tilde{X}$ and $Y\tilde{Y}$, however, need not be equal to I, because generally, if $p < \dfrac{n}{m}$, the matrices X and Y do not fill out their spaces).

The multiplicity of the zero eigenvalue is now

$$(n-p) + (m-p) = n + m - 2p.$$

The total number of eigenvalues must become equal to $n+m$, since the matrix S has $n+m$ rows and columns. Consequently, we get for the number of non-zero eigenvalues:

$$n+m-(n+m-2p) = 2p.$$

Why does our diagram display only p instead of $2p$ axes? The reason is that to every solution

$$(\mathbf{x}_i,\, \mathbf{y}_i,\, \lambda_i)$$

a second solution can be constructed, namely

$$(\mathbf{x}_i,\, -\mathbf{y}_i,\, -\lambda_i).$$

Hence all non-zero eigenvalues appear with both plus and minus signs. We will agree to omit all the negative eigenvalues and keep only the positive ones, since the negative eigenvalues do not add anything new to the eigenvalue problem. We can thus characterize p as the *number of positive eigenvalues* for which the shifted eigenvalue problem (18) is solvable. It can assume any value between 1 and the smaller of the two numbers n and m:

$$1 \leqslant p \leqslant \min\{n, m\}. \tag{28}$$

This number p coincides in fact with the 'rank' of the matrix A. We see that this fundamental number, which in the usual algebraic theory of Kronecker and Frobenius is defined on a completely different basis,† enters our eigenvalue problem again as a quantity of decisive importance.

4. The fundamental decomposition theorem

Apart from the two matrices X and Y we construct a diagonal matrix Λ which contains in the diagonal all the positive eigenvalues $\lambda_1, \lambda_2, ..., \lambda_p$, for which the shifted eigenvalue problem (18) is solvable:

$$\Lambda = \begin{bmatrix} \lambda_1 & \cdot & \cdot & \cdot & \cdot & \cdot \\ \cdot & \lambda_2 & \cdot & \cdot & \cdot & \cdot \\ \cdot & \cdot & \cdot & \cdot & \cdot & \cdot \\ \cdot & \cdot & \cdot & \cdot & \cdot & \cdot \\ \cdot & \cdot & \cdot & \cdot & \cdot & \cdot \\ \cdot & \cdot & \cdot & \cdot & \cdot & \lambda_p \end{bmatrix}.$$

Then the principal axis transformation of the matrix S leads to the

† Cf. [11], p. 10.

following fundamental decomposition theorem which holds without any exception for any matrix A which does not vanish identically:

$$A = X \Lambda \tilde{Y}. \tag{29}$$

The most remarkable feature of this decomposition is that the operator A *completely by-passes the zero fields* X_0, Y_0. Only those principal axes of S are needed for the construction of A, which belong to positive eigenvalues. Hence we will call the columns of the matrices X and Y the 'essential axes' of our problem. It will be possible to formulate the entire theory of solving the basic equation (1) in terms of the matrices X and Y, without any reference to the fields X_0, Y_0 which are associated with the eigenvalue $\lambda = 0.$†

5. Solution of the basic equation

We return to our original problem of solving the matrix equation (1). We will, however, change our notation by denoting the right side of the equation by $\mathbf{x}$:

$$A\mathbf{y} = \mathbf{x}. \tag{30}$$

Then, substituting for A the product (29) we obtain

$$X \Lambda \tilde{Y} \mathbf{y} = \mathbf{x}. \tag{31}$$

We transform our variables according to the law

$$\mathbf{x} = X\mathbf{x}', \quad \mathbf{y} = Y\mathbf{y}', \tag{32}$$

---

† Professor A. S. Householder called the author's attention to a paper of E. G. Kogbetliantz, 'Diagonalization of general complex matrices as a new method for solution of linear equations', *Proceedings of the International Congress of Mathematicians*, (North-Holland, Amsterdam, 1954), II, p. 356, which describes a numerical process for the diagonalization of an arbitrary matrix, on the basis of the equation $U^*AV = D$ (which in our notation becomes $\tilde{U}AV = D$ since it is tacitly understood that 'transposition' in the presence of complex elements includes the change of i to $-i$). No reference is made to the 'shifted eigenvalue problem' (18), nor to the decomposition theorem (29), (in which the zero-field is clipped away).

and obtain for the new variables the relation

$$\Lambda \mathbf{y}' = \mathbf{x}', \tag{33}$$

which can be solved in the form

$$\mathbf{y}' = \Lambda^{-1}\mathbf{x}'. \tag{34}$$

This solution is always possible because the diagonal matrix can have no vanishing elements in the diagonal. But premultiplication of (32) by $\tilde{X}$, respectively $\tilde{Y}$, gives

$$\mathbf{x}' = \tilde{X}\mathbf{x}, \quad \mathbf{y}' = \tilde{Y}\mathbf{y}, \tag{35}$$

and thus we obtain $\qquad \mathbf{y} = Y\Lambda^{-1}\tilde{X}\mathbf{x}. \tag{36}$

This solution gives the impression that *every* linear system has a solution and, in fact, a *unique* solution, which can hardly be expected of arbitrarily over-determined or under-determined systems. But actually this solution depended on the assumption (32) which is equivalent to the statement that the vector $\mathbf{x}$ is inside the X-space, the vector $\mathbf{y}$ inside the Y-space. Let us first consider the latter statement.

The operator A operates solely in the subspaces X and Y. We could describe the situation by imagining that in the matrix diagram (25) the fields X, Y are illuminated while the fields X_0, Y_0 remain entirely in the dark. Now the vector $\mathbf{y}$ can have a projection into Y_0 as well as a projection into Y. However, the given equation determines solely the projection into Y but leaves the projection into Y_0 completely undetermined. Under these circumstances it seems natural that we place $\mathbf{y}$ completely into the space Y and leave the determination of the projection into Y_0 to some additional information. The second equation of the system (32) can thus be conceived as a natural *normalization* of our solution, by putting it into the *space of the least number of dimensions* which is able to hold it.

While the second equation of (32) can be conceived as a matter of choice, this is not so with the first equation. The fact that the left side $\mathbf{x}$ is inside the space X, is not a matter of choice but a consequence of the given equation. If the given left side $\mathbf{x}$ does not satisfy this condition, then the given system is self-contradictory and thus unsolvable. We can thus conceive the condition.

$$\mathbf{x} = X\mathbf{x}', \tag{37}$$

as the *compatibility condition* of the given system which is necessary and sufficient for the existence of a solution. Hence we see that the two conditions (32) solve the problem of over-determination and under-determination. The first condition expresses the compatibility of the given system in the case of over-determination, the second condition normalizes the solution in the case of under-determination. The uniqueness of the solution (36) is thus explained.

The compatibility conditions (37) can be expressed in various equivalent forms. We can put it, for example, in the form of an *orthogonality condition*, expressing the orthogonality of the vector **x** to the space X_0:

$$\tilde{X}_0 . \mathbf{x} = 0. \tag{38}$$

This leads to the traditional formulation of the compatibility condition of an arbitrarily given linear system: 'A given linear system of equations is solvable if and only if the right side of the system is orthogonal to every independent solution of the adjoint homogeneous system.'

It is more adequate, however, to avoid any reference to the space X_0, in which the operator A is not active. We can stay completely within the confines of the space X and express the compatibility condition (37) in the form

$$\mathbf{x} = X\tilde{X}\mathbf{x}. \tag{39}$$

But we can go still further and include all the compatibility conditions of an arbitrary linear system into one single *scalar condition*. If the vector **x** falls completely within X, then the length of the vector and the length of its projection into this space become equal:

$$\mathbf{x}^2 = \xi_1^2 + \xi_2^2 + \dots + \xi_p^2, \tag{40}$$

where

$$\xi_i = \mathbf{x}_i . \mathbf{x}. \tag{41}$$

The converse of the theorem is equally true. The *scalar condition* (40) can thus be considered as the *necessary and sufficient condition for the solvability of the system* (30).

Transition into the realm of differential operators. It will now be our aim to translate these algebraic results into the domain of continuous operators, particularly *differential* operators. And here we encounter first of all the following deviations from the algebraic case: (1) The matrix $\tilde{A}$ is defined by a transposition of rows and columns. This operation does not allow a direct interpretation in the realm of differential equations. (2) A differential equation is usually associated with certain *boundary conditions*. We have to find a way of associating the matrix operators A and $\tilde{A}$ with the given differential operator, *plus* the proper boundary conditions. (Infinite domains are excluded from our considerations.)

We find the answer to these questions by considering the so-called 'bilinear identity'

$$\mathbf{x}.A\mathbf{y} - \mathbf{y}.\tilde{A}\mathbf{x} \equiv 0, \tag{42}$$

which holds for arbitrary vectors $\mathbf{x}$ and $\mathbf{y}$. If for a given matrix A we succeed in finding a matrix B which for arbitrary $\mathbf{x}$ and $\mathbf{y}$ satisfies the identity

$$\mathbf{x}.A\mathbf{y} - \mathbf{y}.B\mathbf{x} \equiv 0, \tag{43}$$

then we know that

$$B = \tilde{A}. \tag{44}$$

Hence we can define $\tilde{A}$ as that particular matrix B which satisfies the identity (43).

Now in the theory of linear differential equations the bilinear identity (42) appears in the form of 'Green's identity':

$$\int (v\,Du - u\,\tilde{D}v)\,d\tau = \text{boundary integral}, \tag{45}$$

which is always obtainable by the method of integrating by parts. The notation D refers to an arbitrary given linear differential operator (ordinary or partial), or systems of such operators (such as for example the left sides of the Cauchy–Riemann equations, or Maxwell's equations, etc.).

In addition to the differential equation

$$Du = \rho$$

some more or less stringent *boundary conditions* may be prescribed for u and its derivatives on the boundary of the domain. These conditions may generally be of the homogeneous or inhomogeneous type. We will agree, however, to replace any inhomogeneous boundary condition by the corresponding *homogeneous* condition. This is always possible by the device of replacing the original unknown function $\bar{u}$ by the sum

$$\bar{u} = u_0 + u, \tag{46}$$

where u_0 is chosen as some function which absorbs the given inhomogeneous boundary conditions, without satisfying, however, any differential equation. Then, considering u as the new unknown, we obtain the differential equation

$$Du = \rho - Du_0, \tag{47}$$

plus boundary conditions which are now of the *homogeneous* type; (the right side being equal to *zero*, instead of some prescribed values).†

Together with the given operator Du we will consider the 'adjoint (transposed) operator' $\tilde{D}v$ which is likewise augmented by suitably

---

† After obtaining our result for u it is not difficult to return to the original function $\bar{u}$ and formulate the results in terms of the original inhomogeneous boundary values (with $\rho = 0$ in most cases). The explicit carrying out of the substitution (46) is thus seldom demanded.

chosen boundary conditions. The definition of $\check{D}v$ follows from the expression on the left side of Green's identity (45). The definition of the 'adjoint boundary conditions' follows from a careful study of the boundary integral on the right side of Green's identity (45). We prescribe for v (and its derivatives) the minimum number of boundary conditions which are necessary and sufficient to make the boundary integral vanish at all points of the boundary. These conditions are once more of the homogeneous kind. The more over-determined the original problem was, the more under-determined is the adjoint problem, and vice versa.

If we now consider the 'shifted eigenvalue problem'

$$Du = \lambda v, \quad \check{D}v = \lambda u, \tag{48}$$

where u is subjected to the given, v to the adjoint boundary conditions, this problem has always the right degree of determination and allows an infinity of possible solutions for an infinite—but discrete—set of eigenvalues λ_i, among which we may find $\lambda = 0$ represented with a finite or infinite multiplicity. We *drop* these latter solutions and keep only the solutions which belong to the non-vanishing (and even positive) λ_i. We thus obtain an infinite set of orthogonal (and normalized) eigenfunctions $u_i(\tau)$ and a corresponding infinite set of orthogonal (and normalized) eigenfunctions $v_i(\sigma)$. These two sets of functions operate generally in two separate portions of Hilbert space, although they belong to the same domain of the independent variables. There is, however, a one-to-one correspondence between these two sets of functions, on account of the relation

$$v_i(\tau) = \frac{Du_i(\tau)}{\lambda_i}. \tag{49}$$

6. Solution of the differential equation $Du = v$

We will now consider the solution of the differential equation

$$Du = v, \tag{50}$$

where u is subjected to the given homogeneous boundary conditions. Although we have dropped the eigenfunctions associated with the eigenvalue zero, the remaining functions $u_i(\tau)$ and $v_i(\tau)$ are still sufficiently complete to serve as base vectors for the representation of the functions $u(\tau)$, respectively $v(\tau)$. Hence we can put

$$\left. \begin{array}{l} u(\tau) = \sum\limits_{i=1}^{\infty} c_i u_i(\tau), \\[2mm] v(\tau) = \sum\limits_{i=1}^{\infty} \gamma_i v_i(\tau). \end{array} \right\} \tag{51}$$

The differential equation (50) establishes the following relation between the two sets of coefficients:

$$c_i = \frac{\gamma_i}{\lambda_i}. \tag{52}$$

Our equation (50) is thus solvable by the infinite expansion

$$u(\tau) = \sum_{i=1}^{\infty} \frac{\gamma_i}{\lambda_i} u_i(\tau), \tag{53}$$

where

$$\gamma_i = \int v(\sigma) v_i(\sigma) \, d\sigma. \tag{54}$$

However, in view of the fact that the functions $v_i(\tau)$ are generally *not* complete (since we have dropped the $v_j^{(0)}$ which belong to $\lambda = 0$), it is necessary to test the given function $v(\tau)$ concerning compatibility. The solvability of the equation (50) demands that $v(\tau)$ is completely within the subspace of the $v_i(\tau)$. This means

$$\int v^2(\sigma) \, d\sigma = \sum_{i=1}^{\infty} c_i^2. \tag{55}$$

This 'completeness relation'—which corresponds to the previous algebraic relation (40)—represents the compatibility condition demanded of the right side of the differential equation (50) in the case of an over-determined system.

7. The two kinds of boundary value problems

If we approach our problem from the previously pursued algebraic angle by considering the given differential equation (50) as the limit of an algebraic set of equations, our general expectations will be as follows. In view of the arbitrarily judicious or injudicious choice of boundary conditions we are confronted with a system which may be arbitrarily over-determined or under-determined. As far as under-determination goes, the uncertainty of the solution is eliminated by a natural normal-ization of our solution, viz. by putting the solution into that u-space in which the operator D is activated. As far as over-determination goes, we have to test the right side of the equation whether or not it satisfies the condition that it is completely contained in that v-space in which the operator D is activated. If this condition is not satisfied, no solution is possible. If this condition is satisfied, a unique solution of the given boundary value problem is obtained. We do not see any further com-plications which may arise.

However, limit processes have their own intricacies and we know that unexpected things can happen because in the limit something may occur which did not occur any time during the limit process. For example the limit of a continuous set of functions may be a discontinuous function. A closer analysis reveals that something similar is at work in our problem.

Let us plot the entire λ_i-spectrum on the positive half-line, from zero to infinity. We have omitted the zero eigenvalue as not included by the operator. And yet, the zero eigenvalue may come into evidence in a more subtle manner. It is possible that our eigenvalue spectrum does not start with a certain finite $\lambda_1 = \epsilon$ but that $\lambda = 0$ is a *limit point*, which means that there are infinitely many eigenvalues which, although discrete, come to zero as close as we wish. These eigenvalues cannot be omitted as not belonging to the operator. They do belong to the operator and their presence has a strong influence on the solution of our problem.

Under these circumstances we can put the entire class of possible boundary value problems into two categories, according to the presence or absence of the limit point $\lambda = 0$.

7.1. Boundary value problems of the first kind.

This class of problems is characterized by the condition that $\lambda = 0$ *is not a limit point of the eigenvalue spectrum*. The traditional type of boundary value problems fall into this category. Within this class of problems we distinguish two subgroups:

(a) $\lambda = 0$ *is not included among the eigenvalues of the adjoint operator* $\tilde{D}$. This means that under the adjoint boundary conditions the equation

$$\tilde{D}v = 0 \tag{56}$$

has no non-vanishing solutions.

(b) $\lambda = 0$ *is included among the eigenvalues of the adjoint operator* $\tilde{D}$. This means that the equation (56) (under the adjoint boundary conditions) has a finite or infinite number of non-vanishing solutions.

The case (a). This is distinguished by the property that the given problem is solvable for arbitrarily prescribed $v(\tau)$,† without the demand of specific compatibility conditions. The solution can be given in the form of the infinite expansion (53). But the same solution may also be put in operational form, corresponding to the matrix solution (36) of the algebraic problem:

$$u(\tau) = \int G(\tau, \sigma)\, v(\sigma)\, d\sigma, \tag{57}$$

---

† The expression 'arbitrary' refers to an aribtrary function from the general class of 'functions of bounded variation'.

where $G(\tau, \sigma)$, the 'Green's function', is defined by the following infinite expansion:

$$G(\tau, \sigma) = \sum_{i=1}^{\infty} \frac{u_i(\tau)\, v_i(\sigma)}{\lambda_i}.$$ (58)

This expansion is a natural generalization of the well-known bilinear expansion† of a symmetric kernel function $K(\tau, \sigma) = K(\sigma, \tau)$:

$$K(\tau, \sigma) = \sum_{i=1}^{\infty} \frac{u_i(\tau)\, u_i(\sigma)}{\lambda_i}.$$ (59)

As we know, this expansion is not always convergent, although convergence can be obtained by averaging over an arbitrarily small neighbourhood of the point σ. The same holds of the expansion (58). Moreover, while the expansion (58) itself may diverge, we get a convergent result if it is applied in (57) under the integral sign, integrating term by term.

The function $G(\tau, \sigma)$ may also be characterized by the solution of the differential equation

$$DG(\underline{\tau}, \sigma) = \delta(\underline{\tau}, \sigma),$$ (60)

where $\delta(\tau, \sigma)$ is Dirac's delta function (the underlining of τ indicates that we consider τ as the variable while σ is a mere parameter).

As a special case within the subgroup (a) we can go one step still further and demand that not only the adjoint homogeneous equation (56) but also the given homogeneous equation

$$Du = 0$$ (61)

(under the given boundary conditions) shall have no non-vanishing solutions. Then the solution given in the form (53), or in the alternate form (57), (58), is unique not merely by normalization but unconditionally. In the algebraic sense we now have the limiting case $n = m = p$. This most restricted group of boundary value problems corresponds to Hadamard's 'well-posed' problem.

The case (b). If the equation (56) possesses non-vanishing solutions

$$v_1^{(0)}(\tau), \quad v_2^{(0)}(\tau), \quad \ldots, \quad v_r^{(0)}(\tau)$$ (62)

we come to the next group of boundary value problems. We *exclude* the solutions $v_i^{(0)}(\tau)$ from our system of eigenfunctions $v_i(\tau)$, and thus lose the completeness of our function system (we do the same with the $u_j^{(0)}(\tau)$, if they exist, with respect to the system $u_i(\tau)$).

The solution (53) is once more valid, and it is also possible to give the solution in terms of the Green's function (57), (58). This function is

† Cf. [2], p. 134.

occasionally called the 'generalized Green's function'† because, although
the bilinear expansion (58) retains its form without any change, the
defining differential equation (60) is now to be modified to

$$
\left.
\begin{aligned}
DG(\underline{\tau}, \sigma) &= \delta(\underline{\tau}, \sigma) - \sum_{i=1}^{r} v_i^{(0)}(\underline{\tau})\, v_i^{(0)}(\sigma), \\
\check{D}G(\tau, \underline{\sigma}) &= \delta(\tau, \underline{\sigma}) - \sum_{i=1}^{s} u_i^{(0)}(\tau)\, u_i^{(0)}(\underline{\sigma}).
\end{aligned}
\right\}
\tag{63}
$$

(The numbers r and s—which are completely independent of each other
—may be finite or infinite. If $s = 0$, the solution is unique not only by
normalization, but unconditionally.)

The only important difference between the cases (a) and (b) is that the
function $v(\tau)$ can no longer be chosen freely but has to satisfy the compatibility conditions

$$
\gamma_i^{(0)} = \int v(\sigma)\, v_i^{(0)}(\sigma)\, d\sigma = 0 \quad (i = 1, 2, \ldots, r). \tag{64}
$$

All these conditions are replaceable by one single scalar condition, viz.
the 'completeness relation' (55), which expresses the fact that $v(\tau)$ is
inside that subspace of the function space which is spanned by the
$v_i(\tau)$ alone, without the $v_j^{(0)}(\tau)$.

7.2. Boundary value problems of the second kind.

This class
of boundary value problems is characterized by the property that
$\lambda = 0$ *is a limit point of the eigenvalue spectrum.* These problems fall
outside the realm of the traditional type of boundary value problems.
We encounter them if an elliptic differential equation is characterized
by hyperbolic type of boundary conditions, or vice versa, or if, for
example, the heat conduction equation is characterized by end-values
instead of initial values. The problem indicated in figure (8) falls likewise
into this category.

These problems are solvable but require a more circumspect approach
than that employed in the previous class of problems. The characteristic
feature of the new eigenvalue problem is the *unusual distribution of the
eigenvalues* λ_i. This feature is deeply interwoven with a fundamental
question that concerns the general structure of the function space. The
concept of the 'function space' envisages a Euclidean space of infinitely
many dimensions. The various orthogonal function systems associated
with self-adjoint differential operators can be conceived as various

† Cf. [2], p. 356.

systems of orthogonal base vectors which in principle are all equivalent to each other. And yet the complete homogeneity of a Euclidean space in every direction does not correspond to the actual structure of the function space. In a Euclidean space the sequence in which we arrange our co-ordinate axes is entirely immaterial. In function space a definite *ordering* of the axes is demanded, in view of the fact that the infinite expansion into eigenfunctions has to *converge* to a definite limit $f(\tau)$. 'Convergence' by its very definition means that the terms of high order contribute negligibly small amounts to the expansion. This, however, assumes that we have arranged our functions properly, namely in order of decreasing significance. From where do we obtain such an ordering principle in a Euclidean space which is homogeneous in every direction?

We are used to the Sturm–Liouville type of eigenfunction problems† in which the ordering of the eigenfunctions is quite systematic and determined by the *magnitude of the eigenvalues* λ_i. By putting the λ_i in increasing order: $\lambda_1 \leqslant \lambda_2 \leqslant \lambda_3 \leqslant \ldots$, we obtain a natural ordering principle for the associated eigenfunctions $\phi_1(x), \phi_2(x), \phi_3(x), \ldots$. This principle carries over into the realm of partial differential equations if we deal with boundary value problems of the conventional type. In our present problem, however, we encounter a situation which is not of the conventional type. First of all, if $\lambda = 0$ is a limit point of the eigenvalue spectrum, the ordering of the λ_i in increasing order is no longer possible. But this ordering would not even be justified since it is no longer true that the orthogonal functions of primary importance belong to the smallest eigenvalues. We experience a peculiar 'inversion of eigenvalues', due to which certain eigenfunctions which should appear quite *late* in order of importance, appear in fact *very early* in the λ_i spectrum. For this reason we will speak of a 'residual spectrum', to which we will relegate all the eigenvalues (and associated eigenfunctions) which cluster around the value zero and which in fact represent eigenfunctions of high order.

In view of this situation we will establish the following procedure for a definite ordering of the λ_i. We prescribe a certain arbitrarily small ϵ and put the eigenvalues into two categories:

Group 1: all the eigenvalues $\lambda_i \geqslant \epsilon$, arranged in increasing order: $\lambda_1 \leqslant \lambda_2 \leqslant \lambda_3 \leqslant \ldots$ (together with the corresponding $u_i(\tau), v_i(\tau)$).

Group 2: all the eigenvalues (now denoted by λ_i') for which $\lambda_i' < \epsilon$, arranged in decreasing order: $\lambda_1' \geqslant \lambda_2' \geqslant \lambda_3' \geqslant \ldots$ (together with the corresponding $u_i'(\tau), v_i'(\tau)$).

† Cf. [2], p. 291.

The solution of the equation (50) under the present circumstances occurs once more by the infinite expansions (51) if we pay attention to the proper ordering of the eigenfunctions:

$$v(\tau) = \sum_{i=1}^{\infty} \gamma_i v_1(\tau) + \sum_{j=1}^{\infty} \gamma_j' v_j'(\tau), \tag{65}$$

where
$$\gamma_i = \int v(\sigma) v_i(\sigma) \, d\sigma, \quad \gamma_j' = \int v(\sigma) v_j'(\sigma) \, d\sigma, \tag{66}$$

and
$$u(\tau) = \sum_{i=1}^{\infty} \frac{\gamma_i}{\lambda_i} u_i(\tau) + \sum_{j=1}^{\infty} \frac{\gamma_j'}{\lambda_j'} u_j'(\tau). \tag{67}$$

In contradistinction to the previous type of problems, a solution in terms of the Green's function is no longer possible. We can define the kernel function

$$G(\tau, \sigma) = \sum_{i=1}^{\infty} \frac{u_i(\tau) v_i(\sigma)}{\lambda_i}, \tag{68}$$

but the corresponding function for the primed eigenfunctions does not exist. Hence we can give the solution only *partially* in terms of the Green's function; the addition of an infinite sum, extended over the residual spectrum, cannot be avoided:

$$u(\tau) = \int G(\tau, \sigma) v(\sigma) \, d\sigma + \sum_{i=1}^{\infty} \frac{\gamma_i'}{\lambda_i'} u_i'(\tau). \tag{69}$$

It is exactly this sum which represents the difference between the conventional and the unconventional type of boundary value problems. We observe that the division by very small λ_i' has the consequence that the sum on the right side (69) will generally *not converge*. Hence the function $v(\tau)$ cannot be prescribed arbitrarily. In order that $u(\tau)$ shall be quadratically integrable, it is necessary and sufficient that the following condition shall be satisfied:

$$\sum_{i=1}^{\infty} \left(\frac{\gamma_i'}{\lambda_i'} \right)^2 < \infty. \tag{70}$$

Beyond this condition, however, we will be generally obliged to demand the absolute convergence of the expansion coefficients:

$$\sum_{i=1}^{\infty} \left| \frac{\gamma_i'}{\lambda_i'} \right| < \infty. \tag{71}$$

The convergence conditions (70) and (71) *restrict* the class of functions $v(\tau)$ for which the given boundary value problem is solvable, although this restriction is *less stringent* than that encountered in the class I, case (b) type of problems. There the conditions (64) demanded that all the expansion coefficients $\gamma_i^{(0)}$ which belonged to the eigenfunctions associated with the eigenvalue zero must *vanish*. Now the demand is that the expansion coefficients γ_i' which belong to the eigenfunctions associated with *almost vanishing* eigenvalues, need not be zero but are *sufficiently small*.

The two subgroups (a) and (b) of the previous class of boundary value problems can once more be distinguished, with quite analogous results:

Case (a). *The value zero is not included among the eigenvalues of the adjoint operator* $\tilde{D}$. This case is covered by our foregoing discussions. The function $v(\tau)$ need not satisfy additional compatibility conditions, beyond the convergence conditions (70) and (71).

Case (b). *The value zero is included among the eigenvalues of the adjoint operator* $\tilde{D}$. Here the solvability of the given problem demands that the function $v(\tau)$ shall satisfy the additional completeness relation:

$$\int v^2(\sigma)\,d\sigma = \sum_{i=1}^{\infty}\gamma_i^2 + \sum_{j=1}^{\infty}\gamma_j'^2. \tag{72}$$

8. An alternative treatment

The same results can be obtained by a somewhat different approach in which the residual spectrum is brought in in the form of a *limit process*. We start again with a given ϵ and all the λ_i which are greater or equal to ϵ. They are once more ordered in increasing magnitude and once more we define the Green's function by the expansion (68). We indicate, however, that this function depends on ϵ:

$$G_\epsilon(\tau,\sigma) = \sum_{i=1}^{\infty}\frac{u_i(\tau)\,v_i(\sigma)}{\lambda_i}. \tag{73}$$

We now decrease ϵ to smaller and smaller values which means that $G_\epsilon(\tau,\sigma)$ goes more and more out of bound. However, while $G_\epsilon(\tau,\sigma)$ does not approach any limit, it is possible that the sequence of functions

$$u_\epsilon(\tau) = \int G_\epsilon(\tau,\sigma)\,v(\sigma)\,d\sigma \tag{74}$$

formed with the help of these $G_\epsilon(\tau,\sigma)$, approaches a definite limit:

$$u(\tau) = \lim_{\epsilon\to 0} u_\epsilon(\tau). \tag{75}$$

If this limit does not exist, the function $v(\tau)$ was not chosen from the class of permissible functions and the given problem has no solution. If the limit (75) does exist, the function $v(\tau)$ was given properly and the limit $u(\tau)$ represents the solution of the given boundary value problem.

9. Elastic vibrations

An interesting example of such a 'boundary value problem of the second kind', which at the same time has physical significance, is provided by the problem indicated in figure (8). Here the function $f(z)$ of the complex variable $z = x + iy$ is given along an open boundary only and our aim is to obtain $f(z)$ inside the given domain with the help of the boundary data. By putting

$$f(z) = \frac{\partial \phi}{\partial y} + i \frac{\partial \phi}{\partial x} \qquad (76)$$

we can formulate our problem in the following alternative fashion. 'Given the potential function $\phi(x,y)$ and its normal derivative $\partial \phi/\partial n$ along an arbitrarily small open portion s of the boundary curve S. Find the value of ϕ inside the given domain.' The associated eigenvalue problem is now

$$\Delta u = \lambda v, \quad \Delta v = \lambda u \qquad (77)$$

with the boundary conditions

$$\left. \begin{array}{ll} u = 0, & \dfrac{\partial u}{\partial n} = 0, \quad \text{along } s, \\[2mm] v = 0, & \dfrac{\partial v}{\partial n} = 0, \quad \text{along } S-s. \end{array} \right\} \qquad (78)$$

Now it is exactly this eigenvalue problem which characterizes the *vibrations of an elastic sheet, clamped along s and free along $S-s$.* This problem is well investigated for simple (particularly rectangular and circular) boundaries but only under the assumption that the *full* boundary is clamped. The interesting fact that for any partially open boundary the eigenvalue $\lambda = 0$ is a limit point of the eigenvalue spectrum—i.e. that there exist infinitely many vibrational modes which belong to arbitrarily small frequencies—has (to the author's knowledge) escaped the attention of the research workers in this field. And yet, we can demonstrate the existence of this limit point without any detailed calculations, solely on the basis of a logical argument.

First of all, the examination of the boundary conditions (78) shows that —in view of the complete symmetry of the differential equations (77) with respect to u and v—the following two problems possess exactly the same eigenvalue spectrum: (a) the sheet clamped along s and free otherwise; (b) the sheet clamped along $S-s$ and free otherwise. Let us now assume that $\lambda = 0$ is *not* a limit point of the eigenvalue spectrum. Then there must exist a definite *smallest* vibrational frequency ν_1. If we now enlarge the length of s by a small amount, we have from the standpoint of (a) *tightened* the boundary conditions and thus ν_1 must *increase*, while from the standpoint of the complementary problem (b) we have *weakened* the boundary conditions and thus ν_1 must *decrease*.† These two statements are self-contradictory, which disproves the existence of ν_1 and thus proves the existence of the limit point $\lambda = 0$.

How far can we tighten the boundary conditions before this limit point disappears? We have to clamp in fact the *entire boundary* for this purpose. Even an arbitrarily small unclamped part of the boundary will cause arbitrarily small frequencies. The limit point $\lambda = 0$ disappears suddenly and jumps up to a finite ν_1 in the moment that even the last portion of the free boundary becomes clamped. But can we believe that this will really happen? Can we assume that any clamping mechanism is so perfect that not even an arbitrarily small portion of the sheet might maintain its free mobility? If this question is answered by 'no', then we automatically admit a distribution of vibrational frequencies which is very different from the traditional one. We should then find that an elastic sheet, if struck with a hammer, contains overtones which are exceedingly *low*-pitched. This can be tested in an indirect way by putting the sheet under lateral pressure. This pressure decreases the vibrational frequencies, until a 'critical load' is reached at which the smallest frequency becomes reduced to zero. At that moment the elastic sheet collapses and we experience the phenomenon of 'buckling'.‡

Now it is a well-known fact that the actual critical load at which buckling occurs is by far *lower* than that evaluated theoretically.§ This surprising result is entirely understandable, if we take into account the effect of the 'residual spectrum' which must accompany even the slightest imperfections of the imposed boundary conditions. 'But'— we may be inclined to say—'this implies another absurdity since now buckling could occur under the slightest lateral pressure, which is certainly not the case.' This possibility is prevented, however, by that peculiar 'reversal of eigenvalues' that we have discussed before. A very small eigenvalue does not belong to a vibrational mode of *low* but of *high* order. This means that a very small eigenvalue would demand a rippling of the sheet of such fineness which is physically unrealizable. And thus the 'residual spectrum' in its *physical* manifestation does not start with the frequency zero but with a definite finite frequency which, however, is still much lower than the lowest frequency calculated for the case of perfect clamping.

† Cf. [2], p. 407.   ‡ Cf. [14], chap. IX, pp. 439–497.   § Cf. [14], p. 462.

The explanation suggested here of the reduced elastic stability of thin sheets is open to experimental verification, since imperfect clamping conditions can be experimentally generated and the corresponding critical loads (at least in rough approximation) calculated. The currently accepted explanation is based on very different considerations. The theory proposed by von Karman[15] operates with vibrations of finite amplitudes, while the theory of Donnell[3] is based on large initial deflections, caused by imperfections of the cylindrical shape. In both cases the discrepancy between theory and experiment is conceived as a non-linear effect which leads to exceedingly complex calculations. The present considerations do not go beyond the realm of the classical eigenvalue theory, although attention is called to the fact that the 'reversal of eigenvalues' which must come in operation under these conditions, confronts us with a situation which is not of the conventional type and which cannot be treated by the customary methods.

10. Summary

The present investigation endeavours to establish a common platform for the theory of linear partial differential equations, subjected to boundary conditions which may be chosen in an arbitrarily judicious or injudicious fashion and given in an arbitrarily over-abundant or underabundant number. The problem is approached from the domain of algebra, exploiting the isomorphism which exists between linear differential equations and systems of linear algebraic equations. First the general problem of $n \times m$ linear systems is solved on the basis of an eigenvalue problem which yields a unique solution of the problem if the right side satisfies the proper compatibility conditions. This method is then translated into the domain of differential equations. It is found that a general boundary value problem belongs to one of two categories, depending on the question whether the eigenvalue spectrum excludes or includes the value $\lambda = 0$ as a limit point. The conventional boundary value problems belong to the first category. The eigenvalue method is applicable, however, to both categories and yields the solution on the basis of an expansion into eigenfunctions. The solvability of the problem demands that the data satisfy compatibility conditions—and in the second category certain convergence conditions—which can be explicitly stated. The explicit construction of the solution and the testing of the data concerning solvability presupposes, however, the preliminary solution of the eigenvalue problem, excluding the eigensolutions associated with the eigenvalue zero which are not needed for the construction of the solution and which are dispensable also from the standpoint of the compatibility conditions.

REFERENCES

[1] Bernoulli, D. *Commentarii Academiae Petropolitanae*, 6, 108 (1732).

[2] Courant, R. and Hilbert, D. *Methods of Mathematical Physics*, I. Interscience Publishers, New York, London, 1953.

[3] Donnell, L. H. and Wan, C. C. Effect of imperfections on buckling of thin cylinders and columns under axial compression. *J. appl. Mech.* 17, 73–83 (1950).

[4] Euler, L. *Methodus inveniendi lineas curvas maximi minimive proprietate gaudentes* (1744).

[5] Frazer, R. A., Duncan, W. J. and Collar, A. R. *Elementary Matrices.* Cambridge University Press, 1950.

[6] Fredholm, I. Sur une classe d'équations fonctionelles. *Acta Math.* 27, 365–390 (1903).

[7] Hadamard, J. *Lectures on Cauchy's Problem.* Dover Publications, New York, 1952.

[8] Hellinger, E. and Toeplitz, O. *Integralgleichungen und Gleichungen mit unendlichvielen Unbekannten.* Chelsea Publishing Co., New York, 1953.

[9] Hilbert, D. *Grundzüge einer allgemeinen Theorie der linearen Integralgleichungen.* Chelsea Publishing Co., New York, 1953.

[10] Lagrange, J. *Miscellanea Taurinensia*, 1, pt. 3, 1759.

[11] MacDuffee, C. C. *The Theory of Matrices.* Chelsea Publishing Co., New York, 1946.

[12] Rayleigh, Lord. *The Theory of Sound*, I. Macmillan and Co., London, 1937.

[13] Schmidt, E. Zur Theorie der linearen und nichtlinearen Integralgleichungen. *Math. Ann.* 64, 161–174 (1907).

[14] Timoshenko, S. *Theory of Elastic Stability.* McGraw-Hill, New York and London, 1936.

[15] Kármán, Th. von and Tsien, H. The buckling of thin cylindrical shells under axial compression. *J. aero. Sci.* 8, 303–312 (1941).

Л. С. ПОНТРЯГИН

ОПТИМАЛЬНЫЕ ПРОЦЕССЫ РЕГУЛИРОВАНИЯ

В этом докладе я излагаю результаты, полученные моими учениками В. Г. Болтянским и Р. В. Гамкрелидзе и мною[1, 2, 3].

1. Постановка задачи

Пусть Ω—некоторое топологическое пространство. Будем говорить, что задан *управляемый процесс*, если имеется система обыкновенных дифференциальных уравнений

$$\frac{dx^i}{dt} = f^i(x^1, \ldots, x^n; u) = f^i(\bar{x}; u) \quad (i = 1, \ldots, n), \qquad (1)$$

или в векторной форме:

$$\frac{d\bar{x}}{dt} = \bar{f}(\bar{x}; u), \qquad (2)$$

где $x^1, \ldots, x^n$ — действительные функции времени t, $\bar{x} = (x^1, \ldots, x^n)$ — вектор n-мерного векторного пространства R, $u \in \Omega$, а

$$f^i(\bar{x}; u) \quad (i = 1, \ldots, n)$$

— функции, заданные и непрерывные для всех значений пары $(\bar{x}, u) \in R \times \Omega$. Предполагается также, что частные производные

$$\frac{\partial f^i}{\partial x^j} \quad (i, j = 1, \ldots, n)$$

также определены и непрерывны на всем пространстве $R \times \Omega$.

Для того, чтобы найти решение уравнения (2), определенное на отрезке $t_0 \leqslant t \leqslant t_1$, достаточно указать *функцию $u(t)$ управления* на отрезке $t_0 \leqslant t \leqslant t_1$ и начальное значение $\bar{x}_0$ решения при $t = t_0$. В соответствии с этим мы будем говорить, что задано *управление*

$$U = (u(t), t_0, t_1, \bar{x}_0) \qquad (3)$$

уравнения (2), если задана функция $u(t)$, отрезок $t_0 \leqslant t \leqslant t_1$ ее определения и начальное значение $\bar{x}_0$ решения $\bar{x}(t)$. В дальнейшем будут рассматриваться кусочно непрерывные функции управления $u(t)$, допускающие разрывы первого рода, и *непрерывные* решения уравнения (2). При этом управления $u(t)$ будут предполагаться непрерывными в начальной точке t_0 и полунепрерывными слева, т.е. удовлетворяющими условию $u(t-0) = u(t)$, $t > t_0$. Мы будем говорить, что управление (3) *переводит* точку $\bar{x}_0$ в точку $\bar{x}_1$, если соответствующее решение $\bar{x}(t)$ уравнения (2), удовлетворяющее начальному условию $\bar{x}(t_0) = \bar{x}_0$, удовлетворяет еще конечному условию: $\bar{x}(t_1) = \bar{x}_1$.

Пусть теперь $f^0(x^1, \ldots, x^n; u) = f^0(\bar{x}, u)$ — функция, определенная и непрерывная вместе со своими частными производными

$$\partial f^0/\partial x^j \quad (j = 1, \ldots, n),$$

на всем пространстве $R \times \Omega$. Каждому управлению (3) соответствует тогда число

$$L(U) = \int_{t_0}^{t_1} f^0(\bar{x}(t), u(t))\, dt.$$

Таким образом, L есть функционал управления (3). Управление U будем называть *оптимальным*, если, каково бы ни было управление

$$U^* = (u^*(t), t_0^*, t_1^*, \bar{x}_0),$$

переводящее точку $\bar{x}_0$ в точку $\bar{x}_1$, имеет место неравенство

$$L(U) \leqslant L(U^*).$$

Замечание 1. Если (3) — оптимальное управление уравнения (2), $\bar{x}(t)$ — соответствующее ему решение уравнения (2), а $t_2 < t_3$ — две точки отрезка $t_0 \leqslant t \leqslant t_1$, то $U' = (u(t), t_2, t_3, \bar{x}(t_2))$ есть также оптимальное управление.

Замечание 2. Если (3) — оптимальное управление уравнения (2), переводящее точку $\bar{x}_0$ в точку $\bar{x}_1$, а τ — произвольное число, то

$$U'' = (u(t-\tau), t_0+\tau, t_1+\tau, \bar{x}_0)$$

—также оптимальное управление, переводящее точку $\bar{x}_0$ в $\bar{x}_1$.

Важным частным случаем является тот, когда функция $f^0(\bar{x}; u)$ определяется равенством $\quad f^0(\bar{x}, u) \equiv 1.$ \hfill (4)

В этом случае имеем $\quad L(U) = t_1 - t_0,$

и оптимальность управления U означает *минимальность времени перехода* из положения $\bar{x}_0$ в положение $\bar{x}_1$.

В применениях важен случай, когда Ω является замкнутой областью некоторого r-мерного эвклидова пространства E; тогда $u = (u^1, \ldots, u^r)$, и один управляющий параметр u превращается в систему числовых параметров $u^1, \ldots, u^r$. В случае, когда Ω представляет собой открытое множество пространства E, сформулированная здесь вариационная задача является частным случаем задачи Лагранжа ([4], стр. 225), и основной результат, приводимый ниже (принцип максимума), совпадает с известным критерием Вейер- штрасса. Для приложений важен, однако, случай, когда управ- ляющие параметры удовлетворяют неравенствам, включающим равенства, например $|u^i| \leqslant 1 \quad (i = 1, \ldots, r)$.

В этом случае критерий Вейерштрасса, очевидно, неверен, и приво- димый ниже результат является новым.

2. Необходимые условия оптимальности (принцип максимума)

Для формулировки необходимого условия оптимальности введем в рассмотрение вектор $\tilde{x} = (x^0, x^1, \ldots, x^n)$

$(n + 1)$-мерного эвклидова пространства S и рассмотрим управляемый процесс

$$\frac{dx^i}{dt} = f^i(x^1, \ldots, x^n, u) = f^i(\bar{x}, u) = f^i(\tilde{x}, u) \quad (i = 0, 1, \ldots, n), \qquad (5)$$

или в векторной форме

$$\frac{d\tilde{x}}{dt} = \tilde{f}(\tilde{x}, u), \qquad (6)$$

где $f^0(\bar{x}, u)$ есть функция, которая определяет функционал L. Для того, чтобы, зная управление (3) уравнения (2), получить управление уравнения (6), достаточно, исходя из начального значения

$$\bar{x}_0 = (x_0^1, \ldots, x_0^n),$$

задать начальное значение $\tilde{x}_0$ уравнения (6). Мы определим вектор $\tilde{x}_0$, положив

$$\tilde{x}_0 = (0, x_0^1, \ldots, x_0^n).$$

Этим способом управление (3) уравнения (2) однозначно определяет управление уравнения (6), и мы просто будем считать, что (3) есть управление уравнения (6). Если теперь управление (3) переводит начальное значение $\tilde{x}_0$ уравнения (6) в конечное значение

$$\tilde{x}_1 = (x_0^1, x_1^1, \ldots, x_1^n),$$

то мы имеем $L(U) = x_1^0,$

и этим определяется связь уравнения (6) с формулированной ранее вариационной задачей.

Наряду с контравариантным вектором $\tilde{x}$ пространства S рассмотрим вспомогательный ковариантный вектор

$$\tilde{\psi} = (\psi_0, \ldots, \psi_n)$$

этого пространства и составим функцию

$$K(\tilde{\psi}, \tilde{x}, u) = (\tilde{\psi}, \tilde{f}(\tilde{x}, u))$$

(справа стоит скалярное произведение векторов $\tilde{\psi}$ и $\tilde{f}$).

При фиксированных значениях $\tilde{\psi}$ и $\tilde{x}$ функция K становится функцией параметра u; верхнюю грань значений этой функции обозначим через $N(\tilde{\psi}, \tilde{x})$. Составим, далее, гамильтонову систему уравнений

$$\frac{dx^i}{dt} = \frac{\partial K}{\partial \psi_i} \quad (i = 0, \ldots, n), \tag{7}$$

$$\frac{d\psi_i}{dt} = -\frac{\partial K}{\partial x^i} \quad (i = 0, \ldots, n). \tag{8}$$

Непосредственно видно, что система (7) совпадает с системой (5), система же (8) есть

$$\frac{d\psi_0}{dt} = 0, \quad \frac{d\psi_j}{dt} = -\sum_{i=0}^{n} \psi_i \frac{\partial f^i(\tilde{x}, u)}{\partial x^j} \quad (j = 1, \ldots, n). \tag{9}$$

Теорема 1. Пусть (3) — *оптимальное управление уравнения* (2) *и* $\bar{x}(t)$ — *соответствующее ему решение уравнения* (2). *Дополним вектор* $\bar{x}(t)$ *до вектора* $\tilde{x}(t)$, *положив*

$$x^0(t) = \int_{t_0}^{t} f^0(\bar{x}(t), u(t)) \, dt.$$

Существует тогда такая ненулевая непрерывная вектор-функция $\tilde{\psi}(t)$, *что*

$$K(\tilde{\psi}(t_0), \tilde{x}(t_0), u(t_0)) = 0 \quad (\psi_0(t_0) \leqslant 0), \tag{10}$$

а функции $\quad \tilde{\psi}(t), \quad \tilde{x}(t), \quad u(t)$

составляют решение гамильтоновой системы (7), (8), *причем*

$$K(\tilde{\psi}(t), \tilde{x}(t), u(t)) = N(\tilde{\psi}(t), \tilde{x}(t)); \tag{11}$$

при этом оказывается, что функция $K(\tilde{\psi}(t), \tilde{x}(t), u(t))$ *постоянна, так что*

$$K(\tilde{\psi}(t), \tilde{x}(t), u(t)) \equiv 0. \tag{12}$$

Для формулировки необходимого условия в случае, когда речь идет о минимализации времени (см. (4)), составим гамильтонову функцию

$$H(\bar{\psi}, \bar{x}, u) = (\bar{\psi}, \bar{f}(\bar{x}, u)).$$

При фиксированных значениях $\overline{\psi}$ и $\overline{x}$ функция $H(\overline{\psi}, \overline{x}, u)$ становится функцией параметра u. Верхнюю грань значений этой функции обозначим через $M(\overline{\psi}, \overline{x})$. Составим, далее, гамильтонову систему

$$\frac{dx^j}{dt} = \frac{\partial H}{\partial \psi_j} \quad (j = 1, \ldots, n), \tag{13}$$

$$\frac{d\psi_j}{dt} = -\frac{\partial H}{\partial x^j} \quad (j = 1, \ldots, n). \tag{14}$$

Очевидно, что система (13) совпадает с системой (1), а система (14) есть

$$\frac{d\psi_j}{dt} = -\sum_{k=1}^{n} \psi_k \frac{\partial f^k(\overline{x}, u)}{\partial x^j} \quad (j = 1, \ldots, n). \tag{15}$$

Теорема 2. Пусть (3) —*оптимальное для функционала* (4) *управление уравнения* (2) *и* $\overline{x}(t)$ —*соответствующее этому управлению решение уравнения* (2). *Существует тогда такая ненулевая непрерывная вектор-функция* $\overline{\psi}(t) = (\psi_1(t), \ldots, \psi_n(t))$, *что*

$$H(\overline{\psi}(t_0), \overline{x}(t_0), u(t_0)) \geqslant 0,$$

а функции $\overline{\psi}(t), \quad \overline{x}(t), \quad u(t)$

удовлетворяют гамильтоновой системе уравнений (13), (14), *причем*

$$H(\overline{\psi}(t), \overline{x}(t), u(t)) = M(\overline{\psi}(t), \overline{x}(t)). \tag{16}$$

Оказывается, кроме того, что функция $H(\overline{\psi}(t), \overline{x}(t), u(t))$ *постоянна, так что*

$$H(\overline{\psi}(t), \overline{x}(t), u(t)) \geqslant 0. \tag{17}$$

Теорема 2 непосредственно вытекает из теоремы 1.

Главным содержанием теорем 1 и 2 являются равенства (11) и (16). Поэтому теорема 2, первоначально опубликованная в качестве гипотезы в заметке[1], названа *принципом максимума*. В этом же смысле и теореме 1 естественно присвоить наименование *принципа максимума*.

3. Доказательство принципа максимума (*теоремы 1 и 2*)

Докажем теорему 1. В доказательстве использованы некоторые конструкции Макшейна[5]. Пусть (3) — некоторое управление уравнения (6) и $\tilde{x}(t)$ —соответствующее ему решение уравнения (6). Система уравнений в вариациях для системы (5) вблизи решения $\tilde{x}(t)$ записывается, как известно, в виде

$$\frac{dy^i}{dt} = \sum_{j=0}^{n} \frac{\partial f^i(\tilde{x}(t), u(t))}{\partial x^j} y^j \quad (i = 0, 1, \ldots, n). \tag{18}$$

Записывая решение системы (18) в векторной форме, получаем вектор

$$\tilde{y}(t) = (y^0(t), \dots, y^n(t)).$$

В дальнейшем будут рассматриваться только непрерывные решения $\tilde{y}(t)$. Систему уравнений в вариациях, как известно, можно истолковать следующим образом. Пусть $\tilde{y}_0$ — произвольный вектор пространства S. Зададимся начальным значением†

$$\tilde{x}_0 + \epsilon\tilde{y}_0 + \epsilon O(\epsilon)$$

для решения уравнения (6). Тогда само решение уравнения (6) с этим начальным значением записывается в форме:

$$\tilde{x}(t) + \epsilon\tilde{y}(t) + \epsilon O(\epsilon),$$

где $\tilde{y}(t)$ есть решение системы (18), взятое с начальным значением $\tilde{y}_0$. Мы будем говорить, что решение $\tilde{y}(t)$ системы (18) является *перенесением* вектора $\tilde{y}_0$, заданного в начальной точке $\tilde{x}_0$ траектории $\tilde{x}(t)$, вдоль всей траектории. В том же смысле можно сказать, что решение $\tilde{y}(t)$ является перенесением вектора $\tilde{y}(\tau)$, заданного в точке $\tilde{x}(\tau)$ траектории $\tilde{x}(t)$, вдоль всей траектории.

Наряду с контравариантным вектором $\tilde{y}(t)$, являющимся решением системы (18), рассмотрим ковариантный вектор $\tilde{\psi}(t)$, являющийся решением системы (8). Непосредственно проверяется, что

$$\frac{d}{dt}(\tilde{\psi}(t), \tilde{y}(t)) = 0,$$

так что $\qquad\qquad (\tilde{\psi}(t), \tilde{y}(t)) = \text{const.} \qquad\qquad (19)$

Если истолковывать ковариантный вектор $\tilde{\psi}(t)$ как плоскость, проведенную через точку $\tilde{x}(t)$, то можно сказать, что плоскость $\tilde{\psi}(t)$ является *перенесением* плоскости $\tilde{\psi}(\tau)$, заданной в точке $\tilde{x}(\tau)$ траектории $\tilde{x}(t)$, вдоль всей траектории.

Вариацией управления (3) будем называть управление

$$U^* = U^*(\epsilon, \alpha) = (u^*(t), t_0, t_1 + \alpha\epsilon, \tilde{x}_0),$$

зависящее от параметра ϵ и действительного числа α, определенное для всех достаточно малых положительных значений параметра ϵ и удовлетворяющее следующему условию: Решение $\tilde{x}^*(t)$ уравнения (6), соответствующее управлению U^*, в точке $t = t_1 + \epsilon\alpha$ может быть записано в виде $\qquad \tilde{x}(t_1) + \epsilon\tilde{\delta}(U^*) + \epsilon O(\epsilon),$

где $\tilde{\delta}(U^*)$ не зависит от ϵ.

† В дальнейшем символ $O(\epsilon)$ используется как типическое обозначение для величин, стремящихся к нулю вместе с ϵ.

Семейство Δ вариаций одного и того же управления (3) будем называть *допустимым*, если наряду с каждыми двумя вариациями $U_1^*(\epsilon,\alpha_1)$ и $U_2^*(\epsilon,\alpha_2)$ в нем найдется при любых неотрицательных γ_1, γ_2 третья вариация $U^*(\epsilon,\gamma_1\alpha_1+\gamma_2\alpha_2)$, удовлетворяющая условию

$$\tilde\delta(U^*) = \gamma_1\tilde\delta(U_1^*)+\gamma_2\tilde\delta(U_2^*). \tag{20}$$

Сконструируем теперь специальную вариацию

$$U^*(\epsilon,\alpha) = V(\epsilon,\alpha,\tau,\sigma,u^*),$$

зависящую от точки τ полуинтервала $t_0 < t \leqslant t_1$ (причем при $\alpha < 0$ должно быть $\tau < t_1$), неотрицательного числа σ и точки u^* пространства Ω. Вариацию $V(\epsilon,\alpha,\tau,\sigma,u^*)$ определим, задав функцию $u^*(t)$ соотношениями

$$u^*(t) = \begin{cases} u(t) \text{ при } t_0 \leqslant t \leqslant \tau-\epsilon\sigma, \\ u^* \text{ при } \tau-\epsilon\sigma < t \leqslant \tau, \\ u(t) \text{ при } \tau < t \leqslant t_1, \\ u(t_1) \text{ при } t_1 < t \leqslant t_1+\epsilon\alpha \quad (\text{если } \alpha > 0). \end{cases} \tag{21}$$

Легко построить допустимое семейство Δ, содержащее все вариации типа (21). Это семейство Δ и будет положено в основу дальнейших построений.

Каждой вариации U^* допустимого семейства Δ соответствует вектор $\tilde\delta(U^*)$, выходящий из точки $\tilde x_1$. Совокупность всех этих векторов заполняет выпуклый конус (см. (20)) Π с вершиной в точке $\tilde x_1$. Пусть

$$\tilde v = (-1,0,\ldots,0)$$

— вектор, выходящий из точки $\tilde x_1$ и идущий в направлении отрицательной оси x^0 в пространстве S. Если конус Π содержит конец вектора $\tilde v$ в качестве внутренней точки, то управление U не является оптимальным. Пусть, в самом деле, $U^*\in\Delta$ — та вариация управления U, для которой

$$\tilde\delta(U^*) = \tilde v.$$

Обозначая через $\tilde x_1^*$ точку, в которую переходит точка $\tilde x_0$ при управлении U^*, получаем

$$\tilde x_1^* = \tilde x_1+\epsilon\tilde v+\epsilon O(\epsilon).$$

Расщепляя это равенство на скалярное для нулевой координаты и векторное для остальных координат, получаем

$$L(U^*) = x_1^{0*} = x_1^0-\epsilon+\epsilon O(\epsilon) = L(U)-\epsilon+\epsilon O(\epsilon),$$

$$\bar x_1^* = \bar x_1+\epsilon O(\epsilon).$$

Таким образом, функционал уменьшен на величину порядка ϵ, а конец траектории отличается от желательного на величину $\epsilon O(\epsilon)$. Уточнение этого построения приводит нас к такой вариации $U^* \epsilon \Delta$, для которой конец $\tilde{x}_1^{\sharp}$ траектории $\tilde{x}^{\sharp}$ удовлетворяет точному равенству $\tilde{x}_1^{\sharp} = \tilde{x}_1 - \epsilon \tilde{v}$, а это противоречит предположению об оптимальности управления U.

Итак, предполагая, что управление U оптимально, мы будем считать в дальнейшем, что вектор $\tilde{v}$ не является внутренним для конуса П. Так как конус П выпуклый, то для него существует такая опорная плоскость Г, что сам конус лежит в одном полупространстве (замкнутом), определяемом этой плоскостью, а вектор $\tilde{v}$ — в другом. Обозначая через $\tilde{\psi}_1$ ковариантный вектор, соответствующий плоскости Г, выбранный с надлежащим знаком, мы получаем

$$(\tilde{\psi}_1, \delta(U^*)) \leqslant 0 \quad (U^* \epsilon \Delta), \tag{22}$$

$$(\tilde{\psi}_1, \tilde{v}) \geqslant 0. \tag{23}$$

Из неравенства (23) сразу следует неравенство

$$\psi_{01} \leqslant 0. \tag{24}$$

Обозначим через $\tilde{\psi}(t)$ ковариантный вектор, получающийся перенесением вектора $\tilde{\psi}_1$, заданного в точке $\tilde{x}_1$, вдоль всей траектории $\tilde{x}(t)$. Покажем, что вектор-функция $\tilde{\psi}(t)$ и есть та, существование которой утверждается в теореме 1.

Пусть $V(\epsilon, 0, \tau, \sigma, u^*)$ — произвольная специальная вариация (см. (21)) семейства Δ и $\tilde{x}^*(t)$ — соответствующее ей решение уравнения (6). Простые вычисления дают

$$\tilde{x}^*(\tau) = \tilde{x}(\tau) + \epsilon[\tilde{f}(\tilde{x}(\tau), u^*) - \tilde{f}(\tilde{x}(\tau), u(\tau))] + \epsilon O(\epsilon).$$

Обозначим через $\tilde{y}(t)$ вектор, получающийся из вектора

$$\tilde{y}(\tau) = \tilde{f}(\tilde{x}(\tau), u^*) - \tilde{f}(\tilde{x}(\tau), u(\tau)),$$

заданного в точке $\tilde{x}(\tau)$, путем переноса вдоль траектории $\tilde{x}(t)$. Тогда мы имеем

$$\tilde{x}^*(t_1) = \tilde{x}_1 + \epsilon \tilde{y}(t_1) + \epsilon O(\epsilon).$$

Так как вектор $\tilde{y}(t_1)$ принадлежит конусу П, то в силу неравенства (22) получаем

$$(\tilde{\psi}_1, \tilde{y}(t_1)) \leqslant 0.$$

В силу (19) отсюда получаем

$$(\tilde{\psi}(\tau), \tilde{f}(\tilde{x}(\tau), u^*) - \tilde{f}(\tilde{x}(\tau), u(\tau))) \leqslant 0.$$

Переписывая последнее неравенство в обозначениях функции K, получаем неравенство

$$K(\tilde{\psi}(\tau), \tilde{x}(\tau), u(\tau)) \geqslant K(\tilde{\psi}(\tau), \tilde{x}(\tau), u^*),$$

эквивалентное равенству (11).

Пусть теперь $U^* = V(\epsilon, \alpha, \tau, 0, u^*)$. Решение уравнения (6), соответствующее этому управлению U^*, обозначим через $\tilde{x}^*(t)$. Мы имеем очевидно

$$\tilde{x}^*(t_1 + \alpha\epsilon) = \tilde{x}_1 + \epsilon\tilde{\delta}(U^*) + \epsilon O(\epsilon),$$

где

$$\tilde{\delta}(U^*) = \alpha\tilde{f}(\tilde{x}_1, u(t_1)).$$

Так как вектор $\tilde{\delta}(U^*)$ принадлежит конусу Π, то, в силу неравенства (22), получаем

$$\alpha(\tilde{\psi}_1, \tilde{f}(\tilde{x}_1, u(t_1))) \leqslant 0.$$

Ввиду того, что α есть произвольное действительное число, последнее неравенство возможно лишь при условии

$$(\tilde{\psi}_1, \tilde{f}(\tilde{x}_1, u(t_1))) = 0,$$

т.е. при

$$K(\tilde{\psi}(t_1), \tilde{x}(t_1), u(t_1)) = 0. \tag{25}$$

Докажем, наконец, что функция $K(t) = K(\tilde{\psi}(t), \tilde{x}(t), u(t))$ переменного t постоянна. Пусть $t_0 \leqslant t_2 < t_3 \leqslant t_1$, причем на полуинтервале $t_2 < t \leqslant t_3$ функция $u(t)$ непрерывна. Покажем, что на этом полуинтервале функция $K(t)$ постоянна. Возьмем две произвольные точки τ_0 и τ_1 полуинтервала $t_2 < t \leqslant t_3$. В силу (11) имеем

$$K(\tilde{\psi}(\tau_0), \tilde{x}(\tau_0), u(\tau_0)) - K(\tilde{\psi}(\tau_0), \tilde{x}(\tau_0), u(\tau_1)) \geqslant 0,$$

$$-K(\tilde{\psi}(\tau_1), \tilde{x}(\tau_1), u(\tau_1)) + K(\tilde{\psi}(\tau_1), \tilde{x}(\tau_1), u(\tau_0)) \leqslant 0.$$

Прибавляя к обеим частям этих неравенств разность $K(\tau_1) - K(\tau_0)$, получим неравенства

$$-K(\tilde{\psi}(\tau_0), \tilde{x}(\tau_0), u(\tau_0)) + K(\tilde{\psi}(\tau_1), \tilde{x}(\tau_1), u(\tau_0)) \leqslant K(\tau_1) - K(\tau_0)$$

$$\leqslant K(\tilde{\psi}(\tau_1), \tilde{x}(\tau_1), u(\tau_1)) - K(\tilde{\psi}(\tau_0), \tilde{x}(\tau_0), u(\tau_1)). \tag{26}$$

Далее, так как функция $K(\tilde{\psi}(t), \tilde{x}(t), u(\tau))$ переменного t на отрезке $t_2 < t \leqslant t_3$ непрерывна и имеет производную, равную нулю в силу (7), (8), то крайние члены неравенств (26) исчезают. Таким образом, $K(\tau_1) - K(\tau_0) = 0$, т.е. $K(t) = \text{const}$ на полуинтервале $t_2 < t \leqslant t_3$.

Пусть теперь τ_0 — точка разрыва функции $u(t)$ и $\tau_1 > \tau_0$ —

близкая к τ_0 точка. Если $K(\tau_0) > K(\tau_1)$, то при достаточно малом $\tau_1 - \tau_0$ мы имеем

$$K(\tilde{\psi}(\tau_1), \tilde{x}(\tau_1), u(\tau_0)) > K(\tilde{\psi}(\tau_1), \tilde{x}(\tau_1), u(\tau_1)),$$

что противоречит равенству (11). Если же $K(\tau_0) < K(\tau_1)$, то при достаточно малом $\tau_1 - \tau_0$ мы имеем

$$K(\tilde{\psi}(\tau_0), \tilde{x}(\tau_0), u(\tau_1)) > K(\tilde{\psi}(\tau_0), \tilde{x}(\tau_0), u(\tau_0)),$$

что также противоречит равенству (11). Таким образом,

$$K(\tau_0) = K(\tau_0 + 0).$$

Из доказанного вытекает (см. (25)) справедливость равенства (12) на всем отрезке $t_0 \leqslant t \leqslant t_1$, чем, в частности, доказано первое из соотношений (10). Второе из соотношений (10) следует из неравенства (24) в силу первого уравнения (9).

Итак, теорема 1 полностью доказана.

Замечание к теореме 1. Теорема 1 остается справедливой и в случае, если в качестве допустимых управляющих функций $u(t)$ рассматривать *измеримые* ограниченные функции; при этом равенство (11) для оптимального управления выполняется почти всюду.

4. Оптимальные в смысле быстродействия линейные управления

Важным для приложений и хорошо иллюстрирующим общие результаты примером является линейная управляемая система

$$\frac{dx^i}{dt} = \sum_{j=1}^{n} a_j^i x^j + \sum_{k=1}^{r} b_k^i u^k \quad (i = 1, \ldots, n),$$

где $\bar{u} = (u^1, \ldots, u^r)$ есть точка выпуклого замкнутого ограниченного многогранника Ω, расположенного в линейном пространстве E с координатами $u^1, \ldots, u^r$. В векторном виде эта система может быть записана так:

$$\frac{d\bar{x}}{dt} = A\bar{x} + B\bar{u}, \tag{27}$$

где A — линейный оператор в пространстве R переменных $x^1, \ldots, x^n$, а B — линейный оператор из пространства E в пространство R. Мы будем рассматривать здесь только задачу о минимализации функционала $\int_{t_0}^{t_1} dt$, т.е. задачу минимализации времени перехода.

Для получения некоторых результатов характера единственности мы будем налагать на управляемое уравнение (27) нижеследующие условия (A), (B), роль которых выяснится в дальнейшем:

(А) Пусть $\overline{w}$ — некоторый вектор, имеющий направление какого-либо из ребер многогранника Ω; тогда вектор $B\overline{w}$ не принадлежит никакому истинному подпространству пространства R, инвариантному относительно оператора A; таким образом, векторы

$$B\overline{w}, \quad AB\overline{w}, \quad \ldots, \quad A^{n-1}B\overline{w} \qquad (28)$$

линейно независимы в пространстве R всякий раз, когда $\overline{w}$ есть вектор, имеющий направление одного из ребер многогранника Ω.

(В) Начало координат пространства E является внутренней точкой многогранника Ω.

Функция $H(\overline{\psi}, \overline{x}, \overline{u})$ в нашем случае имеет вид

$$H = (\overline{\psi}, A\overline{x}) + (\overline{\psi}, B\overline{u}), \qquad (29)$$

а система (15) записывается в виде

$$\frac{d\psi_j}{dt} = -\sum_{i=1}^{n} a_j^i \psi_i \quad (j = 1, \ldots, n),$$

или, в векторной форме, $\quad \dfrac{d\overline{\psi}}{dt} = -A^*\overline{\psi}. \qquad (30)$

Очевидно, что функция H, рассматриваемая как функция переменного $\overline{u} \in \Omega$ достигает максимума одновременно с функцией

$$(\overline{\psi}, B\overline{u}).$$

В соответствии с этим обозначим через $P(\overline{\psi})$ максимум функции $(\overline{\psi}, B\overline{u})$, рассматриваемой как функция переменного $\overline{u} \in \Omega$. Из теоремы 2 следует, таким образом, что если

$$U = (\overline{u}(t), t_0, t_1, \overline{x}_0)$$

есть оптимальное управление уравнения (27), то существует такое решение $\overline{\psi}(t)$ уравнения (30), что

$$(\overline{\psi}(t), B\overline{u}(t)) = P(\overline{\psi}(t)). \qquad (31)$$

Так как уравнение (30) не содержит неизвестных функций $\overline{x}(t)$ и $\overline{u}(t)$, то все решения уравнения (30) легко могут быть найдены, и тем самым по условию (31) легко могут быть найдены и все оптимальные управления $\overline{u}(t)$ уравнения (27). Вопрос о том, насколько однозначно условие (31) определяет управление $\overline{u}(t)$ через функцию $\overline{\psi}(t)$, решается нижеследующей теоремой:

Теорема 3. Если выполнено условие (А), *то при заданном нетривиальном решении* $\overline{\psi}(t)$ *уравнения* (30) *соотношение* (31) *однозначно определяет управляющую функцию* $\bar{u}(t)$; *при этом оказывается, что функция* $\bar{u}(t)$ *кусочно постоянна и ее значениями являются лишь вершины многогранника* Ω.

Доказательство. Так как функция

$$(\overline{\psi}(t), B\bar{u}), \qquad (32)$$

рассматриваемая как функция вектора $\bar{u}$, линейна, то она либо постоянна, либо достигает своего максимума на границе многогранника Ω. Это же соображение применимо и к каждой грани многогранника Ω. Таким образом, либо функция (32) достигает своего максимума лишь в одной вершине многогранника Ω, либо же достигает его на целой грани многогранника Ω. Покажем, что в силу условия (А) последнее возможно лишь для конечного числа значений t. Допустим, что функция (32) достигает своего максимума (и, следовательно, постоянна) на некоторой грани Γ многогранника Ω. Пусть $\bar{w}$ — вектор, имеющий направление некоторого ребра грани Γ. В силу постоянства функции (32) на грани Γ имеем

$$(\overline{\psi}(t), B\bar{w}) = 0.$$

Если бы это соотношение имело место для бесконечного множества значений переменного t, то оно выполнялось бы тождественно по t и, дифференцируя его последовательно по t, мы получили бы

$$\left.\begin{array}{r}(\overline{\psi}(t), B\bar{w}) = 0, \\ (A^{*}\overline{\psi}(t), B\bar{w}) = (\overline{\psi}(t), AB\bar{w}) = 0, \\ (A^{*2}\overline{\psi}(t), B\bar{w}) = (\overline{\psi}(t), A^{2}B\bar{w}) = 0, \\ \cdots\cdots\cdots\cdots\cdots\cdots\cdots\cdots\cdots\cdots \\ (A^{*n-1}\overline{\psi}(t), B\bar{w}) = (\overline{\psi}(t), A^{n-1}B\bar{w}) = 0, \end{array}\right\} \qquad (33)$$

а так как в силу условия (А) векторы (28) образуют базис пространства R, то из соотношений (33) следовало бы $\overline{\psi}(t) \equiv 0$, что противоречит предположению о нетривиальности решения $\overline{\psi}(t)$.

5. Теоремы единственности для линейных управлений

Решим уравнение (27) как неоднородное методом вариации постоянных. Для этого обозначим через

$$\overline{\phi}_1(t), \quad \ldots, \quad \overline{\phi}_n(t) \qquad (34)$$

фундаментальную систему решений однородного уравнения

$$\frac{d\bar{x}}{dt} = A\bar{x},$$

удовлетворяющую начальным условиям $\phi_j^i(t_0) = \delta_j^i$, а через

$$\overline{\psi}^1(t), \quad \ldots, \quad \overline{\psi}^n(t)$$

— фундаментальную систему решений однородного уравнения (30), удовлетворяющую начальным $\psi_i^j(t_0) = \delta_i^j$. Будем искать общее решение уравнения (27) в виде

$$\overline{x}(t) = \sum_{i=1}^n \overline{\phi}_i(t)\, c^i(t).$$

Подставляя это решение в уравнение (27), получим

$$\sum_{i=1}^n \overline{\phi}_i(t) \frac{dc^i(t)}{dt} = B\overline{u}(t).$$

Умножая последнее соотношение скалярно на $\overline{\psi}^j$ и учитывая, что $(\overline{\psi}^j(t), \overline{\phi}_i(t)) = \delta_i^j$, получаем

$$\frac{dc^i(t)}{dt} = (\overline{\psi}^i(t), B\overline{u}(t)). \tag{35}$$

Таким образом, решение уравнения (27) при произвольном управлении $U = (\overline{u}(t), t_0, t_1, \overline{x}_0)$ записывается в виде

$$\overline{x}(t) = \sum_{i=1}^n \overline{\phi}_i(t) \left(x_0^i + \int_{t_0}^t (\overline{\psi}^i(t), B\overline{u}(t))\, dt \right). \tag{36}$$

Теорема 4. Допустим, что уравнение (27) *удовлетворяет условию* (А), *и пусть* $U_1 = (\overline{u}_1(t), t_0, t_1, \overline{x}_0), \quad U_2 = (\overline{u}_2(t), t_0, t_2, \overline{x}_0)$

— *два оптимальных управления уравнения* (27), *переводящие точку* $\overline{x}_0$ *в одну и ту же точку* $\overline{x}_1$; *тогда эти управления совпадают*

$$t_1 = t_2, \quad \overline{u}_1(t) \equiv \overline{u}_2(t).$$

Доказательство. Так как оба управления U_1 и U_2 оптимальны, то $t_1 = t_2$, ибо если бы было, например, $t_1 < t_2$, то управление U_2 не было бы оптимальным. Мы имеем, таким образом, равенство

$$\overline{x}_1 = \sum_{i=1}^n \overline{\phi}_i(t_1) \left(x_0^i + \int_{t_0}^{t_1} (\overline{\psi}^i(t), B\overline{u}_1(t))\, dt \right)$$

$$= \sum_{i=1}^n \overline{\phi}_i(t) \left(x_0^i + \int_{t_0}^{t_1} (\overline{\psi}^i(t), B\overline{u}_2(t))\, dt \right).$$

Так как векторы $\overline{\phi}^1(t_1), \ldots, \overline{\phi}^n(t_1)$ линейно независимы, то из последнего равенства следует

$$\int_{t_0}^{t_1} (\overline{\psi}^i(t), B\overline{u}_1(t))\, dt = \int_{t_0}^{t_1} (\overline{\psi}^i(t), B\overline{u}_2(t))\, dt \quad (i = 1, \ldots, n). \tag{37}$$

Оптимальному управлению U_1 в силу теоремы 3 соответствует вектор-функция $\overline{\psi}(t)$, являющаяся решением уравнения (30). Начальное значение этой функции при $t = t_0$ обозначим через

$$\overline{\psi}_0 = (\psi_{10}, \dots, \psi_{n0});$$

тогда решение $\overline{\psi}(t)$ можно записать в виде

$$\overline{\psi}(t) = \sum_{i=1}^{n} \psi_{i0} \overline{\psi}^i(t). \tag{38}$$

Умножая соотношение (37) на ψ_{i0} и суммируя по i, получаем

$$\int_{t_0}^{t} (\overline{\psi}(t), B\overline{u}_1(t))\, dt = \int_{t_0}^{t} (\overline{\psi}(t), B\overline{u}_2(t))\, dt. \tag{39}$$

В силу теоремы 3 функция $\overline{u}_1(t)$ удовлетворяет условию

$$(\overline{\psi}(t), B\overline{u}_1(t)) = P(\overline{\psi}(t))$$

и определяется этим условием однозначно. Если бы функция $\overline{u}_2(t)$ не совпадала с функцией $\overline{u}_1(t)$, то она не удовлетворяла бы условию

$$(\overline{\psi}(t), B\overline{u}_2(t)) \equiv P(\overline{\psi}(t)),$$

и потому функция $(\overline{\psi}(t), B\overline{u}_2(t))$, нигде не превосходя функции $(\overline{\psi}(t), B\overline{u}_1(t))$, на некотором интервале была бы меньше ее. Таким образом, если на отрезке $t_0 \leqslant t \leqslant t_1$ не имеет места тождество $\overline{u}_1(t) \equiv \overline{u}_2(t)$, то равенство (39) невозможно.

Итак, теорема 4 доказана.

Будем называть управление

$$U = (\overline{u}(t), t_0, t_1, \overline{x}_0)$$

экстремальным, если оно удовлетворяет условию (31), где $\overline{\psi}(t)$ — некоторое нетривиальное решение уравнения (30).

Для нахождения всех оптимальных управлений, переводящих точку $\overline{x}_0$ в точку $\overline{x}_1$, можно найти сперва все экстремальные управления, переводящие точку $\overline{x}_0$ в точку $\overline{x}_1$, а затем выбрать из их числа то единственное, которое осуществляет этот переход в кратчайшее время. Возникает вопрос, может ли существовать несколько экстремальных управлений, переводящих точку $\overline{x}_0$ в точку $\overline{x}_1$? Вообще говоря, их может существовать несколько. Нижеследующая теорема указывает важный случай единственности.

Теорема 5. Допустим, что уравнение (27) удовлетворяет условиям (А) *и* (В), *и пусть*

$$U_1 = (\overline{u}_1(t), t_0, t_1, \overline{x}_0), \quad U_2 = (\overline{u}_2(t), t_0, t_2, \overline{x}_0)$$

— *два экстремальных управления, переводящих точку $\bar{x}_0$ в начало координат $x_1 = 0$ пространства R; тогда управления U_1 и U_2 согпадают*

$$t_1 = t_2, \quad \bar{u}_1(t) \equiv \bar{u}_2(t).$$

Доказательство. По предположению, мы имеем равенства

$$\left.\begin{aligned}
\sum_{i=1}^{n} \overline{\phi}_i(t_1) \left(x_0^i + \int_{t_0}^{t_1} (\overline{\psi}^i(t), B\bar{u}_1(t))\, dt \right) = 0, \\
\sum_{i=1}^{n} \overline{\phi}_i(t_2) \left(x_0^i + \int_{t_0}^{t_2} (\overline{\psi}^i(t), B\bar{u}_2(t))\, dt \right) = 0.
\end{aligned}\right\} \tag{40}$$

Так как векторы (34) линейно независимы при любом t, то из равенств (40) следует равенство

$$-x_0^i = \int_{t_0}^{t_1} (\overline{\psi}^i(t), B\bar{u}_1(t))\, dt = \int_{t_0}^{t_2} (\overline{\psi}^i(t), B\bar{u}_2(t))\, dt. \tag{41}$$

Допустим для определенности, что $t_1 > t_2$, и пусть $\overline{\psi}(t)$ — то решение уравнения (30), для которого имеет место тождество

$$(\overline{\psi}(t), B\bar{u}_1(t)) = P(\overline{\psi}(t)),$$

определяющее функцию $\bar{u}_1(t)$. Как и при доказательстве теоремы 4, функцию $\overline{\psi}(t)$ запишем в виде (38). Умножим соотношение (41) на ψ_{i0} и просуммируем по i. Мы получим

$$\int_{t_0}^{t_1} (\overline{\psi}(t), B\bar{u}_1(t))\, dt = \int_{t_0}^{t_2} (\overline{\psi}(t), B\bar{u}_2(t))\, dt.$$

Заметим теперь, что из условия (B) следует

$$P(\overline{\psi}(t)) \geqslant 0. \tag{42}$$

В самом деле, так как ноль является внутренней точкой выпуклого тела Ω, то функция $(\overline{\psi}(t), B\bar{u})$, как функция переменного $\bar{u}$, либо тождественно равна нулю, либо может принимать как отрицательные, так и положительные значения.

В силу (42) мы имеем неравенство

$$\int_{t_0}^{t_1} (\overline{\psi}(t), B\bar{u}_1(t))\, dt \leqslant \int_{t_0}^{t_2} (\overline{\psi}(t), B\bar{u}_2(t))\, dt.$$

Отсюда так же, как и при доказательстве теоремы 4, получаем

$$\bar{u}_1(t) \equiv \bar{u}_2(t) \quad \text{при} \quad t_0 \leqslant t \leqslant t_2.$$

Далее, так как равенство $P(\overline{\psi}(t)) = 0$ может иметь место только для отдельных значений t, то должно быть $t_1 = t_2$.

Итак, теорема 5 доказана.

6. Существование оптимальных управлений для линейных систем

Теорема 6. Если существует хотя бы одно управление уравнения (27), *переводящее точку $\overline{x}_0$ в точку $\overline{x}_1$, то существует и оптимальное управление уравнения* (27), *переводящее точку $\overline{x}_0$ в точку $\overline{x}_1$.*

Доказательство. Совокупность всех управлений вида

$$U = (\overline{u}(t), 0, t, \overline{x}_0), \qquad (43)$$

переводящих точку $\overline{x}_0$ в точку $\overline{x}_1$, обозначим через $\Delta_{\overline{x}_0, \overline{x}_1}$. Каждому управлению (43) соответствует время перехода t. Нижнюю грань всех таких времен при $U \in \Delta_{\overline{x}_0, \overline{x}_1}$ обозначим через t^* и докажем, что существует управление $U^* = (u^*(t), 0, t^*, \overline{x}_0)$, переводящее точку $\overline{x}_0$ в точку $\overline{x}_1$.

Выберем из множества $\Delta_{\overline{x}_0, \overline{x}_1}$ бесконечную последовательность управлений

$$U_k = (\overline{u}_k(t), 0, t_k, \overline{x}_0) \quad (k = 1, 2, \ldots),$$

для которой имеет место равенство

$$\lim_{k \to \infty} t_k = t^*.$$

Очевидно, имеет место равенство

$$\lim_{k \to \infty} \sum_{i=1}^{n} \overline{\phi}_i(t^*) \left(x_0^i + \int_0^{t^*} (\overline{\psi}^i(t), B\overline{u}_k(t)) \, dt \right) = \overline{x}_1. \qquad (44)$$

Рассмотрим гильбертово пространство L_2 всех измеримых функций с интегрируемым квадратом, заданных на отрезке $0 \leqslant t \leqslant t^*$. Управление $\overline{u}_k(t)$ есть вектор-функция; i-ю координату этой функции обозначим через $u_k^i(t)$. Функция $u_k^i(t)$, рассматриваемая на отрезке $0 \leqslant t \leqslant t^*$, принадлежит пространству L_2. Совокупность всех функций $u_k^i(t)$, $k = 1, 2, \ldots$, очевидно принадлежит некоторому шару пространства L_2, и потому из нее можно выбрать слабо сходящуюся подпоследовательность. Мы будем просто считать, что сама последовательность

$$u_1^i(t), \quad u_2^i(t), \quad \ldots, \quad u_k^i(t), \quad \ldots \qquad (45)$$

слабо сходится к некоторой функции $u^i(t)$, $i = 1, \ldots, r$.

Докажем, что вектор-функция

$$\overline{u}^*(t) = (u^1(t), \ldots, u^r(t))$$

почти для всех значений t удовлетворяет условию

$$\bar{u}^*(t) \in \Omega.$$

Пусть

$$b(\bar{u}) = \sum_{i=1}^{r} b_i u^i = b$$

— уравнение гиперплоскости, несущей одну из $(r-1)$-мерных граней многогранника Ω, причем многогранник Ω расположен в полупространстве

$$b(\bar{u}) \leqslant b.$$

Пусть m — множество всех значений t отрезка $[0, t^*]$, для которых $b(\bar{u}^*(t)) > b$, и $v(t)$ — характеристическая функция множества m. Мы имеем тогда

$$\lim_{k \to \infty} \int_0^{t^*} v(t) \left[b(\bar{u}^*(t)) - b(\bar{u}_k(t)) \right] dt = 0,$$

в силу слабой сходимости последовательностей (45), и так как $b(\bar{u}^*(t)) - b(u_k(t)) > 0$ на множестве m, то $\text{mes}(m) = 0$.

Таким образом, изменяя вектор-функцию $\bar{u}^*(t)$ на множестве меры нуль, мы получим новую функцию, которую снова обозначим через $\bar{u}^*(t)$, удовлетворяющую условию $\bar{u}^*(t) \in \Omega, 0 \leqslant t \leqslant t^*$.

Из соотношения (44) в силу слабой сходимости последовательностей (45) следует

$$\sum_{i=1}^{n} \bar{\phi}_i(t^*) \left(x_0^i + \int_0^{t^*} \bar{\psi}(^i(t), B\bar{u}^*(t)) \, dt \right) = \bar{x}_1.$$

Таким образом, $U^* = (\bar{u}^*(t), 0, t^*, \bar{x}_0)$ является измеримым оптимальным управлением, переводящим точку $\bar{x}_0$ в точку $\bar{x}_1$.

В силу замечания к теореме 1, изменяя управление $\bar{u}^*(t)$ на множестве меры нуль, мы можем превратить его в управление, удовлетворяющее принципу максимума, т.е. в нашем случае — условию

$$(\bar{\psi}(t), B\bar{u}^*(t)) = P(\bar{\psi}(t)).$$

Из этого условия, очевидно, вытекает кусочная непрерывность функции $\bar{u}^*(t)$.

Итак, теорема 6 доказана.

Теорема 7. Если уравнение (27) удовлетворяет условиям (A) и (B) и оператор A устойчив, т.е. все его собственные значения имеют отрицательные действительные части, то для каждой точки $\bar{x}_0 \in R$ существует оптимальное управление, переводящее эту точку в начало координат $0 \in R$.

Доказательство. Докажем прежде всего, что существует окрестность V точки 0 в R, каждая точка $\bar{x}_0$ которой может быть при помощи некоторого управления переведена в 0.

Выберем в Ω такой вектор $\bar{v}$, чтобы вектор $-\bar{v}$ принадлежал Ω и чтобы вектор

$$\bar{b} = B\bar{v}$$

не принадлежал ни к какому истинному подпространству пространства R, инвариантному относительно оператора A. В силу условий (A) и (B) такой вектор $\bar{v}$ существует. При достаточно малом положительном ϵ операторы A и $e^{-\epsilon A}$ имеют совпадающие инвариантные подпространства, и потому векторы

$$e^{-\epsilon A}\bar{b}, \quad e^{-2\epsilon A}\bar{b}, \quad \ldots, \quad e^{-n\epsilon A}\bar{b}$$

линейно независимы.

Пусть $\chi(t)$ — произвольная действительная функция, определенная на некотором отрезке $0 \leqslant t \leqslant t_1$ и не превосходящая по модулю единицы; тогда

$$U = (\bar{v}\chi(t), 0, t_1, \bar{x}_0)$$

есть управление уравнения (27) и управление это переводит точку $\bar{x}_0$ в точку (см. (36))

$$\bar{x}_1 = e^{t_1 A}\left(\bar{x}_0 + \int_0^{t_1} e^{-tA}\bar{b}\chi(t)\,dt\right). \tag{46}$$

Выберем теперь функцию $\chi(t)$ зависящей от параметров $\xi^1, \ldots, \xi^n$ таким образом, чтобы точка (46) — обозначим ее через $\bar{x}_1(\bar{x}_0; \xi^1, \ldots, \xi^n)$ — удовлетворяла следующим условиям:

$$\bar{x}_1(0; 0, \ldots, 0) = 0,$$

а функциональный определитель

$$\frac{\partial(x_1^1, \ldots, x_1^n)}{\partial(\xi^1, \ldots, \xi^n)}\bigg|_{x_0=0,\,\xi^1=0,\,\ldots,\,\xi^n=0}$$

отличен от нуля. Построив такую функцию $\chi(t)$, мы докажем, что уравнение $\bar{x}_1(\bar{x}_0; \xi^1, \ldots, \xi^n) = 0$ разрешимо относительно $\xi^1, \ldots, \xi^n$ для всех значений $\bar{x}_0$, принадлежащих некоторой окрестности V начала 0.

Определим прежде всего функцию $\sigma(t, \tau, \xi)$ переменного t, $0 \leqslant t \leqslant t_1$, где $0 < \tau < t_1$, а ξ — параметр. Функция $\sigma(t, \tau, \xi)$, как функция переменного t, равна нулю всюду вне интервала $[\tau, \tau+\xi]$, а на этом интервале она равна $\operatorname{sign}\xi$. Положим теперь

$$\chi(t) = \sum_{k=1}^{n}\sigma(t, k\epsilon, \xi^k).$$

Простые вычисления показывают, что точка $\bar{x}_1(\bar{x}_0; \xi^1, \ldots, \xi^n)$ при этом выборе функции $\chi(t)$ удовлетворяет высказанным условиям.

Пусть теперь $\bar{x}_0$ — произвольная точка пространства R. Пусть она сперва двигается при управлении $\bar{u}(t) \equiv 0$. Так как все собственные значения оператора A имеют отрицательные действительные части, то по истечении некоторого времени точка придет в окрестность V, после чего ее, по доказанному, можно перевести в начало координат. Отсюда, в силу теоремы 6, вытекает существование *оптимального* управления, переводящего точку $\bar{x}_0$ в начало.

Итак, теорема 7 доказана.

7. Синтез линейного оптимального управления

Задача *синтезирования* оптимального управления имеет смысл для произвольной управляемой системы (1), однако здесь я буду трактовать ее только для линейной управляемой системы (27), удовлетворяющей условиям (А) и (В), с устойчивым оператором A. Для такой системы имеют место теоремы существования и единственности (теоремы 7 и 5), благодаря чему задача синтеза является в принципе решенной. Приводимые здесь соображения дают конструктивный метод решения задачи. Осуществление этого метода в каждом конкретном случае требует, однако, ряда построений. Синтезирование оптимального управления линейной системы (27) совершенно другим методом было осуществлено до сих пор лишь для случая одного управляющего параметра (т.е. при $r = 1$) Фельдбаумом[6] при действительных корнях оператора A и Бушоу[7] для случая, когда $n = 2$, а собственные значения оператора A комплексны.

Будем считать, что уравнение (27) удовлетворяет условиям (А) и (В) и имеет устойчивый оператор A. Тогда для каждой точки $\bar{x}_0 \in R$ существует (и притом только одно) оптимальное управление

$$U_{\bar{x}_0} = (\bar{u}_{\bar{x}_0}(t), t_0, t_1, \bar{x}_0), \tag{47}$$

переводящее точку $\bar{x}_0$ в начало координат $0 \in R$. Единственность имеет место конечно с точностью до сдвига времени (см. замечание 2 к постановке задачи). Величина $\bar{u}_{\bar{x}_0}(t_0)$ зависит, таким образом, только от точки $\bar{x}_0$, а не от случайно выбранного начала отсчета времени t_0, и потому можно положить

$$\bar{v}(\bar{x}_0) = \bar{u}_{\bar{x}_0}(t_0).$$

Пусть $\bar{x}(t)$ — решение уравнения (27), соответствующее управлению

(47); тогда $U_{\bar{x}(\tau)} = (\bar{u}_{\bar{x}_0}(t), \tau, t_1, \bar{x}(\tau))$ (см. замечание 1 к постановке задачи), и потому
$$\bar{u}_{\bar{x}_0}(\tau) = \bar{v}(\bar{x}(\tau)).$$

Таким образом,
$$\frac{d}{dt}\bar{x}(t) = A\bar{x}(t) + B\bar{v}(\bar{x}(t)),$$

и мы видим, что решение уравнения

$$\frac{d\bar{x}}{dt} = A\bar{x} + B\bar{v}(\bar{x}) \tag{48}$$

с произвольным начальным условием $\bar{x}(t_0) = \bar{x}_0$ дает закон оптимального движения точки $\bar{x}_0$ в начало координат. В этом смысле функция $\bar{v}(\bar{x})$ *синтезирует* оптимальное управление, переводящее любую точку $\bar{x}_0$ в начало.

Дадим теперь метод построения функции $\bar{v}(\bar{x})$. Пусть $\bar{\psi}(t)$ — то решение уравнения (30), которое в силу теоремы 2 соответствует управлению (47), так что

$$\frac{d\bar{\psi}(t)}{dt} = -A^*\bar{\psi}(t), \tag{49}$$

а функция $\bar{u}_{\bar{x}_0}(t)$ определяется из уравнения

$$(\bar{\psi}(t), B\bar{u}_{\bar{x}_0}(t)) = P(\bar{\psi}(t)). \tag{50}$$

Пусть, далее, $\bar{x}(t)$ — решение уравнения (27), удовлетворяющее начальному условию
$$\bar{x}(t_0) = \bar{x}_0 \tag{51}$$

и конечному условию $\qquad \bar{x}(t_1) = 0, \tag{52}$

так что
$$\frac{d\bar{x}(t)}{dt} = A\bar{x}(t) + B\bar{u}_{\bar{x}_0}(t). \tag{53}$$

Тогда функция $\bar{v}(\bar{x})$ удовлетворяет условию

$$(\bar{\psi}(t_0), B\bar{v}(\bar{x}(t_0))) = P(\bar{\psi}(t_0)). \tag{54}$$

Из теорем существования и единственности следует, что существует, и притом только одна (с точностью до сдвига времени), пара функций $\bar{u}_{\bar{x}_0}(t), \bar{x}(t)$, заданных на отрезке $t_0 \leqslant t \leqslant t_1$ и удовлетворяющих условиям (49)–(53). Ввиду возможности сдвига времени, числа t_0 и t_1 этими условиями не определены однозначно, а число $t_1 - t_0$ определено.

Совершенно не ясно, как искать функции $\bar{u}_{\bar{x}_0}(t), \bar{x}(t)$, удовлетворяющие всем условиям (49)–(53), но легко найти все функции $\bar{u}_{\bar{x}_0}(t), \bar{x}(t)$, удовлетворяющие лишь условиям (49), (50), (52), (53).

Для этого поступим следующим образом: Ввиду возможности произвольного сдвига времени, зафиксируем число t_1, положив $t_1 = 0$. Пусть теперь $\overline{\chi}$ — произвольный ковариантный вектор, отличный от нуля, и $\overline{\psi}(t,\overline{\chi})$ — решение уравнения (49), удовлетворяющее начальному условию

$$\overline{\psi}(0,\overline{\chi}) = \overline{\chi}$$

и определенное при $t \leqslant 0$. Определим, далее, функцию $\overline{u}(t,\overline{\chi})$ из условия

$$(\overline{\psi}(t,\overline{\chi}), B\overline{u}(t,\overline{\chi})) = P(\overline{\psi}(t,\overline{\chi})) \quad (t \leqslant 0),$$

и функцию $\overline{x}(t,\overline{\chi})$ из уравнения

$$\frac{d\overline{x}(t,\overline{\chi})}{dt} = A\overline{x}(t,\overline{\chi}) + B\overline{u}(t,\overline{\chi}).$$

Согласно сказанному выше, функция $\overline{v}(\overline{x})$ определится соотношением

$$(\overline{\psi}(t,\overline{\chi}), B\overline{v}(\overline{x}(t,\overline{\chi}))) = P(\overline{\psi}(t,\overline{\chi})). \tag{55}$$

Из теоремы существования (теорема 7) следует, что точка $\overline{x}(t,\overline{\chi})$ описывает все пространство R, когда t пробегает отрицательные значения, а вектор $\overline{\chi}$ меняется произвольно. Таким образом, соотношение (55) определяет значение функции $\overline{v}(\overline{x})$ для произвольной точки $\overline{x}$ пространства R.

ЦИТИРОВАННАЯ ЛИТЕРАТУРА

[1] Болтянский, В. Г., Гамкрелидзе, Р. В. и Понтрягин, Л. С. К теории оптимальных процессов. *ДАН*, 110, no. 1, 7–10 (1956).
[2] Гамкрелидзе, Р. В. К теории оптимальных процессов в линейных системах. *ДАН*, 116, no. 1, 9–11 (1957).
[3] Болтянский, В. Г. Принцип максимума в теории оптимальных процессов. *ДАН*, 119, no. 6 (1958).
[4] Блисс, Г. А. *Лекции по вариационному исчислению.* ИЛ, М., 1950.
[5] McShane, E. J.. On multipliers for Lagrange problems. *Amer. J. Math.* 61, 809–819 (1939).
[6] Фельдбаум, А. А. *Автомат. и Телемех.* 16, no. 2, 129 (1955).
[7] Bushaw, D. W. Experimental towing tank. *Stevens Institute of Technology, Report 469.* Hoboken, N.Y., 1953.

RATIONAL APPROXIMATIONS TO ALGEBRAIC NUMBERS

By K. F. ROTH

1. Let α be any algebraic irrational number and suppose there are infinitely many rational approximations h/q to α such that

$$\left| \alpha - \frac{h}{q} \right| < \frac{1}{q^\kappa}. \tag{1}$$

I proved in 1955 that this implies $\kappa \leqslant 2$. In this talk I shall try to outline the proof, and to say a few words about some of the possible extensions, and about the limitations of the method.

It is easily seen that there is no loss of generality in supposing that α is an algebraic integer. Accordingly, we shall suppose that α is a root of the polynomial

$$f(x) = x^n + a_1 x^{n-1} + \ldots + a_n \tag{2}$$

with integral coefficients and highest coefficient 1.

2. Previous work on the problem entailed the use of polynomials in two variables. It has long been recognized that further progress would demand the use of polynomials in more than two variables, and that polynomials in a large number of variables would have to be used to obtain the full result. It is not difficult to formulate properties which a polynomial should have in order to be useful for our purpose.

Suppose that $h_1/q_1, \ldots, h_m/q_m$ are rational approximations to α, all satisfying (1). Let $Q(x_1, \ldots, x_m)$ denote a polynomial with integral coefficients, of degree at most r_j in x_j for each j. Then

$$\left| Q\left(\frac{h_1}{q_1}, \ldots, \frac{h_m}{q_m} \right) \right| \geqslant \frac{1}{P}, \tag{3}$$

where $P = q_1^{r_1} \ldots q_m^{r_m}$, provided, of course, that

$$Q\left(\frac{h_1}{q_1}, \ldots, \frac{h_m}{q_m} \right) \neq 0.$$

Let the Taylor expansion of $Q(h_1/q_1, \ldots, h_m/q_m)$ in powers of $(h_1/q_1) - \alpha, \ldots,$ $(h_m/q_m) - \alpha$ be

$$\sum_{i_1} \ldots \sum_{i_m} Q_{i_1, \ldots, i_m} \left(\frac{h_1}{q_1} - \alpha \right)^{i_1} \ldots \left(\frac{h_m}{q_m} - \alpha \right)^{i_m}.$$

Suppose now that Q has the properties:

A: $$\sum_{i_1} \cdots \sum_{i_m} |Q_{i_1,\ldots,i_m}| < P^\Delta,$$

where Δ is small;

B: $Q_{i_1,\ldots,i_m} = 0$ for all $i_1, \ldots, i_m$ satisfying

$$q_1^{i_1} \cdots q_m^{i_m} \leqslant P^\phi \quad (\phi > 0).$$

Then each term in this Taylor expansion with a non-zero coefficient has

$$\left|\frac{h_1}{q_1} - \alpha\right|^{i_1} \cdots \left|\frac{h_m}{q_m} - \alpha\right|^{i_m} < \frac{1}{(q_1^{i_1} \cdots q_m^{i_m})^\kappa} < \frac{1}{P^{\kappa\phi}}.$$

Hence $$\left|Q\left(\frac{h_1}{q_1}, \ldots, \frac{h_m}{q_m}\right)\right| < \frac{1}{P^{\kappa\phi-\Delta}}. \tag{4}$$

Comparison of (3) and (4) yields

$$\kappa < \frac{1+\Delta}{\phi}. \tag{5}$$

We must bear in mind that to obtain (5) we used, in addition to A and B, the condition

C: $$Q\left(\frac{h_1}{q_1}, \ldots, \frac{h_m}{q_m}\right) \neq 0.$$

To prove our theorem we shall establish the existence of a polynomial Q satisfying the conditions A, B, C with ϕ near to $\frac{1}{2}$. For this purpose it will be necessary to take m large and to choose the approximations $h_1/q_1, \ldots, h_m/q_m$ suitably. Only after choosing these approximations do we choose the polynomial Q, and the latter choice depends on the former.

[With $m = 2$ it is only possible to satisfy condition B with ϕ of the order of $n^{-\frac{1}{2}}$ (where n is the degree of α), and this leads to an estimate of the type $\kappa < cn^{\frac{1}{2}}$.]

3. The logical structure of the proof is as follows. We suppose $\kappa > 2$; m is chosen sufficiently large and is fixed throughout. A small positive number δ ($< 1/m$) will be fixed until the end of the proof, when we let $\delta \to 0$. We denote by Δ any function of δ and m such that $\Delta \to 0$ as $\delta \to 0$ for fixed m.

We begin by choosing $h_1/q_1, \ldots, h_m/q_m$ [with $(h_j, q_j) = 1$] from the assumed infinite sequence of approximations to α (satisfying (1)) by first taking q_1 sufficiently large (in terms of m, δ, α), then taking q_2 sufficiently large in terms of q_1, and so on. It will in fact suffice if

$$\frac{\log q_j}{\log q_{j-1}} > \delta^{-1} \quad (j = 2, \ldots, m).$$

Then we choose integers $r_1, ..., r_m$ which are sufficiently large in relation to $q_1, ..., q_m$ and which satisfy

$$q_1^{r_1} \leqslant q_j^{r_j} < q_1^{r_1(1+\frac{1}{10}\delta)}. \tag{6}$$

This presents no difficulty. We note that (6) implies

$$q_1^{mr_1} \leqslant P < q_1^{mr_1(1+\Delta)}. \tag{7}$$

Condition B now takes the form that the Taylor coefficients $Q_{i_1, ..., i_m}$ vanish for all $i_1, ..., i_m$ satisfying

$$\frac{i_1}{r_1} + ... + \frac{i_m}{r_m} < m\phi + \Delta. \tag{8}$$

4. We shall first outline a proof of the existence of a polynomial Q^* satisfying conditions A and B only (with ϕ near to $\frac{1}{2}$). This proof, due essentially to Siegel, is based on the use of Dirichlet's compartment principle. The question of satisfying C as well, which gives rise to the principal difficulty, is deferred until later.

We put $B_1 = q_1^{\delta r_1}$ and consider all polynomials $W(x_1, ..., x_m)$ of degree at most r_j in x_j, having positive integral coefficients, each less than B_1. We try to find two such polynomials W', W'' such that their derivatives of order $i_1, ..., i_m$ are equal when $x_1 = ... = x_m = \alpha$, for all $i_1, ..., i_m$ satisfying (8). Since any such derivative is of the form

$$A_0 + A_1\alpha + ... + A_{n-1}\alpha^{n-1},$$

where $A_0, ..., A_{n-1}$ are integers, one can estimate the number of possibilities for a derivative for given $i_1, ..., i_m$. This number can be shown to be less than $B_1^{n(1+3\delta)}$. The number of polynomials W is about B_1^r, where $r = (r_1 + 1) ... (r_m + 1)$. Thus the number of polynomials W will exceed the number of possible distinct sets of derivatives provided that the number of sets of $i_1, ..., i_m$ satisfying (8), and with no i exceeding the corresponding r, is less than about $r/\{n(1+3\delta)\}$. The number of integer points $(i_1, ..., i_m)$ in the region defined by the above conditions can be shown to be less than $\frac{2}{3}r/n$ if ϕ is chosen so that

$$m\phi + \Delta = \frac{1}{2}m - 3nm^{\frac{1}{2}}. \tag{9}$$

The polynomial $Q^* = W' - W''$ satisfies B, by its definition, and can be shown by a process of simple estimation to satisfy A. Furthermore, on letting $\delta \to 0$ we would obtain

$$\phi = \frac{1}{2} - 3nm^{-\frac{1}{2}},$$

so that ϕ could be assumed to be sufficiently near to $\frac{1}{2}$, since m is large. [It is in order to be able to choose ϕ near to $\frac{1}{2}$, that we must work with polynomials in many variables.]

5. To find a polynomial Q which satisfies A, B and C as well, we seek a derivative

$$Q = \frac{1}{j_1!}\left(\frac{\partial}{\partial x_1}\right)^{j_1} \cdots \frac{1}{j_m!}\left(\frac{\partial}{\partial x_m}\right)^{j_m} Q^*,$$

of not too high an order, of the polynomial Q^* just considered. We want Q not to vanish at $(h_1/q_1, ..., h_m/q_m)$. The 'order' is measured by $j_1/r_1 + \cdots + j_m/r_m$. The replacement of Q^* by Q will involve a weakening of condition B, but provided the order in question is a Δ, this will make no difference on letting $\delta \to 0$. There will also be an effect on condition A, but this turns out to be insignificant. Condition C is the essential requirement now.

The existence of such a derivative, whose order is a Δ, is not easy to establish. One would in fact expect this to cause difficulty, as the choice of Q^* was designed to make Q^* very small at $(h_1/q_1, ..., h_m/q_m)$.

At this stage it is convenient to introduce the notion of the index of a polynomial at a point. We define the index of a polynomial at the point $(\alpha_1, ..., \alpha_m)$ relative to positive parameters $r_1, ..., r_m$ to be the minimal order of derivative (measured as above) which does not vanish at the point $(\alpha_1, ..., \alpha_m)$. In this language, we need to show that the index of Q^* at $(h_1/q_1, ..., h_m/q_m)$ is a Δ.

For polynomials in two variables, two quite different lines of reasoning have been used to obtain upper bounds for the index of Q^* at $(h_1/q_1, ..., h_m/q_m)$. The first, due to Siegel, is algebraic in nature. It is based on the principle that, under certain conditions, the sum of the indices of a polynomial at a finite number of points (not restricted to be rational) is bounded in terms of its degrees in the various variables. Since Q^* satisfies condition B (with an appropriate ϕ), it has an almost maximal index (in a certain sense) at the point $(\alpha, ..., \alpha)$ and at the points obtained by replacing α by its conjugates; and it can be deduced that the index of Q^* is small at any other point. I have been unable to extend this method to polynomials in more than 2 variables.

The second method, due to Schneider, is arithmetic in nature. It is based on the principle that, under certain conditions, the index of a polynomial at a rational point is bounded in terms of the magnitude of its coefficients. Since the coefficients of Q^* are not too large, this leads to a result of the desired kind.

My treatment is based on Schneider's approach and enables me to prove the following lemma.

Principal lemma. Let $0 < \delta < m^{-1}$, and let $r_1, ..., r_m$ be positive integers satisfying

$$r_m > 10\delta^{-1}, \quad \frac{r_{j-1}}{r_j} > \delta^{-1} \quad (j = 2, ..., m).$$

Let $q_1, ..., q_m$ be positive integers satisfying

$$q_1 > c = c(m, \delta), \quad q_j^{r_j} \geqslant q_1^{r_1}.$$

Consider any polynomial R, not identically zero, with integral coefficients of absolute value at most $q_1^{\delta r_1}$ and of degree at most r_j in x_j. Then

$$\operatorname{index} R < 10^m \delta^{(\frac{1}{2})^m},$$

where the index is taken at a point $(h_1/q_1, ..., h_m/q_m)$ relative to $r_1, ..., r_m$; the h's being integers relatively prime to the corresponding q's.

This suffices for the purpose of finding Q; the hypotheses of the lemma are satisfied when $R = Q^*$, and the lemma shows that the index of Q^* at $(h_1/q_1, ..., h_m/q_m)$ is a Δ, as required.

6. The proof of the lemma is self-contained, as indeed it must be, for it uses induction on the number of variables, whereas in the main proof the number m is fixed. Furthermore, the lemma has to be generalized before the induction can be set up.

We consider the class of all polynomials $R(x_1, ..., x_m)$ with integral coefficients, each coefficient being numerically at most B, say, and of degree at most r_j in x_j. We obtain, under certain conditions, an upper bound for the indices of polynomials of this class at a point $(h_1/q_1, ..., h_m/q_m)$ relative to $r_1, ..., r_m$. During the course of the proof, which is by induction on m, it is necessary to consider various different sets of values of the parameters involved. The final estimate is of the type required to establish the lemma.

The case $m = 1$ is simple. Suppose the coefficients of a polynomial $R(x_1)$ are numerically less than B. If θ_1 is the index of R at h_1/q_1 relative to r_1, the polynomial $R(x_1)$ is divisible by

$$\left(x_1 - \frac{h_1}{q_1}\right)^{\theta_1 r_1}$$

It follows from Gauss's theorem on the factorization of polynomials with integral coefficients into polynomials with rational coefficients, that

$$R(x_1) = (q_1 x_1 - h_1)^{\theta_1 r_1} R^*(x_1),$$

where $R^*(x_1)$ is a polynomial with integral coefficients. Hence the coefficient of the highest term in R^* is an integral multiple of $q_1^{\theta_1 r_1}$, so that

$$q_1^{\theta_1 r_1} \leqslant B, \quad \theta_1 \leqslant \frac{\log B}{r_1 \log q_1}.$$

This gives an upper bound of the required type for $m = 1$.

Now suppose that upper bounds of this kind have been obtained for $m = 1, 2, ..., p-1$, where $p \geqslant 2$. We wish to deduce an upper bound for the indices for classes of polynomials in p variables.

For any given polynomial $R(x_1, ..., x_p)$ we consider all representations of the form

$$R = \phi_0(x_p)\, \psi_0(x_1, ..., x_{p-1}) + ... + \phi_{l-1}(x_p)\, \psi_{l-1}(x_1, ..., x_{p-1}), \quad (10)$$

where the ϕ_ν and ψ_ν are polynomials with rational coefficients, subject to the condition that the ϕ_ν and ψ_ν are of degree at most r_j in x_j. Such a representation is possible, e.g. with $l-1 = r_p$ and $\phi_\nu(x_p) = x_p^\nu$. From all such representations we select one for which l is least.

In this representation the polynomials ϕ form a linearly independent set, and so do the ψ's. Thus the Wronskian $W(x_p)$ of the ϕ's is not identically zero, and the same is true of a certain generalized Wronskian $G(x_1, ..., x_{p-1})$ of the ψ's. From (10) and the rule for multiplication of determinants by rows, it follows that

$$G(x_1, ..., x_{p-1})\, W(x_p) = F(x_1, ..., x_p) \quad (11)$$

is a certain determinant whose elements are all of the form

$$R_{j_1, ..., j_p}(x_1, ..., x_p).$$

Since G and W have rational coefficients, there is an equivalent factorization of F in the form

$$F(x_1, ..., x_p) = U(x_1, ..., x_{p-1})\, V(x_p), \quad (12)$$

where U and V have integral coefficients.

If the coefficients of R are assumed to be numerically less than B, this will imply an upper bound for the coefficients of F; and this, in turn, will imply upper bounds for the coefficients of U and V. The induction hypothesis then gives us upper bounds for the indices of U and V at the points $(h_1/q_1, ..., h_{p-1}/q_{p-1})$ and h_p/q_p respectively; and by a multiplicative property of indices, (12) then yields an upper bound for the index of F at $(h_1/q_1, ..., h_p/q_p)$.

On the other hand, F is obtained from R by the operations of differentiation, addition and multiplication; and by using some simple properties of indices relating to these operations, one obtains a lower bound for the index of F in terms of the index of R. Thus the upper bound for the index of F leads to an upper bound for the index of R.

In this way it is possible to set up the induction on m, although the details are somewhat more complicated than they are made to appear above.

This concludes the outline of the proof of our theorem. We note that the proof of the existence of a polynomial Q satisfying conditions A, B, C of § 2 is very indirect, and it would be of considerable interest if such a polynomial could be obtained by a direct construction.

7. The theorem can be generalized and extended in various ways. For example, instead of considering rational approximations to the algebraic number α, we may consider approximations to α by algebraic numbers β (a) lying in a fixed algebraic field, or (b) of fixed degree. In each case the accuracy of the approximation is measured in terms of $H(\beta)$, the maximum absolute value of the rational integral coefficients in the primitive irreducible equation satisfied by β.

The results already found by Siegel can be improved in both cases. In case (a) the best possible result has been obtained.† In case (b), Siegel's result is significant only if the degree of β is not too large compared to the degree of α, and I do not know how to obtain an improvement which does not suffer from a similar limitation.

The theorem can also be extended to p-adic and g-adic number fields, and this has been done by Ridout and Mahler respectively.

Various deductions can be made from the theorem. For example, for a given α and $\kappa > 2$, it is possible to estimate the number of solutions of (1), as has been done by Davenport and myself. This leads to estimates for the number of solutions of certain Diophantine equations.

The method is subject to a severe limitation, however, due to the role played by the selected approximations $h_1/q_1, \ldots, h_m/q_m$. One cannot answer questions of the following type:

(i) Can one give, in terms of α and κ (if $\kappa > 2$), an upper bound for the greatest denominator q among the finite number of solutions h/q of (1)?

(ii) Can one prove that

$$\left| \alpha - \frac{h}{q} \right| < q^{-(2+f(q))}$$

has only a finite number of solutions h/q for some explicit function $f(q)$ such that $f(q) \to 0$ as $q \to \infty$?

---

† See W. J. LeVeque, *Topics in Number Theory*, Addison Wesley, 1956.

Liouville's result
$$\left| \alpha - \frac{h}{q} \right| > c(\alpha)\, q^{-n}$$

remains the only known result of its type for which an explicit value of the constant can be given.

Our method can only throw light on such questions if some assumption is made concerning the 'gaps' between the convergents to α. It would appear that a completely new idea is needed to obtain any information concerning problems of the above type.

One outstanding problem is to obtain a theorem analogous to ours concerning simultaneous approximations to two or more algebraic numbers by rationals of the same denominator. In the case of such simultaneous approximation to two algebraic numbers α_1, α_2 (subject to a suitable independence condition), one would expect the inequalities

$$\left| \alpha_1 - \frac{h_1}{q} \right| < q^{-\kappa}, \quad \left| \alpha_2 - \frac{h_2}{q} \right| < q^{-\kappa},$$

to have at most a finite number of solutions for any $\kappa > \frac{3}{2}$. But practically nothing is known in this connection.

A complete solution of the problem of simultaneous approximations could lead to the complete solution of many others, such as, for example, case (b) of the first problem mentioned in this section.

EXTREMUM PROBLEMS AND VARIATIONAL
METHODS IN CONFORMAL MAPPING

By MENAHEM SCHIFFER

1. Introduction

One fundamental problem in the classical theory of conformal mapping was the study of the various types of canonical domains upon which any domain, arbitrarily given in the complex plane, can be mapped conformally. The first question to be settled was, therefore, the existence of various types of canonical mapping functions. From the beginning, methods of the calculus of variations were applied in order to establish the necessary existence theorems. The role of the Dirichlet principle in the attempted proof of Riemann's mapping theorem for simply connected domains is well known and also the influence of its initial failure upon the critical period of the calculus of variations and upon the development of the powerful modern direct methods in this important branch of analysis. The existence proofs for canonical conformal mappings by means of extremum problems like the Dirichlet principle are so difficult because they characterize the sought mapping function, which is analytic and univalent, as the extremum function in a much wider class of admissible competing functions. The latter class is so large that the main labour in the proof is spent in establishing the existence of an extremum function of the variational problem considered.

The theory of conformal mapping advanced considerably when one started a systematic study of the univalent analytic functions in a given domain; that is, the class of those functions which realize the various conformal mappings of that domain. The main result of this theory is that all univalent functions in a given domain form a normal family. This fact leads easily to the consequence that for each reasonable extremum problem within the family of univalent functions there exists at least one element of the family which attains the extremum considered[20]. On the basis of this theory, very elegant proofs could be derived for the Riemann mapping theorem and for the existence of numerous other canonical mappings. The characteristic difficulty of the new approach, that is to study extremum problems within the family of univalent functions, lies in the fact that the univalent functions form no linear space; hence, it is not at all easy to characterize an extremum function by comparison with its competitors by infinitesimal variation.

In each particular existence proof a special comparison method had to be devised and the essential step of the whole proof was the characterization of the extremum function by this particular variation.

It is possible to develop a systematic infinitesimal calculus within the family of univalent functions. In 1923 Löwner gave a now classical partial differential equation which has as solutions one-parameter families of univalent functions which admit a very simple geometric interpretation[18]. I showed in 1938 that the univalent extremum functions do satisfy in very many cases a first-order differential equation and gave a standard variational procedure for establishing these ordinary differential equations[26]. In the following years, Schaeffer and Spencer applied this variational procedure systematically to the coefficient problem for functions univalent in the unit circle and developed an extensive theory for it[22, 23, 24]. Golusin applied the same variational technique to numerous questions of geometric function theory[7, 8]. The significance of extremum problems for the general theory of conformal mapping is evident. The great number of possible conformal mappings of a given domain precludes the study of all of them; however, important individual mappings can be singled out as solutions of extremum problems and can be described geometrically and analytically just because of their extremum property. The remaining amorphous mass of conformal mappings is subjected to all the inequalities which flow from the solutions of the various extremum problems and is, thus, at least partially characterized.

In the present paper we shall try to give a brief survey of the basic methods of variation within the family of univalent functions. By discussing a few important extremum problems, we will show the flexibility of the technique. It will appear that the variational method provides very often an elegant and useful transformation of the extremum problem but leads sometimes to functional equations whose solution is a deep problem again. It is clear that the field of research described is by no means completely explored and exhausted and that, because of its interest from the point of view of applied as well as of pure mathematics, it deserves the continued attention of mathematicians.

2. Variation of the Green's function

The simplest approach to the calculus of variations for univalent functions seems to lead through the theory of the Green's function of a domain and its variational formula. Let D be a domain in the complex z-plane whose boundary C consists of n closed analytic curves and let

$g(z, \zeta)$ be its Green's function with the source point ζ. We consider the conformal transformation

$$z^*(z) = z + \frac{e^{i\alpha}\rho^2}{z - z_0} \qquad (z_0 \in D, \rho > 0). \tag{1}$$

This mapping is univalent in the domain $|z - z_0| > \rho$; hence, for small enough ρ it will be univalent on C and transform it into a new set C^* of n closed analytic curves which bounds a new domain D^*. We denote the Green's function of D^* by $g^*(z, \zeta)$ and wish to express it in terms of $g(z, \zeta)$. We observe that $\gamma(z, \zeta) = g^*(z^*(z), \zeta^*(\zeta))$ is a harmonic function in the domain D_ρ which is obtained from D by removal of the circle $|z - z_0| < \rho$. $\gamma(z, \zeta)$ has a pole for $z = \zeta$ and vanishes on the boundary curves C of D_ρ. We choose two fixed points ζ and η in D_ρ and apply Green's identity in the form

$$\frac{1}{2\pi} \int_{C+c} \left[\frac{\partial}{\partial n} g(z, \eta) \, \gamma(z, \zeta) - g(z, \eta) \frac{\partial}{\partial n} \gamma(z, \zeta) \right] ds = \gamma(\zeta, \eta) - g(\zeta, \eta). \tag{2}$$

Here, c denotes the small circumference $|z - z_0| = \rho$. Observe now that the integration takes place only over the circumference c since both g and γ vanish on C.

In order to simplify (2), we introduce the analytic functions of z whose real parts are $g(z, \eta)$ and $g^*(z, \zeta)$, respectively, and denote them by $p(z, \eta)$ and $p^*(z, \zeta)$. These functions have logarithmic poles at η or ζ and have also imaginary periods when z circulates around a boundary continuum. It is now easy to express (2) in the form

$$g^*(\zeta^*, \eta^*) - g(\zeta, \eta) = \mathrm{Re} \left\{ \frac{1}{2\pi i} \oint_c p^*(z^*, \zeta^*) \, dp(z, \eta) \right\}. \tag{3}$$

This integral equation for $g^*(\zeta^*, \eta^*)$ in terms of $g(\zeta, \eta)$ must now hold for the most general domain D which possesses a Green's function at all. Indeed, such a domain D may be approximated arbitrarily by domains D_ν with analytic boundaries C_ν for which the identity (3) is valid. If D and D_ν go under the variation (1) into the domains D^* and D_ν^*, then the D_ν^* will likewise approximate D^*. Since (3) holds for all approximating domains and since at ζ, η and on c the Green's functions of D_ν and D_ν^* converge uniformly to the Green's functions of D and D^*, respectively, the formula (3) must remain valid in the limit and is thus generally proved[31].

We may apply Taylor's theorem in the form

$$p^*(z^*, \zeta^*) = p^*(z, \zeta^*) + p^{*\prime}(z, \zeta^*) \frac{e^{i\alpha}\rho^2}{z - z_0} + O(\rho^4), \tag{4}$$

where the residual term $O(\rho^4)$ can be estimated equally for all domains D which contain a fixed subdomain Δ which, in turn, contains the point ζ and the circle c. Thus, inserting (4) into (3) and using the residue theorem we obtain after an easy transformation

$$g^*(\zeta^*, \eta^*) = g(\zeta, \eta) + \text{Re}\left\{e^{i\alpha}\rho^2 p'(z_0, \zeta) p'(z_0, \eta)\right\} + O(\rho^4), \qquad (5)$$

where again $O(\rho^4)$ can be estimated equally as above. Finally, using Taylor's theorem again, we can reduce (5) to[27, 28]

$$g^*(\zeta, \eta) = g(\zeta, \eta) + \text{Re}\left\{e^{i\alpha}\rho^2\left[\,p'(z_0, \zeta) p'(z_0, \eta) - \frac{p'(\zeta, \eta)}{\zeta - z_0} - \frac{p'(\eta, \zeta)}{\eta - z_0}\right]\right\} + O(\rho^4). \qquad (6)$$

In the preceding, we have restricted ourselves to the particular variation (1) for the sake of simple exposition. It is clear that a corresponding formula can be established for each variation $z^* = z + \rho^2 v(z)$, where $v(z)$ is analytic on the boundary C of the varied domain. On the other hand, such a general variation can be approximated arbitrarily by superposition of elementary variations of the type (1). Indeed, for most applications the formulas (1) and (6) are entirely sufficient.

A remarkable transformation of (6) is possible if the boundary C of D is a set of smooth curves. Indeed, we may express (6) in the form

$$g^*(\zeta, \eta) - g(\zeta, \eta) = \text{Re}\left\{e^{i\alpha}\rho^2 \frac{1}{2\pi i}\int_C \frac{p'(z, \zeta) p'(z, \eta)}{z - z_0}\,dz\right\} + O(\rho^4). \qquad (7)$$

We observe that the real part of $p(z, \zeta)$ is the Green's function $g(z, \zeta)$ and that it vanishes, therefore, on C. Let $z' = dz/ds$ denote the tangent vector to C at the point $z(s)$; it is easy to see that

$$p'(z, \zeta) z' = -i \frac{\partial g(z, \zeta)}{\partial n}$$

and hence (7) may be given the real form

$$g(\zeta, \eta) = -\frac{1}{2\pi}\int_C \frac{\partial g(z, \zeta)}{\partial n_z} \frac{\partial g(z, \eta)}{\partial n_z}\,\delta n\, ds \qquad (8)$$

with

$$\delta n = \text{Re}\left\{\frac{1}{iz'}\frac{e^{i\alpha}\rho^2}{z - z_0}\right\}. \qquad (9)$$

Clearly, δn denotes the shift along the interior normal of the boundary point $z \in C$ under the variation (1).

By linear superposition of elementary variations (1), formula (8) can be proved for very general δn-variations of the boundary curves C.

This formula was first given by Hadamard in 1908[11] and has been very frequently used in applied mathematics because of the very intuitive and geometric significance of the normal displacement of the boundary points. We may mention, in particular, Lavrentieff's systematic use of boundary deformations in many problems of fluid dynamics and conformal mapping[16,17].

If D is a simply connected domain there exists a close relationship between the Green's function of D and the univalent function $\phi(z)$ which maps the domain D onto the exterior of the unit circle. In fact, we have

$$g(z, \zeta) = \log \left| \frac{1 - \phi(z)\,\overline{\phi(\zeta)}}{\phi(z) - \phi(\zeta)} \right|. \tag{10}$$

Julia used this interrelation in order to derive from the Hadamard formula (8) a variational formula for univalent functions[15]. This very intuitive and elegant formula, however, cannot be applied directly to the study of extremum problems in the theory of conformal mapping. In fact, one cannot assert a priori that the extremum domain D will possess a boundary C which is smooth enough to admit a variation of the Hadamard–Julia type.

3. Infinitesimal variations and extremum problems

We are now in a position to construct, by means of the fundamental formula (6), in any given domain D, univalent mappings which are arbitrarily close to the identity mapping. We have to assume only that the boundary C of D contains a non-degenerate continuum Γ. Let $D(\Gamma)$ denote the domain of the z-plane which contains the domain D and the point at infinity and which is bounded by Γ; let $g(z, \zeta)$ denote now the Green's function of $D(\Gamma)$. We choose an arbitrary but fixed point $z_0 \in D$ and subject $D(\Gamma)$ to a variation (1) which transforms it into the varied domain $D(\Gamma^*)$ with the Green's function $g^*(z, \zeta)$. The relation between $g^*(z, \zeta)$ and $g(z, \zeta)$ is given by the variational formula (6).

Let $w = \phi(z)$ be univalent in $D(\Gamma)$, normalized at $z = \infty$ by the requirement $\phi'(\infty) = 1$, and let it map $D(\Gamma)$ onto the domain $|w| > 1$. Analogously, we define $w = \phi^*(z)$ with respect to the domain $D(\Gamma^*)$. By virtue of the relation (10), we have obviously

$$g(z, \infty) = \log |\phi(z)|, \quad g^*(z, \infty) = \log |\phi^*(z)|; \tag{11}$$

these relations permit us to connect $\phi^*(z)$ with $\phi(z)$ by use of (6).

The function

$$v(z) = \phi^{*-1}[\phi(z)] \tag{12}$$

is analytic and univalent in $D(\Gamma)$ and hence, *a fortiori*, in D. A simple calculation based on (6) and (11) shows that

$$v(z) = z + e^{i\alpha}\rho^2\left[\frac{1}{z-z_0} - \frac{\phi'(z_0)^2\,\phi(z)}{\phi'(z)\,\phi(z_0)\,[\phi(z)-\phi(z_0)]}\right]$$

$$+ e^{-i\alpha}\rho^2\frac{\overline{\phi'(z_0)^2}\,\phi(z)^2}{\phi'(z)\,\overline{\phi(z_0)}\,[1-\overline{\phi(z_0)}\,\phi(z)]} + O(\rho^4). \tag{13}$$

Since ρ can be made arbitrarily small, we have in (13) the representation for a large class of univalent variations of the domain D considered. We will now show that this set of variations is general enough to characterize the extremum domains for a large class of extremum problems relative to the family of univalent functions.

We shall consider extremum problems of the following type. Let T be a domain in the complex t-plane which contains the point at infinity and which is analytically bounded. We denote by F the family of all analytic functions $f(t)$ in T which are univalent there, have a simple pole at $t = \infty$ and which are normalized by the condition $f'(\infty) = 1$. Let $\phi[f]$ be a real-valued functional defined for all analytic functions $f(t)$ in T. We suppose that $\phi[f]$ is differentiable in the sense that for an arbitrary analytic function $g(t)$ defined in T

$$\phi[f+\epsilon g] = \phi[f] + \mathrm{Re}\,\{\epsilon\psi[f,g]\} + O(\epsilon^2) \tag{14}$$

holds, where ψ is a complex-valued functional of f and g, linear in g. We suppose that the residual term $O(\epsilon^2)$ can be estimated equally for all analytic functions $g(t)$ which are equally bounded in a specified subdomain of T. Thus, we require for $\phi[f]$ the existence of a Gâteaux differential with the above additional specifications.

We assume also that $\phi[f]$ has an upper bound within the family F. Then, in view of the normality of this family, it is easy to show that there must exist functions $f(t) \in T$ for which $\phi[f]$ attains its maximum value within F. We can characterize each extremum function by subjecting it to infinitesimal variations and comparing $\phi[f]$ with the functional values of the varied univalent elements of the family. Indeed, by means of the functions (13) we can construct the competing functions in F

$$f^*(t) = v[f(t)] \cdot v'(\infty)^{-1}, \tag{15}$$

where $z = f(t)$ maps the domain T onto the extremum domain D in the z-plane. An easy calculation yields

$$\phi[f^*] = \phi[f] + \mathrm{Re}\,\{e^{i\alpha}\rho^2 A + e^{-i\alpha}\rho^2 B\} + O(\rho^4), \tag{16}$$

with

$$A = \psi\left[z, \frac{1}{z-z_0} - \frac{\phi'(z_0)^2\,\phi(z)}{\phi'(z)\,\phi(z_0)\,[\phi(z)-\phi(z_0)]}\right]$$

$$B = \psi\left[z, \frac{\overline{\phi'(z_0)}^2\,\phi(z)^2}{\phi'(z)\,\overline{\phi(z_0)}\,[1-\overline{\phi(z_0)}\,\phi(z)]} + \frac{\overline{\phi'(z_0)}^2}{\overline{\phi(z_0)}^2}\,z\right]$$

$$(z=f(t)). \qquad (17)$$

Since the extremum property of f requires $\phi[f^*] \leqslant \phi[f]$ and since ρ and $e^{i\alpha}$ are at our disposal, we can easily conclude $A + \bar{B} = 0$, that is

$$\psi\left[f(t), \frac{1}{f(t)-z_0}\right]\frac{\phi(z_0)^2}{\phi'(z_0)^2} = \overline{\psi\left[z, \frac{\phi(z)}{\phi'(z)} - z\right]}$$

$$+ \psi\left[z, \frac{\phi(z)\,\phi(z_0)}{\phi'(z)\,[\phi(z)-\phi(z_0)]}\right] + \overline{\psi\left[z, \frac{\phi(z)\,\overline{\phi(z_0)}^{-1}}{\phi'(z)\,[\phi(z)-\overline{\phi(z_0)}^{-1}]}\right]}. \qquad (18)$$

Before discussing the consequences of (18), we introduce some more elementary variations in F which will allow us to simplify the result (18). We map the domain $D(\Gamma)$ onto $|w| > 1$ by means of the function $w = \phi(z)$; we then turn this circle into itself by the linear mapping $w_1 = e^{i\epsilon}w$ and return to the z-plane through $\phi^{-1}(w_1)$. Thus, the function

$$v_1(z) = e^{-i\epsilon}\phi^{-1}[e^{i\epsilon}\phi(z)] \qquad (19)$$

is univalent in $D(\Gamma)$ and hence in D. For small ϵ, we have the series development in ϵ

$$v_1(z) = z + i\epsilon\left[\frac{\phi(z)}{\phi'(z)} - z\right] + O(\epsilon^2). \qquad (20)$$

Since $f^*(t) = v_1[f(t)]$ is an admissible competing function in F, we deduce easily from the extremum property of $f(t)$ and from the freedom in the choice of the real parameter ϵ

$$\psi\left[z, \frac{\phi(z)}{\phi'(z)} - z\right] = \text{real}. \qquad (21)$$

Another possible infinitesimal variation is obtained by

$$v_2(z) = (1+\epsilon)^{-1}\phi^{-1}[(1+\epsilon)\,\phi(z)] \quad (\epsilon > 0). \qquad (22)$$

In fact, we may map $D(\Gamma)$ onto $|w| > 1$, magnify the unit circle by a factor $(1+\epsilon)$ and return through $\phi^{-1}(w)$ to the z-plane. The function $f^*(t) = v_2[f(t)]$ lies also in F and from the extremum property of $f(t)$ we deduce by use of (21) the inequality

$$\psi\left[z, \frac{\phi(z)}{\phi'(z)} - z\right] \leqslant 0. \qquad (23)$$

We return now to formula (18) and observe that in view of (21)

$$\lim_{z_0 \to \Gamma} \psi \left[f(t), \frac{1}{f(t) - z_0} \right] \frac{\phi(z_0)^2}{\phi'(z_0)^2} = \text{real.} \qquad (24)$$

In order to simplify the discussion we shall assume that

$$\psi \left[z, \frac{1}{z - z_0} \right] = W(z_0)$$

is a meromorphic function of z_0; this is, indeed, the case in most applications. We put $z = \psi(w)$, where ψ is the inverse function of $w = \phi(z)$, and obtain from (24) the boundary relation

$$\lim_{|w| \to 1} W[\psi(w)] w^2 \psi'(w)^2 = \text{real,} \qquad (25)$$

for the function $\psi(w)$ which is analytic in $|w| > 1$ and maps this circular domain onto $D(\Gamma)$. By the Schwarz reflection principle, the function $W[\psi(w)] w^2 \psi'(w)^2$ can then be continued analytically into the domain $|w| \leqslant 1$. Thus, $\psi(w)$ satisfies a first-order differential equation with analytic coefficients in the entire w-plane. This fact shows that Γ is composed of analytic arcs and the same holds for the boundary C of the extremum domain D: *C is composed of analytic arcs.*

In order to complete the argument we need a last elementary variation. We again map $D(\Gamma)$ onto the domain $|w| > 1$ by means of $\phi(z)$. The function

$$\omega = p(w) = w + \frac{w_0^2}{w} \quad (|w_0| = 1) \qquad (26)$$

maps the circular region $|w| > 1$ onto the ω-plane slit along the segment between the points $-2w_0$ and $+2w_0$. It is then easily seen that, for $\epsilon > 0$,

$$w_1 = p^{-1}[(1 + \epsilon) p(w) + 2\epsilon w_0] = w + \epsilon \frac{w(w + w_0)}{w - w_0} + O(\epsilon^2) \qquad (27)$$

provides a mapping of $|w| > 1$ onto the same circular region from which a small radial segment issuing from the periphery point w_0 has been removed. The function

$$v_3(z) = (1 + \epsilon)^{-1} \phi^{-1} \left[\phi(z) + \epsilon \frac{\phi(z) [\phi(z) + \phi(z_0)]}{\phi(z) - \phi(z_0)} + O(\epsilon^2) \right]$$

$$= z + \epsilon \left[\frac{\phi(z)}{\phi'(z)} \frac{\phi(z) + \phi(z_0)}{\phi(z) - \phi(z_0)} - z \right] + O(\epsilon^2) \quad (z_0 \in \Gamma) \qquad (28)$$

is then normalized at infinity and univalent in D. Hence, $f^*(t) = v_3[f(t)]$ is again a competing function in our extremum problem, whence

$$\text{Re} \left\{ \psi \left[z, \frac{\phi(z)}{\phi'(z)} \frac{\phi(z) + \phi(z_0)}{\phi(z) - \phi(z_0)} - z \right] \right\} \leqslant 0. \qquad (29)$$

But observe that the left side of (29) coincides with the right-hand term of (18) since $z_0 \in \Gamma$. Hence, we have proved

$$\psi\left[f(t), \frac{1}{f(t)-z_0}\right] \frac{\phi(z_0)^2}{\phi'(z_0)^2} \leqslant 0 \quad (z_0 \in \Gamma). \tag{30}$$

Since $|\phi(z)| = 1$ for $z \in \Gamma$ we have $\log \phi(z) =$ imaginary on Γ and, consequently, we can write on each analytic arc of Γ

$$z' \frac{\phi'(z)}{\phi(z)} = \text{imaginary}, \quad z' = \frac{dz}{ds}. \tag{31}$$

Thus, we may express (30) also in the form

$$\psi\left[f(t), \frac{1}{f(t)-z}\right] \left(\frac{dz}{ds}\right)^2 \geqslant 0 \quad \text{on} \quad C. \tag{32}$$

In this final form the characterization of the extremum domain has become independent of the choice of the subcontinuum Γ. The boundary arcs of C are determined by a first-order differential equation involving the meromorphic function $W(z)$ defined above.

Under our assumptions made regarding the functional $\psi[z, 1/(z-z_0)]$ it is also easy to prove that the extremum domain cannot possess exterior points. For, suppose z_0 were an exterior point of an extremum domain D. In this case, the mapping (1) itself would be an admissible univalent variation for ρ small enough and the extremum property of $f(t)$ would imply

$$\text{Re}\left\{e^{i\alpha}\rho^2 \psi\left[z, \frac{1}{z-z_0}\right]\right\} + O(\rho^4) \leqslant 0, \tag{33}$$

whence easily

$$\psi\left[z, \frac{1}{z-z_0}\right] = 0. \tag{34}$$

But if, as supposed, ψ is a specific meromorphic function $W(z_0)$, not identically zero, this result is impossible since (34) would imply by analytic continuation that $W(z_0) \equiv 0$[19]. Thus, we have proved the

Theorem. The extremum domain of the extremum problem $\phi[f] = \max$ within the family F is a slit domain bounded by analytic arcs. Each satisfies the differential equation:

$$\psi\left[f(t), \frac{1}{f(t)-z(\tau)}\right] \left(\frac{dz}{d\tau}\right)^2 = 1, \tag{35}$$

where τ is a properly chosen real curve parameter.

This theorem was proved originally[26] by means of rather deep theorems of measure theory. It can be derived in elementary manner from the variational formula for the Green's function as shown here. It permits now a systematic and unified treatment of numerous extremum problems of conformal mapping. The extremum domain can be determined either by integrating the differential equation (35) for the boundary slits or by solving the differential equation implied by (25) for the functions $\psi(w)$ which map the circular domain $|w| > 1$ onto the domains $D(\Gamma)$. The latter procedure is particularly convenient in the case that the original domain T is simply connected.

4. The coefficient problem

The best studied extremum problem in conformal mapping is without any doubt the coefficient problem for the functions univalent in the unit circle. We consider all power series

$$f(z) = z + a_2 z^2 + \ldots + a_n z^n + \ldots, \tag{36}$$

which converge for $|z| < 1$ and which represent univalent functions. Bieberbach stated the conjecture

$$|a_n| \leqslant n. \tag{37}$$

Since the 'Koebe function'

$$\frac{z}{(1-z)^2} = z + 2z^2 + \ldots + nz^n + \ldots \tag{38}$$

is indeed such a univalent power series, this function would seem to be the solution of an infinity of extremum problems. Because of its simple formulation the conjecture (37) has attracted the attention of many analysts. Bieberbach himself proved (37) in 1916 for $n = 2$ [2]; Löwner proved the case $n = 3$ in 1923 [18] and Garabedian and Schiffer proved the case $n = 4$ in 1955 [5]. These proofs are to be considered as tests for our technique in handling extremum problems of conformal mapping and the main significance of the coefficient problem is indeed that it raises a challenge to our various methods in this field. We want to give a brief survey of variational methods applied in this problem.

We define a sequence of polynomials $P_n(x)$ of degree $(n-1)$ by means of the generating function

$$\frac{f(z)}{1 - xf(z)} = \sum_{n=1}^{\infty} [a_n + P_n(x)] z^n \quad (P_1(x) = 0). \tag{39}$$

We note down the first few polynomials

$$P_2(x) = x, \quad P_3(x) = 2a_2 x + x^2, \quad P_4(x) = (2a_3 + a_2^2) x + 3a_2 x^2 + x^3. \tag{39'}$$

A simple application of the reasoning in the preceding section leads to the following result. Let $f(z)$ be a univalent function which maximizes $|a_n|$; we can make the permissible assumption $a_n > 0$. Then $f(z)$ satisfies the differential equation [27]

$$\frac{z^2 f'(z)^2}{f(z)^2} P_n\left[\frac{1}{f(z)}\right] = \frac{1}{z^{n-1}} + \frac{2a_2}{z^{n-2}} + \frac{3a_3}{z^{n-3}} + \ldots + \frac{(n-1)a_{n-1}}{z}$$

$$+ (n-1)a_n + (n-1)\bar{a}_{n-1}z + \ldots + 3\bar{a}_3 z^{n-3} + 2\bar{a}_2 z^{n-2} + z^{n-1}. \quad (40)$$

The right side as well as the polynomial $P_n(x)$ depends on the coefficients of the unknown function $f(z)$; hence, (40) represents a rather complicated functional equation for the extremum function sought which has been solved until now only in the cases $n \leqslant 4$.

We may attack the functional equation (40) as follows. It is easily shown in all cases $n \leqslant 4$ that the extremum function $w = f(z)$ maps the domain $|z| < 1$ onto the entire w-plane slit along a single analytic arc Γ which runs out to infinity. We consider then the analytic functions

$$w = f(z,t) = e^t[z + a_2(t)z^2 + \ldots + a_n(t)z^n + \ldots], \quad (41)$$

which map $|z| < 1$ onto the w-plane slit along infinite subarcs Γ_t of Γ. We can read off from (40) that Γ satisfies the differential equation

$$\frac{w'(\tau)^2}{w(\tau)^2} P_n\left[\frac{1}{w(\tau)}\right] + 1 = 0 \quad (\tau = \text{real parameter}), \quad (42)$$

and evidently the subarcs Γ_t satisfy precisely the same equation. Using next the Schwarz reflection principle, we can show that the functions $f(z,t)$ satisfy differential equations which are very similar to (40); namely

$$\frac{z^2 f'(z,t)^2}{f(z,t)^2} P_n\left[\frac{1}{f(z,t)}\right] = \sum_{\nu=-(n-1)}^{n-1} A_\nu(t) z^\nu = q(z,t), \quad A_{-\nu}(t) = \overline{A_\nu(t)}. \quad (43)$$

We may transform (43) into

$$\int_{f(z_0,t)}^{f(z,t)} \sqrt{\left[P_n\left(\frac{1}{w}\right)\right]} \frac{dw}{w} = \int_{z_0}^z \sqrt{[q(z,t)]} \frac{dz}{z}. \quad (44)$$

Löwner has shown [18] that the functions $f(z,t)$ which represent the unit circle on a family of slit domains with growing boundary slits Γ_t of the above type satisfy the partial differential equation

$$\frac{\partial f(z,t)}{\partial t} = z \frac{1+\kappa(t)z}{1-\kappa(t)z} \frac{\partial f(z,t)}{\partial z} \quad (\kappa(t) \text{ continuous}, |\kappa| = 1). \quad (45)$$

Thus, differentiating (44) with respect to t and using (43) and (44) we find

$$\sqrt{[q(z,t)]}\frac{1+\kappa z}{1-\kappa z} - \sqrt{[q(z_0,t)]}\frac{1+\kappa z_0}{1-\kappa z_0} = \frac{1}{2}\int_{z_0}^{z} \frac{\partial q(z,t)}{\partial t}\frac{1}{\sqrt{[q(z,t)]}}\frac{dz}{z}. \qquad (46)$$

Differentiating (46) again with respect to z, we find after simple re-arrangement

$$\frac{\partial q(z,t)}{\partial t} = z\frac{1+\kappa z}{1-\kappa z}\frac{\partial q(z,t)}{\partial z} + \frac{4\kappa z}{(1-\kappa z)^2}q(z,t). \qquad (47)$$

On the other hand, $q(z,t)$ is a simple rational function of z as is seen from its definition (43). When we insert its expression into (47) and compare the coefficients of equal powers of z on both sides, we obtain

$$\frac{dA_\nu(t)}{dt} = \nu A_\nu(t) + 2\sum_{\mu=-(n-1)}^{\nu-1}(2\nu-\mu)A_\mu \kappa^{\nu-\mu}. \qquad (48)$$

In order that $A_\nu(t) \equiv 0$ for all $\nu \geqslant n$ it is necessary and sufficient that

$$\sum_{\mu=-(n-1)}^{n-1} A_\mu \kappa^{-\mu} \equiv 0, \qquad \sum_{\mu=-(n-1)}^{n-1} \mu A_\mu \kappa^{-\mu} \equiv 0 \quad \text{identically in } t. \qquad (49)$$

These conditions guarantee also that $A_{-\nu} \equiv \bar{A}_\nu$ is fulfilled for all values of t.

We observe that the equations (48) for $\nu = -1, -2, ..., -(n-1)$ give $(n-1)$ differential equations for the corresponding functions $A_\nu(t)$; their coefficients depend in a very simple manner on $\kappa(t)$. The function $\kappa(t)$, in turn, can be determined from the $A_\nu(t)$ by means of the second equation (49), which can be written in the form

$$\text{Im}\left\{\sum_{\mu=-(n-1)}^{-1}\mu A_\mu \kappa^{-\mu}\right\} = 0. \qquad (50)$$

Thus, $A_{-1}, A_{-2}, ..., A_{-n-1}$ and κ satisfy a well-determined system of ordinary differential equations.

Let us start with the case $n = 3$. The differential system to be considered is

$$\frac{dA_{-2}(t)}{dt} = -2A_{-2}(t), \quad \frac{dA_{-1}(t)}{dt} = -A_{-1}(t), \left.\vphantom{\begin{array}{c}1\\1\end{array}}\right\} \qquad (51)$$
$$\text{Im}\{2A_{-2}\kappa^2 + A_{-1}\kappa\} = 0.$$

We can integrate immediately and find

$$A_{-2}(t) = \alpha_2 e^{-2t}, \quad A_{-1}(t) = \alpha_1 e^{-t}. \qquad (52)$$

Since for $t = 0$ the function $q(z,0)$ coincides with the right side of (40) for $n = 3$, we determine the constants of integration as follows: $\alpha_2 = 1$, $\alpha_1 = 2a_2$. Thus, $\kappa(t)$ satisfies the equation

$$e^{-2t}\kappa^2 + a_2 e^{-t}\kappa = \text{real}. \qquad (53)$$

From the general Löwner theory it is well-known that

$$a_2 = -2 \int_0^\infty \kappa\, e^{-t} dt. \tag{54}$$

We have to utilize now the inequality $|a_3 - a_2^2| \leqslant 1$ which follows from the elementary area theorem. Since we assume $a_3 \geqslant 3$ we can assert $\mathrm{Re}\,\{a_2^2\} \geqslant 2$ and see that the left side of (53) cannot vanish for $0 \leqslant t < \infty$.

We wish to show next that $a_2^2 =$ real in consequence of (53) and (54). Indeed, if a_2^2 were not real, equation (53) would exclude the possibility $\kappa = \pm \,\mathrm{sgn}\, a_2$ and the expression $\mathrm{Im}\,\{\kappa(t)\, a_2\}$ could never change its sign. Consequently

$$\int_0^\infty \mathrm{Im}\,\{\overline{2\kappa(t)}\, a_2\}\, e^{-t} dt = -\,\mathrm{Im}\,\{|a_2|^2\} \tag{55}$$

could not be zero, which yields a contradiction. Thus, $a_2^2 =$ real and in consequence of the area theorem even $a_2^2 > 2$ holds; hence, we conclude $a_2 =$ real. From (53) follows then easily that κ must be real throughout and it can be shown that $a_3 = 3$.

The above proof for $|a_3| \leqslant 3$ is more complicated than Löwner's original proof which made use only of the formula (45). It can, however, be generalized to the problem of a_4 though it becomes in this case still more complicated. The differential system becomes now

$$\frac{dA_{-3}}{dt} = -3A_{-3}, \quad \frac{dA_{-2}}{dt} = -2A_{-2} - 2A_{-3}\kappa, \quad \frac{dA_{-1}}{dt} = -A_{-1} + 2A_{-3}\kappa^2, \Bigg\}$$
$$\mathrm{Im}\,\{3A_{-3}\kappa^3 + 2A_{-2}\kappa^2 + A_{-1}\kappa\} = 0. \tag{56}$$

We find $A_{-3} = \alpha_3\, e^{-3t}$ and, since $A_{-3}(0) = 1$, we have $A_{-3} = e^{-3t}$. We set up

$$A_{-2}(t) = \alpha_2(e^{-t})\, e^{-2t}, \quad A_{-1}(t) = \alpha_1(e^{-t})\, e^{-t}; \tag{57}$$

inserting into (56) and putting $\sigma = e^{-t}$, we arrive at the differential system

$$\frac{d\alpha_2(\sigma)}{d\sigma} = 2\kappa, \quad \frac{d\alpha_1(\sigma)}{d\sigma} = -2\kappa^2\sigma \quad (0 \leqslant \sigma \leqslant 1), \Bigg\}$$
$$\mathrm{Im}\,\{3\kappa^3\sigma^3 + 2\alpha_2(\sigma)\,\kappa^2\sigma^2 + \alpha_1(\sigma)\,\kappa\sigma\} = 0. \tag{58}$$

A simple calculation leads to the boundary conditions

$$\alpha_2(0) = 3a_2, \quad \alpha_1(0) = 2a_3 + a_2^2, \Bigg\}$$
$$\alpha_2(1) = 2a_2, \quad \alpha_1(1) = 3a_3. \tag{59}$$

Those for $\sigma = 1$, $t = 0$ are obvious; those for $\sigma = 0$, $t = \infty$ follow by comparison of coefficients of powers of e^{-t} in (43) and by passage to the limit $t = \infty$.

The differential system (58), together with the boundary conditions (59), represents a typical Sturm–Liouville boundary value problem. We have to start integration of (58) with such initial values $\alpha_1(0)$ and $\alpha_2(0)$ that we end up at the other end of the interval considered with

$$\alpha_1(1) = \tfrac{3}{2}[\alpha_1(0) - \tfrac{1}{9}\alpha_2(0)^2], \quad \alpha_2(1) = \tfrac{2}{3}\alpha_2(0). \tag{60}$$

The difficulty of the problem lies in the non-linear character of the equations and of the boundary conditions. Each possible set $\alpha_1(0)$, $\alpha_2(0)$ determines a set of possible values a_2, a_3. Clearly, $a_2 = 2$, $a_3 = 3$ and $\kappa(\sigma) \equiv -1$ is an admissible solution which leads to the Koebe function (38), the conjectured extremum function.

The question arises now whether the corresponding special values $\alpha_1(0)$, $\alpha_2(0)$ connected with the conjectured extremum function might not be imbedded into a one-parameter family of initial values such that all of them lead to the boundary relations (60). For this purpose, we have to study the variational equations of the system (58) and of the boundary conditions (60). If we denote the derivatives of α_1, α_2 and κ with respect to the parameter by β_1, β_2 and $i\lambda$, we find easily

$$\left.\begin{array}{l} \dfrac{d\beta_1}{d\sigma} = 4i\lambda\sigma, \quad \dfrac{d\beta_2}{d\sigma} = 2i\lambda, \quad \lambda = \dfrac{1}{2p(\sigma)}\,\mathrm{Im}\,\{\beta_1 - 2\beta_2\sigma\}, \\[2mm] \beta_1(1) = \tfrac{3}{2}[\beta_1(0) - \tfrac{4}{3}\beta_2(0)], \quad \beta_2(1) = \tfrac{2}{3}\beta_2(0); \quad p(\sigma) = 8\sigma^2 - 12\sigma + 5. \end{array}\right\} \tag{61}$$

We are thus led to a linear differential system with linear boundary conditions which can be treated by the standard Sturm–Liouville methods.

It is immediately seen from (61) that λ is real and that $\beta_1(\sigma)$ and $\beta_2(\sigma)$ must be pure imaginary. When we introduce the new unknowns

$$u(\sigma) = \mathrm{Im}\,\{\beta_1 - 2\sigma\beta_2\}, \quad v(\sigma) = \mathrm{Im}\,\{\beta_2\}$$

the system (61) simplifies to

$$\frac{du}{d\sigma} = -2v, \quad \frac{dv}{d\sigma} = \frac{1}{p(\sigma)}\,u, \tag{62}$$

with the boundary conditions

$$u(1) = \tfrac{3}{2}u(0) - \tfrac{10}{3}v(0), \quad v(1) = \tfrac{2}{3}v(0). \tag{63}$$

From the differential system we derive by integration by parts the equality

$$2\int_0^1 v^2\,d\sigma = \int_0^1 pv'^2\,d\sigma + \tfrac{20}{9}v(0)^2; \quad v(1) = \tfrac{2}{3}v(0). \tag{64}$$

We may now apply the calculus of variations in order to estimate the ratio

$$\left[\int_0^1 pv'^2 d\sigma + \tfrac{20}{9}v(0)^2\right] \cdot \left[\int_0^1 v^2 d\sigma\right]^{-1} = R[v] \qquad (65)$$

under the given boundary condition on $v(\sigma)$. Even when we replace in (65) the polynomial $p(\sigma)$ by a piecewise constant function which is nowhere larger than $p(\sigma)$ the minimum value of the new ratio, which can now be computed explicitly, comes out to be larger than 2. Hence, a fortiori, we can assert that $R[v] > 2$ for all admissible $v(\sigma)$ and that (64) is impossible. We have thus shown that the solution $a_2 = 2$, $a_3 = 3$, $\kappa(\sigma) \equiv -1$ cannot be imbedded into a one-parameter family of solutions which can be differentiated continuously with respect to this parameter.

By a more careful analysis we may now treat differences of solution systems $\alpha_1(\sigma)$, $\alpha_2(\sigma)$, $\kappa(\sigma)$ instead of differentials. We can then delimit an entire neighborhood of the point $a_2 = 2$, $a_3 = 3$, $\kappa = -1$ in which no other solution point could be located. On the other hand, one can combine the area theorem with various relations between the coefficients of the extremum function which arise from the differential equation (40), in order to estimate the values $|a_2 - 2|$ and $|a_3 - 3|$ in the extremum case. It can be seen by elementary if very tedious calculations that the point a_2, a_3, $\kappa(1)$ must lie precisely in the neighborhood in which 2, 3, -1 is the only solution point. This proves that the Koebe function (38) is, indeed, the extremum function and establishes the inequality $|a_4| \leqslant 4$ for all univalent functions (36).

The actual labor in the proof sketched here lies in the very extensive elementary estimations and could probably be reduced considerably by extending the uniqueness neighborhood through greater attention to the theory of the differential system (58), (59).

It may be remarked, finally, that the Koebe function (38) satisfies the functional equation (40) which characterizes the extremum function for every $n \geqslant 2$. This fact tends, of course, to strengthen the evidence for the Bieberbach conjecture. The following fact should be mentioned, however, in order to caution against too great reliance on this evidence. One may consider the family of functions

$$f(z) = z + b_0 + \frac{b_1}{z} + \frac{b_2}{z^2} + \dots + \frac{b_n}{z^n} + \dots, \qquad (66)$$

which are univalent in the outside $|z| > 1$ of the unit circle and one may ask for $\max |b_n|$. The same variational technique as above yields for the extremum functions $f_n(z)$ of this 'exterior' problem a differential-functional equation which is analogous to (40). It is easy to show that the functions

$$F_n(z) = \left[z^{n+1} + 2 + \frac{1}{z^{n+1}} \right]^{1/(n+1)} = z + \frac{2}{n+1}\frac{1}{z^n} + \dots \qquad (67)$$

belong to the family considered and satisfy the extremum condition for the corresponding $f_n(z)$. For $n = 1$ and $n = 2$ these functions are, indeed, the extremum functions of the exterior coefficient problem. The estimate $|b_1| \leqslant 1$ was discovered together with the area theorem[2] and $|b_2| \leqslant \frac{2}{3}$ was established in 1938 by Golusin[6] and myself[25]. It was conjectured that $|b_n| \leqslant 2/(n+1)$ was the best possible estimate for the nth coefficient for all values of n. However, in 1955 Garabedian and I[4] showed that the precise value of the maximum for $|b_3|$ is not $\frac{1}{2}$ as expected but $\frac{1}{2} + e^{-6}$. Thus, in spite of the fact that the function $F_3(z)$, defined in (67), satisfies the rather restrictive extremum condition, it is not the extremum function $f_3(z)$. Since e^{-6} is a small number, this example shows also how little empirical numerical evidence can be trusted in problems of this kind. Recently, Waadeland[36] has shown that quite generally

$$\max |b_{2k-1}| \geqslant \frac{1}{k}(1 + 2e^{-2[(k+1)/(k-1)]}), \qquad (67')$$

while for $n = 2k$ no counter example to the conjecture $|b_n| \leqslant 2/(n+1)$ seems to be known.

There are, of course, numerous cross-relations between the coefficient problem for univalent functions and the general theory of conformal mapping. Two examples may serve as illustrations. There is a well-known problem in the theory of conformal mapping: given n points in the complex plane, to find a continuum which contains these points and has minimum capacity[10]. From the topology of the extremum continuum, one can derive by an elementary variation the coefficient inequality $|b_2| \leqslant \frac{2}{3}$[25]. Here, the general theory of conformal mapping helped to solve a coefficient problem. Conversely, de Possel[21] formulated a simple extremum problem for the coefficients of univalent functions in a multiply-connected domain and showed that the extremum functions mapped the domain onto a parallel slit domain. Since the existence of an extremum function is assured, an elegant existence proof for an important canonical mapping was thus established.

5. Fredholm eigenvalues

The problem of conformally mapping a given plane domain D can often be reduced to a boundary value problem for the functions harmonic in D. If the boundary C of D is sufficiently smooth, the latter problem can be attacked through the Poincaré–Fredholm integral equation

$$m(z) = \mu(z) + \frac{1}{\pi} \int_C \frac{\partial}{\partial n_\zeta} \left(\log \frac{1}{|z-\zeta|} \right) \mu(\zeta)\, ds_\zeta \quad (z \in C). \tag{68}$$

In order to solve this fundamental integral equation of two-dimensional potential theory one has to consider the corresponding homogeneous integral equation

$$\phi_\nu(z) = \frac{\lambda_\nu}{\pi} \int_C \frac{\partial}{\partial n_\zeta} \left(\log \frac{1}{|z-\zeta|} \right) \phi_\nu(\zeta)\, ds_\zeta \quad (z \in C), \tag{69}$$

its eigenfunctions $\phi_\nu(z)$ and its eigenvalues λ_ν. The eigenvalue $\lambda = 1$ occurs always and has as eigenfunctions a set of easily described functions on C; we shall call this eigenvalue the trivial eigenvalue of the domain. The non-trivial eigenvalues λ_ν satisfy the inequality $|\lambda_\nu| > 1$. It is easily seen that with each non-trivial eigenvalue λ_ν also the value $-\lambda_\nu$ will occur as eigenvalue of (69) with the same multiplicity. We shall restrict ourselves, therefore, to the positive non-trivial eigenvalues λ_ν and assume them ordered in increasing magnitude. These eigenvalues λ_ν are called the Fredholm eigenvalues of the domain D and they are of importance for the potential theory and the function theory of the domain considered.

It is, for example, of great interest to obtain a lower bound for the first eigenvalue λ_1 of a given domain. Such information would enable us to estimate the speed of convergence of the Neumann–Liouville series which solves the basic equation (68). The larger λ_1 can be asserted to be, the easier the numerical work for the solution of the boundary value problems in the potential theory for D. Thus, the λ_ν seem to be a set of functionals of D which deserves a careful study.

The λ_ν are also closely related to the theory of the Hilbert transformation

$$F(z) = \frac{1}{\pi} \iint_D \frac{f(\zeta)}{(\zeta-z)^2} d\tau_\zeta, \tag{70}$$

which carries each analytic function in D into a new analytic function in the same domain. There exists a set of eigenfunctions $w_\nu(z)$ which are analytic in D and which satisfy

$$w_\nu(z) = \frac{\lambda_\nu}{\pi} \iint_D \frac{\overline{w_\nu(\zeta)}}{(\zeta-z)^2} d\tau_\zeta \quad (\lambda_\nu > 1). \tag{71}$$

The eigenvalues λ_ν are precisely the Fredholm eigenvalues defined above. We shall assume the $w_\nu(z)$ to be normalized by the usual convention

$$\iint_D |w_\nu(z)|^2 \, d\tau = 1. \tag{72}$$

The eigenfunctions $w_\nu(z)$ form an orthonormal set of analytic functions in D and play an interesting role in the theory of the kernel function of $D^{[33]}$.

In order to establish a unified theory for the treatment of extremum problems for the functionals λ_ν of D it is necessary to determine the variation of each λ_ν for a variation of the defining domain D. If we assume the variation to be of the special type (1) with $z_0 \in D$ and if λ_ν is non-degenerate, we have

$$\lambda_\nu^* = \lambda_\nu + (1 - \lambda_\nu^2)\,\pi \operatorname{Re}\{e^{i\alpha}\rho^2 w_\nu(z_0)^2\} + O(\rho^4). \tag{73}$$

An analogous, though slightly more complicated, formula can be given for the variation of degenerate eigenvalues.

When one wishes to apply the variational formula (73) to the solution of extremum problems, one runs immediately into a serious difficulty. The entire theory of the Fredholm eigenvalues has been established under certain smoothness conditions for the boundary and one has to be sure that the extremum domain does possess a boundary of this type. One has to introduce a class of domains which possess admissible boundaries and which is compact; within such a class the calculus of variations based on (73) and the theory of extremum problems become possible.

For this purpose, we introduce the concept of *uniformly analytic curves*. A curve is called analytic if it can be obtained as the image of the unit circumference $|z| = 1$ by means of a function $t(z)$ which is analytic and univalent on $|z| = 1$. A set of curves is said to be uniformly analytic with the modulus of uniformity (r, R) (where $r < 1 < R$) if all of them are obtained by means of mapping functions $f(z)$ which are analytic and univalent in the fixed annulus $r \leqslant |z| \leqslant R$. This concept of uniform analyticity seems to be quite useful in the variational theory of domain functionals.

We can now formulate the theorem:

If a simply connected domain is bounded by a curve which is analytic with the modulus (r, R), then its lowest Fredholm eigenvalue λ_1 satisfies the inequality:

$$\lambda_1 \geqslant \frac{r^2 + R^2}{1 + r^2 R^2}. \tag{74}$$

This estimate is the best possible for every modulus (r, R).

Frequently, the boundary curve C of a domain is given in a parametric representation from which the modulus (r, R) can be readily deduced. Thus, the estimate (74) is often convenient to predict the convergence of the Neumann–Liouville series which solve the various boundary value problems in the domain.

We may also connect with a given domain D the Fredholm determinant

$$D(\lambda) = \prod_{\nu=1}^{\infty} \left(1 - \frac{\lambda^2}{\lambda_\nu^2}\right) \tag{75}$$

of the integral equation (68) and consider, for fixed λ, $D(\lambda)$ as a functional of the domain D. The following extremum problem suggests itself: Let D_0 be a given multiply-connected domain; consider all smoothly bounded domains D which are conformally equivalent to it and ask for those domains in this equivalence class which yield the maximum value of $D(\lambda)$.

This problem has been solved in the case $\lambda = 1$. The main difficulty in the investigation was again the non-compactness of the class of domains considered. It could be overcome by considering maximum sequences of domains and their limit domain; all domains of the sequence were subjected to the same variation (1) and from the fact that they formed a maximum sequence it could be shown that their limit domain is analytically bounded. Then, the existence of a maximum domain is easily established and it can be shown that it is bounded by circumferences. We obtain thus a new proof of Schottky's famous circular mapping theorem and also a characterization of this canonical mapping by an extremum property. Methodologically, the proof is of interest since the method of variation is not applied to the extremum domain, whose existence is not yet known, but to the extremum sequence. This procedure seems to be of very great applicability.

The solution of the maximum problem for general $D(\lambda)$ is not yet known and well deserves additional study.

The Fredholm eigenvalues represent an instructive example for the flexibility of the variational method in dealing with extremum problems for rather difficult types of domain functionals. The great formal elegance of the variational formula (73) enabled us to overcome the quite serious difficulties which arise from the fact that these functionals are defined only for a restricted and non-compact class of domains.

6. Further applications

We have restricted ourselves to a few fundamental problems in order to exhibit clearly the basic ideas of the variational method. It may, however, be applied to much more general function-theoretic problems. It can be used in problems of mapping of domains on Riemann surfaces[29, 33] and leads there to existence theorems for various canonical realizations of Riemann domains. It can be applied to the theory of multivalent functions in a given domain[27], their coefficient problems and distortion theorems. Some interest has been devoted to the problem of developing a calculus of variations within important subclasses of the family of univalent functions in the unit circle. Golusin[9] described a method of variations for the subclass of star-like univalent mappings and Hummel[12, 13] gave an even simpler method of this kind. Singh[34] gave a theory of variations for real univalent functions, for bounded univalent functions and other interesting subclasses. Finally, the role should be mentioned which the method of variations could play as a useful tool in the theory of quasi-conformal mappings and of extremal metrics[14].

The variational method is, of course, only one of many powerful methods in the theory of conformal mapping and complex function theory. There are many problems where other methods give the answer more easily and directly. It seems to me, however, that the method of variations is one of the most systematic and widely applicable methods which we possess in this field.

REFERENCES

[1] Bernardi, S. D. A survey of the development of the theory of schlicht functions. *Duke Math. J.* 19, 263–287 (1952).
[2] Bieberbach, L. Über die Koeffizienten derjenigen Potenzreihen, welche eine schlichte Abbildung des Einheitskreises vermitteln. *Berl. Ber.* 940–955 (1916).
[3] Courant, R. *Dirichlet's Principle, Conformal Mapping and Minimal Surfaces.* New York, 1950. Appendix by M. Schiffer.
[4] Garabedian, P. R. and Schiffer, M. A coefficient inequality for schlicht functions. *Ann. Math.* (2), 61, 116–136 (1955).
[5] Garabedian, P. R. and Schiffer, M. A proof of the Bieberbach conjecture for the fourth coefficient. *J. Rat. Mech. Anal.* 4, 427–465 (1955).
[6] Golusin, G. M. Einige Koeffizientenabschätzungen für schlichte Funktionen. *Rec. Math. (Mat. Sbornik)*, 3 (45), 321–330 (1938).
[7] Golusin, G. M. Method of variations in the theory of conformal mapping. *Rec. Math. (Mat. Sbornik)*, 19 (61), 203–236 (1946); 21 (63), 83–117, 119–132 (1947); 29 (71), 455–468 (1951).
[8] Golusin, G. M. *Geometrical Theory of Functions of a Complex Variable.* Moscow, 1952. (German translation: Berlin, 1957.)
[9] Golusin, G. M. A variational method in the theory of analytic functions. *Leningrad Gos. Univ. Uč. Zap.* 144, Ser. Mat. Nauk, 23, 85–101 (1952).
[10] Grötzsch, H. Über ein Variationsproblem der konformen Abbildung. *Leipzig. Ber.* 82, 251–263 (1930).

[11] Hadamard, J. Mémoire sur le problème relatif à l'équilibre des plaques élastiques encastrées. *Acad. Sci. Paris, Mém. Sav. étrangers*, 33 (1908).

[12] Hummel, J. A. The coefficient regions of starlike functions. *Pacif. J. Math.* 7, 1381–1389 (1957).

[13] Hummel, J. A. A variational method for starlike functions. *Proc. Amer. Math. Soc.* 9, 82–87 (1958).

[14] Jenkins, J. A. On the existence of certain general extremal metrics. *Ann. Math.* (2), 66, 440–453 (1957).

[15] Julia, G. Sur une équation aux dérivées fonctionnelles liée à la représentation conforme. *Ann. Éc. Normale* (3), 39, 1–28 (1922).

[16] Lavrentieff, M. A. Sur deux questions extrémales. *Rec. Math. (Mat. Sbornik)*, 41, 157–165 (1934).

[17] Lavrentieff, M. A. Über eine extremale Aufgabe aus der Tragflügeltheorie. *Cent. Aero-Hydrodyn. Inst.* no. 155, 1–40 (1934).

[18] Löwner, K. Untersuchungen über schlichte konforme Abbildungen des Einheitskreises. *Math. Ann.* 89, 103–121 (1923).

[19] Marty, F. Sur le module des coefficients de Maclaurin d'une fonction univalente. *C.R. Acad. Sci., Paris*, 198, 1569–1571 (1934).

[20] Montel, P. *Leçons sur les fonctions univalentes ou multivalentes.* Paris, 1933.

[21] Possel, R. de. Zum Parallelschlitztheorem unendlichvielfach zusammenhängender Gebiete. *Göttinger Nachr.* 199–202 (1931).

[22] Schaeffer, A. C. and Spencer, D. C. The coefficients of schlicht functions. I. *Duke Math. J.* 10, 611–635 (1943); II. *Duke Math. J.* 12, 107–125 (1945); III. *Proc. Nat. Acad. Sci., Wash.*, 32, 111–116 (1946).

[23] Schaeffer, A. C. and Spencer, D. C. A variational method in conformal mapping. *Duke Math. J.* 14, 949–966 (1947).

[24] Schaeffer, A. C. and Spencer, D. C. *Coefficient Regions for Schlicht Functions.* Amer. Math. Soc. Colloquium Publ., vol. 35 (1950).

[25] Schiffer, M. Sur un problème d'extrémum de la représentation conforme. *Bull. Soc. Math. Fr.* 66, 48–55 (1938).

[26] Schiffer, M. A method of variation within the family of simple functions. *Proc. Lond. Math. Soc.* (2), 44, 432–449 (1938).

[27] Schiffer, M. Variation of the Green function and theory of the p-valued functions. *Amer. J. Math.* 65, 341–360 (1943).

[28] Schiffer, M. Hadamard's formula and variation of domain functions. *Amer. J. Math.* 68, 417–448 (1946).

[29] Schiffer, M. Variational methods in the theory of Riemann surfaces. *Contributions to the Theory of Riemann Surfaces*, pp. 15–30. Princeton, 1953.

[30] Schiffer, M. Variation of domain functionals. *Bull. Amer. Math. Soc.* 60, 303–328 (1954).

[31] Schiffer, M. Application of variational methods in the theory of conformal mappings. *Proc. Symp. appl. Math.* 8, 93–113 (1958).

[32] Schiffer, M. The Fredholm eigenvalues of plane domains. *Pacif. J. Math.* 7, 1187–1225 (1957).

[33] Schiffer, M. and Spencer, D. C. *Functionals of Finite Riemann Surfaces.* Princeton, 1954.

[34] Singh, V. Interior variations and some extremal problems for certain classes of univalent functions. *Pacif. J. Math.* 7, 1485–1504 (1957).

[35] Spencer, D. C. Some problems in conformal mapping. *Bull. Amer. Math. Soc.* 53, 417–439 (1947).

[36] Waadeland, H. Über ein Koeffizientenproblem für schlichte Abbildungen des $|\zeta| > 1$. *Norske Vidensk. Selsk. Forh.* 30, 168–170 (1957).

COHOMOLOGY OPERATIONS AND
SYMMETRIC PRODUCTS

By N. E. STEENROD

This lecture is embodied in a set of mimeographed notes entitled 'Cohomology operations and obstructions to extending continuous functions'. These can be obtained by writing to the Department of Mathematics, Fine Hall, Box 708, Princeton, N.J., U.S.A.

LINEARIZATION AND DELINEARIZATION

By G. TEMPLE†

1. Introduction

My terms of reference, as prescribed by our President, are to survey problems of applied mathematics which still challenge the pure mathematician. This is an agreeable exercise for it enables me to range over a wide field, to select such topics as fancy and caprice may dictate, and above all to shun the rigours of precise proof and detailed definition.

The group of problems which I propose to describe belong to that Cinderella of pure mathematics—the study of differential equations. The closely guarded secret of this subject is that it has not yet attained the status and dignity of a science, but still enjoys the freedom and freshness of such a pre-scientific study as natural history compared with botany. The student of differential equations—significantly he has no name or title to rank with the geometer or analyst—is still living at the stage where his main tasks are to collect specimens, to describe them with loving care, and to cultivate them for study under laboratory conditions. The work of classification and systematization has hardly begun.

This is true even of differential equations which belong to the genus technically described as 'ordinary, linear equations'. The morphology of this genus has progressed only as far as equations which possess three or at most four regular singularities. In the case of non-linear equations, Lie's theory of transformation groups has done little but suggest a scheme of classification. An inviting flora of rare equations and exotic problems lies before a botanical excursion into the non-linear field.

I propose today to speak of some linear and non-linear differential equations as they arise in mathematical physics, with an eye to the unsolved analytical problems which they present.

The history of mathematical physics during the last century may be divided into two periods—the linear period and the non-linear period. In those happy far-off times of the linear period, all differential equations were linear and the principle of superposition reigned supreme. In the present distressful times most differential equations are non-linear and no effective general method of solution has yet been proposed. We have, however, two practical expedients—the method of linearization by which non-linear equations are forcibly reduced to an associated, approximate

† Read by Professor E. C. Titchmarsh.

linear form, and the method of delinearization by which the non-linearities are partially restored.

Linearization and delinearization are the main topics of my address, especially in relation to the equations of fluid dynamics, but perhaps it is desirable to illustrate the nature of the problems involved by some trivial examples.

2. Regular and singular perturbations

Consider the ordinary differential equation of the first order

$$du/dx = F(x, u, \alpha),$$

in which α is a small parameter. The classical existence theorem can be easily proved by the use of dominant functions (Goursat[7]). It shows that, if F is an analytic function of x, u and α in the neighbourhood of a point $x = x_0$, $u = u_0$ and $\alpha = \alpha_0$, then the differential equation possesses a solution $u = u(x, \alpha)$, which is analytic in some neighbourhood of the point $x = x_0$, and such that $u_0 = u(x_0, \alpha)$, if α is in some neighbourhood of α_0. In the type of problem which we wish to study we are especially interested in the solution for small values of α, and therefore in the 'reduced equation'

$$du/dx = F(x, u, 0),$$

which, in practice, is often much simpler than the original 'perturbed equation', $du/dx = F(x, u, \alpha)$. The solution of the reduced equation is called the 'basic solution', $f_0(x) = u(x, 0)$. It is clear that the perturbed equation will possess a solution of the form

$$u = f_0(x) + \alpha f_1(x) + \ldots + \alpha^n f_n(x) + \ldots,$$

convergent in some interval $|\alpha| < \rho$, and reducing to u_0 at $x = x_0$, if the function F is analytic in a neighbourhood of $x = x_0$, $u = u_0$, $\alpha = 0$. Also the leading term $f_0(x)$ will then satisfy the reduced equation. In this case the perturbation is said to be 'regular' at (x_0, u_0). But if the function F is not analytic in a neighbourhood of $x = x_0$, $u = u_0$, $\alpha = 0$, the perturbation is said to be 'singular'. The classical existence theorem then applies no longer. This is the interesting case which frequently arises in applied mathematics.

There is one obvious method of dealing with singular perturbations— it is to find a transformation which will result in an equation (or equations), for which the perturbation is regular.

Consider, for example, the trivial equation

$$(x + \alpha) \, du/dx + u = 0,$$

with the reduced equation
$$x\,du/dx + u = 0,$$
and the initial conditions $\quad x = 1, \quad u = 1.$

The full perturbation equation is regular everywhere in the x, u-plane except at $x = 0$. The solution of the perturbed equation is
$$u(x, \alpha) = (1+\alpha)/(x+\alpha),$$
while the basic solution is $\quad f_0(x) = x^{-1}.$

The relation between the basic solution and the perturbed solution is that
$$u(x, \alpha) - f_0(x) = \frac{\alpha(1-x)}{x(x+\alpha)} = O(\alpha),$$
but the approximation indicated by the order term is not uniformly valid for all values of x. In fact it is uniformly valid only in domains which exclude $x = 0$ and $x = -\alpha$.

If, however, we express the equation and solution in inverted form as
$$u\,dx/du + x + \alpha = 0$$
and
$$x(u, \alpha) = -\alpha + (1+\alpha)\,u^{-1},$$
then the perturbation is regular, the basic solution is
$$x(u, 0) = u^{-1},$$
and
$$x(u, \alpha) - x(u, 0) = -\alpha + \alpha u^{-1} = O(\alpha),$$
uniformly in a neighbourhood of $\alpha = 0$.

3. Neighbouring solutions

If $u = u(x, \alpha)$ is an integral curve of
$$du/dx = F(x, u, \alpha),$$
which passes through a point (x_0, u_0) in a region D in which the differential equation is regular, then
$$u(x, \alpha) - u(x, 0) = O(\alpha) \tag{3.1}$$
uniformly in D. But, as the preceding example shows, this is no longer true if D contains points at which the differential equation is singular.

The significance of the relation (3.1) is that the integral curves $u = u(x, \alpha)$ and $u = u(x, 0)$ are 'neighbouring curves', with ordinates differing by $O(\alpha)$ in D. But the preceding example suggests at once a more general concept of neighbourliness.

Elementary geometrical considerations applied to a system of curves

$$\phi(x, u, \alpha) = 0$$

suggest that the curves

$$\Gamma \quad \text{or} \quad \phi(x, u, 0) = 0,$$

and

$$C \quad \text{or} \quad \phi(x, u, \alpha) = 0,$$

should be regarded as 'neighbouring' in a region D, if, with any point (ξ, η) on Γ, we can associate a point (x, u) on C such that

$$x - \xi = O(\alpha),$$

and

$$u - \eta = O(\alpha),$$

uniformly in D. In the preceding example

$$u = \eta, \quad x - \xi = -\alpha(1 - \xi).$$

This then suggests that the whole system of curves

$$\phi(x, u, \alpha) = 0$$

should be regarded as a system of neighbouring curves if they can be represented in the parametric form

$$x = X(z, \alpha), \quad u = U(z, \alpha),$$

where X and U are analytic functions of z and α reducing to ξ and η respectively when $\alpha = 0$.

Since ξ and η are connected by the relation $\phi(\xi, \eta, 0) = 0$ this representation is equivalent to

$$x = \xi + \sum_1^\infty \alpha^n x_n(\xi), \quad u = \eta(\xi) + \sum_1^\infty \alpha^n u_n(\xi),$$

in a region where $d\eta/d\xi$ is bounded.

Although there is no *a priori* reason to assert that the solutions of a given singular perturbation problem must form a system of neighbouring curves, the preceding ideas do provide a powerful and flexible technique for searching for solutions and approximations which are uniform within a region containing singular points.

This technique is due to Lighthill and has received numerous applications in fluid dynamics. It is reminiscent of the method of small perturbations employed by Poincaré[16], but the motivation of Poincaré's work was not any singularity in the perturbation but practical convenience in calculating the period of non-linear oscillations.

4. Uniformization

If the original perturbation equation

$$du/dx = F(x, u, \alpha) \quad (u = u_0 \text{ at } x = x_0),$$

possesses a system of neighbouring solutions

$$x = x(\xi, \alpha), \quad u = \eta(\xi, \alpha),$$

then the equations for $x(\xi, \alpha)$ and $\eta(\xi, \alpha)$ must be regular in a neighbourhood of $\xi = x_0$, $u = \eta(x_0, 0)$. The search for systems of neighbouring solutions therefore depends upon the introduction of a new variable ξ and the replacement of the original equation

$$du/dx = F(x, u, \alpha)$$

by two new equations,

$$dx/d\xi = X(\xi, x, u, \alpha), \quad du/d\xi = U(\xi, x, u, \alpha),$$

regular in α.

This process may be called the 'uniformization' of the original equation, and it is equivalent to the method introduced by Lighthill[13].

Thus a typical equation discussed by Lighthill

$$(x + \alpha u) \, du/dx + q(x) \, u = r(x)$$

possesses the uniformizing equations

$$\frac{du}{d\xi} = r(x) - q(x) \, u,$$

$$\frac{dx}{d\xi} = x + \alpha u.$$

These equations are manifestly analytic in α, and in fact their solutions are precisely those given by Lighthill if we write $\xi = \log z$.

Consider for example, the equation

$$(x + \alpha u) \, du/dx + (2 + x) \, u = 0,$$

with the condition $\quad u = e^{-1} \quad$ at $\quad x = 1.$

The reduced equation $\quad x \, du/dx + (2 + x) \, u = 0$

has the solution $\quad u = x^{-2} e^{-x}.$

To obtain the solution of the perturbed equation which is valid near $x = 0$ we must therefore uniformize by introducing an auxiliary variable.

To facilitate comparison with Lighthill's solution[13] (p. 1190) we write

$$z\,dx/dz = x + \alpha u, \quad z\,du/dz = -(2+x)\,u,$$

with $$x = 1, \quad u = e^{-1} \quad \text{at} \quad z = 1.$$

These equations are analytic in α and possess solutions of the form

$$x = x_0 + \alpha x_1 + \dots, \quad u = u_0 + \alpha u_1 + \dots,$$

where $$x_0 = z, \quad u_0 = z^{-2} e^{-z}, \quad x_1 = z\phi(z),$$

$$u_1 = -z^{-2} e^{-z} \int_1^z \phi(t)\,dt, \quad \phi(z) = \int_1^z s^{-4} e^{-s}\,ds.$$

Hence, near $z = 0$,

$$x = z - \tfrac{1}{3}\alpha z^{-2} + O(\alpha^2/z^4),$$

$$u = z^{-2} - \tfrac{1}{6}\alpha z^{-4} + O(\alpha^2/z^6),$$

and, at $x = 0$, $$u = (3/\alpha)^{\frac{2}{3}} + O(\alpha^{-\frac{1}{3}}).$$

The method of uniformization suggested here systematizes Lighthill's method of expansion in powers of an auxiliary variable. Its main advantage is that it establishes the *existence* of a solution which is analytic in the small parameter, without becoming embroiled in the details of its *computation*.

5. Singular boundary conditions

The singular perturbation equations which arise in fluid dynamics are often of a rather different character from those discussed above. In the first place they are usually of the second order, and in the second place the singularity is not in the equation but in the boundary conditions.

The first difference is of little importance. An equation of the second order can be replaced by a pair of equations of the first order, e.g.

$$F(x, u, v, du/dx, \alpha) = 0, \quad dv/dx = u.$$

The process of uniformization then consists in introducing an auxiliary variable z in such a way that the original system of equations is replaced by a system

$$dx/dz = X(z, x, u, v, \alpha),$$

$$du/dz = U(z, x, u, v, \alpha),$$

$$dv/dz = uX,$$

which is analytic in α.

The second difference is much more significant and a systematic examination of this question is lacking.

A survey of those problems of compressible fluid flow which can be reduced to ordinary differential equations has been given by Lighthill[12]. Some of these require the location of a shock wave and involve singular boundary conditions.

A striking example given by Lighthill[13] refers to the waves produced in still air by the slow uniform expansion of a circular cylinder with radial velocity αa_0, a_0 being the speed of sound in the undisturbed air, and α a small parameter. The velocity potential has the form

$$\phi = a_0^2 t f(x),$$

where t is the time since the cylinder was of zero radius, and $x = r/(a_0 t)$. The disturbed region is bounded externally by a shock wave at $r = M a_0 t$ or $x = M$, and internally by the surface of the cylinder $r = \alpha a_0 t$ or $x = \alpha$. The main problem is to calculate M for small values of α.

Bernoulli's equation gives the local speed of sound in the form

$$a = a_0\{1 - (\gamma - 1)\,(f - xf' + \tfrac{1}{2}f'^2)\}^{\frac{1}{2}},$$

while the potential equation is

$$a^2 \operatorname{div} \operatorname{grad} \phi = \ddot\phi + 2\phi_r\dot\phi_r + \phi_r^2\phi_{rr}.$$

Hence $\{1 - (\gamma - 1)\,(f - xf' + \tfrac{1}{2}f'^2)\}\,(f'' + x^{-1}f') = (x - f')^2 f''.$

The boundary conditions are

(1) at $x = \alpha$, $f'(\alpha) = \alpha$,

(2) at $x = M$, $f(M) = 0$,

and $f'(M) = 2(M - M^{-1})/(\gamma + 1).$

To put the differential equation in standard form we write $f' = u$, $f = v$, whence
$$P\,du/dx + Qu/x = 0, \quad dv/dx = u,$$

where $P = 1 - x^2 + (\gamma + 1)\,xu - (\gamma - 1)\,v - \tfrac{1}{2}(\gamma + 1)\,u^2,$

and $Q = 1 + (\gamma - 1)\,(xu - v - \tfrac{1}{2}u^2).$

If we linearize this equation we find that

$$(1 - x^2)\,du/dx + u/x = 0,$$

whence $|u| = C\,|x^{-2} - 1|^{\frac{1}{2}},$

and $C = \alpha^2(1 - \alpha^2)^{-\frac{1}{2}}.$

It is then obvious that this approximation fails as we approach the upper limit $x = M$. We therefore proceed to uniformize the equation by writing

$$z\,dx/dz = x/Q, \quad z\,du/dz = -u/P,$$

and we construct solutions of the form

$$x = z + \alpha^2 x_1 + \alpha^4 x_2 + \cdots,$$

$$u = \alpha^2 u_0 + \alpha^4 u_1 + \cdots,$$

$$v = \alpha^2 v_0 + \alpha^4 v_1 + \cdots.$$

This preserves the solution of the linearized equation in the leading terms, with
$$u_0 = (z^{-2} - 1)^{\frac{1}{2}}.$$

The solution then follows the lines of Lighthill's argument[13] (p. 1191) and finally yields
$$M = 1 + \tfrac{3}{8}(\gamma + 1)^2 \alpha^4 + \cdots.$$

6. Perturbations which are singular almost everywhere

A specially interesting type of perturbation equation is one which is singular everywhere in the x, u-plane except on a certain curve C, e.g. the equation
$$\alpha\,du/dx = F(x, u).$$

A classical example occurs in the theory of relaxation oscillations of the type studied by van der Pol. Here the perturbation equation can be expressed in the form
$$\alpha u\,du/dx = u - \tfrac{1}{3}u^3 - x = F(x, u).$$

The periodic solution is represented approximately by a closed curve in the x, u-plane, consisting of certain arcs of the curve $F(x, u) = 0$ and of certain straight lines parallel to the x-axis (Stoker[18], p. 128).

The basic equation
$$F = 0 \quad \text{or} \quad x = u - \tfrac{1}{3}u^3$$

provides an approximate solution except near the points where
$$dx/du \equiv 1 - u^2$$

vanishes, i.e. at
$$x = \pm\tfrac{2}{3}, \quad u = \pm 1.$$

Near these points uniformization is easily carried out by employing the Carrier 'two-way stretch'[4],
$$x = \pm\tfrac{2}{3} + \alpha^m \xi, \quad u = \pm 1 + \alpha^n \eta,$$

with suitable exponents m and n, chosen so as to make the resulting equation regular. The simplest choice is

$$m = 1, \quad n = 0,$$

which yields the regular equation

$$(\pm 1 + \eta)\, d\eta/d\xi = \mp \eta^2 - \tfrac{1}{3}\eta^3 - \alpha\xi.$$

7. The thin aerofoil problem

Although a number of interesting and important problems in fluid dynamics involving singular perturbations of *partial* differential equations have been examined and uniformized by Lighthill[13], Carrier[4] and Whitham ([21] and numerous subsequent papers), the theory is in a much less advanced state than the corresponding theory for *ordinary* equations. It therefore seems preferable to give just a few specific examples.

In the first place we consider the problem of a thin two-dimensional symmetric aerofoil (or strut) with profile

$$y = \pm \alpha f(x) \quad (0 \leqslant x \leqslant 1),$$

placed in a uniform stream of incompressible, inviscid fluid with velocity components $(U \cos \alpha,\ U \sin \alpha)$ at infinity[14]. The potential $\alpha\phi$ of the disturbance velocity satisfies the equation

$$\phi_{xx} + \phi_{yy} = 0,$$

and the boundary conditions

$$\phi = O(R^{-1})$$

for large

$$R = (x^2 + y^2)^{\frac{1}{2}},$$

and

$$(U \sin \alpha + \alpha\phi_y) = \pm (U \cos \alpha + \alpha\phi_x)\, \alpha f'(x),$$

on the surface of the aerofoil. Now near the leading edge

$$[f(x)]^2 = c^2 x + O(x^2) \quad (c \neq 0),$$

and

$$f'(x) = O(x^{-\frac{1}{2}}).$$

The surface boundary condition is therefore singular at the leading edge.

The reduced boundary condition is

$$(U + \phi_y) = \pm U f'(x),$$

to be satisfied on the x-axis, $y = 0$, $0 \leqslant x \leqslant 1$, and it is this equation together with the potential equation for ϕ which forms the basis of 'thin aerofoil theory'. The main problem is to improve this approximation without making a completely fresh start.

The boundary conditions can be uniformized by introducing parabolic co-ordinates ξ, η, where

$$x + iy - \tfrac{1}{4}\alpha^2 c^2 = c^2(\xi + i\eta)^2.$$

Then $$x = c^2(\xi^2 - \eta^2 + \tfrac{1}{4}\alpha^2), \quad y = 2c^2\xi\eta,$$

and the parabola $\eta = \tfrac{1}{2}\alpha$ osculates the leading edge section

$$y^2 = \alpha^2 c^2 x + O(x^2).$$

Hence in parabolic co-ordinates the profile $y = \alpha f(x)$ has an equation of the form

$$\eta = \tfrac{1}{2}\alpha + \alpha P(\xi)$$

$$= \tfrac{1}{2}\alpha + \alpha \sum_{n=1}^{\infty} c_n \xi^n,$$

and the exact boundary condition becomes

$$\{-2Uc^2\eta\cos\alpha + 2Uc^2\xi\sin\alpha + \alpha\phi_\eta\}$$
$$= \alpha P'(\xi)\{2Uc^2\xi\cos\alpha + 2Uc^2\eta\sin\alpha + \alpha\phi_\xi\}.$$

This condition is regular, and hence the problem admits a solution of the form

$$\phi = \phi_0(\xi,\eta) + \alpha\phi_1(\xi,\eta) + \dots.$$

8. The boundary layer on a flat plate

Another problem which exemplifies the techniques of both Lighthill and Carrier is that of the steady flow of an incompressible, viscous fluid past a semi-infinite flat plate

$$y = 0, \quad x \geqslant 0,$$

placed parallel to the main stream. The natural units of length, velocity and pressure are ν/U, U and ρU^2, where ν is the kinematic viscosity, U the main stream velocity and ρ the density. In terms of these units the Navier–Stokes equations for the pressure p and the components of fluid velocity are

$$uu_x + vu_y = -p_x + \Delta u,$$
$$uv_x + vv_y = -p_y + \Delta v,$$

where $$\Delta u = u_{xx} + u_{yy}.$$

The boundary conditions are

$$u \to 1, \quad v \to 0 \quad \text{as} \quad x^2 + y^2 \to \infty,$$

except on the flat plate $y = 0$, $x \geqslant 0$ where

$$u = 0, \quad v = 0.$$

There is no parameter in these equations or boundary conditions, but it is known from experiment that derivatives of u and v with respect to y are small compared with derivatives with respect to x, except at the leading edge, $x = 0$, $y = 0$, where presumably the dominant derivative is in the radial direction. These conditions are conveniently expressed in terms of parabolic co-ordinates ξ, η such that

$$x + iy = (\xi + i\eta)^2,$$

or
$$\xi = [\tfrac{1}{2}(r + x)]^{\frac{1}{2}}, \quad \eta = [\tfrac{1}{2}(r - x)]^{\frac{1}{2}},$$

where
$$r = [x^2 + y^2]^{\frac{1}{2}}.$$

The stream function ψ is defined by the equations

$$u = \psi_y, \quad v = -\psi_x,$$

$\psi \sim y = 2\xi\eta$ for large $\xi^2 + \eta^2$ ($\eta \neq 0$!), and itself satisfies the equation

$$\rho^2 \Delta^2 \psi - 4(\xi \Delta \psi_\xi + \eta \Delta \psi_\eta - \Delta \psi)$$
$$= -\rho^2 (\psi_\xi \Delta \psi_\eta - \psi_\eta \Delta \psi_\xi) + 2(\eta \psi_\xi - \xi \psi_\eta)\,\Delta \psi,$$

where $\rho^2 = \xi^2 + \eta^2$.

To identify the dominant terms we write

$$\eta = \epsilon\bar{\eta}, \quad \psi = \epsilon^{-1}\overline{\psi},$$

thus introducing a small parameter ϵ and thus obtaining a regular equation with parameter ϵ. On retaining the terms of lowest order (i.e. those in ϵ^{-5}) we obtain the reduced equation

$$\xi^2 \overline{\psi}^{\mathrm{iv}} = -\xi^2(\overline{\psi}_\xi \overline{\psi}''' - \overline{\psi}' \overline{\psi}''_\xi) - 2\xi \overline{\psi}' \overline{\psi}'',$$

where accents indicate differentiation with respect to η. If we now write

$$\overline{\psi} = \xi f(\xi, \bar{\eta}),$$

we find that by a remarkable and unexpected simplification the function f satisfies the *ordinary* differential equation

$$f^{\mathrm{iv}} = -ff''' - f'f'',$$

which integrates at once, in virtue of the boundary conditions at infinity, to
$$f''' + ff'' = 0.$$

This is the well-known Blasius equation, with the boundary conditions

$$f \sim 2\eta \quad \text{for large } \eta,$$

$$f = 0, \quad f' = 0 \quad \text{at} \quad \eta = 0.$$

The independent variable however is now

$$\eta = r^{\frac{1}{2}} \sin \tfrac{1}{2}\theta,$$

whereas in the classical Blasius problem it is

$$y/x^{\frac{1}{2}} = r^{\frac{1}{2}} \sin \theta / \sqrt{(\cos \theta)}.$$

The preceding analysis is due to Carrier and Lin[5] and there can be no doubt of the superiority of their solution of this problem over the classical solution given by Blasius[2]. A somewhat similar investigation, carried to the next order of approximation has been given by Kuo[11].

9. Accuracy of approximations

The preceding brief accounts of some methods and problems of interest to applied mathematicians will doubtless suggest many questions for the analyst, but the question of outstanding importance is surely that of the accuracy of the approximations obtained. The existence theorems which have been invoked do little more than guarantee the existence of solutions in the form of power series in the perturbation parameter α. The following questions arise at once:

(1) What is the radius of convergence of the power series?

(2) What is the rapidity of convergence?

(3) Is it possible to prescribe an upper bound to the absolute magnitude of the error which is involved in truncating the power series after N terms? And, in particular, can we do this for the 'basic solution' where $N = 1$?

A classical example of this problem is provided by the Blasius equation which is obtained as the 'reduced equation' from the Navier–Stokes equations for the flow of an incompressible, inviscid fluid past a semi-infinite flat plate. In this case, as in so many other physical problems, even the reduced equation is not linear.

The Blasius equation, as obtained in §8, is

$$f''' + ff'' = 0,$$

primes indicating differentiation with respect to η, and a solution is required for the range $0 \leqslant \eta \leqslant \infty$, with the boundary conditions

$$f = 0 \quad \text{and} \quad f' = 0 \quad \text{or} \quad \eta = 0$$

and
$$f' \to 2 \quad \text{as} \quad \eta \to \infty.$$

There is a power series solution (obtained by Blasius[2]) in the form
$$f = c\left\{\frac{(c\eta)^2}{2!} - \frac{(c\eta)^5}{5!} + \frac{11(c\eta)^8}{8!} - \dots\right\},$$
where
$$c^3 = f''(0).$$

Weyl[20] showed that the radius of convergence R of this power series in η satisfies the inequalities
$$18 < c^3 R^3 < 60,$$
and Punnis[17] obtained the closer limits
$$3 \cdot 11 < cR < 3 \cdot 18,$$
by showing that the power series has a simple pole at $\eta = -R$. There is therefore a real problem for the analyst to determine the value of c so as to satisfy the condition $f' \to 2$ as $\eta \to \infty$, although the practical computer has little difficulty in obtaining the approximate value
$$f''(0) = 1 \cdot 328 \dots.$$

Quite another approach to this problem is provided by Weyl's transformation[20] of the differential equation into an integral equation of the form
$$\log F''(\eta) = \Phi(F'') = -\frac{1}{2}\int_0^\eta (\eta - s)^2 F''(s)\, ds,$$
where
$$f(\eta) = cF(c\eta),$$
and, as before,
$$c^3 = f''(0).$$

If an iteration process is specified by the conditions
$$F_0'' = 0, \quad F_{n+1}'' = \Phi(F_n''),$$
then
$$F_0'' < F_1'', \quad F_1'' > F_2'', \quad F_2'' < F_3'', \quad \text{etc.}$$

The sequence $\{F_n''\}$ converges and any two consecutive members form upper and lower bounds to the limit function. Moreover,
$$F_2''(\eta) = \exp\left(-\tfrac{1}{6}\eta^3\right),$$
is found to be an adequate approximation to the limit.

The use of integral equations of Weyl's type has been successfully exploited by Meksyn[15] in numerous papers on boundary layer theory, although the convergence of the iteration process still requires examination. As examples of other important investigations on approximate solutions of partial differential equations we may cite papers by Westphal[19] and Görtler[6].

10. Conclusion

In the preceding paper the name of 'delinearization' has been given to the process whereby we endeavour to return from a linearized approximation to the original non-linear equation. There is, however, another kind of delinearization, which, I venture to predict, will become increasingly important—namely a process whereby an exact linear equation is replaced by an exact non-linear equation. This apparently retrograde step is sometimes advantageous because good approximate solutions may be obtainable more easily for the non-linear equation than for the original linear equation.

One example is provided by the so-called Wentzel–Kramers–Brillouin method of solving the Schrödinger wave equation

$$\epsilon^2 \psi'' + f(x)\, \psi = 0$$

for small values of the parameter ϵ. This method, due to Jeffreys[8, 9], consists in writing

$$\psi = \exp\left\{ i\epsilon^{-1} \int \chi\, dx \right\},$$

and thus obtaining the Riccati equation

$$-\chi^2 + i\epsilon\chi' + f = 0,$$

with the series solution

$$\chi = \chi_0 - i\epsilon\chi_1 + \cdots, \quad \chi_0 = f^{\frac{1}{2}}, \quad \chi_1 = -\tfrac{1}{2}\chi_0'/\chi_0.$$

Another example is derived from the new theory of diffraction problems, suggested by Birkhoff[1] and recently developed by Keller, Lewis and Seckler[10]. Here the wave equation for monochromatic light,

$$\operatorname{div\,grad} u + k^2 u = 0,$$

is solved in the form

$$u \sim e^{ik\psi} \sum_{n=0}^{\infty} (ik)^{-n} v_n,$$

where $\psi, v_0, v_1, \ldots$ satisfy the non-linear equations

$$(\operatorname{grad} \psi)^2 = 1,$$

$$2\operatorname{grad} v_n \cdot \operatorname{grad} \psi + v_n \operatorname{div\,grad} \psi = -\operatorname{div\,grad} v_{n-1} \quad (v_{-1} \equiv 0\,!).$$

These examples do suggest that the eras of linear equations and of linearized non-linear equations may be succeeded by the era of delinearized linear equations.

REFERENCES

[1] Birkhoff, G. D. Quantum mechanics and asymptotic series. *Bull. Amer. Math. Soc.* 39, 681 (1933).

[2] Blasius, H. Grenzschichten in Flüssigkeiten mit kleiner Reibung. *Z. Math. Phys.* 56, 1 (1908).

[3] Carrier, G. F. Boundary layer problems in applied mathematics. *Commun. Pure appl. Math.* 7, 11–17 (1954).

[4] Carrier, G. F. Boundary layer problems in applied mechanics. *Advanc. appl. Mech.* 3, 1–19 (1953).

[5] Carrier, G. F. and Lin, C. C. On the nature of the boundary layer near the leading edge of a flat plate. *Quart. appl. Math.* 6, 63–68 (1948).

[6] Görtler, H. Über die Lösungen nichtlinearer partieller Differential-gleichungen vom Reibungsschichttypus. *Z. angew. Math. Mech.* 30, 265–267 (1950).

[7] Goursat, E. *Cours d'Analyse Mathématique*, tome II, p. 371. Paris, 1911.

[8] Jeffreys, H. On certain approximate solutions of linear differential equations of the second order. *Proc. Lond. Math. Soc.* (2), 23, 428–436 (1925).

[9] Jeffreys, H. Asymptotic solutions of linear differential equations. *Phil. Mag.* (7), 33, 451–456 (1942).

[10] Keller, J. B., Lewis, R. M. and Seckler, B. D. Asymptotic solution of some diffraction problems. *Commun. Pure appl. Math.* 9, 207–265 (1956).

[11] Kuo, Y. H. On the flow of an incompressible viscous fluid past a flat plate at moderate Reynolds number. *J. Math. Phys.* 32, 83–101 (1953).

[12] Lighthill, M. J. The position of the shock wave in certain aerodynamic problems. *Quart. J. Mech. appl. Math.* 1, 309–318, Oxford, 1948.

[13] Lighthill, M. J. A technique for rendering approximate solutions to physical problems uniformly valid. *Phil. Mag.* (7), 40, 1179–1201 (1949).

[14] Lighthill, M. J. A new approach to thin aerofoil theory. *Aeronaut. Quart.* 3, 193–210 (1951).

[15] Meksyn, D. Integration of the laminar boundary layer equation. I. Motion of an elliptic cylinder. Separation. II. Retarded flow along a semi-infinite plane. *Proc. Roy. Soc.* A, 201, 268–278, 279–283 (1950).

[16] Poincaré, H. *Les méthodes nouvelles de la mécanique céleste*, 1, ch. 3. Paris, 1892.

[17] Punnis, B. Zur Differentialgleichung der Plattengrenzschicht von Blasius. *Archiv Math.* 7, 165–171 (1956).

[18] Stoker, J. J. *Non-Linear Vibrations*, p. 128. New York, 1950.

[19] Westphal, H. Zur Abschätzung der Lösungen nichtlinearer parabolischer Differentialgleichungen. *Math. Z.* 51, 690–695 (1949).

[20] Weyl, H. Concerning the differential equations of some boundary layer problems. *Proc. Nat. Acad. Sci., Wash.*, 27, 578–583 (1941).

[21] Whitham, G. B. The behaviour of supersonic flow past a body of revolution, far from the axis. *Proc. Roy. Soc.* A, 201, 89–109 (1950).

DES VARIÉTÉS TRIANGULÉES AUX VARIÉTÉS DIFFÉRENTIABLES

Par R. THOM

1. Généralités sur les groupes de difféomorphismes

Soit V^n une variété différentiable connexe séparée; par différentiable, on entendra toujours r-fois différentiable, où $2 \leqslant r \leqslant \infty$. On désigne par Dif$(V^n)$ le groupe de tous les automorphismes de la variété différentiable V^n, muni de la C^r-topologie (i.e. la topologie définie par la différence sur tout compact des applications et de leurs dérivées partielles jusqu'à l'ordre r). Si V^n est orientable, on se restreindra aux difféomorphismes de degré $+1$.

Tout difféomorphisme f d'une variété V connexe peut être déformé (au sens de la C^r-topologie) en un difféomorphisme g tel que:

(1) g laisse fixe un point donné p de V;

(2) g est tangent à l'identité (jusqu'à l'ordre r) en p;

(3) g se réduit à l'identité sur un voisinage de p.

Les propriétés (1) et (2) sont connues; on en trouvera une démonstration complète, ainsi que de (3), dans un article à paraître de Jean Cerf.

1.1. Les difféomorphismes des boules et des sphères.

Soient Dif(B^{n+1}), Dif(S^n) les groupes de difféomorphismes de degré un de la $(n+1)$-boule et de la n-sphère. Soient $\pi_0(\text{Dif}(B^{n+1}))$, $\pi_0(\text{Dif}(S^n))$ les groupes discrets quotients des groupes précédents par les composantes connexes de l'identité. Milnor a démontré la propriété suivante:

Théorème 1. Les groupes $\pi_0(\text{Dif}(B^{n+1}))$, $\pi_0(\text{Dif}(S^n))$ sont abéliens.

Démonstration: Soient f, g deux difféomorphismes de S^n; d'après la propriété (3), on peut déformer f, g resp. en f_1, g_1 tels que f_1 se réduise à l'identité sur l'hémisphère Nord E^{n+} de S^n, et tel que g_1 se réduise à l'identité sur l'hémisphère Sud E^{n-} de S^n. Dans ces conditions, il est clair que l'on a, en tout point de S^n:

$$f_1 \circ g_1 = g_1 \circ f_1.$$

Par restriction au bord S^n de B^{n+1}, on définit un homomorphisme canonique (en fait, une fibration sur la composante connexe de l'identité): $j:$ Dif$(B^{n+1}) \to$ Dif(S^n); par suite le quotient Dif$(S^n)/j$. Dif(B^{n+1}) est discret, et isomorphe au quotient $\pi_0(\text{Dif}(S^n))/j$. $\pi_0(\text{Dif}(B^{n+1}))$. Ce groupe abélien sera désigné par Γ_{n+1}. On sait que certains de ces groupes ne sont pas nuls (par exemple Γ_7); néanmoins, on ne connaît aucun exemple où le groupe $\pi_0(\mathrm{Dif}\,(B_m))$ n'est pas nul.

2. Subdivisions différentiables d'une variété différentiable

Soit K un complexe simplicial, $|K|$ l'espace topologique sous-jacent. Supposons que $|K|$ soit une variété qu'on peut munir d'une structure différentiable ($\mathscr{S}$). On dire que K est une subdivision *différentiable* de la variété X, si l'application canonique $\sigma^k \to |K|$ de tout simplexe est une application différentiable de rang maximum du k-simplexe euclidien standard dans X. Tout simplexe apparaît ainsi comme un morceau de variété plongée dans X; un voisinage tubulaire normal de σ^k contient les simplexes de l'étoile de σ^k comme sous-variétés plongées; en coupant en un point x de σ^k par un $(n-k)$-plan transverse, on définira l'"étoile transverse' à σ^k, qui est une $(n-k)$-boule différentiable, contenant les sections des simplexes de dimension $> k$ de l'étoile de σ^k comme sous-variétés localement linéaires. Ceci impose que la triangulation proposée soit une 'triangulation de Brouwer' dans la terminologie de Cairns.†

Etant donné le complexe K, donné par exemple par son schéma combinatoire, on se propose de déterminer si la variété $X = |K|$ peut être munie d'une structure différentiable globale contenant K comme subdivision différentiable, et, de plus, de classifier ces structures à l'équivalence près. Pour que le problème ait un sens, il faudra supposer au départ que $|K|$ est une variété, et que K en est une 'subdivision de Brouwer', ce qu'on peut exprimer en disant qu'à tout simplexe σ^r est attachée une carte locale contenant σ^r-comme sous-variété linéaire, et telle que tous les simplexes de l'étoile de σ^r-soient plongés rectilinéairement dans cette carte.

On construira la structure différentiable sur K par récurrence sur les squelettes K^i successifs; une structure différentiable dans un voisinage du 0-squelette K^0 s'obtient en attachant à chaque sommet sa carte locale qui lui est donnée en raison de la propriété de Brouwer de la subdivision. Le problème consistera donc, étant donnée une structure différentiable ($\mathscr{S}$) sur un voisinage du bord $\partial\sigma^{k+1}$ d'un $(k+1)$ simplexe, à étendre (si possible) cette structure à un voisinage de σ^{k+1}. C'est ce problème d'extension locale d'une structure différentiable qu'on va étudier de près.

---

† Ceci n'est pas une restriction; en effet, J. H. C. Whitehead a démontré que toute variété triangulée possède une subdivision qui a la propriété de Brouwer.

3. Extension locale d'une structure différentiable

Etant donnée une variété à bord M de bord V, et une variété à bord M' de bord V' de même dimensions, supposons qu'on ait défini dans les bords V, V' deux sous-variétés à bords $G \subset V$, $G' \subset V'$ (dim. $G = $ dim. $G' = $ dim. V) qui soient difféomorphes; si, dans la réunion $M \cup M'$, on identifie G à G' par ce difféomorphisme, on obtient une nouvelle variété à bord P, dont la structure est bien déterminée à un difféomorphisme près. Cette structure ne varie pas non plus si on remplace le difféomorphisme d'attachement $g: G \to G'$ par un difféomorphisme qui s'en déduit par déformation continue.

C'est ce procédé qu'on emploiera pour étendre la structure $(\mathscr{S})$ du bord S^k de notre $(k+1)$-simplexe σ^{k+1} à l'intérieur de σ^{k+1}; puisque la structure $(\mathscr{S})$ est définie dans un voisinage de $\partial\sigma^{k+1}$, il est possible de définir une variété à bord M diff. pour $(\mathscr{S})$ qui soit un 'voisinage tubulaire' régulier de $\partial\sigma^{k+1}$ pour $(\mathscr{S})$; un tel voisinage contient $\sigma^{k+1} \cap M$ comme sous-variété plongée, et l'intersection $V = \partial M$ par σ^{k+1} est, au moins topologiquement, une k-sphère S^k topologiquement isotope (dans σ^{k+1} un peu agrandi), au bord polyèdral $\partial\sigma^{k+1}$.

On prendra pour G un voisinage tubulaire de S^k dans V. Au simplexe σ^{k+1} est attachée sa 'carte de Brouwer', dans laquelle il est rectilinéairement plongé; on peut, par une construction facile, définir une sphère S^k différentiablement plongée isotope au bord $\partial\sigma^{k+1}$ pour la structure différentiable $(\mathscr{T})$ associée à la carte de Brouwer; on prendra pour M' un voisinage tubulaire de σ^{k+1} pour $(\mathscr{T})$, limité aux plans normaux tombant sur S^k; le voisinage normal de S^k sera la variété à bord G'. On recherchera si G et G' qui sont tous deux des fibrés en $(n-k)$-boules sont difféomorphes; si oui, on appliquera la construction générale; il restera à vérifier que la variété obtenue par identification est bien homéomorphe à un voisinage de σ^{k+1}; ceci résulte du fait que les deux sphères S^k qu'on identifie dans σ^{k+1} sont isotopes au bord polyèdral $\partial\sigma^{k+1}$, et que les plans normaux à σ^{k+1} pour les deux structures $(\mathscr{S})$ et $(\mathscr{T})$ définissent des polyèdres isotopes à l'étoile transverse en tout point x de σ^{k+1}, étoile qui, elle, est intrinsèquement définie.

En fait la variété M' qu'on ajoute est difféomorphe au produit $B^{k+1} \times B^{n-k}$, comme voisinage tubulaire d'une boule; l'identification proposée sera possible si, pour la structure $(\mathscr{S})$, on peut trouver une sphère Σ approchant $\partial\sigma^{k+1}$ dont la structure différentiable est la structure usuelle, et dont le voisinage normal soit trivial. Autrement dit: si l'on remarque que, pour la structure $(\mathscr{S})$, comme pour $(\mathscr{T})$, le simplexe

σ^{k+1} est différentiablement plongé, le problème d'extension de la structure ($\mathscr{S}$) se trouve décomposé en deux problèmes:

(1) extension de la structure 'tangente', restriction de($\mathscr{S}$) au simplexe σ^{k+1};

(2) extension de la structure fibrée des vecteurs normaux de ($\mathscr{S}$) à tout σ^{k+1}.

Le premier problème étant supposé résolu pour le moment, on va s'occuper du second. La première difficulté est de montrer que la structure fibrée normale à Σ pour ($\mathscr{S}$) est celle du produit. On le montre grâce à la remarque suivante: soit σ^n un n-simplexe quelconque de l'étoile de σ^{k+1}; sa section par un plan normal (pour ($\mathscr{S}$)) à σ^{k+1} en x donne un $(n-k-1)$-simplexe (de dimension maximum) de l''étoile transverse en x', simplexe qui a un sommet en x; les vecteurs tangents en x aux arêtes de ce simplexe-section définissent en tout point $x \in \Sigma$ un repère pour le fibré des vecteurs normaux. Comme ce repère est défini pour tout x et varie continuement avec x, il en résulte bien que le fibré normal est trivial.

On pourra donc étendre sans difficulté la structure normale, mais une sérieuse difficulté apparaît: la nouvelle structure obtenue doit contenir les simplexes de l'étoile de σ^{k+1} comme sous-variétés linéaires en tout point. Cette condition est satisfaite pour la structure ($\mathscr{T}$); le sera-t-elle pour ($\mathscr{S}$) étendue? Dans l'identification des fibrés de vecteurs normaux pour ($\mathscr{S}$) et ($\mathscr{T}$), on peut supposer qu'on prend comme repères homologues ceux définis par le simplexe σ^n de l'étoile transverse dont il a été question plus haut; x variant sur Σ, la section de l'étoile transverse par une sphère normale donne naissance à une triangulation géodésique variable de cette sphère, triangulation dont un simplexe de dimension maximum (à savoir le simplexe σ^{n-k} section de σ^n) demeure fixe; le problème de l'extension sera donc possible si l'application de S^k dans l'espace des triangulations géodésiques de la sphère isotopes à une triangulation donnée (avec simplexe fixe) est homotope à zéro. Par une inversion on est ramené à considérer l'espace de tous les automorphismes semi-linéaires d'une subdivision simpliciale du simplexe. On est donc ramené à démontrer le

Théorème 2. L'espace de tous les homéomorphismes semi-linéaires d'une subdivision simpliciale du m-simplexe dans lui-même (qui se réduisent à l'identité sur le bord du simplexe) est asphérique.

Un homéomorphisme semi-linéaire du simplexe s^m dans lui-même est un homéomorphisme de s^m sur lui-même qui est linéaire sur chacun des simplexes d'une triangulation donnée de s^m. Il est clair qu'un tel homéomorphisme est complètement déterminé dès qu'on s'est donné la position des images des sommets; on a par suite sur cet espace une topologie naturelle définie par l'ouvert de $(R^m)^q$ défini par la position dans s^m des q sommets images. Partons d'une subdivision donnée (K) de s^m, et supposons défini l'espace H des homéomorphismes semi-linéaires de K. Supposons maintenant qu'on subdivise l'un des m-simplexes de K, soit (z), par adjonction d'un nouveau sommet y intérieur à (z); dans ces conditions, on obtient une nouvelle subdivision (K'); formons l'espace (H') des homéomorphismes semi-linéaires correspondants; tout élément de (H') provient d'une application de (H) par subdivision du simplexe (z); ceci définit une application de (H') sur (H); pour les espaces représentatifs, la fibre est l'ensemble des positions de y dans (z); c'est donc un espace asphérique. Par suite (H') et (H) ont même type d'homotopie.

Nous avons donc démontré le théorème pour toutes les subdivisions qui se déduisent de la subdivision triviale du m-simplexe par une suite de 'subdivisions coniques' de simplexes de dimension maximum m. Le cas de la subdivision barycentrique classique échappe malheureusement au résultat. Il faudrait donc invoquer ici, pour obtenir le théorème dans toute sa généralité, une forme particulièrement forte de la 'Hauptvermutung'.

Il en résulte donc que, si les étoiles transverses aux simplexes de K satisfont à cette condition, le prolongement de la structure normale sera possible. C'est ce que nous supposerons dorénavant.

4. Prolongement de la structure tangente

Pour que le prolongement de la structure tangente soit possible il suffit que la sphère Σ approximation différentiable de $\partial\sigma^{k+1}$ pour la structure $(\mathscr{S})$ soit isomorphe à la sphère S^k usuelle; de plus c'est une condition nécessaire de prolongement au $(k+1)$-simplexe: car on peut montrer qu'il y a unicité de la structure de l'approximation différentiable du bord $\partial\sigma^{k+1}$, la structure différentiable de l'espace ambiant étant donnée.

Observons d'ailleurs que la sphère Σ possède une subdivision différentiable isomorphe à $\partial\sigma^{k+1}$; soit s^k un de ses simplexes; le complémentaire $\Sigma - s^k$, formé de k simplexes de dimension k, est nécessairement difféomorphe à l'un de ces simplexes de dimension k (on passe d'un simplexe à la réunion des k-simplexes par une suite de $(k-1)$ expansions élémentaires au sens de Whitehead; or, du fait qu'une boule euclidienne a une structure de produit $B^{n+1} \times I$, toute expansion définit un difféomorphisme).

Il en résulte que la sphère Σ s'obtient en accolant deux k-boules (hémisphères) suivant leur bord commun S^{k-1}. Dans ces conditions deux sphères sont ou non difféomorphes suivant que les difféomorphismes d'attachement $S^{k-1} \to S^{k-1}$ sont ou non dans la même composante connexe de $\mathrm{Dif}\,(S^{k-1})$ modulo $\mathrm{Dif}\,(B^k)$. Donc l'ensemble des structures différentiables de la sphère S^k (compatibles avec une subdivision différentiable isomorphe au bord du $(k+1)$-simplexe) s'identifie au groupe abélien Γ_k. Ce groupe définira par suite la nature de l'obstruction rencontrée dans l'extension au $(k+1)$-squelette de la structure différentiable donnée sur le k-squelette.

De plus, au cas où l'obstruction est nulle, la sphère Σ est difféomorphe à S^k et on peut faire l'identification $S^k \to \Sigma$; en modifiant au besoin la paramétrisation de σ^{k+1}, il est possible de modifier l'application d'attachement par un élément quelconque de Γ_{k+1}; les diverses extensions possibles sont donc en correspondance biunivoque avec les éléments du groupe Γ_{k+1}.

Lemme d'addition. Soit σ^{r+1} le $(r+1)$ simplexe standard; toute r-face F^r de σ^{r+1} admet pour bord $\partial\sigma^r = S^{r-1}$; il est possible de modifier la paramétrisation de chaque face F_j en modifiant l'application d'attachement de S^{r-1} sur F_j par un élément γ_j de Γ_r. Le lemme d'addition exprime que pour la sphère S^r obtenue en recollant tous ces simplexes F_j ainsi modifiés, la structure différentiable globale correspond à l'élément $c \in \Gamma_r$ défini par $c = \Sigma(-1)^j \gamma_j$.

En admettant toujours possible l'extension des structures 'normales', on voit que l'extension des structures 'tangentes' se heurte à des obstructions, à valeurs dans les groupes Γ_j. On peut vérifier que ces obstructions ont toutes les propriétés classiques des obstructions de la théorie de l'homotopie; ce sont des cocycles, et la nullité de leur classe est une condition nécessaire et suffisante de prolongement des structures, après une éventuelle déformation sur le squelette convenable.

De plus, il résulte du lemme d'addition que si K' est une subdivision barycentrique de K, les obstructions relatives à K' donnent les obstructions relatives à K dans l'homomorphisme canonique $C(K') \to C(K)$. La construction d'une structure différentiable sur K se présente donc formellement comme la construction d'une section d'un fibré $\hat{K} \to K$, dont la fibre serait un complexe (abstrait) Γ tel que $\pi_i(\Gamma) = \Gamma_i$. Une telle fibration donne naissance à une factorisation de Postnikov

$$\hat{K} \to \ldots \to \hat{K}_n \to \ldots \to \hat{K}_{j+1} \to \hat{K}_j \to \ldots \to K,$$

la fibre de $\hat{K}_{j+1} \to \hat{K}_j$ étant le 'complexe d'Eilenberg MacLane' $K(\Gamma_j, j)$.

Les invariants k_j de cette factorisation apparaissent comme les obstructions successives à la construction d'une section. On ignore si ces invariants sont de nature topologique (ce sont des invariants combinatoires); un exemple dû à Milnor montre que ce ne sont pas des invariants du type d'homotopie.

5. Classification des structures différentiables

Théorème 3. Soit M^{n+1} une variété à bord différentiable homéomorphe au produit $V^n \times I$, où V^n est une variété différentiable telle que le bord de M^{n+1} soit homéomorphe à $(V, 0) \cup (V, 1)$. Soit K une subdivision différentiable de V; supposons que M^{n+1} admette une subdivision différentiable isomorphe au produit $K \times I$. Alors $(V, 0)$ et $(V, 1)$ sont difféomorphes, et M^{n+1} est difféomorphe au produit $V \times I$.

5.1. Indication sur la démonstration. Dans M on considère les hypersurfaces différentiables par morceaux $t =$ constant, $t \in I$; en tout point p de M, l'hypersurface $t = t(p)$ admet un vecteur transverse, à savoir le vecteur tangent à l'image de $p \times I$ dans M; un tel champ est différentiable par morceaux dans M; on pourra donc l'approcher par un champ différentiable H, qui, lui, sera encore transversal aux hypersurfaces $t =$ constant; l'intégration du champ de vecteurs H (convenablement normé) fournira un groupe de difféomorphismes à un paramètre qui fait de M le produit $V \times I$ au sens différentiable (on notera que cette structure de produit ne conserve pas la triangulation $K \times I$ donnée tout d'abord). Soit K une subdivision, qu'on peut munir de deux structures différentiables $(\mathcal{S}^0)$ et $(\mathcal{S}^1)$; supposons qu'on puisse munir le produit $K \times I$ d'une structure différentiable telle que les bords $(K, 0)$ et $(K, 1)$ y soient différentiablement plongés avec les structures $(\mathcal{S}^0)$ et $(\mathcal{S}^1)$ respectivement. Il résulte alors du théorème 3 que les structures $(\mathcal{S}^0)$ et $(\mathcal{S}^1)$ sont isomorphes. Or, dans l'interprétation qui fait d'une structure différentiable une section du fibré $\hat{K}$ de fibre Γ, on voit que les sections correspondant à $(\mathcal{S}^0)$ et $(\mathcal{S}^1)$ sont la restriction d'une section sur $K \times I$, donc des sections homotopes. Ainsi: à des sections homotopes du fibré K correspondent des structures différentiables isomorphes.

Remarque. La classification ainsi obtenue des structures est plus fine que l'automorphisme différentiable pur et simple: on pourrait l'exprimer ainsi: Deux variétés différentiables V, V' munies de subdivisions différentiables isomorphes K, K' sont 'équivalentes', s'il existe un difféomorphisme h de V sur V' tel que les subdivisions K' et $h(K)$ de V' soient 'isotopes' dans V'.

5.2. Quelques conséquences. Si l'on admet le résultat suivant:
Théorème 4. L'espace des automorphismes semi-linéaires d'une subdivision simpliciale du simplexe est contractile,
on peut affirmer:

(i) Toute variété contractile triangulée admet une structure différentiable compatible avec sa triangulation.

(ii) Sur une variété contractile deux structures différentiables compatibles avec la triangulation sont isomorphes.

Si l'on admettait en plus la Hauptvermutung, on en déduirait l'unicité de la structure différentiable sur les boules et l'espace euclidien. En effet, un fibré de base contractile admet toujours des sections, et deux sections sont homotopes.

SOME FUNDAMENTAL PROBLEMS IN STATISTICAL PHYSICS

By G. E. UHLENBECK

1. Introduction

On a similar occasion, in 1950, I had the privilege of discussing some basic problems of statistical mechanics before an audience of mathematicians. I am grateful for this opportunity to give a kind of repeat performance, especially because in the last ten years there has been a revival of interest among physicists and mathematicians in statistical mechanics and as a consequence some real advances are being made. I do not mean to imply that this revival was due to my lecture, but it suffices perhaps to present again a kind of catalogue of unsolved problems, which are mainly of a mathematical nature, and which therefore may be of interest to some of you. It seems to me anyway, that anyone interested in statistical problems or in the theory of probability might profit from a study of statistical physics. The molecular constitution of matter provides 'populations' of all possible varieties, and in addition the statistical regularities are empirically known, since these are the phenomena and laws of macroscopic physics. And they are again of an enormous variety, of which very little is understood in a really fundamental way.

A natural division of the problems of statistical physics is into *equilibrium* and *non-equilibrium* problems. If the system under consideration, consisting say of N molecules with *known* interactions, is in thermal equilibrium with a heat reservoir of temperature T, then the probability W that the system is in some state is given by the canonical distribution:

$$W \sim e^{-(1/kT)E}, \tag{1}$$

where E is the energy of the total system and is a function of the variables defining the state of the system. There is complete agreement about this, and one could say that (1) completely 'solves' all equilibrium problems. In fact, the so-called partition function

$$Z = \sum_i e^{-(1/kT)E_i}, \tag{2}$$

where the sum goes over all states of the system, is related to the free energy Ψ by $\Psi = -kT \ln Z$, and from Ψ most of the observable average values can be derived by simple differentiations. Of course, as Poincaré said, a problem is never solved, only more or less solved, and the equations (1) and (2) really give very little of what one would like to know. Only in a few cases can the sum (2) be actually evaluated; usually one has to use series expansions or successive approximation procedures of which the convergence can almost never be established. Technically one runs into subtle problems of combinatorial analysis, some of which are notoriously difficult. As a result, even such general phenomena as the melting of a solid at a sharply defined melting point, or the existence of a critical point for gases have not been explained in a really fundamental way.

I will refrain from giving further details, mainly because of the lack of a general method of treating sums like (2) which would allow the investigation of the limit properties for $N \to \infty$, which is needed to explain the phase transitions. This does not mean that there is no physical insight in these problems, but it is based on qualitative considerations, or on intuitive and 'uncontrolled' approximations. They are very valuable, and I do not want to slight them, but they rarely give a mathematical foothold and they are therefore perhaps of less interest to you.

I would like to concentrate on the non-equilibrium problems, because in a sense the problem of the approach to equilibrium is the central problem of statistical mechanics—I propose to call it the *problem of Boltzmann*—and because through recent work a general point of view is emerging, which is I think of mathematical interest since it hints at a kind of generalization of the ergodic theory.

2. The Boltzmann equation

I will begin with the classical kinetic theory of gases. Consider the simplest type of molecular system: N point molecules in a vessel (volume V) repelling each other by a known monotonic central potential $\phi(r_{ij})$ between each pair (i, j), which has a finite range r_0, so that $\phi(0) \to \infty$ and $\phi(r_0) = 0$. If the number of molecules is very large, and if r_0 is very small compared with the mean free path λ of the molecules, then the state of the gas is described by the distribution function $f(\mathbf{r}, \mathbf{v}, t)$ giving the number of molecules at time t in the range $d\mathbf{r} d\mathbf{v}$ of the phase space (μ-space) of a molecule. The temporal development of the state of the gas is governed by the famous Boltzmann equation

$$\frac{Df}{Dt} \equiv \frac{\partial f}{\partial t} + \mathbf{v} \cdot \frac{\partial f}{\partial \mathbf{r}} + \mathbf{a} \frac{\partial f}{\partial \mathbf{v}} = \int d\mathbf{v}_1 \int d\Omega g I(g, \theta) [f'f_1' - ff_1], \qquad (3)$$

which expresses the statement that if one moves with the molecules in μ-space the change of f is due to the gains and losses because of the collisions with other molecules. In (3), a is the acceleration due to an *outside* potential $U(\mathbf{r})$; the indices of f indicate the velocity variables only, so that for instance $f' \equiv f(\mathbf{r}, \mathbf{v}', t)$; the four velocity variables refer to the velocities of the binary collision

$$(\mathbf{v}, \mathbf{v}_1) \leftrightarrow (\mathbf{v}', \mathbf{v}'_1); \quad g = |\mathbf{v} - \mathbf{v}_1| = |\mathbf{v}' - \mathbf{v}'_1|$$

is the relative velocity, which in a collision turns over an angle θ in the solid angle $d\Omega$, and finally $I(g, \theta)$ is the differential cross-section which is uniquely determined by the force law $\phi(r)$. Since $\phi(r)$ has a finite range r_0, the total cross-section

$$\sigma = \int d\Omega I(g, \theta) \tag{4}$$

is also finite.

I will postpone the critical discussion of (3), and first concentrate on its mathematical aspects. Since the famous paper of Hilbert[1] in 1910, the Boltzmann equation has received rather little attention from the mathematicians, and it is therefore perhaps worth while to point out some basic problems which are still unsolved.

In the first place, the classical arguments of Boltzmann, by means of the H-theorem, have made it extremely plausible that *any* initial distribution $f(\mathbf{r}, \mathbf{v}, 0)$ will go over for $t \to \infty$ into the equilibrium or Maxwell–Boltzmann distribution:

$$f_0 = A \exp\left\{ -\beta \left[\frac{mv^2}{2} + U(\mathbf{r}) \right] \right\}, \tag{5}$$

where the constants A and β are determined by the total number of particles and the given total energy. These arguments, which occupy a central position in the explanation of the laws of thermodynamics, are quite convincing to the physicist, but it must be admitted that they lack rigor mainly because they presuppose the existence and the unicity of the solution of the initial value problem. Let me put therefore as:

Problem I. Show that with 'appropriate' conditions for $\mathbf{r}, \mathbf{v} \to \infty$ and for 'sufficiently general' outside potentials $U(\mathbf{r})$, the initial value problem for equation (3) has a unique solution, which for $t \to \infty$ approaches (5).

For the spatially uniform case ($U \equiv 0$; f function of $\mathbf{v}$ and t alone) and for elastic spheres ($I(g, \theta) = $ constant) Carleman[2] has given a rigorous proof. The extension to other force laws is, I think, straightforward, but the generalization to the spatially non-uniform case seems far from obvious.

3. The linearized Boltzmann equation

For the case of a small disturbance of the equilibrium state it is natural to put:

$$f = f_0(1 + h(\mathbf{r}, \mathbf{v}, t)),\qquad(6)$$

and to neglect the quadratic terms in h. One then gets the linearized Boltzmann equation

$$\frac{\partial h}{\partial t} = \left(-\mathbf{v}\frac{\partial h}{\partial \mathbf{r}} - \mathbf{a}\frac{\partial h}{\partial \mathbf{v}}\right) + \int d\mathbf{v}_1\int d\Omega\, gI(g,\theta)\,(h' + h_1' - h - h_1)f_{01}$$

$$\equiv S(h) + C(h).\qquad(7)$$

Consider first again the spatially uniform case. To solve the initial value problem, it is clearly of interest to find the eigenfunctions ψ_i and eigenvalues λ_i of the collision operator C, defined by

$$C(\psi_i) = \lambda_i \psi_i.\qquad(8)$$

Some properties are quite obvious. There are five zero eigenvalues, corresponding to the eigenfunctions 1, $\mathbf{v}$, v^2. This is a consequence of the five conservation laws during a collision. All other eigenvalues must be negative, which follows from the inequality:

$$\int d\mathbf{v} f_0 \psi_i C(\psi_i) \leqslant 0.\qquad(9)$$

Because of the isotropy of C, the eigenfunctions must have the form:

$$\psi_i = R_{rl}(v) Y_{lm}(\theta, \phi),$$

using polar co-ordinates in velocity space. In one case (for so-called Maxwell molecules for which $gI(g,\theta) = F(\theta)$) the ψ_i and λ_i have been found explicitly, but in general very little is known. It is very likely that the spectrum is bounded from below if the total cross-section σ is finite, which is the interesting case. Partially surmising, I put as:

Problem II. Show that the linearized collision operator has a discrete eigenvalue spectrum $\lambda_i \leqslant 0$ which is bounded from below.

Clearly, knowing the ψ_i and λ_i, one can solve the initial value problem in the spatially uniform case by developing $h(\mathbf{v}, t)$ in the ψ_i, and one gets

$$h(\mathbf{v}, t) = \sum_i c_i e^{\lambda_i t} \psi_i(\mathbf{v}),\qquad(10)$$

where the c_i are determined from $h(\mathbf{v}, 0)$. Because of $\lambda_i < 0$, (10) shows the relaxation of the disturbance h to zero.

Again the spatially non-uniform case is much more complex, because of the different nature of the operators S and C. I only want to mention one problem, namely the propagation of sound. If the intensity is small, sound is a small disturbance and should therefore be described by (7). Because of the existence of a sound velocity, one might perhaps think that equation (7) has propagation character in the sense that the solution $h(\mathbf{r}, \mathbf{v}, t)$ of the initial value problem at any t depends only on the initial values in a *finite* domain. I do not think that this can be true rigorously, because the sound velocity is *not* a limiting velocity, and there *is*, in fact, no limiting velocity. But in an approximate sense, for sufficiently 'smooth' initial distributions, it must be so. Although rather vague, let me put therefore as:

Problem III. Show that the solutions of the linearized Boltzmann equation have in some approximate sense propagation character.

Dr Wang Chang and I[3] have approached the problem of the sound propagation by seeking solutions of (7) of the form:

$$h = h_0(\mathbf{v})\, e^{i(\omega t - \sigma z)}, \tag{11}$$

corresponding to a wave in the $+z$ direction. Equation (7) becomes, supposing $\mathbf{a} = 0$,

$$i(\omega - \sigma v_z)\, h_0 = C(h_0). \tag{12}$$

Developing h_0 in the eigenfunctions of C leads to an infinite set of homogeneous linear equations. Putting the infinite determinant equal to zero gives a relation between ω and σ, which expresses the dispersion law for the gas. Breaking off the determinant, using only the eigenfunctions ψ_i corresponding to the zero eigenvalues, gives $\omega = V_0 \sigma$, where $V_0 = (5kT/3m)^{\frac{1}{2}}$ is the well-known sound velocity. Taking in successively more and more eigenfunctions shows that the velocity $V = \omega/\sigma_1$, where σ_1 is the real part of σ ($\sigma = \sigma_1 - i\sigma_2$), increases monotonically with increasing ω. There is now, as to be expected, also absorption of the sound wave, given by σ_2, and this also increases with ω. Unfortunately, we were unable to discuss the convergence of this procedure. I will therefore omit the details and only put as:

Problem IV. Find the dispersion law for a disturbance of the form (11), and show that both the velocity and the absorption coefficient go to infinity if $\omega \to \infty$.

The second part of this problem is of course a surmise. It is perhaps of interest to mention that the experiments of Greenspan[4] seem to agree; at the highest frequencies the sound velocity in helium was already three times V_0.

4. The Chapman-Enskog expansion

The usual description of the macroscopic properties of a gas is by means of the hydrodynamic equations, and the question therefore arises how these equations are contained in the Boltzmann equation. This question was answered in the dissertation of Enskog in 1917, which was inspired by Hilbert. Chapman, using another method which goes back to Maxwell, arrived at the same conclusions. I will give a short sketch of this theory, and I want to emphasize the mathematical features, which are very curious and I think quite fundamental. Further details can be found in the monograph of Chapman and Cowling[5].

Let us go back to the Boltzmann equation and the problem of the approach to equilibrium. Physically one expects that this approach proceeds so to say in *two* stages. Because of the collisions any initial distribution will reach very quickly (in a time of the order of the mean free time $t_0 = \lambda/v_{\mathrm{av.}}$) a *local* Maxwell distribution

$$f^{(0)} = n\left(\frac{m}{2\pi kT}\right)^{\frac{3}{2}} \exp\left[-\frac{m(\mathbf{v}-\mathbf{u})^2}{2kT}\right], \qquad (13)$$

where n, $\mathbf{u}$ and T (*the macroscopic variables*) are still functions of $\mathbf{r}$ and t. Note the difference between (13) and the complete equilibrium distribution (5). In the second stage the slow relaxation of the macroscopic variables to their equilibrium values takes place, and it is this stage which we want to follow more in detail. To do this, put

$$f = f^{(0)}[1 + \phi(\mathbf{r}, \mathbf{v}, t)] \qquad (14)$$

in the Boltzmann equation. In the collision term, keep only terms linear in ϕ, while in $\partial f/\partial t$ and in the streaming terms one puts only $f^{(0)}$. Thus one gets the inhomogeneous, linear integral equation

$$\left.\begin{aligned} &\left(\frac{\partial}{\partial t} - S\right) f^{(0)} = J(\phi), \\ &J(\phi) \equiv \int d\mathbf{v}_1 \int d\Omega\, g I(g, \theta)\, (\phi' + \phi_1' - \phi - \phi_1) f^{(0)} f_1^{(0)}. \end{aligned}\right\} \qquad (15)$$

The homogeneous equation $J(\phi) = 0$ has the five solutions 1, $\mathbf{v}$ and v^2, which I will call ϕ_i. In order that (15) has a solution the left-hand side must be orthogonal to the ϕ_i. These solubility conditions turn out to be the ideal fluid or Euler equations for n, $\mathbf{u}$ and T with $p = nkT$. These conditions allow us to express the time derivatives of the macroscopic variables in terms of their spatial derivatives, and one then finds that

$$\left(\frac{\partial}{\partial t} - S\right) f^{(0)} = f^{(0)} \left[\frac{\partial T}{\partial x_\alpha} \frac{U_\alpha}{T} \left(\frac{m}{2kT} U^2 - \frac{5}{2} \right) + \frac{m}{kT} D_{\alpha\beta}(U_\alpha U_\beta - \tfrac{1}{3}\delta_{\alpha\beta}U^2) \right], \quad (16)$$

where $U_i = v_i - u_i$ is the thermal velocity and

$$D_{ij} = \tfrac{1}{2}(\partial u_i/\partial x_j + \partial u_j/\partial x_i)$$

is the rate of deformation tensor. With (16) as the left-hand side, equation (15) has a solution which is also unique if we require that

$$\int \phi_i f^{(0)} \phi \, d\mathbf{v} = 0,$$

which means that the macroscopic variables must be determined from $f^{(0)}$ alone.

With the solution for ϕ one can then calculate the stress tensor P_{ij} and the heat flux q_i, for which one finds the well-known Newton and Fourier laws:

$$\left. \begin{aligned} P_{ij} &= p\delta_{ij} - 2\mu(D_{ij} - \tfrac{1}{3}D_{\alpha\alpha}\delta_{ij}), \\ q_i &= -\nu \frac{\partial T}{\partial x_i}, \end{aligned} \right\} \quad (17)$$

but with *known* values for the viscosity and heat conduction coefficients μ and ν. They are expressed in terms of the cross-section $I(g, \theta)$, and can therefore be computed if the force law between the molecules is known. With (17) one can then 'correct' the Euler equations and one obtains the 'second order' or Navier–Stokes equations. Let me conclude this sketch with a number of remarks.

(1) This is a successive approximation method which can be continued. The development parameter is of the order of the relative change of the macroscopic variables over a mean free path. For instance: $(\lambda/T)\,\mathrm{grad}\,T$. I will call this parameter the *uniformity parameter*. It can also be looked upon as the ratio t_0/Θ of the mean free time t_0 to a macroscopic relaxation time Θ.

(2) One obtains in this manner a solution of the Boltzmann equation of the form

$$f = f^{(0)}(\mathbf{r}, \mathbf{v} \mid n, \mathbf{u}, T) + f^{(1)}(\mathbf{r}, \mathbf{v} \mid n, \mathbf{u}, T) + \ldots, \quad (18)$$

where each approximation $f^{(k)}$ is a function of $\mathbf{r}$, $\mathbf{v}$ and of the macroscopic variables n, $\mathbf{u}$, T and their spatial derivatives; $f^{(k)}$ does *not* depend on the time explicitly. The time dependence is completely governed by the time dependence of the macroscopic variables, and these are determined by the hydrodynamic equations of successive order, which are of the form

$$\frac{\partial u_i}{\partial t} = V_i^{(1)}(\mathbf{r} \mid n, \mathbf{u}, T) + V_i^{(2)}(\mathbf{r} \mid n, \mathbf{u}, T) + ..., \tag{19}$$

and analogously for n and T. Here, again, the successive approximations $V_i^{(k)}$ do *not* depend on the time explicitly.

(3) The curious mathematical feature of the solution is therefore that the whole temporal development is determined by giving the initial values of n, $\mathbf{u}$, T, which are the first five moments (in velocity) of the distribution function f, while from the Boltzmann equation itself follows that one needs the *whole* initial distribution function. Since Hilbert pointed out this feature of the solution one may perhaps call it the Hilbert paradox.

(4) Physically, one must expect of course that an initial distribution in a time of order t_0 relaxes to a solution of the form (18), *whatever* the initial distribution is. Or one can say, that after a time t_0 (initial chaotization period) a *contraction* of the description of the state of the gas is possible, in which the temporal development will be determined by many fewer variables. And the reason that these variables are n, $\mathbf{u}$ and T is clearly because they correspond to the five quantities 1, $\mathbf{v}$ and v^2 which are conserved in a collision. Therefore, the collisions cannot affect the n, $\mathbf{u}$ and T directly. They change with time only 'secularly', and after a while they therefore completely govern the temporal development.

(5) All this, although I think very plausible, is of course not proved! Let me therefore put as:

Problem V. In which precise sense is the solution of the initial value problem of the Boltzmann equation approximated by the Chapman–Enskog expansion (18), and what is the nature of the convergence of this expansion?

5. The general kinetic equation of Bogolyubov

Let me return now to the *derivation* of the Boltzmann equation. It has been clear for a long time that in the usual derivation one uses, besides the laws of mechanics, a probabilistic assumption about the number of binary collisions, the so-called Stosszahl Ansatz [7]. It is this assumption which produces the irreversible behavior of the gas. A more fundamental derivation should, in my opinion, clarify the relation between the mechanical and the probabilistic assumptions, and in addition it should show the limits of validity of the Boltzmann equation and indicate how to generalize the equation, especially for denser gases where triple and higher order collisions can no more be neglected. I think that all this has in principle been achieved by the work of Bogolyubov[6]. Before giving you a short sketch, let me first make some general remarks.

Since probability theory always connects some given probabilities with other, derived probabilities, it is clear that some probability assumptions have to be made. Since the *main* concern is always the development of the system *in time*, I think the *only* proper place for probabilistic assumptions is for the *initial* state of the system. From then on only the laws of mechanics should be used. In addition, one is really only interested in such systems, where the temporal development soon becomes *independent* of the initial assumptions. One wants to know how the initial information gets lost, how the system gets 'mixed up', and one expects that this process will to a great extent be independent of the initial state. I think that only in this sense can probabilistic descriptions for the *whole* mechanical system (the ensembles of Gibbs) be expected to describe the temporal development of a single given system.

Leaving the philosophy, in our case of a system of N particles with known interactions in a volume V, the state of the whole gas must be described by the probability distribution $D_N(x_1, x_2, ..., x_N, t)$ in the phase space (Γ-space) of the gas. D_N is a symmetric function of the $x_1 ... x_N$ and $x_i \equiv \mathbf{r}_i$, $\mathbf{p}_i$ denotes the co-ordinates and momenta of the ith molecule. D_N changes with time according to the Liouville equation

$$\frac{\partial D_N}{\partial t} = \{H_N, D_N\}, \tag{20}$$

where the brackets denote the Poisson brackets, and H_N is the Hamiltonian of the whole gas of N particles. From (20) follows by integration a hierarchy of equations for the partial distribution functions

$$F_s = V^s \int ... \int D_N \, dx_{s+1} ... dx_N,$$

which has been derived by many authors. One finds

$$\frac{\partial F_s}{\partial t} = \{H_s, F_s\} + \frac{1}{v} \int dx_{s+1} \Big\{ \sum_{i=1}^{s} \phi(|\mathbf{r}_i - \mathbf{r}_{s+1}|), F_{s+1} \Big\}, \tag{21}$$

where $v = \lim V/N$ for N and $V \to \infty$.

The Liouville equation (20) or the hierarchy (21), to which it is equivalent, embodies the mechanical assumptions, and is so to say the basic equation of statistical mechanics. For a bounded system one knows (or better one expects) that *any* initial distribution $D_N(x_1, ..., x_N, 0)$, will approach (in the coarse-grained sense) the microcanonical or uniform distribution over the energy surface, which describes the equilibrium state. It seems that for any given process one would have to know $D_N(0)$ and it is often said that $D_N(0)$ should be chosen so as 'to correspond with our initial macroscopic knowledge of the system'. But the authorities are silent about how this should be done, and I thought for a long time that this was an essential gap in the theory. The answer is, I think, that one is interested only in those phenomena which are *independent* of $D_N(0)$. To see how this can come about we have first to look for the basic relaxation times. In our case there are *three* of such times: the time *of* a collision $\tau \sim r_0/v_{av.}$ (r_0 = range of molecular forces), the time *between* collisions $t_0 \sim \lambda/v_{av.}$, and the macroscopic relaxation time

$$\Theta \sim \psi/v_{av.} \operatorname{grad} \psi,$$

if ψ is a macroscopic quantity. In the usual situations and for not too dense gases $\tau \ll t_0 \ll \Theta$.

Now one can expect, following Bogolyubov, that after an initial chaotization time of order τ and for *any* $D_N(0)$ a stage is reached—Bogolyubov calls it the *kinetic stage*—in which the further temporal development of the gas is determined completely by the temporal change of the *first* distribution function $F_1(x, t)$, which in turn is governed by an equation of the form:

$$\frac{\partial F_1}{\partial t} = A(x \mid F_1), \tag{22}$$

where A depends functionally on F_1 but does *not* depend on the time. All the higher distribution functions depend on the time *only* through F_1 and have therefore the form

$$F_s = F_s(x_1 ... x_s \mid F_1). \tag{23}$$

To elucidate, let me again make a number of remarks.

(1) Equation (22) represents a *contraction* of the description of the state of the gas, which is quite analogous to the Chapman–Enskog solution of the Boltzmann equation, which describes the second or *hydrodynamical stage* of the total relaxation process. Just as in the Chapman–Enskog solution one develops in the uniformity parameter,

Bogolyubov develops the functionals A and F_s in powers of $1/v$ (virial development):

$$\left.\begin{aligned}
A(x \mid F_1) &= A_0(x \mid F_1) + \frac{1}{v} A_1(x \mid F_1) + \frac{1}{v^2} A_2(x \mid F_1) + \cdots, \\
F_s &= F_s^{(0)}(x_1 \ldots x_s \mid F_1) + \frac{1}{v} F_s^{(1)}(x_1 \ldots x_s \mid F_1) + \cdots.
\end{aligned}\right\} \quad (24)$$

Just as the uniformity parameter is $\sim t_0/\Theta$, one easily sees that the dimensionless parameter in (24) is $\sim \tau/t_0$.

(2) That in the kinetic stage F_1 is the basic 'secular' variable, which governs the temporal development, is because of the fact that the intermolecular forces do not affect F_1 directly. Only for $s \geqslant 2$ the 'drift' term $\{H_s, F_1\}$ in (21) contains the intermolecular force so that F_s ($s \geqslant 2$) will change quickly in a time of order τ. The equation for $s = 1$, which can be written as

$$\frac{\partial F_1}{\partial t} = -\frac{p_\alpha}{m} \frac{\partial F_1}{\partial r_\alpha} + \frac{1}{v} \int dp_1 \int dr_1 \frac{\partial \phi(|\mathbf{r} - \mathbf{r}_1|)}{\partial r_\alpha} \frac{\partial F_2(x, x_1, t)}{\partial p_\alpha}. \quad (25)$$

contains the intermolecular force ϕ only in the 'collision' term under the integral sign, so that F_1 will change much more slowly.

(3) Bogolyubov actually succeeded in *finding* solutions of the form (22), (23), (24) by a method which is again quite similar to the method of Enskog for the Boltzmann equation. I have only time to say that one finds that $A_0(x \mid F_1)$ is the drift term $S(F_1)$ in the Boltzmann equation; $A_1(x \mid F_1)$ becomes the collision term if the spatial non-uniformity of F_1 over a distance of order r_0 can be neglected; $A_2(x \mid F_1)$ contains as expected the triple collisions, etc.

(4) Just as with the Chapman–Enskog expansion there remains:

Problem VI. In which precise sense is the solution of the initial value problem of the Liouville equation approximated by the Bogolyubov expansion, and what is the nature of the convergence of this expansion? I think it may very well be that the sense of the approximation in this case is quite different from the way the solution of the Boltzmann equation is approximated by the Chapman–Enskog expansion, because of the different nature of the Liouville operator. Perhaps one gets the Bogolyubov expansion only by an averaging or coarse-graining procedure, while the Chapman–Enskog expansion is actually reached asymptotically.

Let me conclude by saying that in my opinion the successive contraction of the description of the temporal change of the state of the system is the essential feature of non-equilibrium statistical mechanics. It is the reason why there is *no* general method for treating the so-called irreversible processes analogous to the partition function method for the equilibrium phenomena. Everything depends on what are the basic relaxation times and on their relative spacing, and for each case the proper contraction of the temporal description has to be investigated.

REFERENCES

[1] Hilbert, D. *Grundzüge der linearen Integralgleichungen*, chap. 22. Leipzig, Teubner Verlag, 1912.
[2] Carleman, T. *Problèmes mathématiques dans la Théorie cinétique des Gaz.* Uppsala, 1957. This is a generalization of the earlier work in *Acta Math.* 60, 91–146 (1932).
[3] Chang, W. and Uhlenbeck, G. E. This work is up to now only published as a report entitled: On the propagation of sound in monatomic gases. *Eng. Res. Inst. Univ. of Mich.* 1952.
[4] Greenspan, M. Propagation of sound in rarified helium. *J. Acoust. Soc. Amer.* 22, 568–571 (1951).
[5] Chapman, S. and Cowling, T. G. *The Mathematical Theory of Non-Uniform Gases.* Cambridge, 1939; 2nd ed. 1953.
[6] Bogolyubov, N. *J. Phys. U.S.S.R.* 10, 265 (1946). This is an excerpt of his book: *Problemy Dinamicheskoi Teorii v Statisticheskoi Fizike.* Moscow, 1946.
[7] See Ehrenfest, P. and Ehrenfest, T. *Enc. der Math. Wiss.* 4, Art. 32.

ENTWICKLUNGSLINIEN IN DER STRUKTUR-THEORIE DER ENDLICHEN GRUPPEN

Von H. WIELANDT

Vor wenigen Jahrzehnten schien die Theorie der Gruppen von endlicher Ordnung sich einem Zustand der Erschöpfung und Unfruchtbarkeit zu nähern. Es fehlte zwar nicht an bedeutenden Problemen; es sei nur an die bis heute nicht bewältigte Aufgabe erinnert, alle einfachen Gruppen endlicher Ordnung zu bestimmen. Aber es fehlte an Methoden, die auch nur einen aussichtsreichen Ansatz zur Behandlung dieser Probleme hätten bieten können. Andererseits kann man auch sagen, daß es nicht an Methoden fehlte; es standen u. a. schöne Methoden von Hölder, Jordan, Frobenius, Burnside und Schur zur Verfügung. Aber es schienen die Probleme erschöpft, die diesen Methoden zugänglich sind.

Die Lage hat sich geändert und zwar von beiden Seiten her. Einerseits sind neue Hilfsmittel entwickelt worden, die mit ihrem kraftvollen, umfangreichen technischen Apparat zu ermutigenden Ergebnissen insbesondere über einfache Gruppen und über Gruppen von Primzahlpotenz-Ordnung geführt haben; auf diese mit modularen Darstellungen und Lie-Ringen zusammenhängenden Methoden soll hier nicht eingegangen werden; das ist 1954 in Amsterdam durch Brauer und auf dem gegenwärtigen Kongress durch Higman und Chevalley geschehen. Es soll vielmehr gezeigt werden, wie auf der anderen Seite die den älteren Methoden zugänglichen Fragenkreise sich in den letzten drei Jahrzehnten erweitert haben. Wir beschränken uns dabei auf zwei zentrale Probleme, die mit den Stichworten *arithmetische Struktur* und *Normalstruktur* bezeichnet werden können.

Die Frage nach der arithmetischen Struktur einer Gruppe G betrifft das Verhalten von G bezüglich gegebener Primzahlen. Man untersucht vor allem Untergruppen, die durch arithmetische Extremaleigenschaften ihrer Ordnung ausgezeichnet sind. Ein bekanntes Beispiel ist der gleich zu erwähnende Satz von Sylow.

Bei der Normalstruktur von G handelt es sich um die Normalteiler von G und ihre Faktorgruppen; allgemeiner untersucht man *Normalreihen*

$$G = G^0 > G^1 > G^2 > \dots > G^l = 1, \tag{1}$$

in denen jede Gruppe normal in der vorangehenden ist, und die zugehörigen Faktorgruppen $\quad F^\lambda = G^{\lambda-1}/G^\lambda.$

Hierher gehört z. B. der klassische Satz von Jordan und Hölder: Die F^λ sind bis auf Isomorphie und Reihenfolge allein durch G bestimmt, wenn die Normalkette eine Kompositionsreihe, d. h. nicht mehr zu verfeinern ist. Insbesondere sind dann die Indizes $|G^{\lambda-1}/G^\lambda|$ eindeutig bestimmt; sie heißen die Kompositionsindizes von G.

Zwischen beiden Fragenkreisen sind in den letzten Jahrzehnten überraschende Wechselbeziehungen entdeckt worden, vor allem durch Philip Hall. Über diese Entwicklung soll im folgenden ein Überblick gegeben werden an Hand einiger einfach zu formulierender, typischer Sätze. Wir beginnen mit Aussagen über die arithmetische Struktur.

1. Sylowsätze

Wir erinnern an den grundlegenden Satz von Sylow (1872):

Die Ordnung der Gruppe G habe die Primfaktorzerlegung $|G| = \Pi p^\alpha$. Dann enthält G wenigstens eine Untergruppe G_p der Ordnung p^α; je zwei solche Untergruppen sind in G konjugiert: $\bar{G}_p = g^{-1}G_p g$. Jede Untergruppe einer Ordnung p^β ist in wenigstens einer der Gruppen G_p enthalten.

Die Gruppen G_p heißen die zur Primzahl p gehörigen Sylowgruppen oder kurz die p-Sylowgruppen von G. Ihre Bedeutung liegt darin, daß in ihrer Struktur und Lage das Verhalten von G bezüglich der Primzahl p zum Ausdruck kommt.

Obwohl die Bedeutung des Satzes von Sylow sofort erkannt wurde, dauerte es mehr als 50 Jahre, bis ein wesentlicher Fortschritt in der durch Sylow eingeschlagenen Richtung erzielt wurde. Und zwar gelang dies 1928 Hall in einer Arbeit, die den Anstoß zu einer heute noch nicht abgeschlossenen Entwicklung gab. Der Fortschritt bestand darin, daß Hall mehrere Primzahlen gleichzeitig in Betracht zog. Es sei ω irgend eine Menge von Primzahlen. Man nennt eine Untergruppe H von G eine ω-*Gruppe*, wenn ihre Ordnung die Gestalt $|H| = \prod_{p \in \omega} p^\beta$ ($\beta \geqq 0$) hat. Es ist dann $|H| \leqq \prod_{p \in \omega} p^\alpha$, wobei α den Exponenten von p in $|G|$ bezeichnet. Wenn in dieser Ungleichung das Gleichheitszeichen gilt, so wollen wir H nach einem üblich gewordenen Sprachgebrauch eine ω-*Hallgruppe* von G nennen. Hallgruppen sind diejenigen Untergruppen, deren Index zu ihrer Ordnung teilerfremd ist.

Für eine gegebene Gruppe G und eine Primzahlmenge ω kann es eintreten, daß G wenigstens eine ω-Hallgruppe enthält, daß ferner je zwei solche Gruppen in G konjugiert sind und daß jede ω-Untergruppe von G in einer ω-Hallgruppe von G enthalten ist. In diesem Fall wollen wir kurz sagen: In G gilt der ω-*Sylowsatz*. Beispielsweise gilt in jeder Gruppe G für jede Primzahl p der p-Sylowsatz.

Besteht ω aus mehr als einer Primzahl, so braucht der ω-Sylowsatz in G nicht zu gelten. Genauer: G braucht keine ω-Hallgruppe zu enthalten (z. B. gibt es in der kleinsten einfachen Gruppe, der Ikosaedergruppe der Ordnung 60, weder eine {2, 5}-Hallgruppe noch eine {3, 5}-Hallgruppe). Wenn es mehrere ω-Hallgruppen gibt, so brauchen sie nicht konjugiert zu sein. Und selbst wenn dies eintritt, braucht nicht jede ω-Untergruppe von G in einer ω-Hallgruppe von G enthalten zu sein; hierfür gibt wieder die Ikosaedergruppe ein Beispiel mit $\omega = \{2, 3\}$.

Aber in auflösbaren Gruppen G gelten alle diese Aussagen. Das ist die Entdeckung von Hall[4]:

Ist G auflösbar und ω beliebig, so gilt in G der ω-Sylowsatz.

Man kann sagen, daß dieser Satz einen Einfluß der Normalstruktur auf die arithmetische Struktur erkennen läßt. Denn die Voraussetzung der Auflösbarkeit bedeutet, daß alle Kompositionsindizes von G Primzahlen sind, m. a. W. daß G eine Normalreihe mit abelschen Faktorgruppen F^λ besitzt.

Der Beweis ist ein einfacher Induktionsschluß ausgehend von dem klassischen Satz von Sylow.

Der Satz von Hall legt den Wunsch nahe, die Voraussetzung der Auflösbarkeit von G abzuschwächen. In dieser Richtung hat Čunihin seit 1943 eine längere Reihe von Untersuchungen veröffentlicht. Sie behandeln die folgende Frage: Es sei eine Normalreihe (1) irgend einer endlichen Gruppe G gegeben, und es sei etwas über ω-Hallgruppen der einzelnen Faktorgruppen F^λ bekannt. Was kann man dann über ω-Hallgruppen von G sagen? Bei diesen Untersuchungen wurde Čunihin auf einen wichtigen neuen Begriff geführt. Er nennt eine Gruppe ω-auflösbar, wenn jeder ihrer Kompositionsfaktoren entweder eine Primzahl aus ω oder durch keine Primzahl aus ω teilbar ist. Mit anderen Worten: Versteht man unter ω' die zu ω komplementäre Menge von Primzahlen, so heißt G dann ω-auflösbar, wenn G eine Normalreihe besitzt, in der jede Faktorgruppe entweder eine ω'-Gruppe oder eine abelsche ω-Gruppe ist. (Eine auflösbare Gruppe ist also ω-auflösbar für jede Primzahlmenge ω.) Čunihin[2] beweist die folgende Verallgemeinerung des Satzes von Hall:

G sei ω-auflösbar. Dann gilt in G sowohl der ω-Sylowsatz wie der ω'-Sylowsatz.

Der Beweis beruht auf einem Satz von Zassenhaus, der aus der Erweiterungstheorie stammt und später erwähnt werden wird.

In einer anderen Richtung als Hall und Čunihin hat Wielandt 1954 den Satz von Sylow erweitert. Es wird nichts über Normalreihen vorausgesetzt, dafür aber die Existenz einer ω-Hallgruppe von spezieller Struktur gefordert[11]:

Wenn G eine nilpotente ω-Hallgruppe enthält, so gilt in G der ω-Sylowsatz.

Dabei heißt eine Gruppe nilpotent, wenn sie das direkte Produkt ihrer Sylowgruppen ist. Diese zunächst vielleicht unangemessen stark erscheinende Voraussetzung kann, wie Baer[1] und andere gezeigt haben, durch verwandte, etwas schwächere Voraussetzungen ersetzt werden. Doch kann man hier nicht mehr viel einsparen; z.B. weiß man, daß die Voraussetzung der Auflösbarkeit der ω-Hallgruppe nicht ausreicht.

Über den ganzen Fragenkreis der ω-Sylowsätze hat Hall 1956 eine inhaltreiche Übersicht gegeben. Das wichtigste seiner neuen Resultate besagt:

Gegeben sei eine Normalreihe (1) von G. In der obersten Faktorgruppe F^1 gelte der ω-Sylowsatz, und die ω-Hallgruppen von F^1 seien auflösbar; jeder andere Faktor F^λ enthalte eine nilpotente ω-Hallgruppe. Dann gilt in G der ω-Sylowsatz.

Dieser Satz umfaßt sowohl den Satz von Wielandt als auch mehrere Ergebnisse von Čunihin. Naheliegende Fragen sind noch offen; z.B.: Gilt der ω-Sylowsatz in G genau dann, wenn er für jeden Faktor F^λ einer Normalreihe von G gilt?

Wir gehen nun zu Aussagen über die Normalstruktur über und beginnen mit der Betrachtung eines einzelnen Normalteilers.

2. Gruppen-Erweiterung

Gegeben seien zwei Gruppen N und F. Wie konstruiert man die Erweiterungen von N mit F, das heißt diejenigen Gruppen G, die N als einen Normalteiler mit der Faktorgruppe $G/N \cong F$ enthalten?

Dieses *Erweiterungsproblem* ist seit Hölder (1895) mehrfach bearbeitet worden. Wir wollen mit der Erörterung einer besonders einfachen Situation beginnen: Die zu untersuchende Erweiterung G soll zerfallen, d.h. ein Komplement zu N enthalten. Ein Komplement C zu N in G ist durch die beiden folgenden Forderungen erklärt:

(i) C ist Untergruppe von G;
(ii) C enthält aus jeder Nebenklasse von N in G genau ein Element.

Wenn es in G ein Komplement C zu N gibt, so ist $C \cong F$, und jedes

Element g von G läßt sich eindeutig in zwei Komponenten aus N und C zerlegen:
$$g = nc \quad (n \in N,\ c \in C).$$

Das Rechnen mit diesen Komponenten gestaltet sich einfach wegen der Normalität von N:
$$g_1 g_2 = (n_1 . c_1 n_2 c_1^{-1})(c_1 c_2).$$

Hiernach kennt man die Multiplikationstafel von G, sobald man die Tafeln von C (d. h. von F) und von N kennt und sobald man außerdem weiß, welche Automorphismen der Normalteiler N bei Ähnlichkeitstransformation mit den Elementen aus C erleidet. Dies führt zu einer durchsichtigen Konstruktion der zerfallenden Erweiterungen.

Es erhebt sich die Frage, wann eine vorgelegte Erweiterung zerfällt und wie man alle Komplemente findet. Zum letzten Punkt ist zu bemerken, daß jede zu einem Komplement konjugierte Untergruppe von G wieder ein Komplement zum selben Normalteiler ist. Wir formulieren daher das *Komplementenproblem*: Gegeben sei ein Normalteiler N von G. Wann enthält G wenigstens ein Komplement C zu N, und wann sind zwei gegebene Komplemente C, $\bar{C}$ zu N in G konjugiert?

Zu diesem Problem hat Gaschütz 1952 einen wichtigen Beitrag geliefert. Sein Satz[3] kann in etwas vervollständigter Form so ausgesprochen werden:

Sei N ein abelscher Normalteiler von G.

(a) *Genau dann gibt es ein Komplement zu N in G, wenn es für jede Primzahl p ein Komplement zu $G_p \cap N$ in G_p gibt. Dabei bedeutet G_p eine p-Sylowgruppe von G.*

(b) *Zwei Komplemente C, $\bar{C}$ von N in G sind genau dann konjugiert in G, wenn für jede Primzahl p die p-Sylowgruppen $\bar{C}_p$ und C_p in G konjugiert sind.*

Dieser Satz führt das Komplementenproblem im Fall eines abelschen Normalteilers auf die Untersuchung von p-Gruppen zurück. Ein Sonderfall des Satzes von Gaschütz war schon 1937 von Zassenhaus gefunden worden. Zassenhaus setzt voraus, daß N eine Hallgruppe von G ist. In diesem Fall sind die Durchschnitte $G_p \cap N$ entweder gleich G_p oder gleich 1, und die Bedingungen von Gaschütz sind stets erfüllt. Es gibt also stets Komplemente, und je zwei Komplemente sind konjugiert. Darüber hinaus hat Zassenhaus bemerkt, daß man durch einen Induktionsschluß die Einschränkung auf abelsche Normalteiler mildern kann. Der vollständige Satz von Zassenhaus[14] lautet:

(a) *Sei N normal in G und $(|N|, |G/N|) = 1$. Dann gibt es in G ein Komplement zu N.*

(b) *Sei N oder G/N auflösbar. Dann sind je zwei Komplemente zu N in G konjugiert.*

Zassenhaus gibt gute Gründe für die Vermutung an, daß die Voraussetzung der Auflösbarkeit überflüssig ist.

Es wäre schön, wenn man beim Satz von Gaschütz die Voraussetzung der Kommutativität von N streichen könnte. Leider ist das nicht möglich. Wie Gegenbeispiele zeigen, wird der Satz schon dann falsch, wenn man Kommutativität durch Nilpotenz ersetzt. Die Aufgabe erscheint wichtig, das Komplementenproblem für nilpotente Normalteiler zu lösen. Wie sich nämlich unten zeigen wird, würde man daraus auch Nutzen für den Fall eines beliebigen Normalteilers ziehen können.

Wir müssen hier eine Bemerkung über den Beweis der Sätze von Zassenhaus und Gaschütz einschalten. Er benutzt die sogenannten Faktorensysteme. Auf diese wird man in folgender Weise geführt: Wenn es in G kein Komplement zu N gibt (wenn also die beiden Forderungen (i) und (ii) nicht gleichzeitig erfüllt werden können), oder wenn man kein Komplement kennt, so kann man doch stets (ii) erfüllen, indem man auf (i) verzichtet. Man braucht nur aus jeder Nebenklasse von N in G einen Vertreter c_ρ auszuwählen ($1 \leqq \rho \leqq |G/N|$). Die Abweichung des gewählten Vertretersystems $C = \{c_\rho\}$ von der Gruppeneigenschaft macht sich dann dadurch bemerkbar, daß in den Gleichungen

$$c_\rho c_\sigma = n_{\rho\sigma} c_{\rho\sigma}$$

Faktoren $n_{\rho\sigma} \neq 1$ auftreten; dabei bedeutet $c_{\rho\sigma}$ den Vertreter der Nebenklasse $N c_\rho c_\sigma$, und es ist $n_{\rho\sigma} \in N$. Derartige *Faktorensysteme* $\{n_{\rho\sigma}\}$ sind in einem Sonderfall (nämlich N im Zentrum von G) 1904 von Schur und in voller Allgemeinheit 1926 von Schreier eingeführt worden. Die Faktorensysteme genügen auf Grund der Assoziativität der Multiplikation in G einer gewissen Funktionalgleichung. Leider ist diese Gleichung nur im Fall der Kommutativität von N handlich genug, um wesentliche Aussagen über die Struktur von G zu liefern. Daran hat auch die im letzten Jahrzehnt erfolgte Einordnung der Faktorensysteme in die Kohomologie-Theorie (für die wir auf Kurosh [9] verweisen) bisher nichts geändert. Eine weitere Schwierigkeit liegt darin, daß die Faktorensysteme durch die erwähnte Funktionalgleichung eng mit der (im allgemeinen schlecht bekannten) Automorphismengruppe von N zusammenhängen und keine unmittelbare Beziehung zu der besser bekannten Struktur von N selbst (etwa den Sylowgruppen und Normalteilern von N) haben.

Zur Umgehung dieser Schwierigkeiten schlug Hall 1940 einen Weg ein, der dem Erweiterungsproblem eine neue Wendung gab. Während Schreier auf die Gruppeneigenschaft des Vertretersystems (Forderung (i)) verzichtet und die Eindeutigkeit (Forderung (ii)) verlangt, geht Hall gerade umgekehrt vor. Mit anderen Worten: Hall sucht in einer Gruppe G mit Normalteiler N nach Untergruppen S mit der Eigenschaft

$$G = NS. \tag{2}$$

Wir wollen eine solche Untergruppe S kurz ein Supplement zu N in G nennen. Es existiert stets ein triviales Supplement, nämlich $S = G$. Wenn ein nicht triviales Supplement existiert, so liefert die (nicht mehr eindeutige) Zerlegung

$$g = ns \quad (n \in N, s \in S)$$

eine Reduktion des Erweiterungsproblems, nämlich von G auf die kleinere Gruppe S. Die Reduktion ist um so stärker, je kleiner S ist, d. h. je kleiner der Durchschnitt $S \cap N = T$ ist; genau wenn $T = 1$ ist, ist S ein Komplement von N. Es kommt also darauf an, kleine Supplemente zu suchen. Hall hat dies unter der zusätzlichen Voraussetzung durchgeführt, daß der Normalteiler N auflösbar ist. Es ist nicht schwer, seinem Gedanken eine allgemein anwendbare Form zu geben. Das geschieht im folgenden.

Angenommen, wir hätten in N ein System Σ von Untergruppen oder Komplexen $J_1, J_2, \ldots, J_r$ mit folgenden Eigenschaften gefunden: Zu jedem Automorphismus τ von N gibt es einen inneren Automorphismus von N, der auf das System die gleiche Wirkung wie τ hat; d. h. zu τ existiert ein $n \in N$ mit $n^{-1}J_\rho n = J_\rho^\tau$ ($\rho = 1, \ldots, r$). Ein solches System Σ nennen wir *intravariant* in N. Ist dann G eine beliebige Erweiterung von N und verstehen wir unter S den Normalisator von Σ in G, also

$$S = \{s \mid s \in G, s^{-1}J_1 s = J_1, \ldots, s^{-1}J_r s = J_r\},$$

so ist S eine Untergruppe von G, die wegen der Intravarianz von Σ die Eigenschaft (2) hat. Der Durchschnitt $S \cap N = T$ ist normal in S, und wegen (2) ist $S/T \cong G/N \cong F$; daher ist S eine Erweiterung von T mit F. Um also alle Erweiterungen von N mit F zu erhalten, hat man gewisse Erweiterungen S der festen Untergruppe T von N mit F zu bilden und diese, kurz gesagt, gemäß (2) mit N zu verschmelzen. Die letzte Aufgabe wird dadurch erleichtert, daß die von S bewirkten Automorphismen von N jede einzelne der Untergruppen J_ρ in sich überführen.

Die Gruppe $T = S \cap N$ ist nichts anderes als der Normalisator des Systems Σ in N. Also wird man Σ so zu bestimmen suchen, daß der Normalisator klein ist und eine durchsichtige Struktur hat. Hier sind die Sylowsätze von Nutzen. Denn wenn in N der ω-Sylowsatz gilt, so ist jede ω-Hallgruppe intravariant in G. Und es ist leicht zu zeigen: Wenn jede einzelne der Gruppen $J_1, J_2, \ldots, J_r$ intravariant in N ist und wenn außerdem ihre Indizes in N paarweise teilerfremd sind, so ist das ganze System $\Sigma = \{J_1, J_2, \ldots, J_r\}$ intravariant in N. Da in dem von Hall betrachteten Fall N auflösbar ist, kann Hall für die J_ρ ein vollständiges System von p'_ρ-Hallgruppen von N nehmen (zu jedem Primteiler p_ρ von $|N|$ eine). Es ist leicht zu sehen, daß die so entstehende Gruppe T stets nilpotent ist. Damit erhält Hall den Satz[6]:

In jeder auflösbaren Gruppe N gibt es eine nilpotente Untergruppe T mit folgender Eigenschaft: Man erhält jede Erweiterung von N durch eine beliebige Gruppe F, indem man T mit F erweitert und das Ergebnis S mit N verschmilzt.

Den Verschmelzungsprozeß erörtert Hall eingehend. Dabei kommen ihm überraschende Eigenschaften der eben konstruierten Gruppe T zu Hilfe. (Dieser bis auf einen inneren Automorphismus von N eindeutig bestimmte *Sylowsystem-Normalisator T* dürfte die wichtigste Untergruppe einer auflösbaren Gruppe N sein, abgesehen von der Kommutatorgruppe.) Besonders einfach wird die Theorie, wenn alle Sylowgruppen des auflösbaren Normalteilers N abelsch sind. Dann wird das Erweiterungsproblem auf den schon ausreichend geklärten Fall eines abelschen Normalteilers zurückgeführt durch den folgenden Satz von Hall[6]:

Ist N ein auflösbarer Normalteiler von G und sind alle Sylowgruppen von N abelsch, so gibt es in G zu der Kommutatorgruppe von N ein Komplement S. Der Durchschnitt $T = S \cap N$ ist abelsch.

Auf diesem Wege haben Hall[6] und Taunt (1949) die Konstruktion und Struktur der auflösbaren Gruppen mit lauter abelschen Sylowgruppen untersucht.

Ist der Normalteiler N nicht auflösbar, so stehen die p'_ρ-Sylowsätze nicht zur Verfügung. Es sei jedoch erwähnt, daß man stets zum Ziel kommt, indem man für Σ irgend ein maximales in N intravariantes System wählt; dessen Normalisator in N ist nämlich, wie aus dem klassischen Satz von Sylow folgt, stets nilpotent. *Der an vorletzter Stelle erwähnte Satz von Hall gilt also sogar für beliebige endliche Gruppen N.* Der skizzierte Beweis legt die Aufgabe nahe, die maximalen intravarianten Systeme in N und ihre Normalisatoren T zu studieren. Natürlich kann man von diesen T nicht alle schönen Eigenschaften erwarten, die Halls Sylowsystem-Normalisator besitzt; dadurch wird die Diskussion des Verschmelzungsprozesses schwieriger werden als im Fall eines auflösbaren Normalteilers. Jedenfalls aber hat sich hier ein Weg zur Reduktion des allgemeinen Erweiterungsproblems auf den Fall nilpotenter Normalteiler eröffnet, der weitere Verfolgung verdient. Zur Lösung des verbleibenden Erweiterungsproblems für nilpotente Normalteiler würde es nützlich sein, die Darstellungen einer gegebenen Gruppe durch Automorphismen von p-Gruppen zu studieren; die Lösung dieser Aufgabe könnte an die Theorie der modularen Darstellungen anknüpfen.

3. Normalreihen

Wie im ersten Teil gezeigt wurde, kann man aus Voraussetzungen über Normalreihen von G Aussagen über die Gültigkeit von Sylow-sätzen in G gewinnen. Es gibt auch Sätze in der umgekehrten Richtung; sie liegen meist tiefer. So hat Hall 1937 die Umkehrung seines Satzes von 1928 bewiesen [5]:

Wenn in G jeder ω-Sylowsatz gilt, dann ist G auflösbar.

Schärfer gilt sogar [5]:

Eine Gruppe ist genau dann auflösbar, wenn sie für jede Primzahl p eine p'-Hallgruppe enthält.

Der Beweis benutzt einen bisher nur mit Hilfe der Gruppencharaktere bewiesenen Satz von Burnside, nach dem jede Gruppe der Ordnung $p^\alpha q^\beta$ auflösbar ist.

Der zuletzt genannte Satz von Hall ist ein wertvolles Hilfsmittel zum Nachweis der Auflösbarkeit von endlichen Gruppen. Eine ähnliche Kennzeichnung der ω-auflösbaren Gruppen ist bisher nicht bekannt. Einfache Gegenbeispiele zeigen, daß sich jedenfalls der oben erwähnte Satz von Čunihin nicht umkehren läßt; wenn in einer Gruppe sowohl der ω-Sylowsatz wie der ω'-Sylowsatz gilt, so braucht sie nicht ω-auflösbar und ω'-auflösbar, d. h. auflösbar schlechthin zu sein. Die Behauptung wird aber richtig, wenn man die Voraussetzung etwas verschärft [13]:

G enthalte eine ω-Hallgruppe und eine ω'-Hallgruppe, die beide nilpotent sind (ω und ω' bedeuten komplementäre Mengen von Primzahlen). Dann ist G auflösbar.

Dieser Satz ist insofern von Interesse, als er nur zwei Hallgruppen als bekannt voraussetzt.

Es ist bemerkenswert, daß Hall und Higman im Zusammenhang mit Untersuchungen zu einem Problem von Burnside gezeigt haben, daß sogar eine einzige p-Sylowgruppe schon einen gewissen Einfluß auf die Normalreihen von G hat, wenigstens wenn die Gruppe p-auflösbar ist, d. h. wenn jeder Kompositionsindex von G entweder gleich p oder teilerfremd zu p ist. Und zwar kann man, kurz gesagt, Kompositions-faktorgruppen der Ordnung p auf wenige Strecken zusammenschieben, wenn die p-Sylowgruppe eine einfache Struktur hat. Es sei nur der einfachste einer ganzen Reihe von Sätzen erwähnt[8]:

G sei p-auflösbar und besitze eine abelsche p-Sylowgruppe G_p. Dann besitzt G eine Normalreihe der Gestalt

$$G > K > L > 1 \quad \text{mit} \quad K/L \cong G_p. \tag{3}$$

Der Anteil von p an der Gruppe G tritt hier in besonders übersicht-licher Weise im Mittelstück einer ausgezeichneten Normalreihe in Erscheinung.

Der letzte Satz führt auf eine allgemeine Frage: Was läßt sich über die Normalreihen einer beliebigen endlichen Gruppe G sagen, wenn man eine p-Sylowgruppe von G kennt? Zur Behandlung dieser Frage hat Wielandt kürzlich eine Methode entwickelt. Es wird die Gesamtheit SG der subnormalen Untergruppen von G betrachtet; das sind die-jenigen Untergruppen von G, die in Normalreihen von G auftreten: die Normalteiler von G, die Normalteiler der Normalteiler von G, und so weiter. Diese subnormalen Untergruppen stehen den normalen über-raschend nahe. Beispielsweise sind sie unter schwachen Zusatzvoraus-setzungen (man könnte fast sagen: im allgemeinen) miteinander als Ganze vertauschbar, und der Durchschnitt ihrer Normalisatoren in G ist niemals 1, wenn $G \neq 1$ ist (Wielandt 1957, 1958). In unserem jetzigen Zusammenhang ist eine andere Eigenschaft von Bedeutung: Durch-schnitt und Erzeugnis von subnormalen Untergruppen sind subnormal. Mit anderen Worten[10]:

SG ist ein Teilverband des Verbands aller Untergruppen von G.

Man kann nun einen unmittelbaren Zusammenhang zwischen Normal-reihen und Sylowgruppen herstellen, indem man eine feste Sylowgruppe G_p von G wählt und jeder subnormalen Untergruppe A von G den Durchschnitt $A \cap G_p = A_p$ zuordnet. Dieser Durchschnitt ist, wie man leicht erkennt, eine p-Sylowgruppe von A. Außerdem gilt aber[12]:

Die Abbildung $A \to A_p$ ist ein Verbandshomomorphismus von SG in SG_p. (Für Hallgruppen gilt der entsprechende Satz.)

Wenn man die Struktur einer Sylowgruppe von G genügend gut kennt, ergeben sich starke Einschränkungen für den Verband SG und damit für die Normalreihen von G. Bisher ist nur der Fall einer zyklischen Sylowgruppe genauer untersucht worden. Es ergibt sich[12]:

Wenn eine p-Sylowgruppe von G zyklisch ist, so ist G entweder p-auflösbar, oder G besitzt nur einen durch p teilbaren Kompositionsindex.

Im ersten Fall hat G nach Hall und Higman eine Normalreihe der einfachen Gestalt (3); im zweiten Fall kann man G passend als p-einfach bezeichnen.

Der erwähnte Verbands-Homomorphismus ist genau dann treu, wenn jeder Kompositionsindex von G durch p teilbar ist. Ist diese Bedingung nicht erfüllt, so genügt es, hinreichend viele verschiedene Primteiler der Ordnung von G heranzuziehen. Man erhält den Satz[12]:

Der Subnormalverband SG *einer endlichen Gruppe* G *ist ein Teilverband des direkten Produkts der Untergruppenverbände der Sylowgruppen von* G *zu den verschiedenen Primzahlen.*

Die im letzten Abschnitt erwähnten Sätze zeigen einen unerwartet starken Einfluß der Hallgruppen auf die Normal-Struktur. Hier scheint ein ergiebiges Feld für weitere Untersuchungen vorzuliegen.

Insgesamt darf man wohl sagen, daß die an klassische Fragen und Methoden anknüpfende Strukturtheorie der endlichen Gruppen nicht mehr arm an angreifbaren, lohnenden Problemen ist. Sie wird auch der allgemeinen Strukturtheorie unendlicher Gruppen noch auf längere Zeit Anregungen geben können.

LITERATUR

[1] Baer, R. Verstreute Untergruppen endlicher Gruppen. *Arch. Math.* 9, 7–17 (1958).

[2] Čunihin, S. A. Über Sylow-Eigenschaften endlicher Gruppen. *Dokl. Akad. Nauk SSSR* (N.S.) 73, 29–32 (1950). (Russisch.)

[3] Gaschütz, W. Zur Erweiterungstheorie der endlichen Gruppen. *J. reine angew. Math.* 190, 93–107 (1952).

[4] Hall, P. A note on soluble groups. *J. Lond. Math. Soc.* 3, 98–105 (1928).

[5] Hall, P. A characteristic property of soluble groups. *J. Lond. Math. Soc.* 12, 198–200 (1937).

[6] Hall, P. The construction of soluble groups. *J. reine angew. Math.* 182, 206–214 (1940).

[7] Hall, P. Theorems like Sylow's. *Proc. Lond. Math. Soc.* (3), 6, 286–304 (1956).

[8] Hall, P. and Higman, G. On the p-length of p-soluble groups and reduction theorems for Burnside's problem. *Proc. Lond. Math. Soc.* (3), 6, 1–42 (1956).

[9] Kurosh, A. G. *The Theory of Groups,* vol. II. Chelsea Publ. Co., New York, 1956.

[10] Wielandt, H. Eine Verallgemeinerung der invarianten Untergruppen. *Math. Z.* 45, 209–244 (1939).

[11] Wielandt, H. Zum Satz von Sylow. *Math. Z.* 60, 407–408 (1954).

[12] Wielandt, H. Sylowgruppen und Kompositions-Struktur. *Abh. Math. Sem. Hamburg,* 22, 215–228 (1958).

[13] Wielandt, H. Über Produkte von nilpotenten Gruppen. *Illinois J. Math.* 2, 611–618 (1959).

[14] Zassenhaus, H. *Lehrbuch der Gruppentheorie,* Bd. I. Teubner, Leipzig und Berlin, 1937.

HALF-HOUR ADDRESSES

COMPLETENESS RESULTS FOR FORMAL SYSTEMS

By E. W. BETH

1. Suppose that, on the basis of the classical sentential logic in $^{-}$ and $\rightarrow$, I wish to test or to establish the validity of Peirce's Law:

$$[(p \rightarrow q) \rightarrow p] \rightarrow p. \qquad (1)$$

The notion of validity can be taken either syntactically or semantically. I start from the semantic conception, and thus I try to construct a regular valuation under which formula (1) is false. The following *semantic tableau* records the successive steps in the construction and shows that its result is negative[1]:

True		False	
		$[(p \rightarrow q) \rightarrow p] \rightarrow p$	(2)
$(p \rightarrow q) \rightarrow p$		p	(3)
	p	$p \rightarrow q$	(4)
p		q	(5)

A similar tableau can be constructed for every formula X of sentential logic. If X happens to be a *logical identity* (that is, a formula which is true under every regular valuation; by a *regular valuation*, I mean a valuation permitted by the truth-tables), then this fact will manifest itself by the *'closure'* of the tableau. If X is *not* a logical identity, then the tableau will not be closed and it will exhibit a regular valuation under which X becomes false.

2. We now observe that the rules of construction and closure for semantic tableaux, although semantically motivated, can be stated in syntactical terms. Thus a closed semantic tableau can be alternatively considered as a formal derivation in a certain Formal System F, which is trivially complete, so to speak.

3. In order to give this derivation a more familiar shape, it can be rewritten as follows:

$$\cdots\cdots\cdots\cdots\cdots\cdots\cdots\cdots\cdots$$

$$
\begin{array}{c|c}
\multicolumn{2}{c}{(p \to q) \to p} \\
\hline
[\overline{p}] & p \\
\cdots\cdots\cdots\cdots & \\
p & \\
q & \\
\cdots\cdots\cdots\cdots & \\
p \to q & \\
\end{array}
$$

$$p$$

$$\cdots\cdots\cdots\cdots\cdots\cdots\cdots\cdots\cdots$$

$$[(p \to q) \to p] \to p$$

In fact, every closed semantic tableau can be rewritten as a formal derivation in Gentzen's system NK, which hence proves to be complete. Likewise, closed semantic tableaux can be rewritten as formal derivations in Gentzen's system LK, and this remark implies the completeness of this system as well.

It takes somewhat more trouble to rewrite the tableau so as to obtain a formal derivation of Peirce's Law, starting from Church's axiom system P_2:

$$p \to (q \to p), \tag{6}$$

$$[p \to (q \to r)] \to [(p \to q) \to (p \to r)], \tag{7}$$

$$(\overline{p} \to \overline{q}) \to (q \to p), \tag{8}$$

by means of substitution and *modus ponens*. By showing that, for every closed semantic tableau, this transcription can be carried out, we prove the completeness of the axiom system P_2.

4. In their original shape, semantic tableaux also suggest *non-regular valuations* by means of which we can establish the relative independence of Church's axioms. For instance, by means of a valuation w, defined as follows:

(i) $w(a) = 2\ [true]$, $w(p) = 0\ [false]$ for each atom p different from a;
(ii) $w(\overline{X})$ is obtained from $w(X)$ as usual;
(iii) $w(a \to b) = 2$, whereas in all other cases $w(X \to Y)$ is obtained from $w(X)$ and $w(Y)$ as usual;

we show that, since:

$$w([a \to (b \to c)] \to [(a \to b) \to (a \to c)]) = 0, \tag{9}$$

axiom (7) is independent with respect to axioms (6) and (8).

A systematic study of non-regular valuations suggests various new problems which, however, I cannot hope to discuss in this paper.

5. It may cause some surprise (although it is in accordance with Kreisel's no-counterexample interpretation[2]) that an entirely similar approach is found to be available in the theory of quantification. Let us draw up the semantic tableau for the formula:

$$\overline{\overline{(x)\,a(x)}} \to \overline{\overline{(y)\,a(y)}}, \tag{10}$$

namely:

True	False
$\overline{\overline{(x)a(x)}}$	$\overline{\overline{(x)a(x)\to(y)a(y)}}$
$(x)a(x)$	$\overline{\overline{(y)a(y)}}$ ←
$a(1)$	$\overline{\overline{(x)a(x)}}$
$\overline{a(1)}$	$\overline{a(1)}$
	$a(1)$

(with *A* bracketing the True column and *B* bracketing the False column)

Some reflection on the method of construction shows that we could dispense with the decomposition of prenex formulas if, instead of formula (10), we considered the formula

$$[(x)\,a(x) \to a(1)] \to \{[\overline{a(1)} \to (y)\,\overline{a(y)}] \to [\overline{(x)\,a(x)} \to (y)\,\overline{a(y)}]\}, \tag{11}$$

which, therefore, is derivable in sentential logic. A little more reflection is sufficient to convince us that, by introducing additional axioms:

$$(z)\,[(x)\,a(x) \to a(z)], \tag{12}$$

$$(z)\,\{[b(z) \to (y)\,b(y)] \to p\} \to p, \tag{13}$$

and stating suitable (and fairly obvious) rules for (generalized) substitution and (generalized) *modus ponens*, we shall obtain a complete axiomatization for the theory of quantification.

Again, non-regular valuations, suggested by suitable semantic tableaux, can be used in proofs of relative independence. For instance, the independence of axiom (12) with respect to axioms (6)–(8) and (13) can be established by means of the following valuation w:

  (i) if U is one of the atoms $p, q, r, \ldots, a(1), b(1), \ldots, u(1,1), \ldots$, then
$$w(U) = 0;$$
  (ii) $w(\overline{X})$ and $w(X \to Y)$ are obtained from $w(X)$ and $w(Y)$ as usual;
  (iii) $w[(v)\,U(v)] = w[U(1)]$, with one exception: $w[(x)\,a(x)] = 2$.

It hardly needs saying that we can again rewrite every closed semantic tableau so as to obtain formal derivations in Gentzen's systems NK and LK.

6. The above method of semantic tableaux is one among several devices which have been offered as substitutes for the more conventional axiomatic treatment of elementary logic. The origin of these devices is found in Herbrand's ideas, and we may point to Gentzen's systems NK and LK and to the Hilbert–Bernays theory of the ε-symbol as early representatives. More recent contributions are those by Craig[3], Guillaume[4], Hintikka[5], Kanger[6], Kripke[7], Quine[8], and Schütte[9].

In an elementary textbook by Basson and O'Connor[10], one finds an '*Indirect Method of Truth-Table Decision*' which corresponds to the construction in §1. However, these authors seem not to realize the implications of this method.

7. An evaluation of the efficiency of these various systems can be based upon the following considerations:
- (i) simplifications in the proofs of more profound metamathematical results;
- (ii) the avoidance of reduction to prenex and other normal forms;
- (iii) the degree to which Gentzen's *subformula principle* is brought into effect;
- (iv) the possibility of an adaptation to the requirements of modal logic, intuitionistic logic, and many-valued logic.

I feel that in these various respects considerable progress has been made as compared to the state of affairs created by Gentzen's work. In demonstrating this progress I shall discuss the method of semantic tableaux; however, I do not wish to imply that the methods developed by Craig, Hintikka, Kanger, or Schütte do not offer the same or similar advantages.

By way of example, let us consider a special case of Herbrand's Theorem. Suppose we wish to establish the validity of the formula:

$$(Ex)\,(Ey)\,(z)\,(t)\,U(x, y, z, t), \tag{14}$$

where $U(x, y, z, t)$ is quantifier-free.

Besides the formula (14), we consider the disjunctions:

$$\mathop{\mathrm{D}}_{j+k \leq P} U(j, k, f(j, k), g(j, k)), \tag{15}$$

where

$$f(j, k) = (j + k - 1) \times (j + k - 2) + 2k, \tag{16}$$

$$g(j, k) = (j + k - 1) \times (j + k - 2) + 2k + 1, \tag{17}$$

for $P = 1, 2, 3, \ldots$ It will be clear that the semantic tableau for formula (14) will be closed if, and only if, for some P, the semantic tableau for formula (15) is closed; thus formula (14) will be derivable if, and only if, for some P, formula (15) is derivable. This is Herbrand's Theorem, as applied to formula (14). From this example, it may be seen to what extent the method of semantic tableaux (i) simplifies the proofs of certain more profound metamathematical results.

Moreover, it is obvious from our discussion that (ii) the restriction of Herbrand's Theorem to formulas in proof-theoretic Skolem normal form serves no other purpose than that of simplifying the description of the index functions $f, g, \ldots$ which are required. Essentially, Herbrand's Theorem applies to *every* formula; it is easy to construct in each case the necessary index functions by means of a semantic tableau.

Let us now consider Gentzen's Subformula Theorem. For Gentzen's system LK, this theorem can be stated as follows: every derivation of a sequent $K \vdash L$ can be replaced by a derivation in which only subformulas of the formulas in K and in L appear.

On the other hand, a derivation of a sequent $K \vdash L$ by means of a semantic tableau *always* has the property that only subformulas of the formulas in K and in L appear in it. Thus it may be said that (iii) the subformula principle is brought into effect even more strongly than it was in Gentzen's own system.

8. Finally, I turn to point (iv); our conclusions in connection with points (ii) and (iii) suggest the possibility of an adaptation of the method of semantic tableaux to other forms of logic.

(iv a) With a view to many-valued logic this possibility is so obvious as to make any further discussion superfluous. However, I may mention the possibility of combining the method of semantic tableaux with an application of the ideas recently expressed by Dreben[11].

(iv b) In connection with modal logic, I may mention recent work by Guillaume[4], Kanger[6] and Kripke[7]; it would carry me too far, however, to go more deeply into their results.

(iv c) In recent work[1], I have extended the method of semantic tableaux to the case of intuitionistic logic. It is true that Kreisel[12] and Kleene[13] have raised certain objections as to the semantic basis of my construction; however, I do not wish to dwell upon this point.

At any rate, my intuitionistic version of the method of semantic tableaux provides an adequate substitute for other extant formalizations of intuitionistic logic, such as Heyting's axiomatization or Gentzen's

systems NJ or LJ. Moreover, it seems to offer the same advantages which
are characteristic of the classical version. To demonstrate these advantages, I may mention the fact that the well-known topological completeness results for intuitionistic sentential logic by Tarski and for intuitionistic elementary logic by Mostowski and Raisowa can be improved in
this respect, that all topological spaces which are needed are subsets of
the Cantor discontinuum.

9. In order not to create an incorrect impression about the advantages
of the method of semantic tableaux, I now wish to mention the metamathematical results which are needed to justify the application of
closed semantic tableaux as formal derivations.

(I) It must be proved that the closure of a semantic tableau does
not depend upon the relative order in which the formulas are submitted
to a decomposition.

(II) It must be proved that, whenever the tableaux for $K' \vdash L', X$
and for $K'', X \vdash L''$ are closed, so is also the tableau for $K', K'' \vdash L', L''$.

(III) It must be proved that, whenever U is a logical identity, the
tableau for the sequent $\phi \vdash U$ is closed.

If we do not insist on having finitary proofs, then very simple proofs
for the results under (I)–(III) are available. However, for (I) and (II),
and for a suitably weakened version of (III), one may reasonably demand
a finitary proof.

As a weakened version of (III) we have, in the classical case, Herbrand's Theorem, the proof of which is extremely simple, as we have seen.
The results under (I) correspond to those by Curry and Kleene on the
permutability of the rules for Gentzen's systems LK and LJ, whereas
the results under (II) correspond to Gentzen's '*Hauptsatz*' for these
systems. Nevertheless, the proofs of the results under (I) and (II) for
closed semantic tableaux turn out to be simple as compared to those of
the corresponding results for Gentzen's systems.

10. In §§ 6 and 7 the method of semantic tableaux, along with other
similar devices, has been presented as a substitute for the more conventional axiomatic treatment of elementary logic. However, it may be
asked if these devices can be justly appreciated if they are considered
from this angle.

Speaking for myself, I may observe that originally I did not introduce
semantic tableaux as a substitute for an axiomatic treatment. This can
be seen from a somewhat earlier publication [14] where, among others, a

Subformula Theorem was proved in connection with an axiomatic treatment instead of a Gentzen-type formalization. The introduction of semantic tableaux was, however, a result of the investigations contained in that publication.

Perhaps the significance of semantic tableaux can be explained as follows. There is always a certain connection between the proof-theory of an interpreted formal system and the corresponding model-theory; this connection finds its expression in a completeness theorem.

On the other hand, there is always a tendency to avoid the application of non-elementary methods in proof-theory, whereas in model-theory we naturally tend to apply non-elementary methods as well. Hence there arises, so to speak, a need for an elementary substitute for model-theory which lends itself to a treatment by means of more elementary and, if possible, by means of finitary methods. It is in order to provide for this need that Herbrand's 'champs infinis', Hintikka's 'model sets', and my own semantic tableaux are introduced.

In this elementary substitute for model-theory the completeness theorem is replaced by Herbrand's Theorem. Conversely, if we pass on from this elementary substitute to the full theory of models, then such results as Gentzen's Subformula Theorem and Craig's Lemma present themselves as strengthened versions of the completeness theorem. This remark may also explain the title of the present address.

REFERENCES

[1] Beth, E. W. Semantic entailment and formal derivability. *Meded. Kon. Ned. Akad. Wetenschappen, Afd. Letterkunde*, N.R., 18, no. 13 (1955). Semantic construction of intuitionistic logic, *ibid.* 19, no. 11 (1956). *The Foundation of Mathematics*, N.-H. Publishing Co., Amsterdam, 1959.

[2] Kreisel, G. On the interpretation of non-finitist proofs. Parts I–II. *J. Symbol. Logic*, 16, 241–267 (1951); 17, 43–58 (1952).

[3] Craig, W. Linear reasoning. A new form of the Herbrand–Gentzen Theorem. *J. Symbol. Logic.* 22, 250–268 (1957). Three uses of the Herbrand–Gentzen theorem in relating model theory and proof theory. *Ibid.* 22, 269–285 (1957). Review of two papers by E. W. Beth and of three papers by K. J. J. Hintikka. *Ibid.* 22, 360–363 (1957).

[4] Guillaume, M. Rapports entre calculs propositionnels modaux et topologie impliqués par certaines extensions de la méthode des tableaux sémantiques. Système de Feys–von Wright. *C.R. Acad. Sci., Paris*, 246, 1140–1142 (1958). Rapports entre calculs propositionnels modaux et topologie impliqués par certaines extensions de la méthode des tableaux sémantiques. Système S4 de Lewis. *Ibid.* 246, 2207–2210 (1958).

[5] Hintikka, K. J. J. A new approach to sentential logic. *Societas Scientiarum Fennica, Commentationes physico-mathematicae*, 17, no. 2, 1–14 (1953). Notes on quantification theory. *Ibid.* 17, no. 12, 1–13 (1955). Two papers on symbolic logic. *Acta Phil. Fenn.* fasc. VIII (1955).

[6] Kanger, S. Provability in logic. *Stockholm Studies in Philosophy*, fasc. I (1957).

[7] Kripke, S. Unpublished paper on the completeness of modal logic.

[8] Quine, W. V. A proof procedure for quantification theory. *J. Symbol. Logic*, 20, 141–149 (1955).

[9] Schütte, K. Ein System des verknüpfenden Schliessens. *Arch. Math. Logik Grundlagenforschung*, 2, 55–67 (1956).

[10] Basson, A. H. and O'Connor, D. J. *Introduction to Symbolic Logic*, 2nd ed. University Tutorial Press Ltd, London, 1957.

[11] Dreben, B. Relation of *m*-valued quantificational logic to 2-valued quantificational logic. *Summaries of talks presented at the Summer Institute of Symbolic Logic in 1957 at Cornell University, sponsored by the American Math. Society under a grant from the National Science Foundation*, 2, 303–304 (1957).

[12] Kreisel, G. Unpublished review of E. W. Beth, *Semantic Construction of Intuitionistic Logic*.

[13] Kleene, S. C. Review of E. W. Beth, *Semantic Construction of Intuitionistic Logic*. *J. Symbol. Logic*, 22, 363–365 (1957).

[14] Beth, E. W. *L'existence en mathématiques*. Gauthier-Villars, Paris, and Nauwelaerts, Louvain, 1956.

ORDINAL LOGICS AND THE CHARACTERIZATION OF INFORMAL CONCEPTS OF PROOF

By G. KREISEL

1. Introduction

By Gödel's (first) incompleteness theorem the informal notion of *arithmetic truth* cannot be formalized in the originally intended sense of 'formalization': there is no recursive enumeration of the true formulae in the notation of classical arithmetic. All one needs here of the concept of truth is that each sentence or its negation is true, and that each theorem is true. On the other hand, mere incompleteness does not generally preclude the formalization of other informal notions, e.g. certain informal methods of proof. For, in general, it is not to be expected that every formula is either provable or refutable by such methods, even if the notation is restricted. This applies to the main subject of the present lecture, namely *finitist proof* in arithmetic as described by Hilbert–Bernays, vol. 1, and also to its subsidiary subject, namely *predicative proof*.

At first sight, another formulation of the incompleteness theorem, also due to Gödel[2], seems to prohibit such a formalization: if $P(\ulcorner A \urcorner)$ is a provability predicate (enumeration of theorems) then, for certain A, $P(\ulcorner A \urcorner) \to A$ is not provable in the system. Yet another formulation is this: for certain $A(n)$, $P(\ulcorner A(0^{(n)}) \urcorner)$ is provable in the system with free variable n, but not $A(n)$. Now consider *finitist proof*: if $P(\ulcorner A \urcorner)$ has been recognized by finitist means to be the provability predicate of a (partial) formalization, say Σ_μ, of finitist mathematics, and $P(\ulcorner A(0^{(n)}) \urcorner)$ has been established by finitist means then, on the intended meaning of free variables, $A(n)$ is finitistically established. In other words, Σ_μ is incomplete and can be extended to Σ_ν in which $A(n)$ is provable. Similar considerations apply to several other informal concepts of proof.

Thus the conclusion from the reformulation of the first incompleteness theorem is this: if the notion of finitist proof is capable of formalization at all, its proof predicate must not be recognizable as such by finitist means.

Remark. The difference between a mere extensional enumeration of the theorems and a provability predicate which can be recognized as such, is familiar from the need for the derivability conditions in the second incompleteness theorem. Thus, if $P(m, n)$ is a proof predicate for a consistent system (S), $\bar{n}$ denotes the negation of n, and $Con\,S$ denotes $(m) \dashv P(m, \ulcorner 0 = 1 \urcorner)$ then $P_1(m, n)$, i.e. $P(m, n)\,\&\,(p)\,[p < m \dashv \dashv P(p, \bar{n})]$ is also a proof predicate, but $Con_1\,S$ is provable in (S) itself.

To give a precise treatment of this idea of recognizing a proof predicate as such we shall consider formal systems whose constants are not only numerical terms and function symbols, but also *proof predicates*. This is independently justified by the accepted sense of *finitist* according to which finitist proofs can themselves be the subject-matter of finitist reasoning; also it is in accordance with Heyting's and Gödel's view of the place of *proof* as an object of a theory of constructivity. Just as function symbols may be introduced only after they have been shown to represent a computation procedure, so proof predicates may be introduced only after they have been shown to represent an extension procedure as above.

Roughly speaking, we propose to characterize finitist proofs by a precisely defined *class* of formal systems, namely the least class of systems Σ_μ containing a certain basic finitist apparatus and closed under the principle: if a proof predicate P_ν is recognized as such in a system Σ_μ of the class then the corresponding system Σ_ν also belongs to the class.

Our main result is that, in a precise sense, the theorems of this class are co-extensive with those of classical number theory when the latter is suitably interpreted. (Below this is shown only for formulae $(Ex)\,A(n, x)$ with primitive recursive A.) Since each of our extensions is finitist this means at least that finitist results include essentially those of classical number theory.

The important problem of establishing the converse is not discussed here. But we note at least the following point: though each extension is finitist, the general extension principle cannot be regarded as finitistically evident since it is framed in terms of the concept of finitist proof which has no place in finitist mathematics; hence at least the obvious formalization of our class is not finitist.

An important tool in this work is the use of ordinal notations for specifying formal systems, first discussed by Turing[6].

2. Ordinal logics

The following information is needed only for a comparison between the present approach and Turing's. Turing uses a narrow and a wide definition of logics based on ordinals.

The narrow one consists of extending some system, e.g. classical number theory Z or *Principia Mathematica* with a recursively enumerable provability predicate $P(\ulcorner A \urcorner)$ ($\ulcorner A \urcorner$ is provable) by all formulae $P(\ulcorner A \urcorner) \to A$, and iterating the procedure. To ensure recursive enumerability of the theorems of each extension, recursive ordinals are used to describe the iteration. In modern language Turing's construction can be formulated as follows: for each $n \in O$, O denoting the class of recursive ordinals, we set up recursion equations relating P_n with P_m, $m <_O n$, of a kind which have a recursively enumerable solution by results of Kleene[3]. In fact, the only consequence of $n \in O$ needed here is that these recursion equations have a unique solution, and so the full degree of undecidability of $n \in O$ is not used.

Feferman[1] has sharpened these results by giving an unambiguous method for constructing such proof predicates, and has shown that, if one starts the iteration with Z, *the class of theorems is just that obtained by adding to Z all true formulae $(x) A(x)$ with A primitive recursive*. In consequence: (i) the provability predicate for the whole iteration is arithmetically definable while the proof predicate is not, since, at least with the obvious coding of proofs, it would allow us to decide $n \in O$; (ii) exactly the same functions are *provably* recursive in the ordinal logics as in Z since, if $\vdash_Z [(x) A(x) \to (x)(Ey) B(x, y)]$ and $(x) A(x)$ is true,

$$\mu_y B(x, y) = \mu_y [B(x, y) \vee \neg A(y)]$$

and the latter is provably recursive in Z.

The wide definition simply associates a formal system with every recursive ordinal, e.g. for each $n \in O$, we add to Z the 'principle of transfinite induction'

$$\frac{\{y <_O n \ \& \ (x) [x <_O y \to A(x)]\} \to A(y)}{y <_O n \to A(y)}.$$

As shown by Wang, Shoenfield and myself[5], every true arithmetic formula is provable in one of these systems even if we require $|n| < \omega^\omega$. Conversely, for each $\alpha < \omega^\omega$, O_α is arithmetic. This is obtained by showing $\deg H_{n_0} < \deg O_{\omega^\omega}$ and then analysing the argument proof-theoretically.

The following differences between Turing's aims and consequently his methods on the one hand and ours on the other seem worth noting.

(1) Turing aimed at completeness, we do not. The results above show conclusively that on the narrow definition we do not get completeness even for two-quantifier formulae with primitive recursive scope and that on the wider definition we get it too easily. However, with respect

to the latter the following interesting problem remains open: to assign non-constructively a unique notation for each recursive ordinal and reopen the problem of completeness.

(2) For Turing the narrow definition represented simply one means of extension: perhaps the most that can be said for it is that at each stage it is stronger than merely adding consistency: e.g.

$$P(\ulcorner \neg \, Con\, Z \urcorner) \to \neg \, Con\, Z$$

cannot be proved in Z from $Con\, Z$ since otherwise $Z \cup \{Con\, Z\}$ could be proved consistent in itself. We have to choose the extension principle in relation to the informal proof predicate under investigation. As it happens it turns out that the extension principle is intimately related to a kind of modal interpretation of Heyting's arithmetic discussed below.

(3) By Feferman's results the choice of starting system is not important for Turing's aim of achieving completeness since no system will do it. We have to choose one which admits of an interpretation as a partial system of the informal concept under discussion.

(4) We cannot use oracles to supply ordinals or, more specifically, to ensure the unique solubility of the recursion equations for the proof predicate, but this has to be proved in one of the earlier systems.

3. Finitist proofs (general description)

The description of our class of systems, and particularly the sketches of proofs, will have to be brief. It is hoped that the full details will be published before too long.

Our variables range over the natural numbers.

We have three kinds of constants: (i) numerical terms $0, \ldots,$ (ii) function symbols with one or more arguments including the successor function, relation symbols $<$ and $=$, and (iii) proof predicates $P_\mu(m, n)$.

Our formulae are prime formulae built up of the constants and variables above in the usual way, (quantifier-free) truth functional combinations of them, existential quantification, where $(Ex)\, A(n, x)$ below usually denotes a string of existential quantifiers

$$(Ex_1) \ldots (Ex_p)\, A(n, x_1, \ldots, x_p),$$

A quantifier-free, and single truth functional combinations of the latter. This process is not iterated in accordance with the finitist requirement that no premises relating to an infinite totality may be used.

The rules of proof are first those of the classical propositional calculus, the transposition rules for the existential quantifier,

$$\frac{(Ex)\,A(x)\vee(Ex)\,B(x)}{(Ex)\,[A(x)\vee B(x)]} \quad \frac{(Ex)\,A(x)\,\&\,(Ey)\,B(y)}{(Ex)\,(Ey)\,[A(x)\,\&\,B(y)]} \quad \frac{\neg(Ex)\,A(x)}{\neg A(n)}$$

$$\frac{(Ex)\,A(x)\rightarrow(Ey)\,B(y)}{(Ey)\,[A(n)\rightarrow B(y)]}$$

and conversely, with the obvious condition on n,

$$\frac{(Ex)\,A(n,x)}{A[n,\tau(n)],\ m<\tau(n)\rightarrow\neg A(n,m)},$$

if $A(n,x)$ quantifier-free, and similarly for strings of x,

$$\frac{A(\alpha)\vee B}{(Ex)\,A(x)\vee B}, \quad \frac{A(n)}{A(\alpha)}.$$

Also the schema of identity, and induction

$$\frac{A(0),\ A(n)\rightarrow A(n')}{A(n)},$$

and the axioms for the successor function. This is the basic finitist apparatus. The existential quantifier is constructive and because of the restriction on the formulae the truth functional interpretation of the logical connectives is not problematical.

Remark. It seems likely that this formulation could be considerably simplified.

In addition to induction, which is unproblematical, we use primitive recursive definitions

$$\phi(0)=\alpha,\quad \phi(n')=\psi[n,\phi(n)], \tag{1}$$

when α and ψ are already introduced (and ϕ not).

(1) can be interpreted in two ways, either as the *existence of a unique function* ϕ satisfying (1) for all n, or as a set of equations from which, for each number $0^{(n)}$, $\phi(0^{(n)})=0^{(m)}$ can be obtained for a unique m from (1) and the computation rules for α and ψ by a *finite number of substitutions*. The difference is clear: e.g.

$$f(n)=2f(n+1)$$

uniquely defines the function $f(n)\equiv 0$ on the first interpretation since $f(n+k)=2^{-k}f(n)$ and f is integer valued, but not on the second. Only the second interpretation has finitist sense. Bernays has established by finitist methods the permissibility of (1).

Remark. It would be more elegant to start off with a pure equation calculus, and an elementary semiotic, and give a formal proof for each primitive recursive definition that it uniquely defines values for $\phi(0^{(n)})$ in a pure equation calculus.

To deal with so-called definitions by transfinite induction, which are used below, we note a

Lemma. Let τ_1, τ_2 be two functions and define a tree as follows: the descendants of a node p are those values of $\tau_1(p)$, $\tau_2(p)$ which are different from 0. Then, if $\lambda(n)$ is a bound for the length of the tree with vertex n, then there is a function f, primitive recursive in λ, τ_1, τ_2 and g, which satisfies

$$f(0) = \alpha, \quad f(n) = g\{n, f[\tau_1(n)], \quad f[\tau_2(n)]\}.$$

Define $F(n, m)$ by induction w.r.t. m:

$$F(0, m) = \alpha.$$

If the tree with vertex n has length $> m$ then $F(n, m) = 0$. If the tree with vertex n has length $\leqslant m$ then

$$F(n, m) = g\{n, F[\tau_1(n), m-1], F[\tau_2(n), m-1]\}.$$

$F[n, \lambda(n)]$ is our function.

Finally, we come to the conditions to be satisfied by a proof predicate. In the definitive version it is essential to give a careful numbering of expressions, particularly of function symbols.

$P_0(m, n)$ is the usual proof predicate for primitive recursive arithmetic, as extended above.

We use some notation for ordinals, e.g. primitive recursive orderings of primitive recursive subsets of the natural numbers.

Remark. It seems plausible that a wide range of alternative notations gives essentially the same results.

As in [1] we regard each system Σ_μ ($\mu = 2^l 3^k$) as made up of the basic finitist apparatus together with a sequence $\lambda n \pi_l(k, n)$ of axioms (the extensions), where l is a number of a primitive recursive relation which has been proved in Σ_μ to be an ordering whose first element is 0, and k is in the field of l.

If a is the successor of 0, $\pi_l(a, n)$ is an enumeration of formulae $\ulcorner A(p) \urcorner$ such that $(Ey) P_0(y, \ulcorner A(0^{(p)}) \urcorner)$ is provable in the basic finitist apparatus, and $P_0(m, n)$ means that m is a proof of n in this system; $P_\mu(m, n)$ means that m is a proof of n in this system from the appropriate set of axioms $\hat{n} \pi_r(s, n)$.

To establish an extension $< l, k >$ in Σ_μ we require that there be a term $\pi_l(k, n)$ in Σ_μ for which the following two statements can be proved in Σ_μ:

(i) For each k and n, a formula $\ulcorner A(p) \urcorner$ has the number $\pi_l(k, n)$ if, and only if, for some $k' < {}_l k$, either $\pi_l(k, n) = \pi_l(k', n')$ for some n', or it follows (in the basic finitist apparatus) from $\hat{n}\pi_l(k', n)$ that $\ulcorner A(0^{(p)}) \urcorner$ follows from $\hat{n}\pi_l(k', n)$ with free variable p, i.e. it is proved in $\Sigma_{<l,k'>}$ that

$$(Ey) P_{<l,k'>} \ (y, \ulcorner A(0^{(p)}) \urcorner).$$

Note that such a proof in Σ_μ provides a term $\tau(k, n)$ specifying k' and the axioms of $\hat{n}\pi_l(k', n)$ actually needed.

(ii) There is a proof in Σ_μ that each tree with vertex $< k, n >$ is finite if the descendants of a node N are the axioms specified by $\tau(k, n)$.

Our proposal is to identify finitist proofs in arithmetic with *the least class of systems Σ_μ containing primitive recursive arithmetic with a constructive existential quantifier, and if $P_l(k, m, n)$ is proved to be a proof predicate in Σ_μ, then $\Sigma_{<l*k>}$ also belongs to the class.* Call this class $\mathcal{H}$.

It is occasionally natural to include variables f for free choice sequences in finitist mathematics, and, if $(Ex) A(f, x)$ has been proved, to introduce a functional $\tau(f)$ with $A[f, \tau(f)]$. It would be desirable to develop our theory for this extended notation, but we have not done so here in order to keep the basic treatment of finitist proof strictly number theoretical. The use of variables for free choice sequences would permit a more elegant formulation of the conditions to be satisfied by a proof predicate, namely that $<_l$ be a well-ordering. Here it is necessary to distinguish between weak well-ordering when we have proved that every descending sequence is finite, i.e. $(Ex)[f(x+1) \nless f(x)]$ and strong well-ordering (needed above) where we have to prove that every descending binary tree defined by a free choice sequence is finite when $f(1)$ is the value of the vertex; $f(2), f(3)$ at the next level; $f(4), \ldots, f(7)$ at the next, etc. It is clear that in a particular finitist system an ordering may be provable to be a weak well-ordering, but not a strong one, since otherwise one would get up to an ϵ-number in each system.

4. Finitist proofs (results)

Our main results are:

(i) Every function $\tau(n)$ of the class $\mathcal{H}$ is provably recursive in classical arithmetic Z.

(ii) If $A(n, m)$ is a quantifier-free formula whose non-logical constants are provably recursive functions of Z and $(Ey) A(n, y)$ is provable in Z then it is also provable in some system of $\mathcal{H}$.

Thus essentially exactly the same formulae in the notation of $\mathscr{H}$ are provable in $\mathscr{H}$ and in classical arithmetic.

(i) is seen easily by constructing a truth definition in Z for each system Σ_μ of $\mathscr{H}$, and observing that, if P_ν can be proved to be a proof predicate in Σ_μ, a truth definition can be defined for Σ_ν in Z too.

Two steps are used to establish (ii).

First, we use the result [4] that exactly the same functions are provably recursive in Z and in Heyting's arithmetic.

Secondly, we show that Heyting's arithmetic may be regarded as the metamathematics of $\mathscr{H}$ in the following precise sense: Gödel's interpretation of Heyting's propositional calculus in modal logic [2] can be extended to Heyting's arithmetic with the additional restriction that the modal operator B ('is provable') is replaced by: is provable in $\mathscr{H}$.

In detail: we leave prime formulae unchanged; if A^*, B^* are the translations of A and B, $(A \& B)^*$ is $A^* \& B^*$, $(A \overrightarrow{\mathsf{v}} B)^*$ is replaced by $(E\mu)$ $[P_\mu$ is a proof predicate and $P_\mu(\ulcorner A_1^{*}\urcorner)] \overrightarrow{\mathsf{v}} (E\mu)$ $[P_\mu$ is a proof predicate and $P_\mu(\ulcorner B_1^*\urcorner)]$, negation is not needed, $[(x)A(x)]^*$ is $(E\mu)$ $[P_\mu$ is a proof predicate and $P_\mu(\ulcorner A_1^*(x)\urcorner)]$, $[(Ex)A(x)]^*$ is $(Ex)A^*(x)$, where A_1^* is obtained from A^* by replacing each *free* variable n in A^* by $0^{(n)}$. Note that however complicated the logical structure of A may be, e.g. however many iterated implications it may cóntain, A^* will be an assertion of the form: a certain concretely specified formula is provable in some Σ_μ. Also note that an assertion $P_\nu\{(E\mu)$ $[P_\mu$ is a proof predicate and $P_\mu(\ulcorner A^*\urcorner)]\}$ means that P_μ is *proved* to be a proof predicate in Σ_ν.

As in Gödel's original work [2] the axioms of the propositional calculus go into true assertions on this interpretation, the rules for quantifiers are straightforward and so are the equality axioms and those for constant functions. The induction axiom requires essentially that the union of systems $\Sigma_{\nu(n)}$ is again a system, namely, suppose

$$P_\tau(\ulcorner A^*(0)\urcorner), \quad P_\tau\{\ulcorner (E\mu) [P_\mu \text{ is a proof predicate and } P_\mu(\ulcorner A^*(0^{(n)})\urcorner)] \rightarrow$$

$$(E\mu')[P_{\mu'} \text{ is a proof predicate and } P_{\mu'}(\ulcorner A^*(0^{(n+1)})\urcorner)]\urcorner\}$$

then μ' may be replaced by $\mu'(n+1)$, where $\mu'(n)$ is a term of Σ, $\mu'(0) = \tau$, and there is a proof that each $\mu'(n)$ is a number of a proof predicate permissible in Σ_τ. What we need now is the union of $\Sigma_{\mu'(n)}$, say Σ_{μ^*}, when we have

$$P_\tau(\ulcorner P_{\mu^*} \text{ is a proof predicate and } P_{\mu^*}(\ulcorner A^*(n)\urcorner)\urcorner).$$

Recalling the condition on proof predicates this is essentially equivalent to saying that the union of well-orderings is well-ordered, which can certainly be established in each of our systems.

The conditions on Gödel's Bp are

$$Bp \to BBp, \quad [Bp \,\&\, B(p \to q)] \to Bq \quad \text{and} \quad Bp \to p.$$

The first two conditions are clearly true for our interpretation, and so is the third by our closure condition.

The translation of $(Ey)A(n, y)$ asserts the existence of a proof of $(Ey)A(n, y)$ in $\mathscr{H}$.

Note incidentally that the application of the classical propositional calculus is much more reasonable in our translation than in general modal logic. For it is not at all clear that it always makes sense to say of a proposition that it is provable without further restriction, while our provability statements are purely existential assertions about decidable relations, just the kind of statements in our basic finitist apparatus.

5. Predicativity

Wang[7] gives a reasonable indication for treating predicative *proof*. The following modifications are needed: (i) To obtain the ordinals needed for the extension it seems best to add free function variables to his systems and extend the notion of term; a number of an ordering of a system is said to be a notation for a (provable) well-ordering $<$, if $(Ex)[f(x+1) \not< f(x)]$ can be proved in the system. (ii) The extension principle is now: if $<$ is proved to be a well-ordering in a system Σ_μ, then the system with types indexed by $<$ is said to be proved in Σ_μ to be permissible (as a predicative proof predicate). Here, too, though each extension is predicative provided $<$ has been recognized by predicative means to be a well-ordering, the general extension principle is not since the concept of predicative proof has no place in predicative mathematics in the present sense.

It should be noted that the new extension principle includes the previous one since now a truth definition can be introduced, and by means of a truth definition: from $P_\mu(\ulcorner A(0^{(n)})\urcorner)$ follows $A(n)$. For, by induction

$$(n)\,P_\mu(\ulcorner A(0^{(n)})\urcorner) \to (n)\,T_\mu(\ulcorner A(0^{(n)})\urcorner)$$

and

$$(n)\,T_\mu(\ulcorner A(0^{(n)})\urcorner) \leftrightarrow T_\mu(\ulcorner (n)\,A(n)\urcorner).$$

I have no information about the least ordinal not obtained by these extensions starting from Z.

The notion of predicative *definability* does not seem to come into the

298 · · · · · · · · · · · · · · · · G. KREISEL

present scheme at all. To fix ideas we consider number theory and a
theory with two types of variables, for natural numbers and functions
(of the natural numbers into the natural numbers). We call number
theory predicative because it has a *unique* minimal model, i.e. a model
no subset of which is also a model. We call predicative those theories of
second order which have a unique minimal model in which the individuals are the natural numbers.

Evidently hyperarithmetic set theories are predicative in this sense.
Non-trivial group theory (with at least two elements) clearly is not
predicative in the present sense because there exist infinitely many non-
isomorphic minimal models. It seems possible that there are set theories
whose unique minimal model properly includes all hyperarithmetic sets.
Naturally, this definition is itself quite impredicative.

6. General remarks

The present study is at the same time a theory of extensions of formal
systems and of a single concept, namely the totality of such extensions.

In its former role it seems to be complementary to the current tendencies of invoking oracular identities or even statistical principles
proposed for machines which are to 'accept' a rule of inference if it
has been successful a certain number of times. Instead we have here
mathematical principles governing extensions, e.g. for getting from a
fragment of finitist mathematics to a larger one. While this principle is
devoid of finitist sense it is, for example, constructive. At least as far
as the actual work involved in extending the whole body of mathematics
is concerned, the extensions of finitist mathematics as seen by a finitist
seem to be a closer model than the proposals for learning machines.
Naturally, if they are a good model it is also understandable why there
is trouble about a mathematical theory of the principles governing
extensions of the whole body of mathematics.

In its latter role the present theory is closer to old-fashioned number
theory and analysis than to modern algebra. For we do not consider all
models of a formal system, but minimal models (satisfying certain
closure conditions).

REFERENCES

[1] Feferman, S. Ordinal logics re-examined, and on the strength of ordinal
logics. *J. Symbol. Logic (Abstr.)*, 23 (1958).
[2] Gödel, K. Eine Interpretation des intuitionistischen Aussagenkalküls.
Ergebnisse eines mathematischen Kolloquiums, 4, 39–40 (1931–32).

[3] Kleene, S. C. On the form of predicates in the theory of constructive ordinals. *Amer. J. Math.* 66, 41–58 (1944).

[4] Kreisel, G. Mathematical significance of consistency proofs. *J. Symbol. Logic*, 23, 155–182 (1958).

[5] Kreisel, G., Schoenfield, J. R. and Hao Wang, *H, O* and *W* (to be published).

[6] Turing, A. M. Systems of logic based on ordinals. *Proc. Lond. Math. Soc.* (2), 45, 161–228 (1939).

[7] Hao Wang. The formalization of mathematics. *J. Symbol. Logic*, 19, 241–266 (1954).

А. А. МАРКОВ†

НЕРАЗРЕШИМОСТЬ ПРОБЛЕМЫ ГОМЕОМОРФИИ

1. Будем называть общей проблемой гомеоморфии проблему разыскания алгорифма, распознающего для любых двух данных полиэдров, гомеоморфны ли они. Полиэдры при этом задаются комбинаторно их триангуляциями, что дает возможность понимать здесь термин 'алгорифм' в его точном смысле, то есть, например, как 'нормализуемый алгорифм'[3].

Наряду с общей проблемой гомеоморфии естественно возникают различные частные проблемы гомеоморфии, относящиеся к полиэдрам того или иного класса. Можно, например, фиксируя натуральное число n, ставить проблему гомеоморфии для полиэдров размерности не выше n. Можно также ставить проблему гомеоморфии для n-мерных многообразий, если условиться в определенном понимании термина 'многообразие'.

Другим естественным ограничением, налагаемым на сравниваемые полиэдры, является фиксация одного из них. При этом возникает проблема гомеоморфии данному полиэдру A, состоящая в разыскании алгорифма, распознающего для любого полиэдра, гомеоморфен ли он полиэдру A.

Некоторые из этих проблем давно решены, например, проблема гомеоморфии 2-мерных многообразий или проблема гомеоморфии данному 2-мерному многообразию. Однако имеет место следующая

Теорема 1. Для всякого натурального числа n, большего трех, можно указать такое n-мерное многообразие M^n, что проблема гомеоморфии многообразий многообразию M^n явится неразрешимой.

Термин 'многообразие' мы понимаем здесь в смысле Пуанкаре[4] и Веблена[5].

Следствие 1. Проблема гомеоморфии n-мерных многообразий неразрешима при $n > 3$.

Следствие 2. Проблема гомеоморфии полиэдров размерности не выше n неразрешима при $n > 3$.

Следствие 3. Общая проблема гомеоморфии неразрешима.

† Read by Professor M. H. A. Newman.

2. Наметим доказательство теоремы 1.

Пусть K^4 — 4-мерный шар в 4-мерном сферическом пространстве S^4; S^3 — его граница; K^3 — 3-мерный шар; I — отрезок числовой прямой $[-1, 1]$;

$$Z = K^3 \times I,$$

где '$\times$' — знак топологического умножения; r — натуральное число. Построим систему r взаимно-однозначных дифференцируемых отображений $\phi_1, \ldots, \phi_r$ пространства Z в S^4, обладающую следующими свойствами:

$$\phi_i Z \cap \phi_j Z = \emptyset \quad (i, j = 1, \ldots, r; \ i \neq j),$$

$$\left. \begin{array}{l} K^4 \cap \phi_i Z \subset S^3 \\ S^3 \cap \phi_i Z = \phi_i(K^3 \times \{-1, 1\}) \end{array} \right\} \quad (i = 1, \ldots, r),$$

где $\emptyset$ означает пустое множество.

Построим полиэдр

$$L_r = K^4 \cup \bigcup_{i=1}^{r} \phi_i Z,$$

иначе говоря, 'приделаем' к 4-мерному шару K^4 r 'ручек'

$$\phi_i Z (i = 1, \ldots, r).$$

В пространстве Z естественным образом определяются прямолинейные отрезки. Условимся обозначать через $[x, y]$ прямолинейный отрезок в Z с концами x и y. То же обозначение будем применять для прямолинейного отрезка в шаре K^4 с концами x и y.

Будем рассматривать $2r$-буквенный алфавит.

$$\Gamma_r = \{\alpha_1^1, \ldots, \alpha_r^1, \alpha_1^{-1}, \ldots, \alpha_r^{-1}\}.$$

Пусть P — слово в алфавите Γ_r. Условимся называть *изображением* слова P всякую простую замкнутую кривую W, получаемую следующим образом.

Если P пусто, то в качестве W может быть взята любая окружность, содержащаяся внутри K^4.

Пусть P непусто и пусть

$$P = \alpha_{i_1}^{\epsilon_1} \ldots \alpha_{i_s}^{\epsilon_s}, \tag{1}$$

где $i_i, \ldots, i_s$ — числа из ряда $1, \ldots, r$ и где $\epsilon_j = \pm 1 \ (j = 1, \ldots, s)$. Возьмем внутри шара K^3 точки $x_1, \ldots, x_s, y_1, \ldots, y_s$ так, чтобы соблюдались условия

$$[(x_j, -\epsilon_j), (y_j, \epsilon_j)] \cap [(x_h, -\epsilon_h), (y_h, \epsilon_h)] = \varnothing$$

$$(j, h = 1, \ldots, s; \; j \neq h; \; i_j = i_h),$$

$$[\phi_{i_j}(y_j, \epsilon_j), \phi_{i_{j+1}}(x_{j+1}, -\epsilon_{j+1})] \cap [\phi_{i_h}(y_h, \epsilon_h), \phi_{i_{h+1}}(x_{h+1}, -\epsilon_{h+1})] = \varnothing$$

$$(j, h = 1, \ldots, s; \; j \neq h),$$

где x_{s+1} означает x_1 и y_{s+1} означает y_1. Это всегда возможно. Положим

$$W = \bigcup_{j=1}^{s} (A_j \cup B_j),$$

где

$$\left.\begin{array}{l} A_j = \phi_{i_j}[(x_j, -\epsilon_j), (y_j, \epsilon_j)] \\ B_j = [\phi_{i_j}(y_j, \epsilon_j), \phi_{i_{j+1}}(x_{j+1}, -\epsilon_{j+1})] \end{array}\right\} \quad (j = 1, \ldots, s).$$

Ясно, что всякое изображение всякого слова в алфавите Γ_r есть кусочно-гладкая простая замкнутая кривая, содержащаяся внутри L_r.

Пусть изображение W непустого слова (1) построено как только что указано. Пусть c означает центр шара K^3. Может быть построено топологическое отображение ψ полиэдра $K^3 \times W$ во внутренность L_r, обладающее следующими свойствами.

T 1. $\psi(c, x) = x \; (x \in W)$.

T 2. ψ дифференцируемо на всяком полиэдре $K^3 \times A_j$.

T 3. ψ дифференцируемо на всяком полиэдре $K^3 \times B_j$.

T 4. $\psi(K^3 \times A_j) \subset \phi_{i_j} Z$.

T 5. $\psi(K^3 \times B_j) \subset K^4$.

Если ψ обладает этими свойствами, то мы будем говорить о внутренности множества $\psi(K^3 + W)$, что она есть *туннель* слова P.

Для пустого слова туннели определяются аналогично с той разницей, что тогда ψ должно быть дифференцируемым отображением полиэдра $K^3 \times W$ во внутренность шара K^4, удовлетворяющим условию T 1.

Нетрудно видеть, что во всякой окрестности образа слова в алфавите Γ_r содержится туннель этого слова.

Пусть теперь $P_1 * \ldots * P_m$ — система слов в алфавите Γ_r. Построим для всякого $i(1 \leqslant i \leqslant m)$ туннель T_i слова P_i, так чтобы соблюдались условия:

$$T_i \cap T_j = \varnothing \quad (i, j = 1, \ldots, m; \; i \neq j),$$

где черта над буквой означает операцию замыкания в S^4.

Построим затем полиэдры

$$J_0 = L_r \setminus \bigcup_{i=1}^{m} T_i,$$

$$J_i = S^4 \setminus T_i \quad (i = 1, \ldots, m),$$

$$H_{-1} = L_r \times \{-1\},$$

$$H_i = J_i \times \{i\} \quad (i = 0, 1, \ldots, m).$$

Наконец, из полиэдров $H_i (i = -1, 0, \ldots, m)$ построим полиэдр M путем следующих отождествлений точек:

(1) отождествляются всякие две точки $(x, 0)$ и $(x, -1)$, где x принадлежит границе L_r в S^4;

(2) отождествляются всякие две точки $(x, 0)$ и (x, i), где x принадлежит границе туннеля T_i в S^4 $(i = 1, \ldots, m)$.

Описанное построение содержит элементы произвола. Однако результат построения — полиэдр M — определяется однозначно с точностью до гомеоморфии исходной системой слов $P_1 * \ldots * P_m$ и числом r.

С другой стороны, и сам этот произвол легко может быть устранен. Результирующий полиэдр мы будем обозначать символом

$$\mathfrak{M}(P_1 * \ldots * P_m * r).$$

Он, как нетрудно видеть, всегда является 4-мерным многообразием. Описанное построение этого полиэдра является некоторым уточнением указанного Зейфертом и Трельфаллем построения 4-мерного многообразия с заданной фундаментальной группой†. Методами, изложенными в их книге[2], доказывается

Лемма 1. Какова бы ни была система слов $P_1 * \ldots * P_m$ в алфавите Γ_r, фундаментальная группа многообразия $\mathfrak{M}(P_1 * \ldots * P_m * r)$ изоморфна группе, определяемой системой соотношений

$$P_i \leftrightarrow \Lambda \quad (i = 1, \ldots, m), \tag{2}$$

между производящими элементами $\alpha_1^1, \ldots, \alpha_r^1$. Буквы $\alpha_1^{-1}, \ldots, \alpha_r^{-1}$ рассматриваются при этом как элементы, обратные элементам $\alpha_1^1, \ldots, \alpha_r^1.$‡

---

† См. [2], стр. 208.

‡ Соответствующее этой группе групповое исчисление (см. [3], стр. 341) в алфавите Γ_r может быть определено системой соотношений, получаемой из системы (2) после присоединения соотношений $\alpha_i^\epsilon \alpha_i^{-\epsilon} \leftrightarrow \Lambda (i = 1, \ldots, r; \epsilon = \pm 1)$.

Далее могут быть доказаны следующие леммы о гомеоморфии многообразий $\mathfrak{M}(P_1 * \ldots * P_m * r)$.

Лемма 2. Многообразия $\mathfrak{M}(P_1 * \ldots * P_m * r)$ *и* $\mathfrak{M}(Q_1 * \ldots * Q_m * r)$ *гомеоморфны, если система слов* $Q_1, \ldots, Q_m$ *получается от системы слов* $P_1 * \ldots * P_m$ *в результате подстановки пустого слова вместо вхождения слова* $\alpha_i^\varepsilon \alpha_i^{-\varepsilon} (i = 1, \ldots, r; \ \varepsilon = \pm 1)$.

Лемма 3. Многообразия $\mathfrak{M}(P_1 * \ldots * P_m * r)$ *и* $\mathfrak{M}(Q_1 * \ldots * Q_m * r)$ *гомеоморфны, если среди чисел* $1, \ldots, m$ *имеется число* i *такое, что* Q_i *есть результат циклической перестановки букв в слове* P_i *и что*

$$Q_j = P_j \tag{3}$$

при $1 \leqslant j \leqslant m$ *и* $j \neq i$.

Лемма 4. Многообразия $\mathfrak{M}(P_1 * \ldots * P_m * r)$ *и* $\mathfrak{M}(Q_1 * \ldots * Q_m * r)$ *гомеоморфны, если среди чисел* $1, \ldots, m$ *имеется число* i *такое, что* Q_i *есть групповое обращение слова* P_i *и что при* $1 \leqslant j \leqslant m$ *и* $j \neq i$ *имеет место равенство* (3).

Групповое обращение слова P в алфавите Γ_r мы определяем здесь как слово, получаемое из P в результате изменения порядка букв на обратный с последующей заменой всякой буквы α_j^ε буквой $\alpha_j^{-\varepsilon}$.

Лемма 5. Многообразия $\mathfrak{M}(P_1 * \ldots * P_m * r)$ *и* $\mathfrak{M}(Q_1 * \ldots * Q_m * r)$ *гомеоморфны, если среди чисел* $1, \ldots, m$ *имеются числа* i *и* h *такие, что* $i \neq h$, *что*

$$Q_i = P_i P_h$$

и что при $1 \leqslant j \leqslant m$ *и* $j \neq i$ *имеет место равенство* (3).

Лемма 6. Многообразия $\mathfrak{M}(*^k \alpha_1^1 * \ldots * \alpha_r^1 * r)$ *и* $\mathfrak{M}(*^k 0)$ *гомеоморфны, каково бы ни было натуральное число* k.

С помощью лемм 2–6 легко может быть доказана

Лемма 7. Если группа с производящими элементами $\alpha_1^1, \ldots, \alpha_r^1$, *определяемая системой соотношений*

$$R_i \leftrightarrow \Lambda \quad (i = 1, \ldots, k), \tag{4}$$

в алфавите Γ_r, *есть единичная группа, то многообразие*

$$\mathfrak{M}(R_1 * \ldots * R_k *^{r+1} r)$$

гомеоморфно многообразию $\mathfrak{M}(*^k 0)$.

С другой стороны из леммы 1 следует

Лемма 8. Если группа, о которой идет речь в лемме 7, не есть единичная группа, то многообразие $\mathfrak{M}(R_1 * \ldots * R_k *^{r+1} r)$ *не гомеоморфно многообразию* $\mathfrak{M}(*^k 0)$.

Фиксируем теперь натуральные числа r и k. Будем рассматривать группы с производящими элементами $\alpha_1^1, \dots, \alpha_r^1$, определяемые всевозможными системами k соотношений (4) в алфавите Γ_r. Будем называть такие группы (r, k)-*группами*. Из лемм 7 и 8 вытекает, что с помощью всякого алгорифма, распознающего для любого многообразия, гомеоморфно ли оно многообразию $\mathfrak{M}(*^k 0)$, может быть построен алгорифм, распознающий для всякой (r, k)-группы, является ли она единичной. Между тем, из построения, проведенного С. И. Адяном в его работе[1],† непосредственно следует, что числа r и k могут быть заданы так, что алгорифм, распознающий единичность (r, k)-группы, окажется невозможным. Возьмем такую пару чисел (r, k) и положим

$$M^4 = \mathfrak{M}(*^k 0).$$

Тогда проблема гомеоморфии многообразий 4-мерному многообразию M окажется неразрешимой.

Нетрудно, наконец, видеть, что для всякого натурального числа n, большего четырех, неразрешима проблема гомеоморфии n-мерному многообразию

$$M^n = M^4 \times S^{n-4},$$

где S^h означает h-мерную сферу.

Этим завершается доказательство теоремы 1.

3. Аналогично проблемам гомеоморфии могут быть поставлены проблемы гомотопической эквивалентности. Их формулировки получаются из формулировок проблем гомеоморфии путем замены слов 'гомеоморфны', 'гомеоморфен', 'гомеоморфно' словами 'гомотопически эквивалентны', 'гомотопически эквивалентен', 'гомотопически эквивалентно'. Но такая замена, очевидно, возможна в леммах 7 и 8. Это дает следующие результаты.

Теорема 2. Для всякого натурального числа n, большего трех, проблема гомотопической эквивалвнтности многообразий многообразию M^n неразрешима.

Следствие 1. Проблема гомотопической эквивалентности n-мерных многообразий неразрешима при $n > 3$.

Следствие 2. Проблема гомотопической эквивалентности полиэдров размерности не выше n неразрешима при $n > 3$.

Следствие 3. Общая проблема гомотопической эквивалентности неразрешима.

† См. также [6].

ЛИТЕРАТУРА

[1] Адян, С. И. *ДАН*, 103, 4, 533–535 (1955).
[2] Зейферт, Г. и Трельфалль, В. *Топология*. Москва-Лёнинград, 1938.
[3] Марков, А. *Теория алгорифмов*. Труды МИ АН СССР, 42 (1954).
[4] Poincaré, H. *R.C. Circ. mat. Palermo*, 13, 285–343 (1899).
[5] Veblen, O. *Analysis Situs*. New York, 1931.
[6] Rabin, M. O. *Ann. Math.* (2), 67, 172–194 (1958).

LIE RING METHODS IN THE THEORY OF FINITE NILPOTENT GROUPS

By GRAHAM HIGMAN

1. Introduction

There are, of course, many connections between Group Theory and the theory of Lie rings, and my title is, perhaps, designed rather to exclude those I shall not deal with, than to define those that I shall. It excludes, for instance, the classical connection between Lie algebras and continuous groups, the applications, due to Mal'cev and others, to torsion-free nilpotent groups, and the construction, by Chevalley and Tits, using Lie algebra methods, of finite analogues to the exceptional simple Lie groups.

There is one other connection, that could hardly have been excluded by title, but which I can no more than mention. This is the so-called Baker–Hausdorff formula, the theorem that, if x, y are non-commuting variables, the terms of the series

$$\log(e^x . e^y) = x + y + \tfrac{1}{2}[x, y] + \tfrac{1}{12}([[y, x], x] + [[x, y], y]) + \ldots$$

are rational combinations of the elements of the Lie ring generated by x and y under the bracket multiplication $[x, y] = xy - yx$. Under suitable restrictions (cf. Lazard[5]) this fact leads to an elegant and precise correspondence between groups and Lie rings. For the sort of thing I have in mind, however, the restrictions are altogether too severe for the method to be of use. They are, in fact, precisely the sort of thing one wants to have in the conclusion of one's theorems, rather than in the hypothesis.

The method I want to discuss is much less precise, but much more generally applicable. It rests on the observation that the formal properties of multiplication and commutation in a group are similar to those of addition and multiplication in a Lie ring. For instance, commutation is almost bilinear:

$$[xy, z] = y^{-1}[x, z] y [y, z];$$

and, corresponding to the Jacobi identity, the product

$$[[x, y], z][[y, z], x][[z, x], y],$$

if not 1, is at least expressible in terms of still more complicated commutators. These facts can be used to associate with a group G a Lie ring L in the following manner.

20-2

First, one chooses a series of normal subgroups of G

$$G = H_1 \supset H_2 \supset H_3 \supset \ldots \supset H_i \supset \ldots$$

with the property that, for all integers i, j, the commutator group $[H_i, H_j]$ is contained in H_{i+j}. For instance, the lower central series will do. Then the additive group of L is the direct sum of the (abelian) factor groups H_i/H_{i+1} of the series, elements of H_i/H_{i+1} being called homogeneous of degree i. Multiplication between homogeneous elements is defined by

$$g_i H_{i+1} \cdot g_j H_{j+1} = [g_i, g_j] H_{i+j+1}.$$

That this is indeed well-defined, and that it can be extended by bilinearity to a multiplication on the whole of L, under which L is a Lie ring, follows from the commutator identities already cited, and others like them.

Obviously, this process can give information about G itself, rather than about a factor group, only if $\bigcap_{i=1}^{\infty} H_i = 1$, which implies that G is at worst ω-nilpotent. If, however, this is so, then we can associate one or more Lie rings L with G, and, generally speaking, reasonable properties of G will translate into reasonable properties of L. Since the structure of L is richer, and in some ways more regular, than that of G, it is reasonable to hope that this translation will make problems easier to solve. I want to illustrate this by discussing two problems, both of which have their origins in the first decade of this century—the golden age of group theory, if ever there was one.

2. Frobenius groups

A Frobenius group is a permutation group on a finite set, no element of which, except the identity, has more than one fixed point. By a well-known theorem of Frobenius, in such a group the elements without fixed points, together with the identity, form a normal subgroup G. But this by no means exhausts what can be said. If H is the subgroup leaving a particular point fixed, then of course H acts as a group of automorphisms of G, and it is easy to see that H acts regularly, that is, no element of H except the identity leaves fixed any element of G except the identity.

The problem then is, what finite groups H act as regular automorphism groups, and what finite groups G admit regular automorphism groups? The answer to the first question, though a bit complicated, is classical. The investigation was begun by Burnside, and his work was completed and corrected by Zassenhaus. But less is known about the answer to the second question. It is conjectured that G is nilpotent, but all that has been proved is that, if not, there must exist a simple G whose Sylow subgroups are large and possess improbable properties.

Evidently, if G has a (finite) regular automorphism group, it has a regular automorphism of order p, for some prime p, so one may particularise the problem, and inquire, for each fixed prime p, what this implies about the structure of G. It is convenient at the same time to drop the assumption that G is finite, though to get a reasonable theorem we must have some condition on G, since a free product of p isomorphic groups certainly has a regular automorphism of order p. However, Neumann[6, 7] has shown that, under appropriate side conditions, a group with a regular automorphism of order p is abelian if $p = 2$, and nilpotent of class at most two if $p = 3$.

It is in attempting to generalize this result that Lie ring methods come in. They cannot be used to prove G nilpotent. They can be used[3] to prove that if G is nilpotent, and has a regular automorphism of order p, there is a bound $k(p)$ for its class, depending only on p.

Naturally, there are two stages in the application. First one must translate the condition on G into a condition on L, and then one must prove a theorem on Lie rings satisfying this condition. The first half works precisely as one would expect. A nilpotent group G with a regular automorphism of order p has an associated Lie ring L of the same class with an induced automorphism which is also regular and of order p. Two warnings are, however, necessary. This result does not extend to ω-nilpotent groups; and, if G is infinite, it may not be possible to take L to be the associated Lie ring formed from the lower central series itself.

The second stage in the application is to prove that there is a function $k(p)$ of p alone such that a Lie ring L with a regular automorphism of order p is nilpotent of class at most $k(p)$. This depends on a rather complicated combinatorial argument which it is not possible even to sketch here, so that I must confine myself instead to making one or two remarks about it.

First, an important step is to embed the given ring in one having as domain of operators the integer ring of the field of the pth roots of unity, and generated by elements on which the automorphism acts as a scalar multiplier. It would be difficult, if not impossible, to carry out this step within G, and it is in this sense that the richer structure of L makes it easier to handle.

Secondly, a much simpler argument proves an analogous theorem for

GRAHAM HIGMAN

associative rings, and yields the value $k(p) = p$, which is easily seen to be best possible. But in the Lie ring case, apart from the first few values $(k(2) = 1, k(3) = 2, k(5) = 6)$ all that is known is that $k(p) \geqslant \frac{1}{4}(p^2 - 1)$.

Note, finally, that the Lie ring theorem is clearcut; the messy side conditions in the group result come in in the translation.

3. Burnside's problem

The second application is, of course, to Burnside's problem, the problem, that is, of what can be said about a group G which satisfies an identical relation $x^n = 1$, and, in particular, whether, if G is finitely generated, it is necessarily finite.

Since our method requires that G is ω-nilpotent, it will apply only to the case of a prime-power exponent, $n = p^r$, say. Furthermore, it will apply only to the restricted Burnside problem, that is, to the question whether, for given k, there is a (finite) upper bound to the orders of the k-generator groups of exponent n.

As before, there are two stages in the application of the method. The first requires us to translate the group theoretical problem into Lie ring terms; the second to prove a theorem about Lie rings. It cannot be said as yet that we know how to accomplish either half of this programme, in general.

Let us look at the first stage first. If G is an ω-nilpotent group of exponent $n = p^r$, what can be said about its associated Lie ring L? We shall assume that L is formed using the lower central series of G; this has the advantage that if G can be generated by k elements, so can L; which is obviously desirable in the present context. Evidently, L is of characteristic n; that is, $nx = 0$ for all x in L.

But, furthermore, L satisfies certain identical relations. Let us explain how these can be obtained.

We begin with a word
$$w_0 = x_1 x_2 \ldots x_t$$
in the free group on generators $a_1, \ldots, a_N$, which is a product of nth powers; and for simplicity we suppose that w_0 is a positive word, so that each x_i is an a_j. We apply to w_0 a commutator collection process. In the first stage of this the letters other than a_1 are collected to the left, in the order in which they occur. In the process, evidently, we shall have to introduce commutators $[a_1, a_i]$, and then $[a_1, a_i, a_j]$, and so on. The upshot is an expression
$$w_0 = y_1 y_2 \ldots y_s z_1 z_2 \ldots z_u$$
in which each y_i is an a_j, $j \neq 1$, and each z_i is a_1, or a commutator con-

taining a_1. Evidently, $y_1 y_2 \ldots y_s$ is the result of putting $a_1 = 1$ in w_0, so that it is a product of nth powers. Hence

$$w_1 = z_1 z_2 \ldots z_u$$

is a product of nth powers. In the second stage, we similarly collect to the left, and drop, those z_i which do not involve a_2. The result is a product w_2 of commutators, all involving both a_1 and a_2, which is also a product of nth powers. Continuing thus, we ultimately obtain a product w_N of commutators, all involving each of $a_1, \ldots, a_N$, which is also a product of nth powers. Now a commutator of weight N in w_N involves a_1 just once, and hence was introduced at the first stage of the collection process. Hence it is of the form $[a_1, a_{2\sigma}, \ldots, a_{N\sigma}]$, where σ is a permutation of $2, \ldots, N$; and it is easy to see that the number k_σ of times this commutator occurs is equal to the number of times $a_1, a_{2\sigma}, \ldots, a_{N\sigma}$ occurs as a subsequence in w_0. Thus

$$\prod_\sigma [a_1, a_{2\sigma}, \ldots, a_{N\sigma}]^{k_\sigma}$$

is a product of nth powers, and commutators of weight greater than N involving each a_i. It is easy to see that

$$\sum_\sigma k_\sigma x_1 x_{2\sigma} \ldots x_{N\sigma} = 0$$

is an identical relation in L.

If, in particular, we take $w_0 = (a_1 a_2 \ldots a_n)^n$, this relation becomes

$$\sum_\sigma x_1 x_{2\sigma} \ldots x_{n\sigma} \equiv 0 \quad (p).$$

From this point, it is necessary to distinguish the case $n = p$ from the general case $n = p^r$, $r > 1$. If $n = p$, the condition we have obtained is precisely the $(p-1)$st Engel condition, $xy^{p-1} = 0$. That is, the associated Lie ring of a group of exponent p is of characteristic p and satisfies the $(p-1)$st Engel condition. As far as I know, we cannot be sure, except for the smallest values of p, that this gives the complete translation of the problem. But, what is far more important, we do now know that it is sufficient. That is, a Lie ring of characteristic p satisfying the $(p-1)$st Engel condition is locally finite. For $p = 2$ this is obvious; for $p = 3$ it was proved essentially by Levi and van der Waerden, though their argument was conducted entirely in the group. The case $p = 5$ is due independently to Kostrikin (see references in [4]) and to me [2]. The general case has been obtained very recently indeed by Kostrikin [4], whose work must be considered the first major break-through in Burnside's problem. It is not, of course, possible here to give any details, though it is perhaps worth observing that the point here is that the behaviour of

L is more regular than that of G. It is not obvious, for instance, that G is an Engel group; certainly it is not true, in general, that for x, y in G,

$$[x, \underbrace{y, \ldots, y}_{p-1}] = 1.$$

For $q = p^r$ $(r > 1)$, the position is much less satisfactory. We know, in fact, that the translation mentioned above (which is now not a genuine Engel condition, but only a linearized one) is not the whole story, nor is it sufficient to prove L locally finite (e.g. the relation obtained lies in the second derived ring $(L^2)^2$, whereas for large enough k, $xy^k = 0$; also, for $q = 4$, the relation holds in the non-soluble ring of dimension 3).

In view of Kostrikin's achievement, it is, perhaps, appropriate to conclude with a word or two about the present state of Burnside's problem. It has for some time been clear that the problem falls into four parts:

(i) the nilpotent part, the restricted problem for (a) prime and (b) prime power exponents;

(ii) the reduction to (i) of the problem as far as it concerns soluble groups (this deals, in particular, with exponents involving just two primes);

(iii) the reduction to (i) and (ii) of the general restricted problem;

(iv) the passage from the restricted problem to the general problem.

Of these parts, (ii) has been dealt with completely in Hall and Higman[1], and Kostrikin has well begun (i). It seems likely that prime-powers will fall to methods of the same general nature. There remain (iii) and (iv). Of these, (iv) is, perhaps, the most characteristic of the problem; (iii) draws attention to our need for information about the finite simple groups, but it scarcely needs Burnside's problem to do that.

REFERENCES

[1] Hall, P. and Higman, G. On the p-lengths of p-soluble groups, and reduction theorems in Burnside's problem. *Proc. Lond. Math. Soc.* (3), 6, 1–42 (1956).

[2] Higman, G. On finite groups of exponent five. *Proc. Camb. Phil. Soc.* 52, 381–390 (1956).

[3] Higman, G. Groups and rings having automorphisms without non-trivial fixed elements. *J. Lond. Math. Soc.* 32, 321–334 (1957).

[4] Кострикин, А. И. О проблеме бернсайда. *Доклады Академии Наук СССР*, N.S., 119, 1081–1084 (1958).

[5] Lazard, M. Sur les groupes nilpotents et les anneaux de Lie. *Ann. Sci. Éc. Norm. Sup.* (3), 71, 101–190 (1954).

[6] Neumann, B. H. On the commutativity of addition. *J. Lond. Math. Soc.* 15, 203–208 (1940).

[7] Neumann, B. H. Groups with automorphisms that leave only the neutral element fixed. *Arch. Math.* 7, 1–5 (1956).

Ю. В. ЛИННИК†

О ПРОБЛЕМЕ ДЕЛИТЕЛЕЙ И РОДСТВЕННЫХ ЕЙ БИНАРНЫХ АДДИТИВНЫХ ПРОБЛЕМАХ

1. Многие английские авторы изучали асимптотическое поведение сумм делителей вида:

$$\sum_{n \leqslant x} \tau_{k_1}(n)\,\tau_k(n+l), \tag{1,1}$$

$$\sum_{\nu \leqslant n} \tau_{k_1}(\nu)\,\tau_k(n-\nu) \tag{1,2}$$

(см. напр. Ингам[5], Эстерман[3], Титчмарш[14], Хооли[4]).

Были выведены асимптотические формулы для $k_1 = 2$; $k = 2, 3$ (случай $k = 3$ был изучен Хооли[4]). Хорошо известно, что функция $\tau(n) = \tau_2(n)$ тесно связана с функцией $U(n) = \sum\limits_{n=u^2+v^2} 1$; их производящие ряды Дирихле довольно сходны с аналитической точки зрения. Ввиду этого, напр., сумма (1, 2) для $k_1 = k = 2$ отвечает диофантову уравнению $u^2 + v^2 + w^2 + t^2 = n$.

Более общие уравнения, содержащие любую заданную кватернарную форму, были изучены Клостерманом[6], Тартаковским[12,13], и в недавней работе Эйхлером[2].

Эти исследования были связаны с теорией гиперкомплексных чисел и применены автором и А. В. Малышевым (см. напр. [7,8,9,10,11]) для развития асимптотической теории представлений чисел тернарными квадратичными формами.

Интересно, что эта теория оказывается полезной при изучении сумм (1, 1), (1, 2) и связанных с ними задач, касающихся формы $au^2 + buv + cv^2$. Именно, теория тернарных квадратичных форм позволяет вывести асимптотические выражения для сумм (1, 1), (1, 2) для случая $k_1 = 2$, $k \geqslant 2$ (любое).

Пользуясь аналогией между $\tau_2(n)$ и $U(n)$, упоминавшейся ранее, можно вывести асимптотические формулы для числа решений уравнения

$$n = Q(u, v) + gN(\mathfrak{a}), \tag{1,3}$$

где $Q(u, v)$ — примитивная бинарная квадратичная форма дискриминанта $d < 0$; $\mathfrak{a}$ — идеал в заданном поле, абелевом над рацио-

† Read in an English translation by Professor H. A. Heilbronn.

нальным полем; $N(\mathfrak{a})$—его норма, g—заданное положительное число взаимно простое к d. Более общее уравнение

$$n = g_1 Q(u, v) + g_2 N(\mathfrak{a})$$

можно трактовать также.

Уравнение (1, 3) можно рассматривать, как обобщение (частичное) уравнения Г. Д. Клостермана $n = au^2 + bv^2 + cw^2 + dt^2$; здесь абелево поле — $k(\sqrt{-1})$ и $c = d = g$.

2. Метод, основанный на теории тернарных квадратичных форм, позволяет вывести асимптотические формулы для сумм (1, 1) и (1, 2) только для случая $k_1 = 2$, $k \geqslant 2$ (любое число). Формулы для общего l довольно сложны; мы укажем лишь простейшую из них (с $l = 1$) (она верна также для $l = -1$).

Теорема 1. При $n \to \infty$, $k \geqslant 2$ имеем

$$\sum_{n \leqslant x} \tau_2(n)\, \tau_k(n+1) \sim k!\, C_{k-1} S_k x (\ln x)^k, \qquad (2,1)$$

где
$$S_k = \sum_{n=1}^{\infty} \frac{\mu(n)}{n^2} \prod_{p \mid n} \left(p \left(1 - \left(1 - \frac{1}{p} \right)^{k-1} \right) \right), \qquad (2,2)$$

$$\left. \begin{array}{c} C_{k-1} = \lim_{Y \to \infty} \dfrac{1}{(\ln Y)^{k-1}} \displaystyle\int \cdots \int \dfrac{dy_1 \ldots dy_{k-1}}{y_1 y_2 \ldots y_{k-1}}, \\[4mm] 1 \leqslant y_1 \leqslant y_2 \leqslant \ldots \leqslant y_{k-1} \leqslant \dfrac{Y}{y_1 y_2 \ldots y_{k-1}}. \end{array} \right\} \qquad (2,3)$$

Разумеется, константу C_{k-1} можно подсчитать явно, как элементарную функцию k.

Более того, мы можем получить асимптотическое разложение для суммы (2, 1) вида:

$$\sum_{n \leqslant x} \tau_2(n)\, \tau_k(n+1) = x P_k(\ln x) + O(x(\ln x)^{\alpha_0}), \qquad (2,4)$$

где $P_k(\ln x)$—полином от $\ln x$ степени k; $\alpha_0 > 0$—константа (в дальнейшем α_i, η_i, ξ_i $(i = 0, 1, 2, \ldots)$—малые положительные константы).

Подобные асимптотические разложения можно получить для общих сумм (1, 1) и (1, 2).

3. Пусть теперь $d < 0$—заданный фундаментальный дискриминант заданной целочисленной примитивной бинарной квадратичной формы $Q(u, v)$ $(d \equiv 0; 1 \pmod 4)$. Пусть $\mathfrak{R}$ заданное алгебраическое поле, абелево над полем рациональных чисел, с дискриминантом $d_{\mathfrak{R}}$, и $g > 0$—целое такое, что $(d, g d_{\mathfrak{R}}) = 1$.

Теорема 2. Всякое достаточно большое число $n > n_0\,(d, g, d_\Re)$ может быть представлено в форме

$$n = Q(u, v) + gN(\mathfrak{a}), \tag{3, 1}$$

где $\mathfrak{a}$ — целый идеал поля $\Re$.

Можно дать асимптотическую формулу для числа решений (3, 1). Заметим, что в случае: $\Re = k(\sqrt{-1})$, $N(\mathfrak{a}) = z^2 + t^2$, и мы получаем частный (хотя и достаточно широкий) случай теоремы Тартаковского о кватернарных формах[13]:

$$n = Q(u, v) + g(z^2 + t^2).$$

Другой частный случай теоремы 2:

Теорема 2′. Для заданного $D > 0$ и $n > n_0(D)$ имеем:

$$n = \Pi_1 + \Pi_2, \tag{3, 2}$$

где Π_1 состоит только из простых чисел $\equiv 1 \,(\mathrm{mod}\,4)$, а Π_2 — только из простых чисел $\equiv 1 \,(\mathrm{mod}\,D)$. Если считать каждый простой делитель Π_1 дважды, каждый простой делитель $\Pi_2 - \phi(D)$ раз, можем получить асимптотическую формулу для числа решений (3, 2).

Если $D = 4$, получаем известную теорему о четырех квадратах.

Укажем теперь асимптотическую формулу для (3, 1) для некоторых простых случаев $\Re$. Хорошо известно, что ввиду теоремы Кронекера-Вебера, дзета-функция $\Re$ выражается через обыкновенные L-ряды Дирихле. Мы рассмотрим следующий случай:

Пусть p_0 — простое число вида: $p_0 = qm + 1$; q задано; существуют степенные вычеты степени q; пусть $\chi_1, \chi_2, \ldots, \chi_q$ будут соответствующие неглавные характеры Дирихле. Образуем ряд

$$A(s) = \sum_{m=1}^{\infty} \frac{a_m}{m^s} = \zeta(s) \prod_{k=1}^{q-1} L(s, \chi_k); \quad \mathrm{Re}\,s > 1 \tag{3, 3}$$

и рассмотрим сумму

$$T(n) = \sum_{n - Q(u, v) = gm} a_m. \tag{3, 4}$$

Эта сумма дает число решений уравнения (3, 1) в частном случае, когда образующий ряд Дирихле для норм идеалов задается (3, 3). Случай (3, 2) тесно связан с $A(s) = \zeta(s) \prod\limits_{k=1}^{\phi(D)-1} L(s, \chi_k)$, где χ_k — неглавные характеры Дирихле для модуля D.

Асимптотическая формула для $T(n)$ довольно сложна. При заданных n и $D < n$, обозначим: $d_1 = (n, D)$; $D = D_1 d_1$; $d_1 = d_{11} d_{12}$, где d_{11} состоит из простых чисел, делящих D_1, а d_{12} взаимно просто к D_1. Обозначаем

$$\Pi(n, D) = \prod_{p|D}\left(1 - \frac{\chi_d(p)}{p}\right) \prod_{p|d_{11}}(1 + \chi_d(p) + \ldots + \chi_d^{a_p}(p))$$

$$\times \prod_{p|d_{12}}\left(1 + \chi_d(p) + \ldots + \chi_d^{a_p-1}(p) + \frac{\chi_d^{a_p}(p)}{1 - \chi_d(p)/p}\right). \quad (3,5)$$

Здесь d — дискриминант формы $Q(u, v)$, a_p — наивысшая степень p, делящая соответствующее число (d_{11} или d_{12}), $\chi_d(n) = (d/n)$ — характер Кронекера, принадлежащий d.

Теорема 3. При $g = 1$

$$T(n) = \frac{2\pi n}{\sqrt{|d|}}\,\mathfrak{S}(n) + O(n^{1-\alpha_1}), \quad (3,6)$$

где $\quad \mathfrak{S}(n) = \sum_{\delta_1, \delta_2, \ldots, \delta_{q-1}=1}^{\infty} \frac{\chi_1(\delta_1) \ldots \chi_{q-1}(\delta_{q-1})}{\delta_1 \delta_2 \ldots \delta_{q-1}} \Pi(n, \delta_1\delta_2\ldots\delta_{q-1}). \quad (3,7)$

Ряд $\mathfrak{S}(n)$ суммируемый по возрастающим $\delta_1\delta_2\ldots\delta_{q-1}$, сгруппированным для одинаковых значений произведения $\delta_1\delta_2\ldots\delta_{q-1}$, как легко видеть, условно сходится, и $\mathfrak{S}(n) > 1/\ln n$.

4. Сформулируем теперь главные леммы. Хорошо известно, что сумма (1, 1) совпадает с числом решений диофантова уравнения

$$x_1 x_2 - y_1 y_2 \ldots y_k = l \quad (4,1)$$

под условием $x_1 x_2 \leqslant l$. Это число можно постепенно свести к $k!\,N(x, -l)$, где $N(x, -l)$ — число решений (4, 1) при условии:

$$1 \leqslant x_1 x_2 \leqslant x; \quad y_1 \leqslant y_2 \leqslant \ldots \leqslant y_k.$$

Далее, $y_1 y_2 \ldots y_{k-1} \leqslant (x+l)^{(k-1)/k}$, и

$$N(x, -l) \sim \sum_{(y_i)} \sum_{\substack{n \equiv -l(\bmod y_1 \ldots y_{k-1}) \\ n \leqslant x}} \tau(n), \quad (4,2)$$

где область суммирования (y_i) есть

$$1 \leqslant y_1 \leqslant y_2 \leqslant \ldots \leqslant y_{k-1} \leqslant \frac{x+l}{y_1 y_2 \ldots y_{k-1}}.$$

Таким образом, чтобы вывести асимптотику (4, 2) достаточно вывести асимптотические формулы для количеств

$$\sum_{\substack{n \equiv -l(\bmod D) \\ n \leqslant x}} \tau(n) \quad (4,3)$$

для $D \leqslant (2x)^{(k-1)/k}$.

Заметим, однако, что мы не нуждаемся в них для всех значений D, но, грубо говоря, для 'почти всех' $D \leqslant (2x)^{(k-1)/k}$, т.к. формула (4, 2) имеет 'усредняющий характер'.

Формула (4, 3) дает число решений уравнения

$$x_1 x_2 - Dy = -l \quad (x_1 x_2 \leqslant x). \qquad (4,4)$$

Изучение этого уравнения приводит к тригонометрическим суммам Клостермана (см. статью автора[10], стр. 22–29 и мемуар С. Хоули[4]). Применяя хорошо известные оценки Андре Вейля для сумм Клостермана, мы можем получить вполне удовлетворительные асимптотические формулы для $D \leqslant x^{\frac{2}{3}-\alpha_0}$ (смысл α_i был объяснен ранее). Но для нашей проблемы мы нуждаемся в $D \leqslant (2x)^{(k-1)/k}$, однако, для 'почти всех' таких D. Это составляет содержание главной леммы 1.

Лемма 1. Пусть

$$A(x, D) = \frac{x \ln x}{D^2}\phi(D)\left(1 - \frac{1}{\ln x}\left(1 - 2c_0 + 2\sum_{p|D}\frac{\ln p}{p-1}\right)\right), \qquad (4,5)$$

где $\phi(D)$ — функция Эйлера и

$$c_0 = -\frac{\Gamma'(1)}{\Gamma(1)}.$$

Пусть $x^{\frac{1}{2}} \leqslant D_1 \leqslant x^{1-\eta_1}$; $D_2 = D_1^{1-\eta_2}$; $l \leqslant x$. Тогда

$$\sum_{\substack{(D,\, l)=1 \\ D_1 \leqslant D \leqslant D_1+D_2}}\left(\sum_{\substack{n \equiv l\,(\mathrm{mod}\,D) \\ n \leqslant x}}\tau(n) - A(x, D)^2\right) = O\left(\frac{x^{2-\eta_3}}{D_1^2}D_2\right). \qquad (4,6)$$

Здесь η_1, η_2 — небольшие константы; $\eta_3 = \eta_3(\eta_1, \eta_2) > 0$. Эта формула дает 'дисперсию' $\tau(n)$ в арифметических прогрессиях; мы применим ее как неравенство Чебышева в теории вероятностей, чтобы доказать, что 'почти для всех' D имеем удовлетворительную асимптотическую формулу для (4, 3). Для $D < \sqrt{x}$ (даже для $D < x^{\frac{2}{3}-\alpha_0}$) мы можем применить формулы, упомянутые выше. Для оставшихся $D > \sqrt{x}$ применяем тривиальную верхнюю оценку.

5. Функция $\tau_2(n)$ отвечает квадратичной форме uv; для квадратичной формы $Q(u, v)$, указанной выше, можно вывести формулы, совершенно аналогичные (4, 6). Определяя функцию $\Pi(n, D)$ посредством (3, 5), имеем

Лемма 2.

$$\sum_{D_1 \leqslant D \leqslant D_1+D_2}\left(\sum_{\substack{Q(u,v) \equiv n\,(\mathrm{mod}\,D) \\ Q(u,v) \leqslant n}}1 - \frac{2\pi n}{\sqrt{|d|}\,D}\Pi(n, D)\right)^2 = O\left(\frac{n^{2-\eta_3}}{D_1^2}D_2\right), \qquad (5,1)$$

где D_1, D_2 имеют то же значение, что и в лемме 1; x надо заменить на n.

Эта 'формула дисперсии' доставляет хорошие асимптотические формулы для 'почти всех' $D > \sqrt{n}$; для $D \leqslant \sqrt{n}$ (даже для $D \leqslant n^{\frac{2}{3}-\alpha_0}$) индивидуальные формулы:

$$\sum_{\substack{Q(u,\,v)\equiv n (\mathrm{mod}\, D) \\ Q(u,\,v)\leqslant n}} 1 = \frac{2\pi n}{D\sqrt{|d|}}\, \Pi(n, D) + O(n^{1-\eta_4}) \qquad (5,2)$$

имеют место в силу свойств сумм Клостермана.

6. Доказательство формул (4, 6) и (5, 1) базируется на мультипликативных свойствах входящих в них функций и теории тернарных квадратичных форм. Начнем с (4, 6).

Пусть (QFR) — множество свободных от квадратов чисел. Пусть $n_i \in (QFR)$ $(i = 1, 2)$; $(n_1, n_2) = \delta$. Тогда имеем, очевидно:

$$\tau(n_1)\, \tau(n_2) = (\tau(\delta))^2\, \tau\left(\frac{n_1 n_2}{\delta}\right).$$

Это приводит к следующему выражению: пусть ν_1, ν_2, δ фиксированы; $D\nu_i + l \leqslant n$ $(i = 1, 2)$; образуем

$$(\tau(\delta))^2 \sum_{s\leqslant n} \mu(s) \sum_{(D)} \tau\left(\frac{D\nu_1 + l}{\delta}\, \frac{D\nu_2 + l}{\delta}\right). \qquad (6,1)$$

Область (D) определена условиями: $0 \leqslant D\nu_i + l \leqslant n$; $D\nu_i + l \in (QFR)$; $\delta \mid D\nu_i + l$; $s \mid D\nu_i + l$; $(D, l) = 1$. Суммирование по s, с $\mu(s)$, вводится для того, чтобы сделать $(D\nu_i + l)/\delta$ взаимно простыми.

Члены суммы (6, 1) с фиксированными δ, s, ν_1, ν_2, как легко видеть, равны числу решений уравнения

$$(D\nu_1\nu_2 - l(\nu_1 + \nu_2))^2 - 4\nu_1\nu_2\delta^2 vw = (l(\nu_1 - \nu_2))^2, \qquad (6,2)$$

т.е. числу представлений квадрата $(l(\nu_1 - \nu_2))^2$ неопределенной тернарной квадратичной формой $u^2 - vw$ при некоторых простых геометрических и конгруэнциальных условиях на

$$u, v, w \,(\text{напр. } v \equiv 0\ (\mathrm{mod}\, 4\nu_1\nu_2\delta^2)).$$

Если δ, s, ν_1, ν_2, сравнительно малы (порядка $O(n^{\alpha_1})$), а D, v, w достаточно велики ($D > n^{1-\alpha_2}$), представление квадрата тернарной квадратичной формой этого типа может быть трактовано методами автора и А. В. Малышева (см. напр. [8, 9]) вполне удовлетворительно.

Для данной формы $u^2 - vw$ даже нет нужды в этих методах, т.к.

уравнение $u^2 - vw = (l(\nu_1 - \nu_2))^2$ может быть сведено к подсчету матриц

$$A = \begin{Vmatrix} \alpha & \beta \\ \gamma & \delta \end{Vmatrix} \quad \text{c } \det(A) = \pm l(\nu_1 - \nu_2)$$

при некоторых геометрических и конгруэнциальных условиях на $\alpha, \beta, \gamma, \delta$. Это, в основном, проблема теории кватернарных форм, очень сходная с задачей, разобранной в[10], стр. 25–29. (Есть также старая работа Г. Кантора[1], относящаяся к подобным вопросам.) Заметим, что δ и s можно считать порядка $O(x^{\xi_1})$ ($\xi_1 > 0$ нужного порядка малости), т.к. для больших δ и s члены сумм (6, 1) могут быть оценены тривиально и оказываются сравнительно малыми. Произведение $\nu_1\nu_2$ тоже должно быть порядка $O(x^{\xi_1})$; это означает, что D должно быть большим ($D > x^{1-\xi_2}$). Естественно, что усреднение по большому интервалу значений D легче, чем по узкому. Нам нужны, однако, $D \geqslant \sqrt{x}$; мы объясним их трактовку ниже.

Результаты о числе представлений очень громоздки и мы опускаем их трактовку. Чтобы приложить их к выводу формулы (4, 6), мы заставляем сперва n пробегать по свободным от квадратов числам и заменяем $A(x, D)$ на соответствующие $A'(x, D)$. Затем открываем скобки в (4, 6) и суммируем по D, прилагая к произведению $\tau(n_1)\tau(n_2)$ указанные выше результаты. 'Линейное выражение', содержащее $\tau(D\nu_1 + l) A'(x, D)$, тоже суммируется по D, и мы получаем (4, 6) для $n \in (QFR)$. Общий случай затем получается легко.

Поскольку нам нужны 'формулы дисперсии' для $\sqrt{x} < D < x^{1-\alpha_1}$, мы можем получить формулы этого типа, беря $D = pD'$; $p \asymp x^\beta$ выбираются большими простыми числами. Прогрессии

$$n \equiv l \,(\mathrm{mod}\, D'p)$$

могут быть разложены затем на прогрессии

$$n \equiv l \,(\mathrm{mod}\, D'p); \; n \equiv \xi \,(\mathrm{mod}\, p); \; \xi = 1, ..., p-1.$$

Изменяя далее $p \asymp x^\beta$, мы можем получить требуемые усредняющие формулы для $n \equiv l \,(\mathrm{mod}\, D')$; $D' > \sqrt{x}$. Тем же методом мы можем получить асимптотическую формулу для суммы

$$\sum_{m \leqslant x} \tau_2(ax^2 + bx + c)$$

с остаточным членом $O(x^{1-\alpha_0})$. 'Формула дисперсии' (5, 1) проверяется аналогично. Простейший случай: $Q(u, v) = u^2 + v^2$. Для

$$U(m) = \sum_{u^2+v^2=m} 1; \; n_i \in (QFR); \; (n_1, n_2) = \delta,$$

получаем
$$U(n_1)\,U(n_2) = \tfrac{1}{4}(U(\delta))^2\,U\left(\frac{n_1 n_2}{\delta}\right).$$

Общая форма $Q(u,v)$ требует более сложных формул и привлечения теории композиции классов. Действуя как в случае $\tau(n)$, получаем уравнение

$$(n(\nu_1-\nu_2))^2 = (D.\,2\nu_1\nu_2 - n(\nu_1+\nu_2))^2 - \delta^2\nu_1\nu_2 Q_1(v,w). \qquad (6,3)$$

Таким образом, соответствующая тернарная форма будет

$$u^2 - aQ_1(v,w);$$

она должна представлять большой квадрат $(n(\nu_1-\nu_2))^2$; $a = \delta^2\nu_1\nu_2$; $Q(v,w)$ — примитивная бинарная квадратичная форма дискриминанта d. Представления подчинены некоторым геометрическим и конгруэнциальным условиям. Асимптотические формулы можно получить, используя методы, упомянутые выше [8,10,11]; остаточные члены хорошие.

Представление больших квадратов и даже чисел, имеющих большие квадратные делители, тернарными квадратичными формами — в основном, проблема теории кватернарных квадратичных форм.

7. Возвратимся теперь к теореме 3 (формула (3, 6)), чтобы пояснить ее связь с 'формулой дисперсии' (5, 1). Заметим, что в (3 3),

$$a_m = \sum_{\delta_1\delta_2\dots\delta_{q-1}|m} \chi_1(\delta_1)\,\chi_2(\delta_2)\dots\chi_{q-1}(\delta_{q-1}).$$

Берем $g = 1$ (только для простоты). Имеем

$$T(n) = \sum_{\delta_1,\,\delta_2,\,\dots,\,\delta_{q-1}} \chi_1(\delta_1)\dots\chi_{q-1}(\delta_{q-1}) \sum_{\delta_1\delta_2\dots\delta_{q-1}|n-Q(u,v)} 1. \qquad (7,1)$$

Пусть $\delta_1 \leqslant \delta_2 \leqslant \dots \leqslant \delta_{q-1}$. Если $\delta_1\delta_2\dots\delta_{q-1} \leqslant n^{(q-2)/(q-1)+\eta_0}$, имеем хорошие асимптотические выражения для $\displaystyle\sum_{\delta_1\delta_2\dots\delta_{q-1}|n-Q(u,v)} 1$ и 'почти всех' $\delta_1\delta_2\dots\delta_{q-1}$ по лемме 2.

Если $\delta_1\delta_2\dots\delta_{q-1} > n^{(q-2)/(q-1)+\eta_0}$, $\delta_1\delta_2\dots\delta_{q-1}\,|\,m$ $(m \leqslant n)$, мы можем отобразить произведение $\chi_1(\delta_1)\dots\chi_{q-1}(\delta_{q-1})$ на произведение

$$\chi_1\overline{\chi}_{q-1}(\delta_1)\,\chi_2\overline{\chi}_{q-1}(\delta_2)\dots\chi_{q-2}\overline{\chi}_{q-1}(\delta_{q-2})\,\chi_{q-1}\left(\frac{m}{\delta_1\delta_2\dots\delta_{q-1}}\right).$$

Далее, $\quad \delta_1\delta_2\dots\delta_{q-2}\dfrac{m}{\delta_1\delta_2\dots\delta_{q-1}} = \dfrac{m}{\delta_{q-1}} \leqslant n^{(q-2)/(q-1)-\eta_0|q-2}$,

т.к. $\delta_{q-1} \geqslant n^{1/(q-1)+\eta_0/(q-2)}$. Таким образом, лемма 2 будет достаточной, чтобы вычислить асимптотику (3, 4).

ЛИТЕРАТУРА

[1] Cantor, G. *De transformatione formarum ternariarum quadraticarum*, Halis Saxonium (Habilitationsschrift) (1869).

[2] Eichler, M. Quaternäre quadratische Formen und die Riemannsche Vermutung für die Kongruenzzetafunktion. *Archiv Math.* 5, 355–366 (1954).

[3] Estermann, T. On the representation of a number as a sum of two products. *Proc. Lond. Math. Soc.* (2), 31, 123–133 (1930).

[4] Hooley, C. An asymptotic formula in the theory of numbers. *Proc. Lond. Math. Soc.* (3), 7, 396–419 (1957).

[5] Ingham, A. E. Some asymptotic formulae in the theory of numbers. *J. Lond. Math. Soc.* 2, 173–182 (1927).

[6] Kloosterman, H. D. On the representation of a number in the form $ax^2 + by^2 + cz^2 + dt^2$. *Acta Math.* 49, 407–464 (1926).

[7] Linnik, yu. V. A general theorem on positive ternary quadratic forms. *Izvestia AN SSSR, seria matem.* 87–110 (1939). (Russian.)

[8] Linnik, yu. V. Quaternions and Cayley's numbers; some applications of quaternion arithmetics. *Uspekhi Matem. Nauk,* iv, 5 (33), 49–98 (1949). (Russian.)

[9] Linnik, yu. V. and Malyshev, A. V. Applications of quaternion arithmetics to the theory of ternary quadratic forms and to the decomposition of numbers into cubes. *Uspekhi Matem. Nauk,* viii, 5, 3–71 (1953). (Russian; there is an American translation.)

[10] Linnik, yu. V. Asymptotic distribution of reduced binary quadratic forms in connection with Lobatschevskian geometry, II. *Vestnik Leningradskogo Universiteta,* 5, 3–32 (1955). (Russian.)

[11] Malyshev, A. V. Asymptotic distribution of integer points on certain ellipsoids. *Izvestia AN SSSR, seria matem.* 21, 457–500 (1957). (Russian.)

[12] Tartakowski, V. A. Expressions pour le nombre des représentations d'un nombre par une forme quadratique positive à plus de trois variables. *C.R. Acad. Sci., Paris,* 186, 1337–1340 (1928).

[13] Tartakowski, V. A. Die Gesamtheit der Zahlen die durch eine positive quadratische Form $F(x_1,...,x_s)$ darstellbar sind. *Izvestia AN SSSR, seria matem.* §§ 111–112, 165–196 (1929).

[14] Titchmarsh, E. C. Some problems in the analytic theory of numbers. *Quart. J. Math.* (Oxford) 13, 129–152 (1942).

SOME FUNDAMENTAL THEOREMS ON ABELIAN FUNCTION FIELDS

By PETER ROQUETTE

1. Introduction

1.1. The 'fundamental theorems' I have in mind are those which are centred around the celebrated finiteness theorem of Mordell and Weil. Let us first recall what this finiteness theorem says:

Let A be an abelian variety which is defined over a number field k. Let A_k be the group of all points of A which are rational in k. Then the finiteness theorem of Mordell and Weil[1] can be stated as follows: A_k, *as a commutative group, is finitely generated.*

That means, there are finitely many points $a_1, ..., a_m$ in A_k such that each point a in A_k can be written in the form

$$a = \sum_{i=1}^{m} n_i a_i$$

with rational integers n_i.

There is a generalization of this theorem which is due to Néron[2]. Namely, Néron has proved that the finiteness theorem does not only hold for a number field as ground field, but also for a much wider class of fields: *In the above theorem, k may be an arbitrary finitely generated field.*

In the following, when referring to the finiteness theorem, we shall always mean this generalization by Néron rather than the original theorem.

1.2. The purpose of this note is, first, to give another formulation and at the same time a further generalization of the finiteness theorem which, as it seems, is just the 'proper' one from the arithmetical point of view (see the class theorem in § 2.4.) Secondly, we shall speak about some new principles and results by means of which the finiteness theorem (and its generalization) can be proved. Since it is impossible in this short note to deal with the whole proof, we shall therefore restrict ourselves to those results which lead to the so-called 'weak' finiteness theorem (or its generalization, the 'weak' class theorem). This 'weak' theorem says that for an integer $n > 1$ the factor group A_k/nA_k is finite; it is the first step towards the proof that A_k itself is finitely generated. It seems to be interesting that this 'weak' theorem is a consequence of theorems of a cohomological nature (see § 3).

2. The class theorem.

2.1. First we shall give a reformulation of the finiteness theorem, presenting it as a statement about divisor classes instead of points of an abelian variety. In doing this we restrict ourselves to the special case where the abelian variety in question is the *jacobian variety* of an algebraic curve (this is no essential restriction).

So let Γ be a projective curve, defined over a finitely generated field k; assume that Γ is normal. Let $K = k(\Gamma)$ be the function field of Γ over k; this is a finitely generated field extension of k of degree of transcendency 1 over k. By the very definition of the jacobian variety A of Γ[3], the group A_k is naturally isomorphic to the group C_0 of Γ-divisor classes of degree 0 of K. This group C_0 can be described as follows:

We consider the set $M = M_\Gamma$ of Γ-prime divisors of K, i.e. of those prime divisors of K which are trivial on k. We form the corresponding divisor group D, defined as the free commutative group whose generators are the elements of M. We assign to each element $u \neq 0$ of K its divisor (u) which counts the zeros and poles of u with their respective multiplicity. Thus we get a homomorphism

$$d\colon K^\times \to D$$

of the multiplicative group $K^\times$ of K into the divisor group D. This map d, fundamental in arithmetic, is called the *divisor map* of K with respect to the set M of prime divisors.

The cokernel of d, i.e. the factor group of D modulo the image $d(K^\times)$, is called the *divisor class group* of K with respect to M. We shall denote it by C or more precisely by C_M whenever we want to indicate the prime divisor set M which we are referring to.

It is well known that the image $d(K^\times)$ is contained in the subgroup D_0 of those divisors which are of degree 0; this fact can also be described by saying that a function $u \neq 0$ in K has as many zeros as it has poles. It follows that the degree of a divisor does not actually depend on the divisor itself but only on its class in the group C. In particular, the divisor classes of degree 0 form a certain subgroup of C which we shall denote by C_0 or, more precisely, by $C_{0,M}$.

It is this group C_0 which, as already said above, is naturally isomorphic to the group A_k of rational points of the jacobian variety A of Γ. Hence it follows by the finiteness theorem that C_0 is finitely generated. But to say that C_0 is finitely generated is the same as to say that C itself is finitely generated, for the factor group C/C_0 is isomorphic to the additive group of rational integers. Hence we get the following statement as another formulation of the finiteness theorem:

Let K be a finitely generated field and Γ be a projective normal curve which is a model of K (over some subfield k of K); denote by M the set of Γ-prime divisors of K.

Then the corresponding divisor class group C_M of K is finitely generated.

2.2. The generalization of this statement which I have in mind says that the same is true for an *arbitrary* model of K (not only for a projective normal curve). Here, the notion of a model of K may be taken either in the usual sense in algebraic geometry (over some subfield k of K as constant field) or in the more general 'arithmetic' sense given to it by Nagata[4] (over some Dedekind domain contained in K). For reasons of brevity, we shall not explain in detail the definition of a model; we only give the definition of what we call a 'prime divisor model' of K which will suffice for our purpose.

So let K be a finitely generated field and M a set of prime divisors (= valuations) of K. We denote by R the intersection of all the valuation rings belonging to M. We say that M is an 'affine' prime divisor model of K, if the following two conditions are satisfied:

(i) each valuation ring belonging to M is a quotient ring of R;

(ii) R is finitely generated either over the rational integers or over some subfield k of K.

Now we define a *prime divisor model* of K as a set M of prime divisors of K which contains an affine prime divisor model M_0 such that $M - M_0$ is finite.

If a prime divisor model M of K is given, there is a natural divisor map $d = d_M$ of $K^\times$ into the corresponding divisor group, and the cokernel of d is called, exactly as above, the divisor class group of K with respect to M.

Now we can state our theorem as follows:

Class theorem. Let M be a prime divisor model of a finitely generated field K. Then the corresponding divisor class group C_M of K is finitely generated.

2.3. It is not very difficult to deduce this class theorem from the finiteness theorem of Mordell–Weil–Néron[5]. The reason why I stated it is that in this form its arithmetic significance becomes clear: it gives us a certain insight into the structure of the divisor class groups of finitely generated fields. Note that K may very well be an algebraic number field; in this case a prime divisor model of K is just a set of prime divisors which contains all but a finite number of prime divisors of K. Hence, in this case, the class theorem reduces to the well-known classical result that the class number is finite[6].

Hence our class theorem may be regarded as the direct generalization of the classical result that the class number in an algebraic number field is finite.

2.4. As a side remark we would like to mention that there is another theorem, dual to the class theorem, which concerns the unit group E_M with respect to a prime divisor model M of K. The unit group is defined dually to the class group as the *kernel* of the divisor map.

Let us call a prime divisor model M of K an 'absolute' model, if it contains an affine model M_0 whose corresponding integral domain R_0 is 'absolutely' finitely generated, i.e. finitely generated over the rational integers (see condition (ii) above). Then the unit theorem reads as follows:

Unit theorem. Let M be an absolute prime divisor model of a finitely generated field K. Then the corresponding unit group E_M of K is finitely generated.

If K is a number field, then this theorem coincides with the classical unit theorem of Dirichlet. The general unit theorem can be easily deduced from this classical statement[5], contrary to the situation for the class theorem.

3. The weak class theorem

3.1. Let M be a prime divisor model of a finitely generated field K. Then the weak class theorem states that for any integer $n > 1$, prime to the characteristic of K, the factor group C_M/nC_M is finite. Of course this is an immediate consequence of the class theorem; however, as already said in the introduction, the first step towards the proof of the class theorem is the weak class theorem.

In this § 3 we shall outline a proof of the weak class theorem.

Let r be the arithmetic dimension of the field K. This is defined as the degree of transcendency of K if the characteristic of K is $\neq 0$, and $1 +$ degree of transcendency of K if the characteristic of K is 0. If $r = 0$ then K is finite and the weak class theorem is trivial. If $r = 1$ then K is either a number field or a function field in one variable over a finite constant field. In this case the weak class theorem is true by the classical theorem on the finiteness of the class number[6]. Hence we may assume $r > 1$ and use induction on r.

Let k be a subfield of K, algebraically closed in K, such that K is separable and of degree of transcendency 1 over k. The key for the induction argument is the following result[5].

Proposition 1. *If the weak class theorem holds for all models of k, and also for all models of K over k, then it holds for all models of K.*

Hence, using the induction assumption, we see that it suffices to prove the weak class theorem for a model Γ of K over k. It is no essential restriction to assume that Γ is projective and normal; let A be the jacobian variety of Γ. As shown in § 2.3, the finiteness of C_Γ/nC_Γ is equivalent to the finiteness of A_k/nA_k. Hence we are back at the proof of the weak finiteness theorem. However, in our present situation we have the induction assumption that the weak class theorem is already proved for the ground field k. This will turn out to be important in the following proof of the weak finiteness theorem.

Let F be the function field of the jacobian variety A over k. Then A defines a prime divisor model of F over k; let us denote by C_A the corresponding divisor class group of F and by $C_{0,A}$ the subgroup of those divisor classes in C_A which are algebraically equivalent to zero. It is well known that the group A_k is isomorphic to the group $C_{0,A}$; such an isomorphism is given by

$$x \to \text{class of } \theta_x - \theta \quad \text{for} \quad x \in A_k,$$

where θ denotes a theta divisor of A, normed so as to be rational in k.

Hence we have to show that $C_{0,A}/nC_{0,A}$ is finite. This will be done, for an arbitrary abelian variety, in the following theorems.

3.2. According to what we have said above, we consider the following situation:

k is a finitely generated field for which the weak class theorem is true;

A is an abelian variety defined over k;

$F = k(A)$ is the function field of A over k;

$C_0 = C_{0,A}$ is the group of divisor classes of F with respect to A which are algebraically equivalent to 0;

n is a given integer > 1, prime to the characteristic of k.

In this situation we have to prove

$$C_0/nC_0 \text{ is finite.}$$

To prove this, it is no essential restriction to assume that all the nth division points of A are rational in k. Let us assume this, and let us denote by G_n the automorphism group of F consisting of the translations by the nth division points of A.

3.3. The group G_n acts in a natural way on the divisor group $D = D_A$ of F with respect to A, in such a way that the divisor map $d = d_A$ is a G_n-homomorphism. Hence G_n acts also on the cokernel of d, which is the divisor class group $C = C_A$, and on the kernel of d, which is the multiplicative group $k^\times$ of the ground field.

Now, general cohomology theory tells us that there is a natural map of the cohomology groups of G_n in the cokernel C into the cohomology groups of G_n in the kernel $k^\times$, increasing the dimension by two:

$$H^i(G_n, C) \to H^{i+2}(G_n, k^\times).$$

This map has proved to be of fundamental importance in various problems of arithmetic, in particular in class field theory. We shall consider it in the dimension $i = 0$.

The zero cohomology group $H^0(G_n, C)$ is essentially the group of those divisor classes which are fixed under G_n. This group contains in particular the group C_0, since C_0 is elementwise fixed under *all* translations. Hence, by restriction to C_0 we get a map

$$h_n: C_0 \to H^2(G_n, k^\times),$$

which we call the Hasse map (with respect to n). It is this map which we are referring to in the following theorems. Note that $H^2(G_n, k^\times)$ is precisely the group of abelian algebras over k with group G_n in the sense of Hasse[7].

3.4. The first theorem we would like to mention is the following:

Theorem 1. *The kernel of* h_n *is equal to* nC_0.

Thus h_n defines an isomorphism of the factor group C_0/nC_0 into $H^2(G_n, k^\times)$.

3.5. The next theorem is about the image of h_n. Let us choose an absolute prime divisor model M of the field k in the sense of § 2.4. We say that a prime p of M is *regular* with respect to A (or with respect to F) if the reduction mod p of A is non-degenerate and an abelian variety over the residue field k mod p. If this is so then p extends uniquely in a natural way to F[8].

It is known that all but a finite number of primes p of M are regular with respect to A. After removing the irregular primes we may therefore assume that *all* primes of M are regular with respect to A. We then call M a *regular* model of k with respect to A.

Theorem 2. If M is regular with respect to A then the image of h_n splits M-divisorially, i.e. it is contained in the kernel of the natural map

$$H^2(G_n, k^\times) \to H^2(G_n, D_M)$$

which is induced by the divisor map $d_M: k^\times \to D_M$.

3.6. The next theorem is about the kernel of the map mentioned in theorem 2. Before stating it let us remember that the factor group C_M/nC_M is finite, according to our assumption that the weak class theorem is true for the ground field k. Furthermore, we know that the factor group E_M/nE_M is also finite; see § 2.4.

Theorem 3. As a consequence of the fact that the groups C_M/nC_M and E_M/nE_M are finite, it follows that the kernel of the natural map

$$H^2(G_n, k^\times) \to H^2(G_n, D_M)$$

is finite too.

For the proof of the theorems 1, 2 and 3 see [9].

Now, taking the theorems 1, 2 and 3 together we see that h_n defines an isomorphism of the group C_0/nC_0 into a finite group; hence C_0/nC_0 is finite too. This proves the weak finiteness theorem.

3.7. *Remark.* In order to deduce from the weak finiteness theorem the finiteness theorem itself one has to apply the classical method of infinite descent together with the theory of heights of points of abelian varieties. For details of a simple proof see my forthcoming paper in the *Journal f. d. reine u. angewandte Mathematik.*

REFERENCES

[1] Mordell, L. J. On the rational solutions of the indeterminate equations of the third and fourth degree. *Proc. Camb. Phil. Soc.* 21, 179–192 (1922). Weil, A. L'arithmétique sur les courbes algébriques. *Acta math.* 52, 281–315 (1928).

[2] Néron, A. Arithmétique et classes de diviseurs sur les variétés algébriques. *Proc. Int. Symp. Tokyo*, 139–154 (1956).

[3] Weil, A. Variétés abéliennes et courbes algébriques. *Actualités sci. industr.* 1064, Paris, 1948.

[4] Nagata, M. A general theory of algebraic geometry over Dedekind domains. *Amer. J. Math.* 78, 78–116 (1956).

[5] Roquette, P. Einheiten und Divisorklassen in endlich erzeugbaren Körpern. *Jahresberichte dtsch. Math. Ver.* 60, 1–21 (1957).

[6] Artin, E. and Whaples, G. Axiomatic characterization of fields by the product formula for valuations. *Bull. Amer. Math. Soc.* 51, 469–492 (1945).

[7] Hasse, H. Existenz und Mannigfaltigkeit abelscher Algebren mit vorgegebener Galoisgruppe über einem Teilkörper des Grundkörpers. *Math. Nachr.* 1, 40–61 (1948).

[8] Chow, W. L. and Lang, S. On the birational equivalence of curves under specialization. *Amer. J. Math.* 79, 649–652 (1957). Lamprecht, E. Zu Eindeutigkeit von Funktionalprimdivisoren. *Archiv Math.* 8, 30–38 (1957).

[9] Roquette, P. Über das Hassesche Klassenkörperzerlegungsgesetz und seine Verallgemeinerung für beliebige abelsche Funktionenkörper. *J. reine angew. Math.* 197, 49–67 (1957).

FONCTIONS AUTOMORPHES ET
CORRESPONDANCES MODULAIRES

Par GORO SHIMURA

Il est bien connu que le groupe modulaire de Siegel est le groupe de transformations pour les périodes des fonctions abéliennes, mais on sait peu de chose des relations entre les fonctions modulaires et les abéliennes, sauf au cas de dimension 1. L'objet de cette conférence est d'énoncer quelques idées et résultats à ce sujet. Nous démontrerons d'abord que les 'modules' des variétés abéliennes polarisées, regardés comme fonctions des périodes, engendrent les fonctions modulaires de Siegel, et que les fonctions modulaires par rapport aux groupes de congruence s'obtiennent à partir des points t sur les variétés abéliennes tels que $qt = 0$ pour un entier q. On peut appliquer le même procédé à d'autres types de fonction automorphe, par exemple, aux fonctions de Hilbert. Ce sont non seulement la généralisation de la fonction $j(\tau)$ ou des '$\wp$-Teilwerte', mais les outils dont on se sert pour attaquer les problèmes arithmétiques des fonctions automorphes. On voit en effet que les opérateurs T_n, introduits par Hecke pour les formes modulaires elliptiques et généralisés par ses disciples pour divers formes automorphes, ne sont autres que les représentations de certaines correspondances algébriques, appelées correspondances modulaires, définies au moyen d'isogénies de variétées abéliennes. En se basant sur cette idée, on acquiert des formules de congruence pour les correspondances modulaires dans le cas d'une certaine classe de fonctions automorphes de dimension 1, que l'on rencontrera à la fin de cet exposé.

Pour commencer, on va rappeler les notions de variété abélienne polarisée et de ses modules (Weil[7], Matsusaka[3], Shimura[6]). Soient A une variété abélienne et X un diviseur sur A. On désignera par $\mathscr{C}(X)$ l'ensemble de tous les diviseurs X' sur A pour lesquels il existe deux entiers positifs m, m' tels que mX soit algébriquement équivalent à $m'X'$. $\mathscr{C}(X)$ s'appellera une *polarisation* de A si $\mathscr{C}(X)$ contient un diviseur ample. On entendra par une *variété abélienne polarisée* une variété abélienne A sur laquelle est donnée une polarisation $\mathscr{C}$, désignée par $(A, \mathscr{C})$. $(A, \mathscr{C})$ est dit isomorphe à une autre variété abélienne polarisée $(A', \mathscr{C}')$ s'il existe un isomorphisme de A sur A' qui envoie $\mathscr{C}$ sur $\mathscr{C}'$. On dira que $(A, \mathscr{C})$ est défini sur un corps k si A est défini sur k et si $\mathscr{C}$ contient un diviseur X rationnel sur k. Ainsi, soit σ un isomorphisme de k sur un corps k'; on désignera par $\mathscr{C}^\sigma$ la polarisation $\mathscr{C}(X^\sigma)$ de A^σ. On peut démontrer qu'il existe un sous-corps K de k jouissant de la propriété suivante: pour qu'un isomorphisme σ de k soit l'identité sur K, il faut et il suffit que $(A, \mathscr{C})$ soit isomorphe à $(A^\sigma, \mathscr{C}^\sigma)$. Si la caractéristique est 0, K est déterminé par cette condition, et s'appellera le *corps de modules* de $(A, \mathscr{C})$. On l'obtient au moyen de points de Chow ainsi qu'il suit. Soient X un diviseur ample dans $\mathscr{C}$ et A' l'image d'un plongement projectif de A donné par X; soient A^* la transformée de A' par une transformation projective générique sur k et z le point de Chow de A^*. Le lieu $\mathscr{F}$ de z sur k ne dépend que de A et de X; on appellera $\mathscr{F}$ la *famille projective* de (A, X). $(A, \mathscr{C})$ est isomorphe à $(A', \mathscr{C}(X'))$ si et seulement si les familles projectives de (A, X) et de (A', X') coïncident, où l'on suppose que les dimensions des systèmes linéaires définis par X et X' soient les mêmes. Il en résulte que le corps de modules de (A, X) est engendré sur $\mathbf{Q}$† par le point de Chow de $\mathscr{F}$, qui peut être donc considéré comme un 'module' de $(A, \mathscr{C})$. On peut définir de même les 'modules' au cas de caractéristique $\neq 0$; mais on n'en s'occupera pas dans cet exposé.

On va maintenant étudier les fonctions modulaires et para-modulaires. Soient $\delta_1, \ldots, \delta_n$ n entiers positifs tels que

$$\delta_1 = 1, \quad \delta_i \mid \delta_{i+1} \quad (1 \leqslant i \leqslant n-1);$$

soient δ la matrice diagonale ayant pour éléments les δ_i, et

$$E = \begin{pmatrix} 0 & -\delta \\ \delta & 0 \end{pmatrix}, \quad F = \begin{pmatrix} 1_n & 0 \\ 0 & \delta \end{pmatrix},$$

où 1_n est la matrice unité de degré n. On désignera par $\Gamma'(\delta)$ le groupe composé des matrices T de degré $2n$ à coefficients entiers telles qu'on ait $^tTET = E$. On voit que le groupe

$$\Gamma(\delta) = \{F^{-1}{}^tT^{-1}F \mid T \in \Gamma'(\delta)\}$$

est un sous-groupe du groupe symplectique. $\Gamma(1_n) = \Gamma'(1_n)$ est le groupe modulaire de Siegel. Les éléments de $\Gamma(\delta)$ opèrent d'une manière ordinaire sur l'espace de Siegel S_n. Soit $D_\delta(z)$ un lattice de l'espace numérique complexe $\mathbf{C}^n$ engendré par les colonnes de la matrice $(z \, \delta)$ sur $\mathbf{Z}$. Si z est un point de S_n, le tore complexe $\mathbf{C}^n/D_\delta(z)$ a une structure de variété abélienne; la matrice alternée E donne une forme de Riemann sur $\mathbf{C}^n/D_\delta(z)$, pour ainsi dire, la matrice $(z \, \delta)$ détermine une structure de

---

† On désignera par $\mathbf{Z}$, $\mathbf{Q}$, $\mathbf{R}$ et $\mathbf{C}$ l'anneau des entiers rationnels, les corps des nombres rationnels, réels et complexes.

variété abélienne polarisée, qui est réalisée comme une variété projective $A_\delta(z)$ par les fonctions thêta correspondant à la matrice mE pour un entier $m \geqslant 3$. Fixons désormais un tel entier, m, et désignons par $\mathscr{F}_\delta(z)$ la famille projective de $A_\delta(z)$ pour chaque z. Tandis que $A_\delta(z)$ dépend du choix d'une base des fonctions thêta, $\mathscr{F}_\delta(z)$ ne dépend que de z (et de m). On vérifie facilement que les $\mathscr{F}_\delta(z)$ pour $z \in S_n$ sont toutes de même dimension. Il existe, de plus, un sous-ensemble analytique Y de S_n de co-dimension 1 et des fonctions méromorphes $\phi_1(z), \ldots, \phi_\lambda(z)$ sur S_n jouissant des propriétés suivantes:

(m 1) les $\mathscr{F}_\delta(z)$ pour $z \in S_n - Y$ sont de même degré;

(m 2) pour chaque $z \in S_n - Y$, $(1, \phi_1(z), \ldots, \phi_\lambda(z))$ donne le point de Chow de la variété $\mathscr{F}_\delta(z)$.

Soient T un élément du groupe $\Gamma'(\delta)$ et $U = F^{-1}{}^t T^{-1} F$. La relation ${}^t TET = E$ entraîne que deux variétés $A_\delta(z)$ et $A_\delta(U(z))$, polarisées par les sections hyperplanes, sont isomorphes; d'où résulte $\mathscr{F}_\delta(z) = \mathscr{F}_\delta(U(z))$. Réciproquement, on peut démontrer que si l'on a $\mathscr{F}_\delta(z) = \mathscr{F}_\delta(z')$, il existe un élément U de $\Gamma(\delta)$ tel que $z' = U(z)$. On déduit de ceci, en vertu du théorème de plongement projectif de l'espace quotient $S_n/\Gamma(\delta)$ (Baily[1], Satake et Cartan[4]), le théorème suivant:

Théorème 1. *Si* $n > 1$, *le corps* $\mathbf{C}(\phi_i)$ *est le corps des fonctions méromorphes sur* S_n *invariantes par* $\Gamma(\delta)$.

En d'autres termes, les fonctions automorphes par rapport à $\Gamma(\delta)$ sont engendrées par les modules des variétés abéliennes polarisées 'de la famille δ'; en effet, d'après ce qu'on a vu plus haut, pour chaque point z' de S_n, le corps $\mathbf{Q}(\phi_i(z'))$ est le corps de modules de la variété $A_\delta(z')$.

Désignons par $L(\delta)$ le corps $\mathbf{C}(\phi_i)$; le corps $L(1_n)$ est le corps des fonctions modulaires de Siegel, même si $n = 1$.

Considérons maintenant les points t sur $A_\delta(z)$ tels qu'on ait $qt = 0$ pour un entier q; ils engendreront les fonctions automorphes par rapport aux groupes de congruence. Pour avoir ce résultat il faut rappeler la notion de variété de Kummer introduite par Weil[7]. Soit G le groupe des automorphismes d'une variété abélienne polarisée $(A, \mathscr{C})$; on sait que G est d'ordre fini. On entendra par une variété de Kummer de $(A, \mathscr{C})$ une variété quotient W de A par rapport à G. Bien entendu, W n'est pas uniquement déterminé; mais on peut construire, en vertu des résultats de Weil[8], une variété de Kummer W dans un espace projectif ainsi qu'une application naturelle h de A sur W satisfaisant aux conditions suivantes:

(W 1) W est défini sur le corps K de modules de $(A, \mathscr{C})$.

(W 2) h est défini sur tout corps de définition pour $(A, \mathscr{C})$ contenant K.

(W 3) Si σ est un isomorphisme d'un corps de définition pour $(A, \mathscr{C})$ contenant K, et si λ est un isomorphisme de $(A, \mathscr{C})$ sur $(A^\sigma, \mathscr{C}^\sigma)$, on a $h = h^\sigma \circ \lambda$.

On construit pour $A_\delta(z)$ un couple (W, h) jouissant des propriétés (W 1–W 3). Soit $\theta(u)$ l'isomorphisme de $\mathbf{C}^n/D_\delta(z)$ sur $A_\delta(z)$ où $u \in \mathbf{C}^n$. Soit b un vecteur de $\mathbf{R}^{2n}$ (une matrice à $2n$ lignes et une colonne). On peut considérer, en gros, le point $h(\theta(\omega_\delta(z) b))$ comme une fonction de z. Il est difficile d'éclaircir la situation pour tout point de S_n, puisqu'on ignore comment fabriquer (W, h) comme fonction de z. De toute façon, nous pouvons obtenir des fonctions méromorphes $\xi_\alpha(z, b)$ sur S_n dont les valeurs en z donnent les coordonnées du point $h(\theta(\omega_\delta(z) b))$ pour 'presque tout' point z de S_n. Ces fonctions satisfont à l'équation

$$(\xi) \qquad \xi_\alpha(U(z), Tb)/\xi_\beta(U(z), Tb) = \xi_\alpha(z, b)/\xi_\beta(z, b),$$

où U est un élément de $\Gamma(\delta)$ et $T = F^{-1}{}^t U^{-1} F$. C'est une conséquence de la propriété (W 3). Soit q un entier positif; on désignera par $\Gamma'(\delta, q)$ le sous-groupe formé des éléments T de $\Gamma'(\delta)$ tels que $T \equiv \pm 1_{2n} \pmod{q}$, et par $\Gamma(\delta, q)$ le sous-groupe $\{F^{-1}{}^t T^{-1} F \mid T \in \Gamma'(\delta, q)\}$ de $\Gamma(\delta)$. Soient a_i ($1 \leqslant i \leqslant q^{2n}$) les vecteurs de $\mathbf{R}^{2n}$ tels que les coordonnées de $q a_i$ soient des entiers non-négatifs $< q$. D'après la relation (ξ), on voit qu'un élément U de $\Gamma(\delta)$ laisse invariantes les fonctions $\xi_\alpha(z, a_i)/\xi_\beta(z, a_i)$ si et seulement si U est contenu dans $\Gamma(\delta, q)$. Il s'ensuit de là le théorème suivant.

Théorème 2. Si $n > 1$, le corps engendré sur $\mathbf{C}$ par les ϕ_i et les

$$\xi_\alpha(z, a_i)/\xi_\beta(z, a_i)$$

est le corps des fonctions méromorphes sur S_n invariantes par $\Gamma(\delta, q)$.

Dans le cas $n = 1$, les fonctions qu'on vient de construire engendrent les fonctions modulaires elliptiques de 'Stufe' q.

Les systèmes $\{A_\delta(z) \mid z \in S_n\}$ sont les plus grands systèmes de variétés abéliennes polarisées; et chaque membre générique de ces systèmes n'a pas de multiplication complexe, c.-à-d. son anneau des endomorphismes est isomorphe à $\mathbf{Z}$. En considérant les variétés abéliennes dont les anneaux d'endomorphismes contiennent un certain anneau donné, on obtient un système de variétés abéliennes polarisées, auquel notre méthode est applicable également. Il vaudrait mieux, dans ce cas, généraliser quelque peu la notion de corps de modules ainsi qu'il suit. On se bornera au cas de caractéristique 0. Soit $\mathfrak{r}$ un anneau; on entendra par une variété abélienne polarisée de type $\mathfrak{r}$, une variété abélienne polarisée $(A, \mathscr{C})$ pour laquelle est donné un isomorphisme η de $\mathfrak{r}$ dans l'anneau des endomorphismes de A; on la désignera par $(A, \mathscr{C}, \eta)$. Un isomorphisme λ de $(A, \mathscr{C})$ sur $(A' \, \mathscr{C}')$ s'appellera un isomorphisme de $(A, \mathscr{C}, \eta)$ sur $(A', \mathscr{C}', \eta')$ si l'on a $\lambda\eta(r) = \eta'(r)\lambda$ pour tout $r \in \mathfrak{r}$. Soit k un corps de définition pour $(A, \mathscr{C})$ par rapport auquel tout élément de $\eta(\mathfrak{r})$ est défini; et soit σ un isomorphisme de k sur un corps k^σ. On obtient alors une variété abélienne polarisée $(A^\sigma, \mathscr{C}^\sigma, \eta^\sigma)$ de type $\mathfrak{r}$, en posant $\eta^\sigma(r) = \eta(r)^\sigma$. On peut démontrer qu'il existe un sous-corps K' de k pour lequel σ est l'identité sur K' si et seulement si $(A, \mathscr{C}, \eta)$ est isomorphe à $(A^\sigma, \mathscr{C}^\sigma, \eta^\sigma)$. On appellera K' le *corps de modules* de $(A, \mathscr{C}, \eta)$; le corps de modules de $(A, \mathscr{C})$ est un sous-corps de K'; ils coïncident si $\mathfrak{r} = \mathbf{Z}$; la réciproque n'est pas nécessairement vrai.

Considérons par exemple le cas de fonctions de Hilbert. Soit $\mathfrak{f}$ un corps totalement réel de degré $n > 1$ sur $\mathbf{Q}$, et soit $\mathfrak{r}$ l'anneau des entiers de $\mathfrak{f}$. On désignera par $a^{(1)}, \ldots, a^{(n)}$ les conjugués de $a \in \mathfrak{f}$. Soit $\mathfrak{a}$ un idéal de $\mathfrak{r}$; désignons par $\Gamma(\mathfrak{a})$ le groupe des transformations $\tau \to (a\tau + b)/(c\tau + d)$ où a, b, c, d sont quatre éléments tels que $a \in \mathfrak{r}$, $b \in \mathfrak{a}$, $c \in \mathfrak{a}^{-1}$, $d \in \mathfrak{r}$ et que $ad - bc = 1$. $\Gamma(\mathfrak{a})$ opère sur l'espace produit H_n de n demi-plans complexes $\mathrm{Im}\,(\tau) > 0$. Soit $(\tau) = (\tau_1, \ldots, \tau_n)$ un point de H_n; soit $D(\tau, \mathfrak{a})$ le lattice de $\mathbf{C}^n$ composé des vecteurs $(a^{(1)}\tau_1 + b^{(1)}, \ldots, a^{(n)}\tau_n + b^{(n)})$ où $a \in \mathfrak{r}$, $b \in \mathfrak{a}$. Le tore $\mathbf{C}^n/D(\tau, \mathfrak{a})$ a une structure de variété abélienne, sur laquelle toute forme de Riemann correspond à un nombre y de $\mathfrak{f}$ tel que $y^{(i)} < 0$ pour tout i; chaque élément a de $\mathfrak{r}$ définit un endomorphisme de $\mathbf{C}^n/D(\tau, \mathfrak{a})$ donné par la matrice diagonale ayant pour éléments les $a^{(i)}$. On obtient ainsi les systèmes de variétés abéliennes polarisées $A(\tau, \mathfrak{a}, y)$ de type $\mathfrak{r}$. Nous pouvons démontrer qu'il existe des fonctions méromorphes $\psi_\nu(\tau)$ sur H_n telles que $\mathbf{Q}(\psi_\nu(\tau'))$ soit le corps de modules de $A(\tau', \mathfrak{a}, y)$ pour presque tout point τ' de H_n. De plus, ces fonctions engendrent sur $\mathbf{C}$ toutes les fonctions méromorphes sur H_n invariantes par $\Gamma(\mathfrak{a})$. Si l'on ne tient pas compte des endomorphismes, on se procure un certain sous-corps de ce corps de fonctions comme 'corps de modules absolus'; pour qu'ils soient les mêmes, il faut et il suffit qu'on ait $(ya)^\sigma \neq ya$ pour tout automorphisme $\sigma \neq 1$ de $\mathfrak{f}$.

On se propose maintenant d'étudier les corps de définition pour les corps de fonctions automorphes. Supposons qu'on ait défini un système $\{A(s)\}$ de variétés abéliennes polarisées de type $\mathfrak{o}$, où $\mathfrak{o}$ est un anneau, dont les membres dépendent d'une manière convenable de points s sur un sous-ensemble S ouvert connexe de $\mathbf{C}^m$; on obtient alors des fonctions méromorphes χ_ν sur S telles que $\mathbf{Q}(\chi_\nu(s'))$ soit le corps de modules de $A(s')$ pour presque tout $s' \in S$. Soit k un sous-corps dénombrable de $\mathbf{C}$. Le système est dit *complet* par rapport à k s'il satisfait à la condition suivante:

Soit s_0 un point de S tel que tout $A(s)$ soit une spécialisation de $A(s_0)$ sur k. Si B est une spécialisation générique de $A(s_0)$ sur k, il existe un point s_1 tel que deux variétés abéliennes polarisées $A(s_1)$ et B de type $\mathfrak{o}$, soient isomorphes.

Notre critérium de corps de définition s'énonce:

Théorème 3. *Si le système* $\{A(s)\}$ *est complet par rapport à un corps dénombrable* k, *le corps* $k(\chi_\nu)$ *est une extension régulière de* k *et l'on a* $\dim_k k(\chi_\nu) = \dim_{\mathbf{C}} \mathbf{C}(\chi_\nu)$.

On vérifie facilement que les systèmes $\{A_\delta(z)\}$ et $\{A(\tau, \mathfrak{a}, y)\}$ sont tous complets par rapport à $\mathbf{Q}$.

Ce théorème est applicable même au cas de domaine fondamentale compact, où l'on ne peut pas se servir de série de Fourier. On sait à titre d'exemple une classe de fonctions automorphes de dimension 1 définie pour la première fois par Poincaré, qui sont en rapport avec une forme quadratique ternaire indéfinie, et qu'on trouve dans le livre de Fricke et Klein. Les recherches sont 'peu développées', comme Eichler a dit, dans l'arithmétique de ces fonctions. On va maintenant s'occuper de cette classe. Soit $\mathfrak{A}$ une algèbre de quaternions sur $\mathbf{Q}$ dont la norme est une forme quadratique indéfinie, et soit $\mathfrak{o}$ un ordre maximal de $\mathfrak{A}$. Comme $\mathfrak{A}$ contient un corps quadratique réel $\mathfrak{K}$, $\mathfrak{A}$ a une représentation de degré 2 à coefficients dans $\mathfrak{K}$, que l'on désignera par M. Soit γ un unité de $\mathfrak{o}$ tel que $\det M(\gamma) = 1$; γ donne une transformation

$$\tau \to \frac{a\tau + b}{c\tau + d}$$

du demi-plan complexe H, où $M(\gamma) = \begin{pmatrix} a & b \\ c & d \end{pmatrix}$. Désignons par $\Gamma(\mathfrak{o})$ le groupe des transformations ainsi obtenues; si $\mathfrak{A}$ n'a pas de diviseur de zéro, le domaine fondamental est compact. Soient τ un point de H et $D(\tau)$ le lattice de $\mathbf{C}^2$ composé des vecteurs $M(\alpha) \begin{pmatrix} \tau \\ 1 \end{pmatrix}$ pour $\alpha \in \mathfrak{o}$. On obtient sur le tore $\mathbf{C}^2/D(\tau)$ une forme de Riemann correspondant à un élément w de $\mathfrak{A}$ tel que w^2 soit un nombre négatif de $\mathbf{Q}$. La matrice $M(\alpha)$ pour $\alpha \in \mathfrak{o}$ donne un endomorphisme de $\mathbf{C}^2/D(\tau)$. On peut ainsi définir un système $\{A(\tau)\}$ de variétés abéliennes polarisées de type $\mathfrak{o}$. Le corps de modules de $A(\tau)$ est donné par les valeurs de certaines fonctions méromorphes $g_i(\tau)$ sur H, qui engendrent toutes les fonctions automorphes par rapport à $\Gamma(\mathfrak{o})$. On vérifie que $\{A(\tau)\}$ est complet par rapport à $\mathbf{Q}$, de sorte que le corps $\mathbf{Q}(g_i)$ est une extension régulière de $\mathbf{Q}$ de dimension 1. Par suite il existe une courbe algébrique définie sur le corps rationnel donnant un modèle du corps des fonctions automorphes par rapport à $\Gamma(\mathfrak{o})$.

Commençons la théorie des correspondances modulaires par l'étude du groupe $\Gamma(\mathfrak{o}, q)$ composé des transformations obtenues à partir des unités $\gamma \in \mathfrak{o}$ tels que $\gamma \equiv \pm 1 \bmod q$, où q est un entier positif. Dans ce but, on modifie, eu égard aux endomorphismes, la définition de variété de Kummer et les propriétés (W 1–W 3). Soient W une variété de Kummer de $A(\tau)$ et h une application de $A(\tau)$ sur W ayant les propriétés modifiées. Les coordonnées du point $h(t)$, où t est un point sur $A(\tau)$ tel que $qt = 0$, regardées comme fonctions de τ, donnent des fonctions méromorphes f_j sur H, qui engendrent, avec les g_i, toutes les fonctions automorphes par rapport à $\Gamma(\mathfrak{o}, q)$. Prenons un point τ_0 sur H tel que tout $A(\tau)$ soit une spécialisation de $A(\tau_0)$ sur $\mathbf{Q}$, et posons $K_q = \mathbf{Q}(g_i(\tau_0), f_j(\tau_0))$. K_q est une extension galoisienne de K_1 dont le groupe de Galois est isomorphe au groupe G des éléments réguliers de l'anneau $\mathfrak{o}/q\mathfrak{o}$. ζ_q étant une racine primitive q-ième d'unité, $\mathbf{Q}(\zeta_q)$ est algébriquement fermé dans K_q; $K_1(\zeta_q)$ correspond à un sous-groupe des éléments α tels que $\det M(\alpha) \equiv 1 \bmod q$. Un élément α donne un automorphisme $\zeta_q \to \zeta_q^m$ sur $\mathbf{Q}(\zeta_q)$, où $m = \det M(\alpha)$. Si q est premier avec le discriminant de l'algèbre $\mathfrak{A}$ (ce que l'on suppose dans ce qui suit), K_q contient un sous-corps K_q' tel que $K_q = K_q'(\zeta_q)$ et $K_q' \cap \mathbf{Q}(\zeta_q) = \mathbf{Q}$; il existe donc une courbe algébrique C_q définie sur $\mathbf{Q}$, dont le corps des fonctions est le corps des fonctions automorphes par rapport à $\Gamma(\mathfrak{o}, q)$. Soit p un nombre premier. Les points u sur $A(\tau_0)$ tels que $pu = 0$ forment un groupe $\mathfrak{g}$ d'ordre p^4 invariant par $\mathfrak{o}$. Il existe exactement $p + 1$ sous-groupes de $\mathfrak{g}$ d'ordre p^2, invariants par $\mathfrak{o}$, qu'on notera par $\mathfrak{g}_1, \ldots, \mathfrak{g}_{p+1}$. Chaque $\mathfrak{g}_\nu$ correspond à un idéal $\mathfrak{o}\alpha_\nu$ de norme p de telle façon qu'il existe un homomorphisme λ_ν de $A(\tau_0)$ sur $A(\tau_\nu)$ dont le noyau est $\mathfrak{g}_\nu$, où

$$\tau_\nu = \frac{a_\nu \tau_0 + b_\nu}{c_\nu \tau_0 + d_\nu}, \quad M(\tau_\nu) = \begin{pmatrix} a_\nu & b_\nu \\ c_\nu & d_\nu \end{pmatrix}.$$

On définit un isomorphisme σ_ν de K_q par $g_i(\tau_0)^{\sigma_\nu} = g_i(\tau_\nu)$ et $h(t)^{\sigma_\nu} = h_\nu(\lambda_\nu t)$ pour $qt = 0$, où h_ν est l'application naturelle de $A(\tau_\nu)$ sur sa variété de Kummer. Soit x un point générique de C_q et soit X_p le lieu de $x \times x^{\sigma_1}$ par rapport à $\mathbf{Q}$; X_p s'appellera la *correspondance modulaire de degré p* sur C_q. Soit P un diviseur premier de p dans une clôture algébrique de K_q; on indiquera par la barre la réduction modulo P. La réduction modulo P donne un homomorphisme de $\mathfrak{g}$ sur le groupe $\bar{\mathfrak{g}}$ des éléments $\bar{u}$ sur $\overline{A(\tau_0)}$ tels que $p\bar{u} = 0$. Comme $\bar{\mathfrak{g}}$ est d'ordre p^2, le noyau de cet homomorphisme est un des $\mathfrak{g}_\nu$, mettons $\mathfrak{g}_1$. On voit alors que le noyau de $\bar{\lambda}_\nu$ est $\bar{\mathfrak{g}}$ ou $\{0\}$ selon que $\nu > 1$ ou $\nu = 1$. Désignons par μ_ν l'homomorphisme de $A(\tau_\nu)$ sur $A(\tau_0)$ tel que $\mu_\nu \lambda_\nu = p$; le noyau des $\bar{\mu}_\nu$ est d'ordre 1 ou p^2 selon que $\nu > 1$ ou $\nu = 1$. On en déduit que $\overline{A(\tau_1)}$ est isomorphe à $\overline{A(\tau_0)}^p$ et que

$\overline{A(\tau_0)}$ est isomorphe à $\overline{A(\tau_\nu)}^p$ pour $\nu > 1$; les homomorphismes $\overline{\lambda}_1$ et $\overline{\mu}_\nu$ pour $\nu > 1$ sont équivalents aux homomorphismes de p-ième puissance. Nous pouvons démontrer, d'après ces relations, deux formules de congruence pour la correspondance modulaire

$$\overline{X}_p = \Pi + \Pi' \circ \overline{Y}_p, \quad \Pi' \circ \overline{Y}_p = \overline{Z}' \circ \Pi' \circ \overline{Z}$$

sur $\overline{C}_q$ pour presque tous p, où Π est la correspondance $\overline{x} \to \overline{x}^p$ sur $\overline{C}_q$, Y_p est la correspondance birationnelle de C_q donné par $h(t) \to h(pt)$, Z est une certaine correspondance birationnelle de C_q et $'$ désigne l'anti-automorphisme de Rosati. Ces formules sont des généralisations de celles qui ont été obtenues pour les fonctions modulaires elliptiques (Eichler[2], Shimura[5]), puisque nos fonctions automorphes coïncident avec les fonctions modulaires elliptiques de 'Stufe' q, si o est l'ensemble des matrices de degré 2 à coefficients entiers. La représentation de X_p par les formes différentielles de première espèce, n'est autre que l'opérateur T_p de Hecke, pour les formes paraboliques de poids 1. Par suite la fonction ζ de la courbe C_q s'exprime sous la forme

$$\zeta(s, C_q) = r(s)\,\zeta(s)\,\zeta(s-1)\,\Phi(s)^{-1}$$

où $\zeta(s)$ est la fonction ζ de Riemann, $r(s)$ est une fonction rationnelle de p^{-s} et $\Phi(s)$ désigne un produit d'Euler du type introduit par Hecke. D'après un résultat de Weil, on constate que les valeurs absolues des racines caractéristiques de l'opérateur T_p pour les formes paraboliques de poids 1 ne dépassent pas $2\sqrt{p}$ pour presque tous les nombres premiers p.

On signale que les formules de congruence sont démontrées pour les correspondances elles-mêmes non seulement pour les classes de correspondances. Ce fait nous semble bien significatif pour les formes automorphes de poids > 1.

On peut définir, de la même manière que ci-dessus, les correspondances modulaires pour les fonctions automorphes de plusieurs variables au moyen des isogénies de variétés abéliennes; on obtiendra alors les formules de congruence pour ces correspondances. Et il est à souhaiter en déduire quelque chose d'intéressant; le conférencier regrette qu'il n'a rien à dire sur ce que signifient ces formules.

BIBLIOGRAPHIE

[1] Baily, W. L. Satake's compactification of V_n. Amer. J. Math. 80, 348–364 (1958).
[2] Eichler, M. Quaternäre quadratische Formen und Riemannsche Vermutung für die Kongruenzzetafunktion. Arch. Math. 5, 355–366 (1954).

[3] Matsusaka, T. Polarized varieties, fields of moduli and generalized Kummer varieties of polarized abelian varieties. *Amer. J. Math.* 80, 45–82 (1958).

[4] Satake, I. et Cartan, H. Exposés 11–17 du *Séminaire H. Cartan*, 10 (1957–58).

[5] Shimura, G. Correspondances modulaires et les fonctions ζ de courbes algébriques. *J. Math. Soc. Japan*, 10, 1–28 (1958).

[6] Shimura, G. Modules des variétés abéliennes polarisées et fonctions modulaires. *Séminaire H. Cartan*, 10 (1957–58).

[7] Weil, A. On the theory of complex multiplication. *Proc. Int. Symp. Alg. number theory*. Tokyo-Nikko, 1955. Tokyo, Science Council of Japan, 9–22 (1956).

[8] Weil, A. The field of definition of a variety. *Amer. J. Math.* 78, 509–524 (1956).

В. И. АРНОЛЬД

НЕКОТОРЫЕ ВОПРОСЫ ПРИБЛИЖЕНИЯ И ПРЕДСТАВЛЕНИЯ ФУНКЦИЙ†

1. Постановка задачи

Пусть f и g — функции двух переменных. Тогда

$$F(x, y, z) = f[x, g(y, z)]$$

функция трех переменных x, y и z. Это пример *суперпозиции*, составленной из функций f и g.

Вообще, *суперпозицией, составленной из данных функций, или суперпозицией данных функций, называется функция, получающаяся при подстановке одних из них в другие вместо аргументов.*

Понятие суперпозиции —одно из основных в анализе. Например, элементарные функции —это, по определению, суперпозиции функций $a(x, y) = x + y$, $b(x, y) = xy$, $c(x, y) = x^y$ и известных функций одного переменного $\ln x$, $\sin x$ и др.

Очевидно, суперпозиция, составленная из функций двух переменных, может быть функцией любого числа переменных. Здесь рассматривается обратный вопрос: какие функции многих переменных являются суперпозициями функций меньшего числа переменных.

Постановка вопроса принадлежит Д. Гильберту. Корни уравнений 5-ой и 6-ой степени, как функции коэффициентов, оказываются суперпозициями функций двух переменных. Для уравнения 7-ой степени не удавалось получить такое представление: дело сводится к функциям трех переменных. Это побудило Гильберта поставить следующую задачу[1] (13 проблема из 'Математических проблем'):

Всякая ли аналитическая функция трех переменных является суперпозицией, составленной из непрерывных функций двух переменных? Является ли корень $x(a, b, c)$ *уравнения*

$$x^7 + ax^3 + bx^2 + cx + 1 = 0$$

суперпозицией непрерывных функций двух переменных?

Следует обратить внимание на классы функций, из которых составляются суперпозиции. Легко видеть[2], что суперпозиция

† Read in English translation by Professor J. L. B. Cooper.

разрывных функций двух переменных может быть при должном их подборе *любой* функцией 3-х переменных.

С другой стороны, имеет место

Теорема 1 (Д. Гильберта). *Существуют аналитические функции трех переменных, не являющиеся суперпозициями бесконечо дифференцируемых функций двух переменных.*

Это может быть объяснено так: число независимых коэффициентов ряда Тэйлора до порядка n у функции трех переменных порядка n^3, а у функции двух переменных порядка n^2. Поэтому, если функция трех переменных есть суперпозиция любого *фиксированного вида* (например, $f[x, g(y, z)]$) бесконечно дифференцируемых функций двух переменных, то между коэффициентами её ряда Тэйлора достаточно высокого порядка должно выполняться некое соотношение, отвечающее виду суперпозиции (для указанного выше простейшего вида достаточно взять коэффициенты при членах до 4-го порядка). Различных видов суперпозиций счетное множество; существует аналитическая функция трех переменных, избегающая все такие соотношения. Она не может быть суперпозицией *аналитических* функций двух переменных никакого вида.

Этим объясняется постановка вопроса о возможности разложения в суперпозиции *непрерывных* функций. Гильберт ожидал, что и здесь такое разложение не всегда возможно.

Для суперпозиций простейших видов это действительно так[2, 3, 4].

2. Суперпозиции гладких функций

Витушкин[5, 4] показал, что если разложение любой гладкой (имеющей p производных) функции трех переменных в суперпозицию функций двух переменных и возможно, то лишь с понижением гладкости в полтора раза.

Рассмотрим класс заданных на единичном n-мерном кубе E^n функций, имеющих все частные производные до порядка p включительно, и пусть все p-ые производные удовлетворяют условию Гёльдера степени $0 < \alpha \leqslant 1.$†

Определение. Класс всех таких функций обозначается $F_{p,\alpha}^n$; n называется *размерностью*, $p+\alpha$ — *гладкостью*, а $(p+\alpha)/n$ — *качеством* функций класса.

Заметим, что, например, гладкость 2 имеют функции, у которых

---

† Функция $f(x_1 \ldots x_n)$ удовлетворяет условию Гёльдера степени α с константой M, если $|f(x) - f(y)| < M \|x - y\|^\alpha$.

первые производные удовлетворяют условию Липшица (условию Гёльдера с $\alpha = 1$).

Теорема 2 (Витушкина). *Существует функция класса $F_{p,\alpha}^n$, не представимая в виде суперпозиции функций лучшего качества гладкости* ≥ 1, *т.е. функций классов $F_{q,\beta}^m$ с* $(q+\beta)/m > (p+\alpha)/n, q \geq 1$.

Например, функции трех переменных гладкости 3 можно надеяться разложить в суперпозицию функций двух переменных лишь гладкости не более 2.

Полученные очень сложным образом с помощью теории многомерных вариаций[4], созданной Витушкиным на базе исследований Кронрода, Адельсона-Вельского, Ландиса и др.[6], эти результаты были затем связаны Колмогоровым[7] с идеями теории передачи информации Шеннона. Возникшие при этом соображения имеют самостоятельное значение, выходящие далеко за пределы проблемы Гильберта, которая стимулировала их развитие (см. также доклад Колмогорова 'Линейная размерность топологических векторных пространств').

3. ϵ-энтропия классов функций

Чтобы указать точку отрезка (0, 1) с точностью 0,001 требуется 3 десятичных знака. Указание точки отрезка с точностью ϵ требует числа десятичных знаков порядка $\lg 1/\epsilon$, а точки n-мерного куба — порядка $n \lg 1/\epsilon$ знаков. Имеет также смысл говорить о 'числе знаков, необходимых для задания с точностью ϵ функций $f \in F$', где F какой-либо класс функций. Это число будет 'минимальным объемом таблицы функции'. Известно, насколько возрастает объем таблицы с ростом числа переменных и как он уменьшается при увеличении гладкости функций (интерполяция высокого порядка!). Оказывается, минимальный объем таблицы функции класса $F_{p,\alpha}^n(C)$ при точности ϵ имеет когда $\epsilon \to 0$ порядок роста $(1/\epsilon)^{n/(p+\alpha)}$. Здесь $F_{p,\alpha}^n(C)$ есть класс функций $f \in F_{p,\alpha}^n$, у которых $|f|$ и абсолютные величины частных производных порядка $\leq p$ не превосходят C и которые удовлетворяют условию Гёльдера с константой C. В теории передачи информации принято считать не десятичные, а двоичные знаки. Поэтому в приводимых ниже точных формулировках появляются двоичные логарифмы.

Пусть X — вполне ограниченное множество в метрическом пространстве R. По определению, при любом $\epsilon > 0$ существует конечное множество точек R таких, что шары радиуса ϵ с центрами в этих точках крывают X (ϵ-сеть в R для X).

Определение. Пусть

$N_\epsilon^R(X)$ — *минимальное число точек в ϵ-сети в R для X.*

$N_\epsilon(X)$ — *минимальное число множеств диаметра 2ϵ покрывающих X.*

$M_\epsilon(X)$ — *максимальное число ϵ-различных сигналов в X, т.е. максимальное число точек X, таких, что шары радиуса ϵ с центрами в этих точках не пересекаются попарно.*

Тогда

$$H_\epsilon^R(X) = \log_2 N_\epsilon^R(X), \quad H_\epsilon(X) = \log_2 N_\epsilon(X), \quad C_\epsilon(X) = \log_2 M_\epsilon(X)$$

называются соответственно ϵ-энтропией X относительно R, ϵ-энтропией X и ϵ-ёмкостью X.

Легко доказать, что

$$H_{2\epsilon}^R(X) \leqslant C_\epsilon(X) \leqslant H_\epsilon(X) \leqslant H_\epsilon^R(X).$$

Если называть *таблицей с точностью ϵ* функции данного класса любой набор 0 и 1, по которому эта функция определяется в каждой точке с точностью ϵ, то число различных таблиц из n знаков будет 2^n. Поэтому естественен результат Витушкина доказавшего, что $H_\epsilon(X)$ в случае, когда X — компакт пространства C† есть минимальный объем таблицы функции класса X с точностью $\epsilon^{[10]}$.

Для ряда важнейших классов установлены оценки $H_\epsilon^{[8,\,9,\,11]}$.

(А) Для класса функций n комплексных переменных, аналитических в области $G = \prod_{i=1}^{n} G_i$ (G_i — область комплексной плоскости), равномерно ограниченных какой-либо константой C и рассматриваемых на $K = \prod_{i=1}^{n} K_i$ ($K_i \subset G_i$ — континуумы)

$$\lim_{\epsilon \to 0} \frac{H_\epsilon}{C(K, G)(\log 1/\epsilon)^{n+1}} = 1;$$

здесь $C(G, K)$ — вычисляемая определенным в [9] регулярным образом геометрическая характеристика G и K.

(В) В$^{[8,\,11]}$ доказана

Теорема 3 (Витушкина и Колмогорова).

$$k\left(\frac{C}{\epsilon}\right)^{n/(p+\alpha)} \leqslant H_\epsilon(F_{p,\alpha}^n(C)) \leqslant K\left(\frac{C}{\epsilon}\right)^{n/(p+\alpha)},$$

где $0 < k \leqslant K < \infty$ не зависящие от ϵ константы.

† Непрерывных функций на компакте с метрикой

$$\rho(f, g) = \max |f - g|.$$

Используя эту формулу, легко доказать при $q \geqslant 1$, что суперпозиция некоторого фиксированного вида из функций качества $(q+\beta)/m$ данного компактного семейства $F_{q,\alpha}^m(C_1)$ не может дать всех функций компактного семейства $F_{p,\alpha}^n(C)$, если качество последних хуже $((p+\alpha)/n < (q+\beta)/m)$.

Грубо говоря, дело в том, что минимальный объем таблицы функции класса $F_{p,\alpha}^n(C)$ порядка $(1/\epsilon)^{n/(p+\alpha)}$, а объем таблиц всех входящих в суперпозицию конкретного вида функций класса $F_{q,\beta}^m(C_1)$ порядка $(1/\epsilon)^{m/(q+\beta)}$. Если разложение всех функций класса $F_{p,\alpha}^n(C)$ в такую суперпозицию функций класса $F_{q,\beta}^m(C_1)$ возможно, то достаточно полная таблица всех входящих в суперпозицию функций заменяет таблицу с точностью ϵ функции-суперпозиции; поэтому $(1/\epsilon)^{n/(p+\alpha)} \leqslant K(1/\epsilon)^{m/(q+\beta)}$, где $0 < K < \infty$ не зависит от ϵ. Значит $n/(p+\alpha) \leqslant m/(q+\beta)$: если разложение и возможно, то в функции не лучшего качества, чем разлагаемые. Применяя известный метод 'накопления гадостей' строим теперь функцию класса $F_{p,\alpha}^n(C)$, не представимую вообще *никакой* суперпозицией функций классов $F_{q,\beta}^m(C_1)$ при всевозможных C_1 и $(q+\beta)/m > (p+\alpha)/n$. Это и есть теорема Витушкина.

4. Суперпозиции непрерывных функций

Однако в формулировке Гильберта речь шла не о гладких, а о непрерывных функциях. В этой области результаты оказались противоположными его гипотезе.

В 1956 г. Колмогоров показал[12], что любая непрерывная функция, заданная на $n \geqslant 3$-х мерном кубе E^n представима в виде

$$f(x_1, ..., x_n) = \sum_{r=1}^{n} h^r[x_n, g_1^r(x_1, ..., x_{n-1}), g_2^r(x_1, ..., x_{n-1})],$$

где функции $n-1$ переменного g и функции 3-х переменных h действительны и непрерывны.

Применяя это разложение много раз, видим, что любая непрерывная функция $n \geqslant 4$-х переменных есть суперпозиция непрерывных функций 3-х переменных.

Доказательство весьма сложно. Основным аппаратом, используемым при этом, является дерево компонент множеств уровня функции, введенное Кронродом[6].

Множеством уровня c функции $f(\mathbf{x})$ называется совокупность всех тех точек $\mathbf{x}$ области определения, для которых $f(\mathbf{x}) = c$.

Компонентой множества уровня называется каждый связный

кусок множества уровня. На черт. 1 множества уровней $0 \leqslant c \leqslant \frac{1}{2}$ состоят из одной компоненты, при $\frac{1}{2} < c \leqslant \frac{2}{3}$ из двух компонент, при $\frac{2}{3} < c \leqslant 1$ —из одной.

Функция есть отображение области определения на область значений. Это отображение может быть представлено как произведение двух:

(1) Отображение области определения на множество компонент множеств уровня —каждой точке ставится в соответствие та компонента, которой она принадлежит.

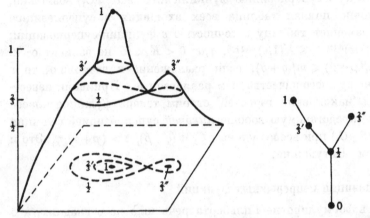

Черт. 1. Множества уровней 0, $\frac{1}{2}$, $\frac{2}{3}$ и 1 обозначены теми же цифрами, что уровни. $\frac{2}{3}{}'$ и $\frac{2}{3}{}''$ – две компоненты множества уровня $\frac{2}{3}$.

Черт. 2. Дерево компонент множеств уровня функции черт. 1. Компоненты обозначены так же, как на черт. 1.

(2) Отображение множества компонент на множество значений: каждой компоненте ставится в соответствие значение, принимаемое на ней функцией.

Пусть область определения —компакт F.

Если функция непрерывна, то в множестве компонент вводится 'естественная топология'. Пусть A—компонента, U—любое открытое множество F содержащее A. Тогда совокупность всех тех компонент множеств уровня, которые пересекаются с U, объявляется окрестностью U_A компоненты A.

Теперь первое отображение оказывается непрерывным монотонным†, а второе—непрерывным с нульмерными прообразами. Отсюда вытекает, что пространство компонент есть локально

_____
† Т.е. прообраз каждой точки связен.

связный односвязный континуум, т.е. дерево [13,14]. Оно и назвается деревом функции.

Связь его с функцией очень простая. Например, функция черт. 1 имеет дерево, гомеоморфное рогатке черт. 2. Число кусков, на которые точка дерева функции разбивает его, равно числу частей, на которые соответствующая компонента множества уровня делит область определения.

Из многочисленных замечательных свойств дерева в [12] используется то, что на плоскости существует универсальное дерево (оно содержит гомеоморфные любому дереву подмножества). О зубчатости функций, входящих в указанную выше суперпозицию и в дальнейшие свидетельствует то, что их деревья универсальны или почти универсальны.

Гильбертова проблема была сформулирована для функций *трех* переменных, и теорема Колмогорова не давала тут ответа. Однако выяснилось [15], что усложняя далее его конструкции и располагая дерево в трехмерном пространстве так, чтобы любая функция на нем представлялась в виде суммы функций координат, можно представить любую непрерывную функцию, заданную на трехмерном кубе, в виде

$$f(x_1, x_2, x_3) = \sum_{i=1}^{3} \sum_{j=1}^{3} h_{ij}[\phi_{ij}(x_1, x_2), x_3],$$

где h и ϕ — действительные и непрерывные функции двух переменных.

Тем самым доказана (вопреки гипотезе Гильберта) возможность представить любую непрерывную функцию $n \geqslant 3$-х переменных в виде суперпозиции непрерывных функций двух переменных.

Наконец, вскоре после этого Колмогорову удалось показать, что справедлива

Теорема 4. Любая функция, непрерывная на n-мерном кубе, представима в виде

$$f(x_1, ..., x_n) = \sum_{i=1}^{2n+1} \chi_i \left[\sum_{j=1}^{n} \phi_{ij}(x_j) \right],$$

где функции χ и ϕ — действительные и непрерывные функции одного переменного.

Таким образом, все непрерывные функции оказались суперпозициями непрерывных функций одного переменного и одной единственной функции двух переменных — сложения.

Метод этой работы элементарнее [12] и [15], и не использует понятия дерева. Доказательство теоремы 4 может быть легко понято по заметке [16].

Функции ϕ_{ij} — стандартные, не зависящие от $f(x_1,\ldots,x_n)$. Построения [16] можно рассматривать поэтому как помещение с помощью стандартных функций

$$\phi_i(x_1,\ldots,x_n)=\sum_{j=1}^{n}\phi_{ij}(x_j)\quad(i=1,\ldots,2n+1)$$

специального гомеоморфа F n-мерного куба E^n в $2n+1$-мерное пространство. Функция $f(x_1,\ldots,x_n)$ индуцирует на F непрерывную функцию $f(\phi_1,\ldots,\phi_{2n+1})$. F обладает тем замечательным свойством, что любая непрерывная функция $f(\phi_1,\ldots,\phi_{2n+1})$ на F представляется в виде суммы функций $\chi_i(\phi_i)$ координат точки F.

5. Задачи

Полученные результаты можно свести в следующую таблицу:

Представляемые Функции \ Используемые Функции	C^m	$q+\beta\to\infty$ $F_{q,\beta}^m$	F_∞^m	A^m
C^n	+	—	—	—
$p+\alpha \downarrow \infty$ $F_{p,\alpha}^n$	+	$q+\beta>(p+\alpha)m/n,\,q\geqslant1$ — $q<1$, или $q+\beta\leqslant(p+\alpha)m/n$?	—	—
F_∞^n	+	?	—	—
A^n	+	?	—	—

Здесь C^n — класс всех непрерывных функций на n-мерном кубе, F_∞^n — всех бесконечно-дифференцируемых, A^n — аналитических; $+$ означает, что все функции класса слева являются суперпозициями функций класса наверху ($m<n$). Отсюда естественно возникает

Задача 1. *Разлагается ли всякая функция класса $F_{p,\alpha}^n$ в суперпозицию функций класса $F_{q,\beta}^m$ при $(q+\beta)/m=(p+\alpha)/n$? при*

$$(q+\beta)/m>(p+\alpha)/n-\varepsilon\quad(\varepsilon>0)\quad(m<n)?$$

Разлагается ли каждая функция класса F_∞^n, A^n в суперпозицию функций $F_{q,\beta}^m$? F_∞^m?

Изучение отдельных видов суперпозиций показывает весьма своеобразные свойства классов функций, представимых в виде суперпозиции данного вида [17]. Отсюда

Задача 2. Найти простейшую суперпозицию функций $m < n$ переменных, в виде которой может быть представлена: (a) данная функция n переменных; (б) данный класс непрерывных функций n переменных; (в) все непрерывные функции n переменных. Исследовать аналогичные вопросы для аппроксимации с произвольной точностью. О более практическом подходе к задачам такого рода см. [18].

Оценка H_ϵ, даваемая теоремой 3, грубая, так как константы k и K остаются неопределенными. Неясно, как они зависят от C, n, p и α; неизвестна асимптотика H_ϵ, т.е. при какой функции $\phi(\epsilon)$ (предположительно — константе)

$$\lim_{\epsilon \to 0} \frac{H_\epsilon}{\phi(\epsilon)\,(1/\epsilon)^{n/(p+\alpha)}} = 1.$$

Трудность этих вопросов становится ясной, если заметить, что в гораздо более простом случае евклидовой метрики им соответствуют задачи о плотнейшей укладке шаров и экономнейшем покрытии пространства шарами.

Задача 3. Улучшить оценки H_ϵ, даваемые теоремой 3. Установить асимптотику $H_\epsilon(F_{p,\alpha}^n(C))$ при $\epsilon \to 0$.

Так как ϵ-энтропия $H_\epsilon(F)$ характеризует минимальный объем таблицы функции класса F с точностью ϵ, знание поведения H_ϵ существенно для оценки различных методов приближенного задания функций, их введения в машины и сохранения в памяти машин [17, 20]. Однако, здесь будет важным знание H_ϵ при малых, но конечных ϵ.

Задача 4. Дать для различных классов ($F_{p,\alpha}^n(C)$ и т. п.) точные оценки H_ϵ при конечных ϵ. Исследовать способы табулирования, при которых объем таблицы приближается к минимальному. Оценить возрастание трудности пользования таблицей при уменьшении её объема.

ЦИТИРОВАННАЯ ЛИТЕРАТУРА

[1] Hilbert, D. *Gesammelte Abhandlungen*, 3, no. 17 (1935).
[2] Полиа, Г. и Сеге, Г. *Задачи и теоремы из анализа*, 1, задачи 119 и 119а.
[3] Арнольд, В. И. О представимости функций двух переменных в виде $\chi[\phi(x)+\psi(y)]$ *УМН*, 12, 2 (74), 119–121 (1956).

[4] Витушкин, А. Г. *О многомерных вариациях*. Гостехиздат, Москва, стр. 201 (1955).

[5] Витушкин, А. Г. К тринадцатой проблеме Гильберта. *ДАН СССР*, 95, 4, 701–704 (1954).

[6] Кронрод, А. С. О функциях двух переменных. *УМН*, 5, в. 1 (35) (1950).

[7] Колмогоров, А. Н. Оценки минимального числа элементов ϵ-сетей в различных функциональных классах и их применение к вопросу о представлении функций нескольких переменных суперпозициями функций меньшего числа переменных. *УМН*, 10, в. 1, стр. 192 (1955).

[8] Колмогоров, А. Н. О некоторых асимптотических характеристиках вполне ограниченных метрических пространств. *ДАН СССР*, 108, 3, 385–387 (1956).

[9] Ерохин, В. Д. *ДАН СССР*, 120, 4 и 5 (1958).

[10] Витушкин, А. Г. Абсолютная ϵ-энтропия метрических пространств. *ДАН СССР*, 117, 5, 745–747 (1957).

[11] Витушкин, А. Г. О наилучших приближениях дифференцируемых и аналитических функций. *ДАН СССР*, 119, 3, 418–420 (1958).

[12] Колмогоров, А. Н. О представлении непрерывных функций нескольких переменных суперпозициями непрерывных функций меньшего числа переменных. *ДАН СССР*, 108, 2, 179–182 (1956).

[13] Kuratowski, C. *Topologie II*, Warszawa-Wrocław, § 46 (1950).

[14] Menger, K. *Kurventheorie*, Berlin-Leipzig, Кар. х (1932).

[15] Арнольд, В. И. О функциях трех переменных, *ДАН СССР*, 114, 4, 679–681 (1957).

[16] Колмогоров, А. Н. О представлении непрерывных функций нескольких переменных в виде суперпозиции непрерывных функций одного переменного и сложения. *ДАН СССР*, 114, 5, 953–956 (1957).

[17] Ли Дя Гон. Представление функции двух переменных в виде $\chi[\phi(x) + \psi(y)]$. *Сухак ка мулли, Математика и физика*, 1, no. 4, 22–28 (1957). (Корейск.)

[18] Шура-Бура, М. Р. Аппроксимация функций многих переменных функциями, каждая из которых зависит от одного переменного. *Вычислительная математика*, Сборник 2, Издательство АН СССР (1957).

[19] Витушкин, А. Г. Некоторые оценки из теории табулирования. *ДАН СССР*, 114, 5 (1957).

[20] Бахвалов, Н. С. О составлении уравнений в конечных разностях при приближенном решении уравнения Лапласа. *ДАН СССР*, 114, 6, 1146–1148 (1957).

SPACES OF RIEMANN SURFACES†

By LIPMAN BERS

This address is a progress report on recent work, partly not yet published, on the classical problem of moduli. Much of this work consists in clarifying and verifying assertions of Teichmüller[23-28] whose bold ideas, though sometimes stated awkwardly and without complete proofs, influenced all recent investigators, as well as the work of Kodaira and Spencer on the higher dimensional case. Following Teichmüller we consider not the space of closed Riemann surfaces of a given genus g but rather an appropriate covering space and certain related spaces. For the sake of brevity the simple and somewhat exceptional cases $g = 0$ and $g = 1$ will be omitted.

Our main technical tools are uniformization theory and the theory of partial differential equations. The problem of moduli has also an algebraico-geometrical aspect, but the topological and analytical methods used here are, of course, restricted to the classical case. On the other hand, they are, in principle, applicable also to open surfaces.

1. Quasiconformal mappings

Let $w = w(x) = u(x, y) + iv(x, y)$ be a homeomorphism of a domain $\mathscr{D}$ in the z-plane onto a domain in the w-plane, and let k be a number such that $0 \leqslant k < 1$; we set $K = (1+k)/(1-k)$. There exist three distinct ways of defining k-quasiconformality of the mapping w.

Definition A (Morrey[17], Caccioppoli[8], Bers and Nirenberg[6]). The derivatives w_x, w_y exist as generalized L_2 derivatives and almost everywhere

$$|w_x + iw_y| \leqslant k |w_x - iw_y|. \tag{1}$$

For a C_1 mapping w with positive Jacobian this is the original definition used by Grötzsch[10-12], Ahlfors[1] and Teichmüller[23]. We recall that w_x is called a generalized L_2 derivative of w if w and w_x are measurable and locally square integrable in $\mathscr{D}$ and $\int w_x \phi \, dx \, dy = -\int w \phi_x \, dx \, dy$ for every C_∞ function ϕ with compact support in $\mathscr{D}$.

Definition B (Ahlfors[2], Pfluger[19], Mori[16]). For every topological rectangle $\mathscr{R} \subset \mathscr{D}$

$$\operatorname{mod} w(\mathscr{R}) \leqslant K \operatorname{mod} \mathscr{R}. \tag{2}$$

† Work performed with the sponsorship of the Office of Ordnance Research, U.S. Army, Contract No. DA–30–069–2153.

We recall that a topological rectangle $\mathscr{R}$ is a conformal image of a closed rectangle $0 \leqslant \xi \leqslant 1$, $0 \leqslant \eta \leqslant m$, and $\mathrm{mod}\,\mathscr{R} = m$.

Definition C (Lavrent'ev[15], Pesin[18], Jenkins[13]). At almost all points z of $\mathscr{D}$

$$\limsup_{r \to 0} \left\{ \max_{|z-\zeta|=r} |w(z)-w(\zeta)| \Big/ \min_{|z-\zeta|=r} |w(z)-w(\zeta)| \right\} \leqslant K.$$

That quasiconformality is a natural concept is shown by the

Equivalence Theorem. *Each of the three definitions A, B, C implies the other two.*

The implication $A \to B$ was proved by Grötzsch for C_1 mappings; his proof extends to the general case in view of the results of Morrey. Mori's work contains implicitly the statements $B \to A$, $B \to C$, cf. Bers[4]. Pesin and Jenkins showed that $C \to A$. (Cf. also Volkoviskiĭ[29], Yujobo[31].)

A k-quasiconformal mapping remains so if followed or preceded by a conformal mapping. Hence we may define a homeomorphism f of a Riemann surface S onto another such surface S' to be quasiconformal if it is so in a neighborhood of every point on S, in terms of local parameters.

2. Beltrami equations

It follows from Definition A that every k-quasiconformal mapping of a plane domain satisfies a Beltrami equation

$$w_x + iw_y = \mu(z)\,(w_x - iw_y) \quad (|\mu| \leqslant k < 1), \tag{3}$$

where μ is a complex-valued measurable function. Conversely, every homeomorphic solution (with generalized L_2 derivatives) of (3) is k-quasiconformal. We recall the geometric meaning of (3): the mapping $z \to w(z)$ is conformal with respect to the metric $ds = |dz + \mu\,d\bar{z}|$.

Let $\mathbf{M}$ denote the set of all bounded measurable functions $\mu(z)$, $|z| < 1$, with $\|\mu\| = \text{true max}\,|\mu(z)| < 1$. We topologize $\mathbf{M}$ by requiring $\mu_j \to \mu$ to mean that $\|\mu_j\| \leqslant k_0 < 1$ and $\mu_j(z) \to \mu(z)$ a.e. For $\mu \in \mathbf{M}$ let w^μ denote a solution of (3) which maps $|z| \leqslant 1$ topologically onto itself leaving the points 1, i, -1 fixed.

Proposition I (Morrey, cf. Bers and Nirenberg[6], Boyarskiĭ[7]). *For $\mu \in \mathbf{M}$, w^μ exists and is unique and every other solution of (3) is an analytic function of w^μ. Also w^μ and $(w^\mu)^{-1}$ satisfy uniform Hölder conditions depending only on $\|\mu\|$.*

Proposition II. Let $\mu \in \mathbf{M}$ depend on several real parameters $t_1, \ldots, t_r$ and *be a function of class* C_ν $(\nu = 0, 1, 2, ..., \infty)$ *of these parameters. For every* z, $|z| \leqslant 1$, $w^\mu(z)$ *is of class* C_ν *as a function of* $t_1, ..., t_r$.

Proposition III. Let $\mu \in \mathbf{M}$ *depend holomorphically on several complex parameters* $s_1, ..., s_r$. *For a sufficiently small* $\epsilon > 0$ *there exists a homeomorphic solution* $w(z)$ *of* (3) *defined for* $|z| < \epsilon$, *such that* $w(z)$ *is a holomorphic function of* $s_1, ..., s_r$.

The proofs of II and III will appear elsewhere.

3. Teichmüller spaces

In what follows conformally equivalent Riemann surfaces are considered identical. Two Riemann surfaces S and S_0 will be called *similar* if there exists a quasiconformal homeomorphism f of S onto S_0. In this case the homotopy class $\mathbf{F}_{S,S_0}$ of f is called allowable and the pair $(S, \mathbf{F}_{S,S_0})$ is called a *marked* Riemann surface. The totality of these forms the *Teichmüller space* $\mathbf{T}(S_0)$. Every allowable class $\mathbf{F}_{S_1,S_0}$ defines in an obvious way a one-to-one mapping (allowable mapping) of $\mathbf{T}(S_1)$ onto $\mathbf{T}(S_0)$. We are interested only in properties invariant under allowable mappings; hence we may identify $\mathbf{T}(S_1)$ with $\mathbf{T}(S_0)$ and call it the Teichmüller space $\mathbf{T}$ determined by a class of similar Riemann surfaces.

A differential of type (p, q) on S_0 is, locally, of the form $\lambda(z)\,dz^p\,d\bar{z}^q$, where z is a local parameter and $\lambda(z)$ a measurable function. Let $m = \mu\,d\bar{z}/dz$ be a differential of type $(-1, 1)$ (*Beltrami differential*). Then $|\mu|$ is a scalar; if $\|m\| = \text{true max} |\mu| < 1$, m is called a proper Beltrami differential. It defines on S_0 a Riemannian metric

$$ds = |dz + \mu\,d\bar{z}|$$

and it follows from I that this metric defines on S_0 a new conformal structure. S_0 with this conformal structure and with the allowable class containing the identity is a marked Riemann surface which we denote by S_0^m. *Every element of* $\mathbf{T}(S_0)$ *is of the form* S_0^m, but $S_0^{m_1} = S_0^{m_2}$ (equality in the sense of marked Riemann surfaces) does not imply that $m_1 = m_2$.

The Teichmüller distance between two elements of $\mathbf{T}(S_0)$, say S_1 and S_2, is defined as $\inf \|m\|$ for all m such that $S_1 = S_2^m$. It defines a topology on $T(S_0)$.

Now let S_0 be neither the sphere, nor the plane, nor the cylinder, nor a closed surface of genus 1. Then we have the representation $S_0 = U/G_0$, where U is the unit disc and G_0 a Fuchsian group (by which we mean here a discrete fixed-point-free group of non-Euclidean motions). G_0 is determined by S_0 uniquely, except that it may be replaced by AG_0A^{-1}, where A is a non-Euclidean motion.

Let $\mathbf{M}_{G_0}$ denote the set of those $\mu \in \mathbf{M}$ which satisfy the functional equation

$$\mu(A(z))\,\overline{A'(z)}/A'(z) = \mu(z) \quad \text{(for } A \in G_0\text{).} \tag{4}$$

Every Beltrami differential on S_0 can be written as $m = \mu(z)\,d\bar{z}/dz$, $|z| < 1$, $\mu \in \mathbf{M}_{G_0}$. For $\mu \in \mathbf{M}_{G_0}$ one verifies (using I) that

$$w^\mu(A(z)) = A^\mu(w^\mu(z)) \quad \text{for} \quad A \in G_0, \tag{5}$$

where A^μ is a non-Euclidean motion. The mapping $A \to A^\mu$ is an isomorphism of G_0 onto a Fuchsian group $G_0^m = w^\mu G_0(w^\mu)^{-1}$. We have that $S_0^m = U/G_0^m$. Thus the study of Teichmüller spaces can be made dependent on the theory of Beltrami equations.

4. The spaces $\mathbf{T}_g$, $\mathbf{T}_{g,n}$, $\mathbf{T}_g^{(n)}$

Let S_0 be a closed Riemann surface of genus g. Every closed surface S of genus g is similar to S_0 and every sense-preserving homeomorphism of S onto S_0 *belongs to an allowable class*. The proof of this is not difficult in view of our definition of quasiconformality. The Teichmüller space of S_0 will be denoted by $\mathbf{T}_g$.

Analogous statements are true if S_0 and S are each obtained by removing n distinct points from a closed Riemann surface of genus g. The corresponding Teichmüller space will be denoted by $\mathbf{T}_{g,n}$.

We shall also consider (for $g > 1$) the space $\mathbf{T}_g^{(n)}$ the elements of which are marked closed Riemann surfaces of genus g on each of which one has distinguished an ordered n-tuple of (not necessarily distinct) points.

There is a natural mapping ϖ of $\mathbf{T}_g^{(n)}$ onto $\mathbf{T}_g$ and the inverse image of a point of $\mathbf{T}_g$ under ϖ is in a one-to-one correspondence with the n-fold product of a Riemann surface by itself. This remark yields a natural way of introducing a topological or differentiable structure in $\mathbf{T}_g^{(n)}$ once we have such a structure in $\mathbf{T}_g$. We denote by $\hat{\mathbf{T}}_g^{(n)}$ the set of points of $\mathbf{T}_g^{(n)}$ corresponding to the choice of n distinct points on a surface. There is a natural mapping π of $\mathbf{T}_{g,n}$ onto $\hat{\mathbf{T}}_g^{(n)}$ which permits one to define a topological or differentiable structure in $\mathbf{T}_{g,n}$ using the corresponding structure of $\mathbf{T}_g^{(n)}$. (π depends upon an arbitrary ordering of the 'removed' points on one element of $\mathbf{T}_{g,n}$.)

5. Embedding of $\mathbf{T}_g$ into $\mathbf{E}_{6g-6}$

In what follows, we consider a fixed g and assume, for the sake of brevity, that $g > 1$. We set $\tau = 3g - 3$.

Let $S_0 = U/G_0$ be a closed surface of genus g. It is known that G_0 consists of the identity and of non-Euclidean translations. Thus every element $A \neq 1$ of G_0 has exactly two fixed points on the unit circle and one can show that two distinct elements have four distinct fixed points. Also, one can choose $2g$ generators $A_j^0, B_j^0, j = 1, 2, \ldots, g$ of G_0 satisfying the relation $\Pi A_j^0 B_j^0 (A_j^0)^{-1} (B_j^0)^{-1} = 1$ (*standard set* of generators). We call a standard set *normalized* if the repelling and attracting fixed points of B_g are 1 and (-1), respectively, and one of the fixed points of A_g is i. We assume that a definite normalized standard generating set of G_0 has been chosen once and for all. (This can always be done, replacing if need be G_0 by $A G_0 A^{-1}$.)

Now set, for some $\mu \in \mathbf{M}_{G_0}$,

$$A_j = w^\mu A_j^0 (w^\mu)^{-1}, \quad B_j = w^\mu B_j^0 (w^\mu)^{-1}. \tag{6}$$

Then $\{A_j, B_j\}$ is a normalized standard set of generators for G^m and hence determines S^m. Moreover, $\{A_j, B_j\}$ depends only on S^m and not on m (for homotopic mappings induce homomorphisms of fundamental groups which differ only by inner automorphisms, and there are canonical mappings of G onto the fundamental group of U/G). Finally, for a standard normalized set, A_g and B_g can be computed from $A_1, B_1, \ldots, B_{g-1}$ by using the relation

$$\prod_{j=1}^{g} A_j B_j A_j^{-1} B_j^{-1} = 1. \tag{7}$$

Each A_j and B_j, $j = 1, \ldots, g-1$ can be represented by 3 real numbers. Thus we can represent every element of $\mathbf{T}(S_0) = \mathbf{T}_g$ by a point in the Euclidean space $\mathbf{E}_{6g-6}$. From now on we identify an element of $\mathbf{T}_g$ with its representative point. Now $\mathbf{T}_g$ appears as a subset of $\mathbf{E}_{6g-6}$, and hence is topologized and even metrized (cf. Siegel[22], Bers[5]).

6. Differentiable structure of $\mathbf{T}_g$

Lemma 1. *Let S_0 be a marked closed Riemann surface of genus g, $\mathbf{m} = (m_1, \ldots, m_\sigma)$ a σ-tuple of Beltrami differentials on S_0, $\boldsymbol{\xi} = (\xi_1, \ldots, \xi_\sigma)$ a point of $\mathbf{E}_\sigma$ of small modulus $|\boldsymbol{\xi}|$. The mapping $\boldsymbol{\xi} \to S^{\boldsymbol{\xi} \cdot \mathbf{m}} = S_0^{\xi_1 m_1 + \cdots + \xi_\sigma m_\sigma}$ of a neighborhood of 0 in $\mathbf{E}_\sigma$ into $\mathbf{E}_{6g-6}$ is C_∞.*

This follows at once from II. In a forthcoming paper Ahlfors and Bers prove that the mapping considered is even real analytic.

A regular quadratic differential Ω on S_0 is locally of the form $\omega(z)\,dz^2$, $\omega(z)$ holomorphic. These Ω's form a complex vector space $\mathbf{Q}_{S_0}$ of dimension $3g - 3 = \tau$ (Riemann–Roch). We note that for any Beltrami differential $m = \mu \, d\bar{z}/dz$ the scalar product

$$(\Omega, m) = \iint_{S_*} \omega(z)\,\mu(z)\,dx\,dy$$

is well defined.

A Beltrami differential m on S_0 will be called *locally trivial* if for real $\epsilon \to 0$

$$|S_0 - S_0^{\epsilon m}| = o(\epsilon). \tag{8}$$

To appreciate this requirement, note that by II we always have

$$|S_0 - S_0^{\epsilon m}| = O(\epsilon).$$

The following result goes back to Teichmüller.

Main Lemma. The Beltrami differential m on S_0 (closed Riemann surface of genus g) is locally trivial if and only if $(\Omega, m) = 0$ for all $\Omega \in \mathbf{Q}_{S_0}$.

Necessity proof (à la Ahlfors). Using II we compute that, for every A in G_0, $A = (\partial A^{\epsilon \mu}/\partial \epsilon)_{\epsilon = 0}$ equals $h(A(z)) - A'(z)\,h(z)$, where

$$h(z) = (\partial w^{\epsilon \mu}(z)/\partial \epsilon)_{\epsilon = 0} \quad \text{and} \quad h_{\bar{z}} = \mu.$$

Equation (8) implies that $A = 0$, i.e. that $h(z)/dz$ is a differential of type $(-1, 0)$ on S_0 and

$$(\Omega, m) = \iint_{S_*} \omega \mu \, dx\,dy = \iint_{S_*} \omega h_{\bar{z}} \, dx\,dy = 0$$

since $\omega_{\bar{z}} \equiv 0$.

Sufficiency proof (à la Weil). Let $\Omega_1, \ldots, \Omega_\tau$ be a complex basis of $\mathbf{Q}_{S_0}$, L some C_∞ density on S_0 (i.e. a differential of type $(1, 1)$, $L = i\lambda \, dz \, d\bar{z}$ with $\lambda \geqslant 0$) and set

$$m_j = \bar{\Omega}_j/L, \quad j = 1, \ldots, \tau, \qquad m_j = -i\bar{\Omega}_{j-\tau}/L, \quad j = \tau + 1, \ldots, 2\tau.$$

Assume that $(\Omega_j, m_0) = 0$ for all j and consider the mapping of $\mathbf{E}_{2\tau+1}$ into $\mathbf{E}_{2\tau}$:

$$(\xi_0, \xi_1, \ldots, \xi_{2\tau}) \to S_0^{\xi_0 m_0 + \xi_1 m_1 + \cdots + \xi_{2\tau} m_{2\tau}}.$$

This mapping has rank $\leqslant 2\tau$ at the origin. Hence there is a $(2\tau + 1)$-tuple $(\xi_0, \ldots, \xi_{2\tau}) \neq (0, \ldots, 0)$ for which $\xi_0 m_0 + \ldots + \xi_{2\tau} m_{2\tau}$ is locally trivial. But then $\xi_1 = \ldots = \xi_{2\tau} = 0$, by the result proved above, so that $\xi_0 \neq 0$ and m_0 is locally trivial.

A real (complex) *Beltrami basis* on S_0 is a basis of the real (complex) factor-space of all Beltrami differentials modulo the locally trivial ones. The preceding argument contains the proof of

Lemma 2. Let $\mathbf{m} = (m_1, \ldots, m_{2\tau})$ be a real Beltrami basis on S_0. The mapping $\boldsymbol{\xi} = (\xi_1, \ldots, \xi_{2\tau}) \to S_0^{\boldsymbol{\xi} \cdot \mathbf{m}}$ has rank 2τ at the origin.

Corollary. $\mathbf{T}_g$ *is an open subset of* $\mathbf{E}_{6g-6}$.

Indeed, S_0 is not distinguished from any other element of $\mathbf{T}_g$.

We have now defined a C_∞ structure in $\mathbf{T}_g$ and hence also in $\mathbf{T}_g^{(n)}$ and in $\mathbf{T}_{g,n}$.

7. Extremal quasiconformal mappings

A *Teichmüller differential* on a closed Riemann surface S_0 of genus g is either 0 or a Beltrami differential of the form $\kappa \bar{\Omega}/|\Omega|$ where Ω is a regular quadratic differential and κ a number, $0 \leqslant \kappa < 1$. These differentials have an extremal property proved correctly in Teichmüller's 1940 paper (cf. also Ahlfors[2], Bers[5]).

Theorem A. If m_0 is a Teichmüller differential and m any other Beltrami differential on S_0 (a closed surface of genus $g > 1$), and if $S_0^m = S_0^{m_0}$, then either $m = m_0$ or $\|m\| > \|m_0\|$.

We can now state the Teichmüller theorem for closed surfaces.

Theorem B. Let S_0 be a marked closed Riemann surface of genus $g > 1$. Every element S_1 of $\mathbf{T}_g = \mathbf{T}(S_0)$ admits the unique representation $S_1 = S_0^m$ where m is a Teichmüller differential.

The theorem means that every homeomorphism of S_0 onto S_1 can be deformed into a unique extremal one, which deviates least from conformality and which can be represented, locally, except near finitely many points, as a conformal mapping followed by a uniform stretching and then by another conformal mapping.

Now let S_0 be a surface obtained from a closed Riemann surface Σ_0 of genus g by removing n distinct points $p_1, \dots, p_n$. A Teichmüller differential on S_0 is either 0 or a Beltrami differential of the form $\kappa \bar{\Omega}/|\Omega|$, where $0 \leqslant \kappa < 1$ and Ω is a quadratic differential which is holomorphic on S except perhaps at the points p_j at which it may have simple poles.

Theorem C. Let S_0 be an element of $\mathbf{T}_{g,n}$. Every other element S_1 of $\mathbf{T}_{g,n}$ admits the unique representation $S_1 = S_0^m$, where m is a Teichmüller differential on S_0.

This Teichmüller theorem can be derived from B (see Ahlfors[2] for details). Theorem A was proved by Teichmüller in [25], another proof is due to Ahlfors[2]. We sketch below the proof in Bers[5]. It differs from Teichmüller's own only technically.

Let $\Omega = (\Omega_1, \dots, \Omega_{6g-6})$ be a real basis of Q_{S_0}. For $\mathbf{x} \in \mathbf{E}_{6g-6}$, $|\mathbf{x}| < 1$, set $\gamma(\mathbf{x}) = S_0^{|\mathbf{x}| \mathbf{x} \cdot \bar{\Omega}/|\mathbf{x} \cdot \Omega|}$ if $\mathbf{x} \neq 0$, $\gamma(0) = S_0$. The mapping $\mathbf{x} \to \gamma(\mathbf{x})$ of $|\mathbf{x}| < 1$ into $\mathbf{T}_g \subset \mathbf{E}_{6g-6}$ is continuous (by II) and one-to-one (by A), hence open and topological (by the theorem on the invariance of domain). We prove that it is onto. For $S_1 \in \mathbf{T}_g$ there is a $\mu \in \mathbf{M}_{G_0}$ with $S_1 = S_0^m$.

For $0 \leqslant t \leqslant 1$ we have that $t\mu \in \mathbf{M}_{G_0}$. Let $\boldsymbol{\theta}$ be the set of those t for which $S_0^{tm} = \gamma(\mathbf{x})$ for some $\mathbf{x}$. Then $\boldsymbol{\theta}$ is open, by the previous result, and contains $t = 0$. We must show that $\boldsymbol{\theta}$ is closed (so that $t = 1$ belongs to it). But by A we have that if $S_0^{tm} = \gamma(\mathbf{x})$, then

$$|\mathbf{x}| \leqslant \|tm\| = t\|m\| \leqslant \|m\| < 1.$$

The closure of $\boldsymbol{\theta}$ follows by the local compactness of $\mathbf{E}_{6g-6}$ and the continuity of γ.

The argument just given also establishes the following results:

Theorem D. $\mathbf{T}_g$ *is a* $(6g-6)$ *cell;* $\mathbf{T}_{g,n}$ *is a* $(6g-6+2n)$ *cell.*

Theorem E. *The Teichmüller metrics in* $\mathbf{T}_g$ *and in* $\mathbf{T}_{g,n}$ *yield the same topology as the embedding of* $\mathbf{T}_g$ *into* $\mathbf{E}_{6g-6}$.

The statement that $\mathbf{T}_g$ is a $(6g-6)$ cell is already contained in the work of Fricke[9]. Fricke's proof is quite different and very difficult to follow.

8. Complex-analytic structure of $\mathbf{T}_g$

The existence of a 'natural' complex analytic structure in $\mathbf{T}_g$ has been asserted by Teichmüller[28]; the first proof was given by Ahlfors[3] after Rauch[21] showed how to introduce complex-analytic co-ordinates in the neighborhood of any point of $\mathbf{T}_g$ which is not a hyperelliptic surface. Other proofs are due to Kodaira–Spencer[14] and to Weil[30]. The proof sketched below gives explicitly a set of co-ordinates near every point of $\mathbf{T}_g$.

Let S_0 be an element of $\mathbf{T}_g$, that is a marked closed Riemann surface of genus g, and $\mathbf{m} = (m_1, \ldots, m_{3g-3})$ a complex Beltrami basis. By Lemmas 1 and 2 the mapping $\mathbf{a} = (a_1, \ldots, a_{3g-3}) \to S_0^{\mathbf{a} \cdot \mathbf{m}}$ is a C_∞ homeomorphism of a neighborhood of the origin of the complex number space $\mathbf{C}_{3g-3}$ onto a neighborhood of S_0. We call the a_j the co-ordinates associated with $\mathbf{m}$.

Theorem F. *The co-ordinates associated with complex Beltrami bases are complex-analytic co-ordinates in* $\mathbf{T}_g$.

It will suffice to prove two statements.

(i) If $\mathbf{m}$ and $\mathbf{n}$ are two complex Beltrami bases on S_0 and the relation $\mathbf{a} = \mathbf{a}(\mathbf{b})$ is defined by the equation $S_0^{\mathbf{a} \cdot \mathbf{m}} = S_0^{\mathbf{b} \cdot \mathbf{n}}$, then $\partial a_i / \partial \bar{b}_j = 0$ at $\mathbf{a} = \mathbf{b} = 0$.

(ii) If $\mathbf{m}$ is a complex Beltrami basis on S_0, and $S_1 = S_0^{\mathbf{c} \cdot \mathbf{m}}$ with $|\mathbf{c}|$ sufficiently small, then there exists a complex Beltrami basis $\mathbf{n}$ on S_1 such that if $\mathbf{a}(\mathbf{b})$ is defined by the equation $S_0^{(\mathbf{c}+\mathbf{a}) \cdot \mathbf{m}} = S_1^{\mathbf{b} \cdot \mathbf{n}}$, then

$$\frac{\partial a_i}{\partial b_j} = \delta_{ij} \quad \text{and} \quad \frac{\partial a_i}{\partial \bar{b}_j} = 0 \quad \text{at} \quad \mathbf{b} = 0.$$

Proof of (i). From our definitions and Lemma 1 we conclude that for any two Beltrami differentials s and t on S_0

$$|S_0^{s+t} - S_0^s| = O(\|t\|) \quad \text{for} \quad s \text{ fixed}, \tag{9}$$

$$|S_0^{es+et} - S_0^{es}| = o(\epsilon) \quad \text{if} \quad t \text{ is locally trivial}. \tag{10}$$

Now we have $\mathbf{n} = Q\mathbf{m} + \mathbf{r}$, where Q is a constant matrix and

$$\mathbf{r} = (r_1, \dots, r_{3g-3}),$$

r_j being locally trivial. It is plainly sufficient to consider the case $Q = I$, i.e. $\mathbf{n} = \mathbf{m} + \mathbf{r}$. Using Lemma 1 and (10), we have that for small $|\mathbf{a}|$ and $|\mathbf{b}|$:

$$|\mathbf{a} - \mathbf{b}| = O(|S_0^{\mathbf{a} \cdot \mathbf{m}} - S_0^{\mathbf{b} \cdot \mathbf{m}}|) = O(|S_0^{\mathbf{b} \cdot \mathbf{m} - \mathbf{b} \cdot \mathbf{r}} - S_0^{\mathbf{b} \cdot \mathbf{m}}|) = o(|\mathbf{b} \cdot \mathbf{r}|) = o(|\mathbf{b}|),$$

so that $\partial a_i / \partial \bar{b}_j = 0$.

Proof of (ii). If s and t are Beltrami differentials on S_0,

$$\lambda_s(t) = \frac{1 + \bar{s}}{1 + s} \frac{t}{1 - |s|^2} \tag{11}$$

is a Beltrami differential on S_0^s and a direct computation based on (9) shows that

$$|S_0^{s+t} - (S_0^s)^{\lambda_s(t)}| = O(\|t\|^2) \quad \text{for} \quad \text{fixed } s. \tag{12}$$

Now set $s = \mathbf{c} \cdot \mathbf{m}$ and $\mathbf{n} = \lambda_s(\mathbf{m})$. Then

$$|\mathbf{a}| = O(|S_0^{s + \mathbf{a} \cdot \mathbf{m}} - S_0^s|) = O(|(S_0^s)^{\mathbf{a} \cdot \mathbf{n}} - S_0^s|) + O(|\mathbf{a}|^2).$$

This shows that $\mathbf{n}$ is a complex Beltrami basis on $S_0^s = S_1$. Next, if $S_0^{s + \mathbf{a} \cdot \mathbf{m}} = S_1^{\mathbf{b} \cdot \mathbf{n}}$, we have, for small $\mathbf{a}$ and $\mathbf{b}$:

$$|\mathbf{a} - \mathbf{b}| = O(|S_1^{\mathbf{a} \cdot \mathbf{n}} - S_1^{\mathbf{b} \cdot \mathbf{n}}|) = O(|S_1^{\mathbf{a} \cdot \mathbf{n}} - S_0^{s + \mathbf{a} \cdot \mathbf{m}}|)$$

$$= O(|(S_0^s)^{\lambda_s(\mathbf{a} \cdot \mathbf{m})} - S_0^{s + \mathbf{a} \cdot \mathbf{m}}|) = O(\|\mathbf{a} \cdot \mathbf{m}\|^2) = O(|\mathbf{a}|^2),$$

whence $\partial a_i / \partial b_j = \delta_{ij}$ and $\partial a_i / \partial \bar{b}_j = 0$.

The space $\mathbf{T}_g$ also has a 'natural' *Hermitian metric* $d\sigma$ defined (by Weil) as follows. For $S_0 \in \mathbf{T}_g$, let L denote the Poincaré density on S_0 (if $S_0 = U/G_0$, then $L = dx\,dy/y^2$). Let $\{\Omega_j\}$ be a complex basis of $\mathbf{Q}_{S_0}$ such that $(\Omega_j, \bar{\Omega}_k/L) = \delta_{jk}$, and set $m_j = \bar{\Omega}_j/L$, $\mathbf{m} = (m_1, \dots, m_r)$. Let a_j be the co-ordinates associated with the complex Beltrami basis $\mathbf{m}$ on S_0; then $d\sigma^2 = \Sigma |da_j|^2$ at the point $S_0 \in \mathbf{T}_g$. Weil proved, by a computation, that the metric $d\sigma^2$ is Kählerian.

9. Complex analytic structure of $\mathbf{T}_g^{(n)}$ and $\mathbf{T}_{g,n}$

The results of this and the following section confirm and extend some of Teichmüller's assertion in [28]. They also show that the complex-analytic structure defined above is natural and coincides with that of Rauch–Ahlfors.

Let S_0 be a marked closed Riemann surface of genus $g > 1$, $p_1, ..., p_n$ points on S_0, not necessarily distinct, $\zeta_1, ..., \zeta_n$ local uniformizers on S_0 with $\zeta_j = 0$ at p_j, $\mathbf{m} = (m_1, ..., m_r)$ a complex Beltrami basis on S_0, $\mathbf{a} = (a_1, ..., a_r)$ a complex vector of small modulus. By a permanent uniformizer near p_j we mean a continuous function $z_j = w_j(a_1, ..., a_r, \zeta_j)$ which vanishes for $\zeta_j = 0$, is holomorphic in $\mathbf{a}$ for fixed ζ_j, and is, for fixed $\mathbf{a}$, a homeomorphic solution of the Beltrami equation in ζ_j with $\mu = \Sigma a_i \mu_i$ (where $m_i = \mu_i(\zeta_j)\,d\bar{\zeta}_j/d\zeta_j$). Clearly, $(a_1, ..., a_r, z_1, ..., z_n)$ are a set of complex co-ordinates, called *distinguished co-ordinates* of a neighborhood of $(S_0, p_1, ..., p_n)$ in $\mathbf{T}_g^{(n)}$. The existence of permanent uniformizers follows from Proposition III.

Theorem G. The distinguished co-ordinates give $\mathbf{T}_g^{(n)}$ a complex-analytic structure.

Assume now that the points $p_1, ..., p_n$ are distinct. By a complex Beltrami basis $\mathbf{m} = (m_1, ..., m_{r+n})$ on $(S_0, p_1, ..., p_n)$ (also called an *extended* Beltrami basis) we mean a basis of the (complex) factor-space of all Beltrami differentials n on S_0 modulo those n for which $(\Omega, n) = 0$ whenever the quadratic differential Ω is regular on S_0 except perhaps for simple poles at the p_j. For $\mathbf{a} = (a_1, ..., a_{r+n})$ small, $(S_0^{\mathbf{a} \cdot \mathbf{m}}, p_1, ..., p_n)$ is a point of $\mathbf{T}_{g,n}$. We call the a_j co-ordinates associated with $\mathbf{m}$.

Theorem H. The co-ordinates associated with extended Beltrami bases are complex-analytic co-ordinates in $\mathbf{T}_{g,n}$.

Theorem I. The natural mappings ϖ (of $\mathbf{T}_g^{(n)}$ onto $\mathbf{T}_g$) and π (of $\mathbf{T}_{g,n}$ into $\mathbf{T}_g^{(n)}$) are holomorphic. ϖ makes $\mathbf{T}_g^{(n)}$ into a complex fibre-space (the fibres being n-fold products of a marked closed Riemann surface by itself). π makes $\mathbf{T}_{g,n}$ into the universal covering space of $\hat{\mathbf{T}}_g^{(n)}$.

The rather simple proofs of Theorems G, H, I are omitted due to lack of space.

10. Meromorphic functions on $\mathbf{T}_g$, $\mathbf{T}_{g,1}$ and $\mathbf{T}_g^{(1)}$

Let us choose a canonical dissection on S_0. This gives us on every marked Riemann surface S similar to S_0 a set of generators $A_1, B_1, ..., A_g, B_g$ for the fundamental group $\pi(S)$ (which are determined but for an inner automorphism and satisfy (7)) and hence also a one-dimensional homology basis. If ϕ is an Abelian differential on S, let (A_j, ϕ), (B_j, ϕ) denote the A_j and B_j periods of ϕ_j respectively. Also, let ω_j denote the Abelian differential of the first kind on S with $(A_i, \phi_j) = \delta_{ij}$, and set $p_{ij} = (B_i, \phi_j)$. Then the p_{ij} are functions on $\mathbf{T}_g$.

We note that S may be considered as a complex analytic sub-manifold of $\mathbf{T}_g^{(1)}$. The ratios $f_{ij} = \omega_i/\omega_j$ are functions on $\mathbf{T}_{g,1}$ and on $\mathbf{T}_g^{(1)}$. Finally, let ω_{ij} $(i \neq j)$ denote the Abelian differential of the third kind on S which has simple poles with residues 1 at the zeros of ω_i and with residues (-1) at the zeros of ω_j. The ratios $f_{ijk} = \omega_{ij}/\omega_k$ are functions on $\mathbf{T}_{g,1}$ and on $\mathbf{T}_g^{(1)}$.

Theorem J. *The p_{ij} are meromorphic on $\mathbf{T}_g$. The f_{ij} and f_{ijk} are meromorphic on $\mathbf{T}_{g,1}$ and on $\mathbf{T}_g^{(1)}$.*

We omit the proof which is based on Proposition II.

Let Φ denote the function field generated by the f_{ij}, f_{ijk}, and $\Phi_0 \subset \Phi$ the field generated by the f_{ij}. It is known that every meromorphic function f on S belongs to Φ (and even to Φ_0, if S is not hyperelliptic). Hence f is a *restriction* of a meromorphic function defined on the whole space $\mathbf{T}_g^{(1)}$.

11. Applications

The following two results can be proved in a few lines using the fact that $\mathbf{T}_g$ is connected.

(a) Let $A_j, B_j, j = 1, \dots, g$, be non-Euclidean motions generating, with the single relation (7), a fixed-point-free Fuchsian group with compact fundamental region. Represent A_j, B_j by (2×2) matrices α_j, β_j. Then $\Pi\alpha_j\beta_j\alpha_j^{-1}\beta_j^{-1} = +I$ (identity matrix) and not $(-I)$. This answers a question of Siegel[22].

(b) Every canonical dissection of a closed Riemann surface $S = U/G$ can be deformed into a dissection which maps into a convex non-Euclidean polygon in U. This was stated, with a different and complicated proof, by Fricke[9].

12. Open questions

Here are some open questions.

(1) The space of (unmarked) Riemann surfaces is the factor-space $\mathbf{T}_g/\Gamma_g$, Γ_g being the so-called mapping class group. Give a precise description of this space.

(2) Does there exist a complex-analytic co-ordinate patch covering $\mathbf{T}_g$? (In other words is $\mathbf{T}_g$ a subset of $\mathbf{C}_{3g-3}$?)

(3) How can the theory sketched above be extended to open surfaces (other than those obtained from a closed surface by removing points or disks)? Our definition of Teichmüller spaces tries to anticipate such an extension.

(4) In particular, if S_0 and S are two similar marked open Riemann surfaces, what is the nature of the extremal quasi-conformal mapping of S onto S_0, and is this mapping unique?

REFERENCES

[1] Ahlfors, L. V. Zur Theorie der Überlagerungsflächen. *Acta Math.* 65, 157–194 (1935).
[2] Ahlfors, L. V. On quasi-conformal mappings. *J. Anal. Math.* 3, 1–58 (1953–4).
[3] Ahlfors, L. V. The complex analytic structure of the space of closed Riemann surfaces. (To appear.)
[4] Bers, L. On a theorem of Mori and the definition of quasiconformality. *Trans. Amer. Math. Soc.* 84, 78–84 (1957).
[5] Bers, L. Quasiconformal mappings and Teichmüller's theorem. (To appear.)
[6] Bers, L. and Nirenberg, L. On a representation theorem for linear elliptic systems with discontinuous coefficients and its applications. *Atti del Convegno Internazionale sulle Equazioni alle derivate parziali, Trieste, 1954*, pp. 111–140 (1955).
[7] Boyarskiĭ, B. V. Homeomorphic solutions of Beltrami systems. *Dokl. Akad. Nauk, SSSR*, 102, 661–664 (1955). (Russian.)
[8] Caccioppoli, R. Fondamenti per una teoria generale delle funzioni pseudo-analitiche di una variable complessa. *R.C. Acc. Naz. Lincei*, (8), 13, 197–204, 321–329 (1952).
[9] Fricke, R. and Klein, F. *Vorlesungen über die Theorie der Automorphen Funktionen*, Band I. B. G. Teubner, Leipzig, 1926.
[10] Grötzsch, H. Über die Verzerrung bei schlichten nichtkonformen Abbildungen und über eine damit zusammenhängende Erweiterung des Picardschen Satzes. *Ber. Verh. sächs. Akad. Lpz.* 80, 503–507 (1928).
[11] Grötzsch, H. Über Verzerrung bei nichtkonformen schlichten Abbildungen mehrfach zusammenhängender Bereiche. *Ber. Verh. sächs. Akad. Lpz.* 82, 69–80 (1930).
[12] Grötzsch, H. Über möglichst konforme Abbildungen von schlichten Bereichen. *Ber. Verh. sächs. Akad. Lpz.* 84, 114–120 (1932).
[13] Jenkins, J. A. A new criterion for quasiconformal mapping. *Ann. Math.* (2), 64, 208–214 (1957).
[14] Kodaira, K. and Spencer, D. C. (To appear.)
[15] Lavrent'ev, M. A. Sur une classe de représentations continues. *Mat. Sbornik*, 42, 407–423 (1935).
[16] Mori, A. On quasi-conformality and pseudo-analyticity. *Trans. Amer. Math. Soc.* 84, 56–77 (1957).
[17] Morrey, C. B. On the solutions of quasi-linear elliptic partial differential equations. *Trans. Amer. Math. Soc.* 43, 126–166 (1938).
[18] Pesin, I. N. Metric properties of quasi-conformal mappings. *Mat. Sbornik*, 40 (82), 281–294 (1956).
[19] Pfluger, A. Quasikonforme Abbildungen und logarithmische Kapazität. *Ann. Inst. Fourier*, 2, 69–80 (1950).
[20] Pfluger, A. Über die Äquivalenz der geometrischen und der analytischen Definition quasikonformer Abbildungen. *Comment. Math. Helv.* 33, 23–33 (1959).

[21] Rauch, H. E. On the transcendental moduli of algebraic Riemann surfaces. *Proc. Nat. Acad. Sci., Wash.*, 41, 42–49 (1955).

[22] Siegel, C. L. Über einige Ungleichungen bei Bewegungsgruppen in der nichteuklidischen Ebene. *Math. Ann.* 133, 127–138 (1957).

[23] Teichmüller, O. Untersuchungen über konforme und quasi-konforme Abbildungen. *Dtsch. Math.* 3, 621–678 (1938).

[24] Teichmüller, O. Extremale quadratische quasikonforme Abbildungen und quadratische Differentiale. *Preuss. Akad. Ber.* 22 (1940).

[25] Teichmüller, O. Bestimmung der extremalen quasikonformen Abbildungen bei geschlossenen Riemannschen Flächen. *Preuss. Akad. Ber.* 4 (1943).

[26] Teichmüller, O. Beweis der analytischen Abhängigkeit des konformen Moduls einer analytischen Ringflächenschar von den Parametern. *Dtsch. Math.* 7, 309–336 (1944).

[27] Teichmüller, O. Ein Verschiebungssatz der quasi-konformen Abbildung. *Dtsch. Math.* 7, 336–343 (1944).

[28] Teichmüller, O. Veränderliche Riemannsche Flächen. *Dtsch. Math.* 7, 344–359 (1944).

[29] Volkoviskiĭ, L. I. *Quasiconformal Mappings.* Lwów, 1954. (Russian.)

[30] Weil, André. Sur les modules des surfaces de Riemann. *Séminaire Bourbaki*, May, 1958.

[31] Yujobo, Zuiman. On absolutely continuous functions of two or more variables in the Tonelli sense and quasiconformal mappings in the A. Mori sense. *Comment. Math. Univ. St Paul*, 4, 67–92 (1955).

DIE RIEMANNSCHEN FLÄCHEN
DER FUNKTIONENTHEORIE
MEHRERER VERÄNDERLICHEN

Von HANS GRAUERT

Seit Riemann betrachtet man mehrdeutige holomorphe Funktionen $w = f(z)$ als eindeutige holomorphe Funktionen auf Riemannschen Flächen R über der abgeschlossenen komplexen z-Ebene. Algebroide Verzweigungspunkte—das sind Punkte, in denen $w = f(z)$ einer pseudo-algebraischen Gleichung $w^k + \sum_{\nu=1}^{k} A_\nu(z) w^{k-\nu} = 0$ genügt—werden dabei in die Betrachtung einbezogen.

In der Funktionentheorie mehrerer Veränderlichen ist man ebenfalls bestrebt, mehrdeutige Funktionen eindeutig zu machen. Dazu können —analog zur klassichen Funktionentheorie—verzweigte Gebilde über dem Raum C^n von n komplexen Veränderlichen benutzt werden. Als 'Riemannsche Fläche' der zweideutigen Funktion $w = \sqrt{(z_1 z_2)}$ bietet sich z. B. die analytische Teilmenge $F = \{(w, z_1, z_2) : w^2 = z_1 z_2\}$ im Raum C^3 der komplexen Zahlentripel (w, z_1, z_2) an, die durch die Projektion $\Phi: (w, z_1, z_2) \to (z_1, z_2)$ über dem Raum C^2 der komplexen Zahlenpaare 'ausgebreitet' liegt. Während aber jeder Punkt einer Riemannschen Fläche der Funktionentheorie einer Veränderlichen uniformisierbar ist, läßt sich zu dem Punkt $0 = (0, 0, 0) \in F$ nicht einmal eine Umgebung $U(0) \subset F$ finden, die man topologisch auf ein Gebiet des C^2 abbilden kann (vgl. [2]): es gibt keine Umgebung von 0, die Träger eines komplexen Koordinatensystems ist. Die heute viel diskutierten komplexen Mannigfaltigkeiten sind darum noch nicht als die wirklichen Verallgemeinerungen der Riemannschen Flächen auf den Fall mehrerer komplexer Veränderlichen anzusehen.

Durch die Ausbreitung der Riemannschen Flächen über der z-Ebene wird für jeden ihrer Punkte P der Begriff der lokalen Ortsuniformisierenden definiert. Eine lokale Ortsuniformisierende in P kann als ein lokales komplexes Koordinatensystem in einer Umgebung $U(P)$ angesehen werden. Eine in (einer Umgebung von) P definierte komplexwertige Funktion heißt holomorph, wenn sie als abhängend von der komplexen Koordinate in U betrachtet eine holomorphe Funktion ist. Hermann Weyl[32] hat von diesen Gedankengängen ausgehend den Aufbau der Riemannschen Fläche abstrakt vollzogen. Nach seiner

Definition ist eine Riemannsche Fläche ein Hausdorffscher Raum R, der durch einen Atlas lokaler komplexer Koordinatensysteme überdeckt ist. In den Durchschnitten der Träger je zweier lokaler Koordinatensysteme bestehen holomorphe Koordinatentransformationen. Da durch den Atlas der Begriff der holomorphen Funktion in Teilbereichen von R eindeutig festgelegt wird, sind auch alle Keime von holomorphen Funktionen wohldefiniert. Die Gesamtheit der holomorphen Funktionskeime auf R ist nach moderner Terminologie eine Garbe $\mathfrak{O}$ von Ringen und eine Untergarbe der Garbe $\mathfrak{C}$ der Keime von stetigen Funktionen auf R. $\mathfrak{O}$ wird durch die komplexen Koordinatensysteme von R gegeben, durch die Garbe $\mathfrak{O}$ werden die holomorphen Funktionen in Teilbereichen von R und mithin die lokalen komplexen Koordinatensysteme von R bestimmt. Es ist also sinnvoll $\mathfrak{O}$ die Strukturgarbe der Riemannschen Fläche R zu nennen.

Nach Cartan heißt ein topologischer Raum X ein (komplex) geringter Raum, wenn X mit einer Strukturgarbe $\mathfrak{S}$ versehen ist, so daß folgendes gilt:

(1) $\mathfrak{S}$ ist eine Untergarbe von $\mathfrak{C}$, der Garbe der Keime von stetigen komplexwertigen Funktionen.

(2) $\mathfrak{S}$ ist eine Garbe von Ringen, d. h. jeder Halm von $\mathfrak{S}$ ist ein Ring.

(3) Die konstanten Funktionskeime sind in $\mathfrak{S}$ enthalten.

Die Schnittflächen in $\mathfrak{S}$ können als spezielle komplexwertige Funktionen angesehen werden. Man nennt sie morphe Funktionen. Offenbar sind zwei geringte Räume X_1 und X_2 als gleich strukturiert (isomorph) anzusehen, wenn es eine topologische Abbildung $\phi: X_1 \to X_2$ gibt, derart, daß die folgende Aussage richtig ist:

Ist $U_2 \subset X_2$ eine beliebige offene Teilmenge, f_2 eine in U_2 komplexwertige Funktion, so ist f_2 genau dann morph, wenn $f_1 = f_2 \circ \phi$ in $U_1 = \phi^{-1}(U_2)$ morph ist.

Jede offene Teilmenge U eines geringten Raumes X, wird durch Beschränkung der Strukturgarbe auf U ebenfalls zu einem geringten Raum. Eine Riemannsche Fläche ist ein geringter Raum, auf dem jeder Punkt eine Umgebung U besitzt, die zu einem Gebiet der z-Ebene isomorph ist.

1. Komplexe Räume

1.1. Es ist zweckmäßig die Argumentbereiche der Funktionen mehrerer Veränderlichen ebenfalls auf einem abstrakten Wege einzuführen. Das führt zu den komplexen Räumen, die als die richtigen Verallgemeinerungen der Riemannschen Flächen einer komplexen Veränderlichen angesehen werden können. Der Begriff des komplexen Raumes wurde von mehreren Autoren auf verschiedene (nicht äquivalente) Weise definiert. In allen Fällen ist jedoch ein komplexer Raum ein geringter Hausdorffscher Raum, der lokal zu gewissen geringten Repräsentantenräumen isomorph ist.†

1.2. Da mehrdeutige holomorphe Funktionen $f(z_1, ..., z_n)$ in algebroiden Punkten über analytische Resultantenmengen verzweigt sind, wählten Behnke und Stein als lokale Repräsentanten *analytische Überlagerungen* von Gebieten $G \subset C^n$. Eine analytische Überlagerung $\mathfrak{Y}$ ist eine endlichblättrige unbegrenzte Pseudoüberlagerung im Sinne von Steenrod, deren Verzweigungspunkte über einer analytischen Menge $A \subset G$, $A \neq G$ liegen.‡ Genauer gesagt ist $\mathfrak{Y}$ ein Tripel (Y, Φ, G), das folgenden Axiomen genügt:

(1) Y ist ein Hausdorffscher Raum, $G \subset C^n$ ein (zusammenhängendes) Teilgebiet, $\Phi: Y \to G$ eine stetige, eigentliche Abbildung.§ Die Urbildmenge jedes Punktes $z_0 \in G$ ist diskret.

(2) Es gibt eine in G analytische Menge $A \neq G$, so daß $\hat{A} = \Phi^{-1}(A)$ in Y nirgends dicht liegt, $\hat{A}$ den Raum Y nirgends zerlegt|| und Φ die offene Teilmenge $\mathring{Y} = Y - \hat{A}$ lokal-topologisch auf $G - A$ abbildet.

Die Punkte $y_0 \in Y$, in denen Φ lokal-topologisch ist, heißen *schlichte Punkte* von Y, die übrigen Punkte werden *Verzweigungspunkte* genannt. Der Begriff der *holomorphen Funktion* in offenen Teilmengen $B \subset Y$ kann wie folgt definiert werden:

Eine komplexwertige Funktion f in B heißt holomorph, wenn

(1) f stetig ist,

(2) f in jedem schlichten Punkt $y_0 \in Y$ in bezug auf die in einer Umgebung $U(y_0)$ definierten Koordinaten $\hat{z}_\nu = z_\nu \circ \Phi$ holomorph ist.

Diese Definition ist sinnvoll, obgleich f in den Verzweigungspunkten nur stetig zu sein braucht. Nach Riemann ist nämlich—in Gebieten $G \subset C^n$—jede stetige, außerhalb einer nirgendsdichten analytischen Menge holomorphe Funktion in ihrem ganzen Definitionsbereich holomorph.

† Natürlich wurde die Definition der komplexen Räume nicht immer in der Terminologie des geringten Raumes gegeben.

‡ Die hiergegebene Definition der analytischen Überlagerung ist eine Vereinfachung der Definition in [2]. Sie wurde in dieser Form zum ersten Male in [9] angegeben. Eine analytische Menge $A \subset G$ ist eine abgeschlossene Teilmenge von G, für die es zu jedem Punkt $z_0 \in A$ eine Umgebung $U(z_0)$ und endlich viele in U holomorphe Funktionen $f_1, ..., f_k$ gibt, so daß $U \cap A = \{z \in U : f_\nu(z) = 0, \nu = 1, ..., k\}$ ist.

§ Eine stetige Abbildung heißt eigentlich, wenn das Urbild jeder kompakten Menge kompakt ist.

|| $\hat{A}$ zerlegt Y nirgends, wenn für jede offene zusammenhängende Menge $Q \subset Y$ die Menge $Q - \hat{A}$ ebenfalls zusammenhängt.

Offenbar bilden die holomorphen Funktionskeime in Y eine Unter-
garbe $\mathfrak{S}$ von $\mathfrak{C}$. $\mathfrak{S}$ ist eine Garbe von Ringen, die die konstanten Funktionskeime enthält. Y ist also ein geringter Raum.

1.3. Cartan und Serre[5, 30] benutzen als lokale Repräsentanenräume
zur Definition der komplexen Räume analytische Teilmengen A von
Teilgebieten G des komplexen Zahlenraumes C^n. In Teilbereichen $B \subset A$
läßt sich der Begriff der *holomorphen Funktion* auf einfache Weise
definieren: Eine in B komplexwertige Funktion f heißt holomorph, wenn
es zu jedem Punkt $z_0 \in B$ eine offene Menge $U \subset G$ mit $z_0 \in U$ gibt, so
daß $f \mid U \cap B$ die Beschränkung einer in U holomorphen Funktion ist.
Natürlich ist jede in B holomorphe Funktion auch stetig. Die holomorphen Funktionskeime machen A wieder zu einem geringten Hausdorffschen Raum.

Weitere Definitionen des komplexen Raumes wurden von Bochner
und Martin[3] und von Chow gegeben. Es kann hier leider nicht näher
darauf eingegangen werden. In einem geringten Raum X, der ein komplexer Raum ist, werden die morphen Funktionen natürlich holomorphe
Funktionen genannt.

1.4. Die komplexen Räume der Definition von Behnke und Stein
(α-*Räume*) als auch die komplexen Räume im Cartan–Serreschen Sinne
(β-*Räume*) treten in der komplexen Analysis in natürlicher Weise auf.
Die analytischen Gebilde der holomorphen Funktionen mehrerer
Veränderlichen sind spezielle α-Räume. Ferner gewinnt man α-Räume
als Quotientenräume bei *analytischen Zerlegungen* von komplexen
Mannigfaltigkeiten[31]. Die β-Räume haben dagegen ihr Vorbild in der
algebraischen Geometrie, in welcher der (projektiv) algebraische Raum
als algebraische Teilmenge des n-dimensionalen komplex-projektiven
Raumes P^n definiert wird. Analog zum algebraischen Fall ist jede
analytische Teilmenge eines β-Raumes wieder ein β-Raum (die gleiche
Aussage gilt für die α-Räume nicht!). Von Cartan[6] wurde auf β-Räumen
die Theorie der kohärenten analytischen Garben entwickelt.† Es
wurden von ihm sehr tiefliegende Resultate hergeleitet. Will man die
gleichen Ergebnisse für α-Räume erzielen, so stößt man auf Schwierigkeiten, die eine direkte Durchführung der Cartanschen Theorie für den
Fall dieser Räume als nicht zweckmäßig erscheinen lassen.

Der Begriff des β-Raumes ist sehr allgemein gehalten. Ist X ein

---

† Die Theorie ist in den Cartanschen Seminarberichten 1951–52 explizit für komplexe Mannigfaltigkeiten durchgeführt. Die Beweise lassen sich jedoch auf den Fall
komplexer Räume übertragen.

beliebiger β-Raum, $x \in X$ ein beliebiger Punkt, so braucht der Ring $\mathfrak{O}_x$ der Keime von holomorphen Funktionen in x z. B. kein Integritätsring zu sein, wie das Beispiel des eindimensionalen komplexen Raumes $X = \{(z_1, z_2) : z_1 z_2 = 0\} \subset C^2$ lehrt. Ein eindimensionaler β-Raum ist also nicht notwendig eine Riemannsche Fläche (was für die eindimensionalen α-Räume natürlich gilt).

Aus diesem Grunde hat Cartan anfangs nur *normale β-Räume* definiert. Ein β-Raum X heißt normal, wenn jeder Halm $\mathfrak{O}_x$ seiner Strukturgarbe $\mathfrak{O}$ ein normaler Ring, d. h. in seinem Quotientenring ganz abgeschlossen ist. In [11] wurde gezeigt:

Satz. Die Klasse der α-Räume stimmt mit der Klasse der normalen β-Räume überein.

Zum Beweise dieses Satzes sei nur angemerkt, daß man mit verhältnismäßig elementaren Mitteln zeigen kann, daß jeder normale β-Raum ein α-Raum ist. Die andere Richtung der Aussage verlangt schwierige Hilfsmittel der Garbentheorie, u. a. auch einen tiefliegenden Satz von K. Oka über die Normalisierung analytischer Mengen (vgl. [24] und [5]). Es muß gezeigt werden, daß man den Träger Y jeder analytischen Überlagerung $\mathfrak{Y} = (Y, \Phi, G)$ lokal als analytische Menge in einem Teilgebiet eines komplexen Zahlenraumes realisieren kann. Zu dem Zwecke müssen auf Y holomorphe Funktionen konstruiert werden.†

1.5. Da die α-Räume spezielle β-Räume sind und eine Theorie möglichst allgemein durchgeführt werden sollte, verstehen wir im folgenden unter einem komplexen Raum stets einen β-Raum. Ein Punkt x eines β-Raumes heißt ein *regulärer* Punkt, wenn es eine Umgebung $U(x)$ gibt, die isomorph zu einem Gebiet $G \subset C^n$ ist. Da sich in einer solchen Umgebung sinnvoll komplexe Koordinaten einführen lassen, ist die Klasse der komplexen Räume, die nur aus regulären Punkten bestehen, mit der Klasse der *komplexen Mannigfaltigkeiten* identisch. Natürlich gibt es sogar normale komplexe Räume X mit nicht-regulären Punkten. Der Raum $X = \{(w, z_1, z_2) : w^2 = z_1 z_2\}$ ist dafür ein Beispiel.

2. Holomorph-vollständige Räume

2.1. Die Riemannschen Flächen der klassischen Funktionentheorie zerfallen in zwei Teilklassen, deren Untersuchung verschiedener Methoden bedarf. Die kompakten Riemannschen Flächen sind isomorph zu (1-dimensionalen) singularitätenfreien algebraischen Teil-

---

† Herr Kawai hat mir brieflich mitgeteilt, daß ihm neuerdings diese Konstruktion mit Hilfe eines Integrals ohne Anwendung von Garbentheorie ebenfalls gelungen ist.

mengen eines n-dimensionalen komplex-projektiven Raumes P^n. Sie können deshalb auf algebraische Weise behandelt werden. Die nichtkompakten Riemannschen Flächen müssen mit transzendenten Mitteln untersucht werden.

In der komplexen Analysis mehrerer Veränderlichen zeigt sich, daß die Theorie der holomorphen Funktionen auf nicht-kompakten komplexen Räumen i. a. in keiner Analogie zur Funktionentheorie auf nicht-kompakten Riemannschen Flächen steht. So gelten z. B. schon für beliebige Gebiete $G \subset C^n$ keine Analoga der Sätze von Mittag– Leffler und Weierstraß: zu einer sinnvoll vorgegebenen Verteilung von Hauptteilen meromorpher Funktionen gibt es i. a. keine meromorphe Funktion, zu sinnvoll vorgegebenen Nullstellenflächen läßt sich nicht immer eine holomorphe Funktion finden. Dagegen sind für die *Holomorphiegebiete* ähnliche Aussagen wie für die nicht-kompakten Riemannschen Flächen gültig. Als Grund, daß diese Aussagen nicht in beliebigen nicht-kompakten komplexen Räumen gelten, stellt sich heraus, daß in X gewöhnlich nicht genügend viele holomorphe Funktionen existieren.

Stein hat deshalb vorgeschlagen, das funktionentheoretische Studium auf die *holomorph-vollständigen Mannigfaltigkeiten* zu beschränken.†
Hier seien gleich allgemeiner *holomorph-vollständige Räume* untersucht:

Definition. Ein komplexer Raum X heißt holomorph-vollständig, wenn er noch zusätzlich zwei Eigenschaften hat:

(1) *Er ist K-vollständig*, d. h. zu jedem Punkt $x_0 \in X$ gibt es endlich viele in X holomorphe Funktionen $f_1, \dots, f_k$, so daß die analytische Menge $A = \{x \in X : f_\nu(x) = f_\nu(x_0), \nu = 1, \dots, k\}$ den Punkt x_0 als isolierten Punkt enthält.

(2) *X ist holomorph-konvex*: zu jeder sich in X nicht häufenden unendlichen Punktfolge x_ν gibt es eine in X holomorphe Funktion f, derart, daß die komplexen Zahlen $f(x_\nu)$ eine unbeschränkte Folge bilden.

Da auf jedem kompakten komplexen Raum jede holomorphe Funktion konstant ist, kann kein K-vollständiger und somit auch kein holomorphvollständiger Raum kompakt sein. Ein Gebiet $G \subset C^n$ ist genau dann ein holomorph-vollständiger Raum, wenn G ein Holomorphiegebiet ist.

2.2. Für holomorph-vollständige Räume gelten die Cartanschen Hauptresultate der Theorie *kohärenter analytischer Garben* ([6], théorèmes

† Nach H. Cartan heißen holomorph-vollständige Mannigfaltigkeiten 'variétés de Stein', d. h. Steinsche Mannigfaltigkeiten. Vgl. [6]. Bei den hier angegebenen Axiomen handelt es sich um eine Vereinfachung der Axiome aus [6]. Die (nicht triviale) Äquivalenz des neuen Axiomensystems mit den ursprünglichen Axiomen wurde in [14] gezeigt.

A und B). In den sehr weitgehenden Cartanschen Sätzen, sind wesentliche zuerst von Oka gewonnene Aussagen enthalten; z. B. folgen aus ihnen Analoga der Sätze von Mittag–Leffler und Weierstraß (als Lösung der sog. Cousinschen Probleme). Die Theorie der *komplex-analytischen Faserbündel* wird über holomorph-vollständigen Räumen besonders einfach. In [15] wurde gezeigt:

Satz I. *Sind* $\Re_1$, $\Re_2$ *zwei topologisch äquivalente komplex-analytische Faserbündel über einem holomorph-vollständigen Raum* $\mathfrak{B}$, *so sind* $\Re_1$ *und* $\Re_2$ *sogar komplex-analytisch äquivalent.*

Satz II. *Zu jedem topologischen komplexen Faserbündel über einem holomorph-vollständigen Raum* $\mathfrak{B}$ *gibt es ein äquivalents komplex-analytisches Bündel.*

Wie die Ergebnisse der Garbentheorie kann man die Sätze I und II als methodisches Hilfsmittel für die komplexe Analysis auf holomorph-vollständigen Räumen verwenden.

2.3. Ein Gebiet $G \subset C^n$ wird meistens durch seinen Rand gegeben. Will man nachweisen, daß G ein Holomorphiegebiet ist, so muß man Eigenschaften des Randes ∂G von G kennen, durch die sich die Holomorphiegebiete charakterisieren lassen. Besonders geeignet sind lokale Randeigenschaften.

Wie schon Levi[19] und Krzoska vermutet haben, ist eine solche Eigenschaft in der *Pseudokonvexität* von ∂G gefunden. Die Pseudokonvexität, die sich für Gebiete mit beliebigen Rand definieren läßt, kann bei Gebieten mit zweimal stetig differenzierbarem Rand durch Differentialungleichungen angegeben werden (analog zur elementaren Konvexität, die man als Krümmungseigenschaft des Randes auffassen kann).

Definition. *Ein Gebiet* $G \subset C^n$ *mit wenigstens zweimal stetig differenzierbarem Rande heiße pseudokonvex, wenn es zu jedem Punkt* $z_0 \in \partial G$ *eine Umgebung* $U(z_0)$ *und eine reell-wertige in* U *zweimal stetig differenzierbare Funktion* ϕ *gibt, so daß* $d\phi \neq 0$, $G \cap U = \{z \in U : \phi(z) < 0\}$ *und die Form* $L(\phi) = \Sigma[(\partial^2\phi)/(\partial z_\nu \partial \bar{z}_\mu)] a_\nu \bar{a}_\mu$ *auf* $U \cap \partial G$ *bedingt positiv semidefinit ist (d. h. positiv semidefinit für alle komplexen Vektoren* $(a_1, ..., a_n)$, *die der Bedingungsgleichung* $\Sigma[(\partial\phi)/(\partial z_\nu)] a_\nu = 0$ *genügen). Läßt sich stets eine Funktion* ϕ *finden, so daß* $L(\phi)$ *auf* $U \cap \partial G$ *bedingt positiv definit ist, so heißt* G *streng pseudokonvex.*

Levi (1911) und Krzoska haben gezeigt, daß jedes Holomorphiegebiet pseudokonvex ist. Die Umkehrung dieser Aussage blieb über vierzig Jahre lang ein offenes Problem (das sog. *Levische Problem*). Schließlich zeigte Oka in den Arbeiten[23, 25]:

(*) *Jedes unverzweigte pseudokonvexe Gebiet G über dem C^n ist holomorph-konvex und ein Holomorphiegebiet.*†

Das Levische Problem ist bis heute noch nicht gelöst für verzweigte und unendliche Gebiete. Doch hat auch hier Oka ein abschließendes Ergebnis angekündigt.

2.4. Der Okasche Beweis von (*) benutzt die Tatsache, daß man jedes pseudokonvexe Gebiet G durch eine aufsteigende Folge $G_\nu \subset G$, $\nu = 1, 2, \ldots$ von streng pseudokonvexen Teilgebieten ausschöpfen kann. Nachdem bewiesen ist, daß alle G_ν Holomorphiegebiete sind, folgt aus einem Satz von Behnke und Stein[1] über die Limiten konvergenter Folgen von Holomorphiegebieten, daß auch $G = \lim G_\nu$ ein Holomorphiegebiet ist.

Die Begriffe des differenzierbaren Randes,‡ der Pseudokonvexität und der strengen Pseudokonvexität lassen sich ebenfalls für relativkompakte Teilgebiete G komplexer Räume X definieren. Mit Hilfe von Garbentheorie ergibt sich[16] als Verallgemeinerung des Okaschen Resultates für streng pseudokonvexe Gebiete $G \subset C^n$:

Satz 1. Jedes streng pseudokonvexe Teilgebiet $G \subset X$ ist holomorph-konvex. Enthält G außerdem keine kompakten analytischen Teilmengen A mit $\dim A > 0$, so ist G ein holomorph-vollständiger Raum.

Natürlich ist kein Gebiet, das kompakte analytische Teilmengen A, $\dim A > 0$, enthält, holomorph-vollständig. Wie einfache Beispiele lehren, gibt es pseudokonvexe relativkompakte Teilgebiete von komplexen Mannigfaltigkeiten, die nicht holomorph-konvex sind. Es ist ein offenes Problem, welche Eigenschaften für die Holomorphiekonvexität notwendig und hinreichend sind.

Man wird bestrebt sein, die holomorph-vollständigen Räume durch leicht nachzuweisende Eigenschaften zu charakterisieren. Für Gebiete $G \subset C^n$ zeigt man auf elementarem Wege[18]: G ist genau dann pseudokonvex, wenn in G die Funktion $-\ln \delta(z)$ plurisubharmonisch ist. Dabei bezeichnet $\delta(z)$ den euklidischen Randabstand der Punkte $z \in G$. Der Begriff der *plurisubharmonischen Funktion* kann sehr allgemein definiert werden. Gewöhnlich ist eine plurisubharmonische Funktion

---

† Für den Fall schlichter Gebiete $G \subset C^n$ wurde dieser Satz 1954 unabhängig voneinander von H. Bremermann[4] und F. Norguet[22] auch gewonnen.

‡ Da komplexe Räume X nicht-uniformisierbare Punkte enthalten können, muß die Differenzierbarkeit auf besondere Weise definiert werden. Weil aber jeder komplexe Raum lokal zu einer analytischen Menge $A \subset G$ isomorph ist, braucht der Begriff der differenzierbaren Funktion nur für solche analytischen Mengen festgesetzt zu werden. Man kann sagen, eine Funktion auf A ist k-mal stetig differenzierbar, wenn sie (lokal) die Beschränkung einer in G k-mal stetig differenzierbaren Funktion ist.

nur halbstetig nach oben. Ihre Werte sind die reellen Zahlen und $-\infty$. Für zweimal stetig differenzierbare reell-wertige Funktionen p läßt sich die Definition jedoch folgendermaßen angeben:

p ist plurisubharmonisch genau dann, wenn die Levische Form $L(p)$ überall (ohne Bedingung) positiv semidefinit ist. Ist $L(p)$ überall positiv definit, so heißt p streng plurisubharmonisch.

Die Begriffe der plurisubharmonischen und der streng plurisubharmonischen Funktion lassen sich in nahe liegender Weise auf komplexen Räumen definieren (vgl. [10]). Mit Hilfe der in einem holomorphvollständigen Raum X holomorphen Funktionen läßt sich leicht eine in X streng plurisubharmonische Funktion $p(x)$ konstruieren, so daß jeder Teilbereich $T(d) = \{x \in X, p(x) < d\}$ mit $d > 0$ relativ-kompakt in X liegt. Ist umgekehrt $p(x)$ eine streng plurisubharmonische Funktion in einem komplexen Raum X, derart, daß jedes $T(d)$ relativ-kompakte Teilmenge von X ist, so sind alle $T(d)$ streng pseudokonvex. Aus Satz 1 und einem Analogon zum Limessatz für Holomorphiegebiete folgt sodann

Satz 2. Ein komplexer Raum X ist ein holomorph-vollständiger Raum dann und nur dann, wenn es in ihm eine streng plurisubharmonische Funktion $p(x)$ gibt, so daß alle offenen Mengen $T(d) = \{x \in X, p(x) < d\} \subset\subset X$ liegen.

2.5. Die Sätze 1 und 2 gestatten mehrere Anwendungen. Aus Satz 2 folgt unmittelbar die Lösung eines Problems von Whitney über die Einbettung *reell-analytischer Mannigfaltigkeiten* in euklidischen Räumen. Whitney hat in Arbeiten [33, 34], die in den Jahren 1936 und 1944 in den *Annals of Mathematics* erschienen, folgenden Satz gezeigt:

Es sei $\mathfrak{R}$ eine n-dimensionale reell-analytische Mannigfaltigkeit mit abzählbarer Topologie. Dann gibt es eine beliebig oft stetig differenzierbare, (lokal) eigentliche, reguläre† eineindeutige Abbildung $\phi\colon \mathfrak{R} \to R^{2n}$ von $\mathfrak{R}$ auf eine reell-analytische Untermannigfaltigkeit $\phi(\mathfrak{R}) = F \subset R^{2n}$.

Als Problem blieb offen, ob man ϕ stets als reell-analytische Abbildung wählen kann (womit $\mathfrak{R}$ als reell-analytische Untermannigfaltigkeit des R^{2n} realisiert wäre). Wie Morrey [21] im Januar dieses Jahres (1958) gezeigt hat, ist diese Frage für den Fall kompakter Mannigfaltigkeiten $\mathfrak{R}$ bejahend zu beantworten. Durch Satz 2 wird nun das Problem im positiven Sinne allgemein gelöst:

Jede reell-analytische Mannigfaltigkeit $\mathfrak{R}$ mit abzählbarer Topologie

---

† 'Regulär' heißt, daß die Funktionalmatrix von ϕ überall in $\mathfrak{R}$ den Rang n hat. F ist also eine singularitätenfreie Fläche im R^{2n}. Whitney hat den angegebenen Satz m. m. gleich für k-mal stetig differenzierbare Mannigfaltigkeiten hergeleitet, $k \geqslant 1$.

läßt sich nämlich *komplexifizieren*: $\mathfrak{R}$ ist reell-analytische Untermannigfaltigkeit einer komplex n-dimensionalen komplexen Mannigfaltigkeit $\mathfrak{M}$ (mit $n = \dim_r \mathfrak{R}$). Mit Hilfe einer 'Teilung der Eins' kann man leicht zeigen:

Bei geeigneter Wahl von $\mathfrak{M}$ gibt es eine in $\mathfrak{M}$ streng plurisubharmonische Funktion p, so daß für alle $T(d) = \{x \in M : p(x) < d\}$ gilt $T(d) \Subset M$.

Nach Satz 2 ist M dann eine Steinsche Mannigfaltigkeit, nach einem Einbettungssatz von Remmert[27] gibt es zu jeder Steinschen Mannigfaltigkeit $\mathfrak{M}$ eine holomorphe, eigentliche, reguläre, eineindeutige Abbildung $\hat{\phi} \colon \mathfrak{M} \to C^k$ (k genügend groß). $\phi' = \hat{\phi} \mid R$ ist dann eine reguläre, eineindeutige, eigentliche, reell-analytische Abbildung $\mathfrak{R} \to C^k \approx R^{2k}$. Die gewünschte reell-analytische Abbildung $\phi \colon \mathfrak{R} \to R^{2n}$ erhält man sodann mit Hilfe eines Approximationssatzes von Whitney[34]†.

Kodaira[17] hat 1953 in der Theorie der Kählermannigfaltigkeiten folgenden grundlegenden Satz bewiesen:

(**) *Eine kompakte komplexe Mannigfaltigkeit $\mathfrak{M}$ ist genau dann eine projektiv algebraische Mannigfaltigkeit, wenn sie eine Hodge'sche Metrik (= Kählersche Metrik mit ganzzahligen Perioden) tragen kann.*

Dieses Resultat verallgemeinert den Satz über die Periodenrelationen $2n$-fach periodischer Funktionen im C^n und kann in der Theorie der algebraischen Mannigfaltigkeiten als ein Analogon des Okaschen Fundamentalsatzes (*) angesehen werden. Es ist deshalb nicht verwunderlich, daß man Satz 1 zum Beweise von (**) verwenden kann. Darüber hinaus folgt aus Satz 1 sogar die Aussage von (**) für den Fall, daß X ein komplexer Raum mit normalen nicht-uniformisierbaren Punkten ist. Als Nebenergebnis dürfte folgende Aussage interessieren:

Ein komplex-analytisches Geradenbündel F über einem kompakten komplexen Raum X ist genau dann negativ (im Sinne von Kodaira), wenn F durch eine eigentliche birationale Transformation aus einer affin-algebraischen Mannigfaltigkeit ensteht.

Weitere Anwendungen unserer Sätze 1 und 2 sind möglich, um Sätze von Enriques, die bislang nur für den Fall algebraischer Räume bekannt waren, auch für beliebige komplexe Räume herzuleiten. Die Enriques'schen Sätze—auf die hier nicht näher eingegangen werden kann—handeln über birationale Transformationen, bei denen kompakte analytische Flächen durch Punkte ersetzt werden.

---

† Ein analoges Resultat läßt sich mit Hilfe von Satz 2 auch für reell-analytische Räume, d. s. reell-analytische Mannigfaltigkeiten mit algebroiden Singularitäten, gewinnen.

3. Kompakte und allgemeinere komplexe Räume

3.1. Zur Untersuchung kompakter komplexer Räume X wird man wie in der klassischen Funktionentheorie zu algebraischen Methoden greifen wollen. Jedoch zeigt sich, daß auf allgemeinen kompakten komplexen Räumen X—außer den konstanten—keine meromorphen Funktionen existieren, also eine Realisierung von X als algebraischer Unterraum des P^k nicht möglich ist.

Weil hat den Begriff des *abstrakten algebraischen Raumes* in die Literatur eingeführt. Diese Räume—komplex-analytisch betrachtet†—können, sofern sie vollständig sind, als spezielle kompakte komplexe Räume X gedeutet werden. Der Körper $K(X)$ der meromorphen Funktionen auf X hat stets den Transzendenzgrad $n = \dim X$ über dem Körper der komplexen Zahlen. Obgleich also viele nicht-konstante meromorphe Funktionen auf X existieren, gibt es schon zweidimensionale normale vollständige algebraische Räume X (mit wenigstens zwei nicht-uniformisierbaren Punkten), die sich nicht als algebraische Unterräume eines P^k realisieren lassen. Nagata hat sogar n-dimensionale kompakte algebraische Mannigfaltigkeiten ($n \geqslant 3$) dieser Art angegeben. Algebraische Räume und Mannigfaltigkeiten, die sich in einem komplex-projektiven Raum einbetten lassen, heißen *projektiv algebraisch*. Wie Chow und Kodaira[8] gezeigt haben, ist jede zweidimensionale komplexe Mannigfaltigkeit mit zwei unabhängigen meromorphen Funktionen in diesem Sinne projektiv algebraisch.

Zur Untersuchung abstrakter wie auch projektiv-algebraischer Räume eignet sich besonders die Theorie *kohärenter algebraischer Garben*, die zuerst in einer Arbeit von Serre[28] definiert wurden. Grothendieck hat mit Hilfe dieser Theorie die Formel von Riemann–Roch–Hirzebruch für projektiv algebraische Mannigfaltigkeiten in einer verallgemeinerten Form auf algebraischer Weise hergeleitet.

3.2. Vollständige projektiv algebraische Mannigfaltigkeiten sind spezielle kompakte Kählersche Mannigfaltigkeiten. Für die Klasse der Kählerschen Mannigfaltigkeiten wurden von Kodaira und Spencer potentialtheoretische Methoden entwickelt. Es wurden Ergebnisse erzielt, die für projektiv algebraische Mannigfaltigkeiten auf algebraischen Wege teils nicht bewiesen werden können, teils noch nicht bewiesen werden konnten.

---

† Die algebraischen Räume der abstrakten algebraischen Geometrie tragen natürlich nur die Zariskitopologie.

Leider ist es bis heute noch nicht gelungen, die Differentialgeometrie und die Potentialtheorie auch bei der Untersuchung von komplexen Räumen mit nicht-uniformisierbaren Punkten auszunutzen. Da diese Räume keine Tangentialbündel besitzen, entstehen wesentliche Schwierigkeiten. Dagegen besteht Hoffnung mit Potentialtheorie gewonnene Sätze durch garbentheoretische Methoden herzuleiten. Die neuen Beweise lassen sich dann i. a. auf den Fall komplexer Räume übertragen. Im § 2 der vorl. Arbeit wurde ein solcher garbentheoretischer Beweis für einen Kodairaschen Satz angegeben. Es wurde gezeigt, daß jede Hodge'sche Mannigfaltigkeit projektiv algebraisch ist.

3.3. Die holomorph-vollständigen und die algebraischen Räume geben zu interessanten funktionentheoretischen Untersuchungen Anlaß, da auf ihnen eine hinreichende Zahl holomorpher bzw. meromorpher Funktionen existiert. Ist X ein holomorph-vollständiger, Y ein algebraischer Raum, so ist das kartesische Produkt $X \times Y$ weder ein holomorph-vollständiger noch ein algebraischer Raum. Der komplexe Raum $X \times Y$ dürfte jedoch das gleiche funktionentheoretische Interesse verdienen wie die Räume X und Y. Aus diesem Grunde wurde in [12] eine Klasse komplexer Räume definiert, die die holomorph-vollständigen Räume und die (projektiv) algebraischen Räume umfaßt:

Definition. Ein komplexer Raum X heißt analytisch-vollständig, wenn er:

(1) *holomorph-konvex ist,*

(2) *es zu jedem Punkt $x_0 \in X$ eine holomorphe Abbildung ϕ von X in einen komplex projektiven Raum P^k gibt, derart, daß x_0 isolierter Punkt der Menge $A = \{x \in X: \phi(x) = \phi(x_0)\}$ ist.*

Wie Remmert[27] gezeigt hat, gibt es zu jedem analytisch-vollständigen Raum X eine *holomorphe Reduktion*, d. h. einen holomorph-vollständigen Raum Y und eine eigentliche holomorphe Abbildung $\rho: X \to Y$. Die Fasern $\rho^{-1}(y_0)$, $y_0 \in Y$, sind projektiv algebraische Räume (zum Beweis vgl. [12]). Jeder analytisch vollständige Raum X kann daher als ein verallgemeinerter analytischer Faserraum mit projektiv algebraischen Fasern über einem holomorph-vollständigen Raum Y aufgefaßt werden.

Bei der Untersuchung analytisch-vollständiger Räume kommt der Abbildung $\rho: X \to Y$ und den *direkten Bildern* $\rho_\nu(\mathfrak{S})$ kohärenter analytischer Garben $\mathfrak{S}$ in X besondere Bedeutung zu. Es zeigt sich, daß alle Bilder $\rho_\nu(\mathfrak{S})$ kohärente analytische Garben in Y sind†. Mit Hilfe

† Vgl. [12]. In [13] wurden die direkten Bilder der kohärenten analytischen Garben bei Produktabbildungen von kartesischen Produkten $Y \times P^n \to Y$ untersucht (Y ein komplexer Raum). Es wurde gezeigt, daß alle direkten Bilder kohärent sind.

dieser Methoden läßt sich der Riemann–Rochsche Satz auf analytisch vollständige Räume verallgemeinern (vgl. [12]).

3.4. Die Erforschung allgemeiner komplexer Räume steckt noch sehr in ihren Anfängen. So sind Fragen wie: 'Wodurch wird ein komplexer Raum holomorph-vollständig, algebraisch oder eine komplexe Mannigfaltigkeit Kählersch?', noch ungeklärt. Zur Lösung dieser Probleme fehlen allgemeine Aussagen und Existenzsätze, die für beliebige komplexe Räume gültig sind. Bislang sind nur wenige solcher Sätze bekannt geworden. Abschließend seien sie hier zusammengestellt:

(1) Es sei X ein kompakter komplexer Raum, S eine kohärente analytische Garbe über X. Dann sind die *Čechschen Kohomologiegruppen* $H^\nu(X, S)$, $\nu = 0, 1, 2, \ldots$ *endlichdimensionale komplexe Vektorräume* (Cartan[5], Cartan und Serre[7]).

(2) Es sei X eine komplexe Mannigfaltigkeit. Dann gilt für X der *Serresche Dualitätssatz* in Bezug auf Kohomologiegruppen mit Koeffizienten in einer freien analytischen Garbe $\mathfrak{S}$ (Serre[29]).

(3) Es sei X eine n-dimensionale komplexe Mannigfaltigkeit, $\mathfrak{S}$ eine kohärente analytische Garbe über X. *Dann ist* $H^\nu(X, \mathfrak{S}) = 0$, $\nu > n$ (Malgrange[20]).

(4) Es sei X ein kompakter komplexer Raum. Dann ist der *Körper der auf X meromorphen Funktionen isomorph* zu einem *algebraischen Funktionenkörper* über C vom Transzendenzgrad k, $k \leqslant \dim X$ (Remmert[26]).

LITERATUR

[1] Behnke, H. and Stein, K. Konvergente Folgen nicht-schlichter Regularitätsbereiche. *Ann. Mat.* (4), 28, 317–326 (1949).
[2] Behnke, H. and Stein, K. Modifikationen komplexer Mannigfaltigkeiten und Riemannscher Gebiete. *Math. Ann.* 124, 1–16 (1951).
[3] Bochner, S. and Martin, W. T. Complex spaces with singularities. *Ann. Math.* (2), 57, 490–516 (1953).
[4] Bremermann, H. Über die Äquivalenz der pseudokonvexen Gebiete und der Holomorphiegebiete im Raum von n komplexen Veränderlichen. *Math. Ann.* 128, 63–91 (1954).
[5] Cartan, H. *Séminaire E.N.S.* 1953–54 (hektographiert).
[6] Cartan, H. Variétés analytiques complexes et cohomologie. *Colloque sur les fonctions de plusieurs variables.* Brussels, 1953.
[7] Cartan, H. and Serre, J. P. Un théorème de finitude concernant les variétés analytiques compactes. *C.R. Acad. Sci., Paris*, 237, 128–130 (1953).
[8] Chow, W. L. and Kodaira, K. On analytic surfaces with two independent meromorphic functions. *Proc. Nat. Acad. Sci., Wash*, 38, 319–325 (1952).

[9] Grauert, H. and Remmert, R. Zur Theorie der Modifikationen. I. Stetige und eigentliche Modifikationen komplexer Räume. *Math. Ann.* 129, 274–296 (1955).

[10] Grauert, H. and Remmert, R. Plurisubharmonische Funktionen in komplexen Räumen. *Math. Z.* 65, 175–194 (1956).

[11] Grauert, H. and Remmert, R. Komplexe Räume. *Math. Ann.* 136, 245–318 (1958).

[12] Grauert, H. and Remmert, R. Espaces analytiquement complets. *C.R. Acad. Sci., Paris,* 245, 882–885 (1957).

[13] Grauert, H. and Remmert, R. Bilder und Urbilder analytischer Garben. *Ann. Math.* (2), 68, 393–443 (1958).

[14] Grauert, H. Charakterisierung der holomorph-vollständigen komplexen Räume. *Math. Ann.* 129, 233–259 (1955).

[15] Grauert, H. Analytische Faserungen über holomorph-vollständigen Räumen. *Math. Ann.* 135, 263–273 (1958).

[16] Grauert, H. On Levi's problem and the imbedding of real-analytic manifolds. *Ann. Math.* (2), 68, 460–472 (1958).

[17] Kodaira, K. On Kähler varieties of the restricted type. *Ann. Math.* (2), 60, 28–48 (1954).

[18] Lelong, P. Fonctions plurisousharmoniques; mesures de Radon associées. Applications aux fonctions analytiques. *Colloque sur les fonctions de plusieurs variables,* 21–40, Brussels, 1953.

[19] Levi, E. E. Studii sui punti singolari essenziali delle funzioni analitiche di due o più variabili complesse. *Ann. Mat.* (3), 17, 61–87 (1910).

[20] Malgrange, B. Faisceaux sur des variétés analytiques réelles. *Bull. Soc. Math. Fr.* 85, 231–237 (1957).

[21] Morrey, Ch. B. The analytic embedding of abstract real-analytic manifolds. *Ann. Math.* (2), 68, 159–201 (1958).

[22] Norguet, F. Sur les domaines d'holomorphie des fonctions uniformes de plusieurs variables complexes. *Bull. Soc. Math. Fr.* 82, 137–159 (1954).

[23] Oka, K. Sur les fonctions analytiques de plusieurs variables. VI. Domaines pseudoconvexes. *Tôhoku Math. J.* 49, 19–52 (1942).

[24] Oka, K. Sur les fonctions analytiques de plusieurs variables. VIII. Lemme fondamental. *J. Math. Soc. Japan,* 3, 204–214, 259–278 (1951).

[25] Oka, K. Sur les fonctions analytiques de plusieurs variables. IX. Domaines finis sans point critique intérieur. *Jap. J. Math.* 23, 97–155 (1954).

[26] Remmert, R. Meromorphe Funktionen in kompakten komplexen Räumen. *Math. Ann.* 132, 277–288 (1956).

[27] Remmert, R. Sur les espaces analytiques holomorphiquement séparables et holomorphiquement convexes. *C.R. Acad. Sci., Paris,* 243, 118–121 (1956).

[28] Serre, J. P. Faisceaux algébriques cohérents. *Ann. Math.* (2), 61, 197–278 (1955).

[29] Serre, J. P. Un théorème de dualité. *Comment. Math. Helv.* 29, 9–26 (1955).

[30] Serre, J. P. Géométrie algébrique et géométrie analytique. *Ann. Inst. Fourier, Grenoble,* 6, 1–42 (1955–56).

[31] Stein, K. Analytische Zerlegungen komplexer Räume. *Math. Ann.* 132, 63–93 (1956).

[32] Weyl, H. *Die Idee der Riemannschen Fläche.* 3. Auflage, Teubner, Stuttgart, 1955.

[33] Whitney, H. Differentiable manifolds. *Ann. Math.* (2), 37, 645–680 (1936).

[34] Whitney, H. The self-intersections of a smooth *n*-manifold in 2*n*-space. *Ann. Math.* (2), 45, 220–246 (1944).

FUNCTIONS OF BOUNDED CHARACTERISTIC
AND LINDELÖFIAN MAPS

By MAURICE HEINS

1. Introduction

The study of the boundary behavior of analytic functions in a modern sense begins with the fundamental and justly celebrated memoir of Fatou[2]. An essential feature of that paper is the exploitation to the fullest of the then novel ideas of Lebesgue. The half century which has elapsed since Fatou's memoir has witnessed the flowering of a large and important chapter of analysis whose concern is the boundary behavior of analytic functions. Our attention will be centered on questions whose parentage may be traced back to the results of Fatou.

Let us recall that the original theorem of Fatou states that a bounded analytic function f in the open unit disk possesses a radial limit p.p. and that the theorem of F. and M. Riesz[13] states that the radial limit function of f vanishes only on a set of measure zero, if $f \not\equiv 0$. The functions of bounded Nevanlinna characteristic in the open unit disk are precisely the non-constant meromorphic functions which admit representation as quotients of bounded analytic functions. Thanks to the original Fatou theorem and the theorem of F. and M. Riesz, the conclusion of the Fatou theorem persists for functions of bounded characteristic.

When we turn to the study of asymptotic values, we find that a marked contrast appears. For a bounded non-constant analytic function in the open unit disk, each asymptotic path tends to a point of the unit circumference and the associated asymptotic value is the Fatou radial limit at this point. On the other hand, while it is still true for a function of bounded characteristic that each asymptotic path tends to a point of the unit circumference, there exist, as examples of Lohwater[9] and Lehto[8] show, functions of bounded characteristic having more than one asymptotic value associated with a given boundary point. A recently announced theorem of Gehring[4] asserts that an *analytic* (pole-free) function of bounded characteristic in the open unit disk admits at most two distinct finite asymptotic values associated with a given boundary point. We shall see later that the situation changes radically for unrestricted functions of bounded characteristic. In fact, there exists a quotient of two Blaschke products whose zeros cluster solely at $z = 1$ which has the property that the set of asymptotic values associated with

$z = 1$ has the power of the continuum.† In the construction which we give of such a function, a decisive role is played by Valiron's example[15] of a meromorphic function of finite order whose set of asymptotic values has the power of the continuum.

However, the realm of pathology is small. In fact, for an arbitrary non-constant meromorphic function f in $\{|z| < 1\}$, the set of points of $\{|z| = 1\}$ with which more than one asymptotic spot[5] is associated is countable. Examples of functions of bounded characteristic where such sets are infinite are readily constructed. We understand that a point η of $\{|z| = 1\}$ is *associated* with an asymptotic spot σ of f provided that $\{\eta\} = \cap \overline{\sigma(\omega)}$, $\omega \in$ domain of σ.

The notion of a function of bounded characteristic may be generalized to conformal maps of Riemann surfaces[6, 12]. The Nevanlinna characteristic function may be adapted to conformal maps of Riemann surfaces in such a manner that the first fundamental theorem and the basic theorems of the Nevanlinna theory associated with it persist in the general theory of conformal maps of Riemann surfaces. The notion of a map of bounded characteristic becomes an important element of the extended theory.

In the present paper we shall confine our attention to Lindelöfian conformal maps (= maps of bounded characteristic) *whose domain is the open unit disk* (meromorphic functions of bounded characteristic are included) and shall apply the methods developed in our papers [5] and [6] to problems concerning the boundary behavior of such maps and the relation between their boundary behavior and their 'covering properties'.‡ By confining our attention to Lindelöfian maps with domain the open unit disk we put at our disposal an extensive apparatus from real function theory. This fact permits us to expect more precise results than one could for arbitrary Lindelöfian maps.

2. Non-negative harmonic functions

Parreau[11] has introduced the notions of a *quasi-bounded* non-negative harmonic function and of a *singular* non-negative harmonic function.

† The author is indebted to Professor F. W. Gehring for pointing out the fact that he carried out a construction along the lines of the second paragraph of §10 of the present paper (F. W. Gehring, The asymptotic values for analytic functions with bounded characteristic, *Oxford Quart. J. Math.* 1958). His construction shows the existence of a function f of bounded characteristic in $\{|z| < 1\}$ having the property that the set of asymptotic values of f associated with paths terminating at $z = 1$ has the power of the continuum. However, he does not establish the more refined result to which the present footnote refers.

‡ For the case of meromorphic functions of bounded characteristic, cf. Lehto [7,8]. I am indebted to Professor Pfluger for raising the question of extending the Fatou theorem to Lindelöfian maps with domain the open unit disk.

These may be defined as follows. A non-negative harmonic function u on a Riemann surface F is termed *quasi-bounded* provided that it is the limit of a monotone non-decreasing sequence of bounded non-negative harmonic functions on F; u is termed *singular* provided that the only non-negative bounded harmonic function on F which is dominated by u is zero.† A non-negative harmonic function on F admits a unique representation as a sum of a quasi-bounded non-negative harmonic function on F and a singular non-negative harmonic function on F. A positive harmonic function u on F is termed *minimal* (Martin[10]) provided that the positive harmonic functions on F dominated by u are proportional to u. If a minimal positive harmonic function is not bounded, it is singular.

μ-maps. Let Ω denote a non-empty open subset of a Riemann surface F which has the property that each point of fr Ω is a point of a continuum contained in fr Ω. Given a non-negative harmonic function u on Ω which vanishes continuously on fr Ω, by $\mu_\Omega(u)$ is meant the least harmonic majorant of the subharmonic function on F which agrees with u on Ω and vanishes elsewhere on F, provided that the subharmonic function in question admits a harmonic majorant; otherwise $\mu_\Omega(u) = +\infty$. Let Q_Ω denote the set of u for which $\mu_\Omega(u) < +\infty$. The restriction of μ_Ω to Q_Ω is a univalent homogeneous additive map of Q_Ω into the set of non-negative harmonic functions on F. Further u ($\in Q_\Omega$) is quasi-bounded (singular) on Ω if and only if $\mu_\Omega(u)$ is quasi-bounded (singular) on F. If Ω is a region, then u ($\in Q_\Omega$) is minimal on Ω if and only if $\mu_\Omega(u)$ is minimal on F. By a *μ-map* we shall understand the restriction of μ_Ω to Q_Ω for some admitted Ω.

3. The Lindelöf principle

Let ϕ denote a conformal map (not necessarily univalent) of a Riemann surface F into a Riemann surface G. Let $n(p; \phi)$ denote the *multiplicity* of ϕ at $p \in F$. Let $\nu_\phi(q)$ denote the *valence* of ϕ at $q \in G$, i.e. $\Sigma_{\phi(p)=q} n(p; \phi)$.

Suppose now that F and G are hyperbolic and that $\mathfrak{G}_F$ and $\mathfrak{G}_G$ are their respective Green's functions. Let

$$S(p,q) = \Sigma_{\phi(r)=q} n(r; \phi) \mathfrak{G}_F(p,r).$$

The *exact form* of the Lindelöf principle asserts that for each $q \in G$,

$$\mathfrak{G}_G(\phi(p),q) = S(p,q) + u_q(p) \quad (p \in F), \tag{3.1}$$

where u_q is a non-negative harmonic function on F. Obviously u_q is

† These concepts are also pertinent to non-negative harmonic functions whose domains are non-empty open subsets of a Riemann surface.

unique. We note that u_q is the greatest harmonic minorant (abbreviated henceforth by 'G.H.M.') of the superharmonic function $p \to \mathfrak{G}_G(\phi(p), q)$. Let v_q denote the quasi-bounded component of u_q and let w_q denote the singular component of u_q. We have the alternatives: $v_q = 0$ for all $q \in G$, or $v_q > 0$ for all $q \in G$. In the first case we say that ϕ is a *map of type-Bl* of F into G, the designation 'Bl' being employed because of the close relation of such maps with Blaschke products. If $u_q = 0$ for all $q \in G$, we say that ϕ is a map of *type-Bl$_1$* of F into G. We observe that regardless of whether ϕ is of type-Bl or not, $w_q = 0$ save for an F_σ of capacity zero [5].

Suppose that F and G are now unrestricted. We say that ϕ is of *type-Bl* (Bl_1) at $q \in G$ provided that there exists a simply connected Jordan region Ω, $q \in \Omega \subset G$, such that $\phi^{-1}(\Omega) \neq \varnothing$ and the restriction of ϕ to each component ω of $\phi^{-1}(\Omega)$ is a map of type-Bl (Bl$_1$) of ω into Ω.

4. Lindelöfian maps

A conformal map ϕ of a hyperbolic Riemann surface F into a hyperbolic Riemann surface G satisfies

$$S(p, q) < +\infty \quad (\phi(p) \neq q). \qquad (4.1)$$

Interest therefore attaches to the study of conformal maps $\phi: F \to G$, where F is hyperbolic but G is unrestricted, which satisfy (4.1). (It is to be noted that in the definition of $S(p, q)$ we need not assume that G is hyperbolic.) Because of the genesis of this condition we term such maps *Lindelöfian*. In terms of the extension of the Nevanlinna theory to conformal maps of Riemann surfaces the Lindelöfian maps are precisely the conformal maps of bounded characteristic.

5. Given distinct points $a, b \in G$, there exists a function u on G which is harmonic save at a and b, which has a normalized positive logarithmic singularity at a and a normalized negative logarithmic singularity at b, and which is bounded in the complement of some relatively compact neighborhood of $\{a, b\}$. (If G is parabolic, then u is determined up to an additive constant.) Given a Lindelöfian map $\phi: F \to G$, it may be concluded [6] from the extended form of Nevanlinna's first fundamental theorem that $u \circ \phi$ *admits a representation of the form*

$$u \circ \phi(p) = S(p, a) - S(p, b) + H(p), \qquad (5.1)$$

where H is the difference of two non-negative harmonic functions on F. We shall see that this representation is the basis of the boundary behavior theorems of the present paper. It is to be observed that H admits a unique representation of the form $P - N$, where P and N are non-negative harmonic functions on F satisfying G.H.M. $\min\{P, N\} = 0$. There is an intimate connection between the harmonic function P and the way in which ϕ covers a neighborhood of a.

In fact, let $\Omega_0 = \{u > 0\}$, let $\omega_0 = \phi^{-1}(\Omega_0)$, and let ϕ_{ω_0} denote the restriction of ϕ to ω_0. If $\omega_0 \neq \varnothing$, then, thanks to (5·1) and simple facts concerning greatest harmonic minorants of superharmonic functions, we are led to

$$\mu_{\omega_0}(\text{G.H.M. } u \circ \phi_{\omega_0}) = P. \tag{5.2}$$

If $\omega_0 = \varnothing$, by convention we take the left-hand side of (5.2) to be zero; it is readily verified that in this case we also have $P = 0$. It is now easy to conclude that, if Ω is an arbitrary Jordan region of G which contains a, $\omega = \phi^{-1}(\Omega)$, and $\mathfrak{G}_\Omega$ is the Green's function of Ω, then

$$\mu_\omega\{\text{G.H.M. } \mathfrak{G}_\Omega(\phi_\omega, a)\} - P = O(1). \tag{5.3}$$

Here ϕ_ω refers to the restriction of ϕ to ω and the obvious convention prevails when $\omega = \varnothing$. From (5.3) we conclude that *the singular component of the first member of the left side of* (5.3) *is independent of* Ω *and is simply the singular component of* P. We denote it by W_a.

A situation in which (5.2) may be exploited advantageously is the following. Suppose that Ω_0 is connected and that the restriction of u to Ω_0 is the Green's function of Ω_0 with pole at a. (This is certainly the case if G is parabolic.) Suppose further that $\omega_0 \neq \varnothing$. *Then P is singular if and only if the restriction of ϕ to each component of ω_0 is a map of type-Bl of that component into Ω_0.* If P is singular, then by (5.2) G.H.M. $u \circ \phi_{\omega_0}$ is singular, and ϕ is readily seen to have the stated property. The converse is also readily established on noting that a μ-map carries a singular non-negative harmonic function into a singular non-negative harmonic function.

To give a specific application of this result, we consider a function f of bounded characteristic in $\{|z| < 1\}$ whose Fatou radial limits are p.p. of modulus one. If f is bounded, a known theorem of Frostman[3] asserts that for all α, $|\alpha| < 1$, save for an F_σ of zero capacity,

$$\frac{f - \alpha}{1 - \bar{\alpha} f} \tag{5.4}$$

is a Blaschke product up to a constant factor of modulus one.† What we shall now see is that, if neither f nor $1/f$ is bounded, then (5.4) *is a*

† The property of w_a cited in §3 constitutes a generalization of Frostman's theorem.

quotient of two Blaschke products up to a constant factor of modulus one, for all α, $|\alpha| < 1$, save for an F_σ of zero capacity. We remark that the case where $1/f$ is bounded is readily reduced to the case where f is bounded.

To establish our assertion we note that with

$$a = 0, \quad b = \infty, \quad u(z) = -\log|z|,$$

P and N are both singular by virtue of the hypotheses on f. Hence the restriction of f to a component of $f^{-1}\{|z| < 1\}$ (or $f^{-1}\{|z| > 1\}$) is of type-Bl relative to $\{|z| < 1\}$ (or $\{|z| > 1\}$ respectively). It now follows from the property of w_q cited in § 3 and (5.2) that save for an exceptional set of α of the described type, the P and N associated with

$$\log\left|\frac{1 - \bar{\alpha}f}{f - \alpha}\right|$$

vanish. The assertion is readily established.

This result admits generalizations to conformal maps of Riemann surfaces. We shall not pursue this question further here.

6. The Fatou theorem

We shall now see that the Fatou theorem persists for Lindelöfian maps $\phi: F \to G$, $F = \{|z| < 1\}$, in the sense that *for almost all ζ on the unit circumference, either ϕ tends to a point of G as z tends to ζ sectorially, or else ϕ tends to the ideal boundary of G as z tends to ζ sectorially.*

We consider $u \circ \phi$ of § 5 having fixed distinct points $a, b \in G$. There exists a meromorphic function f in $\{|z| < 1\}$ satisfying

$$\log|f| = u \circ \phi. \tag{6.1}$$

By (5.1) f is of bounded characteristic, so that the Fatou theorem holds for f. Suppose that f has a sectorial limit at η, $|\eta| = 1$. Suppose that for some δ, $0 < \delta < \frac{1}{2}\pi$, it is not the case that as z tends to η in

$$S_\delta = \{|\arg(1 - \bar{\eta}z)| < \delta\},$$

$\phi(z)$ tends to a point of G or to the ideal boundary. There would then exist a point $q \in G$, a relatively compact open disk Δ of G containing q, and sequences (z_n') and (z_n'') whose members lie in S_δ and which satisfy not only

$$\lim z_n' = \lim z_n'' = \eta,$$

but also

$$\phi(z_n') \in \Delta, \quad \phi(z_n'') \in G - \Delta \quad \text{(all } n\text{)},$$

and

$$\lim \phi(z_n') = q.$$

Let g denote a meromorphic function in Δ for which $\log |g|$ is the restriction of u to Δ. On considering f on the segments $z_n' z_n''$ we are led to conclude that g is constant. This is of course impossible. The extended Fatou theorem follows.

7. Criteria for sectorial and quasi-sectorial limits

We now turn to the examination of sufficient conditions for a point G to be a sectorial (or a quasi-sectorial†) limit of ϕ.

(a) *Suppose that ϕ has a logarithmic ramification over $a \in G$. Then a is a sectorial limit of ϕ.*

More formally, we suppose that the following conditions are fulfilled: (1) for each simply connected Jordan region Ω of G which contains a, $\phi^{-1}(\Omega) \neq \emptyset$, (2) there exists a function σ whose domain consists of the set of such Ω and which satisfies

    (i) $\sigma(\Omega)$ is a component of $\phi^{-1}(\Omega)$,

    (ii) $\Omega_1 \subset \Omega_2$ implies $\sigma(\Omega_1) \subset \sigma(\Omega_2)$,

    (iii) for Ω sufficiently small, $\phi_{\sigma(\Omega)}$, the restriction of ϕ to $\sigma(\Omega)$, is a universal covering of $\Omega - \{a\}$.

Thanks to these conditions, it follows that, for Ω small, $\mathfrak{G}_\Omega(\phi_{\sigma(\Omega)}, a)$ is a minimal positive harmonic function on $\sigma(\Omega)$. From (5.3) it may be concluded that

$$\mu_{\sigma(\Omega)}[\mathfrak{G}_\Omega(\phi_{\sigma(\Omega)}, a)] < +\infty. \tag{7.1}$$

For Ω sufficiently small, the left side of (7.1) is a minimal positive harmonic function in $\{|z| < 1\}$ independent of Ω, say

$$c\Re\left[\frac{\eta + z}{\eta - z}\right] \quad (c > 0, |\eta| = 1).$$

It follows from the Julia–Carathéodory theorem that for given δ, $0 < \delta < \frac{1}{2}\pi$, and for each Ω, the points of the sector S_δ which are sufficiently close to η lie in $\sigma(\Omega)$. We infer that a is the sectorial limit of ϕ at η.

The example

$$f(z) = \exp\left\{\exp\left(\frac{1+z}{1-z}\right)\right\} \quad (|z| < 1),$$

shows that there exists an analytic function in $\{|z| < 1\}$ which possesses a sectorial limit save at 1 and has infinitely many logarithmic ramifications over 0 and yet does not possess the sectorial limit 0. This shows that the Lindelöfian character of ϕ is pertinent to the question considered.

---

† This notion will be defined below in the present section.

Quasi-sectorial limit. We shall say that ϕ has the quasi-sectorial limit $a\,(\in G)$ at η provided that

$$\lim_{r\to 0}\phi[\eta(1-re^{i\theta})] = a \quad (|\theta| < \tfrac{1}{2}\pi), \tag{7.2}$$

save for a set of θ of zero capacity. We have:

(b) If
$$W_a(z) \geqslant c\Re\left[\frac{\eta+z}{\eta-z}\right],$$

where c is a positive constant, then ϕ possesses the quasi-sectorial limit a at η. In fact, it follows from (5.1) that

$$u(\phi(z)) \geqslant c\Re\left[\frac{\eta+z}{\eta-z}\right] - (S(z,b)+N(z)).$$

Now N does not dominate a positive multiple of $\Re[(\eta+z)/(\eta-z)]$ in $\{|z| < 1\}$. It follows from the theorem of Ahlfors and Heins[1] that

$$\lim_{r\to 0} u[\phi(\eta-\eta re^{i\theta})] = +\infty, \quad |\theta| < \tfrac{1}{2}\pi,$$

save for a set of θ of zero capacity. Hence ϕ has the quasi-sectorial limit a at η.

Examples of this phenomenon are readily constructed. We cite the example given by Lehto[8]

$$f(z) = \exp\left\{\frac{1+z}{1-z}\right\}b(z) \quad (|z| < 1), \tag{7.3}$$

where b is a convergent Blaschke product whose zeros are real and cluster solely at $z = 1$.

It is to be observed that the existence of a quasi-sectorial limit of a Lindelöfian map which is not actually a sectorial limit imposes very severe restrictions. In fact, if (7.2) holds even p.p. for $|\theta| < \tfrac{1}{2}\pi$, but a is not the sectorial limit of ϕ at η, then G is conformally equivalent to the extended plane, and for each $b\,(\neq a) \in G$, for some sector

$$S_\delta = \{\eta(1-re^{i\theta})|\,|\theta| < \delta < \tfrac{1}{2}\pi\},$$

$\phi^{-1}(\{b\}) \cap S_\delta$ clusters at η.

8. What information can be drawn from the non-vanishing of W_a? We shall see:

(i) *If G is compact, $W_a > 0$, and $\nu_\phi(b) < +\infty$ for some $b\,(\in G) \neq a$, then a is a radial limit of ϕ.*

(ii) *If G is not compact and $W_a > 0$, then a is a radial limit of ϕ.*†

A special instance of (i) is to be found in Lehto[8]. We shall revert to this result of Lehto below.

Apart from the use of auxiliary harmonic functions on G with assigned singularities and boundary behavior, and the study of the composition of such functions with ϕ, the proof rests principally on the important extension due to Saks[14] of the de la Vallée Poussin decomposition theorem to the case of an unrestricted function of bounded variation of one real variable.

Assertion (i) may be established as follows. We start with the representation (5.1) and note that our hypothesis on b renders $S(z, b)$ innocuous near the unit circumference. Let

$$(2\pi)^{-1} \int_0^{2\pi} \Re \left[\frac{e^{i\theta} + z}{e^{i\theta} - z} \right] d\mu(\theta)$$

be a Poisson–Stieltjes representation for $H(z)$, where μ is a function of bounded variation on each finite interval and satisfies

$$\mu(\theta) = \tfrac{1}{2}[\mu(\theta+) + \mu(\theta-)], \quad \mu(\theta + 2\pi) = \mu(\theta) + [\mu(2\pi) - \mu(0)].$$

We assert that $\mu'(\theta) = +\infty$ for some θ. If this were not the case, then by the de la Vallée Poussin–Saks decomposition theorem, we would be led to conclude that H is of the form $Q - U$, where Q is a quasi-bounded non-negative harmonic function and U is a non-negative harmonic function in $\{|z| < 1\}$. But $W_a \leqslant P \leqslant Q$ so that $W_a = 0$. This is impossible. Hence for some η, $|\eta| = 1$, $\lim_{r \to 1} H(r\eta) = +\infty$. We conclude that $\lim_{r \to 1} \phi(r\eta) = a$.

Part (ii) is similarly treated; the non-compactness of G serves as a substitute for the finite valence hypothesis. The only change that is called for is to replace u by a harmonic function on $G - \{a\}$ which has a normalized positive logarithmic singularity at a and is bounded above outside of each neighborhood of a, and to note that (5.1) persists save for the absence of $-S(p, b)$.

The theorem of Lehto to which reference has been made asserts that a meromorphic function of bounded characteristic in $\{|z| < 1\}$, whose image is dense in the extended plane, omits at most one value outside of the closure of the set of its radial limits. An examination of the proof given by Lehto shows that it suffices, as far as Lehto's theorem is

† Can 'radial' be replaced by 'sectorial'? The fact that we are concerned with Poisson–Stieltjes integrals at points where the generating function has an infinite derivative renders the 'radial' easy; however, the argument used for the existence of a sectorial limit of a Poisson–Stieltjes integral at a point where the generating function has a finite derivative does not appear to be available in the infinite case.

concerned, to appeal to the original de la Vallée Poussin decomposition for continuous functions of bounded variation rather than to the general decomposition theorem.

The result of Lehto may be extended to Lindelöfian maps with domain $\{|z| < 1\}$ and may be given a slightly more refined formulation. Let R denote the closure of the set of radial limits ($\in G$) of ϕ. By (5.2) if $q \in G - R$, then either q belongs to the complement of the closure of the image of ϕ or else ϕ is of type-Bl at q. Let B denote the set of points in $G - R$ at which ϕ is of type-Bl.

If G is not compact, then ϕ is of type-Bl$_1$ at each point of B. This is a consequence of (ii) of the present section. Thus, in particular, on each component of B, ν_ϕ is constant (finite or not).

If G is compact, two possibilities may occur. The *first* is that ν_ϕ is identically infinite on G. There are examples of such ϕ which are not of type-Bl$_1$ at several points of B (e.g. $f(z)/f(-z)$ for f of (7.3); this function is not of type-Bl$_1$ at 0 and ∞ which are both points of B). The *second* is that for some point $q_0 \in G$, we have $\nu_\phi(q_0) < +\infty$. Thanks to (i) of the present section, it follows that *if $q_0 \in G - B$, then ϕ is of type-Bl$_1$ at each point of B; if $q_0 \in B$, then ϕ is of type-Bl$_1$ at each point of $B - \{q_0\}$ and in this case if ϕ is not of type-Bl$_1$ at q_0, ν_ϕ is infinite at each point of $G - \{q_0\}$.* For this latter phenomenon cf. (7.3).

9.   From this point on we shall be concerned with applying the methods described in this paper to problems concerning functions of bounded characteristic.

We first consider a refinement of the theorem of Gehring which we quoted earlier. We note that an *asymptotic spot*[5] of a function of bounded characteristic in $\{|z| < 1\}$ has the property that precisely one point of $\{|z| = 1\}$ adheres to all $\sigma(\omega)$, $\omega \in$ domain of σ. Let $A(\eta)$ denote the number of distinct asymptotic spots σ of f over finite points which have the property that $\{\eta\} = \cap \overline{\sigma(\omega)}$; let $B(\eta)$ similarly denote the number of distinct asymptotic spots σ of f over ∞ which have the property that $\{\eta\} = \cap \overline{\sigma(\omega)}$. The following theorem holds:

For an analytic function f of bounded characteristic in $\{|z| < 1\}$, $A(\eta) \leqslant 2$ and $B(\eta) \leqslant 3$, $|\eta| = 1$. There exists an f for which $A(1) = 2$ and $B(1) = 3$.

We note that, if $A(\eta) \geqslant 3$, then thanks to the classical results of Lindelöf which are employed in the proof of the Denjoy–Carleman–Ahlfors theorem and to (5.2), $u(z)$ being taken as $\log |z|$, we would infer the existence of two distinct asymptotic spots σ_1 and σ_2 over ∞ satisfying $\{\eta\} = \cap \, \overline{\sigma_k(\omega)}$, $k = 1, 2$, and such that for some ω in their common domain not only is it the case that $\sigma_1(\omega) \cap \sigma_2(\omega) = \emptyset$, but also

$$\mu_{\sigma_k(\omega)}[\mathfrak{G}_\omega(f_{\sigma_k(\omega)}, \infty)] = c_k \,\Re\left[\frac{\eta + z}{\eta - z}\right], \qquad (9.1)$$

where $c_k > 0$, $k = 1, 2$. This is impossible since μ-maps carry two admitted positive harmonic functions with *disjoint* domains into two positive harmonic functions the minimum of which has a vanishing greatest harmonic minorant. Hence $A(\eta) \leqslant 2$. The proof that $B(\eta) \leqslant 3$ is similar; it differs only in preliminary details.

To construct an f with $A(1) = 2$, $B(1) = 3$, we may proceed as follows. Let $F(z) = \displaystyle\int_0^z t^{-1} \sin t \, dt$, $\Im z > -1$. Now the positive and negative real axes are asymptotic paths of F and the corresponding asymptotic values are finite and distinct; the positive imaginary axis is an asymptotic path of F and the associated asymptotic value is ∞. The function

$$g(z) = F\left(\frac{2iz}{1-z}\right) \qquad (|z| < 1)$$

is of bounded characteristic.

Next we observe that there exists a function of bounded characteristic in $\{|z| < 1\}$, say h, which has the following two properties: (1) $B(1) = 2$, (2) h tends to zero as $z \to 1$ in $\Delta = \{|z - 2^{-1}| \leqslant 2^{-1}\}$. Such a function h may be constructed as follows. We observe that a positive harmonic function p in $\{|z| < 1\}$ which does not dominate $c\Re[(1+z)/(1-z)]$ for any positive constant c may be approximated uniformly on the intersection of Δ with the open unit disk by a function of the form $-\log|b|$, where b is a Blaschke product whose zeros cluster solely at $z = 1$. If we choose p tending to infinity as $z \to 1$ (in the open unit disk), let w denote an analytic function satisfying $\log|w| = \frac{1}{2}p$, and let b be as above with $p + \log|b| = O(1)$ on the intersection of Δ with the open unit disk, then wb is an admissible h.

It suffices now to take $f = g + wb$.

10. The striking contrast between the analytic and meromorphic cases, in so far as asymptotic values of functions of bounded characteristic are concerned, is brought out by the following example of *a function f of bounded characteristic in $\{|z| < 1\}$ which is the quotient of two Blaschke products whose zeros cluster solely at 1, and which has the property that the* *set of asymptotic values of f associated with z = 1 has the power of the continuum.*

The construction is based on Valiron's example[15] of an even meromorphic function g in the finite plane, which satisfies $T(r; g) = O(r)$, and has the property that there is a non-countable subset $\mathfrak{E}$ of $[0, \pi]$ such that, for $\theta \in \mathfrak{E}$, $\lim_{r \to \infty} g(re^{i\theta})$ exists and is finite, and $\theta \to \lim g(re^{i\theta})$ is univalent on $\mathfrak{E}$. Let $\alpha, \beta \in \mathfrak{E}$, where $\alpha < \beta < \pi$ and a condensation point of $\mathfrak{E}$ lies in (α, β), and let h denote the restriction of g to $\{\alpha < \arg z < \beta\}$. Thanks to the fact that $T(r; g) = O(r)$, h is a Lindelöfian meromorphic function.

There exists a finite point w distinct from $\lim g(re^{i\alpha})$ and $\lim g(re^{i\beta})$ which is a condensation point of

$$\{\lim g(re^{i\theta})| \ \alpha < \theta < \beta, \theta \in \mathfrak{E}\}.$$

It follows that there exists an open circular disk Δ centered at w such that one of the components Ω of $h^{-1}(\Delta)$ has the following properties: (1) every finite frontier point of Ω lies in $\{\alpha < \arg z < \beta\}$ and fr Ω is regular analytic at each such point, (2) the set of asymptotic values in Δ of h restricted to Ω is not countable (and consequently has the power of the continuum since it is an analytic set). Let ρ denote the radius of Δ.

Let Γ denote the component of fr Ω (relative to the finite plane) which separates 0 from Ω, and let Ω^* denote the component of the complement of Γ with respect to the finite plane which contains Ω. Then Ω^* is simply connected and its frontier (relative to the extended plane) is a closed Jordan curve. Let ϕ denote a univalent conformal map of $\{|z| < 1\}$ onto Ω^* which 'carries' 1 into ∞. We introduce

$$f_1 = \rho^{-1}[g \circ \phi - w]$$

and note that f_1 is of bounded characteristic and further has the property that it possesses a limit of modulus one at each point of $\{|z| = 1\}$ other than 1. Further, the set of distinct asymptotic values of f_1 associated with $z = 1$ has the power of the continuum. It follows from § 5 that there exist complex numbers $\alpha, \epsilon, |\alpha| < 1, |\epsilon| = 1$ such that

$$f = \epsilon \frac{f_1 - \alpha}{1 - \bar{\alpha} f_1}$$

is a quotient of Blaschke products. Clearly f satisfies all the imposed requirements.

REFERENCES

[1] Ahlfors, L. and Heins, M. Questions of regularity connected with the Phragmén-Lindelöf principle. *Ann. Math.* (2), 50, 341–346 (1949).

[2] Fatou, P. Séries trigonométriques et séries de Taylor. *Acta Math.* 30, 335–400 (1906).

[3] Frostman, O. Potentiel d'équilibre et capacité des ensembles. Thesis, Lund, 1935.

[4] Gehring, F. W. Asymptotic values for analytic functions of bounded characteristic. *Bull. Amer. math. Soc.* 63, 368, abstract 684t (1957).

[5] Heins, M. On the Lindelöf principle. *Ann. Math.* (2), 61, 440–471 (1955).

[6] Heins, M. Lindelöfian maps. *Ann. Math.* (2), 62, 418–446 (1955).

[7] Lehto, O. Value distribution and boundary behaviour of a function of bounded characteristic and the Riemann surface of its inverse function. *Ann. Acad. Sci. Fenn.* (Ser. A.I.), no. 177 (1954).

[8] Lehto, O. Boundary theorems for analytic functions. *Ann. Acad. Sci. Fenn.* (Ser. A), no. 196 (1955).

[9] Lohwater, A. J. The boundary values of a class of meromorphic functions. *Duke Math. J.* 19, 243–252 (1952).

[10] Martin, R. S. Minimal positive harmonic functions. *Trans. Amer. math. Soc.* 49, 137–172 (1941).

[11] Parreau, M. Sur les moyennes des fonctions harmoniques et analytiques et la classification des surfaces de Riemann. Thesis, Paris, 1952.

[12] Parreau, M. Fonction caractéristique d'une application conforme. *Ann. Fac. Sci. Toulouse*, (4), 19, 175–189 (1956).

[13] Riesz, F. and M. Ueber die Randwerte analytischer Funktionen. *Fourth Scandinavian Math. Congress*. Stockholm, 1916.

[14] Saks, S. *Theory of the Integral*. Warsaw–Lwów, 1937.

[15] Valiron, G. Sur les valeurs asymptotiques de quelques fonctions méromorphes. *R.C. Circ. mat. Palermo*, 49, 1–7 (1925).

PROBLÈMES MIXTES ABSTRAITS

Par J. L. LIONS

1. Position du problème

Soit $\mathfrak{H}$ un espace de Hilbert complexe; $\|f\|$, (f,g) désignent norme et produit scalaire dans cet espace. On donne dans $\mathfrak{H}$ une famille d'opérateurs $A(t)$ *non bornés* (des systèmes différentiels dans les applications), fermés de domaine $D_{A(t)}$ dense dans $\mathfrak{H}$, t (le temps) $\in [0,\mu]$, μ fini pour fixer les idées.

Problème 1.1. Trouver une fonction $u(t)$ une fois continûment différentiable de $[0,\mu]$ dans $\mathfrak{H}$, avec

$$u(t) \in D_{A(t)} \quad (t \in [0,\mu]), \tag{1.1}$$

$$A(t)\,u(t) + u'(t) = f(t) \quad (t \geqslant 0,\ u'(t) = (d/dt)\,u(t)) \tag{1.2}$$

(f est une fonction continue donnée de $[0,\mu]$ dans $\mathfrak{H}$),

$$u(0) = 0. \tag{1.3}$$

Pour les problèmes mixtes dans des ouverts non cylindriques, on peut considérer des sommes mesurables hilbertiennes (cf. no. 2); cf. d'autres méthodes dans [2, 3, 13].

L'étude des solutions des équations du type (1.2) a été inaugurée, semble-t-il dans [1], avec des $A(t)$ indépendants de t et des objectifs différents. C'est désormais une remarque classique [19, 19 a] qu'il est important de remplacer le problème 1.1—on dira *le problème usuel*—par un problème *faible associé*, ce qui peut être fait comme suit: soit $h(t)$ une fonction donnée avec les propriétés: $h(t)$ est une fois continûment différentiable de $[0,\mu]$ dans $\mathfrak{H}$,

$$h(\mu) = 0, \tag{1.4}$$

$$h(t) \in D_{A^*(t)}, \quad A^*(t) = \text{adjoint de } A(t). \tag{1.5}$$

Si u est solution du problème 1.1, on a

$$\int_0^\mu (u(t), A^*(t)\,h(t) - h'(t))\,dt = \int_0^\mu (f(t), h(t))\,dt \tag{1.6}$$

d'où le

Problème 1.2 (*Problème faible associé*). On cherche une fonction u dans l'espace $L^2(0,\mu;\mathfrak{H})$ des fonctions de carré sommable sur $[0,\mu]$ à valeurs dans $\mathfrak{H}$, vérifiant (1.6) pour tout h avec (1.4) et (1.5).

Un cas important est celui où les $A(t)$ ont une partie hermitienne positive; à $A(t)$ est alors attaché un domaine de Friedrichs $V(t)$ et on doit chercher u à valeurs dans $V(t)$. On est ainsi conduit à la formulation du no. suivant:

2. Équation fonctionnelle générale

Les données fondamentales sont deux espaces de Hilbert E_i, $i = 1, 2$; avec $E_i \subset L^2(0, \mu; \mathfrak{H})$ (le signe $\subset$ signifiant inclusion algébrique et topologique), et une forme sesquilinéaire $\Psi(u_1, u_2)$ continue sur $E_1 \times E_2$. On suppose E_i stable par multiplication par $\exp(\lambda t)$, $\lambda \in \mathbf{C}$, et que $\Psi(\exp(-\lambda t) u_1, \exp(\lambda t) u_2) = \Psi(u_1, u_2)$ (cette dernière hypothèse n'est nullement indispensable mais simplifie les énoncés ci-après).

Problème 2.1. Trouver $u \in E_1$ solution de

$$\Psi(u, h) - \int_0^\mu (u(t), h'(t))\, dt = L(h) \qquad (2.1)$$

pour tout h vérifiant

$$h \in E_2, \quad h' \in L^2(0, \mu; \mathfrak{H}), \quad h(\mu) = 0, \qquad (2.2)$$

h' étant pris au sens des distributions sur $[0, \mu]$ à valeurs dans $\mathfrak{H}$ (cf. [16, 17]), $h \to L(h)$ étant une forme semi-linéaire sur l'espace des fonctions h vérifiant (2.2).

On peut considérer des problèmes analogues avec plusieurs variables de temps [12].

Exemple 2.1. $E_1 = L^2(0, \mu; \mathfrak{H})$, $E_2 =$ espace des $u_2(t) \in E_1$ avec $u_2(t) \in D_{A^*(t)}$ presque partout (p.p.), $A^*(t) u_2(t) \in E_1$, muni de sa structure hilbertienne naturelle:

$$\Psi(u_1, u_2) = \int_0^\mu (u_1(t), A^*(t) u_2(t))\, dt.$$

Le problème 2.1 coincide alors avec le problème 1.2.

Exemple 2.2. Soient $\mathfrak{K}_i(t)$ deux familles mesurables hilbertiennes, $\mathfrak{K}_i(t) \subset \mathfrak{H}$; on prendra $E_1 = \int^\oplus \mathfrak{K}_i(t)\, dt$; ceci est utile pour les problèmes mixtes dans des ouverts non cylindriques.

On va donner un critère d'existence d'une solution du problème 2.1. On a besoin pour cela des notions suivantes:

Définition 2.1. Des opérateurs $B(t)$ dans $\mathfrak{H}$ forment une *famille admissible d'auto-adjoints* si

(i) pour chaque $t \in [0, \mu]$, $B(t)$ est auto-adjoint, borné ou non;

(ii) $(B(t)f, f) \geqslant \beta \|f\|^2$, $\beta > 0$, $f \in D_{B(t)}$;

(iii) $(B^{-1}(t)f, g)$ est une fonction une fois continûment différentiable de $[0, \mu]$ dans $\mathfrak{H}$, $f, g \in \mathfrak{H}$, et

$$|((d/dt)\, B^{-1}(t))\, B(t)f,\, B(t)f)| \leqslant \gamma(B(t)f, f), \quad \gamma > 0, f \in D_{B(t)}.$$

Remarque 2.1. Si $B(t)$ est continu, $B(t)f$ étant une fois continûment différentiable de $[0, \mu]$ dans $\mathfrak{H}$ pour tout $f \in \mathfrak{H}$, (iii) disparaît.

La notion suivante est inspirée de Leray[8].

Définition 2.2. La forme $\Psi(u_1, u_2)$ est dite *conditionnellement E_1-elliptique* s'il existe une famille admissible d'auto-adjoints $B(t)$ telle que, pour λ convenable,

$$\operatorname{Re}\Psi(\phi, B(t)\,\phi) + \lambda \int_0^\mu (B(t)\,\phi(t), \phi(t))\, dt \geqslant \alpha \|\phi\|_{E_1}^2 \quad (\alpha > 0), \quad (2.3)$$

pour ϕ vérifiant

$$\phi \in E_1, \quad \phi(t) \in D_{B(t)}\, p.p., \quad B(t)\,\phi \in E_2. \qquad (2.4)$$

Il est utile d'introduire $B^{\frac{1}{2}}(t)$ racine carré positive de $B(t)$, et

$$W(t) = D_{B^{\frac{1}{2}}(t)};$$

on dira que $$f \in L^2(0, \mu;\, W(t))$$

si $$f \in L^2(0, \mu;\, \mathfrak{H}), \quad f(t) \in W(t)\, p.p., \quad B^{\frac{1}{2}}(t)f(t) \in L^2(0, \mu;\, \mathfrak{H});$$

on munit $L^2(0, \mu;\, W(t))$ de sa structure hilbertienne naturelle.

Théorème 2.1. *Si la forme Ψ est conditionnellement E_1-elliptique, et si*

$$L(h) = \int_0^\mu (f(t), h(t))\, dt,\ \textit{où } f \textit{ est donnée dans } L^2(0, \mu;\, W(t)),\ \textit{il existe } u \in E_1$$

vérifiant (2.1) *pour tout h vérifiant* (2.2) *et*

$$B^{-1}(t)\, h \in E_1. \qquad (2.5)$$

On utilisera dans la démonstration le

Lemme 2.1[9]. *Soit F un espace Hilbertien, Φ un sous-espace, muni d'une structure préhilbertienne plus fine que celle induite par F. On donne une forme sesquilinéaire $\Theta(f, \phi)$, avec*

(1) $f \to \Theta(f, \phi)$ *est continue sur F, pour tout $\phi \in \Phi$;*

(2) $|\Theta(\phi, \phi)| \geqslant \delta \|\phi\|_\Phi^2$ $(\delta > 0, \phi \in \Phi)$.

Dans ces conditions, si $\phi \to L(\phi)$ est une forme semi-linéaire continue sur Φ, il existe u dans F vérifiant

$$\Theta(u, \phi) = L(\phi), \quad \textit{pour tout}\quad \phi \in \Phi. \qquad (2.6)$$

On change ensuite u en $\exp(\lambda t)\, u$ dans (2.1); utilisant (2.4), on peut donc se ramener au cas où

$$\operatorname{Re}\Psi(\phi, B(t)\,\phi) \geqslant \alpha \|\phi\|_{E_1}^2 + c \int_0^\mu (B(t)\,\phi(t), \phi(t))\, dt \qquad (2.7)$$

$c > \frac{1}{2}\gamma$. On applique alors le lemme avec $F = E_1 \cap L^2(0, \mu; W(t))$;
$\Phi = $ espace des $\phi \in F$ tels que $\phi(t) \in D_{B(t)} p.p.$ et $B(t)\phi = h$ vérifie (2.2).
On munit Φ de la topologie induite par F. On prend

$$\Theta(u, \phi) = \Psi(u, B(t)\phi) - \int_0^\mu (u(t), (B(t)\phi(t))')\,dt$$

et $\qquad L(\phi) = \int_0^\mu (f(t), B(t)\phi(t))\,dt = \int_0^\mu (B^{\frac{1}{2}}(t)f(t), B^{\frac{1}{2}}(t)\phi(t))\,dt.$

On vérifie que l'on est dans les conditions d'application du lemme,
d'où l'on déduit le théorème.

Notons qu'on obtient u *dans* F.

Remarque 2.2. Plus généralement le théorème 2.1 vaut si $L(h)$ est telle
que $\phi \to L(B(t)\phi)$ soit continue sur l'espace des $\phi = B^{-1}(t)h$, h vérifiant
(2.2) et (2.5), muni de la topologie induite par F.

Remarque 2.3. On dit que l'on *inverse le sens du temps* dans le problème
2.1 si l'on remplace la condition '$h(\mu) = 0$' par '$h(0) = 0$'; désignons
par 'Problème 2.2' le problème correspondant. On dit que le problème
est *réversible* s'il existe une solution (au moins) pour chacun des problèmes 2.1 et 2.2.

Théorème 2.2. *On suppose qu'il existe une famille admissible d'auto-
adjoints $B(t)$ avec*

$$L^2(0, \mu; W(t)) \subset E_1, \tag{2.8}$$

et $\qquad |\mathrm{Re}\,\Psi(\phi, B(t)\phi)| \leqslant \alpha_1 \|\phi\|_{E_1}^2 \quad$ *pour tout ϕ avec* (2.4); $\qquad$ (2.9)

alors le problème est réversible.

3. Applications (I): problèmes de Cauchy

Sur un ouvert Ω de R^n, $\mathfrak{D}$ est l'espace des fonctions indéfiniment
différentiables à support compact (avec la topologie de Schwartz),
$\mathfrak{D}'$ est le dual (distributions sur Ω). On donne trois espaces de Hilbert
de m-uples de distributions:

$$(\mathfrak{D})^m \subset \mathfrak{K} \subset \mathfrak{H} \subset \mathfrak{F} \subset (\mathfrak{D}')^m,$$

chaque espace étant dense dans le suivant.

On donne un système différentiel $\mathscr{A}(t) = \|\mathscr{A}_{ij}(t)\|$ $(i, j = 1, ..., m)$,
$\mathscr{A}_{ij}(t)$ étant un opérateur différentiel en x, à coefficients fonctions une
fois continûment différentiables de $t \in [0, \mu]$ à valeurs dans l'espace des
fonctions indéfiniment différentiables en x (naturellement on peut
raffiner). On suppose que $\qquad \mathscr{A}(t) \in \mathscr{L}(\mathfrak{K}; \mathfrak{F}),$ $\qquad$ (3.1)

la fonction $\mathscr{A}(t)\,u$ étant une fois continûment différentiable dans $[0,\mu]$ à valeurs dans $\mathfrak{F}$, pour tout $u \in \mathfrak{K}$. On suppose qu'il existe des opérateurs $\mathscr{B}(t)$ avec

$$\mathscr{B}(t) \in \mathscr{L}(\mathfrak{H}; \mathfrak{H}'), \langle \mathscr{B}(t)\,u, \overline{u} \rangle \geqslant \beta \|u\|^2 \quad (\beta > 0, \ u \in \mathfrak{H}), \qquad (3.2)$$

$$\mathscr{B}(t) \text{ étant un isomorphisme de } \mathfrak{K} \text{ sur } \mathfrak{F}', \qquad (3.3)$$

$$\mathrm{Re}\,\langle \mathscr{A}(t)\,u, \overline{\mathscr{B}(t)\,u} \rangle + \lambda \,\langle \mathscr{B}(t)\,u, \overline{u} \rangle \geqslant \alpha \,\|u\|_{\mathfrak{K}}^2 \quad (u \in \mathfrak{K}) \qquad (3.4)$$

pour λ convenable, avec en outre les conditions de régularité suivantes: pour $u \in \mathfrak{H}$, $t \to \mathscr{B}(t)\,u$ est une fois continûment différentiable de $[0,\mu]$ dans $\mathfrak{H}'$; pour $v \in \mathfrak{K}$, $\mathscr{B}(t)\,v$ est continue dans $\mathfrak{F}'$; pour $w \in \mathfrak{F}'$, $\mathscr{B}^{-1}(t)\,w$ est continue dans $\mathfrak{K}$.

Théorème 3.1. *Sous les hypothèses* (3.1), $\ldots$, (3.4), *étant donné f dans* $L^2(0,\mu;\mathfrak{F})$, *il existe u dans* $L^2(0,\mu;\mathfrak{K})$, *unique, dépendant continûment de* f, *avec*

$$\mathscr{A}(t)\,u(t) + u'(t) = f(t), \qquad (3.5)$$

$$u(0) = 0. \qquad (3\cdot6)$$

Ce résultat est une généralisation simple de [18] (cf. aussi [15]). L'unicité se démontre directement. Pour l'existence, on se ramène comme suit au théorème 2.1. Si J est l'isomorphisme canonique de $\mathfrak{H}'$ sur $\mathfrak{H}$ on pose

$$B(t) = J\mathscr{B}(t) \in \mathscr{L}(\mathfrak{H}; \mathfrak{H}) \qquad (3.7)$$

et les $B(t)$ forment une famille admissible d'auto-adjoints. On prend maintenant $E_1 = L^2(0,\mu;\mathfrak{K})$, $E_2 =$ espace des $f \in L^2(0,\mu;\mathfrak{H})$ avec $J^{-1}f \in L^2(0,\mu;\mathfrak{H}')$ et $\Psi(u_1, u_2) = \displaystyle\int_0^\mu \langle \mathscr{A}(t)\,u_1(t), \overline{J^{-1}u_2(t)} \rangle\,dt$, le crochet désignant la dualité entre $\mathfrak{F}$ et $\mathfrak{F}'$. On vérifie que l'on est dans les conditions d'application du théorème 2.1.

Exemple 3.1. Grâce à [14], les opérateurs p-paraboliques au sens de Petrowsky entrent dans ce cadre.

Voici maintenant un cas où l'on applique le théorème 2.2. On suppose cette fois que

$$\mathscr{A}(t) \in \mathscr{L}(\mathfrak{H}; \mathfrak{F}) \qquad (3.8)$$

et on suppose qu'il existe des opérateurs $\mathscr{B}(t)$ et $\mathscr{C}(t)$ avec

$$\mathscr{B}(t) \in \mathscr{L}(\mathfrak{H}; \mathfrak{H}'), \quad \mathscr{C}(t) \in \mathscr{L}(\mathfrak{F}; \mathfrak{F}'), \qquad (3.9)$$

$$\langle \mathscr{B}(t)\,u, \overline{u} \rangle \geqslant \beta \|u\|^2, \quad \langle \mathscr{C}(t)\,v, \overline{v} \rangle \geqslant \gamma_1 \|v\|_{\mathfrak{F}}^2 \quad (u \in \mathfrak{H}, \ v \in \mathfrak{F}, \ \beta, \gamma_1 > 0), \qquad (3.10)$$

$$|\mathrm{Re}\,\langle \mathscr{A}(t)\,u, \overline{\mathscr{B}(t)\,u} \rangle| \leqslant \alpha \|u\|^2, \quad (\alpha > 0, \ u \in \mathfrak{H}, \ \mathscr{B}(t)\,u \in (\mathfrak{D})^m), \qquad (3.11)$$

$$|\mathrm{Re}\,\langle \mathscr{C}(t)\,\mathscr{A}(t)\,u, \overline{u} \rangle| \leqslant \alpha_1 \|u\|_{\mathfrak{F}}^2 \quad (\alpha_1 > 0, \ u \in \mathfrak{H}) \qquad (3.12)$$

et les hypothèses de régularité suivantes: pour $f \in \mathfrak{H}$ (resp. $g \in \mathfrak{F}$), $\mathscr{B}(t)f$ (resp. $\mathscr{C}(t)g$) est une fois continûment différentiable à valeurs dans $\mathfrak{H}'$ (resp. $\mathfrak{F}'$).

Théorème 3.2. Sous les hypothèses (3.8), ..., (3.12), *étant donnée f dans* $L^2(0, \mu; \mathfrak{H})$, *il existe u dans* $L^2(0, \mu; \mathfrak{H})$ *unique, avec*

$$\mathscr{A}(t)\,u(t) + u'(t) = f(t), \tag{3.13}$$

$$u(0) = 0. \tag{3.14}$$

Même chose en remplaçant (3.14) *par* $u(\mu) = 0$.

Les opérateurs $\mathscr{B}(t)$ assurent l'existence, les $\mathscr{C}(t)$ l'unicité.

Exemple 3.2. Grâce à [8] les opérateurs hyperboliques au sens de Petrowsky entrent dans ce cadre. Voir aussi [6]. Variantes dans [18].

4. Applications (II): problèmes mixtes

Soient V et H deux espaces de Hilbert, $V \subset H$, V dense dans H; on donne une famille de formes sesquilinéaires $a(t; u, v)$ continues sur V. On suppose

$$a(t; u, v) = \overline{a(t; v, u)}, \quad a(t; v, v) + \lambda \|v\|_H^2 \geqslant \alpha \|v\|_V^2 \quad (\alpha > 0, \ v \in V), \tag{4.1}$$

$a(t; u, v)$ étant une fois continûment différentiable dans $[0, \mu]$ pour tout $u, v \in V$.

Soit t fixé. On désigne par $N(t)$ le sous-espace de V formé des $u \in V$ tels que la forme semilinéaire $v \to a(t; u, v)$ soit continue sur V muni de la topologie induite par H. Alors

$$a(t; u, v) = (\mathscr{A}(t)\,u, v)_H, \tag{4.2}$$

ce qui définit un opérateur $\mathscr{A}(t)$ non borné dans H, de domaine $N(t)$; cet opérateur est auto-adjoint.

On considère maintenant le problème suivant: trouver une fonction $U(t)$ deux fois continûment différentiable dans $[0, \mu]$ à valeurs dans H telle que $U(t) \in N(t)$ pour tout t et

$$\mathscr{A}(t)\,U(t) + U''(t) = F(t) \tag{4.3}$$

où $F(t)$ est continue dans $[0, \mu]$ à valeurs dans H, avec

$$U(0) = U'(0) = 0 \tag{4.4}$$

(pour une situation plus générale, cf. [10, 20]).

On va écrire (4.4) sous forme de système; soit $\mathfrak{H} = H \times H$ avec la structure hilbertienne produit; $f = \{f_1, f_2\}$ désigne un élément de $\mathfrak{H}$. On pose $U(t) = u_1(t)$, $U'(t) = u_2(t)$, $u(t) = \{u_1(t), u_2(t)\}$.

On définit dans $\mathfrak{H}$ l'opérateur $A(t)$ par

$$A(t) u = \{-u_2, \mathscr{A}(t) u_1\} \quad (u_1 \in N(t), \ u_2 \in H). \tag{4.5}$$

Le problème (4.4), (4.3) devient: on cherche une fonction $u(t)$ une fois continûment différentiable de $[0, \mu]$ dans $\mathfrak{H}$, telle que $u(t) \in D_{A(t)}$, avec

$$A(t) u(t) + u'(t) = f(t) = \{0, F(t)\}, \tag{4.6}$$

$$u(0) = 0. \tag{4.7}$$

Notons maintenant que si $u, v \in D_{A(t)}$, on a

$$(A(t) u, v) = -(u_2, v_1)_H + a(t; u_1, v_2) = \Xi(t; u, v). \tag{4.8}$$

Dans ces conditions *une* formulation faible du problème est: on cherche $u \in L^2(0, \mu; V \times H)$, vérifiant

$$\int_0^\mu [\Xi(t; u(t), h(t)) - (u(t), h'(t))]\, dt = \int_0^\mu (f(t), h(t))\, dt, \tag{4.9}$$

pour tout $h \in L^2(0, \mu; H \times V)$, avec $h' \in L^2(0, \mu; \mathfrak{H})$, $h(\mu) = 0$.

Théorème 4.1. Sous les hypothèses (4.1), *il existe une solution unique u de* (4.9); *u dépend continûment de f.*

L'unicité se démontre par adaptation d'une idée de [7].

Pour l'existence, on prend: $E_1 = L^2(0, \mu; V \times H)$, $E_2 = L^2(0, \mu; H \times V)$, et

$$\Psi(u_1, u_2) = \int_0^\mu \Xi(t; u_1(t), u_2(t))\, dt.$$

On va construire une famille $B(t)$ telle que $\Psi(u_1, u_2)$ devienne conditionnellement E-elliptique. Pour cela on choisit un nombre θ réel tel que

$$a(t; v, v) + \theta \|v\|_H^2 \geqslant \alpha \|v\|_V^2 \quad (v \in V); \tag{4.10}$$

on pose alors

$$B(t) u = \{(\mathscr{A}(t) + \theta) u_1, u_2\} \quad (u_1 \in N(t), \ u_2 \in H). \tag{4.11}$$

On vérifie que

$$|\operatorname{Re} \Psi(\phi, B(t) \phi)| \leqslant c \int_0^\mu (B(t) \phi(t), \phi(t))\, dt$$

pour toute fonction ϕ vérifiant (2.4). Ceci démontre le théorème, car $(B(t) u, u)$ définit une norme équivalente à la norme de $V \times H$. Ceci démontre également que le problème (4.9) est réversible.

Remarque 4.1. Dans le cas où $\mathscr{A}(t) = \mathscr{A}$ est indépendant de t, $-A$ est générateur infinitésimal d'un groupe dans $V \times H$ (cf. [22]; cf. aussi pour ce n^o [1 a] et [21]).

5. Solutions usuelles et solutions distributions à valeurs vectorielles

On a cherché dans ce qui précède seulement les solutions de carré sommable à valeurs dans $\mathfrak{H}$ (ou des sous-espaces de $\mathfrak{H}$). On peut également chercher les solutions usuelles des problèmes des nos. 3 et 4—et d'un autre coté, les solutions distributions à valeurs vectorielles.

Solutions usuelles. Renvoyons à [6, 8, 14] pour les problèmes de Cauchy, à [10] et [20] pour les problèmes mixtes.

Solutions distributions à valeurs vectorielles. Il faut que, dans un sens convenable, les opérateurs $A(t)$ dépendent de t de façon indéfiniment différentiable. On cherche alors une distribution **u**, à valeurs dans le domaine de $A(t)$ (ce qui naturellement demande à être précisé lorsque ce domaine dépend de t), à support limité à gauche en t, solution de

$$A(t)\,\mathbf{u} + \mathbf{u}' = \mathbf{T},$$

où **T** est une distribution à valeurs dans $\mathfrak{H}$, à support limité à gauche. Pour inverser le sens du temps, remplacer 'gauche' par 'droite'. Une théorie générale de ces problèmes reste à faire. Dans le cas des problèmes mixtes (no. 4), en supposant $a(t; u, v)$ indéfiniment différentiable en t, ces problèmes sont résolus dans [10, 20] (cf. aussi [12]), par des méthodes différentes; la méthode [20] a l'avantage de donner simultanément les propriétés de régularité en t et les solutions distributions vectorielles (en donnant en outre des renseignements sur la croissance à l'infini). Les méthodes de [10, 12, 20] valent pour les problèmes du no. 3.

Remarque 5.1. On peut essayer de considérer des problèmes non linéaires dans un cadre fonctionnel général. Cf. [11] pour quelques résultats dans ce sens.

Remarque 5.2. Pour certaines classes d'opérateurs $A(t)$ on trouvera une étude de $A(t)\,u + u' = f$, $u(0) = 0$ dans [4] (les résultats de ces auteurs se généralisant aux problèmes posés en distributions à valeurs vectorielles) par des méthodes différentes.

BIBLIOGRAPHIE SOMMAIRE

[1] Bochner, S. et von Neumann, J. On compact solutions of operational differential equations. *Ann. Math.* (2), 36, 255–291 (1935).

[1a] Browder, F. E. Les opérateurs elliptiques et les problèmes mixtes. *C.R. Acad. Sci., Paris*, 246, 1363–1365 (1958).

[2] Duff, G. F. Mixed problems for linear systems of first order equations. *Canad. J. Math.* 10, 127–160 (1958).

[3] Fichera, G. Sulle equazioni differenziali lineari ellittico-paraboliche del secondo ordine. *Atti Acad. Naz. Lincei* (8), 5, 3–30 (1956).

[4] Foiaş, C., Gussi, G. and Poenaru, V. Generalized solutions of a quasi-linear differential equation in a Banach space. *Dokl. Akad. Nauk SSSR*, 119, 884–887 (1958).

[5] Friedrichs, K. O. Symmetric hyperbolic linear differential equations. *Commun. Pure Appl. Math.* 7, 345–393 (1954).

[6] Gårding, L. Solution directe du problème de Cauchy pour les équations hyperboliques. *Colloque Nancy*, 71–90 (1956).

[7] Ladyženskaya, O. A. *Les problèmes mixtes pour les équations hyperboliques.* Moscou, 1953. (En Russe.)

[8] Leray, J. *Lectures on Hyperbolic Equations with Variable Coefficients.* Princeton Inst. for Adv. Study, 1952.

[9] Lions, J. L. Problèmes mixtes pour opérateurs paraboliques. *C.R. Acad. Sci., Paris*, 242, 3028–3030 (1956).

[10] Lions, J. L. Boundary value problems. *Tech. Rep.* Lawrence (1957).

[11] Lions, J. L. Sur certains problèmes mixtes quasi-linéaires, (I), (II). *C.R. Acad. Sci., Paris*, 246, 1644–1647, 1796–1799 (1958).

[12] Lions, J. L. Sur certaines équations aux dérivées partielles à coefficients opérateurs non bornés. *J. Analyse Math.* 6, 333–355 (1958).

[13] Magenes, E. Il problema della derivata obliqua regolare per le equazioni lineari ellittico-paraboliche del secondo ordine in *m* variabili. *R.C. Mat. Univ. Roma*, 16, 363–414 (1957).

[14] Mizohata, S. Le problème de Cauchy pour les équations paraboliques. *J. Math. Soc. Japan*, 8, 269–299 (1956).

[15] Nagumo, M. On linear hyperbolic system of partial differential equations in the whole space. *Proc. Japan Acad.* 32, 703–706 (1956).

[16] Schwartz, L. *Théorie des distributions*, I et II. Paris, Hermann, 1950 et 1951.

[17] Schwartz, L. Théorie des distributions à valeurs vectorielles. *Ann. Inst. Fourier*, 1ère partie, VII, 1–151 (1957); 2ème partie, VIII (1958).

[18] Shirota, T. On Cauchy problem for linear partial differential equations with variable coefficients. *Osaka Math. J.* 9, 43–60 (1957).

[19] Soboleff, S. L. Méthode nouvelle à résoudre le problème de Cauchy pour les équations linéaires hyperboliques. *Mat. Sbornik*, 1 (43), 39–71 (1936).

[19a] Soboleff, S. L. Sur la presque périodicité des solutions de l'équation des ondes, I, II, III. *Dokl. Akad. Nauk SSSR*, 48, 542–545, 618–620 (1945); 49, 12–16 (1945).

[20] Trèves, F. *Relations de domination entre opérateurs différentiels.* Thèse, Paris, Juin 1958. (À paraître aux *Acta Math.*)

[21] Vishik, I. M. Le problème de Cauchy avec des opérateurs comme coefficients. *Mat. Sbornik*, 39 (81), 51–148 (1956).

[22] Yosida, K. An operator theoretical integration of the wave equation. *J. Math. Soc. Japan*, 8, 79–92 (1956).

Д. Е. МЕНЬШОВ

О СХОДИМОСТИ ТРИГОНОМЕТРИЧЕСКИХ РЯДОВ

Теория тригонометрических рядов представляет одну из основных частей анализа. Эта теория стала очень интенсивно развиваться в начале 20-го столетия, когда Лебег дал новое определение интеграла, значительно обобщающее интеграл Римана.

Рассмотрим произвольный тригонометрический ряд

$$\frac{a_0}{2} + \sum_{n=1}^{\infty} (a_n \cos nx + b_n \sin nx). \tag{1}$$

Этот ряд называется рядом Фурье, если его коэффициенты a_n и b_n определяются по известным формулам Фурье. Если интегралы, входящие в формулы Фурье берутся в смысле Лебега, то ряд (1) называется рядом Фурье-Лебега.

Следует отметить, что разделение тригонометрических рядов на ряды Фурье и ряды не Фурье не является вполне определенным. В самом деле, после определения интеграла Лебега появилось много различных определений интеграла, которые оказались необходимыми для различных вопросов анализа. К этим определениям относятся хорошо известные определения, принадлежащие Данжуа (Denjoy), и другие определения интеграла.

После введения каждого из этих определений некоторые тригонометрические ряды оказывались рядами Фурье в смысле этого нового определения интеграла.

Некоторые из определений интеграла были специально введены для того, чтобы превратить в ряды Фурье тригонометрические ряды, которые по тем или другим причинам было естественно рассматривать как ряды Фурье. Приведем два примера.

Предположим, что тригонометрический ряд (1) сходится всюду к конечной функции $f(x)$. Уже давно было известно, что в данном случае сумма ряда, т.е. $f(x)$, не обязательно должна быть интегрируемой по Лебегу. Тогда возник вопрос, каким образом можно обобщить определение интеграла, чтобы в этом случае тригонометрический ряд оказался рядом Фурье от своей суммы. Такое определение интеграла было дано Данжуа в 1922 г. [1]

Приведем еще один пример. Тригонометрическим рядом, сопряженным с рядом (1) называется ряд вида

$$\sum_{n=1}^{\infty} (-b_n \cos nx + a_n \sin nx). \tag{2}$$

Как известно, ряд (2) есть мнимая часть того степенного ряда, у которого действительная часть совпадает с рядом (1).

Существуют ряды Фурье-Лебега вида (1), для которых сопряженный ряд (2) не является рядом Фурье-Лебега. Тогда возник следующий вопрос:

Для какого определения интеграла любой тригонометрический ряд, сопряженный с рядом Фурье-Лебега, является рядом Фурье в смысле этого нового определения. Как оказалось, таким интегралом является так называемый A-интеграл, рассмотренный впервые Тичмаршем (Titchmarsh)[2] и детально изученный Ульяновым[3]. Однако остается не решенным следующий вопрос.

Пусть ряд (1) есть ряд Фурье в смысле A-интеграла. Будет ли сопряженный ряд (2) также рядом Фурье в смысле A-интеграла.

Можно поставить более общий вопрос, состоящий в следующем. Найти такое определение интеграла, чтобы оба тригонометрических ряда (1) и (2) или одновременно были рядами Фурье в смысле этого определения, или же одновременно не были такими рядами Фурье.

Этот вопрос был поставлен А. Колмогоровым, но для интегралов более общих, чем интеграл Лебега ответа на него до сих пор не было дано.

В дальнейшем мы будем рассматривать ряды Фурье-Лебега, а также произвольные тригонометрические ряды, не останавливаясь специально на рядах Фурье, которые являются обобщениями рядов Фурье-Лебега.

Литература, относящаяся к вопросу о сходимости тригонометрических рядов, очень обширна, и в коротком докладе невозможно изложить все, даже наиболее существенные результаты из этой области. Поэтому мне придется ограничиться рассмотрением только некоторых вопросов, причем выбор этих вопросов будет в значительной мере обусловлен личными интересами докладчика. Докладчик в первую очередь рассмотрит те результаты из теории тригонометрических рядов, которые непосредственно примыкают к его работам.

Рассмотрим сперва расходящиеся тригонометрические ряды.

В 1911 г. Лузин[4] дал пример тригонометрического ряда с коэф-

фициентами, стремящимися к нулю, который расходится почти
всюду. В 1912 г. Штейнхауз (Steinhaus)[5] усилил этот результат,
построив пример тригонометрического ряда с коэффициентами,
стремящимися к нулю, который расходится в каждой точке. В 1926 г.
Колмогоров[6] еще более усилил этот результат; а именно, А.
Колмогоров дал пример ряда Фурье-Лебега, который расходится в
каждой точке.

В примере А. Колмогорова сопряженный ряд к расходящемуся
ряду Фурье-Лебега не является рядом Фурье-Лебега. Харди
(Hardy), Рогозинский (Rogosinski)[7] и Сунуоти (Sunouchi)[8] несколько
видоизменили пример А. Колмогорова и построили пример ряда
Фурье-Лебега, для которого сопряженный ряд также является
рядом Фурье-Лебега, причем оба ряда расходятся почти всюду.
Однако в примере Харди, Рогозинского и Сунуоти ряды расходятся
только почти всюду. Остается открытым вопрос, можно ли определить два сопряженных ряда Фурье-Лебега, каждый из которых
расходится во всех точках.

Можно также поставить вопрос об определении ряда Фурье-
Лебега, который сходится в точках одного множества и расходится
в точках дополнительного множества.

В связи с этим вопросом Целлер (Zeller)[9] доказал следующую
теорему. Для любого множества E типа F_σ, лежащего на $[0, 2\pi]$
можно определить ряд Фурье-Лебега, который сходится в каждой
точке множества E и расходится в каждой точке дополнения CE к
множеству E.

При этом, в примере Целлера в каждой точке множества E ряд
Фурье неограниченно расходится, т.е. последовательность частных
сумм этого ряда не ограничена.

Известно, что множество точек неограниченной расходимости
тригонометрического ряда есть множество типа G_δ. В таком случае,
результат Целлера в известном смысле является окончательным,
так как в его примере CE может быть произвольным множеством
типа G_δ.

Результат Целлера является обобщением результата Стечкина[10]
который доказал аналогичную теорему, не предполагая, однако,
что тригонометрический ряд есть ряд Фурье-Лебега.

Рассмотрим теперь ряды Фурье от непрерывных функций.
Известно, что такие ряды могут расходится на множествах мощности
континуума меры нуль. Однако неизвестно, могут ли такие ряды
расходиться на множествах положительной меры. Точно также,

неизвестно, существуют ли ряды Фурье от функций с суммируемым квадратом, которые расходятся на множествах положительной меры.

В случае непрерывной функции можно ставить вопрос об улучшении сходимости ее ряда Фурье путем изменения этой функции на множестве малой меры.

В этом направлении получен следующий результат:

Любую непрерывную функцию можно изменить на множестве сколь угодно малой меры таким образом, что для полученной новой функции ряд Фурье будет сходиться равномерно на всей оси x (Меньшов)[11].

Предыдущая теорема была получена докладчиком при решении вопроса об изображении измеримых функций сходящимися тригонометрическими рядами. Этот вопрос был поставлен Н. Лузиным еще в 1915г. Оказалось, что на этот вопрос получается положительный ответ, если предположить, что изображаемая функция конечна почти всюду; а именно, справедлива следующая теорема:

Любую измеримую фуикцию $f(x)$, конечную почти всюду на сегменте $[0, 2\pi]$, можно представить тригонометрическим рядом, сходящимся к ней почти всюду на этом сегменте (Д. Меньшов)[12].

Эту теорему можно обобщить, если рассмотреть вопрос об определении тригонометрического ряда по его верхней и нижней сумме. Функцию $f_1(x)$ мы будем называть *верхней суммой или верхним пределом* тригонометрического ряда (1), если $f_1(x)$ есть верхний предел частных сумм этого ряда. Аналогично определяется *нижняя сумма или нижний предел* тригонометрического ряда.

Можно доказать следующую теорему.

Пусть $f_1(x)$ и $f_2(x)$ — измеримые функции, конечные и удовлетворяющие неравенству

$$f_2(x) \leqslant f_1(x) \qquad\qquad (3)$$

почти всюду на сегменте $[0, 2\pi]$. Тогда существует тригонометрический ряд (1) с коэффициентами, стремящимися к нулю, для которого $f_1(x)$ и $f_2(x)$ являются соответственно верхней и нижней суммой почти всюду на $[0, 2\pi]$ (Меньшов)[13].

Эта теорема остается справедливой, если предположить, что на множестве положительной меры $f_1(x) = +\infty$, $f_2(x) = -\infty$, а почти всюду в остальных точках выполняются прежние условия.

Формулированная теорема не является окончательной; а именно: остается открытым вопрос, будет ли справедлива предыдущая теорема, если предположить, что на множестве положительной меры $f_1(x) = +\infty$ и $f_2(x) > -\infty$ или $f_1(x) < +\infty$, $f_2(x) = -\infty$.

В частности, остается открытым вопрос, существует ли тригонометрический ряд, сходящийся к $+\infty$ на множестве положительной меры. Этот вопрос был поставлен Н. Лузиным свыше сорока лет тому назад, но до сих пор остается не решенным.

Большой интерес представляет также теорема об одновременной сходимости с точностью до множества меры нуль данного тригонометрического ряда (1) и сопряженного ему ряда (2). Эта теорема была доказана Плеснером[14] и, независимо, Зигмундом (Zygmund) и Марцинкевичем (Marcinkiewicz)[15].

Точная формулировка теоремы такова:

Если тригонометрический ряд (1) сходится на некотором множестве положительной меры, то сопряженный ему ряд (2) сходится почти всюду на этом множестве.

Зигмунд и Марцинкевич получили даже более общий результат, распространив предыдущую теорему на методы суммирования Чезаро (Cesàro) порядка $\alpha > -1$. Кроме того, они установили взаимоотношение между верхними и нижними суммами данного тригонометрического ряда (1) и сопряженного ему ряда (2).

Этот результат они получили как следствие из теоремы о верхних и нижних пределах Чезаровских сумм для двух сопряженных тригонометрических рядов (1) и (2). Для случая верхней и нижней суммы точная формулировка теоремы такова:

Пусть $S(x)$ и $s(x)$ будут соответственно верхней и нижней суммами ряда (1), а $\bar{S}(x)$ и $\bar{s}(x)$ пусть будут аналогичными суммами ряда (2). Тогда, если $S(x)$ и $s(x)$ конечны на некотором множестве E положительной меры, то почти всюду на E $\bar{S}(x)$ и $\bar{s}(x)$ также конечны и удовлетворяют условию

$$\bar{S}(x) - \bar{s}(x) = S(x) - s(x).$$

В связи с вопросом о сходимости тригонометрических рядов возникает еще следующая задача. Даже в том случае, когда тригонометрический ряд расходится почти всюду, у него может существовать последовательность частных сумм $S_{n_k}(x)$ с возрастающими номерами $n_1, n_2, \dots, n_k, \dots$ которая сходится на множестве положительной меры.

Возьмем какой-нибудь тригонометрический ряд (1) и предположим, что

$$\phi(x, E) = \lim_{k \to \infty} S_{n_k}(x) \tag{4}$$

почти всюду на множестве E положительной меры, лежащем на сегменте $[0, 2\pi]$, где $n_1 < n_2 < \dots < n_k < \dots$. Мы скажем тогда, что

$\phi(x, E)$ есть *предельная функция* ряда (1) на множестве E. При этом мы не исключаем возможности того, что $\phi(x, E) = +\infty$ или $-\infty$ на множестве положительной меры.

Рассмотрим теперь некоторое множество $M = \{\phi(x, E)\}$ измеримых функций $\phi(x, E)$, каждая из которых определена почти всюду на соответствующем множестве E, причем

$$\operatorname{mes} E > 0, \quad E \subset [0, 2\pi] \tag{5}$$

для любой функции из множества M. При этом множества E могут быть различными для различных функций $\phi(x, E)$.

Возникает следующий вопрос. Найти необходимое и достаточное условие для того, чтобы M было множеством *всех* предельных функций некоторого тригонометрического ряда (1).

Рассмотрим сперва более простую задачу. Возьмем какое-нибудь определенное множество E_0 положительной меры, лежащее на сегменте $[0, 2\pi]$, и рассмотрим некоторое множество $M_0 = \{f(x)\}$ измеримых функций $f(x)$, каждая из которых определена почти всюду на множестве E_0. Спрашивается, каково необходимое и достаточное условие для того, чтобы множество M_0 было множеством всех предельных функций некоторого тригонометрического ряда на данном множестве E_0.

Чтобы ответить на этот вопрос, введем следующие определения.

Определение 1. Мы скажем, что функция $\psi(x)$, определенная почти всюду на E_0, есть *предельный элемент* множества M_0 в смысле сходимости почти всюду на E_0, если существует последовательность функций $f_n(x)$, $n = 1, 2, \ldots$, принадлежащих множеству M_0, которая сходится к $\psi(x)$ почти всюду на E_0.

Определение 2. Мы будем называть множество M_0 *замкнутым* в смысле сходимости почти всюду на E_0, если оно содержит все свои предельные элементы в смысле сходимости почти всюду на E_0.

Справедлива следующая теорема.

Пусть множество $M_0 = \{f(x)\}$ определяется так же, как и раньше. Для того, чтобы M_0 было множеством всех предельных функций некоторого тригонометрического ряда на данном множестве E_0, необходимо и достаточно, чтобы M_0 было замкнутым в смысле сходимости почти всюду на E_0 (Меньшов)[16].

При этом, если M_0 замкнуто в смысле сходимости почти всюду на E_0, то тригонометрический ряд, о котором идет речь в предидущей теореме, можно определить так, чтобы его коэффициенты a_n и b_n стремились к нулю при $n \to \infty$.

Можно также найти необходимое и достаточное условие для того, чтобы множество $M = \{\phi(x, E)\}$ было множеством *всех* предельных функций некоторого тригонометрического ряда. Здесь уже множества E, соответствующие различным функциям $\phi(x, E)$, могут быть различными. Это условие состоит в том, что множество M должно быть замкнутым в некотором новом смысле.

Чтобы формулировать это условие введем несколько определений.

Определение 3. Возьмем последовательность функций

$$\phi_n(x, E_n) \quad (n = 1, 2, \ldots), \tag{6}$$

каждая из которых определена почти всюду на соответствующем множестве E_n.

Положим
$$E = \lim_{n \to \infty} E_n$$

и предположим, что множество E и некоторое другое множество E' удовлетворяют условиям

$$\operatorname{mes} E > 0, \quad \operatorname{mes} E' > 0, \quad \operatorname{mes}(E' - E) = 0. \tag{7}$$

(В частности, третье условие (7) выполняется, если $E' \subset E$.)

Мы скажем, что функция $\phi(x, E')$, определенная почти всюду на множестве E', есть предельный элемент в широком смысле последовательности (6), если

$$\lim_{n = \infty} \phi_n(x, E_n) = \phi(x, E')$$

почти всюду на E'.

Легко видеть, что предельный элемент в широком смысле последовательности функций определяется не однозначно.

Определение 4. Возьмем какое-нибудь множество $M = \{\phi(x, E)\}$ функций $\phi(x, E)$, каждая из которых определена почти всюду на соответствующем множестве E, $\operatorname{mes} E > 0$. Мы будем называть функцию $f(x, E')$, определенную почти всюду на множестве E' положительной меры, *предельным элементом в широком смысле* множества M, если существует последовательность функций $\phi_n(x, E_n)$, принадлежащих множеству M, для которой $f(x, E')$ есть предельный элемент в широком смысле.

Определение 5. Пусть множество M удовлетворяет тем же условиям, как и раньше. Мы скажем, что M замкнуто в узком смысле, если оно содержит все свои предельные элементы в широком смысле. Можно доказать следующую теорему.

Пусть $M = \{\phi(x, E)\}$ есть множество измеримых функций, каждая из которых определена почти всюду на соответствующем множестве E положительной меры, $E \subset [0, 2\pi]$.

Для того, чтобы M было множеством всех предельных функций некоторого тригонометрического ряда, необходимо и достаточно, чтобы M было замкнутым в узком смысле (Меньшов)[17].

При этом, если M — замкнуто в узком смысле, то соответствующий тригонометрический ряд можно определить так, чтобы его коэффициенты стремились к нулю с возрастанием номера.

Из моего доклада следует, что в теории тригонометрических рядов целый ряд вопросов, относящихся к сходимости этих рядов, остается не решенным. По-видимому, решение большей части этих вопросов представляет серьезные трудности. Не исключена возможность, что решение некоторых из этих вопросов связано с арифметическими свойствами тригонометрических рядов. Во всяком случае, таково положение дела в вопросе о единственности разложения в тригонометрический ряд. За недостатком времени я не мог затронуть этого вопроса в настоящем докладе. В теории единственности разложения в тригонометрический ряд рассматриваются так называемые U-и M-множества. Множество E называется U-множеством, если у любого тригонометрического ряда, сходящегося к нулю вне этого множества, все коэффициеиты равны нулю. В противном случае множество E называется M-множеством.

Из работ ряда авторов следует, что свойство множества меры нуль быть U или M в значительной мере зависит от арифметической природы этих множеств. Это следует из работ Бари[18], Зигмунда (Zygmund) и Салема (Salem)[19], Шапиро-Пятецкого[20] и других авторов.

БИБЛИОГРАФИЯ

[1] Denjoy, A. Calcul des coefficients d'une série trigonométrique convergente quelconque dont la somme est donnée. *C.R. Acad. Sci.*, *Paris*, 172, 1218–1221 (1921).

[2] Titchmarsh, E. C. On conjugate functions. *Proc. Lond. Math. Soc.* (2), 29, 49–80 (1929).

[3] Ульянов, П. Интеграл и сопряженные функции. *Ученые записки МГУ, математика*, 8, 130–157 (1956).

[4] Lusin, N. Über eine Potenzreihe. *R.C. Circ. mat. Palermo*, 32, 386–390 (1911).

[5] Steinhaus, H. Une série trigonométrique partout divergente. *C.R. Soc. Sci.*, *Varsovie*, pp. 219–229 (1912).

[6] Kolmogoroff, A. Une série de Fourier-Lebesgue divergente partout. *C.R. Acad. Sci.*, *Paris*, 183, 1927–1928 (1926).

[7] Hardy, G. and Rogosinski, W. *Fourier Series*, Cambridge, 1944.

[8] Sunouchi, G. A Fourier series which belongs to the class H divergent almost everywhere. *Kōdai Math. seminar reports*, 1, 27–28 (1953).

[9] Zeller, K. Über Konvergenzmengen von Fourierreihen. *Arch. Math.* 6, N4, 335–340 (1955).

[10] Стечкин, С. О сходимости и расходимости тригонометрических рядов. *УМН*, У1, no. 2, 148–149 (1951).

[11] Меньшов, Д. О равномерной сходимости тригонометрических рядов. *ДАН СССР*, 32, 245–246 (1941).

Menchoff, D. Sur la convergence uniforme des séries de Fourier. *Математический сборник*, 11 (53), 67–96 (1942).

[12] Меньшов, Д. Об изображении измеримых функций тригонометрическими рядами, *ДАН СССР*, 26, 222–224 (1940).

Menchoff, D. Sur la représentation des fonctions mesurables par des séries trigonométriques. *Математический сборник*, 9 (51), 667–692 (1941).

[13] Меньшов, Д. О пределах неопределенности тригонометрических рядов. *ДАН СССР*, 74, no. 2, 181–184 (1950).

Меньшов, Д. О пределах неопределенности рядов Фурье. *Математический сборник*, 30 (72), 601–650 (1952).

[14] Плеснер, А. О сопряженных тригонометрических рядах. *ДАН СССР*, 4 (9), no. 6–7 (75–76), 235–237 (1935).

[15] Marcinkiewicz, J. and Zygmund, A. On the differentiability of functions and summability of trigonometrical series. *Fundamenta Mathematicae*, 26, 1–43 (1936).

[16] Меньшов, Д. О пределах последовательностей частных сумм тригонометрических рядов. *ДАН СССР*, 106, no. 5, 777–780 (1956).

[17] Меньшов, Д. О предельных функциях тригонометрического ряда. *ДАН СССР*, 114, no. 3, 476–478 (1957).

[18] Bary, N. Sur la nature diophantique du problème d'unicité du développement trigonométrique. *C.R. Acad. Sci.*, *Paris*, 202, 1901–1903 (1936).

Sur le rôle des lois diophantiques dans le problème d'unicité de développement trigonométrique. *Математический сборник*, 2 (44), 699–724 (1937).

[19] Salem, R. and Zygmund, A. Sur un théorème de Piatetçki-Shapiro. *C.R. Acad. Sci.*, *Paris*, 240, no. 21, 2040–2042 (1955).

[20] Шапиро-Пятецкий, И. К проблеме единственности разложения функции в тригонометрический ряд. *Ученые записки МГУ*, 165, 7, 39 (1954).

HILBERT ALGEBRAS

By S. MINAKSHISUNDARAM

1. Since the publication of the papers 'On Rings of Operators' by F. J. Murray and J. von Neumann, there has been an increasing interest among mathematicians in the structure of topological algebras. Today I wish to discuss some properties of a class of algebras called Hilbert algebras.

Hilbert algebras are defined by means of the following axioms:

A: A is a * algebra over the field of complex numbers in addition to being an inner product space: that is to say

A_1: A is a linear vector space over the field C of complex numbers.

A_2: A is a ring where multiplication is associative and distributive with addition, but not necessarily commutative.

A_3: There exists an involution operation denoted by *, which associates to each element $a \in$ A, a unique element $a^* \in$ A such that:

(α) $(a^*)^* = a$,

(β) $(\lambda a)^* = \bar{\lambda} a^*$, $\lambda \in C$, $\bar{\lambda}$ complex conjugate of λ,

(γ) $(a+b)^* = a^* + b^*$,

(δ) $(ab)^* = b^* a^*$,

(ϵ) $xx^* \neq 0$ unless $x = 0$.

A_4: A is an inner product space: that is, to each pair of elements a, b in A we can associate a scalar denoted by (a, b) which is linear in a and conjugate linear in b, satisfying the following conditions:

(α) $(\lambda a + \mu b, c) = \lambda(a, b) + \mu(a, c)$, $\lambda, \mu \in C$,

(β) $(a, b) = \overline{(b, a)}$,

(γ) $(a, a) \geqslant 0$ for all a in A and is zero if and only if $a = 0$,

(δ) $(ab, c) = (a, cb^*) = (b, a^*c)$,

(ϵ) $(a, b) = (b^*, a^*)$.

We shall consistently use the following notations:

$\|x\| = \sqrt{(x, x)}$, the positive value of the square root being taken.

Greek letters will denote complex numbers and Roman letters (small type) will denote elements of the algebra A or its completion.

Now A may or may not be complete under the Hilbert norm topology. If A is already complete it is natural to assume that multiplication is continuous with respect to both the variables; in which case we have

Ambrose's H*-algebra whose structure is well known [cf. Loomis, *An Introduction to Abstract Harmonic Analysis*].

If A is not complete, we denote by $\hbar$ the completed Hilbert space. Then, following classical procedure, the mappings $x \to ax$ and $x \to xa$ can be defined for all x in $\hbar$ and a in A so that they are continuous for fixed $a \in A$, throughout $\hbar$. But we cannot say anything about the existence of xy or yx when both x and y are in $\hbar$ but outside A. So we add the following axiom B to axiom A.

B. If A is not complete, the operators $x \to ax$ and $x \to xa$ are so extended to $\hbar$ that they are closed linear operators in $\hbar$, and these operators are bounded if and only if $a \in A$.

Axiom B implies that the product xy certainly exists if at least one of the two elements x, y belongs to A. As an immediate consequence of these axioms one readily proves that the * operation is a bounded operation which can be extended all over $\hbar$, so that to each element $x \in \hbar$ there corresponds x^* so that $\|x\| = \|x^*\|$ and axiom A_3 is fulfilled. Further the linear operators $x \to ax$, $x \to xa$, $x \to a^*x$, $x \to xa^*$, for $a \in A$ have the same bounds and their common bound, called the *uniform norm* of a, is denoted by $\|\|a\|\|$.

$$\|\|a\|\| = \underset{x \in \hbar}{\text{l.u.b.}} \frac{\|ax\|}{\|x\|}.$$

With respect to the uniform norm, A will be in general an incomplete Banach space, unless A contains the unit of multiplication.

There is still another axiom which is of interest, though not essential for our discussion, namely:

C. Whenever a and b are two elements in $\hbar$, if there exists a third element c and a Cauchy sequence b_n, such that

$$b_n \to b$$

and $(ab_n, x) \to (c, x)$ for every x,

then and only then ab exists and $c = ab$.

This axiom assures us that the adjoint of the linear operator $x \to ax$ is $x \to xa^*$ for any $a \in \hbar$. It is obvious that axiom C is of significance only when $a \in \hbar$ and is outside A.

2. The set A together with its completion $\hbar$ satisfying axioms A and B is called a Hilbert algebra. Our purpose is to discuss the structure of this algebra. We introduce a few definitions and notations which we need.

(a) An element $x \in \hbar$ is called *self-adjoint* if $x = x^*$.

(b) An element x is said to be *positive* if $x = x^*$ and $(xy, y) \geqslant 0$ for every $y \in \mathbf{A}$.

(c) An element e is called an *idempotent* if e^2 exists and $e^2 = e$. If e is also self-adjoint we call e a *self-adjoint idempotent*.

One easily verifies that a self-adjoint idempotent is invariably an element in $\mathbf{A}$.

(d) If S is a subset of $\hbar$, we denote by $[S]$ its linear closure.

(e) A closed linear manifold $\mathscr{M} \subset \hbar$ is called a *right ideal manifold*, if whenever $x \in \mathscr{M}$, $[x\mathbf{A}] \subset \mathscr{M}$. A similar definition holds for *left ideal manifolds* and (two sided) *ideal manifolds*. The right ideal manifold $[a\mathbf{A}]$ is called a *principal right ideal manifold* and a the *generating element* of it.

One easily verifies that $[e\mathbf{A}] = [e\hbar] = e\hbar$.

(f) The Hilbert algebra $\{\mathbf{A}, \hbar\}$ is called *simple* if $\hbar$ does not contain any (two-sided) ideal manifolds other than the trivial ones. The algebra is said to be *abelian* if multiplication is commutative.

(g) There are two extensions of the algebra $\mathbf{A}$ to the algebra of linear operators in $\hbar$ viz. those linear operators L which satisfy the condition

$$L(xy) = L(x)\, y,$$

the mapping $x \to ax$ being denoted by $L_a(x)$, and those operators R which satisfy the condition

$$R(xy) = xR(y),$$

the mapping $x \to xa$ being denoted by $R_a(x)$. The totality of operators of the type L will be denoted by $\mathscr{L}(\mathbf{A})$ and those of the type R will be denoted by $\mathscr{R}(\mathbf{A})$.

One verifies that the commutant of $\mathscr{L}$ is $\mathscr{R}$ and of $\mathscr{R}$ is $\mathscr{L}$.

That in the algebra $\mathbf{A}$ there are sufficiently many self-adjoint idempotents can be proved, in the usual way, by taking positive elements a in $\mathbf{A}$, polynomials $p(a)$, and their limits, in the Hilbert norm.

3. If the algebra $\mathbf{A}$ is abelian then $\hbar$ will be isometrically isomorphic to square integrable functions on a locally compact Hausdorff space S with a measure ν, multiplication being defined as the ordinary multiplication of two functions in S. $\mathbf{A}$ will be isomorphic to bounded functions vanishing at ∞. This can be proved either using Gelfand's isomorphism theorem for commutative Banach algebras [cf. Dixmier, *Les algèbres d'opérateurs dans l'espace hilbertien*, 1957] or by observing that the self-adjoint idempotents in $\mathbf{A}$ form a measure algebra and using Stone's theorem [Halmos, *Measure Theory*]. If $\mathbf{A}$ contains the unit of multiplication, S will be compact.

4. If the algebra **A** is simple, that is to say there are no (two sided) ideal manifolds in h other than $\{0\}$ and h, then (and only then) $\mathscr{L}(\mathbf{A})$ and $\mathscr{R}(\mathbf{A})$ will be factors. The self-adjoint idempotents in **A** will satisfy the axioms of continuous geometry of von Neumann, in the sense that there exists an equivalence relation among them and the classes of equivalent idempotents form a well-ordered set with a dimension number attached to the self-adjoint idempotent e, viz. $d(e) = \|e\|^2$. There are only four possible cases, viz. the four types of factors of von Neumann I_n (n a finite positive integer), II_1, I_∞, II_∞. If **A** contains a unit and is simple, the types I_n, II_1 will occur, and the types I_∞ and II_∞ will occur when **A** does not contain a unit. I_n and I_∞ are called discrete while II_1 and II_∞ are continuous types. I_n and I_∞ will occur if the simple algebra **A** contains minimal idempotents, and II_1 and II_∞ otherwise.

It is verified easily that I_n is isomorphic to the algebra of matrices of order n with complex numbers as their elements, and I_∞ to the totality of bounded linear operators in a Hilbert space. The structure of II_∞ can be determined if we know the structure of II_1.

The structure of a simple algebra of type II_1 seems to be difficult and intriguing. Von Neumann has constructed two types of examples of an algebra of type II_1. If **A** is a simple algebra of type II_1 then **A** contains a unit, whose norm we shall assume to be 1, and there are no minimal self-adjoint idempotents. It is likely that the structure of **A** can be determined in terms of its maximal abelian subalgebras. It is quite likely the maximal abelian subalgebras of a simple algebra are unitarily equivalent. In the case of I_n the diagonal matrices form a maximal abelian subalgebra and it is well known that all maximal abelian subalgebras are unitarily equivalent to the subalgebra of diagonal matrices. In the case of an algebra **A** of type II_1, let **B** be a maximal abelian subalgebra and $\bar{\mathbf{B}}$ its Hilbert completion. If we observe that the dimension numbers of the self-adjoint idempotents in **B** take all values between 0 and 1, then one can easily show that all maximal subalgebras are unitarily equivalent.

5. If the algebra **A** is neither simple nor abelian then one has to study the structure of **A** in relation to its centre. Let **C** denote the centre of **A** and let **Z** be its Hilbert completion. A self-adjoint idempotent e in the centre is said to be minimal if it cannot be expressed as sum of two idempotents in the centre. If e is minimal $e\mathbf{A}$ is a simple algebra with eh as its Hilbert completion. **C** being abelian, there exists a locally compact (or compact) Hausdorff space S and a measure ν so that **Z** is isomorphic to L_2 functions on S. The atomic elements in S will correspond to minimal idempotents in C, so that A can be expressed as a direct sum of two algebras A_1 and A_2, that is

$$A = A_1 + A_2,$$

$$\hbar = \hbar_1 + \hbar_2, \quad [A_1] = \hbar_1, \ [A_2] = \hbar_2,$$

$\hbar_1$ and $\hbar_2$ being orthogonal manifolds, where A_1 is a direct sum of simple algebras and $\hbar_2$ contains no minimal ideal manifolds. Further decomposition of $\hbar_2$ is possible along the lines of von Neumann's reduction theory, but we shall not discuss it here.

SPECTRAL SETS AND NORMAL
DILATIONS OF OPERATORS†

By BÉLA SZ.-NAGY

1. Spectral sets

In all that follows 'operator' will mean a linear operator on Hilbert space, everywhere defined and bounded. The notion of *spectral set* is due to von Neumann[5]: a set S of points in the plane of complex numbers is a spectral set of the operator T if, for any rational function $r(z)$ which is bounded on S, the operator $r(T)$ exists and the inequality

$$\|r(T)\| \leqslant \sup_{z \in S} |r(z)| \tag{1}$$

holds; the condition of the existence of $r(T)$ is equivalent to the condition that the *spectrum* of T be contained in the closure of S.

It follows from classical spectral theory that, for a normal operator T, the spectrum itself is a spectral set. For non-normal T, the spectrum is in general not large enough to be a spectral set. The introduction of the concept of spectral sets is motivated just by the fact that one can build up, for functions on a spectral set S of a not necessarily normal operator T, a functional calculus which is analogous in some respects to the functional calculus for normal operators. The map $r(z) \to r(T)$ from functions to operators extends namely, by continuity implied by (1), to all S-*analytic* functions, i.e. to those functions $u(z)$ defined on S, which are uniform limits, on S, of bounded sequences of rational functions. These functions form an algebra, closed with respect to uniform convergence, and the map $u(z) \to u(T)$ is an algebra homomorphism for which also the *metric* relation (1) is valid. (It is just this metric relation that distinguishes this functional calculus essentially from the functional calculus of Riesz–Dunford, based on contour integration in the resolvent set.)

Some more properties follow easily: if S' is the set of values of the S-analytic function $z' = u(z)$ taken on S, then S' is a spectral set of the operator $T' = u(T)$, and for any S'-analytic function $v(z')$ the function $v \circ u(z) = v(u(z))$ is S-analytic, and we have $v \circ u(T) = v(T')$.

The usefulness of the theory depends on whether one can find spectral sets for a given operator. It is obvious that any set containing a spectral

† Read by Professor P. R. Halmos.

set as a subset is itself a spectral set, but the meet of two spectral sets need not be again a spectral set. Fundamental is the following theorem of von Neumann[5] on *contraction operators* T, i.e. for which $\|T\| \leqslant 1$:

Theorem 1. The unit disc $S_0 = \{z: |z| \leqslant 1\}$ is a spectral set for any contraction operator T.

Using linear maps of S_0 one gets as easy consequences criteria that the set $\{z: |z-a| \leqslant r\}$, or $\{z: |z-a| \geqslant r\}$, or $\operatorname{Re} z \geqslant 0$ be a spectral set of a given operator T. These are, respectively, $\|T-aI\| \leqslant r$, $\|(T-aI)^{-1}\| \leqslant r^{-1}$, and $\operatorname{Re} T = \frac{1}{2}(T+T^*) \geqslant O$. It would be very desirable to obtain further simple criteria for spectral sets.

The original proof of von Neumann of Theorem 1 uses a theorem of Schur on analytic functions; a proof given by Heinz[4] uses the Poisson integral formula; another proof, given by Sz.-Nagy[12] reduces the problem to the more simple case of unitary operators (see § 4).

The definition of spectral sets works also for Banach space operators. But Foiaş[1] has proved that Theorem 1 is valid only for Hilbert space.

For any spectral set S of the operator T, which is bounded, closed, and whose boundary F is a Jordan curve (briefly: a *bounded Jordan set*), the class of S-analytic functions coincides, by a theorem of Walsh, with the class $\overline{\mathcal{O}}(S)$ of functions which are continuous on S, and holomorphic in the interior of S. In this case, a more refined functional calculus has been built up by Foiaş[2]. Essentially by taking the real parts of holomorphic functions and the corresponding operators, he extends the map $u(z) \to u(T)$ to the class $\mathscr{A}(T)$ of (real or complex valued) functions which are continuous on S, and harmonic in the interior of S. Thus extended, the map does not remain multiplicative, but it has the monotonic property: $u(z) \leqslant v(z)$ implies $u(T) \leqslant v(T)$; the metric property (1) is also true. Since any continuous function on F can be continued uniquely to a function in $\mathscr{A}(T)$, so we have in reality a map of the continuous functions on F, to operators. Passing to limits by monotone sequences, this map can be extended to broader classes of functions on F, in particular to the characteristic functions $\omega(\beta, z)$ of Borel subsets of F. The corresponding operators $\omega(\beta; T) \geqslant 0$ form the so-called *harmonic spectral measure* of T with respect to S. The properties, and the use in functional calculus, of this 'measure' are studied in detail in Foiaş's paper[2].

2. Positive definite functions

There is still another possibility of building up a functional calculus for general operators, with metric properties such as (1), namely by deducing it directly from the functional calculus for normal operators. This will be treated in the following sections.

The fundamental concepts here are those of *dilations* and *projections* of operators. If T is an operator on Hilbert space H, and $\mathbf{T}$ is an operator on Hilbert space $\mathbf{H}$ containing H as a (not necessarily proper) subspace, we say that $\mathbf{T}$ is a dilation of T or, equivalently, that T is a projection of $\mathbf{T}$, in signs

$$T = \mathrm{pr}_H \, \mathbf{T} \quad \text{or simply} \quad T = \mathrm{pr} \, \mathbf{T},$$

if we have, for any $h \in H$, $\quad Th = \mathbf{PT}h,$

where $\mathbf{P}$ denotes the operator of orthogonal projection onto the subspace H of $\mathbf{H}$. (These terms are due to Halmos[3] and Sz.-Nagy[14], respectively; Halmos says 'compression' instead of 'projection'.)

An important pertaining theorem deals with operator-valued functions $T(s)$ on a *group* G, which are *positive definite* in the sense that, for any H-vector-valued function $h(s)$ such that $h(s) \neq 0$ only for a finite set of elements $s \in G$, we have

$$\sum_{s \in G} \sum_{t \in G} (T(t^{-1}s) \, h(s), \, h(t)) \geqslant 0. \tag{2}$$

Particular positive definite operator-valued functions on G are those for which $T(t^{-1}s) = T^*(t) \, T(s)$; in case $T(e) = I$ (e denoting the unit element of G) these operators $T(s)$ are thus unitary and form a representation of the group G (i.e. we have $T(s) \, T(t) = T(st)$). The theorem in question reads as follows:

Theorem 2. For any positive definite H-operator-valued function T on the group G, with $T(e) = I$, there exists a representation of G by unitary operators $\mathbf{U}(s)$ on some Hilbert space $\mathbf{H}$ ($\supseteq H$), such that

(a) $\qquad\qquad\qquad T(s) = \mathrm{pr} \, \mathbf{U}(s) \quad (s \in G),$

(b) *the elements of the form $\mathbf{U}(s) \, h$ ($s \in G$, $h \in H$) span $\mathbf{H}$. The 'structure' $\{\mathbf{H}, \mathbf{U}(s), H\}$ is then determined up to isomorphism. If G is a topological group, and $T(s)$ is a weakly continuous function on G, then $\mathbf{U}(s)$ is weakly (and so also strongly) continuous too.*

This is a straightforward generalization of a theorem of Gelfand and Raikov on complex scalar-valued positive definite functions on groups, and was proved first by Neumark[7]. Later Sz.-Nagy found independently essentially the same proof[13], and generalized it[14] to *semi-groups* G with unit element e and with an *involution* $s \to s^*$:

$$(st)^* = t^*s^*, \quad s^{**} = s, \quad e^* = e.$$

The role played in a group by s^{-1} is played here by s^*; the operator-valued function $D(s)$ is a representation of G if it satisfies the conditions

$$D(e) = I, \quad D(st) = D(s) D(t), \quad D(s^*) = [D(s)]^*;$$

for a *commutative* semi-group G all $D(s)$ are therefore *normal* operators.

Let B_λ be a self-adjoint operator on Hilbert space H, which is a non-decreasing, right-continuous function of the real parameter λ, with $\lim_{\lambda \to -\infty} B_\lambda = O$, $\lim_{\lambda \to +\infty} B_\lambda = I$; call B_λ an *operator distribution function* (o.d.f.). Particular o.d.f.'s are the *spectral families* (called also 'resolutions of the identity'): this is the case if the operators B_λ are (orthogonal) projections.

Now each o.d.f. B_λ gives rise, by the formula

$$(T(s)f, g) = \int_{-\infty}^{\infty} e^{is\lambda} d(B_\lambda f, g) \quad (-\infty < s < \infty),$$

to an operator $T(s)$, which is a weakly continuous positive definite function on the additive group of reals. By Theorem 2, $T(s)$ is the projection of a weakly continuous unitary representation $U(s)$ of this group. By Stone's theorem, we have

$$U(s) = \int_{-\infty}^{\infty} e^{is\lambda} d\mathbf{E}_\lambda,$$

and conclude that B_λ is the projection of the spectral family $\mathbf{E}_\lambda$. This proves the following very useful theorem of Neumark[6, 7]:

Theorem 3. Any operator distribution function B_λ on Hilbert space H is the projection of some spectral family $\mathbf{E}_\lambda$ on a suitable Hilbert space $\mathbf{H} (\supseteq H)$.

An alternative proof may be obtained from Sz.-Nagy's generalization of Theorem 2 to semi-groups G with involution; it is namely easy to see that (putting $B_{-\infty} = O$, $B_{+\infty} = I$) B_λ is a positive definite function on the extended system of reals λ ($-\infty \leqslant \lambda \leqslant \infty$), made to be a semi-group with involution by defining

$$\lambda \circ \mu = \min \{\lambda, \mu\}, \quad \lambda^* = \lambda \quad (e = +\infty).$$

Further applications of Theorem 2 and of its generalization to semi-groups are given in [14]: one obtains in this way a generalized form of Neumark's theorem to additive operator-valued set functions[8]; a characterization of 'subnormal' operators due to Halmos[3] (an operator is subnormal if it has a normal extension on some larger Hilbert space); a theorem on the operator moment problem; and theorems on unitary dilations of contraction operators. The last-mentioned theorems are important for our present subject, so we shall treat them in more detail.

3. Unitary dilations of contraction operators

The starting-point here is the observation that, for any contraction operator T on Hilbert space H, the operator $T(n)$ defined by T^n for $n = 0, 1, 2, \ldots$, and by $(T^*)^{-n}$ for $n = -1, -2, \ldots$, is a positive definite function on the additive group of integers n, i.e.

$$\sum_n \sum_m (T(n-m) h_n, h_m) \geqslant 0$$

($h_n = 0$ for almost all n). This sum is namely equal to

$$\lim_{r \to 1-0} \frac{1}{2\pi} \int_0^{2\pi} (K(r, \phi) h(\phi), h(\phi)) \, d\phi,$$

where $h(\phi) = \sum_n e^{-in\phi} h_n$, and

$$K(r, \phi) = \sum_n r^{|n|} e^{in\phi} T(n) = 2 \operatorname{Re} \left(\tfrac{1}{2} I + \sum_1^\infty z^n T^n \right)$$

$$= \operatorname{Re} (I + zT)(I - zT)^{-1} \geqslant 0 \quad (z = re^{i\phi}, \ 0 \leqslant r < 1);$$

the positivity of $K(r, \phi)$ follows by the relation

$$(K(r, \phi) f, f) = \operatorname{Re} ((I + zT) g, (I - zT) g) = \|g\|^2 - r^2 \|Tg\|^2 \geqslant 0,$$

for arbitrary $f \in H$ and $g = (I - zT)^{-1} f$.

Thus, applying Theorem 2, we get the result (Sz.-Nagy[12, 13]):

Theorem 4. For any contraction operator T on Hilbert space H there exists a unitary operator $\mathbf{U}$ on some Hilbert space $\mathbf{H}$ ($\supseteq H$) such that

$$T^n = \operatorname{pr} \mathbf{U}^n \quad (n = 0, 1, 2, \ldots), \tag{3}$$

and $\mathbf{H}$ is spanned by the elements $\mathbf{U}^n h$ ($n = 0, \pm 1, \pm 2, \ldots$; $h \in H$). The 'structure' $\{\mathbf{H}, \mathbf{U}, H\}$ is then determined up to isomorphism.

If T, T' are two contraction operators such that T is permutable with T' and with T'^*, thus also T^* with T' (we say that T and T' are *doubly permutable*), then $T(n, m) = T(n) T'(m) = T'(m) T(n)$ is a positive definite function on the additive group of vectors (n, m) with integer components. This follows as above, making use now of double Fourier series, and observing that the positive operator functions $K(r, \phi)$, $K'(r', \phi')$, attached to T and T', respectively, are permutable, and their product is therefore also positive. Theorem 2 then implies that

$$T(n, m) = \operatorname{pr} \mathbf{U}^n \mathbf{U}'^m \quad (n, m = 0, \pm 1, \pm 2, \ldots),$$

and in particular

$$T^n T'^m = \operatorname{pr} \mathbf{U}^n \mathbf{U}'^m \quad (n, m = 0, 1, 2, \ldots), \tag{4}$$

with *permutable* unitary **U**, **U′**. A similar result holds for any finite or infinite set of doubly permutable contraction operators.

The problem whether *simple* permutability of T, T' suffices in order that the representation (4) be possible with *permutable* **U**, **U′**, is not yet fully settled. The problem is equivalent to that of finding an operator-valued function $X(n, m)$ which is positive definite on the group of the vectors (n, m), and reduces to $T^n T'^m$ for $n, m \geqslant 0$. A necessary condition for positive definiteness is that $X(-n, -m) = X^*(n, m)$, so one can dispose only of the values of $X(n, -m)$ for $n, m \geqslant 1$. Brehmer, in his Potsdam thesis, showed that the definition $X(n, -m) = (T'^*)^m T^n$ meets the requirements under certain additional conditions, such as the condition $\|T\|^2 + \|T'\|^2 \leqslant 1$, or that T be an arbitrary contraction operator, while T' be an isometric operator. He obtained analogous results for any finite or countable set of permutable contraction operators.

Theorem 4 has a continuous counterpart (Sz.-Nagy[12, 13]):

Theorem 5. For any weakly continuous one-parameter semi-group $T(s)$ *($s \geqslant 0$) of contraction operators on Hilbert space* H *there exists a weakly (thus also strongly) continuous one parameter group* **U**(s) *($-\infty < s < \infty$) of unitary operators on a Hilbert space* **H** *($\supseteq H$) such that*

$$T(s) = \operatorname{pr} \mathbf{U}(s) \quad (s \geqslant 0),$$

and **H** *is spanned by the elements* **U**$(s)h$ *($-\infty < s < \infty$, $h \in H$); the 'structure'* $\{\mathbf{H}, \mathbf{U}(s), H\}$ *is then defined up to isomorphism.*

This follows from Theorem 2 if we remark that the operator-valued function $T(s)$, when continued to negative values of s by putting $T(s) = T^*(-s)$, is a (weakly continuous) positive definite function on the additive group of reals, i.e.

$$\sum_n \sum_m \left(T(s_n - s_m) h_n, h_m\right) \geqslant 0$$

for any finite set of real numbers s_n and vectors $h_n \in H$. For commensurable s_n this inequality follows from what has just been said for the powers of a single contraction operator, and this implies the general case by continuity.

The fact that any contraction operator has a unitary dilation was proved in 1950 by Halmos[3] by a simple matrix construction. (For isometric instead of unitary dilations this was proved already by Julia in his *Comptes Rendus* Notes 1944.) However, the unitary operator constructed by Halmos does not satisfy the relation simultaneously for all n. After Sz.-Nagy published Theorem 4 in 1953, Schäffer[9] succeeded in generalizing Halmos's matrix construction so as to yield a unitary $\mathbf{U}$ satisfying (3) for all n. However, this construction seems inadequate to be applied to the problem of two or more permutable T's or to one parameter semi-groups. As a matter of fact, the $\mathbf{U}$'s attached by his construction to different T's on H, are all defined on the same space $\mathbf{H}$, but are not necessarily permutable even if the T's are doubly permutable.

Call the operator $\mathbf{U}$ attached to the contraction operator T in the sense of Theorem 4, and the semi-group $\mathbf{U}(s)$ attached to $T(s)$ in the sense of Theorem 5, the *strong unitary dilations* of T and of $T(s)$, respectively.

There are some—rather loose—spectral relations between a contraction operator T and its strong unitary dilation $\mathbf{U}$:

(i) If T itself is unitary, then $\mathbf{U} = T$.

(ii) If T is not unitary, then the spectrum of $\mathbf{U}$ covers the whole unit circle ([11], Th. 2; or [16], cor. 2.2.).

(iii) T and $\mathbf{U}$ have the same eigenvalues on the unit circle, and the corresponding eigenvectors are the same for both ([16], Th. 1).

(iv) The strong unitary dilations of proper contraction operators (i.e. with $\|T\| < 1$) are all unitarily equivalent to the orthogonal sum of $\mathfrak{d}$ replicas of the unitary operator $Uf(\phi) = e^{i\phi}f(\phi)$ on $L^2(0, 2\pi)$, where $\mathfrak{d}$ denotes the dimension number of the Hilbert space H. (This has been proved by Schreiber[10] with the restriction $\mathfrak{d} \leqslant \aleph_0$, and by Sz.-Nagy[15] in the general case.)

Using the Hille–Yosida theorem in a form specialized to Hilbert space one can characterize the one-parameter semi-groups $T(s)$ of Theorem 5 by the fact that their infinitesimal generator A satisfies the condition: $T = (A+I)/(A-I)$ is a contraction operator not having the eigenvalue 1 (see [15], [16]). Call T the infinitesimal *cogenerator* of the semi-group $T(s)$. Any contraction operator not having the eigenvalue 1 is the infinitesimal cogenerator of exactly one such semi-group. There is a continuous analogue of (iv):

(v) The strong unitary dilations $\mathbf{U}(s)$ of those one-parameter semi-groups $T(s)$ whose infinitesimal cogenerator T is a proper contraction operator, are all unitarily equivalent to the orthogonal sum of $\mathfrak{d}$ replicas of the one parameter unitary semi-group $U(s)f(x) = e^{isx}f(x)$ on $L^2(-\infty, \infty)$; $\mathfrak{d}$ is the dimension number of the space H (Sz.-Nagy[15]).

4. Functional calculus for general operators

If $\mathbf{U}$ is the strong unitary dilation of the contraction operator T, then we have obviously

$$u(T) = \mathrm{pr}\, u(\mathbf{U}) \tag{5}$$

for any function $u(z) = c_0 + c_1 z + c_2 z^2 + \dots$, holomorphic on a domain containing the unit disc S_0 in its interior. Thus we have

$$\|u(T)\| \leqslant \|u(\mathbf{U})\| \leqslant \sup_{|z|=1} |u(z)|, \qquad (6)$$

the second inequality being a consequence of the spectral representation of the functions of unitary operators, i.e. of the formula

$$u(\mathbf{U}) = \int_0^{2\pi} u(e^{i\theta}) \, d\mathbf{E}_\theta, \qquad (7)$$

where $\mathbf{E}_\theta$ denotes the spectral family of $\mathbf{U}$.

Applying this result in particular to rational functions having all their singularities outside S_0, we get a proof of Theorem 1 in § 1.

Moreover, these formulae enable us to derive a functional calculus for T from that of $\mathbf{U}$; namely we can define $u(T)$ by (5) and (7) for any function $u(z)$ for which $u(e^{i\theta})$ is defined almost everywhere with respect to the spectral measure generated by $\mathbf{E}_\theta$, is bounded, and measurable with respect to $\mathbf{E}_\theta$. However, in order to preserve also the *multiplicative* property of the map $u(z) \to u(T)$, some restriction of the generality is necessary: one possibility is to consider only those functions $u(z)$ which are bounded and holomorphic in the interior of S_0 and whose radial limits $u(e^{i\theta}) = \lim_{r \to 1-} u(re^{i\theta})$ exist almost everywhere with respect to $\mathbf{E}_\theta$. By virtue of (iii), § 3, these conditions are fulfilled in particular for the functions, holomorphic and bounded in the interior of S_0, whose radial limits exist at every point of the unit circle with the possible exception of a finite or denumerable set of points, *no one of which is an eigenvalue of T*. Denote the class of these functions by $\mathcal{O}_T(S_0)$.

The resulting functional calculus is studied in detail by Sz.-Nagy and Foiaş in [16]. In particular, the metric relation (6) holds true for the class $\mathcal{O}_T(S_0)$, with $\sup_{|z|<1} |u(z)|$ on the right-hand side.

As an application they consider a contraction operator for which 1 is not an eigenvalue, and the functions

$$u_s(z) = \exp\left(s \frac{z+1}{z-1}\right) \quad (s \geqslant 0),$$

which belong to $\mathcal{O}_T(S_0)$; the operators $T_s = u_s(T)$ then form exactly the one-parameter semi-group of contraction operators whose infinitesimal cogenerator is equal to the given operator T. This result opens a new way to the study of such semi-groups. For example, assertion (v) in § 3, on semi-groups, appears as a consequence of assertion (iv) on a single operator. The infinitesimal cogenerator of the strong unitary dilation $U(s)$ of $T(s)$ is equal to the strong unitary dilation $\mathbf{U}$ of the infinitesimal cogenerator T of $T(s)$.

Independently, Schreiber[11] has proposed essentially the same functional calculus. As definition he uses the formula

$$u(T) = \int_0^{2\pi} u(e^{i\theta})\, dB_\theta,$$

which results from (5) and (7) by putting $B_\theta = \operatorname{pr} \mathbf{E}_\theta$; B_θ is an operator distribution function, uniquely determined by the equations

$$T^n = \int_0^{2\pi} e^{in\theta}\, dB_\theta \quad (n = 0, 1, 2, \ldots).$$

If T is normal, B_θ can be calculated from the spectral family of T; in this case—and only in this case—T is permutable with all B_θ's.

Sz.-Nagy and Foiaş[16] carry over their results also to the more general case when the role of a contraction operator and the unit disc S_0 is given to an *arbitrary operator* T on H, and to an arbitrary bounded Jordan set S which is a *spectral set* of T (cf. § 1). Let $z \to z_0 = s(z)$ be a conformal mapping of the interior of S on the interior of S_0, continued to the boundaries so as to be a homeomorphic mapping of S on S_0. Let $z_0 \to z = \overset{-1}{s}(z_0)$ be its inverse. Then $s(z)$ is an S-analytic function, $T_0 = s(T)$ exists in the sense of § 1, and T_0 has the spectral set $S_0 = s(S)$. Thus T_0 is a contraction operator (put $r(z) = z$ in (1)). Let now $\mathbf{U}_0$ be the strong unitary dilation of T_0 on the Hilbert space $\mathbf{H}$. Then $\mathbf{N} = \overset{-1}{s}(\mathbf{U}_0)$ is a *normal* operator on $\mathbf{H}$ whose spectrum is the image, by the map $\overset{-1}{s}$, of the spectrum of $\mathbf{U}_0$, and thus lies on the boundary of the set S. We have in particular

$$T^n = (\overset{-1}{s}(T_0))^n = \overset{-1}{s^n}(T_0) = \operatorname{pr} \overset{-1}{s^n}(\mathbf{U}_0) = [\operatorname{pr} \overset{-1}{s}(\mathbf{U}_0)]^n = \operatorname{pr} \mathbf{N}^n$$

$(n = 0, 1, 2, \ldots)$, and the elements $\mathbf{N}^n h$ and $\mathbf{N}^{*n} h$ $(n = 0, 1, 2, \ldots; h \in H)$ span $\mathbf{H}$. This proves the following generalization of Theorem 4 (uniqueness may be proved directly, or by recursion to the uniqueness as asserted in Theorem 4):

Theorem 6. If the bounded Jordan set S is a spectral set of the operator T on Hilbert space H, then there exists a normal operator $\mathbf{N}$ on a Hilbert space $\mathbf{H}$ ($\supseteq H$) such that

(a) *the spectrum of $\mathbf{N}$ lies on the boundary of S,*

(b) $T^n = \operatorname{pr} \mathbf{N}^n$ $(n = 0, 1, 2, \ldots)$,

(c) **H** *is spanned by the elements* $\mathbf{N}^n h$ *and* $\mathbf{N}^{*n} h$ $(n = 0, 1, 2, ...; h \in H)$. *The 'structure'* $\{\mathbf{H}, \mathbf{N}, H\}$ *is determined by these requirements up to isomorphism.*

To get interesting applications one should remember, for example, that *any* bounded Jordan set, containing the unit disk S_0 as a subset, is a spectral set of *every* contraction operator.

On the basis of Theorem 6 it is then possible to build up a functional calculus for T *with respect to* S by deriving it from the functional calculus for the normal operator $\mathbf{N}$, in a similar way as it was done above for contraction operators. An alternative way is to *deduce* this functional calculus from that already existing for contraction operators, namely by putting by definition

$$u(T) = u \circ s^{-1}(T_0)$$

whenever $u \circ s^{-1} \in \mathcal{O}_{T_0}(S_0)$.

The harmonic spectral measure of T with respect to S (cf. § 1) is nothing else than the spectral measure of the normal operator $\mathbf{N}$. Many other results of the paper [2] appear thus in a more general setting.

REFERENCES

[1] Foiaş, C. Sur certains théorèmes de J. von Neumann concernant les ensembles spectraux. *Acta Sci. Math.* 18, 15–20 (1957).

[2] Foiaş, C. La mesure harmonique-spectrale et la théorie spectrale des opérateurs généraux d'un espace de Hilbert. *Bull. Soc. Math. Fr.* 85, 263–282 (1958).

[3] Halmos, P. R. Normal dilations and extensions of operators. *Summa bras. math.* 2, 125–134 (1950).

[4] Heinz, E. Ein v. Neumannscher Satz über beschränkte Operatoren im Hilbertschen Raum. *Göttinger Nachr.* pp. 5–6 (1952).

[5] Neumann, J. von. Eine Spektraltheorie für allgemeine Operatoren eines unitären Raumes. *Math. Nachr.* 4, 258–281 (1951).

[6] Neumark, M. Spectral functions of a symmetric operator. *Bull. Acad. Sci. U.R.S.S.*, Sér. Math., 4, 277–318 (1940). (Russian with English summary.)

[7] Neumark, M. Positive definite operator functions on a commutative group. *Bull. Acad. Sci. U.R.S.S.*, Sér. Math. 7, 237–244 (1943). (Russian with English summary.)

[8] Neumark, M. On a representation of additive operator set functions. *C.R. (Dokl.) Akad. Sci. U.R.S.S.* 41, 359–361 (1943).

[9] Schäffer, J. J. On unitary dilations of contractions. *Proc. Amer. Math. Soc.* 6, 322 (1955).

[10] Schreiber, M. Unitary dilations of operators. *Duke Math. J.* 23, 579–594 (1956).

[11] Schreiber, M. A functional calculus for general operators in Hilbert space. *Trans. Amer. Math. Soc.* 87, 108–118 (1958).

[12] Sz.-Nagy, B. Sur les contractions de l'espace de Hilbert. *Acta Sci. Math.* 15, 87–92 (1953).

[13] Sz.-Nagy, B. Transformations de l'espace de Hilbert, fonctions de type positif sur un groupe. *Acta Sci. Math.* 15, 104–114 (1954).

[14] Sz.-Nagy, B. Prolongements des transformations de l'espace de Hilbert qui sortent de cet espace. *Appendice au livre 'Leçons d'analyse fonctionnelle' par F. Riesz et B. Sz.-Nagy.* Budapest, 1955.

[15] Sz.-Nagy, B. Sur les contractions de l'espace de Hilbert. II. *Acta Sci. Math.* 18, 1–14 (1957).

[16] Sz.-Nagy, B. and Foiaş, C. Sur les contractions de l'espace de Hilbert. III. *Acta Sci. Math.* 19, 26–45 (1958).

AN APPLICATION OF THE MORSE THEORY TO THE TOPOLOGY OF LIE GROUPS

By RAOUL BOTT†

1. Introduction

In [3] Hans Samelson and I applied the Morse theory to the homology of symmetric spaces. Today I would like to report very briefly on a direct application of this theory to the stable homotopy of the classical groups[1], and to point out some unsolved questions.

2. The Freudenthal theorem for symmetric spaces

Our primary interest is in the classical compact groups; nevertheless, it is essential for our method to consider the larger family of compact symmetric spaces.

A compact homogeneous Riemannian manifold M is called symmetric if it admits an 'inverse operation', i.e. if M admits an isometry, keeping a point $P \in M$ fixed, and whose differential at P is -1.

These geometric generalizations of the compact groups seem to be the class of spaces to which the Morse theory is most applicable. The reason is that, on such a space, conjugate points have global implications:

2.1. *If P and Q are conjugate of degree k along the geodesic s, then s is contained in a k-manifold of geodesics joining P to Q.*

In general this proposition is only infinitesimally true.

The principal step towards the solution of our problem is the following Freudenthal-type theorem. Let M be a compact symmetric space and let $u = (P, Q)$ be a pair of points on M. Let $\Omega_u M$ denote the piecewise regular paths from P to Q on M, topologized as in [4], and set $S_u M$ equal to the set of geodesic segments in $\Omega_u M$. The set of geodesics of minimal length in $S_u M$ is denoted by M^u. I like to think of the step from M to M^u as an antisuspension; for instance, if M is the n-sphere S_n, and u is a pair of antipodes, then $M^u = S_{n-1}$.

Each $s \in S_u M$ has an index, $\lambda(s)$, equal to the number of conjugate points of P in the interior of s. We write $|u|$ for the *least positive number* occurring among the integers $\lambda(s)$; $s \in S_u M$. Finally, the composition $\pi_k(M^u) \to \pi_k(\Omega_u M) \to \pi_{k+1}(M)$ will be denoted by u_*.

† The author holds an A. P. Sloan fellowship.

Theorem I. Let M be a compact symmetric space, and let $u = (P, Q)$ be a pair of points on M. Then:

2.2. $u_* : \pi_k(M^u) \to \pi_{k+1}(M)$ *is bijective for* $0 < k < |u| - 1$.

2.3. M^u *is again a symmetric space.*

By virtue of (2.3), the process of antisuspending can be iterated, and we will call a sequence of symmetric spaces

$$M_1 \to M_2 \to M_3 \to \dots,$$

a u-sequence if each M_i is some component of M^u_{i+1} for some choice of u on M_{i+1}. For instance,

$$S_n \to S_{n+1} \to S_{n+2} \to \dots$$

is an unending u-sequence for which Theorem I yields the usual Freudenthal suspension theorem. The following are two new examples of u-sequences:

(A) $U_n \to U_{2n}/U_n \times U_n \to U_{2n} \to \dots,$

(B) $SO_n \to O_{2n}/O_n \times O_n \to U_{2n}/O_{2n} \to Sp_{2n}/U_{2n} \to Sp_{2n}$
$ \to Sp_{4n}/Sp_{2n} \times Sp_{2n} \to U_{8n}/Sp_{4n} \to SO_{16n}/U_{8n} \to SO_{16n} \to \dots.$

Here we have used the usual notation for the classical groups and certain of their homogenous spaces. Thus O_n denotes the group of $n \times n$ orthogonal matrices, U_n the unitary ones, and Sp_n the symplectic ones.

By a more or less explicit computation it can be shown that:

Theorem II. At each step of the sequences (A) and (B), the integer $|u|$ tends to $+\infty$ with n.

On the other hand, it is well known that π_k of each of the spaces occurring in (A) or (B) becomes independent of n for $n \gg k$. Hence Theorem I together with Theorem II yields recursion relations for these stable values of π_k. In particular one has the following corollary:

Corollary. The stable homotopy of the classical groups satisfies the relation

2.4.
$$\left.\begin{array}{l} \pi_k(U) = \pi_{k+2}(U) \\ \pi_k(O) = \pi_{k+4}(Sp) \\ \pi_k(Sp) = \pi_{k+4}(O) \end{array}\right\} \quad (k = 0, 1, 2, \dots).$$

Here we have suppressed the index n to denote stable homotopy.

The explicit computation of these groups is now a simple matter. One obtains 0, Z; for the period of $\pi_*(U)$ and Z_2, Z_2, 0, Z, 0, 0, 0, Z, for the period of $\pi_*(O)$.

Just a word about the proof. One obtains (2.2) analogously to the usual Freudenthal theorem. By means of the Morse theory one constructs a C.W. model for $\Omega_u M$ which consists of M^u with cells of dimension greater than or equal to $|u|$ attached. This is done in two steps. First, a model is found for the subsets of $\Omega_u M$, consisting of the loops of length less than a given number a. This model is a smooth manifold with boundary. On these models a function closely related to the length function takes on its absolute minimum on a homeomorphic image of M^u. Next, by using (2.1), the effect of the other critical points can be estimated. Now the desired C.W. model for $\Omega_u M$ is obtained by applying the Morse theory on manifolds.

The second part of Theorem I is obtained by a quite elementary argument concerning the midpoint of a minimal geodesic on a symmetric space. Theorem II is a routine computation in view of the fundamental conjugacy theorems of Cartan. By virtue of these, it is sufficient to study the geodesics joining P to Q on a maximal flat torus of M, and these are surveyed rather easily.

The result (2.4) was announced in [1], where a somewhat different proof was sketched. The present point of view seems more concise, and is the one which is adopted in a forthcoming paper.

3. Remarks and problems

3.1. In a sense Theorem I states that a first approximation to $\Omega_u M$ is given by M^u. By looking at the longer geodesics in $S_u M$, one can obtain better approximations. In general the second approximation is obtained by attaching a vector-bundle, ξ, to M^u. In the case of a sphere the attaching map of ξ to M^u becomes trivial after a single suspension, and it is rather natural to ask for the proper generalization of this fact.

3.2. In [3], the additive structure of $H^*(\Omega_u M; Z_2)$ is completely determined when M is a compact symmetric space. In fact we construct a gradation preserving isomorphism of $(S_u M)^*$ onto $H^*(\Omega_u M; Z_2)$, where $(S_u M)^*$ denotes the Z_2 module generated by the points of $S_u M$ ($u = (P, Q)$ a *general* pair of points in M !) and graded by the index. On the other hand we have no general description of $H^*(\Omega_u M; Z_p)$ for p an odd prime. It also seems likely that $\Omega_u M$ has no odd torsion.

The structure of $H^*(\Omega_u M; Z)$ as a Hopf-algebra is described in [2] when M is a group. A corresponding description in general is at present not even known mod 2.

REFERENCES

[1] Bott, R. The stable homotopy of the classical groups. *Proc. Nat. Acad. Sci.*, *Wash.*, 43, 933–935 (1957).

[2] Bott, R. The space of loops on a Lie group. To appear in the *Mich. J. Math.*

[3] Bott, R. and Samelson, H. Application of the Morse theory to the topology of symmetric spaces. To appear in the *Amer. J. Math.*

[4] Seifert, H. and Threlfall, W. *Variationsrechnung im Grossen.* B. G. Teubner, 1938.

ON SOME PROBLEMS CONNECTED WITH THE TOPOLOGY OF MANIFOLDS

By A. KOSIŃSKI

1. When a topologist has a topological property which applies to a space as a whole the first thing he does is to localize it, i.e. to consider the space in which every point has arbitrarily small neighbourhoods with the property P. In that way one obtains spaces which are more and more regular, therefore more adapted to an application of some special method.

In 1934 Borsuk applied another method to get a narrower class of spaces starting with a general notion. This method can be roughly described as follows. Instead of supposing that every point has a neighbourhood with the property P we shall suppose now that every point has an arbitrarily small neighbourhood such that its *complement* has P. If that is true for every point of a space M we shall say that M has the property P *generally* and this method will be described as the generalization of the property P. Of course this is only the general scheme and some modifications imposed by the nature of P may be necessary.

It turned out that generalized notions are rather strong ones, i.e. the classes of spaces defined by them are narrow.

By the generalization of the property 'to be acyclic in all dimensions and with any group of coefficients' Borsuk obtained a class of spaces (so-called H-spheroidal spaces) which resemble Euclidean spheres, and the spaces obtained by generalization of the notion of an absolute retract (spheroidal spaces) were hoped for some time to be spheres[3]. Actually it is not so: it may be shown that the suspension of any one of the Poincaré spheres is spheroidal although it fails to be a manifold. Nevertheless, the H-spheroidal spaces of dimension $\leqslant 2$ are spheres; and the question as to whether a 3-dimensional spheroidal manifold is a sphere remains open, and it may be easily proved that it is equivalent to the Poincaré hypothesis.

Borsuk's H-spheroidal spaces were obtained by generalization of the notion of a space acyclic in all dimensions and with respect to any group of coefficients. It may be supposed that some interesting spaces could be obtained by the generalization of acyclicity in some dimensions or with respect to a special group of coefficients only.

The first case to be considered is that of $H_0(X) = 0$, i.e. X is connected. If we consider compact connected and locally connected spaces which possess this property generally (in the sense that for every $p \in X$, $X - p$ is connected), then we find the well-known notion of a cyclic element introduced by Whyburn. Generalizing it again, i.e. considering spaces such that the complement of every point is not disconnected by any point, we get simply spaces which are not disconnected by any pair of points. Repeating this process we get spaces which are not disconnected by any finite set of points. We will call them *strongly cyclic elements* (s.c.e.) and *true s.c.e.* if they contain more than one point.

Most interesting results connected with Whyburn's cyclic elements lie in the possibility of decomposition of a locally connected (l.c.) continuum into cyclic elements. There is no such decomposition into s.c.e. in the general case, but it exists if we suppose the space to be 1-l.c.

It turned out that it is possible to form a very satisfactory theory of decomposition of 1-l.c. continua into s.c.e. This decomposition resembles that into cyclic elements. For instance, true s.c.e. are countable and intersect in finite sets of points which are local separating points. Furthermore, they are 1-l.c. continua containing only a finite number of local separating points, they are not disconnected by any 0-dimensional subset, and through every finite subset of a true s.c.e. passes a simple closed curve. It may also be proved that true s.c.e. contain for every two points a curve $\Theta_{a,b}^n$ (that is the sum of n arcs L_i with extremities a, b and such that $L_i \cap L_j = a \cup b$ for $i \neq j$), and that they are maximal with respect to this property.

This is what we know about the generalization of connectedness. Little is known about the next case, that of $H_1(X) = 0$. If we consider homology with rational coefficients, this reduces (in case of l.c. spaces) to the study of the generalization of the unicoherence property. It is easy to see that a 2-dimensional polyhedron which is generally unicoherent is unicoherent and 'closed', that is each of its 2-simplexes appears in a 2-cycle with a non-zero coefficient. But this is not the sufficient condition to be generally unicoherent.

It can perhaps be expected that spaces obtained by generalization of vanishing of higher homology groups may serve to obtain a decomposition analogous to that into cyclic elements.

The principal tool in handling the generalized homological notions is the Mayer–Vietoris exact sequence. Since no analogue exists in homotopy theory, little is known about the generalization of homotopical notions. It may be hoped, however, that by an application of some recent results, e.g. the triad homotopy groups of Blakers and Massey, one could get some information about them.

2. It is clear that a manifold has 'generally' the same homotopy type. However, if one looks for a topological characterization of, say, 2-manifolds by a direct generalization of some homological or homotopical notions, some difficulties arise. For instance the property $H_1(X) =$ infinite cyclic is enjoyed generally by the square (because we can suitably choose neighbourhoods), and also by the space obtained from a torus by pinching a meridian to a point, and also by the projective 2-space. These difficulties led to the following construction:

Let X be a compact space and F a closed subspace. We shall consider the following 'general' property P_k of $X - F$:

(P_k)   every point $p \in X - F$ has arbitrarily small neighbourhoods U such that $H_k(F) \to H_k(X - U)$ is onto.

(Čech homology with coefficients so chosen as to assure the exactness of the Mayer–Vietoris sequence.)

It can be shown that, if $X - F$ satisfies P_k for $k = n$, $n - 1$ where $n = \dim(X - F)$, then in $X - F$ there holds the theorem about the invariance of domain, the sets that disconnect $X - F$ are characterized by non-trivial $(n-1)$-homology group, and the n-dimensional subsets, and only these, contain subsets open in $X - F$. These are strong properties and if $n < 3$ then $X - F$ must be locally Euclidean.

Now, we may localize the property P_k by supposing, for instance, that every point of a space X has arbitrarily small neighbourhoods U such that the pair $(U, \mathrm{fr}\,(U))$ satisfies P_k. Spaces so obtained, with suitably chosen k, will behave like manifolds and in dimensions < 3 will actually be manifolds. One kind of such spaces was considered in [4], they were called r-spaces and their definition is as follows: a compact, connected, finite dimensional space is called an r-space if any point has arbitrarily small neighbourhoods U with the property (R):

(R)   for every $q \in U$, $\mathrm{fr}\,(U)$ is a deformation retract of $\overline{U} - q$.

Therefore the pair $(\overline{U}, \mathrm{fr}\,(\overline{U}))$ satisfies P_k for all k and it follows that r-spaces have all properties mentioned above, they are manifolds if they are of dimension < 3 and if they are polyhedra of dimension 3. If they are polyhedra then they are homology manifolds; even more: the boundary complex of every point is of the homotopy type of the n-sphere.

It is not known whether r-spaces are manifolds. It is probable that the space B of a certain decomposition of S_3 constructed by Bing is an r-space although it is not a manifold (see [1]). On the other hand, it may be proved that the 5-dimensional space obtained from the Poincaré space by repeating the suspension twice is an r-space although probably it is not a manifold.

3. Up to this moment the property (R) served to obtain very strong regularity conditions. However, specializing this notion a little one may obtain some notions which permit consideration of a more general class of spaces than that of manifolds. For this purpose we shall introduce the notion of a *relative r-point*. We shall say that p is a relative r-point of X rel. to a subset A of X, if p has arbitrarily small neighbourhoods satisfying (R), but only for q belonging to A. The r-points of X rel. to X will be called absolute r-points. For instance, interior points of a square are absolute r-points while the boundary points, though not absolute r-points, are r-points of the square relative to its boundary. With the aid of this notion we can obtain an invariant characterization of 2-manifolds with boundary, 2-dimensional pseudo-manifolds and also of 2-dimensional polyhedra[5].

Let us consider this last characterization in some detail. For every n-dimensional compact space K we introduce a sequence of subspaces as follows:

K_1 will be the set of those absolute r-points p of K in which $\dim_p K = n$;

K_2 will be the set of r-points of K rel. $K - K_1$;

......

K_i, $i > 2$, will be the set of r-points of K rel. $K - \bigcup_{j < i} K_j$.

Now let us introduce the following inductive definition: a 0-dimensional space will be called an r-polyhedron if it is finite. An n-dimensional compact space, $n > 0$, will be called an n-dimensional r-polyhedron if it is an ANR and if the above sets K_i satisfy the conditions:

(a) $K = \bigcup K_i$;

(b) $K - \bigcup_{j \leqslant i} K_j$ is an r-polyhedron of dimension $\leqslant n - i$.

The main theorem about r-polyhedra is that an r-polyhedron of dimension < 3 can be triangulated.

Let us observe incidentally that the above characterization of 2-dimensional polyhedra allows us to prove the following theorem: if $A \times B$ is a polyhedron and $\dim A < 3$ then A is a polytope[6]. It was formerly supposed that the restriction $\dim A < 3$ is due only to the lack of topological characterization of polyhedra of dimension > 2. But the above-mentioned space B of Bing has the property that $B \times S_1$ is topologically equivalent to $S_3 \times S_1$,[2].

4. Now let us go back to the theorem on the invariance of domain and of cuttings. It may be formulated as follows:

Let $M \subset X$ and let $f: M \to X$. If

(a) M is the Euclidean n-space,

(b) f is a homeomorphism,

then $f(M)$ is open (disconnects X) if and only if M is (does).

It is quite natural to ask what are the possible generalizations of the condition (a). Our results above concerning r-spaces may be considered as partial answers to that problem. But it may also be asked about the possible generalization of the condition (b), that is what class of mappings broader than that of homeomorphisms preserve open sets, disconnecting sets, etc. There are some results in that direction too. Sitnikov considered mappings $f: G \to S_n$ where G is an open subset of S_n with the property that, for every $y \in f(G)$, $\delta[f^{-1}(y)] \to 0$ if $\rho[f^{-1}(y), \mathrm{fr}\,(G)] \to 0$, and proved[8] that in this case $f(G)$ is open and has the same homology properties as G. It would be interesting to prove Sitnikov's theorems in a more direct manner, avoiding the use of triangulations. Many other results could follow.

It is not known whether A and $f(A)$ are homeomorphic if f is a Sitnikov mapping. It would be interesting to know other kinds of mappings characterized by simple geometric properties which preserve open sets although the image is not necessarily homeomorphic to the space.

Concerning the invariance of cuttings it is natural to expect that mappings that preserve in some way the boundary of M will preserve the cutting properties. It may be proved for instance that if A, B are compact subsets of S_n, $f: (A, \mathrm{fr}\,(A)) \to (B, \mathrm{fr}\,(B))$ and $f(\mathrm{int}\,(A)) = \mathrm{int}\,(B)$, and if f maps homeomorphically $\mathrm{fr}\,(A)$ onto $\mathrm{fr}\,(B)$ then f induces an isomorphism of the homology sequence of $(A, \mathrm{fr}\,(A))$ onto that of $(B, \mathrm{fr}\,(B))$. (Some other results in this direction are gathered in [7].) In that case A and B are not necessarily homeomorphic.

It seems probable that the above theorem holds with homological conditions only, that is $f(\mathrm{int}\,(A)) = \mathrm{int}\,(B)$ and $f \mid \mathrm{fr}\,(A)$ induces an isomorphism of the homology groups of $\mathrm{fr}\,(A)$ onto those of $\mathrm{fr}\,(B)$.

Some information about mappings which preserve the cutting properties may also be obtained using Borsuk's characterization of sets which disconnect E^n by the existence of essential mappings onto S^{n-1}.

REFERENCES

[1] Bing, R. H. A decomposition of E^3 into points and tame arcs such that the decomposition space is topologically different from E^3. *Ann. Math.* (2), 65, 484–500 (1957).

[2] Bing, R. H. The cartesian product of a certain non-manifold and a line is E^4. *Bull. Amer. Math. Soc.* 64, 82–84 (1958).

[3] Borsuk, K. Über sphäroidale und H-sphäroidale Raume. *Mat. Sbornik*, 1 (43), 643–660 (1936).

[4] Kosiński, A. On manifolds and r-spaces. *Fund. Math.* 42, 111–124 (1955).

[5] Kosiński, A. A topological characterization of 2-polytopes. *Bull. Acad. Pol. Sci.* (Cl. III), 2, 321–323 (1954).

[6] Kosiński, A. On 2-dimensional topological divisors of polytopes. *Bull. Acad. Pol. Sci.* (Cl. III), 2, 325–328 (1954).

[7] Kosiński, A. On mappings which satisfy certain conditions on boundary. *Bull. Acad. Pol. Sci.* (Cl. III), 4, 335–340 (1956).

[8] Sitnikov, K. A. On continuous mappings of open sets of Euclidean space. *Mat. Sbornik*, 31 (73), 439–458 (1952). (In Russian.)

THE THEORY OF THREE-DIMENSIONAL
MANIFOLDS SINCE 1950†

By C. D. PAPAKYRIAKOPOULOS

The spaces of the branch under consideration are 3-dimensional manifolds, and its problems will be understood by the sequel. The restriction
to the dimension 3 is needed, for the time being, because of the difficulty
of the problems. As an indication of this difficulty it may serve that,
though some of the theorems explained below were well-known problems
and conjectures, formulated some decades ago, nevertheless it was
only during the past eight years that their proof was attained. During
this last period new techniques have been developed. It is hoped that
the branch will continue to grow and for dimensions > 3.

An *n-manifold* (*n*-dimensional manifold, $n \leqslant 3$) is a connected separable metric space, each of whose points has a closed neighborhood
homeomorphic to a closed *n*-cell. So we consider both manifolds with
boundary and manifolds without boundary. (Our definition is a bit
different from that of Bing ([2], p. 145, [3]; p. 456) and Moise ([13], V, p. 96).)
A *closed n*-manifold is a compact *n*-manifold without boundary.

We say that a topological space is *triangulable* if it admits a simplicial
subdivision, in the sense of Seifert–Threlfall ([18], p. 42). A homeomorphism
f of a simplicial complex K_1 into another K_2 is called *semi-linear* if there
are simplicial subdivisions K_1' and K_2' of them, such that each simplex
of K_1' is mapped by f linearly onto some simplex of K_2'. The image $f(K_1')$
is called a *polyhedron* in K_2.

In 1950 appeared the paper of Graeub, where a number of theorems
about semi-linear homeomorphisms are proved and the following
theorem, refining *Alexander's theorem*, is proved too ([7], p. 224, Satz 1).
Another proof of the following theorem was given by Moise ([13], II,
p. 172, Theorem 1).

(1) *If S^2 is a polyhedral 2-sphere in euclidean 3-space E^3, then there
exists a semi-linear homeomorphism of E^3 onto itself, throwing S^2 onto the
boundary of a rectilinear 3-simplex of E^3.*

In 1951 appeared the first of the series of Moise's papers. The main
result in this paper is the following *separation theorem* ([13], I, p. 506,
Theorem 1).

(2) *Let R be a set in E^3, which is homeomorphic to the cartesian product of*

† Read by Professor J. H. C. Whitehead.

TP

a closed 2-manifold N and a closed interval. Then there exists a polyhedral 2-manifold P homeomorphic to N, such that $R - P$ is the union of two disjoint open subsets of R, each of which contains one of the two components of the boundary of R.

The above follows from a certain theorem ([13], I, p. 506, Theorem 0), which is the key theorem in Moise's work, and whose proof is rather geometric. In this proof is made use of a certain upper semi-continuous collection in R, and at several crucial places use is made of Vietoris's homology theory, especially Pontrjagin's duality theorem.

Two simplicial complexes K_1 and K_2 are called *combinatorially equivalent* if there are simplicial subdivisions K_1' and K_2' of them, and a one-to-one correspondence between the simplexes of K_1' and those of K_2' preserving incidence relations. If the closed star of every vertex of a simplicial complex K is the image of a rectilinear 3-simplex of E^3 under a semi-linear homeomorphism, we then say that K is a *combinatorial 3-manifold*.

In 1952 appeared the fifth paper of Moise, which is his main one. There, using Alexander's theorem (1), Pontrjagin's duality, and a rather geometric lemma 'On the fitting together of homeomorphisms' ([13], V, p. 101), he proves the following ([13], V, p. 97, Theorem 1).

(3) *Every triangulated 3-manifold is a combinatorial 3-manifold.*

Then using (2), Alexander's theorem (1), and the lemma 'On the fitting together of homeomorphisms' he proves the following *approximation theorem* ([13], V, p. 97, Theorem 2).

(4) *Let M and M' be triangulated 3-manifolds without boundary, let U be an open subset of M, let f be a homeomorphism throwing U into M', and let ϕ be a continuous positive function defined on U. Then there exists a homeomorphism f' throwing U into M', such that f' is semi-linear over every finite polyhedron in U, and for each $p \in U$ the distance*

$$\rho(f(p), f'(p)) < \phi(p).$$

From (4) follows easily the *triangulability* (5) and *Hauptvermutung* (6) for 3-manifolds *without boundary* ([13], V, p. 96, Theorems 3 and 4).

(5) *Every 3-manifold is triangulable.*

(6) *If the simplicial complexes M_1 and M_2 are homeomorphic 3-manifolds, then they are combinatorially equivalent.*

Finally from (4) follows easily the following ([13], V, p. 97, Theorem 7).

(7) *The separation theorem (2) holds if we replace E^3 by any triangulated 3-manifold M.*

Let Q be a set in a triangulated 3-manifold M. We say that Q is *tamely imbedded* in M if there is a homeomorphism of M onto itself that throws Q onto a polyhedron, otherwise we say that Q is *wildly imbedded* in M. We say that Q is *locally tamely imbedded* in M if for each point p of Q there is an open neighborhood U of p in M and a homeomorphism f_p of $\overline{U}$ onto a polyhedron in M, such that $f_p(\overline{U} \cap K)$ is a polyhedron.

In 1954 appeared the paper of Bing on the taming of locally tame sets[2]. This paper is based on Moise's fifth paper. In Bing's paper was proved for the first time the triangulability (5) and the Hauptvermutung (6) for any 3-manifold *with boundary* ([2], p. 151, Theorem 5, and p. 150, Corollary 1). Actually the triangulability is proved in the following form.

(8) *Given any triangulation of the boundary of M, there exists a subdivision of it which can be extended to a triangulation of M.*

Bing's paper contains also the following ([2], p. 154, Theorem 8, and p. 157, Theorem 9).

(9) *If Q is a locally tame closed subset of a triangulated 3-manifold, then M has a triangulation under which Q is a polyhedron.*

(10) *Each locally tame closed subset Q of a triangulated 3-manifold M is tame.*

The proof of (10) is based on that of (9). As far as the proof of (9) is concerned, Bing proceeds thus: he first proves (9) in the special cases where Q is a 2-manifold and an arc and then, using these results, (8) and (4), he proves (9).

The theorems (9) and (10) have been preceded by some theorems of Moise, of the same kind but special ones ([13], II, p. 172, Theorem 2; V, p. 97, Theorem 5; VII, p. 403, Theorems 1 and 2).

In 1954 appeared also the eighth paper of Moise, which contains some results similar to those of Bing, mentioned above ([13], VIII, p. 159, Theorems (9.1) and (9.2)).

A surface, i.e. 2-manifold, N in a triangulated 3-manifold M is called *locally polyhedral* at a point p if there is an open set U in N containing p, such that $\overline{U}$ is a finite polyhedron.

In 1957 appeared the paper of Bing, where the following *approximation theorem for surfaces* is proved ([3], p. 478, Theorem 7).

(11) *Let N be a 2-manifold, in a triangulated 3-manifold M, and let ϕ be a non-negative continuous function on N. Then there exists a 2-manifold N', and a homeomorphism f of N onto N', such that N' is locally polyhedral at $f(p)$ if $\phi(p) > 0$, and $\rho(p,f(p)) \leqslant \phi(p)$, for any $p \in N$.*

This theorem plays a fundamental role in Bing's work, and its proof is geometric. He proves it first for the case where $N = S^2$ and $M = E^3$, and then he generalizes his method to any N and M. At a crucial place in his argument he makes use of a well-known theorem of Moore, concerning upper semi-continuous decompositions of E^2.

This is a suitable place to mention that the theorem of Moore does not generalize to E^3, as Bing showed with a counter example ([5], p. 484).

Using (11) Bing has proved, in a paper to appear soon[4], the following *approximation theorem for polyhedra.*

(12) *Let M be a triangulated 3-manifold, let P be a closed set in M, which is the homeomorphic image of a simplicial k-complex ($k \leqslant 3$), and let ϕ be a positive continuous function defined on P. Then there exists a homeomorphism f of P onto a polyhedron P' in M, such that*

$$\rho(p, f(p)) < \phi(p) \quad \text{for any } p \in P.$$

Using (12) for $k = 2$, Bing proves the triangulability (5) for *any* 3-manifold. Moreover, using (12) for $k = 2$, and Alexander's theorem (1), he proves the following theorem, from which follows the Hauptvermutung (6) for *any* 3-manifold.

(13) *Let M be a 3-manifold triangulated in two ways, and let ϕ be a continuous positive function defined on M. Then there exists a semi-linear homeomorphism f of M under the first triangulation onto M under the second one, such that $\rho(p, f(p)) < \phi(p)$ for any $p \in M$.*

Theorem (7) follows immediately from (11). Moreover, in (7) the imbedding of the closed 2-manifold N in M is locally tame, and therefore tame by (10), while in (11) the imbedding of the arbitrary 2-manifold N in M is not necessarily tame, it may be *wild*. We also emphasize that in (12) the imbedding of P in M may be *wild*. However, we mentioned all these theorems so that the reader may see how things have been developed gradually.

In a 1957 paper of Munkres[14], it is proved that 'locally triangulable spaces are triangulable, for dimensions $\leqslant 3$'. The spaces under consideration are more general than manifolds. His method makes use of some results of Bing and Moise.

Sanderson, in his thesis[17], approximated isotopic deformations of E^3 by semi-linear ones. His method makes use of Moise's work.

Several theorems have been proved concerning 'the taming of closed sets', especially of curves and surfaces, in 3-manifolds. These things are delicate, they require special definitions, and we refer the reader to Harrold, Griffith and Posey[10], Griffith[8], and especially to the detailed exposition of Harrold[9].

By a polygonal *knot* in a triangulated 3-manifold M we mean a simple closed polygonal curve in M. The knot is called *unknotted* in M if it is the boundary of a polyhedral disk (without singularities, i.e. self-intersections) in M.

In 1957 appeared my paper where the following, known as *Dehn's lemma*, is proved ([15], p. 1).

(14) *Let M be a triangulated 3-manifold, and in M let D be a disk with singularities, having as boundary the polygonal knot C, and such that no point of C is singular. Then C is unknotted in M.*

In the same paper of mine the following *sphere theorem* is proved, but under certain 'additional hypotheses' ([15], p. 1).

(15) *Let M be an orientable triangulated 3-manifold, such that $\pi_2(M) \neq 0$. Then there exists a polyhedral 2-sphere (without singularities) S^2 in M, such that S^2 is non-contractible in M.*

The proofs of (14) and the 'qualified' (15), explained in my paper, are parallel to one another. We make use of covering spaces, and at a crucial place we use a theorem due to Seifert ([18], p. 223, Satz IV). We also use a geometric operation introduced by Dehn, the 'Umschaltung'. The methods are geometric and algebraic ([16], nos. 8, 9 and 19). The sphere theorem does not hold for non-orientable 3-manifolds, as the example $S^1 \times P^2$ shows, where P^2 is the real projective plane.

A consequence of Dehn's lemma (14) is the following ([15], p. 19, Theorem (28.1)).

(16) *A polygonal knot C in S^3 is unknotted if and only if $\pi_1(S^3 - C)$ is free cyclic.*

From the sphere theorem (15) follows the theorem below, which provides us with a solution of a problem of Whitehead ([15], p. 18, Theorem (26.1)).

(17) *A non-empty proper open connected subset U of S^3 is aspherical if, and only if, any 2-sphere in U bounds a 3-cell belonging to U.*

An immediate consequence of (17) is the following theorem, which solves a problem of Eilenberg ([15], p. 19, Theorem (26.2)).

(18) *If F is a non-empty proper closed subset of S^3, then each component of $S^3 - F$ is aspherical.*

From (18) follows immediately the *asphericity of knots*, i.e. that the complement of any polygonal knot in S^3 is aspherical. This was proved in a 1956 paper of Aumann[1], in the special case where the knot is 'alternating'.

Using the sphere theorem (15), and a theorem due to Smith, we proved the following ([15], p. 22, Corollary (31.7)).

(19) *If U is an open connected subset of an orientable 3-manifold M, such that $\pi_1(M)$ is a free group, then $\pi_1(U)$ has no element of finite order.*

In the special case where $M = S^3$, (19) provides us with a solution of a conjecture due to Hopf.

Shortly after the appearance of my paper appeared a paper of Homma[11], where he proved Dehn's lemma for the special case $M = S^3$, by a method different from mine.

In a forthcoming paper of Whitehead and Shapiro[20] a simplified proof of Dehn's lemma is given. There is proved also an extension of the lemma, for Dehn disks with more than one boundary curve ([15], p. 24, Problem 2; [20], p. 174, Theorem (1.1)).

As we have mentioned above, this author proved the sphere theorem (15) under certain 'additional hypotheses', i.e. he proved a 'qualified' sphere theorem. Whitehead, modifying this author's proof, succeeded in proving the sphere theorem in full generality ([19], p. 161, Theorem (1.1); [16], no. 9). In the same paper Whitehead, using the sphere theorem, proves the following ([19], p. 165, Theorem (3.6)).

(20) *Let M be a compact triangulated 3-manifold. Then there exists a finite number of elements $a_1, ..., a_r$ $(r \geqslant 0)$ of $\pi_2(M)$, which are represented by disjoint non-contractible polyhedral 2-spheres (without singularities) in M, such that $a_1, ..., a_r$ are $\pi_1(M)$-generators of $\pi_2(M)$, i.e. $\pi_2(M)$ has to be considered as a group with operators the elements of $\pi_1(M)$, in the well-known way.*

One of the main problems in 3-manifolds is the *classification problem* of closed 3-manifolds, i.e. to define an infinite sequence of closed 3-manifolds $M_1, M_2, M_3, ...$, such that any two of these are not homeomorphic, but any closed 3-manifold is homeomorphic with one of them ([16], §§1 and 5).

Let M', M'' be two *oriented* triangulated closed 3-manifolds, and let E', E'' be the interiors of two polyhedral 3-cells in M', M'' respectively. Matching the boundaries of $M' - E'$ and $M'' - E''$, in the proper way, we obtain a new oriented closed 3-manifold M, called the *composition of* M' and M'', and denoted by $M = M' \sharp M''$. *Homotopy sphere* means a simply connected closed 3-manifold. The oriented closed 3-manifolds M_1 and M_2 are called *congruent* if there exist oriented homotopy spheres S_1 and S_2, and appropriate triangulations, such that $M_1 \sharp S_1$ and $M_2 \sharp S_2$ are *isomorphic* (i.e. there is an orientation preserving homeomorphism). An oriented closed 3-manifold M, which is not a homotopy sphere, is called *decomposable* if M is congruent to $M' \sharp M''$, where neither M' nor M'' is a homotopy sphere. M is called *indecomposable* if it is not decomposable.

Using the sphere theorem (15), Milnor proved the following theorems[12].

(21) *Every oriented closed 3-manifold, which is not a homotopy sphere, is isomorphic to a composition of indecomposable 3-manifolds, which are unique up to order and congruence.*

(22) *Every indecomposable 3-manifold is either congruent to an oriented $S^1 \times S^2$, or is aspherical, or has a non-trivial finite fundamental group.*

From (22), the problem naturally arising is: what are the oriented closed 3-manifolds, which are either aspherical, or whose fundamental group is non-trivial and finite? From (21) the problem arising is: what are the homotopy spheres? The answer to the last question is given by the following well-known *Poincaré conjecture.*

(23) *Any homotopy sphere is a 3-sphere.*

This conjecture is still an open question; however, the following has been proved by Bing ([6], Theorem 1).

(24) *A closed 3-manifold is a 3-sphere if each loop in M lies in a homeomorphic image of a 3-cell in M.*

The theorems (21) and (22) have been preceded by the following theorem of mine ([15], p. 23, Theorem (32.1)).

(25) *If the Poincaré conjecture is true, then any orientable closed 3-manifold, whose fundamental group is a free group on h (> 0) free generators, is the composition of h copies of $S^1 \times S^2$.*

These are some of the many theorems on 3-manifolds proved during the past eight years.

REFERENCES

[1] Aumann, R. J. Asphericity of alternating knots. *Ann. Math.* (2), 64, 374–392 (1956).
[2] Bing, R. H. Locally tame sets are tame. *Ann. Math.* (2), 59, 145–158 (1954).
[3] Bing, R. H. Approximating surfaces with polyhedral ones. *Ann. Math.* (2), 65, 456–483 (1957).
[4] Bing, R. H. An alternative proof that 3-manifolds can be triangulated. *Ann. Math.* (2), 69, 37–65 (1959).
[5] Bing, R. H. A decomposition of E^3 into points and tame arcs such that the decomposition space is topologically different from E^3. *Ann. Math.* (2), 65, 484–500 (1957).
[6] Bing, R. H. Necessary and sufficient conditions that a 3-manifold be S^3. *Ann. Math.* (2), 68, 17–37 (1958).
[7] Graeub, W. Die semilinearen Abbildungen. *S. B. Heidelberg Akad. Wiss. (Math.-nat. Kl.)*, 205–272 (1950).
[8] Griffith, H. C. The enclosing of cells in 3-space by simple closed surfaces. *Trans. Amer. Math. Soc.* 81, 25–48 (1956).
[9] Harrold, O. G. Locally tame curves and surfaces in three-dimensional manifolds. *Bull. Amer. Math. Soc.* 63, 293–305 (1957).
[10] Harrold, O. G., Griffith, H. C. and Posey, E. E. A characterization of tame curves in three-space. *Trans. Amer. Math. Soc.* 79, 12–34 (1955).

[11] Homma, T. On Dehn's lemma for S^3. *Yokohama Math. J.* 5, 223–244 (1957).

[12] Milnor, J. W. A unique decomposition theorem for 3-manifolds. (To appear).

[13] Moise, E. E. Affine structures in 3-manifolds. I. *Ann. Math.* (2), 54, 506–533 (1951); II, *ibid.* 55, 172–176 (1952); V, *ibid.* 56, 96–114 (1952); VII, *ibid.* 58, 403–408 (1953); VIII, *ibid.* 59, 159–170 (1954).

[14] Munkres, J. The triangulation of locally triangulable spaces. *Acta Math.* 97, 67–93 (1957).

[15] Papakyriakopoulos, C. D. On Dehn's lemma and the asphericity of knots. *Ann. Math.* (2), 66, 1–26 (1957).

[16] Papakyriakopoulos, C. D. Some problems on 3-dimensional manifolds. *Bull. Amer. Math. Soc.* 64, 317–335 (1958).

[17] Sanderson, D. E. Isotopic deformations in 3-manifolds. Thesis, 1953. University of Wisconsin.

[18] Seifert, H. and Threlfall, W. *Lehrbuch der Topologie.* Teubner, Leipzig–Berlin, 1934.

[19] Whitehead, J. H. C. On 2-spheres in 3-manifolds. *Bull. Amer. Math. Soc.* 64, 161–166 (1958).

[20] Whitehead, J. H. C. and Shapiro, A. S. A proof and extension of Dehn's lemma. *Bull. Amer. Math. Soc.* 64, 174–178 (1958).

DIFFERENTIAL GEOMETRY AND
INTEGRAL GEOMETRY

By SHIING-SHEN CHERN

Integral geometry, started by the English geometer M. W. Crofton, has received recently important developments through the works of W. Blaschke, L. A. Santaló, and others. Generally speaking, its principal aim is to study the relations between the measures which can be attached to a given variety. It is my purpose in the present paper to discuss the services it can render to some problems in differential geometry.

1. Measure of the spherical image of a closed submanifold in Euclidean space

A submanifold of dimension n in a Euclidean space E^{n+N} of dimension $n+N$ is given by an abstract differentiable manifold M^n of dimension n and a differentiable map $x\colon M^n \to E^{n+N}$, whose Jacobian matrix has everywhere the rank n. We say that M^n is imbedded, if the map x is one–one, i.e. if $x(M^n)$ does not intersect itself. All the unit normal vectors of $x(M^n)$ form a bundle of spheres of dimension $N-1$ over M^n and constitute a manifold B_ν of dimension $n+N-1$. If O is a fixed point of E^{n+N} and S_0 the unit hypersphere with origin O, there is a mapping $T\colon B_\nu \to S_0$ which maps a unit normal vector of M^n into the end-point of the unit vector through O and parallel to it. T is a generalization of the normal mapping of Gauss in the theory of surfaces.

Suppose from now on that M^n is compact. Then B_ν is also compact, and we define as the *total curvature* of $x(M^n)$ the volume of the image $T(B_\nu)$ divided by the volume of S_0 itself, each point of $T(B_\nu)$ counted a number of times equal to the number of points mapped into it. This total curvature we will denote by $T_x(M^n)$. It is in a sense a measure of the curvedness of the submanifold. For a closed space curve, for instance, its total curvature is, up to a constant factor, the integral of the absolute value of the curvature.

Concerning the total curvature Lashof and I [3, 4] proved the following theorems:

(1) The total curvature $T_x(M^n)$ is greater than or equal to the sum of the Betti numbers of M^n relative to an arbitrary coefficient field. As a corollary it follows that $T_x(M^n) \geqslant 2$, a result which can be derived

directly by an elementary argument on the maxima and minima of the co-ordinate functions on M^n.

(2) If $T_x(M^n) < 3$, M^n is homeomorphic to a sphere. (This result was proved also by Milnor[8].)

(3) $x(M^n)$ is a convex hypersurface imbedded in a subspace of dimension $n+1$, if and only if $T_x(M^n) = 2$.

A basic reason for these theorems is the existence of the large number of co-ordinate functions on M^n. Morse's critical point theory then furnishes one of the essential tools in the proofs.

For a differentiable manifold abstractly given, one is led to study the immersions for which $T_x(M^n)$ is as small as possible. Two questions naturally arise: (a) What is the minimum value of $T_x(M^n)$, expressed in terms of M^n itself only, for all possible immersions x? (b) To characterize the immersions x for which the total curvature attains this minimum value. Theorem 3 answers these questions for the case when M^n is homeomorphic to a sphere.

For general compact manifolds very little is known about the two questions. Theorem 1 implies that $T(M^n) = \min_x T_x(M^n)$ is greater than or equal to the sum of the Betti numbers mod 2 of M^n. There are sufficient indications to support the conjecture that $T(M^n)$ is equal to the minimum number of cells by which M^n can be subdivided into a cell complex, but the truth of this remains undecided.

As for Question (b), a conjecture of N. H. Kuiper says that if M^n is immersed in E^{n+N} with the minimum total curvature $T(M^n)$, then $N \leqslant \frac{1}{2}n(n+1)$. Some necessary conditions are known when M^n is a hypersurface ($N = 1$), and has minimum total curvature.

A simple case is when M^n is a closed space curve ($n = 1$, $N = 2$). Then Theorem 2 can be sharpened to the following form (Fary[5] and Milnor[8]): A closed space curve with total curvature < 4 is unknotted. The theorem thus gives a simple necessary condition for a knot in space.

Another application is the following consequence of the above Theorem 3: a closed surface immersed in ordinary Euclidean space with Gaussian curvature $K \geqslant 0$ is an imbedded convex surface. Under the stronger assumption that $K > 0$ the conclusion follows from a well-known argument of Hadamard. It may be of interest to remark that a similar statement is not valid in higher dimensions; there are examples of non-convex closed hypersurfaces in a Euclidean space of four or higher dimensions whose Gauss–Kronecker curvature is everywhere non-negative.

2. Measure of the image of a complex analytic mapping

Entirely analogous to the theory of submanifolds in Euclidean space is that of complex analytic submanifolds in a complex projective space. Let M_n be a complex manifold of (complex) dimension n and $Z: M_n \to P_{n+N}$ be a complex analytic mapping of M_n into the complex projective space P_{n+N} of dimension $n + N$. The study of such mappings includes as particular cases various classical theories. In fact, if M_n is compact, $Z(M_n)$ is an algebraic variety. If M_n is the complex Euclidean line E_1 (or the Gaussian plane, as it is commonly called), the complex analytic mapping $Z: E_1 \to P_{1+N}$ defines a meromorphic curve in the sense of H. Weyl, J. Weyl and Ahlfors. In particular, the notion of the complex analytic mapping $Z: E_1 \to P_1$ is identical with that of a meromorphic function defined in the Gaussian plane.

Starting with the classical theorem of Picard, a main problem in such investigations is the determination of the maximum size of the set of linear spaces of dimension N, which will be disjoint with the image $Z(M_n)$. For meromorphic curves a satisfactory solution is provided by the following theorem of E. Borel: Suppose that the meromorphic curve is non-degenerate (i.e. that it does not lie in a hyperplane of P_{1+N}). Given $N + 3$ hyperplanes in general position, the image $Z(E_1)$ meets one of them. Obviously this theorem contains as a particular case the theorem of Picard that an entire function in the Gaussian plane omits at most one value.

That the theory is mainly geometrical can be justified by the following generalization of Borel's theorem, which follows easily from results of Ahlfors: Let $Z: E_1 \to P_{1+N}$ be a non-degenerate meromorphic curve. Given $\binom{N+2}{k+1} + 1$ linear spaces of dimension $N - k$ in general position, $0 \leqslant k \leqslant N$, one of them must meet an osculating linear space of dimension k of the curve.

In the establishment of these and related results, integral geometry plays a role on at least two occasions. Although the theorems relate only to the incidence of the curve with the linear subspaces, it is necessary to use the elliptic Hermitian metric in P_{n+N}. Then, for compact M_n, $Z(M_n)$ has a finite volume and this volume is, up to a numerical factor, equal to the order of the algebraic variety. This identification of volume and order, of sufficient interest in the compact case, will be of paramount importance in the case when M_n is non-compact. For then the notion of order does not exist, while the volume does. As it turns out, the volume does fulfil many of the functions of the order.

Since a non-compact manifold will be exhausted by a sequence of expanding polyhedra with boundaries, we are led to the study of a complex analytic mapping $Z: M_n \to P_{n+N}$, where M_n is compact and is with or without boundary. The first problem is the following: Given a generic linear space L of complementary dimension N, to determine the difference between the number of points of intersection of L and $Z(M_n)$, each counted with its proper multiplicity, and the volume of $Z(M_n)$. This problem was solved by Levine[7], who expressed the difference as an integral over the boundary ∂M_n of M_n. His result can be stated as follows:

Let $Z = (z_0, z_1, ..., z_{n+N}) \neq 0$ be a homogeneous co-ordinate vector of P_{n+N}, so that Z and $\lambda Z = (\lambda z_0, \lambda z_1, ..., \lambda z_{n+N})$, where λ is a non-zero complex number, define the same point. For Z and $W = (w_0, w_1, ..., w_{n+N})$ we introduce the Hermitian scalar product

$$(Z, W) = \overline{(W, Z)} = \sum_{k=0}^{n+N} z_k \overline{w}_k. \tag{1}$$

The linear space L of dimension N can be defined by the equations

$$l_i \equiv (Z, A_i) = 0 \quad (1 \leqslant i \leqslant n), \tag{2}$$

where we suppose $(A_i, A_j) = \delta_{ij}, 1 \leqslant i, j \leqslant n$. Then, for $\zeta \in M_n$, the function

$$u(\zeta, L) = \frac{\|L\|}{|Z|} \leqslant 1 \quad (|Z| = +(Z_1 Z)^{\frac{1}{2}}, \|L\| = +(l_1 \bar{l}_1 + ... + l_n \bar{l}_n)^{\frac{1}{2}}), \tag{3}$$

where $Z = Z(\zeta)$ is a homogeneous co-ordinate vector of the image point of ζ, is a well-defined real-valued function in M_n, and vanishes, if and only if $Z(\zeta) \in L$. Similarly, we define the exterior differential forms

$$\left. \begin{aligned} \Phi &= \frac{i}{\pi} d'd'' \log \|L\|, \\ \Psi &= \frac{i}{\pi} d'd'' \log |Z|, \end{aligned} \right\} \tag{4}$$

and

$$\Lambda = \frac{1}{2\pi i}(d' - d'') \log u \wedge \sum_{0 \leqslant k \leqslant n-1} \Phi^k \wedge \Psi^{n-1-k}. \tag{5}$$

Then we have the formula

$$v(M_n) - n(M_n, L) = -\int_{\partial M_n} \Lambda, \tag{6}$$

where $n(M_n, L)$ is the number of points common to L and $Z(M_n)$, counted with their multiplicities, and $v(M_n)$ is the volume of $Z(M_n)$, suitably normalized. It follows in particular that, if M_n is without boundary and if $Z(M_n)$ is non-degenerate, so that $v(M_n) \neq 0$, then $Z(M_n)$ meets every linear space of dimension N in P_{n+N}.

Perhaps the first example of a non-compact complex manifold is the complex Euclidean space E_n of dimension n. Let $\zeta_1, ..., \zeta_n$ be the co-ordinates in E_n. We will exhaust E_n by the domains $M(r)$:

$$\zeta_1 \bar{\zeta}_1 + ... + \zeta_n \bar{\zeta}_n \leqslant r^2, \tag{7}$$

as $r \to \infty$. This seems to be the most natural exhaustion, because if we compactify E_n by adding a hyperplane π at infinity, the complement of $M(r)$ in E_n will form a tubular neighborhood about π.

Consider first the classical case of a meromorphic function $Z: E_1 \to P_1$. Let $v(r)$ be the volume of the image of $M(r)$. For a generic point $L \in P_1$ let $n(r, L)$ be the number of times L is covered by $Z(M(r))$. Then (6) can be written

$$v(r) - n(r, L) = -\frac{1}{2\pi} \int_0^{2\pi} r \frac{\partial \log u}{\partial r} d\theta \quad (\zeta_1 = re^{i\theta}). \tag{8}$$

This induces us to put

$$T(r) = \int_{r_0}^r \frac{v(t) \, dt}{t}, \quad N(r, L) = \int_{r_0}^r \frac{n(t, L) \, dt}{t} \quad (r_0 > 0). \tag{9}$$

By integrating (8) with respect to r, we get

$$T(r) - N(r, L) = -\frac{1}{2\pi} \int_0^{2\pi} \log u \, d\theta \Big|_{r_0}^r. \tag{10}$$

This is the so-called first main theorem in the theory of meromorphic functions. Our introduction of the order function $T(r)$ is exactly the way it was introduced by Shimizu–Ahlfors.

Since the first main theorem involves a generic point L of P_1, it is natural to integrate it over P_1. If we perform the integration with the invariant density dL, we shall get a formula of the Crofton type

$$T(r) = \int_{L \in P_1} N(r, L) \, dL, \tag{11}$$

which implies that the average of the right-hand side of (10) is zero. On the other hand, we derive from the first main theorem the fundamental inequality

$$T(r) - N(r, L) > \text{const.} \tag{12}$$

If we integrate this inequality over a non-invariant density, it is easy to get the theorem that the complement of the image set $Z(E_1)$ in P_1 has measure zero. An idea initiated by F. Nevanlinna and simplified by Ahlfors[1] consists in the use of a density with singularities. It is the integration of (12) relative to such a density that leads to a proof of the Picard–Borel theorem.

In the case of complex analytic mappings $Z: E_2 \to P_2$ there are known examples which show that the complement of the image $Z(E_2)$ may contain open subsets of P_2. We shall give a brief discussion of the proper restrictions on the mapping Z in order that general statements can be made. In fact, the first main theorem on complex analytic functions has the following generalization:

Let $v(r)$ be the volume of the image of $M(r)$ and, for a generic point $L \in P_2$, let $n(r, L)$ be the number of times L is covered by $Z(M(r))$. Let

$$T(r) = \int_{r_0}^{r} \frac{v(t)\,dt}{t^3}, \quad N(r, L) = \int_{r_0}^{r} \frac{n(t, L)\,dt}{t^3} \quad (r_0 > 0). \tag{13}$$

Then we have the inequality

$$T(r) - N(r, L) > \text{const} - S(r, L), \tag{14}$$

where
$$\left.\begin{aligned} S(r, L) &= -\frac{2}{\pi^2} \int_{r_0}^{r} \frac{dt}{t} \int (v_{11} + v_{22}) \log u \, dV \geqslant 0, \\ v_{kk} &= \frac{\partial^2}{\partial \zeta_k \partial \bar{\zeta}_k} \log\left(|Z| \cdot \|L\|\right) \quad (k = 1, 2), \end{aligned}\right\} \tag{15}$$

the integration being over the volume element dV of the unit hypersphere in E_n. It is clear from this inequality that in order to have a statement on the image set $Z(E_n)$ we must have $T(r) \to \infty$ as $r \to \infty$. The latter is automatic in the 1-dimensional case, but is an additional assumption in the 2-dimensional case. In fact, the well-known examples of Fatou–Bieberbach do not have this property. If, moreover,

$$S(r, L) = o(T(r)), \tag{16}$$

then we have the theorem that $Z(E_2)$ omits at most a set of measure zero.

The assumption (16) is unsatisfactory in the sense that it involves the generic point $L \in P_2$. The expression for $S(r, L)$ suggests that a 'mixed order function' should be introduced. In fact, let Ω and Ω_0 be the associated two-forms of P_2 and E_2 respectively. Then

$$\int_{M(r)} Z^*(\Omega) \wedge \Omega_0 = v_1(r), \tag{17}$$

where $Z^*(\Omega)$ is the inverse image of Ω under the mapping Z, is a mixed volume of the domain $M(r)$. Put

$$S(r) = \int_{r_0}^{r} \frac{v_1(t)\,dt}{t^3}. \tag{18}$$

It is conceivable that condition (16) can be replaced by a condition on the relative growth of $T(r)$ and $S(r)$.

It seems to me that these problems on complex analytic mappings deserve much further study.

3. Integral formulae and rigidity theorems

I believe my discussion of relations between differential and integral geometry will leave a big gap, if I do not touch on the role that integral formulae play in the proofs of rigidity or uniqueness theorems. Perhaps the most well-known example of such considerations is Herglotz's proof of the uniqueness of Weyl's problem. In spite of these important applications it should be of independent interest to derive integral formulae for compact immersed submanifolds for their own sake. A little analytic manipulation shows that there are few such formulae, unless the latter are allowed to involve other geometrical elements in the space, such as fixed points, fixed linear subspaces, fixed directions, etc. The reason is simple: For an immersed submanifold $x: M^n \to E^{n+N}$, the co-ordinate vector $x(p)$, $p \in M^n$, depends on the choice of the origin.

The simplest case is that of a strictly convex hypersurface $x: M^n \to E^{n+1}$. Naturally we orient it so that the Gauss–Kronecker curvature is everywhere > 0. Since the normal mapping of the hypersurface $\Sigma = x(M^n)$ into the unit hypersphere S_0 about the origin is one–one and has a non-zero Jacobian everywhere, the hypersurface can be defined by $x: S_0 \to \Sigma \subset E^{n+1}$, where x maps a point ξ of S_0 into the point of Σ having ξ as the unit normal vector.

To get rigidity theorems suppose $x': S_0 \to \Sigma'$ is a second strictly convex hypersurface. It is then possible to write down a number of globally defined exterior differential forms on S_0. For our purpose we shall restrict ourselves to the following:

$$\left. \begin{aligned} A_{rs} &= (x, \xi, d\xi, \ldots, d\xi, dx, \ldots, dx, dx', \ldots, dx'), \\ A'_{rs} &= (x', \xi, d\xi, \ldots, d\xi, dx, \ldots, dx, dx', \ldots, dx'). \end{aligned} \right\} \quad (19)$$

Each of these expressions is a determinant of order $n + 1$, whose rows are the components of the respective vectors or vector-valued differential forms, with the convention that in the expansion of the determinant the multiplication of differential forms is in the sense of exterior multiplication. The subscript r refers to the number of entries dx and the subscript s that of the entries dx'. Since A_{rs} and A'_{rs} are globally defined on S_0, their integrals over S_0 are zero.

The integral formulae so obtained can be expressed in a more geometrical form as follows: Let $\text{III} = d\xi^2$ be the fundamental form of S_0, and let

$$\text{II} = -dx\,d\xi, \quad \text{II}' = -dx'\,d\xi \tag{20}$$

be the second fundamental forms of Σ, Σ' respectively. Let $\Delta(y, y')$ be the determinant of the ordinary quadratic differential form $y\text{II} + y'\text{II}' + \text{III}$ relative to a local co-ordinate system, so that $\Delta(y, y')/\Delta(0, 0)$ is independent of the choice of the local co-ordinate system. Let

$$\frac{\Delta(y, y')}{\Delta(0, 0)} = \sum_{0 \leqslant r+s \leqslant n} \frac{n!}{r!\,s!\,(n-r-s)!} y^r y'^s P_{rs}, \tag{21}$$

where P_{rs} are mixed invariants of Σ, Σ'. In particular, P_{l0}, P_{0l} are, up to numerical factors, the lth elementary symmetric functions of the principal radii of curvature of Σ, Σ' respectively. Then our integral formulae can be written

$$\int_{S_0} (pP_{rs} - P_{r+1,s})\,dV = 0, \quad \int_{S_0} (p'P_{rs} - P_{r,s+1})\,dV = 0, \tag{22}$$

where dV is the volume element of S_0 and p, p' are the support functions of Σ, Σ' respectively. An important consequence of (22) consists of the formulae

$$\int pP_{0l}\,dV = \int p'P_{1,l-1}\,dV, \quad \int pP_{l-1,1}\,dV = \int p'P_{l0}\,dV \quad (l \geqslant 1), \tag{23}$$

which give

$$2\int p(P_{0l} - P_{l-1,1})\,dV = \int \{p'(P_{1,l-1} - P_{l0}) - p(P_{l-1,1} - P_{0l})\}\,dV. \tag{24}$$

It is important to observe that the right-hand side of (24) is antisymmetric in the hypersurfaces Σ, Σ'.

Formula (24) reduces to a purely algebraic problem the proof of the following uniqueness theorem of Minkowski, A. D. Alexandroff[2], and Fenchel and Jessen[6]: If two closed strictly convex hypersurfaces are such that at points with parallel normals, the lth (for fixed $l \geqslant 2$) elementary symmetric functions of the principal radii of curvature have the same value, then they differ from each other by a translation.

The theorem is also true for $l = 1$, but it will have a different (and simpler) proof.

The algebraic lemma needed has been communicated to me by L. Gårding, as a consequence of his work on hyperbolic polynomials. It can be stated as follows:

Let (λ_{ik}) be an $n \times n$ symmetric matrix, and let

$$\det(\delta_{ik} + y\lambda_{ik}) = \sum_{0 \leqslant r \leqslant n} P_r(\lambda) y^r. \tag{25}$$

Let $P_r(\lambda^{(1)}, ..., \lambda^{(r)})$ be the completely polarized form of $P_r(\lambda)$, so that $\underbrace{P_r(\lambda, ..., \lambda)}_{r} = P_r(\lambda)$. Then, for $r \geqslant 2$ and for positive definite matrices $(\lambda_{ik}^{(1)}), ..., (\lambda_{ik}^{(r)})$, the following inequality is valid:

$$P_r(\lambda^{(1)}, ..., \lambda^{(r)}) \geqslant P_r(\lambda^{(1)})^{1/r} ... P_r(\lambda^{(r)})^{1/r}. \tag{26}$$

Equality holds, if and only if the r matrices are pairwise proportional.

The uniqueness theorem then follows immediately from the lemma and the integral formula (24). For the hypothesis says that $P_{0l} = P_{l0}$. From (26) it follows that $P_{l-1,1} - P_{0l} \geqslant 0$. By (24) this is possible only when $P_{l-1,1} - P_{0l} = 0$. Again by the lemma it follows that the second fundamental forms of the hypersurfaces are equal.

So far as I am aware, it is not known whether a similar uniqueness theorem is valid, if the lth elementary symmetric function of the principal curvatures is prescribed as a function of the normal vector. Alexandroff proved that a closed convex surface in ordinary Euclidean space is defined up to a translation, if its mean curvature is a given function of the normal. His proof made use of a maximum principle. It would be interesting if this theorem can be proved by using integral formulae.

REFERENCES

[1] Ahlfors, L. The theory of meromorphic curves. *Acta Soc. Sci. fenn.*, A, 3, 1–31 (1941).
[2] Alexandroff, A. D. Zur Theorie der gemischten Volumina von konvexen Körpern. (Russian.) *Mat. Sbornik*, 2 (44), 947–972, 1205–1238; 3 (45), 27–46, 227–251 (1937–38).
[3] Chern, S. and Lashof, R. K. On the total curvature of immersed manifolds. *Amer. J. Math.* 79, 302–318 (1957).
[4] Chern, S. and Lashof, R. K. On the total curvature of immersed manifolds. II. *Michigan Math. J.* 5, 5–12 (1958).
[5] Fary, I. Sur la courbure totale d'une courbe gauche faisant un nœud. *Bull. Soc. Math. Fr.* 77, 128–138 (1949).
[6] Fenchel, W. and Jessen, B. Mengenfunktionen und konvexe Körper. *K. danske vidensk. Selsk. (Math.-fysiske Medd.)*, 16, 1–31 (1938).
[7] Levine, Harold I. Contributions to the theory of analytic mappings of complex manifolds into projective space. University of Chicago thesis, 1958.
[8] Milnor, J. W. On the total curvature of knots. *Ann. Math.* (2), 52, 248–257 (1950); also Princeton thesis (unpublished).

THE POLARIZATION OF ALGEBRAIC VARIETIES, AND SOME OF ITS APPLICATIONS

By T. MATSUSAKA

Let V^m be a complete non-singular variety.[†] There are three kinds of equivalence relations of V-divisors, linear, algebraic and numerical equivalence, which we assume are well known. Let G_l, G_a, G_n be respectively the set of V-divisors which are linearly, algebraically and numerically equivalent to 0. These form subgroups of the group G of V-divisors. We say that a divisor X on V is *linearly effective*, if the complete linear system $\mathfrak{L}(X)$ determined by X gives a projective embedding of V. Let $\mathfrak{X}$ be a non-empty set of positive V-divisors such that if X is in $\mathfrak{X}$, a positive V-divisor Y is in $\mathfrak{X}$ if and only if $mX \equiv sY \mod G_n$ for a pair (m, s) of positive integers. It is clear that $\mathfrak{X}$ is determined uniquely by one of the divisors contained in it. When $\mathfrak{X}$ contains at least one linearly effective divisor, then we call the pair $(V, \mathfrak{X})$ a *polarized variety*, polarized by the set $\mathfrak{X}$. A polarized variety $(V, \mathfrak{X})$ is an algebraic variety plus an additional structure determined by $\mathfrak{X}$. When V is a curve, we can put on V one and only one structure of polarization. The same is true in general if $G/G_n \cong Z$ and if V admits a projective embedding. The notion of polarization was first introduced by Weil (cf. Weil[9]) and studied by the present writer (cf. Matsusaka[2]).

1. The group of automorphisms

Let V, V' be two complete non-singular varieties with sets of structures E, E'. An everywhere biregular birational transformation f of V onto V' is said to be an *isomorphism* of (V, E) and (V', E') if $f(E) = E'$. When $V = V'$, $E = E'$, we say that f is an *automorphism*.

It is well known that when V is a curve, the set of automorphisms of it forms an algebraic group and, in particular, when the genus is greater than 1, it is actually a finite group. It is also well known that when V is an Abelian variety of dimension 1, the set of automorphisms of it forms a finite group. But when dim $V > 1$, the situation is different. In fact, there is a non-singular algebraic surface in a projective space such that its group of automorphisms is an infinite group but is not an algebraic group. These facts depend on the fact that a curve carries its uniquely

---

† We follow the same terminology and conventions as in Weil[6,7].

determined polarization, but when dim $V \geqslant 2$, the same is no longer
true in general. In fact we have the following theorem (cf. Matsusaka[2]):

*Let $(V, \mathfrak{X})$ be a polarized variety, and let $\mathfrak{G}$ be the set of automorphisms
of it, then $\mathfrak{G}$ is an algebraic group and the connected component $\mathfrak{G}_0$ of the
neutral element of $\mathfrak{G}$ is an extension of a subgroup of the Picard variety of
V by a linear group.*

Also we can show that *when $\mathfrak{G}'$ is the set of all automorphisms of V,
then $\mathfrak{G}_0$ is the largest connected algebraic group in $\mathfrak{G}'$.* As an immediate
corollary of the above theorem, we see that *the set of automorphisms of a
polarized Abelian variety is a finite group* (cf. Weil[9]; Matsusaka[2]).

It is seen that $\mathfrak{G}_0$ is a normal subgroup of $\mathfrak{G}'$ and, as far as those
examples which the author knows are concerned, $\mathfrak{G}'/\mathfrak{G}_0$ is a finitely
generated group.

2. Equivalent projective embeddings; Moduli

Let $(V, \mathfrak{X})$ be a polarized variety and let X, X' be two linearly effective
divisors in $\mathfrak{X}$. Then X and X' determine uniquely projective embeddings
f, f' of V up to projective transformations (here we consider only those
projective embeddings which come from the linearly independent base
of $\mathfrak{L}(X), \mathfrak{L}(X')$). We say that f and f' are *equivalent embeddings* in the
broad sense. Moreover, when $X \equiv X' \bmod G_a$, we say that they are
equivalent in the strict sense.

Let $\mathfrak{S}$ be a subset of $\mathfrak{X}$; then by $\mathfrak{P}(V, \mathfrak{S})$ we understand the set of
projective varieties $f(V)$, where f is a projective embedding determined
by a linearly effective divisor contained in $\mathfrak{S}$. $\mathfrak{X}$ can be written as
$\bigcup \mathfrak{A}(X)$ where $\mathfrak{A}(X)$ is the coset of G_a containing X, and except for a
finite set, say $\mathfrak{A}(X_1), ..., \mathfrak{A}(X_s)$, every component $\mathfrak{A}(X)$ is an irreducible
algebraic family consisting of linearly effective divisors such that all
divisors in it determine complete linear systems of the same dimension.
From this we can deduce that $\mathfrak{P}(V, \mathfrak{A}(X))$, for such $\mathfrak{A}(X)$, *has a structure
of an open algebraic variety in a projective space, having the smallest field
of definition* (this can be done conveniently in terms of Chow-forms).
Put $\mathfrak{X}* = \mathfrak{X} - (X_1, ..., X_s)$. Then $\mathfrak{P}(V, \mathfrak{X}*) = \bigcup_{X \in \mathfrak{X}*} \mathfrak{P}(V, \mathfrak{A}(X))$ and *there is
a smallest field K such that*

(i) *one of the $\mathfrak{P}(V, \mathfrak{A}(X))$ in the expression for $\mathfrak{P}(V, \mathfrak{X}*)$ is defined
over K,*

(ii) *every other $\mathfrak{P}(V, \mathfrak{A}(X))$ in the expression for $\mathfrak{P}(V, \mathfrak{X}*)$ is defined over
a separably generated extension of K, provided that the order of G_n/G_a is
prime to the characteristic.* We define this field K as the *field of moduli*

of $(V, \mathfrak{X})$. In the case of characteristic 0, K is the smallest field such that $(V^\alpha, \mathfrak{X}^\alpha)$ is isomorphic to $(V, \mathfrak{X})$, where α is any automorphism of the universal domain over K, as Shimura remarked.

In the case when V is an Abelian variety of dimension 1, the field of moduli K is generated over the prime field by the corresponding values of modular functions. The corresponding fact seems to be true in general when V is an Abelian variety and $\mathfrak{X}$ contains a divisor determined by the unimodular principal matrix of the Riemann matrix defining V. For discussions of (ii), cf. Matsusaka[2].

3. Torelli's theorem

It is known that even if two curves Γ and Γ' have isomorphic Jacobian varieties, they may not be birationally equivalent to each other. Let J and J' be Jacobian varieties of Γ and Γ', and let θ and θ' be canonical divisors on J and J' with respect to Γ and Γ'. θ and θ' determine on J and J' polarizations (cf. Weil[8]), which we call canonical polarizations with respect to Γ and Γ'. Torelli's theorem asserts that Γ *and* Γ' *are birationally equivalent to each other if and only if the canonically polarized* J *and* J' *are isomorphic*. This theorem was first proved by Torelli (cf. Torelli[5]), and recently proofs were given by Weil, Andreotti and myself which are valid over fields of any characteristic (cf. Weil[10]; Andreotti[1]; Matsusaka[3]).

4. A characterization of canonically polarized Jacobian varieties

Let $(A, \mathfrak{X})$ be a polarized Abelian variety. When A is of dimension 2, $(A, \mathfrak{X})$ is a canonically polarized Jacobian variety if $\mathfrak{X}$ contains an irreducible divisor X such that $\deg(X . X_u) = 2$ (cf. Weil[10]). The same fact seems to be true for 3-dimensional Abelian varieties. On the other hand, the above theorem is a special case of the following numerical criterion:

Let $(A, \mathfrak{X})$ *be a polarized Abelian variety of dimension* n; *then it is a canonically polarized Jacobian variety if and only if* $\mathfrak{X}$ *contains an irreducible divisor* X *such that*
$$\text{(i)} \quad (X_{u_1} \ldots X_{u_{n-1}})$$
is numerically equivalent to $(n-1)!\, C$, *where* C *is a positive 1-cycle on* A, *and that*
$$\text{(ii)} \quad \deg(X_{u_1} \ldots X_{u_n}) = n!;$$

moreover, when this is so, C *is irreducible,* A *is the Jacobian variety of* C, C *is canonically embedded in* A, *and* X *is a canonical divisor on* A *with respect to* C (cf. Matsusaka[4]).

This theorem was extended to the case when X is reducible by W. Hoyt in his Chicago thesis, and he also studied a structure of a polarized Abelian variety which is a specialization of a canonically polarized Jacobian variety.

REFERENCES

[1] Andreotti, A. On a theorem of Torelli. *Amer. J. Math.* 80, 801–828 (1958).
[2] Matsusaka, T. Polarized varieties, fields of moduli and generalized Kummer varieties of polarized Abelian varieties. *Amer. J. Math.* 80, 45–82 (1958).
[3] Matsusaka, T. Torelli's theorem. (To appear.)
[4] Matsusaka, T. A characterization of a canonically polarized Jacobian variety. (To appear.)
[5] Torelli, R. Sulle varietà di Jacobi. *R.C. Acc. Naz. Lincei*, (5), 22, 98–103, 437–441 (1913).
[6] Weil, A. *Foundations of Algebraic Geometry.* American Mathematical Society Colloquium Publications, vol. 29.
[7] Weil, A. *Variétés Abéliennes et Courbes Algébriques.* Actualités Scientifiques et Industrielles, no. 1064 (1948).
[8] Weil, A. Projective embeddings of Abelian varieties. *Algebraic Geometry and Topology.* (A symposium in honor of S. Lefschetz.) Princeton Mathematical series.
[9] Weil, A. On the theory of complex multiplication. *Proc. Int. Symp. Algebraic Number Theory* (1955).
[10] Weil, A. Zum Beweis des Torellischen Satzes. *Nachr. Akad. Wiss. Göttingen* (*Mathematisch-Physikalische Klasse*), Nr. 2 (1957).

BERNOULLI NUMBERS, HOMOTOPY GROUPS, AND A THEOREM OF ROHLIN

By JOHN W. MILNOR AND MICHEL A. KERVAIRE

A homomorphism $J\colon \pi_{k-1}(\mathrm{SO}_m) \to \pi_{m+k-1}(S^m)$ from the homotopy groups of rotation groups to the homotopy groups of spheres has been defined by H. Hopf and G. W. Whitehead[16]. This homomorphism plays an important role in the study of differentiable manifolds. We will study its relation to one particular problem: the question of possible Pontrjagin numbers of an 'almost parallelizable' manifold.

Definition. A connected differentiable manifold M^k with base point x_0 is *almost parallelizable* if $M^k - x_0$ is parallelizable. If M^k is imbedded in a high-dimensional Euclidean space R^{m+k} ($m \geqslant k+1$) then this is equivalent to the condition that the normal bundle ν, restricted to $M^k - x_0$, be trivial (compare the argument given by Whitehead[17], or Kervaire ([5], §8)).

The following theorem was proved by Rohlin in 1952 (see Rohlin[11,12], Kervaire[6]).

Theorem (Rohlin). *Let M^4 be a compact oriented differentiable 4-manifold with Stiefel–Whitney class w_2 equal to zero. Then the Pontrjagin number $p_1[M^4]$ is divisible by 48.*

Rohlin's proof may be sketched as follows. It may be assumed that M^4 is a connected manifold imbedded in R^{m+4}, $m \geqslant 5$.

Step 1. It is shown that M^4 is almost parallelizable.

Let f be a cross-section of the normal SO_m-bundle ν restricted to $M^4 - x_0$. The obstruction to extending f is an element

$$\mathfrak{o}(\nu, f) \in H^4(M^4; \pi_3(\mathrm{SO}_m)) \approx \pi_3(\mathrm{SO}_m).$$

Step 2. It is shown that $J\mathfrak{o}(\nu, f) = 0$.

Since J carries the infinite cyclic group $\pi_3(\mathrm{SO}_m)$ onto the cyclic group $\pi_{m+3}(S^m)$ of order 24, this implies that $\mathfrak{o}(\nu, f)$ is divisible by 24. Now identify the group $\pi_3(\mathrm{SO}_m)$ with the integers.

Step 3. It is shown that the Pontrjagin class $p_1(\nu)$ is equal to $\pm 2\mathfrak{o}(\nu, f)$.

Since by Whitney duality $p_1(\nu) = -p_1$ (tangent bundle), it follows that $p_1[M^4]$ is divisible by 48.

The first step in this argument does not generalize to higher dimensions. However Step 2, the assertion that $J\mathfrak{o}(\nu, f) = 0$, generalizes immediately. In fact we have:

Lemma 1. Let $\alpha \in \pi_{k-1}(SO_m)$; then $J\alpha = 0$ if and only if there exists an almost parallelizable manifold $M^k \subset R^{m+k}$ and a cross-section f of the induced normal SO_m-bundle ν over $M^k - x_0$ such that $\alpha = \mathfrak{o}(\nu, f)$.

Step 3 can be replaced by the following. Identify the group $\pi_{4n-1}(SO_m)$, $m > 4n$, with the integers (compare Bott[2]). Define a_n to be equal to 2 for n odd and 1 for n even.

Lemma 2. Let ξ be a stable SO_m-bundle over a complex K (dim $K < m$), and let f be a cross-section of ξ restricted to the skeleton $K^{(4n-1)}$. Then the obstruction class $\mathfrak{o}(\xi, f) \in H^{4n}(K; \pi_{4n-1}(SO_m))$ is related to the Pontrjagin class $p_n(\xi)$ by the identity $p_n(\xi) = \pm a_n . (2n-1)! \, \mathfrak{o}(\xi, f)$.

Combining Lemmas 1 and 2, we obtain the following theorems.

Define j_n as the order of the finite cyclic group $J\pi_{4n-1}(SO_m)$ in the stable range $m > 4n$.

Theorem 1. The Pontrjagin number $p_n[M^{4n}]$ of an almost parallelizable $4n$-manifold is divisible by $j_n a_n (2n-1)!$.

(For $n = 1$, this gives Rohlin's assertion, since $j_1 = 24$, $a_1 = 2$.)

Proof. This follows since $\mathfrak{o}(\nu, f)$ must be divisible by j_n.

Conversely:

Theorem 2. There exists an almost parallelizable manifold M_0^{4n} with

$$p_n[M_0^{4n}] = j_n a_n . (2n-1)!.$$

The proof is clear.

Proof of Lemma 1. Given an imbedding $i: V^{k-1} \to R^{m+k-1}$ of a compact differentiable manifold V^{k-1} into Euclidean space, and given a cross-section f of the normal SO_m-bundle over V^{k-1}, a well-known procedure due to Thom associates with i and f a sphere mapping $\phi: S^{m+k-1} \to S^m$ (compare Kervaire[5], p. 223).

The map ϕ is homotopic to zero if and only if there exists a bounded manifold Q^k with boundary V^{k-1} imbedded in R^{m+k} on one side of R^{m+k-1} such that:

(i) the restriction to V^{k-1} of the imbedding of Q^k is the given imbedding of V^{k-1} in R^{m+k-1};

(ii) Q^k meets R^{m+k-1} orthogonally so that the restriction to V^{k-1} of the normal bundle of Q^k is just the normal bundle of V^{k-1} in R^{m+k-1}; and

(iii) the cross-section f can be extended throughout Q^k as a cross-section f' of the normal SO_m-bundle.

These facts follow from Thom [15], ch. I, § 2 and Lemmas IV, 5, IV.5'.

To obtain Lemma 1 above, take $V^{k-1} = S^{k-1}$ and take $i(S^{k-1})$ to be the unit sphere in $R^k \subset R^{m+k-1}$. Since the normal m-plane at each point of $i(S^{k-1})$ in R^{m+k-1} admits a natural basis (consisting of the radius vector followed by the vectors of a basis for R^{m+k-1}/R^k), the cross-section f

provides a mapping $a: S^{k-1} \to \mathrm{SO}_m$. Let $\alpha \in \pi_{k-1}(\mathrm{SO}_m)$ be its homotopy class. It is easily seen (compare Kervaire[7], §1.8) that the map $\phi: S^{m+k-1} \to S^m$ associated with i and f represents $J\alpha$ up to sign.

If $J\alpha = 0$ then there exists a bounded manifold $Q^k \subset R^{m+k}$ satisfying conditions (i), (ii) and (iii). Let $M^k \subset R^{m+k}$ denote the unbounded manifold obtained from Q^k by adjoining a k-dimensional hemisphere, which lies on the other side of R^{m+k-1} and has the same boundary $i(S^{k-1})$. Since the normal bundle ν restricted to Q^k has a cross-section f', it follows that M^k is almost parallelizable. Clearly the obstruction class $\mathfrak{o}(\nu, f')$ is equal to α.

Conversely, let M^k be a manifold imbedded in R^{m+k} and let f be a cross-section of the normal bundle ν restricted to $M^k - x_0$. After modifying this imbedding by a diffeomorphism of R^{m+k} we may assume that some neighborhood of x_0 in M^k is a hemisphere lying on one side of the hyperplane R^{m+k-1}, and that the rest of M^k lies on the other side. Removing this neighborhood we obtain a bounded manifold $Q^k \subset R^{m+k}$ just as above, having the unit sphere $S^{k-1} \subset R^k \subset R^{m+k-1}$ as boundary. The cross-section f restricted to S^{k-1} gives rise to a map $a: S^{k-1} \to \mathrm{SO}_m$ which represents the homotopy class $\mathfrak{o}(\nu, f)$. The argument above shows that $J\mathfrak{o}(\nu, f) = 0$; which completes the proof of Lemma 1.

Remark. Lemma 1 could also be proved using the interpretation of J given in Milnor[9].

Proof of Lemma 2. (Compare Kervaire[8].) The SO_m-bundle ξ induces a U_m-bundle ξ' and hence a $\mathrm{U}_m/\mathrm{U}_{2n-1}$-bundle ξ''. Similarly, the partial cross-section f induces partial cross-sections f' and f''. By definition the obstruction class $\mathfrak{o}(\xi'', f'')$ is equal to the Chern class $c_{2n}(\xi')$ and hence to the Pontrjagin class $\pm p_n(\xi)$. Therefore $p_n(\xi)$ equals $\pm q_* h_* \mathfrak{o}(\xi, f)$, where

$$h: \pi_{4n-1}(\mathrm{SO}_m) \to \pi_{4n-1}(\mathrm{U}_m) \quad \text{and} \quad q: \pi_{4n-1}(\mathrm{U}_m) \to \pi_{4n-1}(\mathrm{U}_m/\mathrm{U}_{2n-1})$$

are the natural homomorphisms and h_*, q_* are the homomorphisms in the cohomology of K induced by the coefficient homomorphisms h, q.

Using the following computations of Bott[2]:

$$\pi_{4n-1}(\mathrm{U}_m) \approx Z, \quad \pi_{4n-1}(\mathrm{U}_m/\mathrm{SO}_m) \approx Z_{a_n}, \quad \pi_{4n-2}(\mathrm{SO}_m) = 0,$$

it follows that h carries a generator into a_n times a generator. Similarly, using the fact that

$$\pi_{4n-2}(\mathrm{U}_{2n-1}) \approx Z_{(2n-1)!} \quad (\text{see } [3]) \quad \text{and} \quad \pi_{4n-2}(\mathrm{U}_m) = 0,$$

it follows that q carries a generator into $(2n-1)!$ times a generator. Therefore $p_n(\xi) = \pm a_n(2n-1)! \, \mathfrak{o}(\xi, f)$. This completes the proof of Lemma 2.

Hirzebruch's index theorem[4] states that the index $I(M^{4n})$ of any $4n$-manifold is equal to

$$2^{2n}(2^{2n-1}-1)\,B_n p_n[M^{4n}]/(2n)! + \text{(terms involving lower}$$
$$\text{Pontrjagin classes)}.$$

Here B_n denotes the nth Bernoulli number. For an almost parallelizable manifold the lower Pontrjagin classes are zero. Therefore

Corollary. The index $I(M_0^{4n})$ *is equal to* $2^{2n-1}(2^{2n-1}-1)\,B_n j_n a_n/n$; *and the index of any almost parallelizable* $4n$-*manifold is a multiple of this number.*

The fact that $I(M_0^{4n})$ is an integer can be used to estimate the number j_n (compare Milnor[9]). However, a sharper estimate, which includes the prime 2, can be obtained as follows, using a new generalization of Rohlin's theorem.

Borel and Hirzebruch ([1], §§ 23.1 and 25.4) define a rational number

$$\hat{A}[M^{4n}] = -B_n p_n[M^{4n}]/2(2n)! + \text{(terms involving } p_1, ..., p_{n-1});$$

and prove that the denominator of $\hat{A}[M^{4n}]$ is a power of 2.

Theorem 3. If the Stiefel–Whitney class w_2 *of* M^{4n} *is zero then* $\hat{A}[M^{4n}]$ *is actually an integer.*†

The proof will be given in a subsequent paper by Borel and Hirzebruch. It is based on the methods of [1], together with the assertion that the Todd genus of a generalized almost complex manifold is an integer (Milnor[10]).

Applying this theorem to the manifold M_0^{4n} of Theorem 2 it follows that $B_n j_n a_n/4n$ is an integer. Therefore:

Theorem 4. The order j_n *of the stable group* $J\pi_{4n-1}(\mathrm{SO}_m)$ *is a multiple of the denominator of the rational number* $B_n a_n/4n$.

As examples, for $n = 1, 2, 3$, the number $B_n a_n/4n$ is equal to $1/12$, $1/240$, and $1/252$ respectively. Since $\pi_{m+7}(S^m)$ is cyclic of order 240, it follows that $j_2 = 240$. Since $\pi_{m+11}(S^m)$ is cyclic of order 504, it follows that j_3 is either 252 or 504. It may be conjectured that j_n is always equal to the denominator of $B_n/4n$.

The theorems of von Staudt[13,14] can be used to compute such denominators (compare Milnor[9]).

Lemma 3. The denominator of $B_n/2n$ *can be described as follows. A prime power* p^{i+1} *divides this denominator if and only if*

$$2n \equiv 0 \mod p^i(p-1).$$

† See note at the end of the paper.

Combining Lemma 3 with Theorem 4, we see that the stable homotopy groups of spheres contain elements of arbitrary finite order. In fact:

Corollary. If $2n$ is a multiple of the Euler Φ function $\Phi(r)$, then the stable group $\pi_{m+4n-1}(S^m)$ contains an element of order r.

REFERENCES

[1] Borel, A. and Hirzebruch, F. Characteristic classes and homogeneous spaces. *Amer. J. Math.* 80, 458–538 (1958).

[2] Bott, R. The stable homotopy of the classical groups. *Proc. Nat. Acad. Sci., Wash.*, 43, 933–935 (1957).

[3] Bott, R. and Milnor, J. W. On the parallelizability of the spheres. *Bull. Amer. Math. Soc.* 64, 87–89 (1958).

[4] Hirzebruch, F. *Neue topologische Methoden in der algebraischen Geometrie.* Springer, 1956.

[5] Kervaire, M. A. Courbure intégrale généralisée et homotopie. *Math. Ann.* 131, 219–252 (1956).

[6] Kervaire, M. A. On the Pontryagin classes of certain $\mathbf{SO}(n)$-bundles over manifolds. *Amer. J. Math.* 80, 632–638 (1958).

[7] Kervaire, M. A. An interpretation of G. Whitehead's generalization of Hopf's invariant. *Ann. Math.* (2), 69, 345–365 (1959).

[8] Kervaire, M. A. A note on obstructions and characteristic classes. (In preparation.)

[9] Milnor, J. W. On the Whitehead homomorphism J. *Bull. Amer. Math. Soc.* 64, 79–82 (1958).

[10] Milnor, J. W. On the cobordism ring Ω^*, and a complex analogue. (In preparation.)

[11] Rohlin, V. A. Classification of mappings of S^{n+3} onto S^n. (Russian.) *Dokl. Akad. Nauk, SSSR*, 81, 19–22 (1951).

[12] Rohlin, V. A. New results in the theory of 4-dimensional manifolds. (Russian.) *Dokl. Akad. Nauk, SSSR*, 84, 221–224 (1952).

[13] von Staudt. *J. Math.* 21, 372–374 (1840).

[14] von Staudt. *De numeris Bernoullianis commentatio.* Erlangen, 1845.

[15] Thom, R. Quelques propriétés globales des variétés différentiables. *Comment. math. Helvet.* 28, 17–86 (1954).

[16] Whitehead, G. W. On the homotopy groups of spheres and rotation groups. *Ann. Math.* (2), 43, 634–640 (1942).

[17] Whitehead, J. H. C. On the homotopy type of manifolds. *Ann. Math.* (2), 41, 825–832 (1940).

[*Added in proof.*] For the case n odd, Hirzebruch has since sharpened Theorem 3, showing that $\hat{A}[M^{4n}]$ is an even integer. Thus the factor a_n in Theorem 4 can be cancelled.

ON THE FOURTEENTH PROBLEM OF HILBERT

By MASAYOSHI NAGATA

The purpose of the present paper is to show that the answer to the 14th problem of Hilbert[1] is negative, even in the following restricted case, which may be called the original 14th problem of Hilbert:

Let G be a subgroup of the full linear group of the polynomial ring in indeterminates $x_1, \ldots, x_n$ over a field k, and let $\mathfrak{o}$ be the set of elements of $k[x_1, \ldots, x_n]$ which are invariant under G. Is $\mathfrak{o}$ finitely generated?

Our construction of a counter-example is independent of the characteristic of k, and k can be the field of complex numbers.

1. The construction of a counter-example

Let $\{a_{ij}\}$ $(i = 1, 2, 3;\ j = 1, 2, \ldots, 16)$ be algebraically independent elements over the prime field π of arbitrary characteristic, and let k be a field containing the a_{ij}. Let V be the vector space of dimension 16 over k and let V^* be the set of vectors in V which are orthogonal to the vectors $(a_{i1}, a_{i2}, \ldots, a_{i16})$ $(i = 1, 2, 3)$. (V^* is a subspace of dimension 13.)

Let $x_1, \ldots, x_{16}, t_1, \ldots, t_{16}$ be algebraically independent elements over k and let G be the set of linear transformations σ such that (i) $\sigma(t_i) = t_i$ for any i and (ii) $\sigma(x_i) = x_i + b_i t_i$ with $(b_1, \ldots, b_{16}) \in V^*$. Then:

The set $\mathfrak{o}$ of elements of $k[x_1, \ldots, x_{16},\ t_1, \ldots, t_{16}]$ which are invariant under G is not finitely generated.

2. A lemma on plane curves

In order to prove the example, we need the following lemma on plane curves:

Fundamental lemma. Let $P_1, \ldots, P_{16}$ be independent generic points of the projective plane S over the prime field π. For any curve C of degree d, the sum of the multiplicities of P_i on C is less than $4d$.

Proof. Assume that there exists a curve C of degree d such that $\Sigma m_i \geqslant 4d$, where m_i is the multiplicity of P_i on C. Since the P_i are independent generic points, the P_i can be specialized to any permutation of the P_i and therefore we see that there exists a curve of degree d' such that the multiplicity of the P_i is equal to m for every i and $d' \leqslant 4m$. Therefore it is sufficient to prove the following lemma (which is equivalent to the fundamental lemma):

Lemma. For any natural number m, there is no curve of degree $4m$ which goes through every P_i with multiplicity at least m.

Proof. Let C_4, C_3 and C_3' be independent generic curves of degree 4, 3 and 3 respectively in S. Let $Q_1, \ldots, Q_8$ be 8 points among $C_3 . C_3'$. Then the Q_i are independent generic points of S over $\pi(C_4)$. Let $Q_1^*, \ldots, Q_8^*$ be independent generic points of C_4 over $\pi(C_4)$, let C_3^* and $C_3^{*'}$ be general curves going through the Q_i^* and let $R_1^*, \ldots, R_8^*$ be such that

$$C_3^* . C_4 = \Sigma Q_i^* + \Sigma_1^4 R_j^*, \quad C_3^{*'} . C_4 = \Sigma Q_i^* + \Sigma_5^8 R_j^*.$$

We consider a specialization

$$(Q_1, \ldots, Q_8, C_3, C_3', C_4) \to (Q_1^*, \ldots, Q_8^*, C_3^*, C_3^{*'}, C_4)$$

over π. We take $R_1, \ldots, R_8$ so that (1) $\Sigma_1^4 R_i \subseteq C_3 . C_4$, $\Sigma_5^8 R_i \subseteq C_4 . C_3'$, and (2) the R_i are specialized to the R_i^* by the specialization considered above.

Assume that for some m there exists a curve of degree $4m$ which goes through every P_i with multiplicity at least m. Then we see that there exists a curve E of degree $4m$ which goes through the Q_i and the R_j with multiplicity at least m. We shall show that E must contain C_3 as a component. Assume the contrary. Then $C_3 . E$ contains $\Sigma m Q_i + \Sigma_1^4 m R_j$. Since $\deg C_3 . E = 12m$, we see that $C_3 . E = \Sigma m Q_i + \Sigma_1^4 m R_j$, which gives a contradiction because $Q_1, \ldots, Q_8$, $R_1, \ldots, R_4$ are independent generic points of C_3, and C_3 is of positive genus. Thus E must contain C_3 as a component. Similarly, E must contain C_3' as a component. Then $E' = E - C_3 - C_3'$ is a curve of degree $4(m-1) - 2$ which goes through the R_j with multiplicity at least $m-1$ and goes through the Q_i with multiplicity at least $m-2$. Specializing the Q_i and R_j to the Q_i^* and R_j^*, we see that there exists a curve E^* of degree $4t - 2$ $(t = m - 1)$ which goes through the Q_j^* with multiplicity at least $t-1$ and the R_i^* with multiplicity at least t. Assume that E^* does not contain C_4 as a component. Since $C_4 . E^*$ contains $\Sigma(t-1)Q_i^* + \Sigma t R_j^*$ and since $\deg C_4 . E^* = 16t - 8$, we see that $C_4 . E^* = \Sigma(t-1)Q_i^* + \Sigma t R_j^*$.

Since
$$C_4 . (C_3^* + C_3^{*'}) = \Sigma 2 Q_i^* + \Sigma R_j^*,$$

we have
$$C_4 . (t C_3^* + t C_3^{*'} - E^*) = \Sigma(t+1) Q_i^*.$$

Since the Q_i^* are independent generic points of C_4 and since C_4 is of positive genus, we have a contradiction. Thus E^* must contain C_4 as a component. Then $E^* - C_4$ is in the same situation with $t-1$ instead of t. Therefore, by induction, we have a contradiction (observe that if $t = 1$, then $\deg E^* = 2 < \deg C_4$ and E^* cannot contain C_4). Thus the lemma is proved.

3. The proof of the example

Set $u = t_1 \ldots t_{16}, v_i = u/t_i, w_i = v_i x_i, y_j = \Sigma a_{ji} w_i \ (i = 1, \ldots, 16; j = 1, 2, 3)$. Then $y_1, y_2, y_3, t_1, \ldots, t_{16}$ are invariant under G. Since

we have
$$k[w_1, \ldots, w_{16}] = k[y_1, y_2, y_3, w_4, \ldots, w_{16}],$$

$$k(x_1, \ldots, x_{16}, t_1, \ldots, t_{16}) = k(y_1, y_2, y_3, x_4, \ldots, x_{16}, t_1, \ldots, t_{16}),$$

and G is the set of linear transformations σ of

$$k[y_1, y_2, y_3, x_4, \ldots, x_{16}, t_1, \ldots, t_{16}]$$

such that $\sigma(y_i) = y_i$, $\sigma(t_j) = t_j$ and $\sigma(x_i) = x_i + b_i t_i$ with arbitrary elements $b_4, \ldots, b_{16}$ of k. Therefore we see that the invariant subfield of G is $k(y_1, y_2, y_3, t_1, \ldots, t_{16})$. Thus

(1) $\mathfrak{o} = k[x_1, \ldots, x_{16}, t_1, \ldots, t_{16}] \cap k(y_1, y_2, y_3, t_1, \ldots, t_{16})$.

Since y_1, y_2, y_3 are algebraically independent over k, we can regard $\mathfrak{H} = k[y_1, y_2, y_3]$ as a homogeneous co-ordinate ring of the projective plane S. Let P_i be the point on S with co-ordinates (a_{1i}, a_{2i}, a_{3i}). Then the P_i are independent generic points of S over the prime field π. Let $\mathfrak{p}_i$ be the homogeneous prime ideal of P_i and set $\mathfrak{a}(n_1, \ldots, n_{16}) = \bigcap_i \mathfrak{p}_i^{n'_i}$, $n'_i = \max(0, n_i)$ (for arbitrary integers n_i). We shall show that

(2) $\mathfrak{o}$ is the set of elements of the form $\Sigma a(n_1, \ldots, n_{16}) t_1^{-n_1} \ldots t_{16}^{-n_{16}}$ with $a(n_1, \ldots, n_{16}) \in \mathfrak{a}(n_1, \ldots, n_{16})$ (finite sum).

Proof. Since $k[w_1, \ldots, w_{16}] = k[y_1, y_2, y_3, w_4, \ldots, w_{16}]$, we have

$$k[x_1, \ldots, x_{16}, t_1, \ldots, t_{16}, 1/t_1, \ldots, 1/t_{16}]$$

$$= k[y_1, y_2, y_3, x_4, \ldots, x_{16}, t_1, \ldots, t_{16}, 1/t_1, \ldots, 1/t_{16}].$$

The intersection of this last ring with $k(y_1, y_2, y_3, t_1, \ldots, t_{16})$ is equal to $k[y_1, y_2, y_3, t_1, \ldots, t_{16}, 1/t_1, \ldots, 1/t_{16}]$. Since this last ring contains $\mathfrak{o}$ (by virtue of (1) above), it follows that for any element c of $\mathfrak{o}$ there exists an integer s such that cu^s is in $k[y_1, y_2, y_3, t_1, \ldots, t_{16}]$. Therefore

(i) *Any element of $\mathfrak{o}$ can be expressed in the form*

$$\Sigma a(n_1, \ldots, n_{16}) t_1^{-n_1} \ldots t_{16}^{-n_{16}}$$

(finite sum, n_j may be negative) with $a(n_1, \ldots, n_{16}) \in \mathfrak{H}$.

Let $\mathfrak{v}_i$ be the valuation ring $k[x_1, \ldots, x_{16}, t_1, \ldots, t_{16}]_{(t_i)}$ and let v_i be the normalized valuation defined by $\mathfrak{v}_i$. $\mathfrak{o}$ is contained in every $\mathfrak{v}_i$. It is obvious that

(ii) *If an element f of $\mathfrak{H}$ has value not less than $m \ (\geqslant 0)$ under v_i, then f is divisible by t_i^m in* $k[x_1, \ldots, x_{16}, t_1, \ldots, t_{16}]$.

$\mathfrak{p}_i$ is generated by $z_i = a_{3i} y_1 - a_{1i} y_3$ and $z'_i = a_{3i} y_2 - a_{2i} y_3$. Obviously, $v_i(z_i) = v_i(z'_i) = 1$. Furthermore, y_i, z_i/t_i and z'_i/t_i are algebraically independent modulo the maximal ideal $\mathfrak{m}_i$ of $\mathfrak{v}_i$ over k. Therefore z_i/z_i' is transcendental over $k(y_3$ modulo $\mathfrak{m}_i)$. Hence

$$\mathfrak{H}[z_i/z_i']\mathfrak{n}_i \quad (\mathfrak{n}_i = \mathfrak{m}_i \cap \mathfrak{H}[z_i/z_i'])$$

is a valuation ring dominated by $\mathfrak{v}_i$. Therefore we have

(iii) *An element f of $\mathfrak{H}$ has value not less than m $(\geqslant 0)$ under v_i if and only if $f \in \mathfrak{p}_i^m$.*

Since $y_3, z_i/t_i, z_i'/t_i, t_1, \ldots, t_{i-1}, t_{i+1}, \ldots, t_{16}$ modulo $\mathfrak{m}_i$ are algebraically independent over k, we see that

(iv) *If a finite number of $a(n_1, \ldots, n_{16})$ are elements of $\mathfrak{H}$, then*

$$v_i(\Sigma a(n_1, \ldots, n_{16}) t_1^{-n_1} \ldots t_{16}^{-n_{16}}) = \min(v(a(n_1, \ldots, n_{16}) t_1^{-n_1} \ldots t_{16}^{-n_{16}})).$$

Now, (ii) and (iii) show that if $a(n_1, \ldots, n_{16})$ are in $\mathfrak{a}(n_1, \ldots, n_{16})$, then $\Sigma a(n_1, \ldots, n_{16}) t_1^{-n_1} \ldots t_{16}^{-n_{16}}$ (finite sum) is in $\mathfrak{v}$ (by virtue of (1)). Conversely, (i) and (iv) show that if c is an element of $\mathfrak{v}$ then c is of the form stated in (2). Thus (2) is proved completely.

Now we shall prove that $\mathfrak{v}$ is not finitely generated. Assume the contrary. Then there are a finite number of homogeneous forms $h(n_1, \ldots, n_{16})$ (in $\mathfrak{H}$) such that $h(n_1, \ldots, n_{16}) \in \mathfrak{a}(n_1, \ldots, n_{16})$ and $\mathfrak{v}$ is generated by the $h(n_1, \ldots, n_{16}) t_1^{-n_1} \ldots t_{16}^{-n_{16}}$ (each $n_i \geqslant 0$ and $\Sigma n_i > 0$) over $\mathfrak{H}[t_1, \ldots, t_{16}]$. Let r be the minimum of $\deg h(n_1, \ldots, n_{16})/\Sigma n_i$. By the fundamental lemma, we have $r > \frac{1}{4}$. Let r' be any rational number such that $r > r' > \frac{1}{4}$. Let m be a sufficiently large natural number such that $16r'm$ is an integer. Since $r' > \frac{1}{4}$, there exists a curve of degree $16r'm$ which goes through every P_i with multiplicity at least m, i.e. there is a homogeneous form h of degree $16mr'$ which is in $\mathfrak{a}(m, \ldots, m)$. Then $ht_1^{-m} \ldots t_{16}^{-m}$ is in $\mathfrak{v}$ and therefore $ht_1^{-m} \ldots t_{16}^{-m}$ can be expressed in a polynomial in the $h(n_1, \ldots, n_{16}) t_1^{-n_1} \ldots t_{16}^{-n_{16}}$ and therefore $\deg h/16m = r'$ cannot be less than r, which is a contradiction. Therefore $\mathfrak{v}$ is not finitely generated.

Remark. $k[x_1, \ldots, x_{16}, t_1, \ldots, t_{16}] \cap k(y_1, y_2, y_3, u)$ is not finitely generated, which gives another counter-example to the 14th problem of Hilbert.

REFERENCES

[1] Hilbert, D. Mathematische Probleme. *Archiv Math. Phys.* (3), 1, 44–63, 213–237 (1901).

[2] Zariski, O. Interprétations algébrico-géometriques du quatorzième problème de Hilbert. *Bull. Sci. Math.* (2) 78, 155–168 (1954).

[3] Nagata, M. A treatise on the 14th problem of Hilbert. *Mem. Coll. Sci. Univ. Kyoto*, A, 30, 57–70 (1956–57). Addition and correction to it, pp. 197–200.

[4] Rees, D. On a problem of Zariski. *Illinois J. Math.* 2, 145–149 (1958).

GEOMETRIC ASPECTS OF FORMAL DIFFERENTIAL OPERATIONS ON TENSOR FIELDS

By ALBERT NIJENHUIS

1. The concept of differential concomitant of tensor fields

We consider a manifold X of class C^∞, equipped with the set of all possible co-ordinate systems compatible with this structure. The symbol A shall denote this set; so that to each $\alpha \in A$ there belongs an open set U_α in X and co-ordinate functions $x_\alpha^1, \dots, x_\alpha^n$ on U_α, satisfying the usual conditions.

In each of the U_α, a C^∞ tensor field S has components with respect to $(x_\alpha^1, \dots, x_\alpha^n)$, which we denote, symbolically, by $S^{(\alpha)}$. Thus $S^{(\alpha)}$ stands for a certain set of real-valued C^∞ functions defined on U_α. The symbol $\partial^r S^{(\alpha)}$ denotes the set of partial derivatives of the functions $S^{(\alpha)}$ with respect to $(x_\alpha^1, \dots, x_\alpha^n)$ of orders from zero up to and including r.

Let the symbol G denote a finite set of functions of several variables. A tensor field T on X is said to be a *differential concomitant* of tensor fields $S_1, \dots, S_k$ of orders at most $r_1, \dots, r_k$ if there is a G such that, for each $\alpha \in A$, we have

$$T^{(\alpha)} = G(\partial^{r_1} S_1^{(\alpha)}, \dots, \partial^{r_k} S_k^{(\alpha)}). \tag{1}$$

In a similar fashion one can define differential concomitants involving a connection or other geometric objects.

In order that a G give rise to differential concomitants it has to satisfy a great many conditions, not only with regard to its domain of definition, but also a number of functional relations. The formula (1) can hold only if G commutes with co-ordinate transformations. Denote by $\lambda_{(\alpha)}^{(\beta)}$ the transformations of the component functions $S^{(\alpha)}$, as follows: $S^{(\beta)} = \lambda_{(\alpha)}^{(\beta)} S^{(\alpha)}$; then we must have

$$\lambda_{(\alpha)}^{(\beta)} G(\partial^{r_1} S_1^{(\alpha)}, \dots, \partial^{r_k} S_k^{(\alpha)}) = G(\partial^{r_1} \lambda_{(\alpha)}^{(\beta)} S_1^{(\alpha)}, \dots, \partial^{r_k} \lambda_{(\alpha)}^{(\beta)} S_k^{(\alpha)}). \tag{2}$$

In what follows we give an intrinsic approach to differential concomitants, and discuss some known special cases.

2. Natural bundles

Our concept of natural bundle is a generalization of that of tensor bundle, bundle of connections, bundle of jets of tensor fields (in the sense

of Ehresmann[3]), bundle of geometric objects (in the sense of Haantjes–
Laman[9]), and can easily be further extended to include also the approach
of Kuiper and Yano[13]. It should be pointed out here that the set $\{\partial^r S^{(\alpha)}\}$
represents a cross-section in a bundle of rth order jets of tensor fields.

A fiber bundle[21] over X, with bundle space B, projection π, and
fiber F, is called a *natural bundle* if

(1) the bundle space B and the fiber F are manifolds, and π is differentiable;

(2) with every diffeomorphism (differentiable homeomorphism with
differentiable inverse) $f: U \to X$ of an open set U of X into X there is
associated a differentiable mapping $f_B: \pi^{-1}(U) \to B$ such that

(a) f_B sends fibers into fibers by admissible maps[21]; if $f(x) = y$, then
$f_B(F_x) = F_y$, or equivalently: $\pi \circ f_B = f \circ \pi$ on $\pi^{-1}(U)$;

(b) if V is an open subset of U, then the 'lifting' of the restriction
$f \mid V$ is the restriction to $\pi^{-1}(V)$ of the lifting f_B: $(f \mid V)_B = f_B \mid \pi^{-1}(V)$;

(c) if 1_X denotes the identity map on X, then its lifting is the identity
map on B: $(1_X)_B = 1_B$;

(d) if f and g are diffeomorphisms, and $f \circ g$ is meaningful, then
$(f \circ g)_B = f_B \circ g_B$;

(3) all admissible maps of any fiber F_x into itself can be obtained as
the restriction to F_x of liftings f_B of diffeomorphisms f with $f(x) = x$.

A natural bundle is *linear* if

(4) the fibers are vector spaces.

A natural bundle is of *order* at most r if

(5) $f_B \mid F_x$ is the identity map whenever f is such that every differentiable path $c: R \to X$ with $c(0) = x$ has contact of order r at x with its
transform $f \circ c$.

Tensor bundles are linear, and of order 1. The most important special
case is the tangent bundle $\mathcal{T}(X)$, whose fiber over x is denoted by $\mathcal{T}_x(X)$.
Conversely, every linear bundle of order 1 is a Whitney sum of tensor
bundles; i.e. its fibers are the direct sums of tensor spaces associated
with the tangent space. The bundle of connections is of order 2.

If a natural bundle B is of order s, its rth *prolongation*, the bundle $\partial^r B$
of rth order jets of sections in B, is a natural bundle of order $r + s$. The
bundle ∂B of first-order jets has a simple geometric interpretation. The
tangent space $\mathcal{T}_b(B)$ has the vector subspace $\mathcal{T}_b(F_{\pi(b)})$. A vector subspace H_b of $\mathcal{T}_b(B)$ is called horizontal if $\mathcal{T}_b(B)$ is the direct sum of
$\mathcal{T}_b(F_{\pi(b)})$ and H_b. The bundle space of ∂B can then be identified with the
collection of all horizontal planes H_b for all $b \in B$. The projection map is
determined by $H_b \to \pi(b)$. Every section $S: X \to B$ in B is a differentiable

mapping, and induces a mapping S_* on the tangent bundle $\mathscr{T}(X)$ to $\mathscr{T}(B)$. Thus *the section S in B leads to a section ∂S in ∂B*, by

$$\partial S(x) = S_*(\mathscr{T}_x(X)),$$

since, indeed, $S_*(\mathscr{T}_x(X))$ is a horizontal space in $\mathscr{T}_{S(x)}(B)$.

For the higher bundles $\partial^s B$, $s > 1$, the situation is largely similar, but more involved. It should be noted, for instance, that $\partial^2 B$ is a proper sub-bundle of $\partial(\partial B)$.

Thus, prolongations of natural bundles are natural bundles. Also, Whitney sums of natural bundles are natural bundles. Finally, a sub-bundle B' of a natural bundle B (same base space; B' a submanifold of B the given bundle space) is again natural, provided all image sets $f_B(\pi^{-1}(U) \cap B')$ belong to B'.

3. Natural mappings

If B and B' are two natural bundles over X, a mapping $G \colon B \to B'$ is called *natural* if the following conditions are satisfied:

(1) G sends the fiber F_x of B into the fiber F'_x of B': or $\pi' \circ G = \pi$;

(2) if $f \colon U \to X$ is a diffeomorphism (U open in X), then for all points of $\pi^{-1}(U)$, $G \circ f_B = f_{B'} \circ G$.

A great many operations in tensor analysis are natural mappings. In fact, in many cases the statement that an operation is an 'invariant' one, means exactly that one deals with a natural mapping. For example, the contraction of tensors over one covariant and one contravariant index corresponds to a natural mapping of the bundle $\mathscr{T}^p_q(X)$ of tensors, p times contravariant, q times covariant, into $\mathscr{T}^{p-1}_{q-1}(X)$. The forming of the tensor product of tensors corresponds to a natural mapping of the Whitney sum of two tensor bundles into a third one:

$$\mathscr{T}^p_q(X) \oplus \mathscr{T}^r_s(X) \to \mathscr{T}^{p+r}_{q+s}(X).$$

More significant here is the case where the bundle B is a natural sub-bundle of a Whitney sum of prolongations of orders $r_1, \ldots, r_k$ of tensor bundles $B_1, \ldots, B_k$; and where B' is a tensor bundle. Then G is a *tensor concomitant* mapping.

When B is all of $\partial^{r_1} B_1 \oplus \ldots \oplus \partial^{r_k} B_k$, we have the situation of differential concomitants as mentioned in the beginning. Whenever B contains a full $\partial^{r_i} B_i$ as a Whitney summand, it is meaningful to ask if a natural mapping is linear in this summand, because $\partial^{r_i} B_i$ is a linear bundle if B_i is one.

30

Connections enter into consideration by studying mappings in which the (natural) bundle $\Gamma(X)$ of affine connections is involved. For example, the forming of the covariant derivative of vector fields corresponds to a natural mapping $\partial \mathscr{T}(X) \oplus \Gamma(X) \to \mathscr{T}_1^1(X)$. The Christoffel symbols $\left\{ \begin{matrix} \kappa \\ \lambda\mu \end{matrix} \right\}$ arise from a natural mapping $\partial \mathscr{S}_2(X) \to \Gamma(X)$, if $\mathscr{S}_2(X)$ denotes the natural bundle of symmetric covariant tensors of degree two, with non-zero determinant. The Riemann–Christoffel tensor arises from natural mappings

$$\partial\Gamma(X) \to \mathscr{T}_3^1(X) \quad \text{and} \quad \partial^2 \mathscr{S}_2(X) \to \mathscr{T}_3^1(X).$$

4. Examples and applications of differential concomitants

Among the natural mappings there are a number of 'trivial' ones, such as $\partial^{r+1}B \to \partial^r B$, and the usual algebraic operations: mappings of zero-order prolongations of tensor bundles.

The number of known non-trivial tensor concomitant mappings is strictly limited. Well known are the Lie bracket $[u, v]$ of vector fields, which corresponds to $\partial \mathscr{T}(X) \oplus \partial \mathscr{T}(X) \to \mathscr{T}(X)$; the exterior derivative $d\omega$ of a differential form ω, which corresponds to $\partial \mathscr{T}_{[p]}(X) \to \mathscr{T}_{[p+1]}(X)$ (square brackets indicate skew-symmetry); the Lie derivative $[u, S]$ of the tensor field S with respect to a vector field u, corresponding to

$$\partial \mathscr{T}(X) \oplus \partial \mathscr{T}_q^p(X) \to \mathscr{T}_q^p(X);$$

and the Riemann–Christoffel map $\partial^2 \mathscr{S}_2(X) \to \mathscr{T}_3^1(X)$. Of these examples only the last one is not linear.

Less well-known linear ones are:

(a) The Schouten invariants $[P, Q]$ of 1940[18]; where P and Q are both contravariant tensors, of degrees p and q respectively; and both are either symmetric or skew-symmetric (round brackets indicate symmetry):

$$\left. \begin{aligned} \partial \mathscr{T}^{(p)}(X) \oplus \partial \mathscr{T}^{(q)}(X) &\to \mathscr{T}^{(p+q-1)}(X), \\ \partial \mathscr{T}^{[p]}(X) \oplus \partial \mathscr{T}^{[q]}(X) &\to \mathscr{T}^{[p+q-1]}(X). \end{aligned} \right\} \tag{3}$$

The former is related to the Poisson bracket of functions on the cotangent bundle[16]; the latter is determined by the fact that if

$$P = u_1 \wedge \ldots \wedge u_p \quad \text{and} \quad Q = v_1 \wedge \ldots \wedge v_q,$$

then

$$[P, Q] = \sum_{i=1}^{p} \sum_{j=1}^{q} (-1)^{i+j} [u_i, v_j] \wedge u_1 \wedge \ldots \hat{u}_i \ldots \wedge u_p \wedge v_1 \wedge \ldots \hat{v}_j \ldots \wedge v_q. \tag{4}$$

Both generalize the Lie bracket of vector fields when $p = q = 1$; and if only $p = 1$, we have Lie derivatives.

(b) A Schouten invariant of 1954 ([20] see p. 117, Exercise II.11.1). If $\mathfrak{X}$ denotes densities of degree $+1$, then this one is of the type

$$\mathcal{T}_q^p(X) \oplus \mathfrak{X}_p^q(X) \to \mathfrak{X}_1(X).$$

It is related to Lagrangians, and has not yet been studied very far.

(c) An operation on so-called vector forms, tensors of the type $\mathcal{T}_{[p]}^1(X)$, $p \geqslant 0$. It is of the formal nature $\partial \mathcal{T}_{[p]}^1(X) \oplus \partial \mathcal{T}_{[q]}^1(X) \to \mathcal{T}_{[p+q]}^1(X)$; see [16]. It is in intimate relation with the theory of derivations on the ring of differential forms on a manifold [5]. Best known and oldest [15] is the case when $p = q = 1$, so we have $\partial \mathcal{T}_1^1(X) \oplus \partial \mathcal{T}_1^1(X) \to \mathcal{T}_{[2]}^1(X)$. If h and k are sections in $\mathcal{T}_1^1(X)$, thus representing fields of linear transformations in the tangent bundle; and if u and v are arbitrary tangent vector fields, then the differential concomitant $[h, k]$ is determined by

$$[h, k](u, v) = [hu, kv] + [ku, hv] - h[u, kv] - h[ku, v]$$
$$- k[u, hv] - k[hu, v] + hk[u, v] + kh[u, v]. \quad (5)$$

It is clear that if $h = k = J$ is an almost-complex structure in the sense of Ehresmann [2], then $\frac{1}{8}[J, J]$ is the torsion tensor of Eckmann–Frölicher [1, 4]. This latter one is of the type $\partial \mathcal{J}(X) \to \mathcal{T}_{[2]}^1(X)$, if $\mathcal{J}(X)$ is the bundle of tensors $J \in \mathcal{T}_1^1(X)$ with $JJ = -1$; but in the way it was originally defined this mapping was not quadratic. A result by Newlander and Nirenberg [14] implies that a C^∞ almost-complex structure is complex-analytic if and only if $[J, J] = 0$. A study of Tonolo on integrability of planes spanned by principal axes of tensor fields of type $\mathcal{S}_2(X)$ in a Riemannian space has culminated in a theorem by Haantjes [10] on integrability of planes spanned by eigenvectors of tensor fields h of type $\mathcal{T}_1^1(X)$, in which $[h, h]$ plays an essential part (see [5] for further references).

The most important applications of operation (c) have, so far, been to complex manifolds. If L is a vector form whose values lie in the summand $\mathcal{T}'(X)$ of the complexified tangent bundle, spanned by $\partial/\partial z^1, \ldots, \partial/\partial z^n$ ($z^1, \ldots, z^n$ complex co-ordinates), then $i[J, L] = 2\mathcal{D}''L$, where $\mathcal{D}''L$ denotes the exterior derivative in the theory of complex vector bundles [8]. A Jacobi-type identity for the operation (c) leads not only to cohomology groups $H^p(X, \Theta_q)$, where Θ_q is the sheaf of germs of holomorphic vector q-forms; but also to mappings [6]:

$$H^p(X, \Theta_q) + H^{p'}(X, \Theta_{q'}) \to H^{p+p'}(X, \Theta_{q+q'}). \quad (6)$$

A family J_t, $0 \leqslant t \leqslant 1$, of complex structures satisfies $[J_t, J_t] = 0$, whence $[J_t, dJ_t/dt] = 0$; and dJ_t/dt determines an element of $H^1(X, \Theta_0(J_t))$. If X is compact, and if for all t the group $H^1(X, \Theta_0(J_t))$ vanishes, then the structures J_t, $0 \leqslant t \leqslant 1$ are all equivalent[7]. If the group vanishes for $t = 0$, then by a theorem of Kodaira and Spencer[11] it also vanishes for t near zero, so that under small deformations the complex structure is sent into equivalent ones. A recent result of Kodaira, Nirenberg and Spencer[12] shows that if $H^2(X, \Theta_0(J)) = 0$, then to every element α of $H^1(X, \Theta_0(J))$ there belongs a family J_t of complex structures with $J = J_0$ whose $(dJ_t/dt)_{t=0}$ determines the element α. The parameter t can be taken as a complex variable, with J_t holomorphic in t.

5. Final remarks

5.1. Instead of considering the diffeomorphisms $f: U \to X$ (U open in X), one can consider an arbitrary transitive pseudo-group of mappings f in a topological space. Kuiper and Yano[13] considered pseudo-groups of diffeomorphisms in manifolds.

5.2. It seems to be somewhat easier to find tensor concomitant mappings $B \to B'$ if B is a proper natural sub-bundle of a Whitney sum $\partial^{r_1} B_1 \oplus \dots \oplus \partial^{r_k} B_k$ of prolongations of tensor bundles. An early example is given by Schouten[19], of the form $B \subset \partial \mathscr{T}_0^2(X) \oplus \partial \mathscr{T}_1^1(X)$. The components of the tensors u and h must be related by $h_\mu{}^\rho u^{\mu\lambda} = h_\mu{}^\lambda u^{\rho\mu}$. There are numerous tensor concomitant mappings in the theory of almost-complex manifolds. Many are linear in all entries except the one for J, the almost-complex structure tensor. A remarkable example is Walker's torsional derivative[22]. Several other examples are found in [8].

5.3. The theory of the special concomitants $[u, v]$, $[u, S]$, $d\omega$, and (a), (c) above can be vastly generalized to a purely algebraic theory associated with commutative algebras with unit. A brief sketch of such a purely formal and abstract approach has been announced[17]. The theory is non-void only for algebras with a non-trivial Lie algebra of derivations.

REFERENCES

[1] Eckmann, B. and Frölicher, A. Sur l'intégrabilité des structures presque complexes. *C.R. Acad. Sci., Paris*, 232, 2284–2286 (1951).

[2] Ehresmann, C. Sur les variétés presque complexes. *Proc. Int. Congr. Mathematicians 1950*, 2, 412–419 (1950).

[3] Ehresmann, C. Les prolongements d'une variété différentiable. *Atti IV Congr. U.M.I. Taormina*, 1950.

DIFFERENTIAL OPERATIONS ON TENSOR FIELDS **469**

[4] Frölicher, A. Zur Differentialgeometrie der komplexen Strukturen. *Math. Ann.* 129, 50–95 (1955).

[5] Frölicher, A. and Nijenhuis, A. Theory of vector-valued differential forms. I. Derivations in the graded ring of differential forms. *Proc. K. Akad. Wet. Amst.* A 59 = *Indag. Math.* 18, 338–359 (1956).

[6] Frölicher, A. and Nijenhuis, A. Some new cohomology invariants for complex manifolds. *Proc. K. Akad. Wet. Amst.* A 58 = *Indag. Math.* 17, 540–564 (1956).

[7] Frölicher, A. and Nijenhuis, A. A theorem on stability of complex structures. *Proc. Nat. Acad. Sci., Wash.*, 43, 239–241 (1957).

[8] Frölicher, A. and Nijenhuis, A. Theory of vector-valued differential forms. II. Almost-complex structures. *Proc. K. Akad. Wet. Amst.* A 61 = *Indag. Math.* 20, 414–429 (1958).

[9] Haantjes, J. and Laman, G. On the definition of geometric objects. I, II. *Proc. K. Akad. Wet. Amst.* A 56 = *Indag. Math.* 15, 208–222 (1953).

[10] Haantjes, J. On X_m-forming sets of eigenvectors. *Proc. K. Akad. Wet. Amst.* A 58 = *Indag. Math.* 17, 158–162 (1955).

[11] Kodaira, K. and Spencer, D. C. On the variation of almost-complex structure. *Algebraic Geometry and Topology*. Princeton, 1957.

[12] Kodaira, K., Nirenberg, L. and Spencer, D. C. On the existence of deformations of complex analytic structures. *Ann. Math.* (2), 68, 450–459 (1958).

[13] Kuiper, N. H. and Yano, K. On geometric objects and Lie groups of transformations. *Proc. K. Akad. Wet. Amst.* A 58 = *Indag. Math.* 17, 411–420 (1955).

[14] Newlander, A. and Nirenberg, L. Complex analytic coordinates in almost-complex manifolds. *Ann. Math.* (2), 65, 391–404 (1957).

[15] Nijenhuis, A. X_{n-1}-forming sets of eigenvectors. *Proc. K. Akad. Wet. Amst.* A 54 = *Indag. Math.* 13, 200–212 (1951).

[16] Nijenhuis, A. Jacobi-type identities for bilinear differential concomitants of certain tensor fields. *Proc. K. Akad. Wet. Amst.* A 58 = *Indag. Math.* 17, 390–403 (1955).

[17] Nijenhuis, A. Differential-geometric operations on rings. *Bull. Amer. Math. Soc.* 63, Abstr. 575, p. 288 (1957).

[18] Schouten, J. A. Ueber Differentialkomitanten zweier kontravarianter Grössen. *Proc. K. Akad. Wet. Amst.* 43, 3–6 (1940).

[19] Schouten, J. A. On the differential operators of first order in tensor calculus. *Convegno di Geom. Diff.* 1953. Rome, 1954.

[20] Schouten, J. A. *Ricci Calculus*. Berlin, 1954.

[21] Steenrod, N. *The Topology of Fiber Bundles*. Princeton, 1951.

[22] Walker, A. G. Dérivation torsionnelle et seconde torsion pour une structure presque complexe. *C.R. Acad. Sci., Paris*, 245, 1213–1215 (1957).

RELATIONS D'ÉQUIVALENCE EN GÉOMÉTRIE ALGÉBRIQUE

Par PIERRE SAMUEL

1. Relations d'équivalence adéquates

Soit V une variété algébrique que, pour simplifier, nous supposerons projective et non-singulière. Etant donnés deux cycles X, Y sur V, le produit d'intersection $X.Y$ n'est pas toujours défini, ce qui a pour conséquence que les cycles ne forment pas un anneau. Pour remédier à cet état de choses, il est utile de remplacer les cycles eux-mêmes par des classes de cycles modulo une relation d'équivalence convenable. D'autre part le groupe des cycles sur V a une structure assez compliquée; il est commode d'y définir des sous-groupes et des groupes quotients plus faciles à étudier, et ceux-ci permettent de définir d'intéressants invariants de V (l'irrégularité, les nombres ρ et σ, etc.). Enfin l'équivalence linéaire des diviseurs a, comme chacun sait, une importance considérable.

Pour être maniable et intéressante, une relation d'équivalence entre cycles doit vérifier un certain nombre de propriétés. Nous dirons qu'une relation d'équivalence, notée $\sim$, est *adéquate* si elle est définie entre cycles de toute variété projective non-singulière, et si elle vérifie les conditions suivantes:

(RA$_{\mathrm{I}}$) *Elle est compatible avec l'addition des cycles.* Autrement dit la restriction de $\sim$ au groupe $\mathfrak{G}^{(d)}(V)$ des cycles de codimension d sur V coïncide avec la relation de congruence modulo un sous-groupe bien déterminé $\mathfrak{G}_e^{(d)}(V)$.

(RA$_{\mathrm{II}}$) *Etant donnés un cycle X sur V et des sous-variétés W_j de V en nombre fini, il existe un cycle $X' \sim X$ tel que $X.W_j$ soit défini quel que soit j.*

(RA$_{\mathrm{III}}$) *Etant donnés deux variétés* (projectives non singulières comme toujours) *V et W, un cycle $X \sim 0$ sur V, et un cycle Z sur $V \times W$ tel que $Z(X) = pr_W((X \times W).Z)$ soit défini, alors $Z(X)$ est ~ 0 sur W.*

Remarque. De nombreuses relations d'équivalence adéquates vérifient aussi:

(RA$_{\mathrm{IV}}$) *Si un cycle X est ~ 0 sur V, et si (V', X') est une spécialisation de (V, X) sur un corps k (V' étant supposée irréductible et non-singulière), alors X' est ~ 0 sur V'.*

Dans les sept propositions faciles qui suivent, $\sim$ désigne une relation d'équivalence adéquate, V et W des variétés projectives non-singulières.

Proposition 1. *Soient* V, W *deux variétés*, X *un cycle sur* V. *Si* $X \sim 0$, *on a* $X \times W \sim 0$ *sur* $V \times W$.

Notons en effet G le graphe de pr_V dans $V \times (V \times W)$, c'est-à-dire l'ensemble des points $(x, (x, y))$. On a $G(X) = X \times W$, d'où notre assertion par $(\mathrm{RA}_{\mathrm{III}})$.

Proposition 2. *Soient* X, Y *deux cycles sur* V *tels que* $X . Y$ *soit défini. Si* $X \sim 0$ *sur* V, *alors* $X . Y \sim 0$ *sur* V.

Soient en effet D la diagonale de $V \times V$ et Z le cycle $(V \times Y) . D$ sur $V \times V$ (cycle qui est toujours défini). Comme $(X \times Y) . D$ est défini, il en est de même, par associativité, de $(X \times V) . (V \times Y) . D = (X \times V) . Z$. Donc $Z(X)$ est défini, et évidemment égal à $X . Y$. D'où notre assertion par $(\mathrm{RA}_{\mathrm{III}})$.

Proposition 3. *Soit* X *un cycle sur* $V \times W$. *Si* $X \sim 0$, *alors* $\mathrm{pr}_V(X) \sim 0$ *sur* V.

Soit en effet H le graphe de pr_V dans $(V \times W) \times V$ (ensemble des points $((x, y), x)$). On a $\mathrm{pr}_V(X) = H(X)$, d'où notre assertion.

Remarque. On voit aussitôt que les prop. 1, 2, 3 entraînent $(\mathrm{RA}_{\mathrm{III}})$.

Proposition 4. *Soient* X *un cycle sur* V *et* W *une sous-variété non-singulière de* V *telle que* $X . W$ *soit défini. Si* $X \sim 0$ *sur* V, *alors* $X . W \sim 0$ *sur* W.

En effet, en notant I le graphe, dans $V \times W$, de l'application identique de W dans V, on a $W . X = \mathrm{pr}_W((X \times W) . I)$. D'où notre assertion par $(\mathrm{RA}_{\mathrm{III}})$.

Proposition 5. *Soient* W *une sous-variété non-singulière de* V, *et* Y *un cycle sur* W. *Si on a* $Y \sim 0$ *sur* W, *alors* $Y \sim 0$ *sur* V.

En effet, avec les notations précédentes, on a $Y = \mathrm{pr}_V((V \times Y) . I)$.

Remarque. La réciproque est fausse.

Proposition 6. *Les classes de cycles sur* V *pour la relation* $\sim$ *forment un anneau commutatif pour l'addition et la multiplication déduite du produit d'intersection. Cet anneau est gradué par la codimension. La classe de* V *est élément unité.*

En effet, d'après $(\mathrm{RA}_{\mathrm{I}})$, les classes de cycles forment un groupe gradué $\mathfrak{E}(V)$. Etant données deux classes ξ et η, $(\mathrm{RA}_{\mathrm{II}})$ montre qu'il existe des représentants X de ξ et Y de η tels que $X . Y$ soit défini. Si X' et Y' sont d'autres représentants de ξ et η tels que $X' . Y'$ soit défini, on choisit $(\mathrm{RA}_{\mathrm{II}})$ un représentant Y'' de η tel que $X . Y''$ et $X' . Y''$ soient définis; comme

$$X' . Y' - X' . Y'' = X' . (Y' - Y'')$$

et

$$X . Y - X' . Y'' = X . (Y - Y'') + (X - X') . Y'',$$

on a $X' . Y' \sim X' . Y'' \sim X . Y$ (prop. 2); ceci montre que la classe de

472 PIERRE SAMUEL

$X.Y$ ne dépend que de ξ et η. On la note $\xi\eta$. Ceci définit une multi-plication sur $\mathfrak{E}(V)$. Le fait que $\mathfrak{E}(V)$ est un anneau gradué par la co-dimension et admettant la classe de V pour élément unité est alors conséquence immédiate des propriétés classiques des intersections.

Remarque. Si on se donne aussi une relation d'équivalence adéquate $\equiv$ *moins fine* que $\sim$, les classes modulo $\sim$ des cycles $\equiv 0$ sur V forment un *idéal* homogène de l'anneau $\mathfrak{E}(V)$ (prop. 2 appliquée à $\equiv$).

Nous noterons désormais $\mathfrak{E}(V)$ l'anneau des classes de cycles sur V pour la relation d'équivalence $\sim$.

Proposition 7. Soit Z un cycle sur $V \times W$. Alors $X \to Z(X)$ définit un homomorphisme du groupe additif de $\mathfrak{E}(V)$ dans celui de $\mathfrak{E}(W)$; cet homo-morphisme ne dépend que de la classe de Z dans $\mathfrak{E}(V \times W)$. Lorsque Z est le graphe d'un morphisme f de W dans V, c'est un homomorphisme pour les structures d'anneaux.

Il résulte en effet du lemme 2, no. 1, de[6], et de $(\mathrm{RA_{II}})$ que, dans toute classe $\xi \in \mathfrak{E}(V)$, il existe un cycle X tel que $(X \times W).Z$, donc $Z(X)$, soit défini. La classe de $Z(X)$ ne dépend que de ξ (par $(\mathrm{RA_{III}})$), d'où une application de $\mathfrak{E}(V)$ dans $\mathfrak{E}(W)$, qui est évidemment un homomorphisme additif. Si Z' est un cycle sur $V \times W$ tel que $Z' \sim Z$, on choisit (ce qui est loisible) un représentant X de ξ tel que $(X \times W).Z$ et $(X \times W).Z'$ soient définis; on a alors $(X \times W).Z \sim (X \times W).Z'$ (prop. 2), d'où $Z(X) \sim Z'(X)$ (prop. 3); ceci démontre notre seconde assertion.

Supposons enfin que Z soit le graphe d'un morphisme f de W dans V; on a alors $Z(X) = f^{-1}(X)$ lorsque ce cycle est défini. Nous avons à démontrer que, étant données deux classes ξ, $\eta \in \mathfrak{E}(V)$, il existe des représentants X, Y de ces classes tels que $f^{-1}(X.Y) = f^{-1}(X).f^{-1}(Y)$ (les deux membres étant définis). Comme plus haut prenons $X \in \xi$ tel que $(X \times W).Z$ soit défini. Il s'agit alors de trouver $Y \in \eta$ tel que $Y.X$, $(Y \times W).Z$ et $(Y \times W).(X \times W).Z$ soient définis; il suffit pour cela que $(Y \times W).((X \times W).Z)$ soit défini, et ceci est réalisable comme plus haut (en remplaçant Z par $(X \times W).Z$). Ceci étant, l'hypothèse montre que la restriction de pr_W à Z est un isomorphisme de Z sur W; donc Z est une variété non-singulière. On a donc

$$(X \times W).Z.(Y \times W) = ((X \times W).Z)._Z((Y \times W).Z)$$

(le $._Z$ désignant le produit d'intersection sur Z) $= (X.Y \times W).Z$. En appliquant l'isomorphisme pr_W de Z sur W, on obtient

$$Z(X).Z(Y) = Z(X.Y),$$

c'est-à-dire $f^{-1}(X.Y) = f^{-1}(X).f^{-1}(Y)$. C.Q.F.D.

Remarque. La dernière assertion ne subsiste pas lorsque Z est seulement le graphe d'une application rationnelle f de W dans V (exemple simple: W est un plan, V une transformée quadratique de W).

Avec les notations de la prop. 7, l'homomorphisme additif de $\mathfrak{E}(V)$ dans $\mathfrak{E}(W)$ sera noté Z_*; si Z est le graphe d'un morphisme f de W dans V, nous écrirons f^* au lieu de Z_*. Il est facile de voir que $V \to \mathfrak{E}(V)$ est un *foncteur contravariant* pour la catégorie des variétés projectives non-singulières et des morphismes de variétés.

Remarque. Si W est une sous-variété non singulière de V, l'injection $i : W \to V$ définit un homomorphisme i^* de l'anneau $\mathfrak{E}(V)$ dans l'anneau $\mathfrak{E}(W)$. Ainsi $\mathfrak{E}(W)$ est un *module* sur $\mathfrak{E}(V)$.

2. Exemples de relations d'équivalence adéquates

2.1. Equivalence rationnelle.

Rappelons[6] qu'un cycle X sur une variété projective non-singulière V est dit *rationnellement èquivalent à* 0 s'il existe une variété unirationnelle (resp. une droite projective, un espace projectif, un produit de droites projectives) T, un cycle Z de $T \times V$ et deux points a, b de T tels que $Z(a)$ et $Z(b)$ soient définis et que $X = Z(b) - Z(a)$. La relation '$X - Y$ est rationnellement équivalent à 0' entre cycles X, Y de V est une relation d'équivalence adéquate (*ibid.*).†

Notons la propriété extrémale suivante:

Proposition 8. *L'équivalence rationnelle est la plus fine des relations d'équivalence adéquates.*

Soit en effet $\sim$ une relation d'équivalence adéquate. Il suffit, d'après $(\mathrm{RA_{III}})$, de montrer qu'on a $(a) \sim (b)$ pour deux points quelconques a, b de la droite projective P_1. Or, étant donné $a \in P_1$, il existe d'après $(\mathrm{RA_{II}})$ un cycle $Y = \sum_i n_i(b_i)$ sur P_1 tel que $Y \sim (a)$ et que (a). Y soit défini; ceci veut dire $b_i \neq a$ pour tout i. Etant donné $b \neq a$, il existe une application rationnelle (donc un morphisme) de P_1 sur P_1 telle que $f(a) = a$, $f(b_i) = b$ pour tout i (par exemple, en prenant a à l'infini, le polynôme $x \to b + \prod_i (x - b_i)$). Donc, d'après $(\mathrm{RA_{III}})$, on a $(a) \sim n(b)$ (où $n = \sum_i n_i$). Comme il existe un morphisme g de P_1 dans P_1 tel que $g(a) = b$ et $g(b) = b$, on a $(b) \sim n(b)$; d'où $(a) \sim (b)$ par transitivité. C.Q.F.D.

Remarque. La propriété $(\mathrm{RA_{IV}})$ de spécialisation est vraie pour l'équivalence rationnelle ([6], th. 4).

---

† Des lacunes dans la démonstration de $(\mathrm{RA_{II}})$ ont été comblées par C. Chevalley.

2.2. Equivalence algébrique. Rappelons[8] qu'un cycle X sur une variété projective non singulière V est dit *algébriquement équivalent à* 0 s'il existe une variété (resp. une courbe, une variété abélienne) T, un cycle Z de $T \times V$ et deux points a, b de T tels que $Z(a)$ et $Z(b)$ soient définis et que $X = Z(b) - Z(a)$. Il résulte de [8] et des raisonnements faits dans [6] que la relation '$X - Y$ est algébriquement équivalent à 0' est une relation d'équivalence adéquate (pour (RA$_{II}$) on constate que l'équivalence algébrique est moins fine que l'équivalence rationnelle). La propriété (RA$_{IV}$) est évidemment vraie.

2.3. Equivalence numérique. Etant donné un cycle Z de dimension 0, nous noterons deg(Z) son degré. On dit que deux cycles X, X' (sur une variété projective non singulière V) sont numériquement équivalents s'ils ont même dimension et si, pour tout cycle Y de dimension complémentaire tel que $X.Y$ et $X'.Y$ soient définis, on a

$$\deg(X.Y) = \deg(X'.Y).$$

Comme, d'après le principe de conservation du nombre, on peut remplacer Y par un cycle algébriquement équivalent, cette relation est bien une relation d'équivalence vérifiant (RA$_I$). Le principe de conservation du nombre montre aussi qu'elle est moins fine que l'équivalence algébrique, donc qu'elle vérifie (RA$_{II}$). Au lieu de vérifier (RA$_{III}$) nous vérifierons que les conclusions des prop. 1, 2, 3 sont vraies: pour les prop. 1 et 3 ceci résulte de la formule de projection et du fait qu'un cycle de dimension 0 sur un produit a même degré que ses projections; pour la prop. 2 on applique la formule d'associativité. La propriété (RA$_{IV}$) est vraie.

2.4. Equivalence du carré et du n-cube. Soit n un entier > 1. On dit qu'un cycle X sur une variété projective non-singulière V est n-*cubique* s'il existe n variétés de paramètres T_i, $2n$ points a_{ij} ($j = 1, 2$, $a_{ij} \in T_i$) et un cycle Z sur $T_1 \times T_2 \times \ldots \times T_n \times V$ tels que les 2^n cycles

$$Z(a_{1,j(1)}, \ldots, a_{n,j(n)}) = \mathrm{pr}_V((a_{1,j(1)} \times \ldots \times a_{n,j(n)} \times V).Z)$$

soient définis et que l'on ait

$$X = \sum_{j \in (1,2)^n} Z(a_{1,j(1)}, \ldots, a_{n,j(n)}) (-1)^{j(1)+\cdots+j(n)}. \tag{1}$$

Nous allons voir que, pour tout n, la relation '$X - Y$ est n-cubique' est une relation d'équivalence adéquate; nous l'appellerons l'*équivalence n-cubique*. Comme tout cycle n-cubique est différence de deux cycles

$(n-1)$-cubiques, l'équivalence n-cubique est plus fine que l'équivalence $(n-1)$-cubique. Pour $n=1$ on obtient l'équivalence algébrique; pour $n=2$ nous parlerons de cycles *carrés* et d'équivalence *carrée*. Pour vérifier qu'il s'agit bien d'équivalence adéquates, nous nous bornerons, pour alléger les formules, au cas $n=2$ (le cas général est analogue).

(RA_I). Soient X et X_1 des cycles carrés de même dimension sur V; il nous suffit de montrer que $X+X_1$ est carré. Notons T, T', T_1, T'_1 des variétés de paramètres, a et b (resp. a' et b', a_1 et b_1, a'_1 et b'_1) des points de T (resp. T', T_1, T'_1) tels que

$$X = Z(a,a') - Z(a,b') - Z(b,a') + Z(b,b'),$$

$$X_1 = Z_1(a_1,a'_1) - Z_1(a_1,b'_1) - Z_1(b_1,a'_1) + Z_1(b_1,b'_1),$$

où Z est un cycle sur $V \times T \times T'$, et Z_1 un cycle sur $T_1 \times T'_1 \times V$. Considérons, sur $T_1 \times T'_1 \times V \times T \times T'$, le cycle

$$Y = (Z_1 \times T \times T') + (T_1 \times T'_1 \times Z).$$

On vérifie qu'on a

$$Y(a_1,a'_1,a,a') - Y(a_1,b'_1,a,b') - Y(b_1,a'_1,b,a') + Y(b_1,b'_1,b,b') = X+X_1;$$

donc $X+X_1$ est un cycle carré.

(RA_{II}). Il nous suffira de montrer que tout cycle rationnellement équivalent à 0 est carré. Soit donc X un cycle sur V tel que

$$X = Z(b) - Z(a),$$

Z étant un cycle sur $V \times P_1$ et a et b deux points de P_1 que nous supposerons à distance finie. Soit T le graphe, dans $P_1 \times (P_1 \times P_1)$, de l'application $(x,y) \to (x-a)y/(b-a)$ de $P_1 \times P_1$ dans P_1. Posons

$$Y = Z \circ T = \mathrm{pr}_V \times {}_{(P_1 \times P_1)}((Z \times (P_1 \times P_1)) . (V \times T)).$$

Si $Z(T(x,y))$ est défini, il est égal à $Y(x,y)$ $(x,y \in P_1)$: en effet

$$(Z \times (P_1 \times P_1)) . (V \times T) . (V \times P_1 \times (x,y))$$

est alors défini, donc égal à

$$(Z \times (P_1 \times P_1)) . (V \times T(x,y) \times (x,y)) = (Z . (V \times T(x,y))) \times (x,y).$$

Comme $\quad T(a,a) = T(a,b) = 0, \quad T(b,a) = a, \quad T(b,b) = b,$
on a

$$Y(a,a) - Y(a,b) - Y(b,a) + Y(b,b) = Z(0) - Z(0) - Z(a) + Z(b) = X;$$

ceci montre que X est carré. C.Q.F.D.

(RA_{III}). Comme ci-dessus vérifions que les conclusions des prop. 1, 2, 3 sont vraies.

(1) Si $X = Z(a, a') - Z(a, b') - Z(b, a') + Z(b, b')$ est un cycle carré sur V (Z étant un cycle sur $T \times T' \times V$), on a évidemment, en posant

$$Y = Z \times W, \quad X \times W = Y(a, a') - Y(a, b') - Y(b, a') + Y(b, b');$$

donc $X \times W$ est carré.

(2) Si X est carré comme dans (1), et si X' est un cycle sur V tel que $X.X'$ soit défini, nous choisissons un cycle X'' rationnellement équivalent à X' tel que $Z(a, a').X''$, $Z(a, b').X''$, $Z(b, a').X''$, $Z(b, b').X''$ soient définis. Alors $Y = (T \times T' \times X'').Z$ est défini, et on a

$$Y(a, a') = Z(a, a').X''$$

et trois autres formules analogues. D'où

$$Y(a, a') - Y(a, b') - Y(b, a') + Y(b, b') = X.X'',$$

ce qui montre que $X.X''$ est carré. Comme $X.X'$ lui est rationnellement équivalent, il est aussi carré d'après ce qui a été vu dans la vérification de (RA$_{\mathrm{II}}$).

(3) Soit X un cycle carré sur $V \times W$; posons $X = Z(a, a') - \dots$, Z étant un cycle sur $T \times T' \times V \times W$. En posant

$$Y = \mathrm{pr}_{T \times T' \times V}(Z), \quad \text{on a} \quad Y(a, a') = \mathrm{pr}_V(Z(a, a'))$$

d'après la formule de projection. Donc $\mathrm{pr}_V(X)$ est carré.

Remarques. (1) Comme dans le cas de l'équivalence algébrique[8] on peut se limiter au cas où les variétés de paramètres T_i sont des courbes (resp. des variétés abéliennes). Comme, sur $T_1 \times \dots \times T_n$, il existe une courbe T joignant les deux points $(a_{1,1}, \dots, a_{n,1}) = a$ et $(a_{1,2}, \dots, a_{n,2}) = b$, on peut se limiter au cas où toutes les T_i sont égales à une même courbe T, les points $a_{i,1}$ (resp. $a_{i,2}$) étant tous égaux. De même avec des variétés abéliennes.

(2) Comme pour l'équivalence rationnelle ([6], th. 2) on peut supposer que le cycle Z de la formule (1) est positif.

(3) En prenant des droites (resp. des variétés unirationnelles) pour les T_i, on obtient une équivalence n-cubique 'rationnelle', qui est adéquate, et plus fine que l'équivalence rationnelle. Mais le prop. 8 (ou le raisonnement fait ci-dessus pour (RA$_{\mathrm{II}}$)) montre qu'elle est identique à l'équivalence rationnelle.

Proposition 9. Le produit cartésien et la produit d'intersection de deux (resp. n) cycles algébriquement équivalents à 0 est un cycle carré (resp. n-cubique).

Soient en effet X, X' des cycles sur V, V', T et T' des variétés de paramètres, Z et Z' des cycles sur $T \times V$ et $T' \times V'$, a, b des points de T, a', b' des points de T' tels que

$$X = Z(b) - Z(a) \quad \text{et} \quad X' = Z'(b') - Z'(a').$$

Posons $Y = Z \times Z'$. On a

$$Y(a, a') = \text{pr}_{V \times V'}(((a) \times V \times (a') \times V').(Z \times Z'))$$
$$= \text{pr}_{V \times V'}((a) \times Z(a) \times (a') \times Z'(a')) = Z(a) \times Z'(a').$$

De ceci et de trois autres formules analogues il résulte qu'on a

$$X \times X' = Y(a, a') - Y(a, b') - Y(a', b) + Y(b, b');$$

d'où l'assertion relative au produit cartésien. Pour le produit d'intersection, on utilise le procédé du produit et de la diagonale et on applique (RA$_{\text{III}}$).

2.5. Equivalence abélienne. Soient V une variété projective non-singulière, $\mathfrak{G}^{(d)}(V) = \mathfrak{G}$ le groupe des cycles de codimension d sur V, et A une variété abélienne. Nous dirons qu'un homomorphisme h de $\mathfrak{G}$ dans A est *rationnel* s'il existe un corps k de définition de V et A tel que

(HR$_{\text{I}}$). *Si $X \in \mathfrak{G}$ est rationnel sur $k' \supset k$, $h(X)$ est un point rationnel sur k';*

(HR$_{\text{II}}$). *Si $X, X' \in \mathfrak{G}$ sont deux cycles tels que X' soit une spécialisation de X sur un corps $k' \supset k$, alors $(X', h(X'))$ est une spécialisation de $(X, h(X))$ sur k'.*

Il revient au même de dire que la restriction de h à chaque système algébrique de cycles est une application rationnelle. Nous dirons qu'un cycle $X \in \mathfrak{G}$ est *abélien* si on a $h(X) = 0$ pour toute variété abélienne A et tout homomorphisme rationnel h de $\mathfrak{G}$ dans A. La relation '$X - X'$ est abélien' est une relation d'équivalence *adéquate*. En effet (RA$_{\text{I}}$) est évident, et (RA$_{\text{II}}$) résulte du fait que, puisque toute application rationnelle d'une droite dans une variété abélienne est constante, tout cycle rationnellement équivalent à 0 est abélien. Reste à vérifier (RA$_{\text{III}}$).

Considérons pour cela un cycle abélien $X \in \mathfrak{G}^{(d)}(V)$, un cycle Z sur $V \times W$ tel que $Z(X)$ soit défini, et un homomorphisme rationnel h de $\mathfrak{G}^{(d')}(W)$ dans une variété abélienne A ($d' = \dim Z(X)$). Pour tout $X' \in \mathfrak{G}^{(d)}(V)$, choisissons un cycle X'_1 rationnellement équivalent à X' tel que $Z(X'_1)$ soit défini, et posons $u(X') = h(Z(X'_1))$; on voit facilement que $u(X')$ ne dépend pas du choix de X'_1 et que u est un homomorphisme de $\mathfrak{G}^{(d)}(V)$ dans A. Prenons pour k un corps infini de définition de V, W,

Z et h. Si X' est rationnel sur $k' \supset k$, la méthode des cônes ([6], lemme 3) montre qu'on peut prendre X'_1 rationnel sur k'; donc $u(X') = h(Z(X'_1))$ est rationnel sur k', et (HR$_\mathrm{I}$) est vérifié par u. Soit enfin $\mathfrak{S}$ un système algébrique irréductible contenu dans $\mathfrak{G}^{(d)}(V)$; si $Z(X_g)$ n'est pas défini pour un élément générique X_g de $\mathfrak{S}$ sur k', il existe X'_g rationnellement équivalent à X_g tel que $Z(X'_g)$ soit défini, et la méthode des cônes montre qu'on peut supposer que X_g est une spécialisation de X'_g sur k'; soit $\mathfrak{S}'$ le lieu de X'_g sur k'; alors $\mathfrak{S}$ peut être considéré comme plongé dans $\mathfrak{S}'$; d'autre part $X'_g \to Z(X'_g)$ définit une application rationnelle f de $\mathfrak{S}'$ dans le lieu de $Z(X'_g)$ sur k', et, sur le domaine de définition f, on a $u = h \circ f$; ainsi la restriction de u à $\mathfrak{S}'$ est une application rationnelle dans A, donc aussi la restriction de u à $\mathfrak{S}$. Ceci démontre que u est un homomorphisme rationnel de $\mathfrak{G}^{(d)}(V)$ dans A. Comme X est abélien, on a $u(X) = 0$, d'où $h(Z(X)) = 0$; ceci montre que $Z(X)$ est abélien et démontre (RA$_\mathrm{III}$). La relation '$X - X'$ est abélien' s'appelle l'*équivalence abélienne*.

Remarques. (1) Il est évident que l'équivalence abélienne vérifie (RA$_\mathrm{IV}$).

(2) Etant donnés une variété V et un entier d, on peut se demander s'il existe une variété abélienne A et un homomorphisme rationnel h de $\mathfrak{G}^{(d)}(V)$ dans A qui soient *universels* pour les homomorphismes rationnels de $\mathfrak{G}^{(d)}(V)$. La réponse est affirmative en dimension 0 (variété d'Albanese) et en codimension 1 (variété de Picard).

(3) Voici un exemple d'homomorphisme rationnel. Soit W une sous-variété de V; pour tout cycle X sur V de dimension complémentaire à celle de W et tel que $X.W$ soit défini, on pose $h(X) = S(f(X.W))$, f étant l'application canonique de V (ou de W) dans sa variété d'Albanese; lorsque $X.W$ n'est pas défini, on définit $h(X)$ par équivalence rationnelle.

Proposition 10. L'équivalence abélienne est plus fine que l'équivalence algébrique.

Il suffit de démontrer que, si X est un cycle non algébriquement équivalent à 0 sur V, il existe un homomorphisme rationnel h de $\mathfrak{G}(V)$ dans une variété abélienne A tel que $h(X) \neq 0$. Pour cela notons $\mathfrak{G}_a$ le groupe des cycles algébriquement équivalents à 0 sur V de dimension $\dim(X)$ et démontrons le lemme suivant:

Lemme. Soit h un homomorphisme de $\mathfrak{G}_a$ dans une variété abélienne $A \neq (0)$ vérifiant (HR$_\mathrm{I}$) et (HR$_\mathrm{II}$). Alors h se prolonge en un homomorphisme rationnel h' de $\mathfrak{G}$ dans A tel que $h'(X) \neq 0$.

En effet on peut supposer que le corps k intervenant dans (HR$_\mathrm{I}$) et (HR$_\mathrm{II}$) est algébriquement clos et que X est rationnel sur k. Notons $(\mathfrak{G})_k$ le groupe des cycles rationnels sur k. Comme le groupe $(A)_k$ des points de A qui sont rationnels sur k est divisible, la restriction h_1 de h à $(\mathfrak{G})_k \cap \mathfrak{G}_a$ se prolonge en un homomorphisme h_1' de $(\mathfrak{G})_k$ dans A_k, et on peut supposer que $h_1'(X)$ est $\neq 0$ puisque $(A)_k$ contient des points d'ordre infini et des points d'ordre fini de tous ordres. Comme, d'après la théorie de Chow–van der Waerden tout système irréductible maximal de cycles positifs sur V est défini sur k et contient donc des cycles rationnels sur k, toute classe de $\mathfrak{G}$ modulo $\mathfrak{G}_a$ contient des cycles rationnels sur k; autrement dit on a $\mathfrak{G} = \mathfrak{G}_a + (\mathfrak{G})_k$; comme les homomorphismes h et h_1' coïncident sur $\mathfrak{G}_a \cap (\mathfrak{G})_k$, il en résulte qu'il existe un homomorphisme h' et un seul de $\mathfrak{G}$ dans A qui prolonge h et h_1'. Reste à montrer qu'il vérifie $(\mathrm{HR_I})$ et $(\mathrm{HR_{II}})$. Si Y est un cycle rationnel sur $k' \supset k$, on prend un représentant Y_1 rationnel sur k de la classe de Y mod. $\mathfrak{G}_a$; alors $h'(Y) = h(Y - Y_1) + h_1'(Y_1)$ est rationnel sur k' puisque k' contient k. D'autre part, si $Y \to Y'$ est une spécialisation sur $k' \supset k$, Y et Y' appartiennent à la même classe mod. $\mathfrak{G}_a$; prenons un représentant Y_1 rationnel sur k de cette classe; alors $Y' - Y_1$ est une spécialisation de $Y - Y_1$ sur k', et (d'après $(\mathrm{HR_{II}})$ appliqué à h) se prolonge donc en $h(Y - Y_1) \to h(Y' - Y_1)$ c'est-à-dire en $h'(Y) \to h'(Y')$ puisque $h_1'(Y_1)$ est rationnel sur k donc sur k'. C.Q.F.D.

Remarque. Le lemme montre que, dans la définition des cycles abéliens, on aurait pu se restreindre aux homomorphismes de $\mathfrak{G}_a$ dans les variétés abéliennes vérifiant $(\mathrm{HR_I})$ et $(\mathrm{HR_{II}})$.

Proposition 11. *L'équivalence du carré est plus fine que l'équivalence abélienne.*

En effet soient $X = Z(a, a') - Z(a, b') - Z(b, a') + Z(b, b')$ un cycle carré sur V (Z étant un cycle sur $T \times T' \times V$), et h un homomorphisme rationnel de $\mathfrak{G}(V)$ dans une variété abélienne A. Pour $(x, x') \in T \times T'$, $(x, x') \to h(Z(x, x'))$ est une application rationnelle de $T \times T'$ dans A. Il existe donc ([7], cor. du th. 7) deux applications rationnelles $f : T \to A$ et $f' : T' \to A$ tels que $h(Z(x, x')) = f(x) + f'(x')$. D'où $h(X) = 0$ par un calcul immédiat.

2.6. Pseudo-équivalence. Soit $\sim$ une équivalence adéquate. La relation 'il existe un entier $q \neq 0$ tel que $qX' \sim qX$' s'appelle la *pseudo-équivalence* associée à $\sim$; dans les exemples on parlera plutôt de l'équivalence *pseudo-rationnelle* ou *pseudo-abélienne*, etc. On vérifie aussitôt que c'est là une relation d'équivalence *adéquate*.

Remarque. D'après le principe de conservation du nombre l'équivalence pseudo-algébrique est plus fine que l'équivalence numérique, et l'équivalence pseudo-numérique est identique à l'équivalence numérique.

2.7. Equivalence triviale. On appelle équivalence *triviale* celle où tous les cycles sont équivalents à 0. Elle est évidemment adéquate. C'est la moins fine de toutes.

Proposition 12. L'ensemble des équivalences adéquates est réticulé.

Soient en effet $\mathfrak{G}_1$ et $\mathfrak{G}_2$ les groupes correspondant à deux équivalences adéquates données. Il suffit de montrer que les groupes $\mathfrak{G}_1 \cap \mathfrak{G}_2$ et $\mathfrak{G}_1 + \mathfrak{G}_2$ définissent des équivalences adéquates. La condition (RA$_\mathrm{I}$) est évidente, ainsi que (RA$_\mathrm{II}$) pour $\mathfrak{G}_1 + \mathfrak{G}_2$ et (RA$_\mathrm{III}$) pour $\mathfrak{G}_1 \cap \mathfrak{G}_2$. La prop. 8 montre que (RA$_\mathrm{II}$) est vérifiée par $\mathfrak{G}_1 \cap \mathfrak{G}_2$. Enfin, si $X \in \mathfrak{G}_1 + \mathfrak{G}_2$, écrivons $X = X_1 + X_2$ avec $X_1 \in \mathfrak{G}_1$ et $X_2 \in \mathfrak{G}_2$; si Z est un cycle sur $V \times W$ tel que $Z(X)$ soit défini, prenons X_1' rationnellement équivalent à X_1 tel que $Z(X_1')$ soit défini; on a alors $X = X_1' + (X_2 + X_1 - X_1')$ avec $X_1' \in \mathfrak{G}_1$ et $X_2 + X_1 - X_1' \in \mathfrak{G}_2$ (prop. 8); comme $Z(X)$ et $Z(X_1')$ sont définis, il en est de même de $Z(X_2 + X_1 - X_1')$; donc $Z(X)$ appartient au groupe correspondant à $\mathfrak{G}_1 + \mathfrak{G}_2$. C.Q.F.D.

Cas de la codimension 1.

En codimension 1 l'équivalence abélienne coïncide avec l'équivalence rationnelle (appelée alors 'équivalence linéaire'), donc avec les équivalences du carré et du n-cube[8]. Si on note $\mathfrak{G}_l$(resp. $\mathfrak{G}_a$, $\mathfrak{G}_n$, $\mathfrak{G}$) le groupe des cycles de codimension 1 sur V qui sont linéairement (resp. algébriquement, numériquement, trivialement) équivalents à 0, $\mathfrak{G}_a/\mathfrak{G}_l$ est une variété abélienne (variété de Picard[2]), $\mathfrak{G}/\mathfrak{G}_a$ est un groupe abélien de type fini (Néron-Severi[4]), et $\mathfrak{G}_n/\mathfrak{G}_a$ est son sous-groupe de torsion[3]. Ce dernier résultat veut dire que, en codimension 1, l'équivalence numérique coïncide avec l'équivalence pseudo-algébrique. Les énoncés analogues en codimension quelconque constituent d'intéressantes conjectures.

3. Etude de certaines transformées monoïdales

Nous nous proposons de comparer ici les anneaux des classes de cycles sur une variété V et sur certaines transformées monoïdales ([9], no. 11) de V. On suppose V projective et non-singulière, de dimension n. Soient W une sous-variété non-singulière *de dimension $n-2$* de V, et V' la transformée de V par la transformation monoïdale de centre W; rappelons[9] que l'application canonique h de V' sur V est un morphisme birationnel, que V' est projective non-singulière, que h^{-1} est régulière en tout point de $V - W$, que l'image réciproque (au sens ensembliste) W' de W par h est une variété de dimension $n-1$, et que la restriction de h à W' définit une fibration de W' sur W par des droites projectives; les points de W' au-dessus d'un point donné y de W correspondent biunivoquement aux variétés linéaires de dimension $n-1$ contenant la variété linéaire tangente à W en y et contenues dans la variété linéaire tangente à V en y.

Etant donnée une sous-variété U de W, nous noterons $j(U)$ la sous-variété de W' image réciproque (au sens ensembliste) de U par h; elle est fibrée sur U par des droites projectives; on a

$$\text{codim}_{V'}(j(U)) = \text{codim}_W(U) + 1. \tag{2}$$

Nous étendons j par linéarité aux cycles sur W, puis aux classes de cycles.

Soit X un cycle sur V dont aucune composante n'est contenue dans W; comme toutes les composantes de $\text{Supp}(X) \cap W$ sont de dimension $\dim(X)-1$ ou $\dim(X)-2$, l'ensemble $h^{-1}(\text{Supp}(X))$ a toutes ses composantes de même dimension que X (puisque W' est non-singulière); donc le cycle $h^{-1}(X)$ est défini. Notons $i(X)$ le cycle sur V' correspondant à X par l'isomorphisme de $V'-W'$ sur $V-W$, et X_1 la composante de dimension $\dim(X)-1$ (c'est-à-dire d'excès 1) du cycle intersection excédentaire $W \top X$ (on a $X_1 = 0$ si $W.X$ est défini); nous utiliserons les notations $i(X)$ et X_1 dans la suite. On a les relations

$$h^{-1}(X) = i(X) + j(X_1), \tag{3}$$

$$h(h^{-1}(X)) = X. \tag{4}$$

D'autre part, pour tout cycle X' sur V', $h(X')$ est défini. De plus, si $X'.W'$ est défini (c'est-à-dire si aucune composante de X' n'est contenue dans W'), aucune composante du cycle $h(X')$ n'est contenue dans W, et $h^{-1}(h(X'))$ est défini; on voit facilement qu'il existe un cycle X_1 sur W tel que

$$X' - h(h^{-1}(X')) = j(X_1). \tag{5}$$

Nous désignerons désormais par $\sim$ une relation d'équivalence adéquate, et par $\mathfrak{E}(V)$ l'anneau des classes de cycles sur V modulo $\sim$.

Lemme 1. Notons α' la classe de W' dans $\mathfrak{E}(V')$. Pour tout $\eta \in \mathfrak{E}(W)$, on a

$$h^*(j(\eta).\alpha') = -\eta.$$

Par linéarité nous sommes ramenés au cas où η est la classe d'une sous-variété Y de W. Dans l'espace projectif P_N de V, soit C un cône suffisamment général de base Y, de dimension $\dim(Y)+N-n+1$, et coupant transversalement W en Y; alors $C.V$ est défini; en posant $C.V = L$, on a $\dim(L) = \dim(Y)+1$, et toutes les composantes de $L \cap W$ sont propres sur V, à l'exception de Y qui est d'excès 1 et de multiplicité 1; donc, avec les notations de (3), on a $h^{-1}(L) = i(L) + j(Y)$. Soit M un cycle sur V, intersectant proprement W, et tel que $M \sim L$; on a $h^{-1}(M) = i(M)$. Comme $L \sim M$, il vient $h^{-1}(L) \sim h^{-1}(M)$, d'où

$$j(Y) \sim i(M) - i(L).$$

Les cycles $i(M).W'$ et $i(L).W'$ sont définis et de dimension dim (Y); comme $h(\mathrm{Supp}\,(i(M).W')) = M \cap W$ est de dimension dim $(Y) - 1$, on a $h(i(M).W') = 0$. D'autre part $h(\mathrm{Supp}\,(i(L).W')) = L \cap W$ admet Y pour unique composante de dimension maximum; donc $h(i(L).W')$ est un multiple de Y; or, étant donné un corps de définition k de tout ce qui précède, il y a, au dessus d'un point générique y de Y sur k, un point et un seul de $i(L) \cap W'$, et ce point est rationnel sur $k(y)$ puisqu'il correspond à la variété linéaire engendré par les variétés linéaires tangentes à W et à L en y; comme, de plus, $i(L)$ et W' sont transversales, il en résulte qu'on a $h(i(L).W') = Y$. D'où aussitôt notre assertion. C.Q.F.D.

Ceci étant nous allons pouvoir étudier la structure de $\mathfrak{E}(V')$. Pour être plus complets, nous allons aussi considérer une relation d'équivalence adéquate $\equiv$, moins fine que $\sim$; nous noterons $\mathfrak{G}_r$ et $\mathfrak{G}_a$ les groupes correspondant à $\sim$ et $\equiv$, et $\mathfrak{E}_a$ le quotient $\mathfrak{G}_a / \mathfrak{G}_r$ (pour obtenir l'anneau $\mathfrak{E}$, il suffira de prendre pour $\equiv$ l'équivalence triviale). Considérons les homomorphismes:

$$\mathfrak{E}_a(V) \underset{h^*}{\to} \mathfrak{E}_a(V') \underset{h_*}{\to} \mathfrak{E}_a(V). \tag{6}$$

D'après (3) on a $h_* \circ h^* = 1$, donc $\mathfrak{E}_a(V)$ s'identifie à un facteur direct de $\mathfrak{E}_a(V')$. Notons $\mathfrak{N}$ son supplémentaire, c'est-à-dire le noyau de h_*. L'homomorphisme j applique $\mathfrak{E}_a(W)$ dans $\mathfrak{N}$. Nous allons montrer que cette application est une *bijection*:

(a) *Surjectivité*. Prenons $\xi' \in \mathfrak{N}$, et un représentant X' de ξ' tel que $X'.W'$ soit défini. Comme $h(X') \sim 0$, on a $h^{-1}(h(X')) \sim 0$, d'où l'existence d'un cycle X_1 sur W tel que $X' \sim j(X_1)$ (cf. (4)). On a alors $j(X_1) \equiv 0$, d'où, par application du lemme à $\equiv$, $X_1 \equiv 0$ sur W. Ceci démontre la surjectivité.

(b) *Injectivité*. Soit Y un cycle sur W tel que $j(Y) \sim 0$ *sur* V'. Le lemme montre qu'on a $Y \sim 0$ sur W. D'où l'injectivité.

Nous avons donc démontré le résultat suivant:

Proposition 13. Avec les notations précédentes, le groupe additif $\mathfrak{E}_a(V')$, gradué par la codimension, est canoniquement isomorphe au produit $\mathfrak{E}_a(V) \times \mathfrak{E}_a(W)$, où $\mathfrak{E}_a(V)$ a sa graduation naturelle et $\mathfrak{E}_a(W)$ la graduation obtenue en augmentant les codimensions d'une unité.

Exemple. Prenons $V = P_3$ et pour W une courbe non singulière de P_3

de genre $\neq 0$; $\equiv$ sera l'équivalence algébrique et $\sim$ l'équivalence rationnelle. En dimension 1, on a $\mathfrak{E}_a(V) = (0)$, tandis que $\mathfrak{E}_a(V')$ est isomorphe
à la jacobienne de W. Ceci montre que $\mathfrak{E}_a$ n'est pas un invariant birationnel absolu en codimension > 1 (en codimension 1, $\mathfrak{E}_a$ est la variété
de Picard et est un invariant birationnel absolu).

Nous prendrons désormais pour $\equiv$ l'équivalence triviale (et écrirons
donc $\mathfrak{E}$ au lieu de $\mathfrak{E}_a$). Nous allons donner quelques renseignements sur
la *structure multiplicative* de $\mathfrak{E}(V')$:

(1) Nous avons déjà vu que h^* identifie $\mathfrak{E}(V)$ à un *sous-anneau*
unitaire de $\mathfrak{E}(V')$.

(2) Considérons deux classes de cycles, ξ sur V et η sur W, et calculons
$h^*(\xi)j(\eta)$. Pour celà prenons des représentants X de ξ et Y de η tels
que $X.Y$ soit défini; alors $h^{-1}(X).j(Y)$ est défini et a même support que
$j(X.Y)$. Or, si C est une composante de $X.Y$ en laquelle les cycles X et
Y sont transversaux, il en est de même de $h^{-1}(X)$ et de $j(Y)$ en $j(C)$. La
formule $h^{-1}(X).j(Y) = j(X.Y)$ est donc vraie si X et Y sont transversaux en toutes leurs composantes. On a donc

$$h^*(\xi)j(\eta) = j(\xi\eta) \tag{7}$$

chaque fois que ξ est la classe d'une section plane de V (dans n'importe
quel plongement projectif de cette variété); or, comme V est non-
singulière, tout diviseur sur V est différence de deux sections hyperplanes
(dans deux plongements différents); donc (6) est vraie lorsque ξ est de
codimension 1, et par conséquent lorsqu'il est somme de produits de
classes de codimension 1. Il semble probable que (6) soit vraie en
général.

(3) Pour calculer $j(\eta_1)j(\eta_2)$ $(\eta_1, \eta_2 \in \mathfrak{E}(W))$, nous aurons besoin de
connaître la classe α'^2 (où α' désigne la classe de W'); pour $\eta \in \mathfrak{E}(W)$,
nous poserons $c(\eta) = \alpha'j(\eta)$ (rappelons que le lemme nous a montré
que $h(c(\eta)) = -\eta$). Notons que $\alpha'^2 = c(1)$. La 'formule d'induction'
$X'._{V'}Z = (X'._{V'}W')._{W'}Z$ (où X' est un cycle V' et Z un cycle sur W')
montre qu'on a $c(\eta) = \alpha'j(\eta) = (\alpha'^2)._{W'}j(\eta)$ d'où

$$c(\eta) = c(1)._{W'}j(\eta) \quad (\eta \in \mathfrak{E}(W)). \tag{8}$$

Il résulte du lemme que $c(1)$ est l'opposé de la classe d'une section
rationnelle du fibré W'.

Ceci étant la formula d'induction montre encore qu'on a

$$j(\eta_1)j(\eta_2) = c(\eta_1)._{W'}j(\eta_2) = c(1)._{W'}j(\eta_1)._{W'}j(\eta_2) = c(1)._{W'}j(\eta_1\eta_2).$$

D'où, pour η_1 et $\eta_2 \in \mathfrak{E}(W)$, on a

$$j(\eta_1)j(\eta_2) = c(1)._{W'}j(\eta_1\eta_2) = c(\eta_1\eta_2). \tag{9}$$

Remarque. Des résultats plus précis et plus généraux sur les transformées monoïdales ont été obtenus indépendamment par M. A. Grothendieck.

4. Problèmes et résultats sur les cycles de dimension 0.

Il semble raisonnable de commencer l'étude des rapports entre les diverses relations d'équivalence définies en § 2 par le cas des cycles de dimension 0: en effet nous disposons, d'une part, grâce aux travaux de F. Severi, d'un riche matériel expérimental concernant les groupes de points sur une surface; on peut d'autre part espérer démontrer un jour des 'critères d'équivalence' permettant une réduction à la dimension 0.

Pas besoin d'insister sur l'équivalence algébrique ni sur l'équivalence abélienne; rappelons que le groupe de cette dernière est le 'noyau d'Albanese'.

4.1. Equivalence rationnelle. On peut se demander si l'ensemble des points d'une variété V qui sont rationnellement équivalents à un point donné P_0 de V est *fermé* (pour la topologie de Zariski). Il est facile de voir qu'il est stable par spécialisation (puisque l'équivalence rationnelle vérifie (RA_{IV})). Problème voisin: l'ensemble des points de V qu'on peut relier à P_0 par une suite finie et connexe de courbes rationnelles (tracées sur V) est-il fermé? coïncide-t-il avec l'ensemble des points rationnellement équivalent à P_0? Une variété dont deux points quelconques sont connectables par une courbe rationnelle est-elle unirationnelle? Une surface dont tous les points sont rationnellement équivalents est évidemment régulière; vérifie-t-elle $p_g = 0$?; est-elle rationnelle?

Le seul résultat que connaisse l'auteur et qui puisse avoir une certaine utilité dans ce genre de questions est le suivant:

Proposition 14. *Notons* $\sim$ *l'équivalence rationnelle. Si X et X' sont deux cycles positifs sur V, rationnels sur un corps k de définition de V et tels que $X \sim X'$, alors il existe un cycle positif L sur V rationnel sur k, un entier $q \geqslant 1$, un cycle Z positif sur $P_1 \times V$ et rationnel sur k, et deux points a, b de P_1 rationnels sur k tels que*

$$Z(a) = qX + L, \quad Z(b) = qX' + L. \tag{10}$$

En effet il existe un cycle positif Y sur $P_1 \times V$ et deux points a, b rationnels sur k de P_1 (par exemple 0 et ∞) tels que $X - X' = Y(b) - Y(a)$. Par spécialisation de Y sur k, on peut supposer Y algébrique sur k; soient $Y_1, \ldots, Y_q$ ses conjugués (éventuellement répétés); posons

$$Z_1 = Y_1 + \ldots + Y_q.$$

On a $$Z_1(b) - Z_1(a) = q(X' - X).$$

Or, en écrivant $X = X_0 + X_1$, $X' = X_0 + X_1'$, où X_1 et X_1' sont des cycles positifs sans composante commune, la relation

$$Z_1(b) - Z_1(a) = qX_1' - qX_1$$

montre qu'il existe un cycle positif L_1 tel que $qX_1 + L_1 = Z_1(a)$ et $qX_1' + L_1 = Z_1(b)$. Alors le cycle $Z = Z_1 + (P_1 \times L_1)$ est rationnel sur k, et $L = L_1 + qX_0$ vérifie (10). C.Q.F.D.

4.2. Equivalence du carré. Nous avons vu que l'équivalence du carré est plus fine que l'équivalence abélienne. On peut se demander si elles coïncident. En dimension 0 on a deux résultats partiels:

Proposition 15. Sur une variété abélienne V, tout cycle X de dimension 0 et de degré 0 qui est abéliennement équivalent à 0 (c'est-à-dire tel que $S(X) = 0$, avec les notations de[7]), est un cycle carré.

Posons $X = \Sigma n_i(x_i)$ avec $x_i \in V$; on a $\Sigma n_i x_i = 0$ par hypothèse. Notons que, pour $a, a', b, b' \in V$, le cycle

$$(a + a') - (a + b') - (b + a') + (b + b')$$

est carré. Procédons par récurrence sur $n = \Sigma |n_i|$, qui est pair puisque $\Sigma n_i = 0$. Notre assertion est triviale pour $n = 0$ et $n = 2$. Pour $n \geq 4$, mettons en évidence, dans X, deux termes munis du signe $+$ et un terme muni du signe $-$, et écrivons $X = (x) + (y) - (z) + X'$. Le cycle $X_1 = (x) + (y) - (z) - (x + y - z)$ est carré (prendre, ci-dessus, $a = x$, $a' = 0$, $b' = z - x$, $b = x + y - z$). D'autre part $X - X_1 = (x + y - z) + X'$ est abéliennement équivalent à 0, donc carré d'après l'hypothèse de récurrence. C.Q.F.D.

L'auteur ignore si la prop. 15 se généralise aux équivalence n-cubiques.

Proposition 16. Si V et V' sont deux variétés telles que l'équivalence du carré coïncide en dimension 0 avec l'équivalence abélienne, il en est de même de $V \times V'$.

Fixons en effet des points a de V et a' de V'. Notons $\sim$ l'équivalence du carré. Pour $x \in V$ et $x' \in V'$, on a

$$(x, x') \sim (a, x') + (x, a') - (a, a').$$

Donc, pour tout cycle X de dimension 0 sur $V \times V'$, il existe des cycles Y sur V et Y' sur V', tous deux de dimension 0, tels que

$$X \sim (a) \times Y' + Y \times (a').$$

Si X est de degré 0, on peut prendre Y et Y' de degré 0, et on a alors $\mathrm{pr}_V(X) \sim Y$ et $\mathrm{pr}_{V'}(X) \sim Y'$. Si X est abéliennement équivalent à 0, il en est donc de même de Y et Y', qui, par conséquent, sont des cycles carrés en vertu de l'hypothèse. Ainsi X est un cycle carré. C.Q.F.D.

Remarques. (1) Notons $\mathfrak{C}_c^{(0)}(V)$ le groupe des classes de cycles de dimension 0 sur V modulo l'équivalence carrée. Le raisonnement de la prop. 16 montre que $\mathfrak{C}_c^{(0)}(V \times V')$ est isomorphe à $\mathfrak{C}_c^{(0)}(V) \times \mathfrak{C}_c^{(0)}(V')$.

(2) Soient A une variété abélienne, et V sa *variété de Kummer* (quotient de V par la relation d'équivalence $x + y = 0$). Notons h l'application canonique de A sur V. Etant donnés deux points quelconques a, b de V, soient x, y des points de A tels que $a = h(x)$ et $b = h(y)$. Il existe un point z de A tel que $2z = x + y$. Le cycle $X = (x) + (-y) - (z) + (-z)$ sur A est carré (prop. 15). Comme $h(x) = a$, $h(-y) = h(y) = b$ et $h(z) = h(-z)$, on a $h(X) = (a) - (b)$, ce qui montre que $(a) - (b)$ est un cycle carré. Ainsi, sur une variété de Kummer, *l'équivalence du carré coïncide avec l'équivalence algébrique* en dimension 0. A fortiori ces deux équivalences coïncident avec l'équivalence abélienne; en particulier une variété de Kummer est *régulière* (ce qui est d'ailleurs facile à voir directement).

(N.B. Il est facile de voir que, en dimension 0, le groupe $\mathfrak{C}^{(0)}(V)$ est un invariant birationel absolu; il a donc un sens pour les variétés birationnellement équivalentes à des variétés non-singulières, en particulier pour les surfaces.)

On peut se demander si l'équivalence du carré coïncide avec l'équivalence rationnelle. Pour qu'il en soit ainsi il suffit (d'après (RA$_{III}$)) que, quels que soient les variétés (resp. courbes, variétés abéliennes) T, T' et les points a, b de T et a', b' de T', le cycle

$$(a, a') - (a, b') - (b, a') + (b, b')$$

soit rationnellement équivalent à 0; il suffit même que, pour toute courbe C et tous points a, b de C, le cycle $(a, a) - (a, b) - (b, a) - (b, b)$ sur $C \times C$ soit rationnellement équivalent à 0. Le premier cas non trivial est celui où C est une courbe elliptique; la conjecture équivaut alors à la suivante; sur la variété abélienne $C \times C$, les cycles $(x) + (-x)$ et $2(0)$ sont rationnellement équivalents pour tout point x de $C \times C$; ceci veut dire que, sur la surface de Kummer W de $C \times C$, tous les points sont rationnellement équivalents; comme W contient une infinité dénombrable de courbes rationnelles (à savoir les images des sous-variétés abéliennes de $C \times C$), cette conjecture est conséquence de celle qui dit que les points rationnellement équivalents à un point donné forment un ensemble fermé.

4.3. Sous-variétés représentatives. Soit $\sim$ une relation d'équivalence adéquate. Une sous-variété W d'une variété V est dite *repré-*

sentative (pour $\sim$) si tout point de V est équivalent (pour $\sim$) à un cycle de dimension 0 porté par W. Ceci veut dire que l'homomorphisme canonique $\mathfrak{C}^{(0)}(W) \to \mathfrak{C}^{(0)}(V)$ est *surjectif*. L'existence de sous-variétés représentatives de V permet donc d'avoir des renseignements sur $\mathfrak{C}^{(0)}(V)$.

Etant donnée une variété V, l'existence de courbes sur V représentatives pour l'équivalence rationnelle semble un problème ouvert en dehors des cas triviaux. On peut aussi conjecturer que, pour qu'une variété V admette une courbe représentative (pour l'équivalence rationnelle), il faut et il suffit que le noyau d'Albanese de V coïncide avec le groupe de l'équivalence rationnelle; la condition est suffisante puisque toute variété abélienne admet une courbe génératrice (cf. Matsusaka).

BIBLIOGRAPHIE

[1] Chow, W. L. On equivalence classes of cycles in an algebraic variety. *Ann. Math.* (2), 64, 450–479 (1956).

[2] Matsusaka, T. On the algebraic construction of the Picard variety. I, II. *Jap. J. Math.* 21, 217–235 (1951) et 22, 51–62 (1952).

[3] Matsusaka, T. The criteria for algebraic equivalence and the torsion group. *Amer. J. Math.* 79, 53–66 (1957).

[4] Néron, A. Problèmes arithmétiques et géométriques attachés à la notion du rang d'une courbe algébrique. *Bull. Soc. Math. Fr.* 80, 101–166 (1952).

[5] Samuel, P. *Méthodes d'algèbre abstraite en géométrie algébrique.* Ergebn. Math. N.F., Heft 4, Berlin, Springer, 1955.

[6] Samuel, P. Rational equivalence of arbitrary cycles. *Amer. J. Math.* 78, 383–400 (1956).

[7] Weil, A. *Variétés abéliennes et courbes algébriques.* Paris, Hermann, 1948.

[8] Weil, A. Sur les critères d'équivalence en géométrie algébrique. *Math. Ann.* 128, 95–117 (1954).

[9] Zariski, O. Foundations of a general theory of birational correspondences. *Trans. Amer. Math. Soc.* 53, 490–542 (1943).

ON GALOIS GEOMETRIES

By BENIAMINO SEGRE

1. Introduction

Finite spaces, i.e. finite sets of point-elements on which some geometric structure is superimposed, are of importance in many branches of applied mathematics, e.g. statistics (Bose[6], Seiden[57]). Quite recently, they have been used in various efforts to build up a geometry which would be better adapted to the quantistic theories (Järnefelt[23, 24], Järnefelt and Kustaanheimo[25], Kustaanheimo[27-32]).

The present exposition will dwell chiefly on the geometric aspects in the study of the simplest and more important among finite spaces, the so-called *Galois spaces*, $S_{r,q}$, i.e. the linear spaces—of any dimension r—over a $GF(q)$. These spaces can be characterized as those finite graphic (or projective) spaces where Fano's postulate (existence of at least three distinct points on every line) and Desargues' theorem (on homological triangles) hold, the latter condition being a consequence of the former if $r \geqslant 3$. In every graphic space satisfying these conditions Pappus's theorem also holds, which is tantamount to the Maclagan–Wedderburn theorem affirming the commutativity of every finite field. Attempts at proving this theorem geometrically—by directly showing the validity of Pappus's or equivalent theorems in those spaces—have also been made (cf., for example, Locher[34]), but up to now without success (cf. Artin[1], p. 75). Quite recently I have added a fresh attempt [Segre[54]], with a view to a geometric proof of the Maclagan–Wedderburn theorem through a study of certain properties of non-linear non-commutative geometry, concerning reguli in projective spaces over a skew field and their plane sections.

Many of the ideas and results of algebraic geometry can be adapted to Galois spaces without difficulty, paying attention to the fact that the ground field $GF(q)$ of a $S_{r,q}$ has non-zero characteristic p (where $q = p^h$, for some positive integer h), and that the field is not algebraically closed. Additional questions and properties arise from the finiteness of the field, and some of them are dealt with in § 2.

In § 3 we briefly indicate how certain algebraic subvarieties of $S_{r,q}$ can be characterized by means of very simple properties of a numerical and graphical character.

Finally, §§ 4 and 5 concern the study of *k-sets*, i.e. sets of k points of

$S_{r,q}$, for $r = 2$ and $r \geqslant 3$ respectively; and it has to be noticed that each of them can always be considered, in different ways, as the set of points of an algebraic variety. The theory has been recently developed for certain k-sets of special interest, called k-arcs and k-caps; but very much remains to be done in this direction.

Our exposition—while rather sketchy—will implicitly suggest a number of further questions which, however, we shall pass over for lack of time.

2. Some properties of algebraic varieties

Projective geometry and non-Euclidean geometry of $S_{r,q}$ have many interesting group-theoretic aspects, which were substantially known a long time ago (Jordan[26], Dickson[10]). However, only recently have appropriate geometric terminology and reasoning been employed in order to clarify them, particularly for the smallest values of r and q. Thus the geometry of $S_{r,3}$ has been investigated in detail for $r \leqslant 4$ (Edge[11, 13]), and used for studying the group of the 27 lines of the general cubic surface and its separation into 25 conjugate classes (Edge[16]), as well as other group properties in $S_{5,3}$ (Edge[18]). Also the geometry of $S_{2,q}$ has been described, especially for $q = 5, 7, 11$ (Edge[14, 17]), and suggestive geometric arguments have been devised for establishing the known isomorphism between the linear fractional group $LF(4, 2)$ and the alternating group A_8 (Edge[12]), and between $LF(2, 9)$ and A_6 (Edge[15]).

The number of linear subspaces of $S_{r,q}$ of a given dimension can be easily obtained (Segre[46], n. 159); thus, for instance, the number of points of $S_{r,q}$ is $1 + q + q^2 + \ldots + q^r$. Similar questions may be asked for any algebraic variety of $S_{r,q}$; but the answer is then usually far less simple and not precisely known. Estimates for the number of points lying on certain algebraic varieties have been obtained, and in special cases reasonably simple exact expressions of this number were given, by means of deep and often rather intricate algebraic and analytic arguments (Châtelet[9], Vandiver[64, 65], Hua and Vandiver[19-22], Weil[66], Lang and Weil[33], Carlitz[8]). Here we confine ourselves to quoting the result (Hua and Vandiver[22]) $N = (q-1)[(q-1)^{r-1} + (-1)^r]/q$ concerning the number N of solutions $x \in GF(q)$, with $x_1 x_2 \ldots x_r \neq 0$, of the equation $c_1 x_1^{n_1} + c_2 x_2^{n_2} + \ldots + c_r x_r^{n_r} = 0$, where the c's are given non-zero elements of $GF(q)$, the n's are integers satisfying $0 < n_i < q-1$ such that the numbers $(n_i, q-1)$ are relatively prime in pairs, and $r > 2$.

Some of the questions of the type indicated above can be conveniently treated by means of purely combinatorial and geometric methods. For instance, a well-known algebraic result (Dickson[10]) can be stated by saying that *every irreducible conic of* $S_{2,q}$ *contains precisely* $q+1$ *points*, and this has also been freshly proved by showing directly that in $S_{2,q}$ there are as many irreducible conics as there are conics containing exactly $q+1$ points, both numbers of conics being equal to q^5-q^2 (Segre[53]). Thus the whole theory of quadratic forms over $GF(q)$ can be geometrized (Primrose[39]), and partially new results can be obtained as follows (Segre[55]).

If $r=2s$ is even ($s \geqslant 1$), the non-singular quadrics of $S_{r,q}$ are two by two homographic, and every one of them contains a positive number of linear subspaces $S_{n,q}$ of $S_{r,q}$, of each dimension $n=0,1,...,s-1$, this number being

$$\prod_{i=0}^{n}(q^{s-i}-1)\prod_{i=s-n}^{s}(q^i+1)/\prod_{i=0}^{n}(q^{n-i+1}-1).$$

If $r=2s-1$ is odd, the non-singular quadrics of $S_{r,q}$ fall into two different types, the *hyperbolic* and the *elliptic*, those of the same type being two by two homographic. Every hyperbolic quadric contains a positive number of $S_{n,q}$, of each dimension $n=0,1,...,s-1$, this number being

$$\prod_{i=0}^{n}(q^{s-i}-1)\prod_{i=s-n-1}^{s-1}(q^i+1)/\prod_{i=0}^{n}(q^{n-i+1}-1)$$

(as usual, the spaces of maximum dimension $s-1$ then constitute two different systems, and those of lower dimension a single one). Every elliptic quadric contains no $S_{s-1,q}$ and a positive number of $S_{n,q}$, of each dimension $n=0,1,...,s-2$, this number being

$$\prod_{i=0}^{n}(q^{s-i-1}-1)\prod_{i=s-n}^{s}(q^i+1)/\prod_{i=0}^{n}(q^{n-i+1}-1).$$

It has to be noticed that a non-singular quadric f of $S_{r,q}$ defines a polarity, which is a null system if, and only if, $p=2$ (i.e., if q is even). In this case, the polarity is non-singular if r is odd, while it has one, and only one, singular point O if r is even; the point O is called the *kernel* of f, does not lie on f and is the point of concurrence of all the tangent primes of f. This is but an example of the discrepancies which may appear in algebraic geometry between the case where the characteristic of the ground field has the value $p=2$ and the case $p \neq 2$; further examples have been investigated (Boughon, Nathan and Samuel[7], Segre[50]), and deep reasons for the occurrence of such discrepancies have been given (Segre[55]).

Other special algebraic varieties of a Galois space have been studied, as for instance the rational normal curves of $S_{r,q}$ (Segre[49]), and the cubic surfaces of $S_{3,q}$, with odd q. It has been shown (Rosati[42]) that the equation of the 27 lines and the equation of the 45 tritangent planes of such a surface are always reducible, and the various kinds of reducibility have been classified; moreover, the exact number of points lying on the surface has been obtained in several cases (Rosati[41]). Another interesting result is the coincidence between the locus of the kernels of the conics lying on the Veronesean representing the quadrics of $S_{r,q}$, with even q, and the Grassmannian of the lines of $S_{r,q}$ (Tallini).

An upper bound for the number of the points of an algebraic curve will be given later (§4).

3. Characterization of certain algebraic varieties

While every irreducible conic of $S_{2,q}$ consists of a set of $q+1$ points of $S_{2,q}$ (§2) no three of which are collinear, it has been proved that *the converse is true if, and only if, q is odd* (Segre[47, 48]) *or if $q = 2, 4$* (Segre[51]). Another similar result is that, if q is odd, any set of $q+1$ points of $S_{3,q}$ no four of which are in a plane consists of the points of a twisted cubic curve (Segre[49]). These very simple but rather unexpected results have been the starting-point of further investigations, some of which we shall now summarize.

In $S_{r,q}$, where q is odd and $r \geqslant 3$, let us consider any set of k points such that every line of $S_{r,q}$ containing three distinct points of the set consists entirely of points of the set. It has been proved (Tallini[58, 59]) that, if $1 + q + \ldots + q^{r-1} \leqslant k < 1 + q + \ldots + q^r$, then the set can only be the set of all the points of one of the following algebraic varieties: (i) the variety consisting of a prime and a $S_{t,q}$ of $S_{r,q}$ ($t = -1, 0, 1, \ldots, r-1$); (ii) a non-singular quadric in a space of even dimension, or a quadric cone projecting such a quadric from its vertex; (iii) a non-singular hyperbolic quadric in a space of odd dimension, or a quadric cone projecting such a quadric from its vertex. Also the case of q even has been treated (*l.c.*). Moreover, similar characterizations have been obtained for non-singular elliptic quadrics in spaces of odd dimension and their projecting cones (Tallini[60]), as well as for certain cubic surfaces of $S_{3,q}$ with odd q (Tallini[63]).

Finally, the Veronese surface of $S_{5,q}$ can be characterized as follows (Tallini[62]). In a $S_{r,q}$, where $r \geqslant 5$ and q is odd, consider any set of $k \geqslant q^2 + q + 1$ planes two by two incident, such that no three of these have a common point; then $r = 5$, $k = q^2 + q + 1$, and the set consists precisely of the tangent planes of a Veronese surface.

In the sequel we shall give some results improving those stated at the beginning of the present § 3. By using them suitably, similar improvements of the remaining results of this § 3 could be easily deduced.

4. On planar k-sets

With every k-set of $S_{2,q}$ we can associate, for any $n = 1, 2, \ldots$, an integer N_n given by the maximum number of points of the k-set which lie on some algebraic curve of order n. The consideration of N_n is significant only for the smallest values of n, precisely when $\frac{1}{2}n(n+3) < k$, and then we have $\frac{1}{2}n(n+3) \leqslant N_n \leqslant k$. Not much is known about these characters N_n; for instance, it has been proved (Segre[52]) that, *if $N_1 = 2$ and $N_2 \geqslant \frac{1}{2}q + 2$, then $N_2 = k$*, i.e. the k-set lies entirely on a conic, *with only one possible exception for even q, when it can happen that $N_2 = k - 1$*, the k-set consisting then of $k - 1$ points of a conic and of the latter's kernel.

It may be noticed that the character N_1 can be defined—more generally—for k-sets belonging to any projective plane of order q, namely to a graphic finite plane, Π_q say, each line of which contains $q + 1$ points. Of very particular importance are the k-sets of Π_q having $N_1 = 2$ (i.e. containing no triplet of collinear points), which are called simply k-*arcs*. It can be shown (Qvist[40]) that, for no k-arc of Π_q may k be greater than $q + 1$ or $q + 2$ according as q is odd or even, these maximum values of k being in fact reached by some k-arcs, called *ovals* (Segre[47, 48]).

The consideration and study of ovals are important also on planes Π_q which are non-Desarguesian, i.e. non-Galoisian (Ostrom[37]), as can be anticipated from the fact that, *when q is odd, every oval of $S_{2,q}$ is simply a conic* (§ 3). When $q = 2^h$ is even, we obtain an oval in a Galois plane $S_{2,q}$ by adding the kernel to the $q + 1$ points of a conic of $S_{2,q}$ (§ 2); this is then the only way of obtaining an oval if $h = 1, 2, 3$ (i.e. $q = 2, 4, 8$), but there are always ovals not obtainable in this manner if $h > 3$, *with a single possible exception for $h = 6$* (Segre[51]). A complete classification of ovals in a plane $S_{2,q}$, with q even, remains however to be done.

The results stated above concerning the maximum value of k, for k-arcs of Π_q, can be extended as follows (Barlotti[4]): *if, for a k-set of Π_q, the character $N_1 = n$ is such that $2 < n \leqslant q$, then $k \leqslant (n-1)q + n$ or $k \leqslant (n-1)q + n - 2$ according as q is or is not divisible by n.*

It may be noticed that, if a k-set of $S_{2,q}$ lies on an algebraic curve of order $n \leqslant q$ free from rectilinear components, then the k-set has $N_1 \leqslant n$. The last result can in this case be refined, by proving (Segre[55]) that:

The number k of points of $S_{2,q}$ lying on an algebraic curve of order $n \leqslant q$ of $S_{2,q}$ satisfies the limitation $k \leqslant nq + 1$, where the equality sign *occurs if, and only if, the curve consists of n distinct lines of a pencil. Moreover, if the curve does not break up into n lines, then*

$$k \leqslant (q+1)(n-1) - [\tfrac{1}{2}n] + 1,$$

where the equality sign may occur only if n is even or if the curve has some rectilinear component.

We shall now confine ourselves to the study of the k-arcs of Π_q or, in particular, of $S_{2,q}$. Such a k-arc, K, is said to be *incomplete* or *complete* according as there is or there is not a $(k+1)$-arc containing it (for instance, every oval is manifestly complete); moreover, a line of the plane is said to be *external* to K or a *tangent*, or a *secant* of K according as it has 0, or 1, or 2 distinct points in common with K. Clearly, every point of K lies on $k-1$ secant lines and so on

$$t = q - k + 2$$

distinct tangents (each of which is said *to touch K at the point*). To any point of the plane not lying on K we can attach an integer i, called the *index* of the point and satisfying the limitations

$$0 \leqslant i \leqslant [\tfrac{1}{2}k],$$

given by the number of distinct secants of K containing it. Denoting by c_i the number ($\geqslant 0$) of points of the plane not lying on K and having index i, K is obviously complete if, and only if, $c_0 = 0$; more generally, if $c_0 = c_1 = \ldots = c_{\alpha-1} = 0$, but $c_\alpha \neq 0$, K is said to be *complete of index α.*

The projective characters c_i, as well as other integers attached in a similar way to K, are connected by a system of Diophantine equations; and a simple discussion of this system gives necessary conditions in order that K be complete, possibly of a given index (Segre[55], Sce[44, 45]). The question of obtaining sufficient conditions is much more difficult, and has been studied only on Galois planes and in special cases.

For instance, it has been shown (Segre[49], Tallini[61]) that no q-arc of $S_{2,q}$ can be complete; on the contrary (Lombardo-Radice[35]), if $q \equiv 3 \pmod 4$, there certainly exist some complete $\tfrac{1}{2}(q+5)$-arcs in any $S_{2,q}$. Moreover (Segre[55]), while the necessary arithmetical conditions for completeness are, e.g. satisfied for $q = 13$, $k = 7$, there exists no complete 7-arc in $S_{2,13}$; finally (Segre[51, 55]), for the first non-trivial values of q:

$$q = 7, \quad q = 8, \quad q = 9,$$

the exact values of k, such that there exists some complete k-arc in $S_{2,q}$ are

$$k = 6, 8, \quad k = 6, 10, \quad k = 6, 7, 8, 10$$

respectively. The known complete k-arcs which are not ovals have been obtained by different methods (see also Scafati[43]), one of which (Lunelli and Sce[36]) has required the use of an electronic calculating machine; and many of those k-arcs are endowed with remarkable groupal properties.

Other results on completeness have been deduced from the following theorem (Segre[55]), established in its turn by means of certain considerations of algebraic geometry having a much wider applicability.

Theorem. If K denotes any k-arc of $S_{2,q}$ for which (putting as above $t = q - k + 2$) $t > 0$, then it is possible to associate with K an algebraic envelope of lines of $S_{2,q}$, Γ say, containing no pencil of lines with the centre on K as a component, and such that:

if q is even, Γ has class t, and the t lines of Γ issuing from any point of K coincide with the t distinct tangents of K at the same point;

if q is odd, Γ has class $2t$, without being an envelope of class t counted twice, and the $2t$ lines of Γ issuing from any point of K coincide with the t distinct tangents of K at the same point, each counted twice.

In the first of the two cases considered in the theorem, the envelope Γ has therefore class t and contains as elements each of the

$$kt = t(q - t + 2)$$

tangents of K. From the dual of a previous result we see that, if q is sufficiently large with respect to t, the envelope Γ must contain some pencil of lines as a component; the centre of such a pencil does not lie on K and, by aggregating it to K, we obtain a $(k+1)$-arc containing K, so that K is certainly incomplete. The argument now sketched, suitably completed, shows for instance that:

If q is even and $t = 1, 2, 3, 4$, there exists no complete $(q - t + 2)$-arc in $S_{2,q}$, save for just one exception, given by the complete 6-arcs of $S_{2,8}$.

The argument can be further developed, and also suitably modified so as to adapt to the second case of the theorem (q odd); thus new results can be established, whose qualitative content is expressed by the following theorem.

If a denotes any non-negative constant and $k \geqslant q - a$, where q is sufficiently large with respect to a, then every k-arc of $S_{2,q}$ is contained in one and only one oval (i.e. a conic, if q is odd), and is therefore incomplete if it is not an oval.

5. On spatial k-sets

The notion of k-arc can be extended to higher spaces, by considering in $S_{r,q}$ sets of k points any $s + 1$ distinct of which are linearly independent

$(2 \leqslant s \leqslant r)$; such a set will be denoted by $k_{r,q}^s$, and the maximum of k for given r, q, s will be indicated by $m_{r,q}^s$. Of particular importance are the cases when $s = r$ or $s = 2 < r$: then the k-set will be called a k-arc or a k-cap of $S_{r,q}$ respectively.

If c is an integer satisfying $1 \leqslant c \leqslant r-2$, $c < k$, then the projection of any $k - c$ points of a k-arc of $S_{r,q}$, from the $S_{c-1,q}$ joining the remaining c points, onto a $S_{r-c,q}$ skew to $S_{c-1,q}$, is clearly a $(k-c)$-arc of $S_{r-c,q}$. By applying the last theorem of § 4 and this remark, we obtain (Segre[55]) that:

If q is odd and sufficiently large with respect to r, then $m_{r,q}^r = q+1$, the k-arcs with maximum k ($= q+1$) being the rational normal curves of $S_{r,q}$. If, moreover, q is sufficiently large with respect to $q - k$, then every k-arc of $S_{r,q}$ is contained in a rational normal curve of $S_{r,q}$.

The characters m of the k-caps have a special significance in statistics. For them, the following results have been proved:

$$m_{3,q}^2 = q^2 + 1 \qquad \text{if } q \text{ is odd,}$$
$$q^2 + 1 \leqslant m_{3,q}^2 \leqslant q^2 + q + 2 \qquad \text{if } q \text{ is even,} \qquad \text{(Bose[6]).}$$
$$m_{r,2}^2 = 2^r \qquad \text{for any } r \geqslant 2,$$

$$m_{3,4}^2 = 17 \qquad \text{(Seiden[57], Barlotti[2]).}$$

$$m_{r,q}^2 \leqslant q^{r-1} - (q-5)\sum_{i=0}^{r-4} q^i + 1 \qquad \text{if } q \text{ is odd and } r \geqslant 4 \quad \text{(Barlotti[5]).}$$

$$m_{4,q}^2 \geqslant 2q^2 + 2(q+1)[\tfrac{1}{4}q] - 6q - 2 \qquad \text{(Segre[55]).}$$

Moreover, it has been established (Barlotti[2]) that:

If q is odd, the k-caps of $S_{3,q}$ having maximum k ($= q^2+1$) coincide with the sets of points of an elliptic quadric of $S_{3,q}$.

If $q > 17$, this property is a particular case of the following (given by Segre[55], and also including another result by Barlotti[3]):

If q is odd and $k \geqslant q^2 - q + 19$, every k-cap of $S_{3,q}$ lies in an elliptic quadric of $S_{3,q}$.

Stronger improvements can be derived from the last theorems of § 4. Thus, for instance (Segre[55]),

If a denotes any positive constant, and q is odd and sufficiently large with respect to a, then any k-cap of $S_{3,q}$ for which $k \geqslant q^2 - aq$ is contained in an elliptic quadric of $S_{3,q}$; moreover, under the same conditions for q and if $r \geqslant 4$, for every k-cap of $S_{r,q}$ we have $k < q^{r-1} - aq^{r-2}$.

Also the case when q is even has been studied further, even if not so extensively as the case when q is odd. First of all, one of the results by Bose previously quoted has been completed by proving (Barlotti[2]) that, *if $q > 2$ is even, then $m_{3,q}^2 = q^2 + 1$.* Moreover, it has been shown that, for even q (e.g. for $q = 8$), *there exist $(q^2 + 1)$-caps of $S_{3,q}$ which are not quadrics*; however, *any such $(q^2 + 1)$-cap defines a null polarity*, exactly in the same way as in the case of a quadric (cf. Segre[56]).

We conclude by remarking that the sets $k_{r,q}^s$, considered above, are in their turn generalized by the k-sets of $S_{r,q}$ such that, if they contain any $(s+1)$-subset of linearly dependent points, then the space joining such a subset wholly consists of points of the k-set. Every k-set of this type defines a non-negative index δ, and can therefore be indicated by $k_{r,q}^{s,\delta}$, δ being the maximum dimension of the subspaces of $S_{r,q}$ lying entirely in the k-set (clearly, the $k_{r,q}^{s;0}$'s thus coincide with the $k_{r,q}^s$'s). Up to now, only the $k_{r,q}^{2;\delta}$'s (or k-caps of index δ) have been studied (Tallini[59]); and we have already given some of the results concerning them.

REFERENCES

[1] Artin, A. *Geometric Algebra*. Interscience Publ., New York, 1957.

[2] Barlotti, A. Un'estensione del teorema di Segre–Kustaanheimo. *Boll. Un. mat. ital.* (3), 10, 498–506 (1955).

[3] Barlotti, A. Un'osservazione sulle k-calotte degli spazi lineari finiti di dimensione tre. *Boll. Un. mat. ital.* (3), 11, 248–252 (1956).

[4] Barlotti, A. Sui $\{k; n\}$-archi di un piano lineare finito. *Boll. Un. mat. ital.* (3), 11, 553–556 (1956).

[5] Barlotti, A. Una limitazione superiore per il numero di punti appartenenti a una calotta $\mathscr{C}(k, 0)$ di uno spazio lineare finito. *Boll. Un. mat. ital.* (3), 12, 67–70 (1957).

[6] Bose, R. C. Mathematical theory of factorial design. *Sankhyā*, 8, 107–166 (1947).

[7] Boughon, P., Nathan, J. and Samuel, P. Courbes planes en caractéristique 2. *Bull. Soc. math. Fr.* 83, 275–278 (1955).

[8] Carlitz, L. A note on nonsingular forms in a finite field. *Proc. Amer. Math. Soc.* 7, 27–29 (1956).

[9] Châtelet, F. Les courbes de genre 1 dans un champ de Galois. *C.R. Acad. Sci., Paris*, 224, 1616–1618 (1947).

[10] Dickson, L. E. *Linear Groups with an Exposition of the Galois Field Theory*. Teubner, Leipzig, 1901.

[11] Edge, W. L. Geometry in three dimensions over $GF(3)$. *Proc. Roy. Soc. A*, 222, 262–286 (1953).

[12] Edge, W. L. The geometry of the linear fractional group $LF(4, 2)$. *Proc. Lond. Math. Soc.* (3), 4, 317–342 (1954).

[13] Edge, W. L. Line geometry in three dimensions over $GF(3)$, and the allied geometry of quadrics in four and five dimensions. *Proc. Roy. Soc. A*, 228, 129–146 (1955).

[14] Edge, W. L. 31-point Geometry. *Math. Gaz.* 39, 113–121 (1955).
[15] Edge, W. L. The isomorphism between $LF(2, 3^2)$ and $\mathscr{A}_6$. *J. Lond. Math. Soc.* 30, 172–185 (1955).
[16] Edge, W. L. The conjugate classes of the cubic surface group in an orthogonal representation. *Proc. Roy. Soc.* A, 233, 126–146 (1955).
[17] Edge, W. L. Conics and orthogonal projectivities in a finite plane. *Canad. J. Math.* 8, 362–382 (1956).
[18] Edge, W. L. The geometry of an orthogonal group in six variables. *Proc. Lond. Math. Soc.* (3), 8, 416–446 (1958).
[19] Hua, L. K. and Vandiver, H. S. On the existence of solutions of certain equations in a finite field. *Proc. Nat. Acad. Sci., Wash.*, 34, 258–263 (1948).
[20] Hua, L. K. and Vandiver, H. S. Characters over certain types of rings with applications to the theory of equations in a finite field. *Proc. Nat. Acad. Sci., Wash.*, 35, 94–99 (1949).
[21] Hua, L. K. and Vandiver, H. S. On the number of solutions of some trinomial equations in a finite field. *Proc. Nat. Acad. Sci., Wash.*, 35, 477–481 (1949).
[22] Hua, L. K. and Vandiver, H. S. On the nature of the solutions of certain equations in a finite field. *Proc. Nat. Acad. Sci., Wash.*, 35, 481–487 (1949).
[23] Järnefelt, G. A plane geometry with a finite number of elements. *Veröf. finn. geod. Inst.* n. 36 (1949).
[24] Järnefelt, G. Reflections on a finite approximation to Euclidean geometry. Physical and astronomical prospects. *Ann. Acad. Sci. Fenn.*, ser. A, 1, n. 96 (1951).
[25] Järnefelt, G. and Kustaanheimo, P. An observation on finite geometries. *Comptes Rendus XI Congr. Math. Scand.* 166–182. Trondheim, 1949.
[26] Jordan, C. *Traité des substitutions et des équations algébriques.* Paris, 1870.
[27] Kustaanheimo, P. A note on a finite approximation of the Euclidean plane geometry. *Soc. Sci. Fenn. Comm. Phys.-Math.* 15, 19 (1950).
[28] Kustaanheimo, P. On the fundamental prime of a finite world. *Ann. Acad. Sci. Fenn.*, ser. A, 1, n. 129 (1952).
[29] Kustaanheimo, P. Über die Versuche, ein logisch finites Weltbild aufzubauen. *Actes congr. int. Un. int. Phil. Sci.* 2, 60–65. Zürich, 1954.
[30] Kustaanheimo, P. On the relation of order in geometries over a Galois field. *Soc. Sci. Fenn. Comm. Phys.-Math.* 20, 8 (1957).
[31] Kustaanheimo, P. On the relation of congruence in finite geometries. *R.C. Mat. pura appl.* 16, 286–291 (1957).
[32] Kustaanheimo, P. On the relation of order in finite geometries. *R.C. Mat. pura appl.* 16, 292–296 (1957).
[33] Lang, S. and Weil, A. Number of points of varieties in finite fields. *Amer. J. Math.* 76, 819–827 (1954).
[34] Locher, L. Struktur der Axiome der projectiven Geometrie. *C.R. Congr. Int. Oslo 1936*, 2, 167–168 (1937).
[35] Lombardo-Radice, L. Sul problema dei k-archi completi di $S_{2,q}$. *Boll. Un. mat. ital.* (3), 11, 178–181 (1956).
[36] Lunelli, L. and Sce, M. Sulla ricerca dei k-archi completi mediante calcolatrice elettronica. *R.C. Convegno Reticoli geom. proi.* 81–86. Palermo, 1957.
[37] Ostrom, T. G. Ovals, dualities, and Desargues's theorem. *Canad. J. Math.* 7, 417–431 (1955).

498   BENIAMINO SEGRE

[38] Panella, G. F. Caratterizzazione delle quadriche di uno spazio (tridimen-
    sionale) lineare sopra un corpo finito. *Boll. Un. mat. ital.* (3), 10, 507–513
    (1955).

[39] Primrose, E. J. F. Quadrics in finite geometries. *Proc. Camb. Phil. Soc.*
    47, 299–304 (1951).

[40] Qvist, B. Some remarks concerning curves of the second degree in a finite
    plane. *Ann. Acad. Sci. Fenn.* ser. A, I, no. 134 (1952).

[41] Rosati, L. A. Sul numero dei punti di una superficie cubica in uno spazio
    lineare finito. *Boll. Un. mat. ital.* (3), 11, 412–418 (1956).

[42] Rosati, L. A. L'equazione delle 27 rette della superficie cubica generale
    in un corpo finito (Note I, II). *Boll. Un. mat. ital.* (3), 12, 612–626 (1957);
    (3), 13, 84–99 (1958).

[43] Scafati, M. Sui 6-archi completi di un piano lineare $S_{2,8}$. *R.C. Conv.
    Reticoli geom. proi.* 128–132. Palermo, 1957.

[44] Sce, M. Sui k_q-archi di indice h. *R.C. Conv. Reticoli geom. proi.* 133–135.
    Palermo, 1957.

[45] Sce, M. Sulla completezza degli archi nei piani proiettivi finiti. *R.C. Acc.
    Naz. Lincei,* (8), 25, 43–51 (1958).

[46] Segre, B. *Lezioni di geometria moderna.* I. Zanichelli, Bologna,
    1948.

[47] Segre, B. Sulle ovali dei piani lineari finiti. *R.C. Acc. Naz. Lincei,* (8), 17,
    141–142 (1954).

[48] Segre, B. Ovals in a finite projective plane. *Canad. J. Math.* 7, 414–416
    (1955).

[49] Segre, B. Curve razionali normali e k-archi negli spazi finiti. *Ann. Mat.*
    (4), 39, 357–379 (1955).

[50] Segre, B. Intorno alla geometria sopra un corpo di caratteristica due.
    Rev. Fac. Sci. de l'Univ. d'Istanbul (A), 21, 97–123 (1956).

[51] Segre, B. Sui k-archi nei piani finiti di caratteristica due. *Rev. Math.
    pures appl.* 2, 283–294 (1957).

[52] Segre, B. Sulle geometrie proiettive finite. *R.C. Conv. Reticoli geom. proi.*
    46–61. Palermo, 1957.

[53] Segre, B. Sulla geometria sopra un campo a caratteristica. *Archimede,*
    10, 53–60 (1958).

[54] Segre, B. Fondamenti di geometria non lineare sopra un corpo sghembo.
    R.C. circ. mat. Palermo, (2), 7, 81–122 (1958).

[55] Segre, B. Le geometrie di Galois. (Due to appear in *Ann. Mat.*)

[56] Segre, B. On complete caps and ovaloids in three-dimensional Galois
    spaces of characteristic two. (Due to appear in *Acta Arithm.*)

[57] Seiden, E. A theorem in finite projective geometry and an application to
    statistics. *Proc. Amer. Math. Soc.* 1, 282–286 (1950).

[58] Tallini, G. Sulle k-calotte degli spazi lineari finiti (Note I, II). *R.C. Acc.
    Naz. Lincei,* (8), 20, 311–317, 442–446 (1956).

[59] Tallini, G. Sulle k-calotte di uno spazio lineare finito. *Ann. Mat.* (4), 42,
    119–164 (1956).

[60] Tallini, G. Caratterizzazione grafica delle quadriche ellittiche negli spazi
    finiti. *R.C. Mat. pura appl.* 16, 328–352 (1957).

[61] Tallini, G. Sui q-archi di un piano lineare finito di caratteristica $p=2$.
    R.C. Acc. Naz. Lincei, (8), 23, 242–245 (1957).

[62] Tallini, G. Una proprietà grafica caratteristica della superficie di Veronese
    negli spazi finiti (Note I, II). *R.C. Acc. Naz. Lincei,* (8), 24, 19–23,
    135–138 (1958).

[63] Tallini, G. Caratterizzazione grafica di certe superficie cubiche di $S_{3,q}$. (Due to appear in *R.C. Acc. Naz. Lincei.*)

[64] Vandiver, H. S. Limits for the number of solutions of certain general types of equations in a finite field. *Proc. Nat. Acad. Sci., Wash.*, 33, 236–242 (1947).

[65] Vandiver, H. S. Quadratic relations involving the numbers of solutions of certain types of equations in a finite field. *Proc. Nat. Acad. Sci., Wash.*, 35, 681–685 (1949).

[66] Weil, A. Number of solutions of equations in finite fields. *Bull. Amer. Math. Soc.* 55, 497–508 (1949).

SOME GEOMETRICAL ASPECTS OF
COSET SPACES OF LIE GROUPS

By HSIEN-CHUNG WANG

1. Introduction

Let M be a differentiable manifold with a geometrical structure J. This J, for example, can be a Riemannian metric, a projective connection, a complex structure or a G-structure in the sense of Chern. The manifold M will be called J-*homogeneous* if there exists a group G of automorphisms of $\{M, J\}$ transitive on M. For a given J, it is of interest to classify all the J-homogeneous spaces. Suppose M to be J-homogeneous. Let us choose a point x of M, and denote by G_x the isotropic subgroup of G at x. In the natural manner, we can identify M with the space G/G_x of left cosets. The classification problem usually consists of three parts: (I) to determine whether the group of all automorphisms of $\{M, J\}$ is a Lie group; (II) if G is a Lie group, to express the structure J and its local invariants in terms of some tensors over a certain vector space related to the Lie algebras of G and G_x; (III) to classify all the pairs (G, G_x) such that the tensor corresponding to J really exists.

Problem I has been discussed by various authors, and the answer is affirmative when $\{M, J\}$ is a Riemannian space (Myers–Steenrod), a bounded domain (H. Cartan), a compact complex manifold (Bochner–Montgomery) or a manifold with a Cartan connection (S. Kobayashi). We note that the last case includes all the geometrical structures which determine a unique Cartan connection. As for Problem II, E. Cartan gave a general and effective method ([6], pp. 293–299) which can be applied to nearly all the known J's. We shall illustrate this in the next section. Problem III is a purely algebraic one, and in general can be very complicated. Nevertheless, if we assume M to be compact and simply-connected, then G_x is connected and by a theorem of Montgomery[12], we can choose G to be compact and even semi-simple. On account of our thorough knowledge of compact Lie groups (connected) and their representations, the classification can be actually carried out in most cases.

2. Two examples

In this section, we shall give two examples concerning Problem II. The first is more or less trivial, while the second illustrates the method of

Cartan in modern terms. Throughout this section, M denotes a differentiable manifold with a geometrical structure J, G a transitive Lie group of automorphisms of $\{M, J\}$, G_x the isotropic subgroup of G at a point x of M, and $p\colon G \to M$ the projection defined by $p(g) = g(x)$, $g \in G$.

2.1. J *is a tensor field over* M. Let V_x denote the tangent space of M at x. Then the value J_x of J at x is a tensor over V_x. Any k of G_x induces an automorphism $\dot{k}\colon V_x \to V_x$ which can be extended in a unique way to an automorphism of the tensor algebra over V_x. Since J is invariant under G, J_x is invariant under all $\dot{k}$ for $k \in G_x$. Conversely, G being transitive, any tensor over V_x invariant under all $\dot{k}$ ($k \in G_x$) can be extended to a G-invariant tensor field over M. Thus we have a one-to-one correspondence between G-invariant tensor fields over M and tensors over V_x which are left unaltered by all $\dot{k}$ with $k \in G_x$. The latter has a simple interpretation in terms of G and G_x. In fact, let $\mathbf{G}$, $\mathbf{G}_x$ be the Lie algebras of G, G_x respectively, and E the ·quotient vector space $\mathbf{G}$ (mod $\mathbf{G}_x$). We shall always identify the Lie algebra of a Lie group with the tangent space of the group at the identity. The projection $p\colon G \to M$ induces an onto linear map $p\colon \mathbf{G} \to V_x$ with $\mathbf{G}_x$ as its kernel, and so gives an isomorphism $\bar{p}\colon E \to V_x$. For $k \in G_x$, Ad k carries $\mathbf{G}$ into $\mathbf{G}$ and $\mathbf{G}_x$ into $\mathbf{G}_x$, and thus induces an automorphism $\mathrm{Ad}_E k\colon E \to E$. We see immediately that the diagram is commutative.

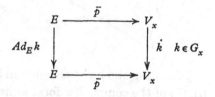

In other words, if we identify E with V_x by p, then $\mathrm{Ad}_E k$ coincides with $\dot{k}$. We have then

'*There is a one-to-one correspondence between* G-*invariant tensor fields over* $M = G/G_x$ *and the* $\mathrm{Ad}_E G_x$-*invariant tensors over* E.'

2.2. J *is a linear connection.* Consider the bundle $\mathscr{B}$ of frames over M. This is a differentiable principal bundle with group $H = GL(n, R)$, $n = \dim M$. For any $b \in \mathscr{B}$, we define $\eta_b\colon H \to \mathscr{B}$ by $\eta_b(\sigma) = b\sigma$, $\sigma \in H$. This map carries H homeomorphically onto the fibre containing b. A linear connection on M can be regarded as an **H**-valued (**H** = Lie algebra of H) Pfaffian form ω over $\mathscr{B}$ such that: (i) for each b, $\eta_b^* \omega$ is the left Maurer–Cartan form of H, and (ii) for each σ of H and each tangent vector $\mathbf{b}$ of $\mathscr{B}$, $\omega(\mathbf{b}\sigma) = \mathrm{Ad}\,(\sigma^{-1})\,\omega(\mathbf{b})$.

Now let G be a transitive Lie group of automorphisms of $\{M, J\}$, where J is a linear connection with connection form ω. In the natural manner, G acts differentiably on $\mathscr{B}$. As transformations of $\mathscr{B}$, elements of G and elements of H commute, and moreover $g^*\omega = \omega$, $g \in G$. Choose a point b_0 of $\mathscr{B}$ whose projection on M is x, and define $\pi\colon G \to \mathscr{B}$ by $\pi(g) = g(b_0)$, $g \in G$. The form $\varpi = \pi^*\omega$ is then an H-valued Pfaffian form over G. From the commutativity of the diagram, where L_g denotes the left translation associated with g, it follows that $L_g^*\varpi = \varpi$, i.e. ϖ is left invariant. We shall see that ϖ satisfies two other conditions.

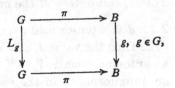

For $k \in G_x$, $k(b_0)$ and b_0 belong to the same fibre, and so there exists a unique $\sigma \in H$ such that $k(b_0) = b_0\sigma$. Thus we get a map $\phi\colon G_x \to H$ defined by $k(b_0) = b_0\phi(k)$, $k \in G_x$. Since H and G commute elementwise, ϕ is a homomorphism. In fact, up to a natural identification, ϕ coincides with the representation Ad_E of G_x. For simplicity, let us use the same letter ϕ to denote the Lie algebra homomorphism: $\mathbf{G}_x \to \mathbf{H}$ induced by the group homomorphism $\phi\colon G_x \to H$. From the commutativity of the diagrams

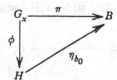   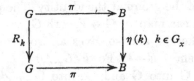

(R_k denoting the right translation associated with k) and the properties (i), (ii) of the connection form ω mentioned above, we verify easily that

$$\mathrm{Ad}\,\phi(k)\,\varpi = R_k^*\varpi = (\mathrm{Ad}\,k)^*\varpi, \quad \phi(\mathbf{k}) = \varpi(\mathbf{k}) \quad (k \in G_x,\ \mathbf{k} \in \mathbf{G}_x). \quad \text{(A)}$$

Summing up, we know that $\varpi = \pi^*\omega$ is an H-valued Pfaffian form over G which is invariant under left translations and satisfies conditions (A). Conversely, given such a Pfaffian form over G, we can construct a G-invariant linear connection over M in the natural manner. Since ϖ is left invariant, it can be interpreted simply as a linear map: $\mathbf{G} \to \mathbf{H}$, and thus

Theorem. There is a one-to-one correspondence between G-invariant linear connections over M and the linear maps $\Phi\colon \mathbf{G} \to \mathbf{H}$ such that

(a) Φ *coincides with* ϕ *when restricted to* $\mathbf{G}_x$, *and*

(b) $(\mathrm{Ad}\,\phi(k))\,\Phi(\mathbf{g}) = \Phi((\mathrm{Ad}\,k)\,(\mathbf{g}))$, $k \in G_x$, $\mathbf{g} \in \mathbf{G}$.

Here $\phi\colon G_x \to H$ *can be regarded as the homomorphism obtained from the representation* Ad_E *of* G_x *on* E.

The curvature and torsion forms as well as the holonomy algebra of the connection can be explicitly written out in terms of Φ[15].

3. Compact complex homogeneous manifolds

Now let us take up the complex homogeneous manifolds and discuss them in detail. Suppose G to be a complex Lie group and L a complex subgroup of G (i.e. the underlying space of L is closed and is a complex submanifold of G). Then the coset space G/L has a natural complex structure which is left invariant by the action of G. This complex manifold will be called the *complex coset space G/L*. Obviously, it is homogeneous.

Suppose M to be a compact complex homogeneous manifold, A the group of all analytic† homeomorphisms of M, and L the isotropic subgroup of A at a point x of M. Bochner and Montgomery[1] proved that A has a complex structure with respect to which (a) G becomes a complex Lie group and (b) the transformation map $A \times M \to M$ defined by $(g, y) \to g(y)$ is analytic, where $g \in A$, $y \in M$. It follows that the isotropic subgroup L is a complex subgroup of A, and M is analytically homeomorphic with the complex coset space A/L. Thus the class of compact complex homogeneous manifolds coincides with the class of compact complex coset spaces. The classification of the latter is largely an algebraic problem. When the subgroup L has infinitely many connected components, the algebraic problem becomes extremely complicated. To avoid this, we assume M to be simply-connected and hence L must be connected.

For simplicity, let us call *C-spaces* the compact and simply-connected complex homogeneous spaces.

Now let M be a C-space and A the group of all analytic homeomorphisms of M. By a theorem of Montgomery[12], the maximal compact subgroup K of A is transitive on M. Since M is simply-connected, the semi-simple part of K, and hence a maximal semi-simple subgroup G of A, must be transitive on M. We note that A and G are complex Lie groups while K is not. Hence M can be written as a complex coset space G/L of a semi-simple complex Lie group.

Let $\mathbf{G}$ be the Lie algebra (over the field of complex numbers) of G, and $\mathbf{H}$ a Cartan subalgebra. Since G is semi-simple, we know that the subgroup H generated by $\mathbf{H}$ in G is analytically isomorphic with the direct product of l multiplicative groups C^* of non-zero complex numbers, where l is the rank of G. An element $h \in \mathbf{H}$ will be called *rational* if the characteristic roots of $\mathrm{Ad}\, h$ are all rational numbers. This condition is equivalent to saying that $\exp(mih)$ is the identity for some positive

† By analytic, we shall always mean complex analytic.

integer m. A base of H is called *rational* if it consists of only rational elements. Let $h_1, \ldots, h_l$ be a rational base of H, and α, β be two roots. From Weyl's results, we know that α, β take rational values on the h's. We shall say that $\alpha > \beta$ if there exists an integer $s < l$ such that $\alpha(h_j) = \beta(h_j)$ for $j = 1, 2, \ldots, s$ and $\alpha(h_{s+1}) > \beta(h_{s+1})$. A root is called *positive* if it is bigger than the 0 root.

The following lemma is basic for our further discussions:

Basic Lemma. Let G be a complex semi-simple Lie group, L a connected complex subgroup of G, and $\mathbf{G}$, $\mathbf{L}$ the Lie algebras of G, L respectively. The complex coset space G/L is compact if and only if there exists a Cartan subalgebra $\mathbf{H}$ of $\mathbf{G}$, a linear subspace $\mathbf{H}_1$ of $\mathbf{H}$, and a base

$$k_1, \ldots, k_u, \quad h_{b+1}, h_{b+2}, \ldots, h_l, \quad e_{\alpha_1}, \ldots, e_{\alpha_m}, e_{-\alpha_1}, \ldots, e_{-\alpha_s} \quad (s < m)$$

of $\mathbf{L}$ such that (i) $b = 2u$, $h_j \in \mathbf{H}$, $k_j \in \mathbf{H}$, the h's are rational while no complex linear combination of the k's is rational, (ii) the e's are root vectors, and $\alpha_1, \ldots, \alpha_m$ exhaust all the positive roots with respect to a certain rational base of $\mathbf{H}$, and (iii) $\{\pm \alpha_1, \ldots, \pm \alpha_s\}$ is the totality of roots which annihilate the subspace $\mathbf{H}_1$.

It can be easily verified that the condition (i) is equivalent to the condition (i'): *The factor group $H/(L \cap H)$ is compact, where H denotes the subgroup generated by $\mathbf{H}$ in G.*

Let us write a C-space M in the form G/L with G complex semi-simple. Then L has the properties as in the Lemma. We see immediately that $\mathbf{L} + \mathbf{H}$ is the normalizer of $\mathbf{L}$, and the quotient $(\mathbf{L} + \mathbf{H})/\mathbf{L}$ is abelian. Hence the subgroup L_1 generated by $\mathbf{L} + \mathbf{H}$ in G is a connected complex subgroup of G, and the quotient L_1/L is a complex toral group of $2u$ real dimensions. Therefore G/L is a complex analytic principal bundle over G/L_1 with the complex torus L_1/L as the group. Since L_1 is connected and contains a Cartan subgroup of G, the compact complex coset space G/L_1 must be simply-connected and has positive Euler characteristic. Thus we have proved

Theorem 1. Each C-space M is an analytic principal bundle, with a complex torus as the group, over a C-space M_1 having positive Euler characteristic.

As we shall see later, this analytic principal bundle structure over M is unique (i.e. independent of the choice of G). In fact, the structure group must be the identity component of the centre of the group of all analytic homeomorphisms of M.

Now suppose that a C-space M is Kählerian. From Theorem 1, M is a complex torus bundle over a C-space M_1 with positive Euler characteristic. Since M is Kählerian, the fibre T^{2u} is non-homologous to zero in M. On account of the simply-connectedness of M, T^{2u} must be a point, and $M = M_1$. Hence

Theorem 2[11]. *If a C-space is Kählerian, then it has positive Euler characteristic.*

This theorem has the following converse:

Theorem 3[7, 4, 13]. *If a C-space M has positive Euler characteristic, then it is algebraic.*

This can be proved by using the fact that if G/L is a C-space of positive Euler characteristic with G complex semi-simple, then L is its own normalizer.

Goto proved, furthermore, that M is a rational variety.

4. Underlying topological manifold of a C-space

We shall introduce a class of real manifolds and show that it coincides with the class of the underlying manifolds of C-spaces. In fact, let K be a compact semi-simple Lie group, and X a connected subgroup whose semi-simple part coincides with the semi-simple part of the centralizer of a toral subgroup of K. If dim K/X is even, we shall call K/X a Q-*space*. We note that rank (K)–rank (X) is always even and is a topological invariant of the space K/X.

Now let $M = G/L$ be a C-space with G complex semi-simple, K a maximal compact subgroup of G, and $X = L \cap K$. By Montgomery's theorem[12], M is homeomorphic with $K/(K \cap L)$. Using the Basic Lemma, we can easily verify that $K/(K \cap L)$ is a Q-space.

Now we shall establish the converse in a strong form. First let us write down the following lemma, whose proof is quite direct:

Lemma. Let C^{*2u} be the complex Lie group which is analytically isomorphic with the direct product of $2u$ multiplicative groups of non-zero complex numbers. Given any complex torus T^{2u} of $2u$ real dimensions (note that for a fixed u, there are infinitely many essentially different such tori), there exists a complex subgroup J† of C^{*2u} such that C^{*2u}/J is analytically isomorphic with T^{2u}.

Let $M' = K/X$ be a Q-space. Here K is a compact semi-simple Lie group, X is a connected subgroup whose semi-simple part U coincides with the semi-simple part of the centralizer Z of a toral subgroup T_1 in K. Moreover, dim K/X is even. It is our purpose to construct various homogeneous complex structures over M'. In what follows, real and complex Lie algebras will appear simultaneously. To distinguish them, we shall

† J is analytically isomorphic with the u-dimensional complex vector space.

put an upper index '0' in all the real Lie algebras. Denote by $\mathbf{K}^0$, $\mathbf{X}^0$, $\mathbf{T}_1^0$, $\mathbf{U}^0$ the Lie algebras (over reals) of K, X, T_1, U respectively, and by $\mathbf{G} = \mathbf{K}^0 \otimes C$ the complex form of $\mathbf{K}^0$. Let G be the complex semi-simple Lie group corresponding to $\mathbf{G}$ and containing K as its maximal compact subgroup. Choose a Cartan subalgebra $\mathbf{H}$ of G such that $\mathbf{T}_1^0 \subset \mathbf{H}$ and that $\mathbf{K}^0$ is spanned (over reals) by

$$i(e_\alpha + e_{-\alpha}), \quad e_\alpha - e_{-\alpha}, \quad i[e_\alpha, e_{-\alpha}],$$

where e_α runs over all the *normalized* root vectors[16] of $\mathbf{G}$ with respect to $\mathbf{H}$. Since T_1, U, X are compact and $\mathbf{T}_1^0 \subset \mathbf{H}$, it can be proved[14] that $\mathbf{H}$ has a rational base

$$h_1, ..., h_b, \quad h_{b+1}, ..., h_a, \quad h_{a+1}, ..., h_l$$

with the following properties:

(A) $\mathbf{U}^0$ is spanned over the reals by

$$\{ih_{a+1}, ..., ih_l, e_\beta - e_{-\beta}, i(e_\beta + e_{-\beta}) : \beta(H_1) = 0\}, \quad \mathbf{H}_1 = \mathbf{T}_1 \otimes C;$$

(B) $\mathbf{T}_1^0 = Rih_1 + Rih_2 + ... + Rih_b + Rih_{b+1} + ... + Rih_l$, R = field of real numbers;

(C) $\mathbf{X}^0 = \mathbf{U}^0 + Rih_{b+1} + ... + Rih_a$.

Let us order the roots by means of this rational base $h_1, h_2, ..., h_l$. Denote by $\alpha_1, ..., \alpha_s$ the totality of positive roots which annihilate $\mathbf{H}_1$ and by $\alpha_{s+1}, \alpha_{s+2}, ..., \alpha_m$ the remaining positive roots of G. On account of (A) and (B), we know that $\alpha_j > \alpha_k$ for $j > s$, $k \leqslant s$, and so the complex linear subspace

$$\mathbf{P} = Ch_{b+1} + ... + Ch_l + Ce_{\alpha_1} + ... + Ce_{\alpha_m} + Ce_{-\alpha_1} + ... + Ce_{-\alpha_s}$$

forms a complex subalgebra of $\mathbf{G}$ where C denotes the field of complex numbers. Let $\mathbf{H}_0 = Ch_1 + ... + Ch_b$. Since the h's are rational $\mathbf{H}_0$ generates in G a complex subgroup H_0 which is analytically isomorphic with the direct product of b multiplicative groups of non-zero complex numbers. By definition of Q-spaces, dim K/X is even, so rank (K)–rank (X) is even and thus we can write $b = 2u$. From the lemma at the beginning of this section, given any complex torus T^{2u}, we can find a complex subgroup J of H_0 such that H_0/J is analytically isomorphic with T^{2u}. Denoting the Lie algebra of J by $\mathbf{J}$, we see immediately that the complex subalgebra $\mathbf{L} = \mathbf{P} + \mathbf{J}$ of $\mathbf{G}$ satisfies the conditions (ii) and (iii) in the Basic Lemma. Let L be the subgroup generated by $\mathbf{L}$ in G. Then $H/(H \cap L) \approx H_0/J$ is compact. This tells us that L is closed in G, and then, by the Basic Lemma, G/L is compact. Since G/L has finite fundamental group (because L is connected and G semi-simple), K acts transitively on G/L with isotropic subgroup $L \cap K$ which coincides with X. Hence the underlying topological manifold of G/L is our given Q-space M'.

Let L_1 be the subgroup generated by $H + L$ in G. It is closed and the quotient group L_1/L is analytically isomorphic with H_0/J which is the pre-assigned T^{2u}. It follows that G/L is an analytic T^{2u} bundle over G/L_1 which is evidently a C-space with positive Euler characteristic. Summing up the results, we have

Theorem 4. *The underlying topological manifold of a C-space is a Q-space. Conversely, given a simply-connected Q-space K/X and any complex torus T^{2u} with $2u = \text{rank}(K) - \text{rank}(X)$,† the real manifold K/X has a complex structure to make it a C-space M, and moreover, this C-space M is an analytic principal bundle over a Kählerian C-space with the pre-assigned T^{2u} as the fibre.*

5. Group of all analytic homeomorphisms of a C-space

Let M be a C-space and A the identity-component of the group of all analytic homeomorphisms of M. It can be proved that[14]

Theorem 5. *The complex Lie group A is locally the direct product of a complex semi-simple Lie group and a complex vector group.*

Let Z^{2v} denote the identity-component of the centre of A, where $2v$ denotes the real dimension. A base $E_1, E_2, ..., E_v$ of the complex Lie algebra of Z^{2v} can be regarded as v complex analytic vector fields over M. Since Z^{2v} is the centre of A, these vector fields $E_1, ..., E_v$ must be linearly independent everywhere. It follows that if M is Kählerian, it has positive Euler characteristic, and hence $v = 0$, i.e., A has a discrete centre.

Now suppose that M has an analytic principal fibre structure with base M_1 and with a complex torus T^{2u} as the group. Since T^{2u} acts on M analytically, we can regard it as a subgroup of A. By the maximal modulus principle, every element of A merely permutes the fibres, and therefore T^{2u} is a normal subgroup of A, whence $T^{2u} \subset Z^{2v}$. If, moreover, the base M_1 is Kählerian, then $T^{2u} = Z^{2v}$, for otherwise Z^{2v} would give a nowhere-zero analytic vector field on M_1, which contradicts the fact that M_1 has positive Euler characteristic. Combining this with Theorem 1, we have

Corollary 1. *The connected centre Z^{2v} of the group A is a complex torus. The action of Z^{2v} on M gives an analytic principal fibring of M with Kählerian base space, and this is the only analytic fibring of M with this property.*

† Note here that u is a topological invariant of Q, and is independent of the way of expressing Q as a coset space.

Now we come to the homogeneous complex structures over a given real compact manifold Q. If Q has positive Euler characteristic, then there exist only a finite number of pairs (G, L), where G is complex semi-simple, and L is a connected complex subgroup of G such that Q is homeomorphic with G/L. Therefore, Q has at most a finite number of homogeneous complex structures. Borel and Hirzebruch[3] gave an effective method of determining all the homogeneous complex structures over such a Q. For real manifolds with zero Euler characteristic, we have from Theorem 4 and Corollary 2, the following

Theorem 6. Let $Q = K/X$ be a Q-space wth zero Euler characteristic, and $2u = \text{rank}(K) - \text{rank}(X)$. To each complex torus T^{2u} there corresponds a homogeneous complex structure over Q. Two such complex structures are equivalent if and only if the corresponding complex tori are analytically homeomorphic. Therefore there are infinitely many inequivalent homogeneous complex structures over Q.

6. Global analytic invariants of C-spaces

Let $M = G/L$ be a C-space with complex G. Recently Bott[5] studied the complex vector bundle over M associated with the principal bundle $L \to G \to M$. Abundant results are obtained. In particular, a definite formula for the Dolbeault groups of M is given. When M is, moreover, Kählerian, Borel and Hirzebruch[3] computed its various characteristic classes.

7. Homogeneous bounded domains

Among the non-compact homogeneous complex manifolds, the homogeneous bounded domains are of particular interest. We know that a bounded domain D has an intrinsic Kählerian metric (Bergman metric), and the group of all analytic homeomorphisms is a Lie group, but never a complex Lie group. E. Cartan has determined all the symmetric bounded domains. Borel[2] and Koszul[10] proved independently that if a bounded domain admits a transitive, semi-simple Lie group G of automorphisms, then it must be symmetric. Recently, Hano[8] proved the same result under the much weaker condition that G is unimodular.

REFERENCES

[1] Bochner, S. and Montgomery, D. Groups of analytic manifolds. *Ann. Math.*
    (2), 48, 659–669 (1947).
[2] Borel, A. Kählerian coset spaces of semi-simple Lie groups. *Proc. Nat.*
    Acad. Sci., Wash., 40, 1147–1151 (1954).

[3] Borel, A. and Hirzebruch, F. Characteristic classes and homogeneous spaces. *Amer. J. Math.* 80, 458–538 (1958).

[4] Borel, A. and Weil, A. Représentations linéaires et espaces homogènes Kählériens des groupes de Lie compacts. *Séminaire Bourbaki*, 1954 (Exposé par J.-P. Serre).

[5] Bott, R. Homogeneous vector bundles. *Ann. Math.* (2) 66, 203–248 (1957).

[6] Cartan, E. *Leçons sur la géométrie des espaces de Riemann*, 2e ed. Paris, 1946.

[7] Goto, M. On algebraic homogeneous spaces. *Amer. J. Math.* 76, 811–818 (1954).

[8] Hano, J. On Kählerian homogeneous spaces of unimodular Lie groups. *Amer. J. Math.* 79, 885–900 (1957).

[9] Hawley, N. S. Complex bundles with abelian groups. *Pacific J. Math.* 6, 65–82 (1956).

[10] Koszul, J. L. Sur la forme hermitienne canonique des espaces homogènes complexes. *Canad. J. Math.* 7, 562–576 (1955).

[11] Lichnerowicz, A. Espaces homogènes Kählériens. *Coll. Géom. Diff., Strasbourg*, 171–184 (1953).

[12] Montgomery, D. Simply-connected homogeneous spaces. *Proc. Amer. Math. Soc.* 1, 467–469 (1950).

[13] Tits, J. Sur certaines classes d'espaces homogènes de groupes de Lie. *Mém. Acad. R. Belg.* 29 (1955).

[14] Wang, H. C. Closed manifolds with homogeneous complex structure. *Amer. J. Math.* 76, 1–32 (1954).

[15] Wang, H. C. On invariant connections over a principal fibre bundle. *Nagoya Math. J.* 13, 1–20 (1958).

[16] Weyl, H. Theorie der Darstellung kontinuierlicher halb-einfacher Gruppen durch lineare Transformationen. II, III. *Math. Z.* 24, 328–395 (1925).

CONTINUOUS PARAMETER
MARKOV CHAINS†

By KAI-LAI CHUNG

The name given in the title is an abbreviation of 'Markov processes with continuous time parameter, denumerable state space and stationary transition probabilities'. This theory is to the discrete parameter theory as functions of a real variable are to infinite sequences. New concepts and problems arise which have no counterpart in the latter theory. Owing to the sharply defined nature of the process, these problems are capable of precise and definitive solutions, and the methodology used well illustrates the general notions of stochastic processes. It is possible that the results obtained in this case will serve as a guide in the study of more general processes. The theory has contacts with that of martingales and of semi-groups which have been encouraging and may become flourishing. For lack of space the developments from the standpoint of semi-groups or systems of differential equations cannot be discussed here.

Terms and notation not explained below follow more or less standard usage such as in [11].

Let $(\Omega, \mathfrak{F}, \mathbf{P})$ be a probability triple where $(\mathfrak{F}, \mathbf{P})$ is complete; $\mathbf{T} = [0, \infty)$, $\mathbf{T}^0 = (0, \infty)$, $\mathfrak{B}$ the usual Borel field on $\mathbf{T}$. Let $\{x_t, t \in \mathbf{T}\}$ be a Markov chain with the minimal state space $\mathbf{I}$, the stationary transition matrix $\{(p_{ij})\}$ and an arbitrary fixed initial distribution. By the minimal state space we mean the smallest denumerable set (of real numbers) such that $\mathbf{P}\{x_t(\omega) \in \mathbf{I}\} = 1$ for every $t \in \mathbf{T}$. The transition matrix is characterized by the following properties: for every $i, j \in \mathbf{I}$, $s, t \in \mathbf{T}^0$:

$$p_{ij}(t) \geqslant 0, \quad \sum_{j \in \mathbf{I}} p_{ij}(t) = 1, \quad p_{ij}(s+t) = \sum_{k \in \mathbf{I}} p_{ik}(s) p_{kj}(t). \tag{1}$$

The last of these relations is the semi-group property. In order to have a separable and measurable Markov chain with the given $\mathbf{I}$ and $\{(p_{ij})\}$ it is sufficient (and essentially necessary) that

$$\lim_{t \downarrow 0} p_{ii}(t) = 1 \tag{2}$$

for every $i \in \mathbf{I}$. In this case we define $p_{ij}(0) = \delta_{ij}$. Each p_{ij} is then uniformly continuous in $\mathbf{T}$. We shall confine ourselves to this case. We may suppose, by going to a *standard modification*, that the process is separable

† This paper was prepared with the partial support of the Office of Scientific Research of the United States Air Force.

(relative to the class of closed sets) and measurable. The basic sample function properties, due largely to Doob[9] and Lévy[17], can then be deduced. We cannot detail these properties but, as a consequence of them, there is a specific *version* (or *realization*) which has desirable properties. To obtain this let us take **I** to be the set of positive integers and compactify it by adjoining one fictitious state ∞, in the usual manner. Let us call a function f on **T** *lower right semi-continuous* with respect to a denumerable set R dense in T if $f(t) = \lim_{R \ni s \downarrow t} f(s)$ for every t. Then there is a version of the given Markov chain such that

(i) $x(\cdot, \cdot)$ is measurable $\mathfrak{B} \times \mathfrak{F}$, or Borel measurable;

(ii) each sample function $x(\cdot, \omega)$ is lower right semi-continuous with respect to any R.

Other properties which are valid for almost all sample functions may be further imposed; we need not elaborate them here. Clearly (i) implies measurability and (ii) implies well-separability, namely separability with respect to any denumerable set dense in **T**. We mention that, despite (ii), it is possible that for all ω, the t-set $S_\infty(\omega)$ where $x(t, \omega) = \infty$ is everywhere dense in **T** (see [14]).

It has been known for some time that an important concept in the study of general Markov processes is the so-called strong Markov property (see [3, 12, 22]). It turns out that the version specified above has this property, which we now proceed to describe (in a slightly restricted form). Let $\mathfrak{F}_t$ be the Borel subfield of $\mathfrak{F}$ generated by $\{x_s, 0 \leqslant s \leqslant t\}$ and augmented by all sets of probability zero. A random variable α with domain of finiteness Δ_α is said to be *optional* (or 'stopping time' or 'independent of the future') if for every t we have

$$\{\omega : \alpha(\omega) < t\} \in \mathfrak{F}_t.$$

The collection of sets Λ in $\mathfrak{F}$ such that $\Lambda \cap \{\omega : \alpha(\omega) < t\} \in \mathfrak{F}_t$ is a Borel field $\mathfrak{F}_\alpha$ called the *pre-α field*. The process $\{\xi_t, t \in \mathbf{T}\}$ on the reduced triple $(\Delta_\alpha, \Delta_\alpha \mathfrak{F}, \mathbf{P}(\cdot \mid \Delta_\alpha))$ where

$$\xi(t, \omega) = x(\alpha(\omega) + t, \omega)$$

is called the *post-α process* and the augmented Borel field generated by this process the *post-α field* $\mathfrak{F}'_\alpha$. Observe that if α is optional then so is $\alpha + t$ for each $t > 0$. For the sake of brevity we shall suppose that $\Delta_\alpha = \Omega$ in the following. The following assertions, collectively referred to as the strong Markov property here, are true for every optional α.

(a) For each $t \in \mathbf{T}^0$, ξ_t is finite (i.e. $\xi_t \in \mathbf{I}$) with probability one.

(b) The post-α process is a Markov chain in $\mathbf{T}^0$ whose state space and transition matrix are restrictions of those of the given Markov chain.

(c) For each $t \in T$, the pre-$(\alpha+t)$ and post-$(\alpha+t)$ fields are conditionally independent given ξ_t, wherever the latter is finite (hence in particular almost everywhere if $t \in T^0$, by (a)). Thus if $\Lambda \in \mathfrak{F}_{\alpha+t}$ and $\Lambda' \in \mathfrak{F}'_{\alpha+t}$ then on the set $\{\omega : \xi_t(\omega) \in I\}$ we have

$$\mathbf{P}\{\Lambda\Lambda' \mid \xi_t\} = \mathbf{P}\{\Lambda \mid \xi_t\}\,\mathbf{P}\{\Lambda' \mid \xi_t\}.$$

(d) The post-α process is well-separable and Borel measurable as it stands.

Furthermore, let us consider for each given $\Lambda \in \mathfrak{F}_\alpha$ and $s \leqslant t$, the conditional probability

$$\mathbf{P}\{x(t,\omega) = j \mid \Lambda; \alpha = s\} = r_j(s, t \mid \Lambda) \tag{3}$$

defined for almost all s according to the measure induced by α on $\mathfrak{B}$. The following additional assertions are true.

(e) For each $j \in I$ and almost all s according to the α-measure: the function $r_j(s, \cdot \mid \Lambda)$ satisfies conditions analogous to (1) and is continuous in (s, ∞); and we have

$$\mathbf{P}\{\xi(t,\omega) = x(\alpha(\omega)+t, \omega) = j \mid \Lambda; \alpha = s\} = r_j(s, s+t \mid \Lambda). \tag{4}$$

(f) The pre-α and post-α fields are absolutely independent if and only if $r_j(s, t \mid \Lambda)$ as a function of the pair (s, t) is a function of $t - s$, for each Λ and j.

Let us observe, comparing (3) and (4), that the assertion (e) is a nontrivial substitution property for the conditional probability.

A preliminary view of the above assertions, together with a justification of the name 'strong Markov property', may be obtained by considering the particular case $\alpha = $ constant. In this case the assertions (a)–(d) become the defining properties of $\{x_t, t \in I\}$, while (e) reduces to the continuity of each p_{ij}. This simple observation implies the truth of (a)–(e) if α is denumerably-valued and shows that for a discrete parameter Markov chain the corresponding assertions hold almost trivially. Proofs of the above assertions except (d), in somewhat more precise terms, are given in [7]. Similar results which overlap these are given by Yushkevič[22]†; for another proof of (a), (b) and part of (c) see Austin[2].

The essence of the strong Markov property may be briefly stated as follows: The ordinary Markov property valid at a fixed time t remains valid at a variable time α chosen according to the evolution of the process but without prevision of the future. The classical illustration is that of a gambler who chooses his turn of playing according to a

---

† His assertion involving another random variable $\geqslant \alpha$ and measurable $\mathfrak{F}_\alpha$ follows easily from (e).

gambling system which he has devised without the aid of prescience. Similar concepts for a martingale have been developed by Doob[11].

Let us discuss some applications of the strong Markov property. It should be remarked that while its invocation is a basic step in each of the following cases further work is needed to establish the results to be mentioned.

(A) The simplest case of an optional random variable is the first entrance time into a given state. This satisfies (f) and its judicious use yields various 'decomposition formulas'. For example, let $_Hp_{ij}(t)$ denote the transition probability from i to j in time t 'under the taboo H' (namely, before entering any state in H), and $_{k,H}p_{ij}(t)$ the analogous probability where H is replaced by the union of H and k; $_HF_{ik}$ the first entrance time distribution from i to k under the taboo H. The intuitive meaning of the following formula is clear: if $k \notin H$, we have

$$_Hp_{ij}(t) = {}_{k,H}p_{ij}(t) + \int_0^t {}_Hp_{kj}(t-s) \, d \, {}_HF_{ik}(s);$$

but its rigorous proof requires the strong Markov property, in particular (e). Specialization of H to one state leads to ratio limit theorems of the Doeblin type concerning $\int_0^t p_{jj}(s) \, ds \Big/ \int_0^t p_{ii}(s) \, ds$ as $t \to \infty$; see [8].

Next, let us recall that the state i is called *stable* or *instantaneous* according as $q_i = -p'_{ii}(0)$, which always exists, is finite or infinite. Let $i \neq j$ and consider in a recurrent class (see [8]) the successive returns to i via j (the intervention of j is necessary only if i is instantaneous). These return times partition the time axis T into independent blocks to which Doeblin's method of treating a functional of the Markov chain can be applied. In this way the classical limit theorems, like the laws of large numbers, the law of iterated logarithm and the central limit theorem, can be easily extended. For the discrete case see [4], where there are some errors in the proofs which can be corrected (see the last footnote in [8]).

Finally, Kolmogorov's example[16], in which there is exactly one instantaneous state, can be analysed probabilistically by use of certain entrances into this state and taboo probabilities. It can be shown as a consequence that the construction of sample functions of this process given by Kendall and Reuter[15] with semi-group methods is indeed the unique one. Namely, the version specified above having Kolmogorov's transition matrix must have the properties implied by the Kendall–Reuter construction.

(B) Let the process start at a stable state i and consider the first exit time from i (see [6]). This is optional and the condition in (f) is again satisfied. Denoting for $\Lambda = \Omega$ the corresponding $r_j(s, s+t)$ by $r_{ij}(t)$, which is continuous in t by (e), we have readily

$$p_{ij}(t) = \delta_{ij} e^{-q_i t} + q_i \int_0^t e^{-q_i(t-s)} r_{ij}(s) \, ds.$$

This integral representation implies the existence of a continuous derivative $p'_{ij}(t) = q_i[r_{ij}(t) - p_{ij}(t)]$ and various complements including an interpretation of Kolmogorov's first (backward) system of differential equations. (The second (forward) system can be dually treated and falls under (A) above.) This gives a probabilistic proof of a result which was first established by analytic means by Austin[1]. A proof similar to the one sketched here was announced by Yushkevič but has not yet appeared.

The independence of the pre-exit and post-exit fields implies the fundamental observation due to Lévy[17] that the lengths of the stable intervals are independently distributed; the separability of the post-exit process asserted in (d) then yields the negative exponential distributions for these lengths; see [5].

(C) Let the process start at an instantaneous state i and put

$$S_i(\omega) = \{t : x(t, \omega) = i\}, \quad \mu_i(t, \omega) = \mu[S_i(\omega) \cap (0, t)],$$

where μ is the Borel–Lebesgue measure on $\mathfrak{B}$. Then for each s, the random variable α_s defined by

$$\alpha_s(\omega) = \inf\{t : \mu_i(t, \omega) > s\}$$

is optional. In words, α_s is the first time when the total amount of time spent in the state i exceeds s. This idea, which is a partial analogue of the exit time from a stable state discussed under (B), is due to Lévy[17,18,19]. Lévy makes use of the more general device of counting time only on a selected set of states, thereby annihilating the remaining states together with the time spent in them. This idea remains to be fully exploited. As a simple example, if $i \neq j$, then the total time spent in i before entering j has the negative exponential distribution $1 - e^{-a_{ij}t}$, where

$$a_{ij} = \int_0^\infty {}_jp_{ii}(t) \, dt$$

in our previous 'taboo' notation.

(D) In this last application we touch upon a chapter of the theory of continuous parameter Markov chains which has yet to be written. It is to be observed that the strong Markov property fails on the set where $\xi_0 = \infty$. While the assertions (a)–(d) are always valid for $t \in T^0$, our information is inadequate as the critical time α is approached from the right if the sample function values approach ∞ there. This failure may be formally attributed to the crude one-point compactification we have adopted, which does not distinguish between the various modes of approaching ∞ that ought to correspond to distinct adjoined fictitious states rather than the one and only ∞. From this point of view the main task, called the 'boundary problem' by certain authors, is the proper compactification of the minimal state space I so as to restore the strong Markov property on the set where $\xi_0 \notin I$ and to induce the appropriate boundary behavior (as in potential theory). Without loss of generality we may suppose that this set has probability one. For fixed $j \in I$ and $t_0 \in T^0$ let us consider the process $\{\eta_t, 0 \leqslant t \leqslant t_0\}$, where $\eta(t, \omega) = p_{\xi(t, \omega), j}(t_0 - t)$. Since $\{\xi_t\}$ is a Markov chain by (b) the new process $\{\eta_t\}$ is easily seen to be a martingale. Applying the martingale convergence theorem we see that $\lim_{t \downarrow 0} \eta(t, \omega)$ exists and is finite with probability one, and the limit has certain gratifying properties. The idea of considering this sort of martingale is due to Doob[9], and the present application to the post-α process will undoubtedly play a role in the compactification problem. For other formulations of the boundary problem see Feller[13], Reuter[21], and Ray[20].

The preceding discussion is centered around the strong Markov property as a convenient rallying point. Lest the impression should have been made that there was nothing else to be done I should like to conclude my discussion by mentioning some other problems not directly connected with the above.

A very natural circle of problems concerns the analytical properties (not to say characterization) of the elements of a transition matrix defined by (1) and (2). These may be regarded as problems in pure analysis. For example, it is still an open problem whether $p'_{ij}(t)$ exists if $t > 0$ and both i and j are instantaneous.† The solution of such a problem would be the more interesting if probabilistic significance is found. In this connection Jurkat[14] has observed that the differentiability results discussed under (B) hold even if the second condition in (1) is omitted, the condition (2) being assumed of course. The following even more primitive and probabilistically meaningful result is only a few weeks old: each p_{ij} is either identically zero or never zero. The original proof of this result, due to Austin, makes ingenious use of the strong Markov property. It is almost 'unfortunate' that a simplification has been found by myself which uses only the separability and measurability of an associated process. This is not the only example where a purely analytic

† *(Added in proof.)* D. Ornstein has now proved that for every i and j, $p'_{ij}(t)$ exists and is continuous for $t > 0$.

and simple-sounding proposition has so far been proved only by properly probabilistic methods.† It follows from this result, as observed by Austin, that if all states communicate then each of the Kolmogorov systems holds as soon as one of its equations holds for one value of the argument.

Another circle of problems is the approximation of the continuous parameter chain $\mathfrak{C} = \{x_t, t \in \mathbf{T}\}$ by its *discrete skeletons* $\mathfrak{C}_s = \{x_{ns}, n \in \mathbf{N}\}$ where $s > 0$ and $\mathbf{N}$ is the sequence of non-negative integers. In what sense and how well do the skeletons $\mathfrak{C}_s$ approximate $\mathfrak{C}$ as $s \downarrow 0$? This does not appear to be as simple a matter as might be expected. To cite a specific example: let m_{ij} denote the mean first entrance time (or return time if $i = j$) from i to j in $\mathfrak{C}$, and let $m_{ij}(s)$ denote the analogous quantity in $\mathfrak{C}_s$. The well-known theorem that $\lim_{t \to \infty} p_{ii}(t)$ exists (see [17]) implies that if i is stable then $m_{ii}(s) = q_i m_{ii}$ for every s. If i and j are distinct states in a positive (or strongly ergodic) class then it can be shown that $\lim_{s \downarrow 0} sm_{ij}(s) = m_{ij}$ by a rather devious method. But I do not know what the situation is with moments of higher order. We may also mention the open problem of characterizing a discrete parameter Markov chain which can be imbedded in a continuous parameter one, namely which is a skeleton of the latter.

Finally, let me mention an annoying kind of problem. Various models of Markov chains can be easily described by so-to-speak word-pictures but the rigorous verification that they are indeed Markovian is often laborious. The well-known construction by Doob[10] is an example. Other examples are given by Lévy[17] of which one (his example II.10.5) may be roughly described as follows. Consider first the infinite *descending escalator* such that from the state $i + 1$ one necessarily goes into the state i while the mean sojourn times in all the states form a convergent series, the process terminating at the state 1. This is a Markov chain in $\mathbf{T}^0$ and one need only hitch it on to a new state at the beginning to obtain a Markov chain in $\mathbf{T}$. In fact, the resulting process is the second example given by Kolmogorov[16], which like the first one mentioned under (A) above has been analysed in detail by Kendall and Reuter[15]. Now modify this scheme by allowing, upon leaving each step, the alternative of either entering the next lower step or starting all over again from the (infinite) top of the escalator. By proper choice of the probabilities of the alternatives it is possible to jump to and return from infinity a nondenumerably infinite number of times. It seems 'intuitively obvious' that the resulting process is still Markovian, but if so why does it elude a simple proof?

† *(Added in proof.)* D. Ornstein has now found an analytical proof of this result.

REFERENCES

[1] Austin, D. G. On the existence of derivatives of Markoff transition probability functions. *Proc. Nat. Acad. Sci., Wash.*, 41, 224–226 (1955).
[2] Austin, D. G. A new proof of the strong Markov theorem of Chung. *Proc. Nat. Acad. Sci., Wash.*, 44, 575–578 (1958).
[3] Blumenthal, R. M. An extended Markov property. *Trans. Amer. Math. Soc.* 85, 52–72 (1957).
[4] Chung, K. L. Contributions to the theory of Markov chains. II. *Trans. Amer. Math. Soc.* 76, 397–419 (1954).
[5] Chung, K. L. Foundations of the theory of continuous parameter Markov chains. *Proceedings Third Berkeley Symposium on Mathematical Statistics and Probability*, 2, 29–40 (1955).
[6] Chung, K. L. Some new developments in Markov chains. *Trans. Amer. Math. Soc.* 81, 195–210 (1956).
[7] Chung, K. L. On a basic property of Markov chains. *Ann. Math.* (2), 68, 126–149 (1958).
[8] Chung, K. L. Some aspects of continuous parameter Markov chains. *Publ. Inst. Statist. Univ. Paris*, 6, 271–287 (1957).
[9] Doob, J. L. Topics in the theory of Markoff chains. *Trans. Amer. Math. Soc.* 52, 37–64 (1942).
[10] Doob, J. L. Markoff chains—denumerable case. *Trans. Amer. Math. Soc.* 58, 455–473 (1945).
[11] Doob, J. L. *Stochastic Processes*. New York, 1953.
[12] Dynkin, E. and Yushkevič, A. Strong Markov processes. *Theory of Probability and Its Applications*, 1, 149–155 (1956).
[13] Feller, W. On boundaries and lateral conditions for the Kolmogorov differential equations. *Ann. Math.* (2), 65, 527–570 (1957).
[14] Feller, W. and McKean, H. A diffusion equivalent to a countable Markov chain. *Proc. Nat. Acad. Sci., Wash.*, 42, 351–354 (1956).
[15] Kendall, D. G. and Reuter, G. E. H. Some pathological Markov processes with a denumerable infinity of states and the associated semi-groups of operators on *l*. *Proceedings International Congress of Mathematicians, Amsterdam*, 3, 377–415 (1954).
[16] Kolmogorov, A. N. On some problems concerning the differentiability of the transition probabilities in a temporally homogeneous Markov process having a denumerable set of states. *Učenye Zapiski (Matem.) Moskov. Gos. Univ.* (4), 148, 53–59 (1951).
[17] Lévy, P. Systèmes markoviens et stationnaires; cas dénombrables. *Ann. Sci. Éc. Norm.* (3), 68, 327–381 (1951).
[18] Lévy, P. Compléments à l'étude des processus de Markoff. *Ann. Sci. Éc. Norm.* (3), 69, 26–212 (1952).
[19] Lévy, P. Processus markoviens et stationnaires du cinquième type (infinité dénombrable des états possibles, paramètre continu). *C.R. Acad. Sci., Paris*, 236, 1630–1632 (1953).
[20] Ray. D. (To appear.)
[21] Reuter, G. E. H. Denumerable Markov processes and the associated contraction semi-groups on *l*. *Acta Math.* 97, 1–46 (1957).
[22] Yushkevič, A. On strong Markov processes. *Theory of Probability and Its Applications*, 2, 187–213 (1957).

Б. В. ГНЕДЕНКО

О ПРЕДЕЛЬНЫХ ТЕОРЕМАХ ТЕОРИИ ВЕРОЯТНОСТЕЙ

Я счастлив прочесть доклад на настоящем конгрессе и хотел бы прежде всего сердечно поблагодарить слушателей за оказанное мне внимание, а оргкомитет конгресса — за приглашение и предоставленную возможность выступления.

Среди разнообразных и сильно разветвившихся за последние годы направлений исследований теории вероятностей классические задачи суммирования, а также их естественное развитие, продолжают занимать значительное место. В сороковые годы, после блестящих работ Линдеберга, П. Леви, С. Н. Бернштейна, Г. Крамера, А. Я. Хинчина, А. Н. Колмогорова, В. Феллера, Г. М. Бавли, В. Дёблина и многих других учёных, казалось, что все основные вопросы теории суммирования уже разрешены. Однако последующее развитие событий показало, что даже хорошо изученные проблемы способны при иных к ним подходах вызвать к жизни новые увлекательные постановки задач и глубокие исследования. В результате появились исследования Ессеена, Донскера, Ю. В. Линника, М. Лоева, Ю. В. Прохорова, А. Н. Колмогорова, А. В. Скорохода, С. Х. Сираждинова, Р. Л. Добрушина и ряда других, открывших новые пути в старых, казалось бы исхоженных областях теории вероятностей. Более того в некотором смысле теория суммирования вновь становится одной из центральных линий развития теории вероятностей. И здесь снова, как это постоянно происходит в истории математики, новые проблемы подсказаны не столько естественным внутренним развитием науки, сколько практической жизнью в самых разнообразных её проявлениях.

В связи со сказанным естественно вспомнить слова одного из крупнейших геометров прошлого века —П. Л. Чебышева: 'Сближение теории с практикой даёт самые благотворные результаты, и не одна только практика от этого выигрывает; сами науки развиваются под влиянием её; она открывает им новые предметы для исследования, или новые стороны в предметах давно известных.... Она предлагает вопросы существенно новые для науки и, таким образом, вызывает на изыскание совершенно новых методов. Если теория много выигрывает от новых приложений старой методы или от новых развитий её, то она еще более приобретает открытием новых метод, и в этом случае наука находит себе верного руководителя в практике.'

Вопросы, которые выдвигали в 18 веке демография, практика страхования, обработка результатов наблюдений, привели Я. Бернулли, Муавра, Лапласа к первым предельным теоремам для сумм случайных величин. Позднее, в значительной степени задачи теории стрельбы, послужили Пуассону путеводной нитью при обобщении закона больших чисел и теорем Муавра-Лапласа, а также открытии распределения, носящего теперь его имя. Известно, что исследования Коши по теории суммирования вызваны развитием теории ошибок наблюдений. Развитие статистической физики, а позднее математических методов в биологии, выдвинуло перед теорией вероятностей новые интересные вопросы. Значительная часть этих вопросов относилась к изучению поведения сумм большого числа случайных величин. Математическая статистика с первых дней своего существования не только широко использовала уже готовые результаты теории суммирования, но и постоянно наталкивала на новые изыскания. Много своеобразных вопросов выдвигают перед теорией суммирования экономические и технические науки, а также непосредственная производственная деятельность.

Об одном из таких заказов производственников мне хотелось бы рассказать сейчас несколько подробнее. Недавно ко мне обратились за консультациями специалисты по расчёту электрических сетей промышленных предприятий. Вопрос ими был поставлен так: имеется большое число потребителей электрического тока — станков, сварочных аппаратов или нагревательных приборов. Каждый из этих приборов включается в сеть в случайные моменты времени и в процессе использования затрачивает не вполне определенную, а меняющуюся случайным образом мощность. Для примера экскаватор в карьере при черпании затрачивает энергию, колеблющуюся в зависимости от плотности породы, далее при переносе выбранной породы требуется иная мощность, которая вновь меняется при высыпании из ковша и возвращении пустого ковша в исходное положение. Очевидно, что общая потребляемая мощность является суммой потребляемых мощностей каждым из приборов. Мы таким образом оказались перед следующей задачей: имеется последовательность случайных процессов

$$\xi_1(t), \ \xi_2(t), \ \ldots, \ \xi_n(t),$$

каждый из которых в некотором смысле мал по сравнению с суммой остальных. Спрашивается что можно сказать о суммарном процессе

$$\zeta_n(t) = \sum_{k=1}^{n} \xi_k(t),$$

если число слагаемых велико? Понятно, что так поставленная задача еще неопределенна, требуются дополнительные уточнения. Однако ясно, что техника выдвигает перед математикой необходимость развития теории предельных теорем для сумм случайных процессов.

Интересно отметить, что вскоре после специалистов электриков с аналогичным вопросом обратились ко мне инженеры, занимающиеся расчетом газовых сетей. Но ведь такие же вопросы возникают также в водопроводных, теплофикационных, канализационных сетях и при расчете потоков машин на автострадах. Собственно тот же вопрос возникает и во многих физических теориях, в теории массового обслуживания и в математической статистике.

Мне хотелось бы указать на еще одно чисто математическое использование теории суммирования, начало которого связано с именем Э. Бореля. Я имею в виду построение новой главы теории чисел, связанной как в формулировках результатов, так и в методах их получения с теорией вероятностей. В последние годы тут получены многочисленные результаты. Мне нет необходимости на них останавливаться подробнее, т.к. этому посвящен специальный доклад проф. А. Реньи. Единственно, что я позволю себе это упомянуть о систематическом использовании этой теорией всего арсенала результатов теории суммирования случайных величин.

Мы видим, что предельные теоремы теории вероятностей играют в развитии современной науки значительную роль. Естественно спросить себя чем это вызвано, какие общие причины приводят к тому, что к одним и тем же проблемам приводят и естественные науки, и технические проблемы, и математическая статистика? Мне представляется, что в известном смысле предельные теоремы особенно полно отражают специфическую особенность тех явлений, которые изучает теория вероятностей. Эти явления обусловлены тем, что они проистекают под влиянием большого числа причин, причем каждая причина в отдельности оказывает лишь ничтожное влияние на течение явления. Более того нам неизвестны, как правило, и сами эти причины, по крайней мере в достаточной мере. Закономерности же, которые возникают, определены тем, что они являются результатом действия огромного количества причин. Собственно эта мысль была высказана еще давно великим Лалпасом.

Я остановлюсь теперь на обзоре тех исследований, которые мне представляются наиболее значительными и попутно укажу на некоторые задачи, которые еще далеки от завершения.

Суммирование независимых случайных величин, если только слагаемые малы в определенном смысле, может приводить, как известно, только к безгранично делимым распределениям. Однако функции распределения сумм не всегда сходятся к предельному распределению. Можно ли, тем не менее, утверждать, что в каком то смысле функции распределения сумм бесконечно малых независимых слагаемых всегда сближаются с безгранично делимыми распределениями? Положительный ответ на этот вопрос недавно был получен Колмогоровым[1]. Более того оказалось, что если суммируются одинаковораспределенные независимые слагаемые, то можно найти и оценку быстроты сближения. Пусть $\phi_n(x)$ означает функцию распределения суммы $s_n = \sum\limits_{k=1}^{n} \xi_k$. Тогда можно указать такое безгранично делимое распределение $\psi_n(x)$, что

$$|\phi_n(x) - \psi_n(x)| \leqslant \frac{c}{\sqrt[5]{n}}. \tag{1}$$

c — постоянное.

В частном случае схемы Бернулли и выбора $\psi_n(x)$ из класса состоящего из нормальных и пуассоновых распределений, Прохоров обнаружил, что имеет место более точное неравенство, а именно

$$|\phi_n(x) - \psi_n(x)| \leqslant \frac{c}{\sqrt[3]{n}}. \tag{2}$$

Из устного сообщения Прохорова известно, что одному из его учеников удалось показать, что в неравенстве (2) $\sqrt[3]{n}$ можно заменить на $\sqrt{n}$, если только выбирать $\psi_n(x)$ из всего класса безгранично делимых распределений. Естественно выяснить можно ли соответствующим подбором распределений $\psi_n(x)$ улучшить оценку (1)?

Понятно, что только что рассмотренная задача является естественным обобщением классической задачи оценки быстроты сходимости функций распределения нормированных сумм к нормальному распределению. Исследования П. Л. Чебышева и А. М. Ляпунова недавно были далеко продвинуты Эссеном. После его известной работы 1945 г. в которой им была найдена оценка

$$\sqrt{n} \sup_x |\phi_n(x) - \phi(x)| \leqslant C\rho_3 \tag{3}$$

$(1/\sqrt{(2\pi)} \leqslant C \leqslant 2,890)$ для случая одинаково распределенных слагаемых, обладающих третьим моментом, возник вопрос об определении наименьшей константы C, удовлетворяющей неравенству (3) при любых n. Второй вопрос возник о разыскании наименьшего k для которого при любых $F(x)$ удовлетворялось бы неравенство

$$\lim_{n \to \infty} \sqrt{n} \sup_x |\phi_n(x) - \phi(x)| \leqslant k\rho_3. \tag{4}$$

Одно время казалось, что по крайней мере в одном из этих случаев, а может быть и в обоих искомые константы окажутся равными $1/\sqrt{(2\pi)}$. В недавней работе Эссеена[3] показано, что

$$k = (3 + \sqrt{10})/(6\sqrt{(2\pi)}).$$

Значение минимума константы C в неравенстве (3), как мне известно, еще не найдено. Подобные вопросы не разрешены еще ни для других предельных распределений, ни для многомерного случая.

Нет возможности коснуться даже мельком описания прекрасных результатов последнего времени относящихся к асимптотическим разложениям, оценке вероятностей больших уклонений и многим другим направлениям исследований. Я остановлю внимание только на последние успехи в формулировке и доказательстве локальных предельных теорем. Мы ограничимся при этом лишь случаем целочисленных слагаемых. Введем обозначения

$$P\{\xi_k = j\} = p_{kj}, \quad M\xi_k = a_k, \quad D\xi_k = b_k^2, \quad S_n = \xi_1 + \xi_2 + \ldots + \xi_n,$$

$$P_n(m) = P\{S_n = m\}, \quad A_n = MS_n, \quad B_n^2 = DS_n.$$

Мы будем говорить, что имеет место локальная предельная теорема в усиленной форме, если соотношение

$$\lim_{n \to \infty} \sup_m \left| B_n P_n(m) - \frac{1}{\sqrt{(2\pi)}} \exp\left[-\frac{(m - A_n)^2}{2B_n^2} \right] \right| = 0$$

выполняется, как для заданной последовательности случайных величин, так и для любой другой, которая может быть получена из нее путем изменения конечного числа членов. Такую форму локальной теоремы пришлось ввести в силу того, что были указаны примеры, когда локальная теорема имела место в зависимости от того принималось во внимание или нет одно единственное слагаемое.

Прохоров[4] доказал, что если (1) $|\xi_k| \leqslant K$, где K — постоянное, (2) $p_{n0} \geqslant p_{nj}$ при всех n, (3) $B_n \to \infty$ при $n \to \infty$, то локальная предельная теорема в усиленной форме имеет место тогда и только тогда, когда общий наибольший делитель тех j, для которых

$$\sum_{n=1}^{\infty} p_{nj} = \infty,$$

равен 1.

Розанов[5] показал, что если выполнено (3), то для применимости усиленной л.т. необходимо чтобы при любом $h \geqslant 2$

$$\prod_{k=1}^{\infty} [\max_{0 \leqslant m \leqslant h} P\{\xi_k \equiv m(\mathrm{mod}\, h)\}] = 0. \tag{5}$$

Если вдобавок равномерно по k при $N \to \infty$

$$b_k^{-2} \sum_{|j-a_k| \leqslant N} (j - a_k)^2 p_{kj} \to 1, \tag{6}$$

то (5) является и достаточным условием.

Наконец, если к указанным двум предположениям добавить условие (2) теоремы Прохорова, то для применимости локальной теоремы в усиленной форме необходимо и достаточно, чтобы для любого множества целых чисел, для которого

$$\prod_{k=1}^{\infty} P\{\xi_k \in A\} > 0,$$

общий наибольший делитель равнялся 1.

Петров[6] указал помимо некоторых иных достаточных условий также оценки быстроты сходимости в случае локальной теоремы.

Я не останавливаюсь на формулировке многочисленных задач, которые остаются нерешенными, относящихся к локальным теоремам.

Класс предельных распределений для сумм зависимых величин должен, естественно, включать в себя все безгранично делимые распределения. Понятно, что сама задача о разыскании возможных предельных распределений для сумм каким то образом зависимых слагаемых не имеет смысла. Для того, чтобы она приобрела интерес, необходимо указать на характер зависимости. В последнее время задача определения класса возможных предельных распределений рассматривалась почти исключительно для слагаемых, связанных в цепь Маркова. Но даже при этом ограничении пока еще рано говорить об окончательных результатах. Пока имеется большое число интересных, но сравнительно частных результатов. До сих пор еще не выяснен ни объем, ни характерные особенности класса возможных предельных распределений даже для однородных цепей Маркова. Подобные вопросы для слагаемых, являющихся членами стационарно связанных последовательностей, только начинают ставиться. В то же время все эти вопросы представляют значительный интерес как общетеоретический, так и прикладной, в частности для математической статистики.

Для последовательностей величин, связанных однородной цепью Маркова имеются работы Феллера[7], Дынкина[8], Нагаева[9]. В первой из названных работ рассматривались цепи со счетным, а в остальных с произвольным множеством состояний. Нагаевым доказана такая теорема: Если при некотором k вероятность перехода из состояния x в множество состояний A удовлетворяет неравенству

$$\sup_{x, y, A} |P^{(k)}(x, A) - P^{(k)}(y, A)| = \delta < 1,$$

то функции распределения сумм

$$s_n = \frac{1}{B_n} \sum_{k=1}^{n} f(\xi_k) - A_n,$$

где A_n и B_n — постоянные, а $f(x)$ — вещественная, измеримая по введенной мере функция, могут сходиться только к устойчивым распределениям.

Предельные распределения для последовательности серий случайных величин, связанных цепной зависимостью, получены лишь в частных предположениях. Наиболее законченный результат здесь принадлежит Добрушину[10]. Он рассмотрел цепи, принимающие лишь два состояния и нашел как все возможные предельные распределения, так и условия сходимости к ним. Понятно, что более ранние результаты Купмэна вошли естественным образом в окончательную теорему Добрушина. Распространение этих теорем на случай цепей Маркова с произвольным конечным числом состояний рассматривалось в работе А. А. Ильяшенко. Его результаты, однако, далеки от окончательности.

Особенно большое число исследований в последние годы было посвящено выяснению условий, при выполнении которых функции распределения сумм связанных величин сходятся к нормальному распределению. Совершенно законченный вид получила эта задача для величин, связанных однородной цепной зависимостью и принимающих конечное число состояний. Связное изложение можно найти в небольшой монографии Сираждинова[11].

Существенный сдвиг получен также в изучении неоднородных цепей Маркова. Теперь можно считать, что некоторые из проблем, возникших еще в трудах А. А. Маркова и С. Н. Бернштейна, получили достаточно окончательное решение. Мы не будем указывать на успехи отдельных ученых и в первую очередь Ю. В. Линника, Н. А. Сапогова, т.к. наиболее законченные формулировки результатов содержатся в недавней статье Добрушина[12]. Мы приведем оттуда формулировку одной теоремы. Пусть $P_i^{(n)}(x, A)$ обозначает вероятность перехода из состояния $\xi_{i-1}^{(n)} = x$ в состояние $\xi_i^{(n)} = z \in A$. Обозначим

$$\alpha_i^{(n)} = 1 - \sup_{x, y, A} \left| P_i^{(n)}(x, A) - P_i^{(n)}(y, A) \right| \quad (\alpha^{(n)} = \min_i \alpha_i^{(n)}).$$

Если при некоторых $c > 0$ и $C < \infty$

$$c \leqslant D\xi_k^{(n)} \leqslant C,$$

а также
$$\alpha^{(n)} n^{\frac{1}{3}} \to \infty$$

и последовательность распределений сумм $\bar{s}_n$

$$s_n = \sum_{k=1}^{n} \xi_k^{(n)}, \quad \bar{s}_n = \frac{s_n - Ms_n}{\sqrt{(Ds_n)}}$$

сходится к собственному предельному распределению, то это распределение безгранично делимо. Если кроме того все величины ограничены одной константой, то предельное распределение существует и нормально.

Формулировка достаточно общих локальных теорем для неоднородных цепей Маркова была дана Линником[13] и Статулявичусом[14]. Интересные теоремы относительно асимптотических разложений были даны Статулявичусом[15] и Гихман[16].

Существенный толчок в развитии предельных теорем для сумм был дан работой Донскера[17]. Собственно основные идеи работы Донскера являются развитием одного давнего результата А. Н. Колмогорова. Представление о теореме Донскера, получившей наименование теоремы инвариантности, мы дадим на базе более сильного позднейшего результата Прохорова[18]. Пусть имеется последовательность серий случайнх величин $\xi_{n1}, \xi_{n2}, \dots, \xi_{nk_n}$, подчиненных условиям бесконечной малости: при каждом $\epsilon > 0$

$$\max_{1 \leqslant k \leqslant k_n} P\{|\xi_{nk}| > \epsilon\} \to 0 \quad (n \to \infty)$$

и предположениям

$$M\xi_{nk} = 0, \quad b_{nk}^2 = D\xi_{nk} > 0, \quad \left(\sum_{k=1}^{k_n} b_{nk}^2 = 1\right).$$

Положим

$$\zeta_{n0} = 0, \quad \zeta_{nk} = \sum_{j=1}^{k} \xi_{nj} \text{ при } 1 \leqslant k \leqslant k_n, \; t_{nk} = D\zeta_{nk}$$

и построим ломаную $\zeta_n(t)$ с вершинами в точках (t_{nk}, ζ_{nk}). Пусть P_n означает соответствующее этой ломаной распределение вероятностей и W — винеровское распределение. Для сходимости P_n к W необходимо и достаточно чтобы при каждом $\tau > 0$ выполнялось условие Линдеберга

$$\lim_{n \to \infty} \sum_{k=1}^{k_n} \int_{|x| > \tau} x^2 dF_{nk}(x) = 0.$$

Отсюда вытекает, что для каждого функционала, который почти непрерывен в равномерной метрике, имеет место сходимость

$$P\{V(\zeta_n(t)) < x\} \to P\{V(\zeta(t)) < x\}.$$

Случай, когда предельное распределение отлично от винеровского, был рассмотрен в работах Гихман[19], Скорохода[20], Прохорова[18], Биллингслея[21] и Кимма[22]. Дальнейшие обобщения касались распространения принципа инвариантности на случай зависимых слагаемых. В первую очередь рассматривались величины, связанные цепной зависимостью. Мы должны здесь отметить работы Гихман[19], Биллингслея[21], Скорохода[23].

Результаты, о которых только что была речь, конечно, имеют самое непосредственное отношение к той проблеме, которая была обрисована в начале доклада. Теперь одной из самых важных задач, стоящих перед теорией суммирования является изучение асимптотических разложений (в том числе и для распределений функционалов), а также быстроты сходимости к распределению предельного процесса. Пока в этом направлении сделано очень мало. Пожалуй, самый сильный результат здесь принадлежит Прохорову ([18], гл. 4).

В настоящем докладе нет возможности даже в общих чертах коснуться распространению теории суммирования на случайные объекты более сложной природы —группы, полугруппы и пр. В то же время я убежден в большом значении этих исследований хотя бы по той причине, что они способствуют проникновению алгебраических методов в теорию вероятностей. Такое же проникновение методов других математических дисциплин в теорию вероятностей будет значительно способствовать её развитию и, в частности, будет содействовать освобождению многих доказательств от значительных аналитических трудностей. У меня нет возможности остановиться также на развитии предельных теорем для сумм в предположении, что выполнены некоторые дополнительные условия. Важность таких результатов для приложений, в том числе и для математической статистики, хорошо известна.

ЛИТЕРАТУРА

[1] Колмогоров, А. Н. Две равномерные предельные теоремы для сумм независимых слагаемых. *Теория вероятностей и её применения*, т. 1, вып. 4, 426–436 (1956).

[2] Прохоров, Ю. В. Асимптотическое поведение биномиального распределения. *Успехи математических наук*, т. 8, вып. 3, 136–142 (1953).

[3] Esseen, C. G. A moment inequality with an application to the central limit theorem. *Scand. Aktuarietidskrift*, H. 3–4, 160–170 (1956).

[4] Прохоров, Ю. В. О локальной предельной теореме для решетчатых распределений. *ДАН СССР*, т. 98, no. 4, 535–538 (1954).

[5] Розанов, Ю. А. О локальной предельной теореме для решетчатых распределений. *Теория вероятностей и её применения*, т. 2, вып. 2, 275–280 (1957).

[6] Петров, В. В. Локальная теорема для решетчатых распределений. *ДАН СССР*, т. 115, no. 1, 49–52 (1957).

[7] Feller, W. Fluctuation theory of recurrent events. *Trans. Amer. Math. Soc.* 67, 98–119 (1949).

[8] Дынкин, Е. Б. О некоторых предельных теоремах для цепей Маркова. *Украинский математический журнал*, т. 6, no. 1, 21–27 (1954).

[9] Нагаев, С. В. Некоторые предельные теоремы для однородных цепей Маркова. *Теория вероятностей и её применения*, т. 2, вып. 4, 389–416 (1957).

[10] Добрушин, Р. Л. Предельные теоремы для цепи Маркова из двух состояний. *Известия Акад. наук СССР, серия матем.* 17, 291–330 (1953).

[11] Сираждинов, С. Х. Предельные теоремы для однородных цепей Маркова. *Изд. Акад. наук Узбекской ССР* (1955).

[12] Добрушин, Р. Л. Центральная предельная теорема для неоднородных цепей Маркова. *Теория вероятностей и её применения*, т. 1, вып. 1, 72–89; вып. 4, 365–425 (1956).

[13] Линник, Ю. В. и Сапогов, Н. А. Многомерные интегральные и локальные законы для неоднородных цепей Маркова. *Известия Академии наук СССР*, 19, no. 6, 533–566 (1949).

[14] Статулявичус, В. А. О локальной предельной теореме для неоднородных цепей Маркова. *ДАН СССР*, т. 107, no. 4, 516–519 (1956).

[15] Статулявичус, В. А. Асимптотическое разложение для неоднородных цепей Маркова. *ДАН СССР*, т. 112, no. 2, 206 (1957).

[16] Гихман, И. И. Об одной асимптотической теореме для сумм малых случайных слагаемых. *Труды института матем. и механики Акад. наук Узбекской ССР*, вып. 10, часть 1, 36–43 (1953).

[17] Donsker, M. D. An invariance principle for certain probability limit theorems. *Mem. Amer. Math. Soc.* 6, 1–12 (1951).

[18] Прохоров, Ю. В. Сходимость случайных процессов и предельные теоремы теории вероятностей. *Теория вероят. и её применения*, т. 1, вып. 2, 177–238 (1956).

[19] Гихман, И. И. Об одной теореме А. Н. Колмогорова. *Матем. сборник Киев. университета*, 12, no. 7, 75–94 (1953).

[20] Скороход, А. В. О связи между последовательностью сумм независимых случайных величин и однородными случайными процессами с независимыми приращениями. *ДАН СССР*, т. 104, no. 3, 364–367 (1955).

[21] Billingsley, P. The invariance principle for dependent random variables. *Trans. Amer. Math. Soc.* 83, 250–268 (1956).

[22] Kimme, E. On the convergence of sequences of stochastic processes. *Trans. Amer. Math. Soc.* 84, 208–229 (1957).

[23] Скороход, А. В. Об одном классе предельных теорем для цепей Маркова. *ДАН СССР*, т. 106, no. 5, 781–784 (1956).

PROBABILISTIC METHODS IN
NUMBER THEORY

By A. RÉNYI

1. Introduction

Probability theory was created to describe random mass-phenomena. Since the appearance in 1933 of the fundamental book [1] of Kolmogoroff, however, probability theory has become an abstract, axiomatic theory, and as such is capable of other interpretations too. Thus methods and results of probability theory may be applied as tools in any other branch of mathematics. Many important applications of probabilistic methods have been made in number theory. There exist excellent previous surveys of these results (see [2, 3, 4]); these surveys contain also many references to the literature. In the present paper I should like to give an account of some recent results, obtained since the appearance of the surveys mentioned.

I do not aim at completeness, and shall mention mainly such results as have some connection with my own work, done partly in collaboration with others, especially with Erdős, to whom I am indebted for kindly agreeing to include in the present paper some yet unpublished results of our collaboration.

Erdős and the author of the present paper are, following a suggestion of Doob, preparing a monograph on 'Probabilistic methods in number theory' to appear in the series 'Ergebnisse der Mathematik' published by the Springer Verlag. This monograph will contain a full bibliography of the subject.

2. Additive number theoretical functions

A real valued function $f(n)$ defined for all natural numbers $n = 1, 2, \ldots$ is called *additive* if
$$f(nm) = f(n) + f(m), \tag{1}$$
provided that $(n, m) = 1$, where (n, m) denotes the greatest common divisor of n and m. Typical additive functions are: the function $V(n)$ denoting the number of all prime factors of n; the function $U(n)$ denoting the number of different prime factors of n; the function $\log d(n)$ where $d(n)$ denotes the number of divisors of n. If $f(n)$ has besides (1) the property that $f(p^\alpha) = f(p)$ if p is a prime and $\alpha = 2, 3, \ldots, f(n)$ is called *strongly additive*. If $f(n)$ is such that (1) holds for all n and $m, f(n)$ is called

34   T P

absolutely additive. Clearly $U(n)$ is strongly additive and $V(n)$ absolutely additive, but $\log d(n)$ has neither of these properties.

The distribution of values of additive number theoretic functions has been investigated in detail. To express the results we may make use of the terminology of conditional probability spaces (see [5,6,7]). Let Ω be the set of all natural numbers, Ω_N the set of the first N natural numbers. Let $\mathscr{A}$ denote the set of all subsets of Ω and $\mathscr{B}$ the set of all finite and non-empty subsets of Ω; we denote by $v(A)$ the number of elements of $A \in \mathscr{A}$. We denote by AB the intersection of the sets A and B, and put for $A \in \mathscr{A}$, $B \in \mathscr{B}$

$$P(A \mid B) = \frac{v(AB)}{v(B)}.$$

Then clearly $[\Omega, \mathscr{A}, \mathscr{B}, P]$ is a conditional probability space in the sense of [5] and [6]. All results concerning the distribution of values of additive number theoretical functions can be conveniently expressed in terms of this conditional probability space.

The first fundamental result concerning additive number theoretical functions was the theorem of Erdős and Kac[8]. For the sake of brevity we shall formulate their result only for the function $V(n)$. If $S(n)$ is a proposition concerning the natural number n, we denote the set of those n for which this proposition is valid also by $S(n)$. The theorem of Erdős and Kac contains as a special case the assertion that

$$\lim_{N \to +\infty} P(V(n) - \log\log n < x\sqrt{(\log\log n)} \mid \Omega_N) = \Phi(x), \qquad (2)$$

where
$$\Phi(x) = \frac{1}{\sqrt{(2\pi)}} \int_{-\infty}^{x} e^{-\frac{1}{2}u^2} du \qquad (3)$$

is the distribution function of the normal probability law. Thus the distribution of values of $V(n)$ on Ω_N tends for $N \to +\infty$ to the normal distribution. A similar statement holds for a broad class of additive number theoretical functions; the most general results have been obtained by Kubilius (see [4]). The method of proof used by Erdős and Kac, as well as by Kubilius, was a combination of probabilistic methods (theorems on the limiting distribution of sums of independent random variables) with elementary number theoretical results (the sieve of Viggo Brun). LeVeque[9] and Kubilius[4] proved certain improvements of (2) too, by estimating the remainder term. LeVeque conjectured that the best result is

$$P(V(n) - \log\log n < x\sqrt{(\log\log n)} \mid \Omega_N) = \Phi(x) + O\left(\frac{1}{\sqrt{(\log\log N)}}\right) \qquad (4)$$

uniformly for $-\infty < x < +\infty$. Other proofs of (3) have been given by Delange[10] and Halberstam[11] using the method of moments.

Recently we succeeded with Turán[12] in proving the conjecture (4) of LeVeque. The method used in [12] was a combination of an analytical approach,† working with standard tools of analytical number theory (such as the zeta-function of Riemann and contour-integration) with the method of characteristic functions. This method can be applied for other functions than $V(n)$ too, and to other related problems.

3. The probabilistic generalization of the large sieve of Linnik

Linnik[14] discovered in 1941 an ingenious new method, called by him 'the large sieve'. I generalized this method in 1947, and used it in the proof of the following theorem (see [15]), being a step towards the unsolved hypothesis of Goldbach: there exists a constant K such that every natural number n can be represented in the form $n = p + P$ where p is a prime and $V(P) \leqslant K$. For another application of the large sieve see [16]. Later on, in 1948–9, I realized that the large sieve is a special case of a general theorem of probability theory (see [17,18,19]). This theorem, in a recently obtained improved form (see [40]), can be stated as follows:

Let $\xi_1, \xi_2, \ldots, \xi_n, \ldots$ be random variables each having a distribution of the discrete type. Let x_{nk} ($k = 1, 2, \ldots$) denote the values taken on by ξ_n with positive probability; let us denote by A_{nk} the event $\xi_n = x_{nk}$ and by $P(A_{nk})$ the probability of this event. Let us denote by $\phi(\xi_n, \xi_m)$ the mean square contingency of ξ_n and ξ_m ($n \neq m$) as defined by Pearson (see [20]), i.e. put

$$\phi(\xi_n, \xi_m) = \left(\sum_k \sum_l \frac{(P(A_{nk} A_{ml}) - P(A_{nk}) P(A_{ml}))^2}{P(A_{nk}) P(A_{ml})} \right)^{\frac{1}{2}}. \tag{5}$$

We call the sequence ξ_n *weakly dependent with bound B* if for any sequence x_n of real numbers such that $\sum_n x_n^2 < +\infty$ we have

$$\left| \sum_{\substack{n \ m \\ n \neq m}} \phi(\xi_n, \xi_m) x_n x_m \right| \leqslant B \sum_n x_n^2. \tag{6}$$

Let $M(\eta)$ and $D^2(\eta)$ denote respectively the mean value and variance of the random variable η, $M(\eta \,|\, \xi)$ the conditional mean value of η with respect to a fixed value of ξ, and $D_\xi^2(\eta)$ the variance of the random variable $M(\eta \,|\, \xi)$. Denote by $\Theta_\xi(\eta)$ the correlation ratio of η on ξ, as defined by Pearson (see [20]), i.e. put

$$\Theta_\xi^2(\eta) = \frac{D_\xi^2(\eta)}{D^2(\eta)}. \tag{7}$$

† This approach has already been applied to the investigation of the distribution of values of additive number theoretical functions in 1934 in the paper[13] of Turán.

Our theorem asserts that if η is an arbitrary random variable having finite second moment $M(\eta^2)$, and if $\{\xi_n\}$ is a sequence of weakly dependent random variables with bound B, then

$$\sum_n \Theta^2_{\xi_n}(\eta) \leqslant (1+B). \tag{8}$$

The application of this theorem to number theory makes it possible to prove that an arbitrary sufficiently dense sequence of integers $\leqslant N$ is 'almost' uniformly distributed among 'almost' all residue classes with respect to 'almost' all primes $p \leqslant N^\alpha$, where $\alpha < \frac{1}{2}$. For the exact meaning of the term 'almost' (occurring three times in the above sentence, and having a different meaning on each occasion) we refer to [18] and [19].

4. Statistical properties of the digits in various representations of real numbers

The history of the subject of this paragraph goes back to Gauss, whose conjecture on continued fractions was proved in 1928 by Kuzmin[21]. The starting-point of recent investigations is the classical result of Borel[22], according to which the limiting frequencies of all possible digits in the q-adic representation of a real number x are equal to one another $\left(\text{and thus equal to } \dfrac{1}{q}\right)$ for almost all x. Some years ago I generalized this theorem by considering, instead of q-adic expansions, general *Cantor's series*[23]. Let $q_n \geqslant 2$ $(n = 1, 2, ...)$ be an arbitrary sequence of positive integers. Then every real number x $(0 \leqslant x \leqslant 1)$ can be represented in the form

$$x = \sum_{n=1}^{\infty} \frac{\epsilon_n(x)}{q_1 q_2 \cdots q_n}, \tag{9}$$

where the 'digit' $\epsilon_n(x)$ may take on the values $0, 1, 2, ..., q_n - 1$ $(n = 1, 2, ...)$. The digits $\epsilon_n(x)$, considered as random variables on the probability space $[\Omega, \mathscr{A}, P]$, where Ω is the interval $(0, 1)$, $\mathscr{A}$ the set of all measurable subsets of Ω and P the ordinary Lebesgue measure, are independent, and $\epsilon_n(x)$ takes on each of the values $0, 1, ..., q_n - 1$ with probability $1/q_n$. Now it was shown in [23] that if $N_n(k, x)$ denotes the frequency of the number k $(k = 0, 1, ...)$ among the first n digits in the representation (9) of x, then

$$\lim_{n \to \infty} \frac{N_n(k, x)}{\sum\limits_{\substack{j=1 \\ q_j > k}}^{n} \dfrac{1}{q_j}} = 1 \tag{10}$$

for almost all x, provided that

$$\lim_{n \to \infty} \sum_{\substack{j=1 \\ q_j > k}}^{n} \frac{1}{q_j} = +\infty.$$

Equation (10) implies that, if the series $\Sigma(1/q_n)$ is divergent and $q_n \to +\infty$, then for any pair k, l of non-negative integers we have

$$\lim_{n \to \infty} \frac{N_n(k, x)}{N_n(l, x)} = 1. \tag{11}$$

Thus in this case all digits $k = 0, 1, 2, \ldots$ occur in the limit in a certain sense equally frequently among the digits in the representation (9) of almost every real number x.

Recently we have obtained, with Erdős, the following results. Let us put

$$M_n(x) = \max_{(k)} N_n(k, x), \tag{12}$$

i.e. let $M_n(x)$ denote the frequency of the most frequent number among the first n digits in (9). Then if, putting $Q_n = \sum_{k=1}^{n} 1/q_k$, we have $\lim_{n \to +\infty} \frac{Q_n}{\log n} \to +\infty$, then for almost all x

$$\lim_{n \to \infty} \frac{M_n(x)}{Q_n} = 1. \tag{13}$$

On the other hand, in case $q_n/n \to +\infty$, and $Q_n \to +\infty$ we have for almost all x

$$\lim_{n \to \infty} \frac{M_n(x)}{Q_n} = +\infty \tag{14}$$

Borel's theorem has been generalized by Raikov[24] as follows: If $g(x)$ is integrable in the interval $(0, 1)$ and periodic with period 1, then for almost all x $(0 \leqslant x \leqslant 1)$ and for all integers $q \geqslant 2$ we have

$$\lim_{n \to \infty} \frac{1}{n} \sum_{k=0}^{n-1} g(q^k x) = \int_0^1 g(t) \, dt. \tag{15}$$

This theorem, as has been remarked by Riesz[25], is a special case of the individual ergodic theorem. An analogous result for continued fractions has been obtained by Ryll-Nardzewski[26]. Let x denote a real number $(0 < x < 1)$ and consider its continued fraction

$$x = \cfrac{1}{\epsilon_1(x) + \cfrac{1}{\epsilon_2(x) + \cfrac{1}{\epsilon_3(x) + \ldots}}}, \tag{16}$$

where the digits $\epsilon_n(x)$ are positive integers. Put $r_0(x) = x$ and

$$r_n(x) = \cfrac{1}{\epsilon_{n+1}(x) + \cfrac{1}{\epsilon_{n+2}(x) + \cfrac{1}{\epsilon_{n+3}(x) + \ldots}}} \qquad (n = 1, 2, \ldots), \qquad (17)$$

i.e. $r_n(x)$ is the nth remainder of the continued fraction (16). Then if $g(x)$ is L-integrable in $[0, 1]$ the theorem of Ryll-Nardzewski asserts that, for almost every x,

$$\lim_{n \to \infty} \frac{1}{n} \sum_{k=0}^{n-1} g(r_k(x)) = \frac{1}{\log 2} \int_0^1 \frac{g(t)}{1+t} \, dt. \qquad (18)$$

The fundamental idea of Ryll-Nardzewski was the following: he introduced the measure

$$v(E) = \frac{1}{\log 2} \int_E \frac{dt}{1+t}, \qquad (19)$$

where E is a measurable subset of $(0, 1)$. Now the transformation† $Tx = (1/x)$ of the interval $(0, 1)$ leaves the measure $v(E)$ invariant; on the other hand $r_n(x) = T^n x$ $(n = 1, 2, \ldots)$. Thus (18) follows also from the individual ergodic theorem.

The q-adic expansion and the continued fraction expansion (16) are both special cases of a general class of representations of real numbers x $(0 < x < 1)$ in the form (called the 'f-expansion' of x)

$$x = f(\epsilon_1(x) + f(\epsilon_2(x) + f(\epsilon_3(x) + \ldots))), \qquad (20)$$

where the monotonic function $f(x)$ has to satisfy certain conditions and the digits $\epsilon_n(x)$ are non-negative integers, determined by the following algorithm: Let $x = \phi(y)$ denote the inverse function of $y = f(x)$, determine the sequence $r_n(x)$ by the recursion

$$r_0(x) = x, \quad r_{n+1}(x) = (\phi(r_n(x)) \qquad (21)$$

and put $\qquad\qquad\qquad \epsilon_{n+1}(x) = [\phi(r_n(x))]. \qquad (22)$

(Here (z) denotes again the fractional part and $[z]$ the integral part of z.)

The q-adic expansion is obtained from (20) as a special case if $f(x) = x/q$ for $0 \leqslant x \leqslant q$ and the continued fraction expansion if $f(x) = 1/x$ for $x \geqslant 1$.

f-expansions with decreasing $f(x)$ have been considered previously by Bissinger[27] and with increasing $f(x)$ by Everett[28]. f-expansions (both with increasing and decreasing f) have been considered under more general conditions by the author of the present paper in [29], where a general theorem on the distribution of digits of a general f-expansion is

† Here and in what follows (z) denotes the fractional part of z.

given, which includes as special cases the theorems of Raikov and Ryll-Nardzewski. The proof of the above-mentioned theorem is based on the ergodic theorem of Dunford and Miller. I mention only the following special case: Let $\beta > 1$ be an arbitrary real number; then every real number x ($0 \leqslant x \leqslant 1$) can be represented in the form

$$x = \sum_{n=1}^{\infty} \frac{\epsilon_n(x)}{\beta^n}, \tag{23}$$

where the digits $\epsilon_n(x)$ may take on the values $0, 1, ..., [\beta]$ and are determined by the algorithm (20)–(22) with $f(x) = x/\beta$ ($0 \leqslant x \leqslant \beta$). There exists a measure $v_\beta(E)$, defined on the interval $(0, 1)$, which is equivalent to the Lebesgue measure, and which is invariant with respect to the transformation $Tx = (\beta x)$. Thus we have for any $g(x)$ which is L-integrable on $[0, 1]$, for almost every x,

$$\lim_{n \to \infty} \frac{1}{n} \sum_{k=0}^{n-1} g(r_k(x)) = \int_0^1 g(x)\, dv_\beta, \tag{24}$$

where $r_k(x)$ is defined by (21). Clearly, if β is an integer, v_β is the ordinary Lebesgue measure and (24) reduces to the theorem of Raikov. I have not succeeded in determining explicitly the measure v_β except for some special (algebraic) values of β. E.g. if β is the only positive root of the equation

$$\beta^n = \beta^{n-1} + 1 \quad (n \geqslant 1 \text{ integral}) \tag{25}$$

then

$$v_\beta(E) = \int_E h_\beta(x)\, dx, \tag{26}$$

where

$$h_\beta(x) = \begin{cases} \lambda & \text{for} \quad 0 < x < \dfrac{1}{\beta^{n-1}}, \\[2mm] \dfrac{\lambda}{\beta^k} & \text{for} \quad \dfrac{1}{\beta^{n-k}} < x < \dfrac{1}{\beta^{n-k-1}} \quad (k = 1, 2, ..., n-1), \end{cases} \tag{27}$$

where

$$\lambda = \frac{\beta}{(\beta - 1)\,(n(\beta - 1) + 1)}. \tag{28}$$

It follows from (24) that, denoting by p_0 and p_1 the limits of the relative frequencies of the digits 0 and 1 in the expansion (23) for almost all x (as (25) implies $1 < \beta < 2$ these are the only possible values of the digits $\epsilon_n(x)$), then

$$p_0 = \frac{(n-1)\,(\beta - 1) + 1}{n(\beta - 1) + 1} \quad \text{and} \quad p_1 = \frac{\beta - 1}{n(\beta - 1) + 1}. \tag{29}$$

In particular, for $n = 1$, $\beta = 2$ we obtain the well-known special case of Borel's theorem that in the dyadic representation of almost all numbers both digits 0 and 1 have the limiting frequency $\frac{1}{2}$.

536   A. RÉNYI

Let us consider now the representation of the real number x $(0 < x < 1)$ in the form of Engel's series

$$x = \frac{1}{q_1} + \frac{1}{q_1 q_2} + \ldots + \frac{1}{q_1 q_2 \cdots q_n} + \ldots, \qquad (30)$$

where the q_n are integers, $q_{n+1} \geqslant q_n$. The probabilistic theory of Engel's series has been considered by Borel[30] and Lévy[31].

Borel has announced, without proof, that for almost all x

$$\lim_{n \to \infty} \sqrt[n]{q_n} = e. \qquad (31)$$

A proof of (31) has been given by Lévy, who also proved that if we define the probability $P(E)$ of a measurable subset E of the interval $[0, 1]$ to be equal to the Lebesgue measure of E, we have

$$\lim_{n \to \infty} P\left(\frac{\log q_n - n}{\sqrt{n}} < y\right) = \Phi(y). \qquad (32)$$

In a recent paper[32] by Erdős, Szüsz and the author, new and simple proofs of these theorems and of some other results have been given, based on the fact that the sequence q_n, considered as a sequence of random variables in the above-mentioned probabilistic interpretation, forms a homogeneous Markov chain, with the transition probabilities

$$P(q_n = k \mid q_{n-1} = j) = \frac{j-1}{k(k-1)} \quad \text{for} \quad k \geqslant j. \qquad (33)$$

By using a theorem of the author on mixing sets[33] it can be shown (see [34]) that (32) can be replaced by the more general relation

$$\lim_{n \to \infty} Q\left(\frac{\log q_n - n}{\sqrt{n}} < y\right) = \Phi(y), \qquad (34)$$

which is valid whenever Q is a measure which is absolutely continuous with respect to the Lebesgue measure.

In [32] similar results have been obtained for Sylvester's series

$$x = \frac{1}{Q_1} + \frac{1}{Q_2} + \ldots + \frac{1}{Q_n} + \ldots \qquad (35)$$

(where the Q_n are integers, $Q_{n+1} \geqslant Q_n(Q_n - 1) + 1$), e.g. it has been shown that $\lim_{n \to \infty} Q_n^{1/2^n} = L(x)$ exists for almost all x, but we were unable to determine the limit $L(x)$ explicitly. In [35] similar results have been obtained for Cantor's products

$$x = \prod_{n=1}^{\infty} \left(1 - \frac{1}{Q_n}\right). \qquad (36)$$

5. Problems of additive number theory

In this section I should like to give an account of some yet unpublished results on the additive properties of random sequences of integers, obtained by Erdős and the author of the present paper. Some results in this direction have been announced without proof in [36]. Random sequences of integers $v = (v_1, v_2, ..., v_n, ...)$ are defined as follows: Let A_n denote for each n ($n = 1, 2, ...$) the event that n belongs to the random sequence v, and let us suppose that $P(A_n) = p_n$ is given. Let us suppose further that the events A_n are independent. By these hypotheses a probability measure is determined on a σ-algebra $\mathscr{A}$ of subsets of the space Ω consisting of all possible sequences of integers. $\mathscr{A}$ is defined as the least σ-algebra which contains all subsets A of Ω defined by fixing for a finite number of integers whether it belongs to the sequence v or not.

Now let $\psi_2(N)$ denote the number of representations of the natural number N in the form $N = v_i + v_j$ ($i \leqslant j$), where v_i and v_j are elements of the random sequence v. Then $\psi_2(N)$ is a random variable.

We have obtained, for example, the following result. If $p_n = cn^{-\frac{1}{2}}$ then the sequence $S_k^{(2)}$ of integers N, for which $\psi_2(N) = k$, has for almost all sequences v the density $d_k = (\lambda^k e^{-\lambda})/k!$ ($k = 0, 1, ...$), with $\lambda = \frac{1}{2}c^2\pi$, i.e. the distribution $\{d_k\}$ is of Poisson's type.

More generally, if $\psi_r(N)$ denotes the number of representations of the natural number N in the form $N = v_{i_1} + v_{i_2} + ... + v_{i_r}$ ($i_1 \leqslant i_2 \leqslant ... \leqslant i_r$), and $p_n = c/n^{1-(1/r)}$ ($n = 1, 2, ...$), then denoting by $S_k^{(r)}$ the sequence of integers N for which $\psi_r(N) = k$, $S_k^{(r)}$ has for almost all sequences v the density $d_k = (\lambda^k e^{-\lambda})/k!$ ($k = 0, 1, ...$), where

$$\lambda = \frac{1}{r!}\left[c\Gamma\left(\frac{1}{r}\right)\right]^r \quad (r = 2, 3, ...).$$

We considered with Erdős also the distribution of differences of elements of a random sequence of integers. Other applications of probabilistic methods to problems of additive number theory are discussed in [36].

6. Some further applications in number theory

In this section we mention, without going into details, some other lines of research. Linnik[37] (see also [4] for further literature) has obtained interesting results by applying the theory of Markov chains to the arithmetic of quaternions. His results are of great importance in the theory of representation of integers by means of ternary quadratic forms. By the same method Linnik has proved[38] that the points (x, y, z) with integral co-ordinates which lie on the spherical surface

$$x^2 + y^2 + z^2 = m$$

are in the limit for $m \to +\infty$ uniformly distributed on this surface (where $m \equiv 1$ or $2 \bmod 4$ or $m \equiv 3 \bmod 8$).

Probabilistic methods have also been applied in the theory of diophantine approximation; see, for example, the paper[39] of Erdős and the author.

I am convinced that in spite of the wide variety of problems to which results or methods of probability theory have already been applied with success, only a small part of the possibilities of such an approach has yet been exhausted, and there will be a rapid development in this direction in the years to come. This remark applies to chapters of mathematics other than number theory too.

REFERENCES

[1] Kolmogoroff, A. Grundbegriffe der Wahrscheinlichkeitsrechnung. *Ergebnisse der Math.* Springer, Berlin, 1933.

[2] Kac, M. Probability methods in some problems of analysis and number theory. *Bull. Amer. Math. Soc.* 55, 641–665 (1949).

[3] Erdős, P. On the distribution of values of additive arithmetical functions. *Proceedings of the International Congress of Mathematicians*, Amsterdam, 1954.

[4] Кубилюс, И. П. Вероятностные методы в теории чисел. *Успехи Математических Наук*, 11, 31–66 (1956).

[5] Rényi, A. On a new axiomatic foundation of the theory of probability. *Proceedings of the International Congress of Mathematicians*, Amsterdam, 1954.

[6] Rényi, A. On a new axiomatic theory of probability. *Acta Math. Acad. Sci. Hung.* 6, 285–335 (1955).

[7] Rényi, A. On conditional probability spaces generated by a dimensionally ordered set of measures. *Теория вероятностей и ее применения*, 1, 61–71 (1956).

[8] Erdős, P. and Kac, M. The Gaussian law of errors in the theory of additive number-theoretical functions. *Amer. J. Math.* 62, 738–742 (1940).

[9] LeVeque, W. J. On the size of certain number-theoretic functions. *Trans. Amer. Math. Soc.* 66, 440–463 (1949).

[10] Delange, H. Sur le nombre des diviseurs premiers de n. *C.R. Acad. Sci.*, Paris, 237, 542–544 (1953).

[11] Halberstam, H. On the distribution of additive number-theoretic functions. *J. Lond. Math. Soc.* 30, 43–53 (1955).

[12] Rényi, A. and Turán, P. On a theorem of Erdős and Kac. *Acta Arithmetica*, 1, 71–84 (1957).

[13] Turán, P. Az egész számok primosztóinak számáról. *Matematikai és Fizikai Lapok*, 41, 103–130 (1934).

[14] Linnik, Yu. V. The large sieve. *C.R. Acad. Sci.*, *U.R.S.S.*, 30, 292–294 (1941).

[15] Rényi, A. О представлении четных чисел в виде суммы простого и почти простого числа. *Известия Академии Наук СССР, сер. мат.* 12, 57–78 (1948).

[16] Bateman, P. T., Chowla, S. and Erdős, P. Remarks on the size of $L(1, \chi)$. *Publicationes Math. Debrecen*, 1, 165–182 (1950).

[17] Rényi, A. On the large sieve of Yu. V. Linnik. *Compositio Mathematica*, 8, 68–75 (1956).

[18] Rényi, A. Un nouveau théorème concernant les fonctions indépendantes et ses applications à la théorie des nombres. *J. Math. pures et appl.* (9), 28, 137–149 (1949).

[19] Rényi, A. Sur un théorème général de probabilité. *Ann. Inst. Fourier*, 1, 43–52 (1949).

[20] Cramér, H. *Mathematical Methods of Statistics*. Princeton, 1946.

[21] Kuzmin, R. O. Sur un problème de Gauss. *Atti del Congresso Internazionale dei Matematici, Bologna*, 6, 83–89 (1928).

[22] Borel, E. Sur les probabilités dénombrables et leurs applications arithmétiques. *Rendiconti del Circolo Mat. di Palermo*, 26, 247–271 (1909).

[23] Rényi, A. A számjegyek eloszlása valós számok Cantor-féle előállításaiban. *Matematikai Lapok*, 7, 77–100 (1956).

[24] Raikov, D. On some arithmetical properties of summable functions. *Mat. Sbornik*, 1 (43), 377–384 (1936).

[25] Riesz, F. Az ergodikus elméletről. *Matematikai és Fizikai Lapok*, 1943.

[26] Ryll-Nardzewski, C. On the ergodic theorems. II. Ergodic theory of continued fractions. *Studia Math.* 12, 74–79 (1951).

[27] Bissinger, B. H. A generalization of continued fractions. *Bull. Amer. Math. Soc.* 50, 868–876 (1944).

[28] Everett, C. J. Representations for real numbers. *Bull. Amer. Math. Soc.* 52, 861–869 (1946).

[29] Rényi, A. Representations for real numbers and their ergodic properties. *Acta Math. Acad. Sci. Hung.* 8, 477–493 (1957).

[30] Borel, É. Sur les développements unitaires normaux. *C.R. Acad. Sci., Paris*, 225, 51 (1947).

[31] Lévy, P. Remarques sur un théorème de M. Émile Borel. *C.R. Acad. Sci., Paris*, 225, 918–919 (1947).

[32] Erdős, P., Szüsz, P. and Rényi, A. On Engel's and Sylvester's series. *Ann. Univ. L. Eötvös* (Sect. Math.) 1, 7–32 (1958).

[33] Rényi, A. On mixing sequences of sets. *Acta Math. Acad. Sci. Hung.* 9, 215–228 (1958).

[34] Rényi, A. and Révész, P. On mixing sequences of random variables. *Acta Math. Acad. Sci. Hung.* 9, 389–393 (1958).

[35] Rényi, A. On Cantor's products. *Colloquium Math.* 6, 135–139 (1958).

[36] Erdős, P. Problems and results in additive number theory. *Colloque sur la théorie des nombres*, 127–137. Brussels, 1955.

[37] Линник, Ю. В. Применение теории цепей Маркова в арифметике кватернионов. *Успехи Мат. Наук*, 9, 203–210 (1954).

[38] Линник, Ю. В. Асимптотическое распределение целых точек на сфере. *Доклады Академии Наук СССР*, 96, 909–912 (1954).

[39] Erdős, P. and Rényi, A. A probabilistic approach to problems of diophantine approximation. *Illinois J. Math.* 1, 303–315 (1957).

[40] Rényi, A. On the probabilistic generalization of the large sieve of Linnik. *Publications of the Mathematical Institute of the Hungarian Academy of Sciences*, 3, 199–206 (1958).

540

RECENT TENDENCIES IN THE
FOUNDATIONS OF STATISTICS

By LEONARD J. SAVAGE†

1. Introduction

This is an expository talk directed mainly at any non-statisticians who may have wandered in. It is important to address the non-specialists at a congress like this, to help maintain the bonds between the diversifying branches of mathematics. In this particular talk, a restraint on technicalities will have the added advantage of helping experts and the speaker keep their feet on the ground.

The foundations of statistics are controversial, changing, and subtle. Therefore, try though I shall to be fair and clear, you must keep yourselves especially aware that you are hearing mainly the present views of a single person, imperfectly expressed.

The foundations of statistics are a part of the foundations of science in its widest sense. Their study is not mathematics in principle, and by no means all the important contributions to them have been made by mathematicians. But the use of some mathematical techniques is inevitable in the study of a quantitative subject. Still more, mathematical training and outlook have led, and will surely continue to lead, to important advances in the foundations of statistics. The relation is reciprocal in that mathematics is sometimes stimulated by the foundations, as it is by the other theoretical aspects of statistics.

The reference to *recent* tendencies in my title has a continuum of possible meanings, and in fact various parts of the talk will refer to tendencies of the present century, of the period since World War II, and of the last few years or even months.

2. Meanings of 'statistics'

I begin by outlining some meanings that have been given to the word 'statistics', not to enter into an argument that would be out of place here (and perhaps anywhere) as to what the word ought to mean, but to indicate the subject of this talk and to set the stage for it.

Etymologically, 'statistics' refers to numerical data about the state.

† This work was supported in part by the Office of Naval Research through a contract with the Department of Statistics of the University of Chicago, and in part by grants to the author by the Jóhn Simon Guggenheim Memorial Foundation and by the Ford Foundation.

Even today there are many professional statisticians to whom 'statistics' means effectively demography in a more or less extensive sense—the compilation and interpretation of census data, economic statistics, or vital statistics (records of births, deaths, and illness).

For many of us, however, the word has drifted far from its original meaning and come to refer to quantitative thinking about uncertainty as it affects scientific and other investigations. It is this meaning, suggested by 'inductive' or 'statistical inference', that 'statistics' has for us here. This subject goes back historically to at least the early eighteenth century, when James Bernoulli, and a little later Thomas Bayes, made great contributions to it. It was pushed forward in the nineteenth century by Laplace, Gauss, and others, and it has been subject to a fervor of activity since the early twenties of this century, when it received great impetus from the work of R. A. Fisher.

In physics, 'statistics' usually pertains to probability without special reference to the problem of inference but with emphasis on large aggregates.

3. Inductive inference and inductive behavior

One of the most important trends of the past few decades is a shift in formulation of the central problem of statistics from the problem of what to say or believe in the face of uncertainty to the problem of what to do. It would be hard and unrewarding to seek out the very beginning of this economic movement, as I shall call it because of its emphasis on the costs of wrong decision. It goes back at least as far as Gauss[7], but Neyman brought it forward with particular explicitness in 1938[8], coining the expression 'inductive behavior' in contrast to 'inductive inference'. Wald took up the theme with energy and enthusiasm, exploring it in great detail and stimulating many others to do so during his own life and after his untimely death.

That many important and interesting problems concerned with uncertainty are economic in nature is clear and undisputed. Going much further, some of us believe that economic models are of great value throughout the whole of statistics. This is controversial, and it is maintained, especially by Barnard[1, 3] and Fisher[5, 6] that the methods and ideas appropriate to frankly economic problems are not appropriate to the problems of science, the problems of finding out the truth for the truth's sake. Fisher says in a particularly pungent way that science ought not to be confused with the sordid business of either the market place or the five-year planners' bureau[5].

Admittedly a close relation between frankly economic problems and more academic ones is not obvious, or even thoroughly demonstrable, but some case can be made for it. To illustrate, in practical problems of point estimation, there are certain systematic reasons why the penalty for mis-estimation is often nearly proportional to the square of the error. These same reasons are, to say the least, suggestive even for problems of pure science—a precedent for this idea can be seen in Gauss[7]. More generally, it should be kept in mind that science does have goals and that mistakes made in approaching them do entail costs, however subtle and abstract.

There seems to me nothing at the present time to substitute for the hope that an economic theory of decision in the face of uncertainty will be a valuable guide for the whole problem of inference. If there is an important kind of inference problem that cannot properly be discussed in economic terms, no one yet seems able to state these problems with enough precision so that they can be analyzed and solved. In brief, the economic outlook seems to me of great promise for the whole of statistics, though it is not necessarily the last word. We should continue to explore and use it with hope and discretion and with an eye open for new ideas.

One thing that has been said about the putative distinction between scientific and economic problems is that the scientific inference to be drawn from given data is unique and universal, whereas the economic conclusions change with circumstances, such as values and opportunities[3]. I myself believe that the idea of a universal summary of data—that is, the likelihood-ratio function or some effective substitute—is valid and important, but the idea of such a summary does not for me rest on any distinction between science and business.

4. Objectivism and subjectivism

It was for a long time generally believed that all uncertainties could be measured by probabilities, and a few of us today believe that this view, which has recently been very unpopular, must soon again come into its own. It was part of the creed of the great renaissance of statistics in the second quarter of the century that only special uncertainties associated with gambling apparatus and the like were measurable by probabilities and that other uncertainties would have to be analyzed and dealt with in some other ways. This renunciation swept away the classical framework for inference, built on Bayes's theorem, and thereby created many new problems. There was especially the problem of finding new meanings to important-sounding questions that had been rendered nonsensical by the renunciation. The situation was a fertile and stimulating one. Many new ideas directed at filling the gaps were introduced. Some of these ideas are apparently of lasting value, but some of them (such as confidence limits in their current formulation or tests of narrow hypotheses) may not be. In any event, the over-all program has not yet been even nearly successful, nor do I think it ever can be.

Statisticians have always recognized that subjective judgments of fact (as well as of value) necessarily play a role in statistical practice. First, much personal, that is subjective, judgment is obviously required to decide what kind of an experiment is the promising one to perform, and on what scale. There are, therefore, subjective aspects to the essential statistical activity of designing experiments and other investigations. Again, it has long been recognized that the user of statistics, in analyzing data, must make a subjective choice among available operating characteristic curves and the like. To be sure, the minimax theory can be seen as an attempt almost to eliminate all judgments but those of value from both design and analysis, but few if any would contend that there has been more than the formal appearance of success here.

A certain subjective theory of probability formulated by Ramsey[9] and later and more extensively by de Finetti[4] promises great advantages for statistics. Contrary to what the word 'subjective' seems to connote to many, the theory is not mysterious or particularly unoperational. It gives, a few of us believe, a consistent, workable, and unifying analysis for all problems about the interpretation of the theory of probability, a much contested subject. It unifies the treatment of uncertainties, measuring them all by probabilities and emphasizing that they depend not only on patterns of information but on the opinions of individual people. Experience seems to me to show that this theory provides a better framework for understanding both the objective and the subjective aspects of statistics than we have heretofore had.

5. Does it matter?

As is often said, and with much truth, the explicit study of the foundations of a subject is usually of relatively little practical importance, for common sense and experience over the course of time develop a science more securely than it could possibly be built up by direct application of abstract principles. None the less, I believe that present-day discussions about inference and behavior, about subjectivism and objectivism are

544   LEONARD J. SAVAGE

stimulating practical advances in statistics. The evidences of this are widely scattered, but I shall mention only two examples.

First, it is becoming increasingly accepted that, once an experiment has been done, any analysis or other reaction to the experiment ought to depend on the likelihood-ratio function and on it alone, without any further regard to how the experiment was actually planned or performed. I believe that this doctrine, which contradicts much that was recently most firmly established in statistical theory and practice, is basically correct and that it will soon greatly simplify and strengthen statistics. Let me not falsify history by intimating that appreciation of the likelihood-ratio function as much more than is ordinarily understood by a 'sufficient statistic' originated in the economic outlook and subjectivism. Actually, it was, so far as I know, begun by Barnard[2] and Fisher[6], and quite apart from these ideas. None the less, the economic outlook and the subjectivistic theory of probability lend strong support to the likelihood-ratio doctrine and promise to hasten its acceptance and exploitation.

Secondly, David Wallace has recently obtained a valuable new insight into the much vexed Behrens–Fisher problem by reconsidering it from the point of view of subjective probability.

REFERENCES

[1] Barnard, G. A. Sequential tests in industrial statistics. *J. R. Statist. Soc.* (Suppl.), 8, 1–26 (1946).
[2] Barnard, G. A. A review of 'Sequential Analysis' by Abraham Wald. *J. Amer. Statist. Ass.* 42, 658–664 (1947).
[3] Barnard, G. A. Simplified decision functions. *Biometrika*, 41, 241–251 (1954).
[4] de Finetti, Bruno. La prévision: ses lois logiques, ses sources subjectives. *Ann. Inst. Poincaré*, 7, 1–68 (1937).
[5] Fisher, Sir Ronald A. Statistical methods and scientific induction. *J. R. Statist. Soc.* (B), 17, 69–78 (1955).
[6] Fisher, Sir Ronald A. *Statistical Methods and Scientific Inference.* Oliver and Boyd, Edinburgh, 1956.
[7] Gauss, Carl Friedrich. *Abhandlungen zur Methode der kleinsten Quadrate von Carl Friedrich Gauss.* Berlin, 1887. (Translation from Latin by A. Borsch and P. Simon.)
[8] Neyman, Jerzy. L'estimation statistique, traitée comme un problème classique de probabilité. Pp. 25–57 of *Actualités scientifiques et industrielles*, no. 739. Hermann et Cie., Paris, 1938.
[9] Ramsey, Frank P. *The Foundations of Mathematics and Other Logical Essays.* Kegan Paul, London, 1931.
[10] Savage, Leonard J. *The Foundations of Statistics.* John Wiley and Sons, New York, 1954.

DISCRETE VARIABLE METHODS
IN NUMERICAL ANALYSIS

By D. H. LEHMER

The term Numerical Analysis is of comparatively recent origin and owes its origin to the rapid growth of electronic computing in the last dozen years. To be sure, Gauss's investigations into the method of least squares, for example, would be called numerical analysis today and his results are generally applicable to computation whether it be done by hand, or by analogue or digital equipment. However, the most significant machine development of the recent past and future has been not in the analogue but in the digital or discrete variable field, and this type of equipment brings with it opportunities and handicaps connected with its characteristic discontinuous behavior. It is my purpose to point out and illustrate these opportunities and handicaps as we see them today.

In the first place, most numerical analysis problems are set up in imitation of a physical situation based on continuity or infinitesimal analysis. To obtain numerical answers, finite difference methods are employed with attendant difficulties which were recognized by Leibnitz and Newton, not to mention Archimedes. It is not my purpose to expatiate on the questions of what we now call truncation error, important as these questions are. Their consideration is made in the same system as that of the infinitesimal analysis on which it is based, namely the real number system. Digital computers do not operate with real numbers. Very frequently they attempt to imitate the rational operations of the real number system and, still more crudely, the various limit processes of analysis. The desire or need to imitate infinitesimal analysis has influenced the design of computers but, even from the beginning, it was understood that a digital computer must be capable of other modes of operation if it is to process its own program of calculation. Newer machines have acquired an even larger repertory of these logical commands not only because they are needed for program processing but because they are genuinely useful for problem solvings, and in some cases, it must be admitted, because they are cheap to realize electronically.

The number system of the digital computer is, after all, a finite one with limitations in time and space. Its effective use involves a subtle compromise between time and space and the requirements of the problem. Thus, in ordinary coding, the associative law of multiplication: $a(bc) = (ab)c$ (certainly one of the cornerstones of the real number system) does not hold because of round off. If it is absolutely essential that this law hold, special coding requiring more space and more time is needed. Even the sum of two numbers may not be given correctly by the machine's addition command because of overflow. The implication

$$ax = b \to x = b/a$$

is not always valid for the machine. This is not because a might be zero but because of the 'inexactitude' of the finite division operation of the machine.

Small wonder that mathematicians prefer not to worry about these small departures from real analysis. We are not surprised either when, after ordering millions of these operations to be performed in a few minutes time, the mathematician is not happy at the look of the output. Nor is this difficulty to be met by more use of floating arithmetic, either programmed or built in, with even less precision in each computer word. While we may guard against overflow at the left, information may disappear towards the right via 'underflow'.

A good deal of interest has been shown recently in automatic error analysis in which the machine's output is an interval rather than a number. This may be satisfactory for some problems and some answers, but often the analysis of the error is too conservative and so the interval is too large to be of much use. There is often a good case for the use of double precision, i.e. two computer words for each number, and little or no error analysis. Certainly a good deal depends on one's overall knowledge of the problem, its stability and the bounds for its variables, a knowledge which may be acquired from immediately previous experience.

There is, of course, a simple way to avoid most of these difficulties, namely to cease our imitation of the real number system and to deal only with genuine discrete variable problems.

Many sorts of number theory problems are quite naturally explored by digital computers, and in most of these problems operations other than addition and multiplication play dominant roles.

Less familiar problems of a discrete variable type are to be found in Combinatorial Analysis. Some of these rate low in theoretical interest but high in practicality as, for example, sorting and collating problems. Less utilitarian are problems involving set inclusion. Here it may be of interest to contrast again the infinite and finite attitudes. Many so-called well-defined sets of real numbers are so ill-defined that it is well-nigh impossible to say whether a given number belongs to the set or not, nor does this fact interfere much with the theoretical discussion of such a set. As soon as we consider finite sets, we can say that the problem is trivial since it involves only a finite number of trials or we can ask what are the efficient ways of deciding whether a number belongs to the set or not. Of course the notion of efficiency here is taken with respect to the capabilities of a given machine to perform certain operations. The operation of multiplication, we can be sure, is not one of these. Lest we might think that all such problems are straightforward, we could consider the set of all the 455052512 prime numbers $< 10^{10}$ and ask whether 2001791881 belongs to the set or not. Obviously there are many ways of handling this problem of set inclusion with widely varying efficiencies.

Other combinatorial problems, useful in management and scheduling, include problems of optimal choice and optimal arrangement. Still more elaborate applications are being made in chemistry and crystallography. Subroutines that can be used to advantage here include the methodical or random selection of k words from a population of n words, and the methodical or random permutation of a set of n words.

Related to these problems are the more prolix programs arising from problems in abstract algebra, group theory, and finite geometries.

Finally, one may cite problems in symbolic logic, chess playing and theorem proving as extreme examples of abstraction in discrete variable techniques.

It almost goes without saying that the usual library of subroutines and the various automatic algebraic coding routines are of little help in carrying out these problems.

Instead of trying to avoid real variable calculations one can try to apply to them some of the discrete variable techniques that we have learned. In fact these methods may be actually forced upon us when we come to formulate in detail a problem which at first sight seems not to involve discrete variables at all.

As an example of such a problem I should like to describe some of the features of a question raised ten years ago by Hardy and Littlewood on inequalities involving integrals. The question is the following one.

Let the monotone sequence of real numbers

$$1 \geqslant a_1 \geqslant a_2 \geqslant \ldots \geqslant 0$$

be the coefficients of the convergent cosine series

$$f^*(\theta) = 1 + a_1 \cos \theta + a_2 \cos 2\theta + \ldots.$$

548   D. H. LEHMER

Further, let $c_1,\quad c_2,\quad c_3,\quad \dots$

be real numbers such that the sequence

$$|c_1|,\quad |c_2|,\quad |c_3|,\quad \dots$$

is a rearrangement of the sequence of a's and let

$$f(\theta) = 1 + c_1 \cos\theta + c_2 \cos 2\theta + \dots.$$

The question is: Can we conclude that

$$\int_0^\pi |f^*(\theta)|\, d\theta \le \int_0^\pi |f(\theta)|\, d\theta ? \tag{1}$$

In other words, to make

$$\int_0^\pi |f(\theta)|\, d\theta$$

a minimum should one replace the coefficients of $f(\theta)$ by their absolute values and arrange to have the large coefficients on the terms of low frequency? The answer to this question is *no* and was given recently by the IBM 701 computer at the University of California. To show that the answer is no, means, of course, to exhibit two examples of functions f for one of which the inequality (1) is true and for the other of which it is false. As a matter of fact (1) is true for almost every choice of f so that the difficulty reduces to finding an f for which (1) is false.

Integrals of the absolute value of a function occur with great frequency in analysis as norms and upper bounds but almost never are these integrals computed. In the present case we must evaluate the two integrals in (1) with sufficient accuracy to make sure that the inequality fails.

One unattractive way of doing this would be to locate fairly accurately the various roots of the function lying in $(0, \pi)$ and to integrate piecewise between successive roots of odd multiplicity attributing the appropriate sign to the function as one goes. The integration would be termwise and consist of evaluating $\sin k\theta$ for θ a root of f or f^*.

A different method involving less computing would be to resort to mechanical quadrature to compute directly the integral of $|f|$ (and $|f^*|$). In this case all the well-known methods, based on the fact that the function to be integrated is nearly a polynomial, fail because $|f|$ has no derivative when $f = 0$. Hence we are driven to use the crude trapezoidal or mid-point rectangular rule with very fine subdivision of the interval $(0, \pi)$. A simple table of cosines based on this subdivision will then suffice to carry out the computation of the integrals. A simple error analysis shows that this discrete variable approach will suffice, when $(0, \pi)$ is subdivided into 128 subintervals, to give about 6-decimal accuracy.

So far we have replaced integration by summation. Next we replace f by a cosine polynomial, since our discrete variable machine cannot really handle infinite series. In fact we can make the degree, n, of this polynomial one of the main parameters of the calculation. Starting with $n = 2$, since (1) is seen to hold for

$$f(\theta) = 1 - a_1 \cos \theta, \quad f^*(\theta) = 1 + a_1 \cos \theta,$$

we make a more or less exhaustive search for failures of (1) before going from n to $n + 1$.

We come now to the next part of the problem in which we must use discrete variables, namely the selection of coefficients a_k and c_k. This amounts to an exploration of the n-dimensional cube

$$-1 \leqslant c_k \leqslant 1 \quad (k = 1 \, (1) \, n),$$

by means of a grid of lattice points. If h is the mesh size, the number of these points, $(2/h)^n$, is prohibitively large for only moderately small h. Of course many of these points are uninteresting. Thus if

$$\sum_{k=1}^{n} c_k = \sum_{k=1}^{n} a_k \leqslant 1,$$

then $f(\theta)$ and $f^*(\theta)$ remain non-negative over $(0, \pi)$ and the two integrals we are considering become equal to π so that (1) will automatically hold in these cases. Hence the region we are looking for is not near the origin. More generally, if

$$\int_0^{\pi} |f^*(\theta)| \, d\theta = \pi, \tag{2}$$

then $f^*(\theta)$ is non-negative and its companion integral

$$\int_0^{\pi} |f(\theta)| \, d\theta \tag{3}$$

cannot then be less than π. Hence it pays to compute the integral of $|f^*|$ first and to abandon the set of a's and the associated $2^n n!$ sets of c's in case (2) holds. Finally the integral (3) remains unchanged if we change the signs of all of its c's. Hence we may suppose $c_1 > 0$ and eliminate one half of our exploration. Our procedure should then be the following. Having disposed of a set of coefficients a_k we generate a new one and compute the integral of $|f^*(\theta)|$. If (2) does not hold we begin to change the signs of the a's to obtain sets of c's computing (3) in each instance and checking the inequality (1). After all sign changes have been taken care of we begin to permute the a's and return to the sign-changing subroutine. When all $n!$ permutations of the a's have been made, the set of a's is now disposed of and we have come full circle. As soon as all interesting a's have been processed we reduce the mesh size by replacing h by $\frac{1}{2}h$ and repeat the program unless the mesh is now too fine, in which case n is replaced by $n+1$.

This resumé of our program introduces still more discrete variable procedures involved in the following operations:

(a) Generating a new set of a's.

(b) Changing signs in all 2^{n-1} ways.

(c) Permuting the set of a's.

A word or two about each of these operations may suffice to indicate possible procedures.

The operation (a) may be performed recursively by setting $A_k = 2^m a_k$ treating the vector

$$(A_1, A_2, ..., A_n)$$

as a positive integer to the base 2^m of which the a's are the digits. Beginning with the number

$$(2^m, 2^m, ..., 2^m)$$

we diminish the last digit by unity to get the next set of A's. When any digit becomes zero a unit is subtracted from the next left A and all succeeding digits are set equal to this reduced digit. In this simple way we canvass all instances of numbers with monotone decreasing digits.

The operation (b) may be performed by means of a characteristic binary number N of $n-1$ bits, whose kth bit is 0 or 1 according as c_k is equal to a_k or $-a_k$. By using intentional overflow and the instruction $M + M$ replaces M, where initially M is a shifted copy of N, the machine instructs itself as to which c's are negative. Initially we set N at $00 ... 0$, a number of $n-1$ bits that produces no overflows and thus generates $a_1, a_2, ..., a_n$. Next a unit is added to N to give $00 ... 01$ that produces an overflow on the very last step to give $a_1, a_2, ..., -a_n$. Finally we produce $a_1, -a_2, ..., -a_n$ as the last set of c's.

There are at least eight different ways of generating the operation (c), the permutation operation. There is enough other computing in the whole problem to make it unnecessary to choose a permutation subroutine on the basis of speed alone. Furthermore, it is clear that we shall be dealing with very modest values of n since the number of integrals to examine is some moderate fraction of $(2/h)^n n!$. Finally the number of different objects to be permuted is often less than n. Equalities among the a's, which are frequent, produce redundant information and thus waste time. Nevertheless it was considered tolerable, and preferable to a more complicated routine made to avoid duplication of effort.

In the actual calculation the permutation subroutine used is based on the cyclic permutation method of Tompkins and Paige. Very briefly this is a method based on the vector S of $n-1$ integers

$$S: \quad (s_{n-1}, s_{n-2}, \ldots, s_2, s_1),$$

in which s_1 is a binary digit, s_2 is a ternary digit, etc., so that

$$0 \leqslant s_k \leqslant k-1.$$

S may be regarded as a number with a different base assigned to each of its digit positions. Addition is defined in terms of these bases with the appropriate 'carry rule' at each position. The vector

$$(0, 0, \ldots, 0)$$

is made to represent the permutation

$$a_1, \quad a_2, \quad \ldots, \quad a_n.$$

The recursive generation of the next permutation from a given one is done as follows. The last two marks of the given permutation are interchanged and a unit is added to s_1. If there is no carry so that s_1 has changed from 0 to 1 we have our next permutation and its corresponding vector S. If there is a carry to s_2 only, we have an old permutation. The last three marks are then subjected to the cyclic permutation

$$\begin{pmatrix} 3 & 2 & 1 \\ 2 & 1 & 3 \end{pmatrix}.$$

In general, if the addition of a unit forces a carry to s_{k-1} only, then the last k marks are subjected to the cyclic permutation

$$\begin{pmatrix} k & k-1 & \ldots & 2 & 1 \\ k-1 & k-2 & \ldots & 1 & k \end{pmatrix}$$

to produce the next permutation.

It is thus easy to set up a general permutation subroutine for any value of n. However, in retrospect, there is not much that can be done with our whole problem for $n \leqslant 6$. In such cases it is reasonable to devote $5! = 120$ words of the storage to represent all the permutations on 5 marks. Each such word contains 5 subwords that are extracted one at

552   D. H. LEHMER

a time to modify commands for depositing the numbers a_k in the appropriate addresses to form the corresponding permutation.

In conclusion there is one more feature of discrete variable methods that is often encountered, namely the ability to verify the results without reference to the machine that found it. In our example the machine indicated that in the case of $n = 4$ and

$$f(\theta) = 1 - \tfrac{3}{4}\cos\theta + \tfrac{3}{4}\cos 2\theta - \tfrac{1}{2}\cos 3\theta + \tfrac{3}{4}\cos 4\theta,$$

and $\qquad f^*(\theta) = 1 + \tfrac{3}{4}\cos\theta + \tfrac{3}{4}\cos 2\theta + \tfrac{3}{4}\cos 3\theta + \tfrac{1}{2}\cos 4\theta,$

an exception to (1) exists. That this is true can be verified independently by a completely rational argument involving no approximations whatsoever.

In fact, by setting $x = 2\cos\theta$, the functions f and f^* become polynomials in x. It can then be shown that for x in the interval $(-2, 2)$ corresponding to $(0, \pi)$ for θ, the function f does not vanish and $f^*(\theta)$ has a pair of distinct real roots. It follows that

$$\int_0^\pi |f^*(\theta)|\, d\theta > \int_0^\pi f^*(\theta)\, d\theta = \pi = \int_0^\pi f(\theta)\, d\theta = \int_0^\pi |f(\theta)|\, d\theta$$

so that (1) is violated.

It is hoped that these examples have served to indicate what roles the discrete variable may play in present-day Numerical Analysis.

SURVEY OF EXPERIMENTS ON THE
SOLUTION OF LINEAR SYSTEMS

By H. RUTISHAUSER

A detailed report will be published as

'Mitteilung Nr. 8 aus dem Institut für angewandte Mathematik der
ETH', published by Birkhäuser Verlag, Basel, Switzerland in 1959.

ÜBER EINE EUKLID-BEARBEITUNG, DIE DEM ALBERTUS MAGNUS ZUGESCHRIEBEN WIRD

Von J. E. HOFMANN

Im Frühjahr 1944 konnte ich dank der Liebenswürdigkeit des Herausgebers der Werke des Albertus Magnus[1]†, Prälat Prof. B. Geyer, eine wenige Wochen später bei der damaligen Preußischen Akademie der Wissenschaften eingereichte Abhandlung[2] einsehen, die im Herbst 1944 in Satz ging, jedoch wegen Zerstörung des Druckmaterials während der letzten Kriegshandlungen nicht zur Ausgabe gelangte. In dieser sehr interessanten Studie, die nunmehr in revidierter Form[3] unmittelbar vor der Veröffentlichung steht, berichtet Herr Geyer über eine schön geschriebene lateinische Handschrift des XIII. Jahrhunderts[4], die sich in Wien befindet und gemäß einer Notiz des Rubricators[5] am oberen Rand des ersten Blattes als Werk eines Albertus anzusehen ist.

Der Fundort macht es von vorne herein wahrscheinlich, daß es sich um einen Dominikaner handelt, und zwar um Albertus Magnus. Tatsächlich findet sich in den beiden alten Katalogen, worin die den Zeitgenossen bekannten Schriften des großen Mannes aufgeführt werden, ein diesbezüglicher Hinweis[6]. In seiner Abhandlung führt Herr Geyer Stellen aus der Paraphrase zur Aristotelischen *Physik* und aus der Schrift *De sensu et sensato* an, in denen Albertus die Absicht bekundet, eine Geometrie zu schreiben. Weitere Stellen—sie wurden mir freundlicherweise durch Herrn Geyer übermittelt—zeigen volle Vertrautheit mit dem Inhalt der Euklidischen *Elemente*. Ich erwähne einen Hinweis auf *Elemente* I, 32 (Satz vom Außenwinkel und der Winkelsumme im Dreieck)[7], auf die Größenlehre in Buch V[8], auf die Definition der Einheit zu Beginn von Buch VII[9], auf die Lehre vom Kommensurablen und Inkommensurablen in Buch X[10], auf die Behandlung der fünf regelmäßigen Körper in Buch XIII[11] und auf das 'von Assicolaus' verfaßte Buch XV (es stammt von einem Schüler des Isidoros, vielleicht von Damaskios)[12].

Schließlich gibt es in späteren Schriften des Albertus auch direkte Hinweise auf die *Geometrie*. Erwähnt wird unter Bezugnahme auf *Elemente* III, 15–16 der Berührpunkt der Kreistangente[13], ohne genaue Stellenangabe (gemeint ist *Elemente* X, 117) die Inkommensurabilität der Quadratdiagonale hinsichtlich der Seite vermittels des Gegensatzes

† Anmerkungen: S. 562 ff.

von gerade und ungerade[14] und schließlich unter Hinweis auf *Elemente* II, 14 eine Bemerkung über die Verwandlung eines Rechtecks in ein flächengleiches Quadrat[15]. Sie findet sich sinngleich in der Wiener Handschrift als Zusatz unter Bezugnahme auf Aristoteles, *De anima* II[16].

Aus dem Bisherigen geht eindeutig hervor, daß das Wiener Manuskript eine echte Schrift des Albertus enthält. Leider ist sie nicht vollständig; vielmehr reicht sie nur bis zu Buch IV der *Elemente*; wo sich der Rest befindet, ist zur Zeit noch unbekannt.

Ursprünglich war verabredet, daß der verdienstvolle Bonner Mathematiker E. Bessel-Hagen, ein ausgezeichneter Kenner der Mathematikgeschichte, die Herausgabe der Wiener Handschrift übernehmen sollte. Der unter so betrüblichen Umständen eingetretene Tod des hochgeschätzten Mannes hat veranlaßt, daß ich die Edition übernommen habe. Meine Frau hat die Handschrift entziffert. Was ich im folgenden über Eigenart, Charakter und Inhalt des Manuskriptes vorbringe, stützt sich größtenteils auf ihre sehr sorgfältigen Editionsvorbereitungen.

Es handelt sich um eine kommentierte Euklid-Ausgabe, im wesentlichen auf lateinischen Übersetzungen arabischer Fassungen beruhend. Albertus erweist sich wie in seinen sonstigen Schriften so auch hier als ein außerordentlich belesener Berichterstatter, der jeder Einzelfrage sorgfältig nachgeht, Gewährsleute nennt, gelegentlich auch eigene Zusätze beifügt und in Zweifelsfällen mit kritischen Bemerkungen nicht zurückhält. Die Wortlaute der einzelnen Euklidischen Definitionen, Postulate und Axiome und der nachfolgenden Sätze stimmen größtenteils mit jenen der von Campanus überarbeiteten Übersetzung des Adelhard von Bath überein[17]; die Beweise sind fast durchwegs in abweichender Fassung ausgeführt, deren Vorlage noch nicht genau feststeht. Daß Albertus die Übersetzungen des Boëtius und des Adelhard gekannt hat, geht aus einer kritischen Bemerkung zu I, 5—der *elefuga* der wissenschaftlich nicht Ansprechbaren[18]—hervor[19]. Den Wortlaut von I, 6 gibt Albertus 'nach verschiedenen Codices' in viererlei Fassungen wieder[20]. Ein Name fehlt—nämlich Campanus—und das wohl mit gutem Grund: dessen Bearbeitung der Adelhardschen Übersetzung dürfte erst nach der des Albertus entstanden sein, die wohl in der Zeit zwischen 1262 und 1265 niedergeschrieben wurde—in Italien vielleicht, wo sich Albertus längere Zeit am Hofe Urbans IV. aufgehalten haben dürfte, zusammen mit Thomas von Aquin, Wilhelm von Moerbeke und Campanus[21].

Was uns die *Geometrie* des Albertus interessant macht, ist freilich nicht die Wiedergabe der Euklidischen Texte selbst, sondern die Fülle der zugehörigen Erläuterungen aus arabischen Quellen. Erwähnt wird Alfârâbî, auf den sich Albertus auch in philosophischen Fragen so häufig bezieht[22], und zwar im Vorwort, auf das ich unten näher eingehen werde[23], und bei Erwähnung der Definition des Punktes[24] und der Ebene[25] und bei Aufzählung der verschiedenen Arten von Vierecken[26]. Auf weitere arabische Kommentare bezieht sich Albertus an mehreren Stellen; diese Stellen finden sich fast wörtlich wieder im Euklid-Kommentar des Annairîzî, den wir bisher nur in der Übersetzung des Gerhard von Cremona kannten[27]. Es handelt sich um die Definition der geraden Linie[28], der Ebene[29] und des Winkels[30], ferner um den Begriff des *Inventum* (gemeint ist das griechische πόρισμα)[31]. Bei Besprechung von IV, 4 taucht schließlich auch der Name *Anarizus* auf. Es handelt sich um die von Annairîzî hinzugefügte Diskussion der gestaltlich verschiedenen Fälle bei der Konstruktion des Umkreises um ein Dreieck[32].

Fast alles weitere, was Albertus an mathematischen Erläuterungen vorbringt, gleicht der Gerhardschen Übersetzung des Annairîzî, jedoch nur dem Sinne nach; wörtliche Übereinstimmung ist selten vorhanden. Zumeist ist Gerhard etwas ausführlicher, gelegentlich auch Albertus. Wir lesen von den zahlreichen Beiträgen des Heron (anfangs als *Hermydes*, später ausschließlich als *Yrinus* bezeichnet) zu den Definitionen, Axiomen und Postulaten Euklids[33] und von den interessanten Ergänzungen zum Satzgefüge und zur Beweistechnik, vor allem von Herons Versuch, die bei Euklid so häufig auftretende indirekte Schlußweise nach Möglichkeit zu vermeiden und die schleppenden geometrischen Beweise durch Verwendung algebraischer Elemente zu verkürzen. Manche neue Einzelheit wird dabei bekannt; des öftern sind Ergänzungen, die Gerhard nach Annairîzî schlechthin als Zusätze anderer bezeichnet, genauer präzisiert.

Seltener sind die dem Simplikios (als Sambelichios bezeichnet) zugeschriebenen Beiträge; neu ist z. B. die Zuweisung einer Definition für die Ebene[34]. Daß Simplikios bei Albertus als Autor eines mißlungenen Beweisversuches für das Parallelenpostulat bezeichnet wird, ist wohl ein Irrtum; nach Annairîzî[35] ist Geminos der Urheber, Simplikios der Berichterstatter, der das Unzutreffende des Vorgebrachten recht wohl erkennt[36].

Auch Geminos (als *Aganyz* bezeichnet) wird erwähnt, und zwar vor allem im Zusammenhang mit Fragen, die das Parallelenpostulat und seine Folgerungen betreffen[37]. Auf Platon wird im Zusammenhang mit der Definition des Punktes[38] und der Geraden[39] verwiesen, auf Archimedes (als *Assamites* bezeichnet) wegen der wohlbekannten Minimaldefinition der Geraden[40], auf Apollonios wegen der Winkeldefinition[41], auf Poseidonios (als *Aposedonius* bezeichnet) wegen einer Punktdefinition[42] und auf Tâbit ben Qurrah (als *Thabit Benchorat* bezeichnet) wegen des so anschaulichen, gewöhnlich als 'Stuhl der Braut' bezeichneten algebraischen Beweises für den Pythagoreischen Lehrsatz[43]. Bei dieser Gelegenheit bemerkt Albertus—vermutlich handelt es sich um eine eigene Zutat—, man könne leicht vermittels eingefügter Halbkreise konstruieren (Abb. 1 deutet das an), gibt jedoch leider keine Figur dazu.

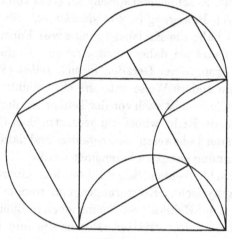

Fig. 1

Nun steht freilich für Albertus nicht das Fachmathematische im Vordergrund, vielmehr einerseits das Philosophische, andererseits das Enzyklopädische; geht es ihm doch darum, den Zeitgenossen einen Überblick über alles wissenschaftlich Bedeutsame zu vermitteln, und das vorzugsweise unter philosophisch-theologischem Aspekt. Diese Tendenz zeigt sich besonders deutlich in der Einleitung, die übrigens vom Autor nicht eigens als solche gekennzeichnet ist[44]. Zu Anfang handelt es sich um die Begriffsbestimmung der Mathematik, innerhalb deren nach den allgemeinen Lehren der Pythagoreer die *quantitas discreta* Gegenstand der Arithmetik (in bezug auf die ganzen Zahlen) beziehungsweise der Musik (in bezug auf deren Verhältnisse) ist; die *quantitas continua* hingegen ist Gegenstand der Geometrie (in bezug auf unbewegliche Figuren) beziehungsweise der Astrologie (in bezug auf Kreisbewegungen). Hier wird im Text auf Aristoteles verwiesen; die Wortlaute beziehen sich auf *Metaphysica* I, 1, 2, 6 und zeigen deutliche Verwandtschaft mit entsprechenden Stellen in des Albertus zugehöriger Paraphrase. Als Erfinder der Geometrie werden, wie üblich, die Ägypter bezeichnet[45]; dann folgt die hübsche Erzählung von Aristipp dem Sokratiker, der bei einem Schiffbruch an unbekanntem Strand (es war der von Rhodos) aus geometrischen Figuren im Sand auf die Existenz freundlicher Menschen schloß[46].

Unter Bezugnahme auf den schon oben erwähnten Kommentar des Alfârâbî zu Euklidischen Sätzen[23] werden Linie, Fläche und Körper als die Grundelemente des Kontinuums bezeichnet, von denen die Geometrie handelt. Anschließend spricht Albertus von der 'Erzeugung' der Geraden durch Bewegung (*motus simplex secundum formam*); im Gegensatz hierzu habe die Kreisbewegung zwei Formen (Konvexität und Konkavität) und sei daher nicht einfach. Einfach sei auch die punktweise Bewegung einer Geraden in sich selbst (Verlängern einer Strecke). Auf jede andere Weise entstehe ein Gebilde mit Länge und Breite usw. Anschließend ist auch von der Teilbarkeit der Grundelemente des Kontinuums die Rede, wobei die geometrischen Gebilde als *regulares*, die Formen der Lebewesen als *irregulares* und damit einer genauen Verhältnisbestimmung entzogen bezeichnet werden.

Mit einem abschließenden Satz[47] beendet Albertus diese seine symbolisch-metaphysischen Erörterungen und wendet sich zur eingehenden Diskussion der Euklidischen Definitionen in Buch I der *Elemente*. Als gewissenhafter Berichterstatter zählt er die ihm bekannt gewordenen Lehrmeinungen auf, wobei auch Stellen aus nicht eigens erwähnten Autoren durchzufühlen sind[48] und der wohlerfahrene Verfasser auch seine eigene Meinung deutlich zu Gehör bringt[49]. Gelegentlich freilich kommen auch mathematische Ungereimtheiten vor; so findet sich in einer Ergänzung zu Definition 17 (Kreisdurchmesser), woselbst bewiesen wird, daß der Durchmesser die Kreisfläche halbiert, eine sinnstörende Diskrepanz zwischen Text und Figur[50].

Auch Worterklärungen treten auf, vor allem bei Übernahme arabischer Fachbezeichnungen; sie haben für die Abhängigkeit von bestimmten Übersetzern große Bedeutung und seien daher eigens aufgezählt[51]. Zunächst erwähne ich *elmuhaym* = al-mu'ayyia (Raute beziehungsweise Rhomboid) und *elymharifa* = al-munḥarif (Trapez)[52], ferner *meguar* = al-miḥwar (Achse)[53], *muchephy* = al-murabba'î (Viereck)[54], *elgyther* = al-quṭr (Durchmesser)[55] und schließlich *dulcarnon* = ḏu'l-qarnain (zweigehörnt) als Fachausdruck für den Pythagoreischen Lehrsatz[56]. Interressant ist außerdem die Bemerkung zum Parallelenpostulat: Albertus weist auf die aus anderer Quelle wohlbekannten Ausführungen des Ptolemaios[57] hin und fügt hinzu, das Ganze befinde sich in dessen *liber theorematum*[58].

Nach Aufzählung von acht Axiomen[59], von denen sich nach Albertus in den ältesten Schriften nur die ersten drei vorfinden[60], folgen zusätzliche Axiome, von denen zwei auch bei Adelhard–Campanus auftreten, jedoch dort in etwas abweichender und weniger geschickter Fassung. Erst nach längeren vorbereitenden Begriffsbestimmungen[61] wendet sich Albertus zum eigentlichen Thema, dem Satzgefüge.

Auch hier stimmt der Aufbau mit Adelhard–Campanus überein, die Kommentierung mit Annairîzî–Gerhard, jedoch zeigen sich wiederum im Wortlaut der Beweise und Erläuterungen erhebliche Abweichungen. Die Figuren sind nicht immer sorgfältig genug konstruiert und häufig zu speziell angelegt; das führt gelegentlich zu Mißverständnissen[62]. Der Text der Vorlagen ist im allgemeinen sorgfältig kopiert; nur an einigen wenigen Stellen, die jedoch nicht unbedingt dem Schreiber zur Last fallen müssen, sind Zeilensprünge nachweisbar, die auch nicht durch Vergleich mit Annairîzî textlich behoben werden können. Ein interessanter Fall liegt bei der Kommentierung von III, 13 vor (Gleichlange Kreissehnen stehen gleichweit vom Mittelpunkt ab; Sehnen gleicher Entfernung vom Mittelpunkt sind gleichlang). Hier bemerkt Heron ergänzend, der Mittelpunkt liege 'zwischen' den beiden Sehnen (d. h. nicht in den kleineren Segmenten, die von den Sehnen mit dem Kreis erzeugt werden, und auf der einen Winkelhalbierenden des durch die Trägergeraden der beiden Sehnen erzeugten Winkels). Den Beweis für gleichlange parallele Sehnen gibt Albertus getreulich nach seiner Vorlage wieder[63]; den nachfolgenden für nicht parallele Sehnen schreibt er ebenfalls ab[64], nimmt jedoch an dem Umstand Anstoß, daß Heron zum Beweis den späteren Satz III, 20 (vom Umfangswinkel) heranzieht, streicht das Ganze durch und begründet sein Verfahren mit dem Zusatz: *Et est non multum valens haec.* Dieses Vorgehen spricht sehr für die von Herrn Geyer[3] außerdem auch auf Grund des Schriftbildes und einer Reihe von Schreibeigentümlichkeiten vertretene Meinung, es handle sich bei der Wiener Handschrift um ein Autograph.

Größere Abweichungen von Campanus im Wortlaut der Sätze finden sich im ersten Buch bei 4, 5, 24, 25, 26 und 40, im zweiten Buch bei 12 und 13 und im dritten Buch bei 30. Jedesmal handelt es sich um ziemlich umfangreiche Sätze mit Nebensatzkonstruktionen. Auf Grund dieser Tatsache wird es vielleicht möglich sein, die wirklich benutzten Vorlagen zu ermitteln.

Zu I, 16 (Im Dreieck ist ein Außenwinkel größer als jeder Gegeninnenwinkel) macht Albertus nicht die allgemeine, sondern eine spezielle Figur, nämlich ein gleichseitiges Dreieck, und verwendet unerlaubterweise deren Eigenschaften auch beim Beweis. An diese Figur anknüpfend, behauptet er auf Grund eines unzulässigen Zirkelschlusses, der Außenwinkel sei sogar gleich der Summe der beiden Gegeninnenwinkel und die Winkelsumme im Dreieck betrage also zwei Rechte. Folglich sei auf diesem Wege I, 32 (Satz vom Außenwinkel und von der Winkelsumme im Dreieck) erwiesen, und das ohne Verwendung des Parallelenpostulats, und es sei seltsam genug, daß weder Heron noch Geminos noch ein anderer auf diesen beachtlichen Sachverhalt gestoßen sei. Hier hat eine handgreifliche Figur zu einem unzureichenden Beweisverfahren verführt.

An anderen Stellen finden sich sehr hervorzuhebende Bemerkungen. So ist es etwa im Fall von I, 24 (In Dreiecken mit zwei entsprechend gleichen Seiten liegt dem größeren Zwischenwinkel die größere Gegenseite gegenüber). Hier ersetzt Albertus (vielleicht im Anschluß an seine Vorlage) den indirekten Beweis Euklids durch einen direkten[65], der sich als vortreffliches Gegenstück neben den von Annairîzî gegebenen[66] stellt. Natürlich weiß Albertus von der heftigen Kritik, die Euklids Beweis für I, 29 (Zwei parallele Gerade werden von einer dritten in gleichen Wechselwinkeln geschnitten, usw.) schon in der Antike gefunden hat. Er führt einen oben[54] erwähnten Beweisversuch des Apollonios und den in [36] und [37] berührten des Geminos vor und fügt einen eigenen hinzu, der sich auf seinen unrichtig bewiesenen Zusatz zu I, 16 stützt.

Im weiteren Verlauf seiner Ausführungen weicht Albertus immer weniger von den Gedankengängen bei Euklid beziehungsweise Annairîzî ab. Hübsch ist die wahrscheinlich selbständige Umgestaltung, die er der Beweisfigur zu II, 8 gibt—jenem Satz, der in moderner Umschrift durch $(2a+b)^2 = 4a(a+b)+b^2$ wiedergegeben werden könnte (Fig. 2). Interessant ist auch die Vorüberlegung, die Albertus dem Beweis für III, 10 (Kreise schneiden sich nur in zwei Punkten) vorausschickt: aus der Vorstellung (secundum imaginationem) gehe hervor, daß sich zwei Kreise auf keinen Fall in einer ungeraden Anzahl von Punkten durchschneiden könnten. Hierin steckt eine Anwendung des Zwischenwertsatzes, der schon seit den Zeiten der Sophisten (Zenon von Elea) so heftig angegriffen wurde—eine Diskussion, die in der Ergänzung des Albertus zum Kontingenzwinkelsatz III, 15 ihren interessanten Niederschlag findet. Für die fragliche Stelle[67] gibt es bei Annairîzî nichts Entsprechendes, wohl aber bei Adelhard–Campanus[68].

Ich habe hier einige Einzelheiten aus der Geometrie des Albertus gegeben, die mir vom Standpunkt der Wissenschaftsgeschichte aus aufschlußreich zu sein scheinen. Zusammenfassend möchte ich sagen, daß wir Albertus als einen auch auf fachmathematischem Gebiet sehr wohlunterrichteten Gelehrten anzusehen haben. Daß ihm gelegentlich Fehler unterlaufen, ist für den größeren Zusammenhang von geringer Bedeutung gegenüber der Fülle des gesicherten Wissens, das Albertus

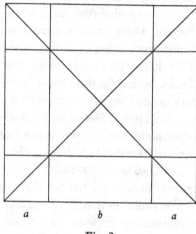

Fig. 2

sowohl an mathematischen Einzelheiten wie an entwicklungsgeschichtlichen Zusammenhängen besitzt—ein Wissen freilich, das bedauerlicherweise nicht auf die nachfolgenden Forschergenerationen weitergewirkt hat. Leider ist das Manuskript nicht vollständig. Das geht—abgesehen von den oben erwähnten Selbstzitaten auf die späteren Teile—auch daraus hervor, daß das letzte der vorgeführten Probleme, nämlich IV, 16 vom regelmäßigen Fünfzehneck, nur unvollständig behandelt ist; es fehlt der dritte Teil und das *Explizit*, das sonst am Ende eines jeden größeren Abschnittes zu finden ist.

Hier noch ein Wort zur Jordanus-Frage. Auf Grund der Albertus-Geometrie muß in Zukunft darauf verzichtet werden, Jordanus Nemorarius mit dem zweiten Dominikaner-General Jordanus Saxo (†1237) zu identifizieren, der bekanntlich Albertus 1223 für seinen Orden gewonnen hat[69]. Wäre der Mathematiker Jordanus der Dominikaner, dann müßte mit Sicherheit angenommen werden, daß Albertus

über die eigentümlichen Definitionen in *De triangulis*[70] etwas zu sagen
hätte. Dem ist aber ni c h t so; Albertus nennt weder den Namen Jordanus noch führt er auch nur ein e dieser Definitionen an. Daraus ist
zu schließen, daß Albertus von dem Mathematiker Jordanus nichts
wußte und daß dieser auf keinen Fall der Dominikaner sein kann. Dies
stimmt auch gut zusammen mit dem Charakter der Jordanischen
Schriften, die nach Aufbau und Inhalt viel besser in die z w e ite Hälfte
des 13. Jahrhunderts passen als in die erste.

Schließlich noch zu Campanus! Daß auch s e in Name bei Albertus
fehlt, lehrt uns, daß die Revision der Adelhardschen Euklid-Übersetzung
noch nicht abgeschlossen, wahrscheinlich noch gar nicht begonnen war.
Diese Übersetzung jedoch, beschränkt auf das, was man. damals für
echt Euklidisch hielt, mit nur wenigen kommentierenden Bemerkungen
behaftet, verdankt ihre Entstehung mit Sicherheit jener Bewegung,
die sich auf jedem Gebiet bemühte, die echten griechischen Texte aus
der Verkleidung überwuchernder Nebenbemerkungen herauszulösen.
Mag das damals noch nicht in jenem Maße gelungen sein, wie es in späterer
Zeit erreicht wurde—fest steht doch, daß erst mit Campanus die eigentliche Neubelebung der mathematischen Wissenschaften im lateinischen
Mittelalter einsetzt, und daß sein Bemühen erfolgreich noch bis zu
Clavius und weit darüber hinaus spürbar ist. Aber durch diesen 'best-
seller' ist auch vieles Wertvolle in den Hintergrund geschoben worden
und schließlich in Vergessenheit geraten, und dazu gehört leider auch
die Albertus-Geometrie, die unser heutiges Wissen über viele Einzel-
fragen der Antike und auch der mittelalterlichen Mathematik wesentlich
bereichert.

ANMERKUNGEN

[1] *Opera omnia*, Münster seit 1951, Aschendorff.

[2] B. Geyer, *Die mathematischen Schriften des Albertus Magnus.*

[3] B. Geyer, *Die mathematischen Schriften des Albertus Magnus*, Angelicum 35, 159–175 (1958).

[4] Wien, Dominikaner-Bibliothek, *cod.* 80/45, fol. 105ᵛ–145ʳ (s. XIII).

[5] *Primus Euclidis cum commento Alberti.*

[6] Im Katalog des Heinrich von Herford: ...[*scripsit*] *expositiones Euclidis, per-spectivae et Almagesti...*; im Stanser Katalog: *Item exposuit Euclidem, perspectivam almagesti* [*sic!*] *et quosdam alios mathematicos.*

[7] In der Physik-Paraphrase I, 2, *trac.* 3; in der 38-bändigen Gesamtausgabe von A. Borgnet, Paris 1890–99 (zukünftig zitiert als AB) III, S. 174.

[8] Paraphrase zu *De anima* I, 2, *tract.* 1, *cap.* 5 (AB V, S. 200a–201b).

[9] Paraphrase zur Metaphysik (AB VI, S. 62b).

[10] Paraphrase zur Metaphysik (AB VI, S. 27a); außerdem *Liber de indivisibilibus* (AB III, S. 465b, 466a, 470a, 471). Dieses Werk schließt sich an die Pseudo-Aristotelische Schrift *De lineis insecabilibus* an, die vielleicht von Theophrast, dem Nachfolger des Aristoteles in der Leitung der peripatetischen Schule, stammt.

[11] Paraphrase zur Metaphysik (AB VI, S. 84a).

[12] Paraphrase zu *De caelo et mundo* I, 3, *tract.* 1, *cap.* 3 (AB IV, S. 243b).

[13] Paraphrase zur Metaphysik (AB VI, S. 148a): *...cum tamen, sicut in XV et XVI tertii Geometriae nostrae determinatum est, linea contingens non nisi secundum punctum contingat ipsum* [= *circulum*].

[14] Paraphrase zur Metaphysik (AB VI, 36a): *Hoc autem a nobis jam in geometricis est demonstratum.*

[15] Paraphrase zur Metaphysik (AB VI, S. 142b).

[16] *Opera*, ed. I. Bekker, S. 413a, 13–20.

[17] Erstdruck Venedig 1482, besorgt von E. Ratdolt. Dieser und die zahlreichen Nachdrucke stimmen mit den zahllosen mittelalterlichen Handschriften des Campanus größtenteils bis auf unwesentliche Textvarianten überein, enthalten jedoch einige kennzeichnende Zusätze, wie etwa die Dreiteilung des Winkels durch Einschiebung am Ende des IV. Buches. Über die Beziehung zwischen Adelhard und Campanus vergleiche H. Weissenborn: 'Die Übersetzungen des Euklid aus dem Arabischen in das Lateinische durch Adelhard von Bath etc.', *Z. Math. Physik*, 25 (1880), hist.-lit. Abt., S. 143–166. Weiteres wertvolles Material über die Adelhardsche Übersetzung und ihre drei Fassungen bringt M. Clagett: 'The Medieval Latin Translations from the Arabic of the Elements of Euclid, with Special Emphasis on the Versions of Adelard of Bath', *Isis*, 44 (1953), Nr. 135–136, S. 16–42. Darnach ist die Vorlage für Campanus die Version II. An deren Satztexte (nicht an die Beweise) hat sich auch Albertus angeschlossen. Die Textproben von Beweisen, die Herr Clagett anführt (dortselbst Fußnote 31), geben keine Auskunft über die möglicherweise von Albertus verwendete Vorlage.

[18] *Ab ele*, fügt Albertus hinzu, *quod est miser, et fugare, quia fugit miserum desidiosum, qui in disciplinalibus non intendit.*

[19] Albertus nimmt Anstoß daran, daß Euklids Text aus zwei Sätzen besteht, die in umgekehrter Reihenfolge bewiesen werden: *et similia faciunt commenta Boëtii et Adelardi.* Diese Stelle lehrt, daß Albertus nicht die Pseudo-Boëtische Geometrie (ed. G. Friedlein, Leipzig, 1867, S. 372–428) gemeint haben kann, woselbst nur der Satz (in anderem Wortlaut als bei Albertus) erwähnt wird (ebenda S. 380), sondern die heute nur mehr in Bruchstücken erhaltene echt Boëtische. Zum Gegenstand: N. Bubnow: *Gerberti opera mathematica*, Berlin, 1899.

[20] Eine davon ist die der Pseudo-Boëtischen Geometrie: [19], S. 380.

[21] Vergleiche Fr. Überweg und B. Geyer, *Die Patristische und Scholastische Philosophie*, 11. Auflage, Berlin, 1928, S. 423.

[22] Vergleiche Überweg–Geyer [21], S. 409 in wörtlicher Wiedergabe des schon in Fr. Überweg–M. Baumgartner, 10. Auflage, Berlin, 1915, S. 468 Ausgeführten.

[23] Hier gibt Albertus Gedanken aus einem in lateinischer Vorlage noch nicht bekannten Kommentar des Alfârâbî zu den Euklidischen Sätzen wieder. Vielleicht handelt es sich um den Text in der Münchner Staatsbibliothek, *Cod. hebr.* 36, Nr. 3 (*fol.* 17b–21b), auf den ich durch Herrn A. P. Juschkewitsch-Moskau hingewiesen wurde.

[24] *Alfarabius vero* [*diffinit*] *sic*: *Punctum est, quod non habet dimensionem quantitatis continuae habentis situm.*

[25] *Superficies plana est, cuius spatium est aequale spatio lineae, quae ipsum comprehendit, aut spatio linearum ipsam comprehendentium.*

[26] *Sed cum Alfarabius introducit Euclidem in libro divisionum....* Anscheinend gibt es also eine Einleitung Alfârâbîs zum Euklidischen *Liber divisionum*; sie ist uns jedoch im Augenblick ebensowenig bekannt wie die zugehörige lateinische Übersetzung.

[27] Ed. M. Curtze, Leipzig, 1899, ²1909 (als Supplement zu Euklid, *Opera omnia*, ed. J. L. Heiberg und H. Menge) nach *Cod.* 569 (DD.IV.19) der Krakauer Universitätsbibliothek. Ich zitiere die Erstausgabe Curtzes stets als AC mit nachfolgender Seiten- und Zeilenzahl. Inzwischen sind weitere Handschriften aufgetaucht, jedoch noch nicht ausgewertet. Herr Clagett gibt in [17], Fußnote 28 an: *Vat. Reg. lat.* 1268 (s. XIII oder XIV), *ff.* 144–205, einen Auszug in Oxford, *Bodl. Digby* 168 (s. XIV), 124ʳ–125ʳ und eine sehr erweiterte Fassung in Paris, *Bibliothèque Nationale* 7215.

[28] *Linea recta est, quae est posita super aequale, quod est inter omnia duo puncta cadentia super ipsam* (AC, S. 5, 19–21).

[29] *Superficies plana est, quae est posita super dimensionem, quae est aequalis ei, quod est inter duas lineas rectas, quae sunt super ipsam* (so auch AC, S. 9, 9–11 mit folgenden Abweichungen: plana *add.* illa, super *vert. in* supra AC). Anschließend wird ähnlich wie in AC, S. 9, 12–13 festgestellt, daß die Ebene die 'kürzeste' Fläche zwischen zwei Geraden ist, und daß hier von der Oberfläche eines Zylinders, einer Pyramide oder einer Kugel keine Rede sein könne (ähnlich: AC, S. 10, 26–8).

[30] *Angulus superficialis* [*est*] *inclinatio duarum linearum in una superficie sibi obviantium non secundum rectitudinem positarum* (AC, S. 11, 4–6). Hier hebt Albertus den Unterschied gegenüber der 'aus dem Griechischen stammenden' Übersetzung hervor: *Angulus planus est duarum linearum alternus contactus, quarum expansio est supra superficiem applicatioque non directa.* Den nämlichen Wortlaut hat auch Adelhard–Campanus. Sollte auch hier eine Bezugnahme auf die echte Boëtische Geometrie [19] vorliegen? Die Pseudoboëtische [19], S. 374 hat einen etwas anderen Text, aber ebenfalls *angulus planus.* Um dieses Wort *planus* geht es hier vor allem.

[31] Bei AC, S. 39, 11–27 findet sich ein viel ausführlicherer Text. Es handelt sich um die endgültige logische Bestimmung eines bereits beim Aufbau der Figur mitverwendeten Elementes, nicht um die Auffindung eines unbekannten Zusammenhanges.

[32] Der Text aus AC, S. 142, 22–145, 24 wird gedankentreu, jedoch nicht wörtlich wiedergegeben.

[33] Ed. J. L. Heiberg, Leipzig, 1912.

[34] AC, S. 10, 6–8: *Superficies plana est, in qua possibile est protrahi ab omni puncto ad omnem punctum lineam rectam.* Dortselbst folgt dann eine lange Erörterung über die Stellung dieser und anderer Definitionen zur Euklidischen. Albertus sagt statt dessen ganz kurz—und das könnte recht wohl seine eigene, von keiner Vorlage beeinflußte Meinung sein: *Sed cuilibet patet, hanc non esse diffinitionem, sed potius signum plani esse ab effectu plani sumptum.*

[35] AC, S. 66, 24–68, 6.

[36] AC, S. 73, 5–31. Albertus hebt den Hauptpunkt, daß nämlich im Grunde nur eine Umstellung der Euklidischen Anordnung vorliegt und die unzulässige Annahme verwendet wird, man könne durch jeden Punkt zu jeder Geraden eine Parallele (in seiner Redeweise: Linie gleichen Abstandes) legen, selbständig und mit vorbildlicher Kürze vor.

[37] Das alles ist in etwas anderer Formulierung auch in AC, S. 26, 11–28; 34, 31–35, 4 und 66, 24–73, 31 enthalten. Vergleiche auch [35] und [36].

[38] *Punctum est unitas habens situm, unitas autem contra punctum non habens situm.* Die Stelle wird wohl aus Aristoteles, *De anima*, III, 6 [16], S. 430b, 20–1 stammen.

[39] AC, S. 6, 25–7.

[40] AC, S. 5, 22–3.

[41] AC, S. 13, 12–14.

[42] AC, S. 3, 23–5.

[43] AC, S. 84, 27–86, 20.

[44] Nähere Einzelheiten, Textproben und eine Handschriftenprobe finden sich in [3].

[45] Vergleiche Joh. Tropfke: *Geschichte der Elementarmathematik* [3]IV, ed. K. Vogel, Berlin, 1939, S. 4, Fußnote 2.

[46] Der Bericht stammt aus Vitruv, *De architectura* VI, 1. Dieses Werk war dem lateinischen Mittelalter schon seit dem 9. Jahrhundert wieder bekannt (z. B. London, *Brit. Mus., Cod. Harl.* 2767).

[47] *Quia autem omnium horum principium est punctum, ab ipso diffinitionum, quae principia quaedam demonstrationum sunt, sumamus exordium.*

[48] So erscheint bei Besprechung der Punktdefinition ein Hinweis auf die drei Erstreckungen des Raumes, die sich auf Aristoteles, *Metaph.* IV, 13 [16], S. 1020a, 3–14 bezieht und im Wortlaut mit der entsprechenden Stelle aus Hugo de S. Victor, *Practica geometriae*, ed. M. Curtze, *Monatshefte Math. Physik*, 8 (1897), S. 193–220, insbesondere 195–6 sehr verwandt ist. Ähnlich drückt sich auch Heron in Definition 135, 3a aus; ed. J. L. Heiberg, Leipzig, 1912, S. 96ff.

[49] So entscheidet sich Albertus in der Frage, ob dem Winkel die Größeneigenschaft zuzuweisen sei oder nicht, wie folgt: Der Winkel ist eine Quantität, aber das Winkelsein eine Qualität.

[50] Der Sachverhalt wird aus der Parallelstelle in AC, S. 20, 14–21, 30 klar; die Zusammenhänge zwischen Text und Abbildungen werden dadurch verständlich, daß man zwei der vorhandenen Teilfiguren gleichzeitig betrachtet und die fraglichen Stücke bald aus der einen, bald aus der anderen Teilfigur entnimmt.

[51] Die Umschriften der arabischen Fachausdrücke und ihre Bedeutung verdanke ich der liebenswürdigen Auskunft von Herrn H. Giesecke vom Institut für Orientforschung an der Deutschen Akademie der Wissenschaften Berlin.

[52] Beide erscheinen in Definition 22 des I. Buches, die von der Einteilung der Vierecke handelt, und werden in der Zweitliteratur häufig erwähnt, weil sie durch die Euklid-Ausgabe des Campanus [17] allgemeine Verbreitung fanden.

[53] Diese Bezeichnung kommt bei der Definition der Geraden vor und findet sich auch in AC, S. 7, 19 im nämlichen Zusammenhang, jedoch fehlt dort die Worterklärung, die Albertus gibt.

[54] Das Wort erscheint zu Beginn der Diskussion um I, 29, die oben erwähnt wurde [36]. Es wird bei Albertus als *sector* übersetzt und als ein von Apollonios stammendes Fachwort bezeichnet, das beim schrägen Schnitt einer (geraden quadratischen) Pyramide verwendet worden sei. Weder in AC noch bei Proklos Diadochos: *In primum Euclidis elementorum librum commentarii*, ed. G. Friedlein, Leipzig, 1873 findet sich eine Parallelstelle.

[55] Das Wort erscheint in der in [15] erwähnten Ergänzung zu II, 14. Albertus verwendet es nicht ganz glücklich: *Et hoc, quod intendunt antiqui, dicentes, ex hoc inveniri posse latus tetragonicum, quod dicunt elgyther, cuiuslibet altera parte longioris.* Hier zeigt sich der Einfluß der oben [17] erwähnten Adelhard-Version II: bei Clagett [17], S. 33 rechts lesen wir nämlich: *Nota quoque, quod hinc inveniri potest latus tetragonicum (quod dicunt elgydar, cuiuslibet parte altera longioris formae).*

[56] Albertus bemerkt hierzu: *Haec figura ab antiquis accepit nomen proprium, quod arabice quidem dicitur dulcarnon, quod latine est cornuta. Quidam tamen ignari virtutis vocabuli dixerunt dulcarnon dictam, quia inventa ea propter utilitatem sui geometrae dulciter canebant. Alii autem rudiores eos tunc dicebant dulces carnes comedisse et ideo sic vocatam. Sola autem prima ratio ab auctoritate habetur.* Beim Beweis zu III, 34 (Sehnensatz) wendet er den Pythagoreischen Lehrsatz an, zitiert aber nicht, wie sonst, entweder wörtlich oder mit I, 46, sondern sagt kurz *per bicornem.* Das ist die wörtliche Übersetzung von *dulcarnon.* Übrigens merkt Regiomontan in der Euklid-Handschrift nach Campanus, die sich in der Nürnberger Stadtbibliothek Cent. VI, 13 vorfindet und bis zu III, 8 von Regiomontan selbst geschrieben wurde, zu I, 46 mit roter Tinte an: *Tunica Francisci.*

[57] So auch in AC, S. 65, 24–66, 10, woselbst zusätzlich auf die von Ptolemaios verwendeten Sätze I, 13, 15 und 18 hingewiesen wird. Dessen Ausführungen kennen wir aus Proklos [54], S. 365–9, deutsch von L. Schönberger, ed. M. Steck, Halle, 1945, S. 418–23.

[58] Diese heute verschollene Schrift scheint den Muslimen noch vorgelegen zu sein. Vielleicht hat sie deren interessante Beiträge zur Parallelenfrage ausgelöst.

[59] Das heutige 9. Axiom: Zwei Gerade umschließen keine Fläche—übrigens ein Einschiebsel der Theonischen Redaktion—erscheint bei Albertus ebenso wie in AC, S. 35, 5–6 und in Adelhard, Version II, siehe Clagett [17], S. 31 rechts, als letztes Postulat. Der Grund für diese Verschiebung ist in der Definition der Postulate als Mitteldinge zwischen den Axiomen und den Sätzen zu suchen (so auch AC, S. 29, 12–13). Gleich AC, S. 35, 7–9 weiß Albertus, daß sich dieses sein letztes Postulat (das er übrigens später niemals also solches zitiert, sondern bestenfalles wörtlich anführt) stets in den Übersetzungen aus dem Griechischen vorfindet, nicht aber in den aus dem Arabischen kommenden. Anschließend gibt er den Beweis der '*moderni*' wieder, den auch AC, S. 35, 10–36, 9 hat. Proklos sagt von diesem Axiom, es sei nur der Geometrie eigen und daher überflüssig, weiß aber noch nichts von einem Beweis: [54], S. 196=[57], S. 304.

[60] Proklos [54], S. 196=[57], S. 304 sagt, die Beschränkung auf nur drei Axiome stamme von Heron. Daraus folgt, dass die Aufzählung in den Heronischen Definitionen [48], Nr. 134, 2, S. 94, woselbst 9 Axiome aufgeführt sind, von Proklos n i c h t gemeint sein kann. Übrigens erscheint bei Heron das vorhin [59] erwähnte 9. Axiom gleichzeitig

auch als 6. Postulat: [48], Nr. 134, 1, S. 94. In den Adelhardschen Versionen ist das Axiom stets unter die Postulate gerechnet.

[61] Das Entsprechende bei AC, S. 38, 14–42, 19 ist zumeist etwas ausführlicher gehalten. Allgemeines Vorbild ist Proklos [54], S. 198–213 = [57], S. 306–16, jedoch sicher nicht direkt, sondern in einer Zwischenbearbeitung, als die etwa Teile aus Heron in Frage kämen; vergleiche [48], Nr. 136, S. 108 ff., 134 ff. und Nr. 137, S. 156 ff., jedoch ist schon von den Arabern das meiste der so vielgestaltigen Neuplatonischen Symbolik abgestreift und in erster Linie das nüchtern-Fachliche exzerpiert worden. Daß die Filiation indirekt ist, läßt sich auch aus vielen anderen Kleinigkeiten schließen, so z. B. aus der Diskussion zu I, 5 (Im gleichschenkligen Dreieck sind die Winkel an der Grundlinie gleich). Hier fehlt sowohl bei Annairîzî wie bei Albertus der Hinweis auf Pappos, dessen einfaches und so überzeugendes Umwendeverfahren bei Proklos [54], S. 249–50 = [57], S. 341 steht und um seiner Überzeugungskraft willen mit Sicherheit gebracht worden wäre, wenn es den Muslimen vorgelegen wäre. Eine Übersicht über die Begriffsbestimmungen findet sich auch in Adelhard, Version III (Oxford, Bodleian Digby 174, fol. 99ʳ–99ᵛ; siehe Clagett [17], S. 34), ferner in der Euklid-Übersetzung des Hermann von Carinthia (Paris, Bibl. Nat., fl. 16646; siehe Clagett, S. 38 links) und des Gerhard von Cremona (Paris, Bibl. Nat., fl. 7216; siehe Clagett, S. 38 rechts).

[62] Über die Stellung des Albertus zur Zeichengenauigkeit lesen wir bei den Begriffsbestimmungen: *Quia tamen tota scientia finem habet theoreticum, ideo theorica est et nominatur. Operatio huius, praxis, non est, ut fiat figura in materia sensibili, sed potius, ut describatur in lineis imaginalibus. Quod multi sciunt, qui in materia figuras facere nesciunt. Et e converso multi figuras proferunt in materia, qui eas in lineis demonstrabilibus nesciunt describere.* Seiner ganzen Geisteshaltung nach zählt sich Albertus zu denen, die größeren Wert auf die theoretische Durchdringung als auf die technische Ausführung legen.

[63] Ähnliche Fassung: AC, S. 125, 33–126, 17.

[64] Ähnliche Fassung, ebenfalls mit Hinweis auf das unsystematische Vorgehen Herons, in AC, S. 126, 18–127, 28.

[65] Ein weniger geschickt durchgeführter direkter Beweis steht bei Proklos [54], S. 342 = [57], S. 402/3.

[66] AC, S. 61, 10–62, 2 entspricht Proklos [54], S. 339 = [57], S. 400.

[67] *Nota autem, quod sophistae impugnant istud theorema, dicentes, omnem quantitatem esse divisibilem in infinitum, ergo angulus contingentiae etiam dividi debet in infinitum. Et adhuc attendam, quod apud geometram nihil dicitur dividi nisi quod linea recta dividitur; et quod linea recta non dividitur, dicitur non habens quantitatem, et ideo [angulus contingentiae] etiam minimus dicitur, quia quantitatem non habet.*

[68] Zunächst wird festgestellt, daß der Kontingenzwinkel nicht durch eine Gerade geteilt werden kann; und zur Begründung: *Ex hac vero, quod non valet ista argumentatio:* ‘*Hoc transit a minori ad maiorem per omnia media, ergo per aequale*’, *nec ista:* ‘*Contingit reperire maius hoc et minus eodem, ergo contingit reperire aequale.*’ Diese Stelle wird wörtlich zitiert in Chr. Clavius: *Euclidis elementorum libri XV*, Erstausgabe Rom 1574, jedoch erst von der 2. Ausgabe Rom 1589 ab, Erläuterung zur Kontingenzwinkelfrage als Zusatz zu Euklid, *Elemente*, III, 16.

[69] Vergleiche Überweg–Geyer [21], S. 403.

[70] Vergleiche M. Curtze: *Jordani Nemorarii Geometria vel de Triangulis libri quattuor*, Mitteilungen des Coppernicus-Vereins 6, Thorn, 1887, S. 3.

SOME PRINCIPLES OF
MATHEMATICAL EDUCATION

By GEORGE KUREPA

In connection with some fundamental new discoveries† the problem of instruction in general, and that of mathematics in particular, will be examined from some new points of view. Nowadays one is aware of the possibility of asking and answering: what the process of teaching is, how the process of teaching is taking place, for one knows many factors involved in the process of teaching. The process of teaching should be considered experimentally, scientifically and by measurements.

The first approximation of an instruction process consists in considering it as an input-output process, in which the input factors, as well as the outfactors, are measurable. Therefore, instruction exploring is becoming a very interesting field for mathematico-statistical considerations.

I would like to stress a point, so as to avoid misunderstandings. The factors or variables involved in the education structure are more numerous than it would seem at first view. We must ask rather what further factors are to be considered than to state axiomatically that the whole instruction process would be based on one or two items. In particular, besides the factors: Teacher–Learner, the environment factors in very various senses (physical [e.g. small or big classrooms], biological [e.g. age of pupils], social [e.g. village, town], etc.) are of paramount importance.

We are going to stress some points of view we consider of great importance.

1. Functional point of view

1.1. Set and function are of vital significance, not only in mathematics but in all sciences and elsewhere. Neither of these two notions has been consciously considered or taught previously. It is important to allow both set and function to be general and not to restrict them unnecessarily. One tends to use function and set in quite restricted and general ways respectively. In this connection it is instructive to mention that the notion of function was introduced several centuries earlier than the notion of set.

† Cf. G. Kurepa, The role of mathematics and mathematicians at the present time (*Proc. Int. Congress of Math. 1954*, Amsterdam, vol. 3, 305–317, in particular §7).

1.2. If we do not wish to compromise the functional standpoint, it is necessary to allow both, function as well as set, to be general; in particular, variables need not be only numbers; variables might be: points, objects, functions (e.g. sequences), propositions, etc., in such a way that the notion of function (mapping, transformation, process) covers also geometrical, mechanical, physical, biological, transformations.

In particular, a function need not be uniform: it would be strange if the inverse of a function were no function; let us recall that complex functions are usually multiform.

The classical restricted notion of function discredited the classical functional principle in instruction of mathematics. Modern considerations and applications of mathematics, in particular of statistics, require, for example, that any finite set K of points in a co-ordinate plane should be considered as the graph of a function; a function in the restricted sense can also be associated with K in a different way, e.g. as the best-fitting function (curve) in the sense of the method of least squares.

1.3. Definition of a function. Let S, S' be an ordered pair of collections (sets, classes, etc.); if one associates with every element of S one or more elements of S' we are speaking of a function on S into S'. If a function on S into S' is denoted by f then for each $x \in S$ one denotes by fx each element of S' that is associated with x. S is called the domain of f and may be denoted Df. The class of all the values fx, x running over S, is called the range or the antidomain of f and may be denoted fS as well as ^-Df. By $\{fx\}$ we denote the set of all the values fx of f in x; consequently, $fS = {}^-Df = \underset{x}{U} \{fx\}$.

It is very useful to consider binary relations R originating in S: to every $x \in S$ one associates any object, point, set, class, denoted Rx. In particular, any function f on S into S' is closely connected with the binary relation $x \to \{fx\}$.

Sometimes it is more convenient to consider a relation $x \to Rx$ and then to consider the corresponding function $x \to fx$, fx running over Rx.

Instead of the relation $x \to Rx$ one speaks of the set-function $x \to Rx$ (abbreviated: S-function).

1.4. Functional principle. Consider any mapping, any associating or function f in any collection S and study what is happening with values fx and f when x is running over S and when S is changing respectively.

In particular, for any S and any subcollection S' of S and any function f on S, the induced function $f \mid S'$ equal to f in S' should be considered ($f \mid S'$ is the induced subfunction of f; f is an overfunction of $f \mid S'$).

2. Active interpenetration principle

2.1. An example. We know that Arithmetic and Geometry arose neither at the same place nor at the same time; they were separated in space as well as in time. The link between them was found consciously only by the discovering of Analytic Geometry. Since then both of them, Arithmetic–Algebra and Geometry, are progressing not only in parallel but also in conjunction and in mutual interpenetration.

2.2. If we consider, moreover, the third fundamental mathematical discipline, mechanics or at least kinematics and its fundamental concept: movement and generalization of movement (mapping, function), we become aware how the conjunctive and simultaneous considerations of the basic triad:

Figure, Number, Function

represent a tremendous step in mathematical and human development. The notions figure (set), number, function originated separately and not at the same time, but they joined one another and are in uninterrupted mutual interpenetration.

2.3. Practical consequence for mathematical education. The arithmetic with algebra and geometry are to be taught as closely together as possible. All three aspects of mathematics concerning (a) number and calculations, (b) figure, schemes, drawings, constructions, sets, ..., (c) movements, agitations, transformations, functions, are to be taught at all steps and all levels including the very beginning of teaching and even the prescholar period in kindergartens.

In particular, it is inconceivable that in many countries the first geometrical considerations are taught as late as in third or even fourth grade or still later. On the contrary, we emphasize the necessity that all three mentioned items should be taught not only in parallel but also in conjunction.

2.4. Considerations of everyday objects, collections, manipulations with them like: collecting, partitioning, transferring, etc., are items occurring everywhere and always. Education should start by combining actions and perceptions of this kind with various other kinds of manifestations and connections like play, speech, descriptions, associations, etc.

2.5. An active interpenetration of methods, domains, topics, is of great importance and usefulness. The more an item X (say a method, topic, individual, community) is linked with various other items, the more vital X is. Unilaterality, restricted exclusiveness on the one hand—joint activity, hybridization, on the other hand are in play. Biology shows how the second items are useful and fruitful. Isolation is a necessary evil, a way only for special purposes for examination and researches but not for successful development.

2.6. Practical consequence. Let us not restrict ourselves to a unique simple method: other methods are to be considered and applied too. In particular, let us not restrict ourselves to rigid Euclidean methods. The analytic method and synthetic method are both to be used: each is inadequate without the other.

2.7. A vital example for active interpenetration consists in considering the following basic items (factors):

Action, Perception, Intelligence, Imagination.

These factors are to work harmoniously in almost every individual or group in dependence upon the biological status (e.g. age of pupils), environment, etc. The more harmoniously they are in developmental respect interrelated and working together, the greater the success will be. In this respect the results of Piaget on genetical psychology and mathematics should be stressed.

2.8. Practical consequence for mathematical education. In the teaching process, the hands are to be active (writing, showing), the tongue, ears, brain, i.e. all organs are more or less in active interdependence and co-operation. Let us remember that for a long time, even in the instruction of geometry, and still more in arithmetic, the factors action and perception were either eliminated or at least neglected.

3. Shifting principle

3.1. As mankind is progressing, the piloting and optimal examples, situations, phenomena, ... for grasping mathematical ideas are occurring more and more in biological, economical, social, ... fields. In particular, in everyday biological, social and other situations, the most fundamental mathematical concepts occur as in a germ: set, class or population, organization or structure, classification, ordination, relation, process, interdependence, sense of evolution, etc. For pupils, they are perceptible, manipulatable, understandable, because the pupils were and are working and living with them. When the idea of a notion is grasped, one proceeds to its representation, definition, generalization, exploration,... e.g. the basic concept of function in the sense of associating occurs everywhere and it is wrong to substitute for it such or such representation or technical tool; the idea of a function is not the same as a representation of the same function.

Example of a function: to every pupil x of a class C associate Cx meaning his parents, sisters, brothers and himself. We are dealing here with a function $x \to Cx$.

3.2. From the practical and philosophical standpoint, it is very important to observe how to the fundamental notions occurring in biology like: classification, ordination, structure (organization), process, ... correspond analogous notions in mathematics using the fundamental interplay:

$$\text{Particular—Some—Any (all)}$$

(Quantifiers interplay).

3.3. In particular, a family, in the usual sense, is a certain organization consisting of a father, mother, children, ...; and there is no isomorphism between any two of these 'structures'; likewise for various other communities of plants, animals, ... The polyvalence of structures is a natural item in biology; in mathematics it was found as late as the last century by invention of groups; in the field of education this is unfortunately not yet introduced.

It is really astonishing that the idea of organization or structure has not been consciously introduced in mathematics earlier.

3.4. In particular, to ecological considerations in biology should correspond similar ones in mathematics. Ecological principles in mathematics and in education are of great importance.

4. Ecological principles

An individual might more reliably be explored, when it is considered jointly with other individuals than when it is considered alone: a family, group or collection property might contribute very much to a better exploring of a single individual. The environments of various kinds of an item should be considered too and jointly with this item.

Usually, a collection is more easily perceptible than an individual or particular case. In this respect there exists more or less normal or optimal size, situation, ... In education processes it is important to set up such more or less optimal situations in order to initiate and provoke so that the pupils (as individuals and as a collection) would perceive, discover, grasp the requested phenomenon. Afterwards follows the discussion, exploration, ... on varying what one wants.

The ecological point of view in mathematics may be considered as a synthesis of the classical functional standpoint and modern statistical ideas and gestalt psychology.

5. Impact of Logic. Quantors. Implication

5.1. So far one used to exaggerate in respect to the rigour of proofs and one tried to prove evident things too. We want to give to pupils interesting and non-obvious items, situations only obtainable by a certain number of justified steps. In such a way the pupils would appreciate the method of proving and deduction.

5.2. One helps the pupil enormously if one accustoms him to speak in the conditional pattern: If A,   then B,
and symbolically $A \Rightarrow B$ (A implies B). In this way, the fundamental logical function: implication is involved automatically.

5.3. Another quite recent acquisition in mathematics and logic consists in the awareness of interplay of logical quantors or quantifiers: some $\exists$ and every (all, each) $\forall$. When we ask how much or how many, the answers should be not only: $1, 2, 3, \ldots$, but also: nothing, some, almost all, all, all up to ϵ, with such and such probability or indetermination, etc. E.g. the convergence of a sequence $a_1, a_2, \ldots$ of numbers means that for some numbers a, n_0 and for every $\epsilon > 0$ and every integer l the relation $l > n_0$ implies $|a_l - a| < \epsilon$.

5.4. Set considerations are very suitable for automatic use of logical quantors: one more reason for using sets.

INDEX

Alexandrov, A. D., 3
Arnold, V. I., 339

Bers, L., 349
Beth, E. W., 281
Bogolyubov, N. N. and Vladimirov, V. S.,
  19
Bott, R., 423

Cartan, H., 33
Chern, S. S., 441
Chevalley, C., 53
Chung, K. L., 510

Feller, W., 69

Gårding, L., 87
Gnedenko, B. V., 518
Grauert, H., 362
Grothendieck, A., 103

Heins, M., 376
Higman, G., 307
Hirzebruch, F., 119
Hofmann, J. E., 554

Kervaire, Michel A., see Milnor, J. W. and
  Kervaire, Michel A.
Kleene, S. C., 137
Kosiński, A., 427
Kreisel, G., 289
Kurepa, G., 567

Lanczos, C., 154
Lehmer, D. H., 545

Linnik, Yu. V., 313
Lions, J. L., 389

Markov, A. A., 300
Matsusaka, T., 450
Menchoff, D. E., 398
Milnor, J. W. and Kervaire, Michel A., 454
Minakshisundaram, S., 407

Nagata, M., 459
Nijenhuis, A., 463

Papakyriakopoulos, C. D. 433
Pontryagin, L. S., 182

Rényi, A., 529
Roquette, P., 322
Roth, K. F., 203

Samuel, P., 470
Savage, L. J., 540
Schiffer, M., 211
Segre, B., 488
Shimura, G., 330
Sz.-Nagy, B., 412

Temple, G., 233
Thom, R., 248

Uhlenbeck, G. E., 256

Vladimirov, V. S., see Bogolyubov, N. N.,
  and Vladimirov, V. S.

Wang, H. C., 500
Wielandt, H., 268

Printed in the United States

By Bookmasters

Printed in the United States
By Bookmasters